Stem Cell Transplantation

Edited by Pier Paolo Piccaluga,
Giuseppe Visani
and Shaimaa Salaheldin Khattab

Published in London, United Kingdom

Stem Cell Transplantation
http://dx.doi.org/10.5772/intechopen.1003323
Edited by Pier Paolo Piccaluga, Giuseppe Visani and Shaimaa Salaheldin Khattab

Contributors
Alejandro Toro-Pedroza, Ali Fayyazpour, Alina Tanase, Amir Mehdizadeh, Ana I. Flores, Andreea Andronesi, Ankur Varma, Anup Kumar Trikannad, Aysu Timuroğlu, Azadeh Kiumarsi, Baris Malbora, Carmino De Souza, Danial Hashemi Karoii, Daniela Cardozo, Deepak Garg, Doria Valentini, Elvira Maličev, Francesc Fernández Avilés, Francisco Jaramillo, Gaurav Verma, Gener Ismail, Ghasem Janbabaei, Hannah E. Doan, Hiroshi Nagasaki, Hongjian Pu, Huayin Zhou, Janina Simončič, Jeane Eliete Laguila Visentainer, José Joaquín Merino, Kenji Takamori, Khadijeh Abbasi, Lavinia Lipan, Lucas Vinícius de Oliveira Ferreira, Luis Gerardo Jiménez Arias, Maria Toumazou, Marios Kozakos, Marko Hawlina, María Gabriela Villalba, Meenakshi Mendiratta, Mihaela Gherghiceanu, Miho Kawata, Mildrey Mosquera, Mitsutoshi Tominaga, Mohammad Biglari, Mohammad Reza Rostami, Mohammed S. Inayat, Mohini Mendiratta, Mortaza Raeisi, Naina Soni, Neslihan Mandacı Şanlı, Nishanth Thalambedu, Parisa Fayyazpour, Pedro Caraballo, Petra Schollmayer, Primož Rožman, Ranjit Kumar Sahoo, Rashmi Rameshwari, Ricardo Martínez-Murillo, Rizaev Jasur Alimdjanovich, Rogério Martins Amorim, Sonia Ruiz Boy, Srividhya Ravichandran, Sruthi Vellanki, Stefania Grasselli, Stefano Botti, Sujata Mohanty, Sumika Toyama, Tahereh Rostami, TingTing Ou, Vasiliki E. Kalodimou, Vasiliki E. Kalodimou, Vincent S. Gallicchio, Wendan Pei, Xingliang Fan, Zahra Hasani Mahforoozmahalleh, Zala Lužnik Marzidovšek, Zhilang Xie, Ziyadullaev Shukhrat Khudayberdiyevich, Álvaro Mondragón

First published in London, United Kingdom, 2025 by IntechOpen
IntechOpen is the global imprint of INTECHOPEN LIMITED, registered in England and Wales, registration number: 11086078, 167-169 Great Portland Street, London, W1W 5PF, United Kingdom

For EU product safety concerns: IN TECH d.o.o., Prolaz Marije Krucifikse Kozulić 3, 51000 Rijeka, Croatia, info@intechopen.com or visit our website at intechopen.com.

British Library Cataloguing-in-Publication Data
A catalogue record for this book is available from the British Library

Stem Cell Transplantation
Edited by Pier Paolo Piccaluga, Giuseppe Visani and Shaimaa Salaheldin Khattab
p. cm.

This title is part of the Biochemistry Book Series, Volume 62
Topic: Cell and Molecular Biology
Series Editor: Andrei Surguchov
Topic Editor: Mary C. Maj
Associate Topic Editor: Felicia Ikolo

Print ISBN 978-0-85466-524-2
Online ISBN 978-0-85466-523-5
eBook (PDF) ISBN 978-0-85466-525-9
ISSN 2632-0983

If disposing of this product, please recycle the paper responsibly.

IntechOpen Book Series

Biochemistry

Volume 62

Aims and Scope of the Series

Biochemistry, the study of chemical transformations occurring within living organisms, impacts all of the life sciences, from molecular crystallography and genetics, to ecology, medicine and population biology. Biochemistry studies macromolecules - proteins, nucleic acids, carbohydrates and lipids –their building blocks, structures, functions and interactions. Much of biochemistry is devoted to enzymes, proteins that catalyze chemical reactions, enzyme structures, mechanisms of action and their roles within cells. Biochemistry also studies small signaling molecules, coenzymes, inhibitors, vitamins and hormones, which play roles in the life process. Biochemical experimentation, besides coopting the methods of classical chemistry, e.g., chromatography, adopted new techniques, e.g., X-ray diffraction, electron microscopy, NMR, radioisotopes, and developed sophisticated microbial genetic tools, e.g., auxotroph mutants and their revertants, fermentation, etc. More recently, biochemistry embraced the 'big data' omics systems.
Initial biochemical studies have been exclusively analytic: dissecting, purifying and examining individual components of a biological system; in exemplary words of Efraim Racker, (1913 –1991) "Don't waste clean thinking on dirty enzymes." Today, however, biochemistry is becoming more agglomerative and comprehensive, setting out to integrate and describe fully a particular biological system. The 'big data' metabolomics can define the complement of small molecules, e.g., in a soil or biofilm sample; proteomics can distinguish all the proteins comprising e.g., serum; metagenomics can identify all the genes in a complex environment e.g., the bovine rumen.
This Biochemistry Series will address both the current research on biomolecules, and the emerging trends with great promise.

Meet the Series Editor

Andrei Surguchov, Ph.D., joined Baylor College of Medicine, Houston, TX, as a faculty member in 1992, where he studied the mechanisms of the genetic control of lipid metabolism. At the University of Utah, his research interests focused on cloning new genes encoding retinal proteins. He studied molecular and cellular mechanisms of neurodegenerative diseases and retinal degeneration at Washington University, St. Louis. Currently, his research focuses on the structure-function relationship of proteins involved in neurodegeneration and ocular diseases. Andrei Surguchov is an Editor-in-Chief at Biochemistry Research International and Associate Editor in several biomedical journals.

Meet the Volume Editors

Pier Paolo Piccaluga is an Associate Professor of Pathology at the Department of Medical and Surgical Sciences at the Bologna University School of Medicine. He is also a co-founder and Executive Physician at the Biobank of Research, IRCCS S. Orsola-Malpighi Hospital in Italy. He has led the Molecular Hematopathology Laboratory for many years. In 2018, he was appointed to teach at Queen Mary University of London (UK) and Jomo Kenyatta University of Agriculture and Technology in Nairobi (Kenya), and since 2023, at the University of Nairobi and the University of Botswana. He is the author of several international publications indexed in the Current Contents (including Nature Medicine, Journal of Clinical Investigation, Journal of Experimental Medicine, Journal of Clinical Oncology, Blood, Lancet Oncology, and Lancet Infectious Diseases, with a total impact factor >1,000, a mean impact factor >6, and an H-index of 62 (ranked among the VIA-Academy Top Italian Scientists). He is the author of several presentations at national and international conferences. Dr. Piccaluga has participated in several clinical trials as a coordinator or sub-investigator and has won multiple awards for his contributions to study and research. He has served as a Principal Investigator (PI) in various research projects funded by nationally and internationally recognized organizations. He is the Editor-in-Chief of the World Journal of Hematology, Digital Medicine and Healthcare Technology, Advances in Precision Medicine, and the Journal of Cancer Biomoleculars and Therapeutics.

Dr. Giuseppe Visani pioneered stem cell transplantation, acute and chronic myeloid leukemia biology and treatment. Currently working as a Director of the Hematology Unit and Transplant Center at AORMN Marche North.

Shaimaa Salaheldin Khattab is an experienced clinical hematologist and hematopathologist with a strong academic and research background. She holds a Master's degree in Clinical Hematology, where her research focused on early predictive markers for septic shock in acute leukemia patients, and an MB Bch from Alexandria University in Egypt. With over a decade of experience, Shaimaa has worked in various prestigious institutions, including Alexandria University Teaching Hospitals and the Medical Research Institute. She has served as an assistant lecturer and a clinical hematology consultant, contributing significantly to both patient care and medical education. Her expertise extends to hematopoietic stem cell transplantation, having worked in specialized centers across Egypt.

Contents

Preface XV

Section 1
Biology and Clinics of Stem Cell Transplantation **1**

Chapter 1 **3**
Adoptive Immune Cell Therapy in Pediatric Solid Tumors
by Tahereh Rostami, Azadeh Kiumarsi, Mohammad Biglari, Mohammad Reza Rostami and Ghasem Janbabaei

Chapter 2 **37**
Allogeneic Stem Cell Transplantation in Developing Countries
by Francisco Jaramillo, Alejandro Toro-Pedroza, Pedro Caraballo, Álvaro Mondragón and Mildrey Mosquera

Chapter 3 **47**
Congenital Neutropenia Associated with Bone Marrow Failure
by Baris Malbora

Chapter 4 **75**
Current Advances in Stem Cell-Based Therapies: Adult Stem Cell Applications
by Neslihan Mandacı Şanlı and Aysu Timuroğlu

Chapter 5 **101**
Effective Techniques in Improving Stem Cell Function and Success in Stem Cell Transplantation
by Parisa Fayyazpour, Amir Mehdizadeh, Khadijeh Abbasi, Ali Fayyazpour and Mortaza Raeisi

Chapter 6 **123**
Ethical Consideration of Stem Cell Transplantation
by Luis Gerardo Jiménez Arias

Chapter 7 **135**
Fundamentals of Stem Cells and Application in Complex Disorders
by Gaurav Verma and Srividhya Ravichandran

Chapter 8 171
Home Care and Pharmacy in Hematopoietic Transplantation
by Sonia Ruiz Boy and Francesc Fernández Avilés

Chapter 9 189
The Role of Long Noncoding RNAs in Gene Expression and Epigenetic Regulation in Spermatogonial Stem Cells and Therapeutic Implications
by Danial Hashemi Karoii and Zahra Hasani Mahforoozmahalleh

Chapter 10 217
The Role of NK Cell, HLA, and Peptide Interaction in Hematopoietic Stem Cell Transplantation
by Daniela Cardozo, Jeane Eliete Laguila Visentainer and Carmino De Souza

Section 2
Mesechymal Stem Cells Applications in Regenerative Medicine 233

Chapter 11 235
Dissecting the Multifaceted Therapeutic Implications of Mesenchymal Stem Cell-Derived Extracellular Vesicles: An Intricate Exploration
by Naina Soni and Rashmi Rameshwari

Chapter 12 253
Mesechymal Stem Cells Applications in Regenerative Medicine and in Biotherapies
by Vasiliki E. Kalodimou

Chapter 13 271
Mesenchymal Stem Cells from Fat: From Differentiation Mechanisms to Biomedical Application in Patients
by José Joaquín Merino, María Gabriela Villalba, Ricardo Martínez-Murillo and Ana I. Flores

Chapter 14 311
Mesenchymal Stem Cell Applications in General Surgery
by Vasiliki E. Kalodimou, Marios Kozakos and Maria Toumazou

Chapter 15 327
The Integral Role of Mesenchymal Stem Cells in Stem Cell Transplantation from Promotion of Stem Cell Engraftment to Immunomodulation
by Mohini Mendiratta, Meenakshi Mendiratta, Deepak Garg, Sujata Mohanty and Ranjit Kumar Sahoo

Chapter 16 365
Schwann-Like Cells Derived from Mesenchymal Stem Cells: Their Potential for Peripheral Nerve Regeneration
by Rogério Martins Amorim and Lucas Vinícius de Oliveira Ferreira

Chapter 17 389
The Function and Effectiveness of MSCs in Preclinical Models of Diseases
by Zhilang Xie, Huayin Zhou, TingTing Ou, Wendan Pei and Xingliang Fan

Chapter 18 427
An *In Vitro* Model for the Study of Itch Using Human iPS Cell-Derived Sensory Neurons
by Sumika Toyama, Mitsutoshi Tominaga and Kenji Takamori

Section 3
Stem Cell Transplant for Non-Neoplastic Diseases **437**

Chapter 19 439
Advancements of Stem Cell Transplantation in Myocardial Infarction and Ischemic Heart Failure
by Rizaev Jasur Alimdjanovich and Ziyadullaev Shukhrat Khudayberdiyevich

Chapter 20 455
Regenerative Medicine for the Hypothalamic-Pituitary Axis: Current Progress and Future Challenges
by Miho Kawata and Hiroshi Nagasaki

Chapter 21 479
Stem Cell Therapy for Stroke: Mechanisms and Functional Recovery
by Hongjian Pu

Chapter 22 491
Stem Cell Use to Treat Dermatological Disorders
by Hannah E. Doan, Mohammed S. Inayat and Vincent S. Gallicchio

Chapter 23 515
Stem Cells for Therapy of Eye Diseases: Current Status
by Zala Lužnik Marzidovšek, Janina Simončič, Petra Schollmayer, Elvira Maličev, Primož Rožman and Marko Hawlina

Section 4
Toxicity of SCT and Cellular Therapies **541**

Chapter 24 543
CART Cell Toxicities and Management
by Nishanth Thalambedu, Sruthi Vellanki, Anup Kumar Trikannad and Ankur Varma

Chapter 25 563
Oral Mucositis Management and Care
by Stefano Botti, Doria Valentini and Stefania Grasselli

Chapter 26 **581**
Renal Complications Following Hematopoietic Stem Cell Transplantation
by Alina Tanase, Andreea Andronesi, Lavinia Lipan, Mihaela Gherghiceanu and Gener Ismail

Preface

The field of stem cell transplantation is continually evolving, offering promising avenues for treating many diseases once deemed incurable. As an editor of this comprehensive volume, "Stem Cell Transplantation", I am excited to present an exploration of recent advancements, challenges, and future directions within this dynamic field. This book delves into the biological, clinical, and ethical dimensions of stem cell research and application, organized into four distinct sections to provide a cohesive and thorough understanding of the subject.

The initial section lays the foundation by examining stem cell transplantation's biological underpinnings and clinical applications. This section begins with exploring adoptive immune cell therapy in pediatric solid tumors, highlighting novel approaches to combat childhood cancers. The chapter on allogeneic stem cell transplantation in developing countries underscores the logistical and clinical challenges, offering insights into innovative solutions tailored for resource-limited settings.

Further, the section addresses congenital neutropenia associated with bone marrow failure, thoroughly examining genetic and therapeutic perspectives. Current advances in adult stem cell applications are explored, showcasing breakthroughs in regenerative medicine. The emphasis on practical techniques for improving stem cell function underscores the importance of optimizing transplantation success rates.

Ethical considerations are pivotal in stem cell research, and this is addressed comprehensively, encouraging a dialogue on balancing scientific progress with moral responsibility. The fundamentals of stem cells and their application in complex disorders are also discussed, providing a broad overview of potential therapeutic landscapes. Additionally, the role of home care and pharmacy in hematopoietic transplantation is analyzed, emphasizing patient-centric care models.

The section concludes with a discussion on the roles of long noncoding RNAs and NK cells in gene expression and immune interactions, offering insights into the molecular intricacies influencing transplantation outcomes.

Mesenchymal stem cells (MSCs) are at the forefront of regenerative medicine, and an entire section of the book delves into their diverse applications. With a focus on MSC-derived extracellular vesicles, the book explores their multifaceted therapeutic potential. Applications in biotherapies and general surgery are detailed, highlighting the versatility of MSCs in clinical settings.

The potential of mesenchymal stem cells from adipose tissue is explored, revealing their biomedical applications and differentiation mechanisms. Innovative approaches, such as Schwann-like cells derived from MSCs for nerve regeneration, are also examined.

The section further investigates the role of MSCs in preclinical disease models and their integral function in stem cell transplantation, from promoting engraftment to modulating the immune response. An intriguing in vitro model for studying itch using iPS cell-derived sensory neurons is also presented, reflecting the innovative research in this area.

The third section addresses the use of stem cell transplantation in treating non-neoplastic diseases. Advancements in myocardial infarction and ischemic heart failure are discussed, providing insights into cardiovascular regeneration. The chapter on regenerative medicine for the hypothalamic-pituitary axis highlights the complexities and future challenges of treating endocrine disorders.

Stem cell therapy for stroke is examined, focusing on mechanisms and functional recovery, while the use of stem cells in treating dermatological disorders and eye therapy is also explored, showcasing the broad therapeutic scope of stem cell applications.

The final section confronts the challenges of toxicity associated with stem cell transplantation and cellular therapies. An in-depth analysis of CART cell toxicities and management strategies is provided, addressing a critical aspect of immunotherapy. Oral mucositis management and renal complications following hematopoietic stem cell transplantation are also discussed, emphasizing the need for comprehensive care strategies to mitigate adverse effects.

The production of this book was made feasible by the involvement of many persons. We sincerely appreciate the authors, researchers, and physicians who devoted their time and expertise to this effort. Their fervor and dedication to progressing the discipline of hematology have been genuinely motivating. We extend our gratitude to the institutions and organizations that offered help and resources, facilitating the completion of this book. Their steadfast dedication to research and education in hematology is crucial for advancing advancement.

Pier Paolo Piccaluga
Associate Professor,
University of Bologna,
Bologna, Italy

Dr. Giuseppe Visani
AORMN Hospital,
Pesaro, Italy

Dr. Shaimaa Salaheldin Khattab
Alexandria University,
Egypt

Section 1

Biology and Clinics of Stem Cell Transplantation

Chapter 1

Adoptive Immune Cell Therapy in Pediatric Solid Tumors

Tahereh Rostami, Azadeh Kiumarsi, Mohammad Biglari, Mohammad Reza Rostami and Ghasem Janbabaei

Abstract

While the prognosis for pediatric solid tumors has improved in recent decades, patients with refractory, metastatic, or recurrent disease continue to face poor outcomes. Innovative therapeutic approaches like adoptive cell therapy (ACT) have emerged as a promising strategy to overcome tumor cell evasion of the immune system. However, targeting solid tumors with ACT has been limited, as tumor cells employ various escape mechanisms, such as downregulating tumor antigens, creating an immunosuppressive microenvironment by secreting anti-inflammatory cytokines and expressing immune checkpoint inhibitors that can suppress immune responses. This review aims to highlight the progress and challenges of immunotherapy for pediatric patients with solid tumors, focusing on new perspectives of ACT in this patient population.

Keywords: Pediatric solid tumors, adoptive cell therapy, immunotherapy, chimeric antigen receptor, natural killer cells

1. Introduction

Recent advancements in immunology and immune-based therapies have presented significant promise in cancer treatment by focusing on tumor-specific antigens. This evolution commenced with the successes of immune checkpoint inhibitors (ICH) and chimeric antigen receptor (CAR) T-cell therapy, leading to approvals from the U.S. Food and Drug Administration (FDA) for the treatment of a wide range of cancers [1].

Adoptive cell therapy (ACT) is a highly personalized immunotherapy [2] that involves harvesting the patient's immune cells, ex vivo expansion, and redirecting them to target cancer cells [3]. The majority of ACT approaches utilize T cells, including CAR-T-cell therapy, T-cell receptor (TCR) gene-modified T cells, and tumor-infiltrating lymphocytes (TILs) which have the potential to act as "living drugs" by expanding, engrafting, and persisting in vivo, leading to enduring remission in a subset of patients [1, 4]. In addition to T cells, the adoptive transfer of natural killer (NK) cells is emerging as a promising strategy in various pediatric solid tumors, which are sensitive to NK cell-mediated cytotoxicity [5].

While ACT has demonstrated curative clinical outcomes in hematologic malignancies, its utilization in solid tumors is still at a nascent stage, with ongoing research efforts aimed at improving its effectiveness [6–9]. This chapter investigates the potential role of ACT in the treatment of pediatric patients with solid tumors. We specifically focus on the associated promises and challenges and explore future research directions in this rapidly evolving field.

2. Cancer immune surveillance

The tumor microenvironment consists of malignant cells, the extracellular matrix (ECM), and non-neoplastic cells such as mesenchymal support cells, endothelial cells, and infiltrated inflammatory immune cells. The interaction between neoplastic and non-neoplastic cells plays a crucial role in determining the fate of the tumor. Studies have demonstrated that a high infiltration of T cells in the tumor microenvironment is linked to a favorable prognosis in many different tumor types, whereas high levels of macrophage infiltration are associated with a poorer prognosis [10, 11].

Malignant cells are identified by macrophages and NK cells as part of the innate immunity, leading to phagocytosis by antigen-presenting cells (APCs) like dendritic cells (DC) [12]. The APCs present tumor-associated antigens (TAAs) to the T-cell receptors (TCRs) on naïve CD8+ T cells via major histocompatibility complex class I (MHC-I), triggering the effector phase of adaptive immunity, ultimately leading to the death of tumor cells [13]. Additionally, the interaction between MHC-II on

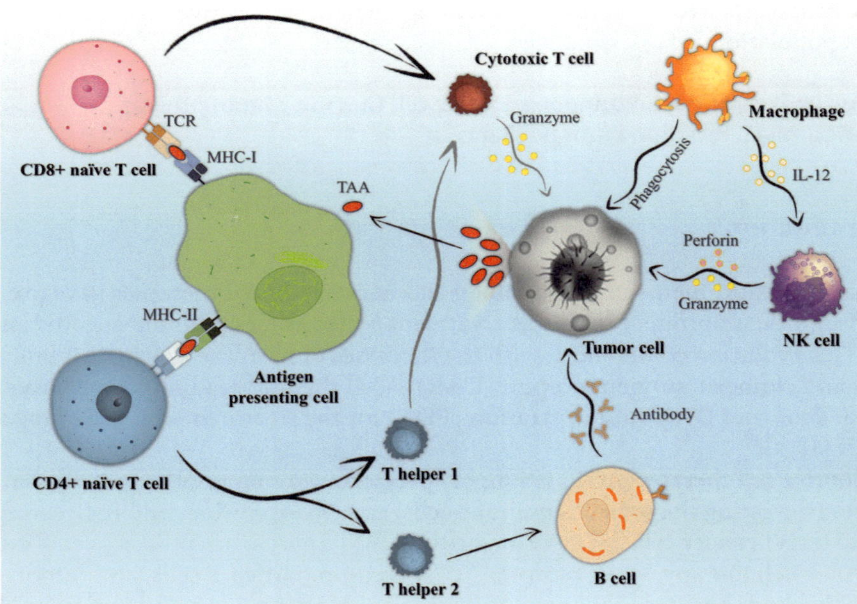

Figure 1.
Tumor immunosurveillance. A particular role of the immune system incorporating various immune cells to recognize and annihilate tumor cells as they develop. This consists of T cells interacting with antigen presenting cells via the major histocompatibility complex molecules, B cells producing antibody, and innate immunity via phagocytosis and cell lysis performed by macrophages and natural killer cells. MHC, major histocompatibility complex; NK, natural killer; TAA, tumor-associated antigen; TCR, T-cell receptor.

malignant cells and TCRs on CD4+ T-helper cells triggers the production of anti-tumor antibodies through B-cell activation [14, 15].

Immunosuppressive cells originated from the tumor microenvironment (TME), which encompass myeloid-derived suppressor cells (MDSCs), tumor-associated macrophages (TAMs), and regulatory T lymphocytes (T-regs), hinder the effector phase by suppressing T-cell responses [16].

Figure 1 represents the concept of immune surveillance against tumor cells.

3. Types of adoptive cell therapy (ACT)

3.1 Tumor-infiltrating lymphocyte (TIL)

Tumor-infiltrating lymphocytes, which include CD4+ helper cells and CD8+ cytotoxic T cells (CTLs), represent a heterogenous population of polyclonal T cells, directed to TAAs [17, 18]. These cells may be extracted from the immunosuppressive tumor microenvironment, which is intrinsically reactive against tumor antigens, activated, expanded ex vivo in the presence of homeostatic cytokines, and then transferred back into the patient to enhance their activity against a heterogeneous repertoire of autologous tumor-associated antigens [19, 20].

TIL therapy, a form of targeted T-cell therapy, has been most widely employed in melanoma patients following a non-myeloablative (NMA) lymphodepletion regimen, which is thought to amplify the antitumor efficacy of TILs by eliminating T-regs and removing homeostatic cytokines that would otherwise dampen effector cell proliferation [21–25].

Even though solid tumors, unlike hematologic malignancies, are made up of various cell populations with significant antigenic heterogeneity, which restricts the effectiveness of antigen-targeted therapies, there has been a rise in multi-center trials examining TIL therapy in patients with solid tumors [26].

The effective potential of TIL therapy to benefit more cancer patients has been constrained by several obstacles.

Tumor-specific neoantigens, which arise from accidental somatic mutations, are effectively targeted by TILs. So, there may be a higher likelihood of tumor clones with a high mutational burden to generate neoantigens and respond to T-cell-based therapies [27–30]. However, the cytolysis of neoantigen-containing tumor clones can result in the emergence of antigen-negative tumor clones that evade recognition by TILs, posing a challenge to the effectiveness of TIL therapy. In addition, tumor cells may downregulate the expression of MHC molecules or other parts of antigen presentation to escape immune surveillance [31].

The low mutational burden in pediatric cancers typically leads to a restricted set of neoantigens that can be identified and targeted by T cells. Consequently, the tumor microenvironment exhibits limited T-cell infiltration, as seen in neuroblastoma with MYCN and ALK (anaplastic lymphoma kinase) amplification, as well as in brain tumors and sarcomas. The TME is a complex system consisting of both cellular and non-cellular components that play crucial roles in tumor progression. The cellular component is made up of various cell types, each with distinct functions and influenced by the cytokines or inhibitory ligands they express. Pediatric solid tumors, in particular, have a highly heterogeneous TME [32].

The TME can impede the infiltration of TILs through mechanisms like generating a dense extracellular matrix, releasing chemokines, or promoting abnormal angiogenesis. Additionally, the TME harbors immunosuppressive cell populations like MDSCs and Tregs that secrete anti-inflammatory cytokines and hinder APCs, further dampening T-cell-mediated immune responses. Ultimately, this environment can lead to T-cell exhaustion, resulting in the loss of their effector functions [27, 33–36].

3.2 Genetically engineered TCR

Conventional T cells recognize antigens via their TCRs [37], comprising α and β chains crucial for defining T-cell antigen specificity [38]. These TCR α and β heterodimers form TCR-CD3 signaling complexes that detect and react to target antigens presented through MHC-dependent pathways [39]. The activation of T cells and subsequent cytolytic destruction of target cells occur upon recognizing the specific peptide displayed on MHC molecules. Considering that tumors originate from self-tissue and are subject to immune tolerance, T cells typically cannot recognize TAAs. However, through genetic engineering, T cells can be modified by inserting genes that encode cell surface receptors capable of detecting TAAs, thereby enabling T-cell recognition and response. Given that T cells possess a distinct TCR tailored to a particular antigenic peptide sequence, engineered T-cell receptors (TCR-T cells) offer an alternative approach to adoptive cell therapy [40–42].

Engineered TCR-T cells provide an advantage by recognizing not only membrane proteins but also intracellular antigens presented by MHC molecules, thereby broadening the scope of tumor neoantigens and therapeutic targets [43].

3.3 Chimeric antigen receptor (CAR) T cell

Chimeric antigen receptor T cells are modified lymphocytes engineered with synthetic receptors comprising three main components:

1. Extracellular domain: this domain is derived from immunoglobulin heavy and light chains and contains a variable single-chain fragment (scFv) that provides specificity for tumor antigens.

2. Transmembrane domain: this domain is connected to the extracellular segment via a spacer.

3. T-cell activation domain: this domain, which is typically derived from the zeta chain of the CD3 complex, triggers downstream signaling pathways.

Additionally, CAR structures often include multiple costimulatory domains, such as CD28 and/or 4-1BB, to enhance the intracellular activity of T cells [44–46].

CAR-T-cell therapy has progressed significantly as its discovery in the late 1980s, with the aim of increasing activation, persistence, proliferation, safety, and efficacy [46]. Five generations of CAR-T cells have been developed, each representing advancements in the field of CAR-T-cell therapy. These generations include the following:

The first generation (1G) of CAR-T cells featured an extracellular domain with a scFv and a cytoplasmic domain containing only a CD3ζ (FcεRIγ) signaling domain, lacking any costimulatory molecules. This generation had limited signaling capacity, leading to suboptimal T-cell activation and weak in vivo persistence of T-cell responses. Due to these limitations, the first generation is now considered obsolete [47–49].

The second generation (2G) of CAR-T cells includes the addition of a costimulatory domain, such as CD28, CD134 (OX-40), or CD137 (4-1BB), to the intracellular domain. This modification results in an improved in vivo life span of CAR-T cells and provides two signals through CD3ζ and costimulatory molecules. This enhancement contributes to better T-cell activation, proliferation, and persistence compared to the first generation [50–57].

The third generation (3G) of CAR-T cells is characterized by the inclusion of multiple signaling domains, typically incorporating CD3ζ and two complementary costimulatory domains, such as CD28 and 4-1BB. CD28 signaling tends to promote faster T-cell expansion and more rapid elimination of tumor cells, while 4-1BB signaling is associated with longer persistence of CAR-T cells within the host. This generation aims to enhance both T-cell activation and persistence compared to previous generations [58].

The fourth generation (4G) of CAR-T cells, also known as T-cell redirected for universal cytokine-mediated killing (TRUCK), universal CAR (UniCAR-T), or armored CAR-T cells, shares similarities with the second generation. However, its distinct feature lies in the intracellular signaling domain, which expresses transgenic proteins such as interleukin 12 (IL-12) after CAR activation. IL-12's antitumor effect is attributed to enhanced activation of cytotoxic T cells and NK cells, as well as the induction of a Th1-type helper T-cell response to targeted tumor tissue. This modification aims to modulate T-cell response and improve the overall efficacy of CAR-T cell therapy by improving their persistence and altering tumor microenvironment characteristics [59–61].

The fifth generation (5G), also known as the next generation, builds upon the second generation by incorporating a truncated cytoplasmic IL-2 receptor β-chain domain with a binding site for the transcription factor STAT3. This alteration enables the initiation of a cytokine stimulation pathway through JAK-STAT3/5, resulting in full T-cell activation and proliferation. The fifth generation, or 5G, offers an improved safety profile and a broader therapeutic range. It promotes a more favorable environment within tumors and restores the immune system post-infusion [62].

A significant drawback of utilizing CAR-T cells is related to the TAAs that can also be present as soluble antigens released from tumors into the surrounding tissue and bloodstream. These soluble antigens have the potential to hinder the interaction between CAR-T cells and cancer cells [63].

3.4 Natural killer (NK) cell

Natural killer cells, essential components of the innate immune system, originate from progenitor cells derived from $CD34^+$ common lymphoid progenitors. With a lifespan estimated at approximately 7–10 days in circulation, NK cells account for 5 to 15% of peripheral blood mononuclear cells (PBMCs). They are characterized by the presence of CD65 and the absence of the T-cell marker CD3. NK cells are classified into two subsets, $CD56^{dim}$ and $CD56^{bright}$, based on the intensity of CD56 fluorescence detected via flow cytometry. In peripheral blood, the predominant NK cells are $CD56^{dim}$ $CD16^+$, which primarily mediate cytotoxicity, whereas, in lymphoid tissues, the majority of NK cells are $CD56^{bright}$ $CD16^-$ that are immature and specialize in cytokine production, with a limited capacity for cytolytic response [64–67].

NK cells identify their target cells through the binding of killer cell immunoglobulin-like receptors (KIRs) to MHC class I molecules, which serve as ligands on the surface of target cells. KIRs can be either activating or inhibitory. When KIRs recognize the appropriate MHC class I molecule, they transmit an inhibitory signal (**Figure 2a**). However, if the MHC class I molecule is missing or mismatched, no inhibitory signal is generated (**Figure 2b**), and the balance tilts toward activation [68–70].

NK cells recognize and target tumors in the absence of self-antigen presentation, making cancer cells with low Human leukocyte antigen (HLA) expression, such as leukemic cells, more susceptible to NK cell-mediated lysis. In contrast, tumor cells with high HLA expression, such as lymphomas and some solid tumors, are more resistant to NK cell-induced killing. When the KIR on an NK cell fails to match the HLA of a target cell, the probability of cytolysis increases, suggesting that allogeneic NK cells with a KIR-HLA mismatch may be effective in eliminating tumor cells resistant to autologous NK cells [71–74].

Antibody-dependent cytotoxicity (ADCC) is another mechanism of adaptive immunity mediated by NK cells (**Figure 2c**). ADCC is initiated by the Fc region of various therapeutic antibodies, including CD16 (or FcγRIIIA), an activating receptor expressed on the surface of NK cells [75, 76].

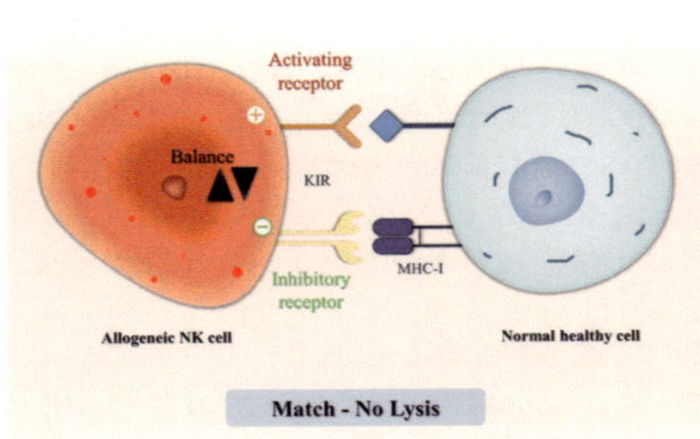

(a)

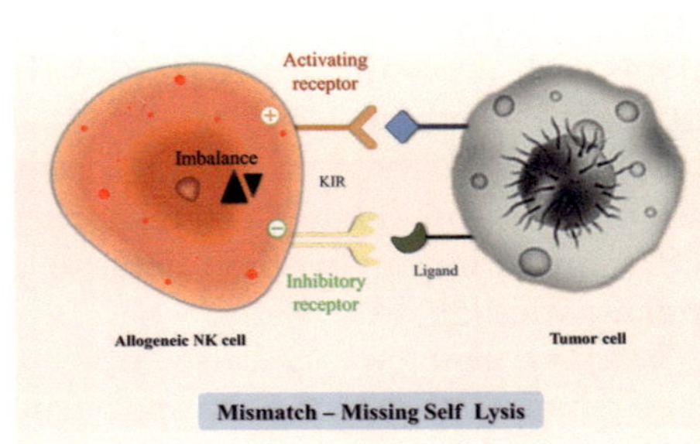

(b)

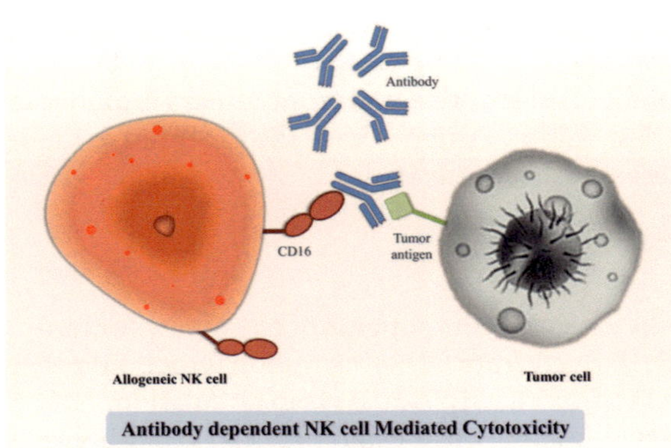

(c)

Figure 2.
The interaction of NK cells with a normal self cell (a) and a tumor cell (b) and antibody-dependent cytotoxicity (ADCC) (c). Natural killer (NK) cells recognize their target via KIR to MHC interaction. KIRs can be either activating (+) or inhibitory (−). When KIRs recognize the appropriate MHC class I molecule, they transmit an inhibitory signal, and if the MHC class I molecule is missing or mismatched, no inhibitory signal is generated, which results in missing self lysis. ADCC is a type of adaptive immunity reaction initiated by the antibody Fc region interaction with a specific tumor antigen and a specific receptor on immune cells like CD16 (FcγRIIIA) on the surface of NK cells.

3.5 CAR NK cell

NK cells possess distinct characteristics that make them a promising alternative to T cells for CAR-based immunotherapy. NK cells lack a TCR and target cancer cells in an HLA-independent manner, which enables them to better recognize MHC-downregulated cancer cells and reduces the risk of graft versus host disease (GvHD) compared to T cells. CAR-T cells identify cancer cells via a CAR-dependent mechanism, while CAR NK cells can recognize target cells through both CAR-dependent and CAR-independent pathways [77]. The CAR-independent killing mechanisms of CAR-NK cells make them more effective than T cells in eradicating minimal residual cancer cells that do not express the antigen targeted by the CAR [78].

CAR-NK cells offer another advantage over CAR-T cells due to their lower expression of programmed death-1 (PD-1), resulting in greater efficacy. However, the relatively short lifespan of NK cells necessitates the administration of booster doses to ensure the elimination of cancer cells [79, 80].

4. Clinical trials of adoptive cell therapy (ACT) in pediatric solid tumors

4.1 Neuroblastoma

Neuroblastoma (NB) is the most common extracranial solid tumor in childhood, originating from the neural crest during embryonic development [81, 82]. It exhibits significant clinical heterogeneity, with around 50% of patients categorized as having high-risk disease according to the International Neuroblastoma Risk Group (INRG) staging system [83, 84]. Despite intensive treatment involving chemotherapy, stem cell transplantation, radiation, and antibody therapy, the prognosis remains poor, with long-term progression-free survival (PFS) rates ranging from 50 to 60%, highlighting the need for novel approaches [85].

4.1.1 Adoptive cell therapy with NK cells

The idea of using NK cells in neuroblastoma treatment emerged from successful clinical trials using monoclonal antibodies targeting NB cells. These trials showed that complete remission or disease stabilization was linked to an increase in NK cell count and activity, including NK cell-mediated ADCC [86–90]. NK cells are known for their ability to target and destroy tumor cells that lack HLA-class I molecules, which are ligands for killer inhibitory receptors. Therefore, metastatic NB cells, which typically have low to absent HLA-class I expression and exhibit various ligands for NK cell activating receptors, are considered ideal targets for NK cell-mediated lysis [91–93].

Numerous ongoing clinical trials are investigating the use of adoptive NK cell infusions in patients with NB.

Federico et al. carried out a pilot study in which patients received treatment with the anti-GD2 (Disialoganglioside) antibody hu14.18K322A, GM-CSF (Granulocyte-macrophage colony-stimulating factor), and IL2, either with or without the infusion of haploidentical parental NK cells. The NK cell infusion was found to be safe and well-tolerated. The trial resulted in an overall response rate of 61.5%, with a complete response/very good partial response rate of 38.5%, and no instances of disease progression were observed. However, the study did not specifically assess the role of NK cells in contributing to therapeutic success [94].

In a phase I clinical trial led by Modak et al., neuroblastoma patients underwent induction chemotherapy before receiving varying doses of haploidentical parental NK cells and the anti-GD2 murine antibody 3F8. The administration of NK cells was found to be safe and well-tolerated. Additionally, patients who exhibited detectable NK cells in their peripheral blood on day 14 achieved complete remission. Furthermore, those who received higher doses of NK cell infusions demonstrated improved progression-free survival compared to individuals who received lower doses ($p = 0.018$) [95].

The potential therapeutic advantages of NK cells in neuroblastoma treatment necessitate thorough assessment in phase III randomized clinical trials. Presently, there are no active phase III/IV clinical trials involving NK cell infusions, with all registered trials on clinicaltrials.gov being phase I/II studies [84]. **Table 1** displays a summary of NK cell immunotherapy clinical trials in patients with NB.

4.1.2 Adoptive cell therapy with CAR-T cells

The effectiveness of CAR-T-cell therapy in solid tumors relies on the capacity of CAR-T cells to infiltrate the tumor site, overcome the immunosuppressive tumor microenvironment, and persist for a long time [84]. While CAR-T cells have exhibited promising results in clinical trials for neuroblastoma, with multiple objective responses observed in initial phase studies, their overall efficacy in neuroblastoma has not been as impressive as in hematologic malignancies [96].

The most extensively investigated tumor-associated antigen in neuroblastoma is GD2, a disialoganglioside that is highly and widely expressed on neuroblastoma tissue and may contribute to the tumor's immune evasion capabilities [96]. Given the success of anti-GD2 monoclonal antibody therapy, GD2 represents a logical target for CAR-T-cell therapy in neuroblastoma [88, 97–99]. Another antigen of interest in neuroblastoma is L1-CAM (L1 cell adhesion molecule), an adhesion molecule that is overexpressed on neuroblastoma cells [100].

In addition to GD2 and L1-CAM, other novel target antigens for CAR-T-cell therapy in neuroblastoma are under investigation. One such target is Glypican 2 (GPC2), a cell surface molecule that is highly expressed in neuroblastoma but has low expression on normal tissue, making it a potentially ideal target for CAR-T-cell-based immunotherapy [101, 102]. Researchers have developed CARs containing heavy chain-only scFvs against GPC2, with 4-1BB and CD3ζ endodomains. These anti-GPC2 CAR-T cells have demonstrated in vitro activity and the ability to clear human neuroblastoma xenografts in vivo [103].

Another emerging target is B7-H3 (CD276), which is highly expressed in various pediatric solid tumors, including neuroblastoma, and its overexpression is associated with a poor prognosis in many cancers. These features have positioned B7-H3 as a promising target for immunotherapeutic approaches, with early-phase clinical trials using monoclonal antibodies showing promising results in neuroblastoma and other malignancies [104–107]. A functional CAR directed against B7-H3 has been engineered, utilizing the scFv derived from MGA271. The efficacy of this CAR is currently under investigation in neuroblastoma [108, 109].

ALK is a potential oncogene [110–112] that is expressed in the central and peripheral nervous system [113]. ALK-targeted CAR-T cells, incorporating a 4-1BB costimulatory domain, have demonstrated in vitro activity. However, these ALK CAR-T cells had limited efficacy in vivo in neuroblastoma xenograft models, which may be attributed to the low threshold of ALK expression required for optimal CAR-T cell activity [114–121].

ClinicalTrials.gov ID	Study title	Status/phase/date of start	Source of NK cells	Description
NCT02573896	A Phase I Dose Escalation Study of Autologous Expanded NK Cells for Immunotherapy of Relapsed Refractory Neuroblastoma With Dinutuximab +/- Lenalidomide	Active, Not recruiting/Phase I/2019	Autologous	To determine the maximum tolerated dose (MTD) of autologous expanded NK cells when combined with standard dosing of dinutuximab and assess the feasibility of adding lenalidomide at the recommended Phase II dose of the expanded NK cells with dinutuximab
NCT03242603	Pilot Study of Anti-GD2 and Expanded, Activated Natural Killer Cell Infusion for Neuroblastoma	Phase I-II/2017	Haploidentical	To determine the safety and efficacy of expanded activated haploidentical NK cells in combination with anti-GD2 (ch14.18/CHO)
NCT05754684	Quadruple Immunotherapy for Pediatric Patients With Relapsed or Refractory Neuroblastoma	Recruiting/Phase II/2022	Haploidentical	To evaluate the efficacy and safety of quadruple immunotherapy with NK cells, anti-GD2 antibody, cytokines (interleukin-2 (IL-2) and granulocyte-macrophage colony stimulating factor (GM-CSF)) and retinoid X receptor gamma (RXRg) agonist spironolactone
NCT01875601	A Phase I Study of Autologous Activated NK Cells +/- rhIL15 in Children and Young Adults With Refractory Solid Tumors	Completed/Phase I/2013	Autologous	Outcome: Stable disease
NCT02100891	Phase II Solid Tumor Immunotherapy Trial Using HLA-Haploidentical Transplant and Donor Natural Killer Cells: The STIR Trial	Completed/Phase II/2013	Haploidentical	HLA-haploidentical HSCT followed by an early, post-transplant infusion of donor NK cells on day +7
NCT02130869	A Pilot Study of Immunotherapy Including Haploidentical NK Cell Infusion Following CD133+ Positively-Selected Autologous Hematopoietic Stem Cells in Children With High Risk Solid Tumors or Lymphomas	Completed/Phase I/2014	Haploidentical	Addition of haploidentical natural killer cell infusion to autologous stem cell transplantation/natural killer cells from a haploidentical family member will be given after high dose chemotherapy and positively selected autologous stem cells/anti-GD2 antibody hu14.18K322A will also be given
NCT04211675	A Phase I/II Safety Lead in Study of Ex-Vivo Expanded Allogeneic Universal Donor TGFβi NK Cell Infusions in Combination With Irinotecan, Temozolomide, and Dinutuximab in Patients With Relapsed or Refractory Neuroblastoma: The Allo - STING Trial	Phase I-II/2022	Recruiting/Allogeneic Universal Donor	Phase 1 will assess the safety and tolerability of universal donor TGFβi NK cell in combination with irinotecan, temozolomide, and dinituximab. Phase 2 of the study will estimate the response to treatment

ClinicalTrials. gov ID	Study title	Status/phase/ date of start	Source of NK cells	Description
NCT01576692	A Safety/Feasibility Trial of the Addition of the Humanized Anti-GD2 Antibody (hu14.18K322A) With and Without Natural Killer Cells to Chemotherapy in Children and Adolescents With Recurrent/Refractory Neuroblastoma	Completed/ Phase I/2012	Haploidentical	A safety/feasibility trial evaluating the combination of a humanized anti-GD2 antibody (HU14.18K322A) with allogeneic NK cells and standard chemotherapy
NCT00877110	Phase I Study of Anti-GD2 3F8 Antibody and Allogeneic Natural Killer Cells for High-Risk Neuroblastoma	Completed/ Phase I/2009	Haploidentical	To see if it is safe and feasible to give chemotherapy, NK cells, and an antibody called 3F8
NCT02650648	Phase I Study of the Humanized Anti-GD2 Antibody Hu3F8 and Allogeneic Natural Killer Cells for High-Risk Neuroblastoma	Active, Not recruiting/Phase I/2016	Haploidentical	Safety and feasibility of combining HLA-mismatched (KIR ligand incompatible) NK cells with hu3F8 in high-risk NB patients Following chemotherapy, patients will be treated in sequential groups with a minimum of three patients/dose of NK cells.
NCT02100891	Phase II Solid Tumor Immunotherapy Trial Using HLA-Haploidentical Transplant and Donor Natural Killer Cells: The STIR Trial	Completed/ Phase II/2013	Haploidentical	Phase 2 cellular and adoptive immunotherapy study using HLA-haploidentical HSCT followed by an early, post-transplant infusion of donor NK cells on Day +7

Table 1.
Summary of NK cell immunotherapy clinical trials for neuroblastoma.

Neural Cell Adhesion Molecule (NCAM/CD56) is a glycoprotein that is overexpressed on neuroblastoma [122]. A second-generation CAR with a CD28 costimulation domain targeting CD56 was able to control tumor burden in a neuroblastoma xenograft model but had only modest effects on survival. Since CD56 is significantly expressed on normal tissue, the potential off-tumor toxicity of this approach will need to be carefully monitored [123]. **Table 2** shows a summary of CAR-T-cell immunotherapy clinical trials in patients with NB.

4.2 Brain tumors

Pediatric central nervous system (CNS) tumors are the second most common childhood malignancy and the most common solid tumor in children with a high mortality rate. They are classified by the World Health Organization (WHO) as low-grade gliomas (LGG), high-grade gliomas (HGG), ependymomas, and embryonal tumors [129, 130].

These tumors continue to be the leading cause of pediatric cancer-related mortality, highlighting the critical necessity for the development of novel treatments [131]. These innovative therapies must be tailored to target malignant cells specifically, reducing off-target cytotoxicity commonly associated with traditional chemotherapeutic agents, while effectively and persistently exerting cytotoxic effects on cancer cells to reduce the risk of recurrence. It seems that CAR-T cells may hold promise in achieving these objectives [132].

Several clinical trials are currently underway to evaluate the efficacy of CAR-T cell therapy in solid tumors, with brain tumors being the most common solid tumor type currently undergoing such clinical testing. These early trials have shown promise in treating glioblastoma (GBM) [133]. However, several key challenges remain in achieving success with CAR-T cell therapy for solid tumors, which are particularly prominent in brain tumors. These include difficulty in trafficking to the tumor site, the presence of an immunosuppressive tumor microenvironment, potential toxicity, and tumor antigen heterogeneity [134]. The unique properties of the blood-brain barrier (BBB) and the blood-tumor barrier (BTB) formed by brain tumors further hinder the delivery of many therapeutics to these tumors, which have heterogeneous perfusion and permeability throughout [135, 136]. **Table 3** displays a summary of CAR-T-cell immunotherapy clinical trials in patients with brain tumors.

4.3 Sarcomas

Soft tissue sarcomas and malignant bone tumors accounting for 20% of solid tumors diagnosed among children and 13% among adolescent and young adults. The most common pediatric sarcomas are osteosarcoma (OS), Ewing sarcoma (EWS), and rhabdomyosarcoma [137].

While the majority of children with localized sarcomas can be successfully treated with conventional approaches like chemotherapy, radiation, and/or surgery, the prognosis for patients with recurrent or metastatic disease remains poor. Additionally, the significant toxicity associated with traditional therapies makes immunotherapy an appealing alternative treatment option [3]. However, due to the diverse nature of sarcomas, identifying optimal TAA that are widely expressed has been challenging.

Osteosarcoma is one of the most common primary malignancies of the bone in children and adolescents. Researchers have explored several potential

ClinicalTrials. gov ID	Study title	Status/phase/date of start	Target	Signaling domains	Response/ toxicity
NCT00085930 [124, 125]	Administration of Peripheral Blood T-Cells and EBV Specific CTLs Transduced to Express GD-2 Specific Chimeric T Cell Receptors to Patients With Neuroblastoma	Active, Not Recruiting/Phase I/2003-04	GD2	CD3ζ	CR in 3/19 patients PR in 1/19 patients No dose limiting toxicity
NCT01822652 [126]	Autologous Activated T-Cells Transduced With A 3rd Generation GD-2 Chimeric Antigen Receptor And iCaspase9 Safety Switch Administered To Patients With Relapsed Or Refractory Neuroblastoma (GRAIN)	Active, Not Recruiting/Phase I/2013-08	GD2	CD28.OX40. CD3ζ	No objective responses No dose limiting toxicity
NCT02311621	A Phase I Feasibility and Safety Study of Cellular Immunotherapy for Recurrent/ Refractory Neuroblastoma Using Autologous T-cells Lentivirally Transduced to Express CD171-specific Chimeric Antigen Receptors	Active, Not Recruiting/Phase I/2014-11-25	L1-CAM	4-1BB. CD3ζ; CD28.4-1BB. CD3ζ	Results not published
NCT02761915 [127]	A Cancer Research UK Phase I Trial of Anti-GD2 Chimeric Antigen Receptor (CAR) Transduced T-cells (1RG-CART) in Patients With Relapsed or Refractory Neuroblastoma	Completed/Phase I/2016-02-29	GD2	CD28. CD3ζ	No response in 6/9 patients PD in 3/9 patients No dose limiting toxicity
NCT02765243 [128]	Anti-GD2 4th Generation Chimeric Antigen Receptor-modified T Cells (4SCAR-GD2) Targeting Refractory and/or Recurrent Neuroblastoma	Completed/Phase I/2016-01-01	GD2	CD28.4-1BB. CD27. CD3ζ	SD in 38% PR in 15% PD in 47% No grade 3 or 4 toxicity
NCT03373097	Phase I-II Study of Anti-GD2 Chimeric Antigen Receptor-Expressing T Cells in Pediatric Patients Affected by High Risk and/or Relapsed/Refractory Neuroblastoma or Other GD2-positive Solid Tumors	Recruiting/Phase I-II/2018-01-05	GD2	CD28.4-1BB. CD3ζ	Results not published
NCT02107963	A Phase I Trial of T Cells Expressing an Anti-GD2 Chimeric Antigen Receptor in Children and Young Adults With GD2+ Solid Tumors; A 3rd generation anti-GD2-CAR (GD2-CAR.OX40.28.z.ICD9)	Completed /Phase I/ 2014-02-28	GD2	OX40.CD28. CD3ζ	Results not published

DOI: http://dx.doi.org/10.5772/intechopen.1007383

ClinicalTrials.gov ID	Study title	Status/phase/date of start	Target	Signaling domains	Response/toxicity
NCT02919046	Single Arm and Multicenter Clinical Trial to Evaluating the Efficacy and Safety of the Chimeric Antigen Receptor T Cell Immunotherapy (CAR-T) for Relapsed or Refractory Neuroblastoma in Children	Recruiting/NA/2016-09	GD2	CD28.OX40. CD3ζ	Results not published
NCT01460901	Phase I Study of Donor Derived, Gene Modified, Multi-virus-specific, Cytotoxic T-Lymphocytes Redirected to GD2 for Relapsed/Refractory Neuroblastoma Post-allo Stem Cell Transplantation With Submyeloblative Conditioning	Completed/Phase I/2012-10	GD2	CD3ζ	Non-complete response No dose limiting toxicity
NCT03721068	A Phase I Study of Autologous Activated T-Cells Expressing a 2nd Generation GD2 Chimeric Antigen Receptor, IL-15, and iCaspase9 Safety Switch Administered To Patients With Relapsed/Refractory Neuroblastoma or Relapsed/Refractory Osteosarcoma	Recruiting/Phase I/2019-02-19	GD2	2G CAR-T	Results not published
NCT04637503	Multi-center Phase I-II Clinical Trial of 4SCAR-T Therapy Targeting GD2, PSMA and CD276 for Treating Neuroblastoma	Recruiting/Phase I-II/2020-11-18	GD2 PSMA CD276 (B7-H3)	4SCAR-T	Results not published
NCT04483778	Phase I Study of B7-H3 CAR-T Cell Immunotherapy for Recurrent/Refractory Solid Tumors in Children and Young Adults	Recruiting/Phase I/2020-07-13	CD276 (B7-H3)	4-1BB. CD3ζ	Results not published
NCT04897321	B7-H3-Specific Chimeric Antigen Receptor Autologous T-Cell Therapy for Pediatric Patients With Solid Tumors (3CAR)	Recruiting/Phase I/2022-07-06	CD276 (B7-H3)	CD28.4-1BB. CD3ζ	Results not published

Source: Refs. [124–128].

Table 2.
Summary of CAR-T cell clinical trials for neuroblastoma.

ClinicalTrials.gov ID	Study title	Status/phase/date of start	Target	Delivery
NCT05298995	Phase I Study of Anti-GD2 Chimeric Antigen Receptor-Expressing T Cells in Pediatric and Young Adult Patients Affected by Relapsed/Refractory Central Nervous System Tumors	Recruiting/Phase I/ 2023-11-09	GD2	Intravenous injections
NCT03638167	Phase I Study of EGFR806 specific CAR-T Cell Locoregional Immunotherapy for EGFR-positive Recurrent or Refractory Pediatric Central Nervous System Tumors	Active, Not recruiting/Phase I /2019-03-19	EGFR	Via an indwelling catheter into the tumor cavity or the ventricular system
NCT03500991	Phase I Study of HER2-Specific CAR-T Cell Locoregional Immunotherapy for HER2 Positive Recurrent/Refractory Pediatric Central Nervous System Tumors	Active, Not recruiting/Phase I/2018-07-26	HER2	Delivered by an indwelling catheter in the tumor resection cavity or ventricular system
NCT04185038	Phase I Study of B7-H3-Specific CAR-T Cell Locoregional Immunotherapy for Diffuse Intrinsic Pontine Glioma/Diffuse Midline Glioma and Recurrent or Refractory Pediatric Central Nervous System Tumors	Recruiting/Phase I/2019-12-11	B7-H3 (CD276)	Via an indwelling catheter into the tumor resection cavity or ventricular system
NCT05768880	Phase I Study of B7-H3, EGFR806, HER2, And IL13-Zetakine (Quad) CAR-T Cell Locoregional Immunotherapy For Pediatric Diffuse Intrinsic Pontine Glioma, Diffuse Midline Glioma, And Recurrent Or Refractory Central Nervous System Tumors	Recruiting/Phase I/2023-05-05	Combinations of B7-H3, EGFR806, HER2, and IL13-zetakine	Via an indwelling catheter into the ventricular system
NCT02442297	Phase I Study of Intracranial Injection of T Cells Expressing HER2-specific Chimeric Antigen Receptors (CAR) in Subjects With HER2-Positive Tumors of the Central Nervous System (iCAR)	Active, Not recruiting/Phase I /2016-04-01	HER2	Intracranial injection
NCT04099797	Phase I Study of Autologous T Lymphocytes Expressing GD2-specific Chimeric Antigen and Constitutively Active IL-7 Receptors for the Treatment of Patients With GD2-expressing Brain Tumors (GAIL-B)	Recruiting/Phase I/2020-02-03	GD2	IV infusions will be combined with infusions directly into the brain through the Ommaya reservoir or programmable VP shunt.
NCT04196413	Phase I Clinical Trial of Autologous GD2 Chimeric Antigen Receptor (CAR) T Cells (GD2CART) for Diffuse Intrinsic Pontine Gliomas (DIPG) and Spinal Diffuse Midline Glioma (DMG)	Recruiting/Phase I/2020-06-04	GD2	Intravenously and intracerebroventricularly

Table 3.
Summary of CAR-T cell clinical trials for brain tumors.

CAR-T-cell therapy targets in osteosarcoma, including HER-2(human epidermal growth factor receptor 2), GD2, B7-H3, IL-11 Ra (Interleukin 11 Receptor Subunit Alpha), IGF1R (type I insulin-like growth factor receptor), and EphA2 (hepatocellular receptor tyrosine kinase class A2). While HER-2 expression is generally low in osteosarcoma [138], a phase I/II trial demonstrated that HER-2-targeted CAR-T cells can effectively penetrate and persist within the tumor [139]. GD2 expression is higher in osteosarcoma compared to neuroblastoma [140], and GD2-directed CAR-T cells were able to destroy GD2-positive osteosarcoma cells in vitro. However, the approach was ineffective in vivo in an osteosarcoma xenograft model, potentially due to the accumulation of MDSCs [141]. The fourth generation of CAR-T cells may be more successful at eliminating GD2-expressing osteosarcoma cells, but the interaction between the CAR-T cells and tumor cells could lead to upregulation of PD-L1 in the tumor and increased PD-1 levels in the CAR-T cells, resulting in CAR-T-cell apoptosis. Reducing PD-L1 expression on the tumor cells may significantly enhance the efficacy of these fourth-generation CAR-T cells [142]. Another target being investigated is B7-H3, and higher doses of B7-H3-directed CAR-T cells have shown more favorable outcomes [143]. A phase I trial is currently underway to determine the maximum tolerated dose of B7-H3-CAR-T cells [144].

Ewing's sarcoma, arises from mesenchymal stem cells, is the second most frequent primary bone tumor after osteosarcoma that occurs predominantly in children and young adults [145]. Targets currently identified for CAR-T-cell therapy in EWS comprise the vascular endothelial growth factor receptor 2 (VEGFR2), IGF1R, receptor tyrosine kinase-like orphan receptor 1 (ROR1), GD2, B7-H3 (CD276), EphA2, and natural-killer group 2D ligands (NKG2D ligands) [146].

Regarding the GD2 target, two clinical trials are underway for treating Ewing sarcoma using GD2 CAR-T cells. The first trial is a phase I study aimed at determining the maximum tolerated dose of GD2-C7R T cells and assessing their persistence in the blood and their impact on cancer [147]. The second trial is a phase II study to evaluate the safety and efficacy of fourth-generation safety-engineered CAR-T cells targeting GD2-positive sarcomas [148].

B7-H3 is another tumor-associated antigen (TAA) that exhibits high expression in solid tumor cells and has been investigated in clinical trials for Ewing sarcoma. A phase I, open-label, non-randomized study has been developed to assess the safety, feasibility, and effectiveness of administering second-generation CAR-T cells targeting B7H3. The study consists of two arms: in Arm A, patients will receive B7H3-specific CAR-T cells alone, while in Arm B, participants will be treated with CAR-T cells targeting both B7H3 and CD19, a surface marker on B lymphocytes. This approach is based on the hypothesis that CD19+ B cells serve as antigen-presenting cells to T cells, supporting the expansion and persistence of CAR-T cells. The CAR-T cells in Arm A contain the EGFRt protein, while those in Arm B contain the HER2tG protein, both of which are utilized for monitoring and eliminating the CAR-T cells if any adverse effects occur [149].

An additional trial is a Phase I clinical study investigating the application of third-generation B7-H3 CAR-T cells in patients aged 21 years or younger, who have relapsed or refractory B7-H3-positive solid tumors, including Ewing sarcoma. The primary objective of the study is to determine the maximum safe dose of B7-H3 CAR-T cells for patients with B7-H3-positive solid tumors [144].

Other targets identified for CAR-T cell therapy in Ewing sarcoma are currently in the preclinical study phase.

5. Barriers to the efficacy of adoptive cell therapy (ACT) for solid tumors

5.1 Heterogeneity of target antigens

Successful ACT relies on an identifiable antigen that is homogeneously expressed by tumor cells, expressed at much higher levels on tumor cells than healthy cells and associated with manageable on-target/off-tumor effects. These characteristics have contributed to the success of ACT trials in hematologic malignancies [7].

However, solid tumors exhibit greater heterogeneity in their antigen profiles, as the development of these malignancies is not always dependent on the clonal expansion of a single aberrantly expressed antigen, as seen in lymphomas or leukemias [150, 151]. Furthermore, intratumoral heterogeneity is another concern in solid tumors, which may contribute to resistance mechanisms and adaptive compensatory responses, posing additional challenges for cellular therapies targeting solid tumor malignancies [152–155].

5.2 T-cell trafficking

For effector T cells to effectively eliminate malignant cells, they must efficiently migrate from the bloodstream to the tumor site. In hematologic malignancies, this migration is usually well-established [156]. However, in solid tumors, this movement can be impeded by a discrepancy between the chemokine receptors on T-cells and the chemokines secreted by the tumors. Tumors frequently release chemokines that draw in inhibitory cells, forming physical obstacles that hinder the infiltration of antigen-specific T-cells [8, 151].

To overcome the challenge of inefficient T-cell trafficking to solid tumors, researchers have engineered TILs and CAR-T cells to express alternative chemokine receptors [157–159]. Another strategy utilizes oncolytic viruses that either are injected directly into the tumor or possess specific tumor-targeting properties to secrete chemokines that attract T cells to the tumor sites. Although local administration of T cells has been investigated, it can present technical complexities and may not be feasible for treating metastatic tumors [160, 161].

5.3 Immunosuppressive tumor microenvironment

The key factors within the tumor microenvironment (TME) that challenge the clinical efficacy of ACT include the following:

- Immunosuppressive molecules, such as transforming growth factor-β (TGF-β) and PD-1

- Immunosuppressive cell types, including MDSCs, TAMs, and Tregs

- Physical barriers, such as stromal components and fibrotic material

These elements of the TME create a highly suppressive milieu that hinders the effectiveness of ACT [3, 162]. **Figure 3** illustrates the role of immunosuppressive cells in facilitating tumor immune evasion.

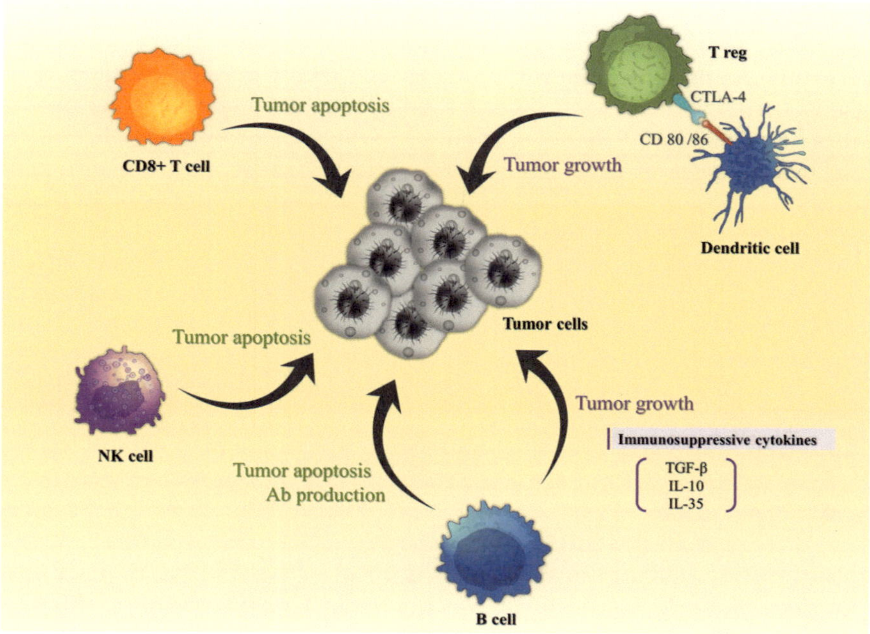

Figure 3.
The role of immunosuppressive cells in facilitating tumor immune evasion. Various cells may induce tumor apoptosis while immune evasion may occur due to immune suppression caused by some cytokines. The interaction of CTLA-4 from regulatory T cells with CD80/86 ligand on dendritic cells also inhibit tumor suppression. Ab, antibody; CTLA-4, cytotoxic T-lymphocyte-associated protein 4; NK, natural killer; TGF, transforming growth factor.

Several novel immunotherapy strategies are being developed to target immune cells within the TME to enhance the body's anti-tumor immune responses:

Blocking immune checkpoints like cytotoxic T-lymphocyte-associated protein 4 (CTLA-4) and PD-1/PD-L1 can relieve the functional inhibition of T cells in the TME.

- Reprogramming TAMs from an immunosuppressive M2 phenotype to an anti-tumor M1 phenotype can lead to reduced immunosuppression and increased CD8+ T-cell responses.
- Dendritic cell-based vaccines can activate T-cell responses by overcoming the inhibition of antigen presentation in the TME.
- Oncolytic viruses injected intratumorally can be engineered to secrete chemokines that attract T cells into the TME.
- Adoptive cell therapies like CAR-T cells can be modified to express chemokine receptors to enhance their trafficking into the tumor.

Targeting the immunosuppressive TME with these strategies aims to reshape the tumor immune microenvironment in a way that allows the body's immune system to more effectively recognize and eliminate cancer cells. This could potentially improve outcomes for patients who are resistant to current immunotherapies [163–165].

6. Conclusion

Recent advancements in adoptive immune cell therapy present a promising avenue for patients battling advanced malignant solid tumors. Despite numerous ongoing clinical trials utilizing ACT in these patients, its widespread clinical application is hindered by challenges in achieving optimal clinical efficacy. Specific characteristics of the pediatric tumor microenvironment and the absence of reliable biomarkers for guiding treatment decisions contribute to these obstacles.

A comprehensive exploration of the composition of the tumor microenvironment (TME) in pediatric solid tumors, along with the identification of tumor-associated antigens (TAAs) that are broadly expressed by tumor cells but not in normal tissues, can play a pivotal role in expanding treatment options, understanding resistance mechanisms, and facilitating the development of innovative functional models for testing therapeutic interventions.

Acknowledgements

The authors would like to thank Mrs. Ashraf Sadat Hosseini, Dr. Mojtaba Azari, Leila Sharifi, Zohreh Shahabi, and Dr. Morteza Azari for contributing to this research. Additionally, we would like to acknowledge the support of the Research Institute for Oncology, Hematology, and Cell Therapy.

Funding

This work was conducted without any funding.

Conflict of interest

The authors declare no conflict of interest.

Appendices and nomenclature

ACT	adoptive cell therapy
ADCC	antibody-dependent cellular cytotoxicity
ALK	anaplastic lymphoma kinase
APC	antigen-presenting cell
BBB	blood-brain barrier
BTB	blood-tumor barrier
CAR	chimeric antigen receptor
CNS	central nervous system
CTL	cytotoxic T-cell
CTLA-4	cytotoxic T-lymphocyte-associated protein 4
DC	dendritic cell
ECM	extracellular matrix
EphA2	hepatocellular receptor tyrosine kinase class A2
EWS	Ewing sarcoma

FDA	Food and Drug Administration
GBM	glioblastoma
GD2	disialoganglioside
GM-CSF	granulocyte-macrophage colony-stimulating factor
GPC2	glypican 2
GvHD	graft versus host disease
HER-2	human epidermal growth factor receptor 2
HGG	high-grade glioma
HLA	human leukocyte antigens
ICH	immune checkpoint inhibitor
IGF1R	type I insulin-like growth factor receptor
IL-11 Ra	interleukin 11 receptor subunit alpha
INRG	International Neuroblastoma Risk Group
KIR	killer cell immunoglobulin-like receptor
L1-CAM	L1 cell adhesion molecule
LGG	low-grade glioma
MDSC	myeloid-derived suppressor cell
MHC	major histocompatibility complex
NB	neuroblastoma
NCAM	neural cell adhesion molecule
NK	natural killer
NKG2D	natural-killer group 2D
NMA	non-myeloablative
OS	osteosarcoma
PBMC	peripheral blood mononuclear cell
PD-1	programmed death-1
PD-L1	programmed death-ligand 1
PFS	progression-free Survival
ROR1	receptor tyrosine kinase-like orphan receptor 1
TAA	tumor-associated Antigen
TAM	tumor-associated macrophage
TCR	T-cell receptor
TGF-β	transforming growth factor-β
TIL	tumor-infiltrating lymphocyte
TME	tumor microenvironment
T-reg	regulatory T lymphocytes
VEGFR2	vascular endothelial growth factor receptor 2
WHO	World Health Organization

Author details

Tahereh Rostami[1*], Azadeh Kiumarsi[2], Mohammad Biglari[3], Mohammad Reza Rostami[1] and Ghasem Janbabaei[1]

1 Hematologic Malignancies Research Center, Research Institute for Oncology, Hematology and Cell Therapy (RIOHCT), Tehran University of Medical Sciences (TUMS), Tehran, Iran

2 Department of Pediatric Hematology–Oncology, Children's Medical Center Hospital, Faculty of Medicine, Tehran University of Medical Sciences (TUMS), Tehran, Iran

3 Hematology, Oncology and Bone Marrow Transplant Research Center, Research Institute for Oncology, Hematology and Cell Therapy (RIOHCT), Tehran University of Medical Sciences (TUMS), Tehran, Iran

*Address all correspondence to: trostami@sina.tums.ac.ir

References

[1] Bear AS, Fraietta JA, Narayan VK, O'Hara M, Haas NB. Adoptive cellular therapy for solid tumors. American Society of Clinical Oncology Educational Book. 2021;**41**:57-65. DOI: 10.1200/EDBK_321115

[2] Zheng W, Ling S, Cao Y, Shao C, Sun X. Combined use of NK cells and radiotherapy in the treatment of solid tumors. Frontiers in Immunology. 2024;**14**:1306534. DOI: 10.3389/fimmu.2023.1306534

[3] Tesfaye M, Savoldo B. Adoptive cell therapy in treating pediatric solid tumors. Current Oncology Reports. 2018;**20**:1-12. DOI: 10.1007/s11912-018-0715-9

[4] Albarrán V, San Román M, Pozas J, Chamorro J, Rosero DI, Guerrero P, et al. Adoptive T cell therapy for solid tumors: Current landscape and future challenges. Frontiers in Immunology. 2024;**15**:1352805. DOI: 10.3389/fimmu.2024.1352805

[5] Cho D, Shook DR, Shimasaki N, Chang Y-H, Fujisaki H, Campana D. Cytotoxicity of activated natural killer cells against pediatric solid tumors. Clinical Cancer Research. 2010;**16**(15):3901-3909. DOI: 10.1158/1078-0432.CCR-10-0735

[6] June CH, O'Connor RS, Kawalekar OU, Ghassemi S, Milone MC. CAR T cell immunotherapy for human cancer. Science. 2018;**359**(6382):1361-1365. DOI: 10.1126/science.aar6711

[7] Schuster SJ, Bishop MR, Tam CS, Waller EK, Borchmann P, McGuirk JP, et al. Tisagenlecleucel in adult relapsed or refractory diffuse large B-cell lymphoma. New England Journal of Medicine. 2019;**380**(1):45-56. DOI: 10.1056/NEJMoa1804980

[8] Neelapu SS, Locke FL, Bartlett NL, Lekakis LJ, Miklos DB, Jacobson CA, et al. Axicabtagene ciloleucel CAR T-cell therapy in refractory large B-cell lymphoma. New England Journal of Medicine. 2017;**377**(26):2531-2544. DOI: 10.1056/NEJMoa1707447

[9] Wang M, Munoz J, Goy A, Locke FL, Jacobson CA, Hill BT, et al. KTE-X19 CAR T-cell therapy in relapsed or refractory mantle-cell lymphoma. New England Journal of Medicine. 2020;**382**(14):1331-1342. DOI: 10.1056/NEJMoa1914347

[10] Gonzalez H, Hagerling C, Werb Z. Roles of the immune system in cancer: From tumor initiation to metastatic progression. Genes & Development. 2018;**32**(19-20):1267-1284. DOI: 10.1101/gad.314617.118

[11] Mantovani A, Allavena P, Sica A, Balkwill F. Cancer-related inflammation. Nature. 2008;**454**(7203):436-444. DOI: 10.1038/nature07205

[12] Chen DS, Mellman I. Oncology meets immunology: The cancer-immunity cycle. Immunity. 2013;**39**(1):1-10. DOI: 10.1016/j.immuni.2013.07.012

[13] Dustin ML. The immunological synapse. Cancer Immunology Research. 2014;**2**(11):1023-1033. DOI: 10.1158/2326-6066.CIR-14-0161

[14] Mills DM, Cambier JC. B lymphocyte activation during cognate interactions with CD4+ T lymphocytes: Molecular dynamics and immunologic consequences. Seminars in Immunology (Academic Press). 1 Dec 2003;**15**(6):325-329

[15] Slaney CY, Kershaw MH, Darcy PK. Trafficking of T cells into tumors. Cancer Research. 2014;**74**(24):7168-7174. DOI: 10.1158/0008-5472.CAN-14-2458

[16] Hadrup S, Donia M, Thor SP. Effector CD4 and CD8 T cells and their role in the tumor microenvironment. Cancer Microenvironment. 2013;**6**:123-133. DOI: 10.1007/s12307-012-0127-6

[17] Wu R, Forget M-A, Chacon J, Bernatchez C, Haymaker C, Chen JQ, et al. Adoptive T-cell therapy using autologous tumor-infiltrating lymphocytes for metastatic melanoma: Current status and future outlook. The Cancer Journal. 2012;**18**(2):160-175. DOI: 10.1097/PPO.0b013e31824d4465

[18] Li F, Li C, Cai X, Xie Z, Zhou L, Cheng B, et al. The association between CD8+ tumor-infiltrating lymphocytes and the clinical outcome of cancer immunotherapy: A systematic review and meta-analysis. EClinicalMedicine. 1 Nov 2021;**41**

[19] Dudley ME, Wunderlich JR, Shelton TE, Even J, Rosenberg SA. Generation of tumor-infiltrating lymphocyte cultures for use in adoptive transfer therapy for melanoma patients. Journal of Immunotherapy. 2003;**26**(4):332-342. DOI: 10.1097/00002371-200307000-00005

[20] Junker N, Kvistborg P, Køllgaard T, Thor Straten P, Andersen MH, Svane IM. Tumor associated antigen specific T-cell populations identified in ex vivo expanded TIL cultures. Cellular Immunology. 2012;**273**(1):1-9. DOI: 10.1016/j.cellimm.2011.12.004

[21] Rosenberg SA, Packard BS, Aebersold PM, Solomon D, Topalian SL, Toy ST, et al. Use of tumor-infiltrating lymphocytes and interleukin-2 in the immunotherapy of patients with metastatic melanoma. New England Journal of Medicine. 1988;**319**(25):1676-1680. DOI: 10.1056/NEJM198812223192527

[22] Dudley ME, Yang JC, Sherry R, Hughes MS, Royal R, Kammula U, et al. Adoptive cell therapy for patients with metastatic melanoma: Evaluation of intensive myeloablative chemoradiation preparative regimens. Journal of Clinical Oncology. 2008;**26**(32):5233. DOI: 10.1200/JCO.2008.16.5449

[23] Rosenberg SA, Yang JC, Sherry RM, Kammula US, Hughes MS, Phan GQ, et al. Durable complete responses in heavily pretreated patients with metastatic melanoma using T-cell transfer immunotherapy. Clinical Cancer Research. 2011;**17**(13):4550-4557. DOI: 10.1158/1078-0432.CCR-11-0116

[24] Gattinoni L, Finkelstein SE, Klebanoff CA, Antony PA, Palmer DC, Spiess PJ, et al. Removal of homeostatic cytokine sinks by lymphodepletion enhances the efficacy of adoptively transferred tumor-specific CD8+ T cells. The Journal of Experimental Medicine. 2005;**202**(7):907-912. DOI: 10.1084/jem.20050732

[25] Antony PA, Piccirillo CA, AL A, Finkelstein SE, Speiss PJ, Surman DR, et al. CD8+ T cell immunity against a tumor/self-antigen is augmented by CD4+ T helper cells and hindered by naturally occurring T regulatory cells. The Journal of Immunology. 2005;**174**(5):2591-2601. DOI: 10.4049/jimmunol.174.5.2591

[26] Zhao Y, Deng J, Rao S, Guo S, Shen J, Du F, et al. Tumor infiltrating lymphocyte (TIL) therapy for solid tumor treatment: Progressions and challenges. Cancers. 2022;**14**(17):4160. DOI: 10.3390/cancers14174160

[27] Qin SS, Melucci AD, Chacon AC, Prieto PA. Adoptive T cell therapy for solid tumors: Pathway to personalized standard of care. Cells. 2021;**10**(4):808. DOI: 10.3390/cells10040808

[28] Tran E, Robbins PF, Rosenberg SA. 'Final common pathway' of human cancer immunotherapy: Targeting random somatic mutations. Nature Immunology. 2017;**18**(3):255-262. DOI: 10.1038/ni.3682

[29] Martincorena I, Campbell PJ. Somatic mutation in cancer and normal cells. Science. 2015;**349**(6255):1483-1489. DOI: 10.1126/science.aab4082

[30] Alexandrov LB, Stratton MR. Mutational signatures: The patterns of somatic mutations hidden in cancer genomes. Current Opinion in Genetics & Development. 2014;**24**:52-60. DOI: 10.1016/j.gde.2013.11.014

[31] Beatty GL, Gladney WL. Immune escape mechanisms as a guide for cancer immunotherapy. Clinical Cancer Research. 2015;**21**(4):687-692. DOI: 10.1158/1078-0432.CCR-14-1860

[32] Belgiovine C, Mebelli K, Raffaele A, De Cicco M, Rotella J, Pedrazzoli P, et al. Pediatric solid cancers: Dissecting the tumor microenvironment to improve the results of clinical immunotherapy. International Journal of Molecular Sciences. 2024;**25**(6):3225. DOI: 10.3390/ijms25063225

[33] van der Woude LL, Gorris MA, Halilovic A, Figdor CG, de Vries IJM. Migrating into the tumor: A roadmap for T cells. Trends in Cancer. 2017;**3**(11):797-808. DOI: 10.1016/j.trecan.2017.09.006

[34] Gabrilovich DI, Ostrand-Rosenberg S, Bronte V. Coordinated regulation of myeloid cells by tumours. Nature Reviews Immunology. 2012;**12**(4):253-268. DOI: 10.1038/nri3175

[35] Ohue Y, Nishikawa H. Regulatory T (Treg) cells in cancer: Can Treg cells be a new therapeutic target? Cancer Science. 2019;**110**(7):2080-2089. DOI: 10.1111/cas.14069

[36] Van der Leun AM, Thommen DS, Schumacher TN. CD8+ T cell states in human cancer: Insights from single-cell analysis. Nature Reviews Cancer. 2020;**20**(4):218-232. DOI: 10.1038/s41568-019-0235-4

[37] Shafer P, Kelly LM, Hoyos V. Cancer therapy with TCR-engineered T cells: Current strategies, challenges, and prospects. Frontiers in Immunology. 2022;**13**:835762. DOI: 10.3389/fimmu.2022.835762

[38] DembiĆ Z, Haas W, Weiss S, McCubrey J, Kiefer H, von Boehmer H, et al. Transfer of specificity by murine α and β T-cell receptor genes. Nature. 1986;**320**(6059):232-238. DOI: 10.1038/320232a0

[39] Sadelain M, Rivière I, Riddell S. Therapeutic T cell engineering. Nature. 2017;**545**(7655):423-431. DOI: 10.1038/nature22395

[40] Groscurth P, Filgueira L. Killing mechanisms of cytotoxic T lymphocytes. Physiology. 1998;**13**(1):17-21. DOI: 10.1152/physiologyonline.1998.13.1.17

[41] Halle S, Halle O, Förster R. Mechanisms and dynamics of T cell-mediated cytotoxicity in vivo. Trends in Immunology. 2017;**38**(6):432-443. DOI: 10.1016/j.it.2017.04.002

[42] Baulu E, Gardet C, Chuvin N, Depil S. TCR-engineered T cell therapy

in solid tumors: State of the art and perspectives. Science Advances. 2023;**9**:eadf3700. DOI: 10.1126/sciadv.adf3700

[43] Sun Y, Li F, Sonnemann H, Jackson KR, Talukder AH, Katailiha AS, et al. Evolution of CD8+ T cell receptor (TCR) engineered therapies for the treatment of cancer. Cells. 2021;**10**(9):2379. DOI: 10.3390/cells10092379

[44] June CH, Sadelain M. Chimeric antigen receptor therapy. New England Journal of Medicine. 2018;**379**(1):64-73. DOI: 10.1056/NEJMra1706169

[45] Snyder A, Makarov V, Merghoub T, Yuan J, Zaretsky JM, Desrichard A, et al. Genetic basis for clinical response to CTLA-4 blockade in melanoma. New England Journal of Medicine. 2014;**371**(23):2189-2199. DOI: 10.1056/NEJMoa1406498

[46] Zhang C, Liu J, Zhong JF, Zhang X. Engineering car-t cells. Biomarker Research. 2017;**5**:1-6. DOI: 10.1186/s40364-017-0102-y

[47] Kulczycka M, Derlatka K, Tasior J, Lejman M, Zawitkowska J. CAR T-cell therapy in children with solid tumors. Journal of Clinical Medicine. 2023;**12**(6):2326. DOI: 10.3390/jcm12062326

[48] Firor AE, Jares A, Ma Y. From humble beginnings to success in the clinic: Chimeric antigen receptor-modified T-cells and implications for immunotherapy. Experimental Biology and Medicine. 2015;**240**(8):1087-1098. DOI: 10.1177/1535370215584936

[49] Smith AJ, Oertle J, Warren D, Prato D. Chimeric antigen receptor (CAR) T cell therapy for malignant cancers: Summary and perspective. Journal of Cellular Immunotherapy. 2016;**2**(2):59-68. DOI: 10.1016/j.jocit.2016.08.001

[50] Imai C, Mihara K, Andreansky M, Nicholson I, Pui C, Geiger T, et al. Chimeric receptors with 4-1BB signaling capacity provoke potent cytotoxicity against acute lymphoblastic leukemia. Leukemia. 2004;**18**(4):676-684. DOI: 10.1038/sj.leu.2403302

[51] Maher J, Brentjens RJ, Gunset G, Rivière I, Sadelain M. Human T-lymphocyte cytotoxicity and proliferation directed by a single chimeric TCRζ/CD28 receptor. Nature Biotechnology. 2002;**20**(1):70-75. DOI: 10.1038/nbt0102-70

[52] Finney HM, Lawson AD, Bebbington CR, Weir ANC. Chimeric receptors providing both primary and costimulatory signaling in T cells from a single gene product. The Journal of Immunology. 1998;**161**(6):2791-2797. DOI: 10.4049/jimmunol.172.1.104

[53] Acuto O, Michel F. CD28-mediated co-stimulation: A quantitative support for TCR signalling. Nature Reviews Immunology. 2003;**3**(12):939-951. DOI: 10.1038/nri1248

[54] Hombach A, Wieczarkowiecz A, Marquardt T, Heuser C, Usai L, Pohl C, et al. Tumor-specific T cell activation by recombinant immunoreceptors: CD3ζ signaling and CD28 costimulation are simultaneously required for efficient IL-2 secretion and can be integrated into one combined CD28/CD3ζ signaling receptor molecule. The Journal of Immunology. 2001;**167**(11):6123-6131. DOI: 10.4049/jimmunol.167.11.6123

[55] Jenkins MK, Burrell E, Ashwell JD. Antigen presentation by resting B cells. Effectiveness at inducing T cell proliferation is determined

by costimulatory signals, not T cell receptor occupancy. Journal of immunology (Baltimore, Md: 1950). 1990;**144**(5):1585-1590. DOI: 10.4049/jimmunol.144.5.1585

[56] Krause A, Guo H-F, Latouche J-B, Tan C, Cheung N-KV, Sadelain M. Antigen-dependent CD28 signaling selectively enhances survival and proliferation in genetically modified activated human primary T lymphocytes. The Journal of Experimental Medicine. 1998;**188**(4):619-626. DOI: 10.1084/jem.188.4.619

[57] Finney HM, Akbar AN, Lawson AD. Activation of resting human primary T cells with chimeric receptors: Costimulation from CD28, inducible costimulator, CD134, and CD137 in series with signals from the TCRζ chain. The Journal of Immunology. 2004;**172**(1):104-113. DOI: 10.4049/jimmunol.172.1.104

[58] Tomasik J, Jasiński M, Basak GW. Next generations of CAR-T cells-new therapeutic opportunities in hematology? Frontiers in Immunology. 2022;**13**:1034707. DOI: 10.3389/fimmu.2022.1034707

[59] Chmielewski M, Kopecky C, Hombach AA, Abken H. IL-12 release by engineered T cells expressing chimeric antigen receptors can effectively muster an antigen-independent macrophage response on tumor cells that have shut down tumor antigen expression. Cancer Research. 2011;**71**(17):5697-5706. DOI: 10.1517/14712598.2015.1046430

[60] Chmielewski M, Abken H. TRUCKs: The fourth generation of CARs. Expert Opinion on Biological Therapy. 2015;**15**(8):1145-1154. DOI: 10.1517/14712598.2015.1046430

[61] Rafiq S, Hackett CS, Brentjens RJ. Engineering strategies to overcome the current roadblocks in CAR T cell therapy. Nature Reviews Clinical Oncology. 2020;**17**(3):147-167. DOI: 10.1038/s41571-019-0297-y

[62] Asmamaw Dejenie T, Tiruneh G, Medhin M, Dessie Terefe G, Tadele Admasu F, Wale Tesega W, et al. Current updates on generations, approvals, and clinical trials of CAR T-cell therapy. Human Vaccines & Immunotherapeutics. 2022;**18**(6):2114254. DOI: 10.1080/21645515.2022.2114254

[63] Kershaw MH, Westwood JA, Darcy PK. Gene-engineered T cells for cancer therapy. Nature Reviews Cancer. 2013;**13**(8):525-541. DOI: 10.1038/nrc3565

[64] Fehniger TA, Cooper MA, Nuovo GJ, Cella M, Facchetti F, Colonna M, et al. CD56bright natural killer cells are present in human lymph nodes and are activated by T cell–derived IL-2: A potential new link between adaptive and innate immunity. Blood, The Journal of the American Society of Hematology. 2003;**101**(8):3052-3057. DOI: 10.1182/blood-2002-09-2876

[65] Ferlazzo G, Thomas D, Lin S-L, Goodman K, Morandi B, Muller WA, et al. The abundant NK cells in human secondary lymphoid tissues require activation to express killer cell Ig-like receptors and become cytolytic. The Journal of Immunology. 2004;**172**(3):1455-1462. DOI: 10.4049/jimmunol.172.3.1455

[66] Yokoyama WM, Kim S, French AR. The dynamic life of natural killer cells. Annual Review of Immunology. 2004;**22**:405-429. DOI: 10.1146/annurev.immunol.22.012703.104711

[67] Campbell KS, Hasegawa J. Natural killer cell biology: An update and future

directions. Journal of Allergy and Clinical Immunology. 2013;**132**(3):536-544. DOI: 10.1016/j.jaci.2013.07.006

[68] Ljunggren H-G, Kärre K. In search of the 'missing self': MHC molecules and NK cell recognition. Immunology Today. 1990;**11**:237-244. DOI: 10.1016/0167-5699(90) 90097-S

[69] Anfossi N, André P, Guia S, Falk CS, Roetynck S, Stewart CA, et al. Human NK cell education by inhibitory receptors for MHC class I. Immunity. 2006;**25**(2):331-342. DOI: 10.1016/j.immuni.2006.06.013

[70] Chiossone L, Dumas P-Y, Vienne M, Vivier E. Natural killer cells and other innate lymphoid cells in cancer. Nature Reviews Immunology. 2018;**18**(11):671-688. DOI: 10.1038/s41577-018-0061-z

[71] Bachanova V, Miller JS. NK cells in therapy of cancer. Critical Reviews in Oncogenesis. 2014;**19**(1-2). DOI: 10.1615/CritRevOncog.2014011091

[72] Ciccone E, Pende D, Viale O, Than A, Di Donato C, Orengo A, et al. Involvement of HLA class I alleles in natural killer (NK) cell-specific functions: Expression of HLA-Cw3 confers selective protection from lysis by alloreactive NK clones displaying a defined specificity (specificity 2). The Journal of Experimental Medicine. 1992;**176**(4):963-971. DOI: 10.1084/jem.176.4.963

[73] Colonna M, Brooks EG, Falco M, Ferrara GB, Strominger JL. Generation of allospecific natural killer cells by stimulation across a polymorphism of HLA-C. Science. 1993;**260**(5111):1121-1124. DOI: 10.1126/science.849355

[74] Moretta A, Tambussi G, Bottino C, Tripodi G, Merli A, Ciccone E, et al. A novel surface antigen expressed by a subset of human CD3-CD16+ natural killer cells. Role in cell activation and regulation of cytolytic function. The Journal of Experimental Medicine. 1990;**171**(3):695-714. DOI: 10.1084/jem.171.3.695

[75] Nimmerjahn F, Ravetch JV. Fcγ receptors as regulators of immune responses. Nature Reviews Immunology. 2008;**8**(1):34-47. DOI: 10.1038/nri2206

[76] Morvan MG, Lanier LL. NK cells and cancer: You can teach innate cells new tricks. Nature Reviews Cancer. 2016;**16**(1):7-19. DOI: 10.1038/nrc.2015.5

[77] Moradi V, Omidkhoda A, Ahmadbeigi N. The paths and challenges of "off-the-shelf" CAR-T cell therapy: An overview of clinical trials. Biomedicine & Pharmacotherapy. 2023;**169**:115888. DOI: 10.1016/j.biopha.2023.115888

[78] Marofi F, Al-Awad AS, Sulaiman Rahman H, Markov A, Abdelbasset WK, Ivanovna Enina Y, et al. Retracted: CAR-NK cell: A new paradigm in tumor immunotherapy. Frontiers in Oncology. 10 Jun 2021;**11**:673276

[79] Pan K, Farrukh H, Chittepu VCSR, Xu H, Pan C-X, Zhu Z. CAR race to cancer immunotherapy: From CAR T, CAR NK to CAR macrophage therapy. Journal of Experimental & Clinical Cancer Research. 2022;**41**(1):119. DOI: 10.1186/s13046-022-02327-z

[80] Alvarez M, Simonetta F, Baker J, Morrison AR, Wenokur AS, Pierini A, et al. Indirect impact of PD-1/PD-L1 blockade on a murine model of NK cell exhaustion. Frontiers in Immunology. 2020;**11**:7. DOI: 10.3389/fimmu.2020.00007

[81] Johnsen JI, Dyberg C, Wickström M. Neuroblastoma—A neural crest derived embryonal malignancy. Frontiers in

Molecular Neuroscience. 2019;**12**:9. DOI: 10.3389/fnmol.2019.00009

[82] Maris JM, Hogarty MD, Bagatell R, Cohn SL. Neuroblastoma. Lancet. 2007;**369**(9579):2106-2120. DOI: 10.1016/S0140-6736(07)60983-0

[83] Irwin MS, Naranjo A, Zhang FF, Cohn SL, London WB, Gastier-Foster JM, et al. Revised neuroblastoma risk classification system: A report from the children's oncology group. Journal of Clinical Oncology. 2021;**39**(29):3229-3241. DOI: 10.1200/JCO.21.00278

[84] Morandi F, Sabatini F, Podestà M, Airoldi I. Immunotherapeutic strategies for neuroblastoma: Present, past and future. Vaccines. 13 Jan 2021;**9**(1):43

[85] Pinto NR, Applebaum MA, Volchenboum SL, Matthay KK, London WB, Ambros PF, et al. Advances in risk classification and treatment strategies for neuroblastoma. Journal of Clinical Oncology. 2015;**33**(27):3008-3017. DOI: 10.1200/JCO.2014.59.4648

[86] Brenner MK, Heslop H, Krance R, Horowitz M, Strother D, Nuchtern J, et al. Phase I study of chemokine and cytokine gene-modified autologous neuroblastoma cells for treatment of relapsed/refractory neuroblastoma using an adenoviral vector. Human Gene Therapy. 2000;**11**(10):1477-1488. DOI: 10.1089/10430340050057549

[87] Handgretinger R, Baader P, Dopfer R, Klingebiel T, Reuland P, Treuner J, et al. A phase I study of neuroblastoma with the anti-ganglioside GD2 antibody 14.G2a. Cancer Immunology, Immunotherapy. 1992;**35**(3):199-204. DOI: 10.1007/BF01756188

[88] Kushner BH, Kramer K, Cheung NK. Phase II trial of the anti-G(D2) monoclonal antibody 3F8 and granulocyte-macrophage colony-stimulating factor for neuroblastoma. Journal of Clinical Oncology. 2001;**19**(22):4189-4194. DOI: 10.1200/JCO.2001.19.22.4189

[89] Ozkaynak MF, Sondel PM, Krailo MD, Gan J, Javorsky B, Reisfeld RA, et al. Phase I study of chimeric human/murine anti-ganglioside G(D2) monoclonal antibody (ch14.18) with granulocyte-macrophage colony-stimulating factor in children with neuroblastoma immediately after hematopoietic stem-cell transplantation: A children's cancer group study. Journal of Clinical Oncology. 2000;**18**(24):4077-4085. DOI: 10.1200/JCO.2000.18.24.4077

[90] Rousseau RF, Haight AE, Hirschmann-Jax C, Yvon ES, Rill DR, Mei Z, et al. Local and systemic effects of an allogeneic tumor cell vaccine combining transgenic human lymphotactin with interleukin-2 in patients with advanced or refractory neuroblastoma. Blood. 2003;**101**(5):1718-1726. DOI: 10.1182/blood-2002-08-2493

[91] Moretta L, Moretta A. Killer immunoglobulin-like receptors. Current Opinion in Immunology. 2004;**16**(5):626-633. DOI: 10.1016/j.coi.2004.07.010

[92] Castriconi R, Dondero A, Corrias MV, Lanino E, Pende D, Moretta L, et al. Natural killer cell-mediated killing of freshly isolated neuroblastoma cells: Critical role of DNAX accessory molecule-1-poliovirus receptor interaction. Cancer Research. 2004;**64**(24):9180-9184. DOI: 10.1158/0008-5472.CAN-04-2682

[93] Castriconi R, Dondero A, Augugliaro R, Cantoni C, Carnemolla B, Sementa AR, et al. Identification of 4Ig-B7-H3 as a neuroblastoma-associated molecule that exerts a protective role from an

NK cell-mediated lysis. Proceedings of the National Academy of Sciences of the United States of America. 2004;**101**(34):12640-12645. DOI: 10.1073/pnas.0405025101

[94] Federico SM, McCarville MB, Shulkin BL, Sondel PM, Hank JA, Hutson P, et al. A pilot trial of humanized anti-GD2 monoclonal antibody (hu14.18K322A) with chemotherapy and natural killer cells in children with recurrent/refractory neuroblastoma. Clinical Cancer Research. 2017;**23**(21):6441-6449. DOI: 10.1158/1078-0432.CCR-17-0379

[95] Modak S, Le Luduec JB, Cheung IY, Goldman DA, Ostrovnaya I, Doubrovina E, et al. Adoptive immunotherapy with haploidentical natural killer cells and anti-GD2 monoclonal antibody m3F8 for resistant neuroblastoma: Results of a phase I study. Oncoimmunology. 2018;**7**(8):e1461305. DOI: 10.1080/2162402X.2018.1461305

[96] Richards RM, Sotillo E, Majzner RG. CAR T cell therapy for neuroblastoma. Frontiers in Immunology. 2018;**9**:2380. DOI: 10.3389/fimmu.2018.02380

[97] Yu AL, Gilman AL, Ozkaynak MF, London WB, Kreissman SG, Chen HX, et al. Anti-GD2 antibody with GM-CSF, interleukin-2, and isotretinoin for neuroblastoma. The New England Journal of Medicine. 2010;**363**(14):1324-1334. DOI: 10.1056/NEJMoa0911123

[98] Cheung NK, Cheung IY, Kramer K, Modak S, Kuk D, Pandit-Taskar N, et al. Key role for myeloid cells: Phase II results of anti-G(D2) antibody 3F8 plus granulocyte-macrophage colony-stimulating factor for chemoresistant osteomedullary neuroblastoma. International Journal of Cancer. 2014;**135**(9):2199-2205. DOI: 10.1002/ijc.28851

[99] Gilman AL, Ozkaynak MF, Matthay KK, Krailo M, Yu AL, Gan J, et al. Phase I study of ch14.18 with granulocyte-macrophage colony-stimulating factor and interleukin-2 in children with neuroblastoma after autologous bone marrow transplantation or stem-cell rescue: A report from the children's oncology group. Journal of Clinical Oncology. 2009;**27**(1):85-91. DOI: 10.1200/JCO.2006.10.3564

[100] Hong H, Stastny M, Brown C, Chang WC, Ostberg JR, Forman SJ, et al. Diverse solid tumors expressing a restricted epitope of L1-CAM can be targeted by chimeric antigen receptor redirected T lymphocytes. Journal of Immunotherapy. 2014;**37**(2):93-104. DOI: 10.1097/CJI.0000000000000018

[101] Bosse KR, Raman P, Zhu Z, Lane M, Martinez D, Heitzeneder S, et al. Identification of GPC2 as an oncoprotein and candidate immunotherapeutic target in high-risk neuroblastoma. Cancer Cell. 2017;**32**(3):295-309.e12. DOI: 10.1016/j.ccell.2017.08.003

[102] Orentas RJ, Yang JJ, Wen X, Wei JS, Mackall CL, Khan J. Identification of cell surface proteins as potential immunotherapy targets in 12 pediatric cancers. Frontiers in Oncology. 2012;**2**:194. DOI: 10.3389/fonc.2012.00194

[103] Li N, Fu H, Hewitt SM, Dimitrov DS, Ho M. Therapeutically targeting glypican-2 via single-domain antibody-based chimeric antigen receptors and immunotoxins in neuroblastoma. Proceedings of the National Academy of Sciences of the United States of America. 2017;**114**(32):E6623-E6e31. DOI: 10.1073/pnas.1706055114

[104] Baral A, Ye HX, Jiang PC, Yao Y, Mao Y. B7-H3 and B7-H1 expression in

cerebral spinal fluid and tumor tissue correlates with the malignancy grade of glioma patients. Oncology Letters. 2014;**8**(3):1195-1201. DOI: 10.3892/ol.2014.2268

[105] Gregorio A, Corrias MV, Castriconi R, Dondero A, Mosconi M, Gambini C, et al. Small round blue cell tumours: Diagnostic and prognostic usefulness of the expression of B7-H3 surface molecule. Histopathology. 2008;**53**(1):73-80. DOI: 10.1111/j.1365-2559.2008.03070.x

[106] Zhao JL, Chen FL, Zhou Q, Pan W, Wang XH, Xu J, et al. B7-H3 protein expression in a murine model of osteosarcoma. Oncology Letters. 2016;**12**(1):383-386. DOI: 10.3892/ol.2016.4675

[107] Zhou Z, Luther N, Ibrahim GM, Hawkins C, Vibhakar R, Handler MH, et al. B7-H3, a potential therapeutic target, is expressed in diffuse intrinsic pontine glioma. Journal of Neuro-Oncology. 2013;**111**(3):257-264. DOI: 10.1007/s11060-012-1021-2

[108] Majzner RG, Theruvath JL, Nellan A, Heitzeneder S, Cui Y, Mount CW, et al. CAR T cells targeting B7-H3, a pan-cancer antigen, demonstrate potent preclinical activity against pediatric solid tumors and brain tumors. Clinical Cancer Research. 2019;**25**(8):2560-2574. DOI: 10.1158/1078-0432.CCR-18-0432

[109] Bocca P, Di Carlo E, Caruana I, Emionite L, Cilli M, De Angelis B, et al. Bevacizumab-mediated tumor vasculature remodelling improves tumor infiltration and antitumor efficacy of GD2-CAR T cells in a human neuroblastoma preclinical model. Oncoimmunology. 2017;**7**(1):e1378843. DOI: 10.1080/2162402X.2017.1378843

[110] Chen Y, Takita J, Choi YL, Kato M, Ohira M, Sanada M, et al. Oncogenic mutations of ALK kinase in neuroblastoma. Nature. 2008;**455**(7215):971-974. DOI: 10.1038/nature07399

[111] George RE, Sanda T, Hanna M, Fröhling S, Luther W 2nd, Zhang J, et al. Activating mutations in ALK provide a therapeutic target in neuroblastoma. Nature. 2008;**455**(7215):975-978. DOI: 10.1038/nature07397

[112] Mossé YP, Laudenslager M, Longo L, Cole KA, Wood A, Attiyeh EF, et al. Identification of ALK as a major familial neuroblastoma predisposition gene. Nature. 2008;**455**(7215):930-935. DOI: 10.1038/nature07261

[113] Iwahara T, Fujimoto J, Wen D, Cupples R, Bucay N, Arakawa T, et al. Molecular characterization of ALK, a receptor tyrosine kinase expressed specifically in the nervous system. Oncogene. 1997;**14**(4):439-449. DOI: 10.1038/sj.onc.1200849

[114] Walker AJ, Majzner RG, Zhang L, Wanhainen K, Long AH, Nguyen SM, et al. Tumor antigen and receptor densities regulate efficacy of a chimeric antigen receptor targeting anaplastic lymphoma kinase. Molecular Therapy. 2017;**25**(9):2189-2201. DOI: 10.1016/j.ymthe.2017.06.008

[115] Arcangeli S, Rotiroti MC, Bardelli M, Simonelli L, Magnani CF, Biondi A, et al. Balance of anti-CD123 chimeric antigen receptor binding affinity and density for the targeting of acute myeloid Leukemia. Molecular Therapy. 2017;**25**(8):1933-1945. DOI: 10.1016/j.ymthe.2017.04.017

[116] Chmielewski M, Hombach A, Heuser C, Adams GP, Abken H. T cell activation by antibody-like immunoreceptors: Increase in affinity

of the single-chain fragment domain above threshold does not increase T cell activation against antigen-positive target cells but decreases selectivity. Journal of Immunology. 2004;**173**(12):7647-7653. DOI: 10.4049/jimmunol.173.12.7647

[117] Watanabe K, Terakura S, Martens AC, van Meerten T, Uchiyama S, Imai M, et al. Target antigen density governs the efficacy of anti-CD20-CD28-CD3 ζ chimeric antigen receptor-modified effector CD8+ T cells. Journal of Immunology. 2015;**194**(3):911-920. DOI: 10.4049/jimmunol.1402346

[118] Hombach AA, Görgens A, Chmielewski M, Murke F, Kimpel J, Giebel B, et al. Superior therapeutic index in lymphoma therapy: CD30(+) CD34(+) hematopoietic stem cells resist a chimeric antigen receptor T-cell attack. Molecular Therapy. 2016;**24**(8):1423-1434. DOI: 10.1038/mt.2016.82

[119] Caruso HG, Hurton LV, Najjar A, Rushworth D, Ang S, Olivares S, et al. Tuning sensitivity of CAR to EGFR density limits recognition of normal tissue while maintaining potent antitumor activity. Cancer Research. 2015;**75**(17):3505-3518. DOI: 10.1158/0008-5472.CAN-15-0139

[120] Turatti F, Figini M, Balladore E, Alberti P, Casalini P, Marks JD, et al. Redirected activity of human antitumor chimeric immune receptors is governed by antigen and receptor expression levels and affinity of interaction. Journal of Immunotherapy. 2007;**30**(7):684-693. DOI: 10.1097/CJI.0b013e3180de5d90

[121] Weijtens ME, Hart EH, Bolhuis RL. Functional balance between T cell chimeric receptor density and tumor associated antigen density: CTL mediated cytolysis and lymphokine production. Gene Therapy. 2000;**7**(1):35-42. DOI: 10.1038/sj.gt.3301051

[122] Wachowiak R, Rawnaq T, Metzger R, Quaas A, Fiegel H, Kähler N, et al. Universal expression of cell adhesion molecule NCAM in neuroblastoma in contrast to L1: Implications for different roles in tumor biology of neuroblastoma? Pediatric Surgery International. 2008;**24**(12):1361-1364. DOI: 10.1007/s00383-008-2264-z

[123] Crossland DL, Denning WL, Ang S, Olivares S, Mi T, Switzer K, et al. Antitumor activity of CD56-chimeric antigen receptor T cells in neuroblastoma and SCLC models. Oncogene. 2018;**37**(27):3686-3697. DOI: 10.1038/s41388-018-0187-2

[124] Pule MA, Savoldo B, Myers GD, Rossig C, Russell HV, Dotti G, et al. Virus-specific T cells engineered to coexpress tumor-specific receptors: persistence and antitumor activity in individuals with neuroblastoma. Nature Medicine. 2008;**14**(11):1264-1270. DOI: 10.1038/nm.1882

[125] Louis CU, Savoldo B, Dotti G, Pule M, Yvon E, Myers GD, et al. Antitumor activity and long-term fate of chimeric antigen receptor–positive T cells in patients with neuroblastoma. Blood, The Journal of the American Society of Hematology. 2011;**118**(23):6050-6056. DOI: 10.1182/blood-2011-05-354449

[126] Heczey A, Louis CU, Savoldo B, Dakhova O, Durett A, Grilley B, et al. CAR T cells administered in combination with lymphodepletion and PD-1 inhibition to patients with neuroblastoma. Molecular Therapy. 2017;**25**(9):2214-2224. DOI: 10.1016/j.ymthe.2017.05.012

[127] Straathof K, Flutter B, Wallace R, Thomas S, Cheung G, Collura A, et al. Abstract CT145: A cancer research UK phase I trial of anti-GD2

chimeric antigen receptor (CAR) transduced T-cells (1RG-CART) in patients with relapsed or refractory neuroblastoma. Cancer Research. 1 Jul 2018;**78**(13_Supplement):CT145

[128] Yang L, Ma X, Liu Y-C, Zhao W, Yu L, Qin M, et al. Chimeric antigen receptor 4SCAR-GD2-modified T cells targeting high-risk and recurrent neuroblastoma: A phase II multi-center trial in China. Blood. 2017;**130**:3335. DOI: 10.1182/blood.V130.Suppl_1.3335.3335

[129] Lutz K, Jünger ST, Messing-Jünger M. Essential management of pediatric brain tumors. Children. 2 Apr 2022;**9**(4):498

[130] Udaka YT, Packer RJ. Pediatric brain tumors. Neurologic Clinics. 2018;**36**(3):533-556. DOI: 10.1016/j.ncl.2018.04.009

[131] Withrow DR, Berrington de Gonzalez A, CJK L, Warren KE, Shiels MS. Trends in pediatric central nervous system tumor incidence in the United States, 1998-2013. Cancer Epidemiology, Biomarkers & Prevention. 2019;**28**(3):522-530. DOI: 10.1158/1055-9965.EPI-18-0784

[132] Patterson JD, Henson JC, Breese RO, Bielamowicz KJ, Rodriguez A. CAR T cell therapy for pediatric brain tumors. Frontiers in Oncology. 2020;**10**:1582. DOI: 10.3389/fonc.2020.01582

[133] Akhavan D, Alizadeh D, Wang D, Weist MR, Shepphird JK, Brown CE. CAR T cells for brain tumors: Lessons learned and road ahead. Immunological Reviews. 2019;**290**(1):60-84. DOI: 10.1111/imr.12773

[134] Ma S, Li X, Wang X, Cheng L, Li Z, Zhang C, et al. Current progress in CAR-T cell therapy for solid tumors. International Journal of Biological Sciences. 2019;**15**(12):2548-2560. DOI: 10.7150/ijbs.34213

[135] Abbott NJ. Blood-brain barrier structure and function and the challenges for CNS drug delivery. Journal of Inherited Metabolic Disease. 2013;**36**(3):437-449. DOI: 10.1007/s10545-013-9608-0

[136] Arvanitis CD, Ferraro GB, Jain RK. The blood-brain barrier and blood-tumour barrier in brain tumours and metastases. Nature Reviews. Cancer. 2020;**20**(1):26-41. DOI: 10.1038/s41568-019-0205-x

[137] Ou JY, Spraker-Perlman H, Dietz AC, Smits-Seemann RR, Kaul S, Kirchhoff AC. Conditional survival of pediatric, adolescent, and young adult soft tissue sarcoma and bone tumor patients. Cancer Epidemiology. 2017;**50**(Pt. A):150-157. DOI: 10.1016/j.canep.2017.08.015

[138] Scotlandi K, Manara MC, Hattinger CM, Benini S, Perdichizzi S, Pasello M, et al. Prognostic and therapeutic relevance of HER2 expression in osteosarcoma and Ewing's sarcoma. European Journal of Cancer. 2005;**41**(9):1349-1361. DOI: 10.1016/j.ejca.2005.03.015

[139] Ahmed N, Brawley VS, Hegde M, Robertson C, Ghazi A, Gerken C, et al. Human epidermal growth factor receptor 2 (HER2)-specific chimeric antigen receptor-modified T cells for the immunotherapy of HER2-positive sarcoma. Journal of Clinical Oncology. 2015;**33**(15):1688-1696. DOI: 10.1200/JCO.2014.58.0225

[140] Roth M, Linkowski M, Tarim J, Piperdi S, Sowers R, Geller D, et al. Ganglioside GD2 as a therapeutic target for antibody-mediated therapy in patients with osteosarcoma. Cancer.

2014;**120**(4):548-554. DOI: 10.1002/cncr.28461

[141] Long AH, Highfill SL, Cui Y, Smith JP, Walker AJ, Ramakrishna S, et al. Reduction of MDSCs with all-trans retinoic acid improves CAR therapy efficacy for sarcomas. Cancer Immunology Research. 2016;**4**(10):869-880. DOI: 10.1158/2326-6066.CIR-15-0230

[142] Chulanetra M, Morchang A, Sayour E, Eldjerou L, Milner R, Lagmay J, et al. GD2 chimeric antigen receptor modified T cells in synergy with sub-toxic level of doxorubicin targeting osteosarcomas. American Journal of Cancer Research. 2020;**10**(2):674-687

[143] Murty S, Labanieh L, Murty T, Gowrishankar G, Haywood T, Alam IS, et al. PET reporter gene imaging and ganciclovir-mediated ablation of chimeric antigen receptor T cells in solid Tumors. Cancer Research. 2020;**80**(21):4731-4740. DOI: 10.1158/0008-5472.CAN-19-3579

[144] ClinicalTrials.gov. Identifier NCT04897321, B7-H3-Specific Chimeric Antigen Receptor Autologous T-Cell Therapy for Pediatric Patients with Solid Tumors (3CAR) [Internet]. Bethesda (MD): National Library of Medicine (US); 2021, 2000. Available from: https://clinicaltrials.gov/ct2/show/NCT04897321

[145] Riggi N, Cironi L, Provero P, Suvà ML, Kaloulis K, Garcia-Echeverria C, et al. Development of Ewing's sarcoma from primary bone marrow-derived mesenchymal progenitor cells. Cancer Research. 2005;**65**(24):11459-11468. DOI: 10.1158/0008-5472.CAN-05-1696

[146] Lin Z, Wu Z, Luo W. A novel treatment for Ewing's sarcoma: Chimeric antigen receptor-T cell therapy. Frontiers in Immunology. 2021;**12**:707211. DOI: 10.3389/fimmu.2021.707211

[147] ClinicalTrials.gov. Identifier NCT03635632, C7R-GD2.CART Cells for Patients with Relapsed or Refractory Neuroblastoma and Other GD2 Positive Cancers (GAIL-N); 2018 Aug 13. [Internet]. Bethesda (MD): National Library of Medicine. (US); 2000. Available from: https://clinicaltrials.gov/ct2/show/NCT03635632

[148] ClinicalTrials.gov. Identifier NCT03356782, Safety and Efficacy Evaluation of 4th Generation Safety-Engineered CAR T Cells Targeting Sarcomas; 2017 Nov 24 [Internet]. Bethesda (MD): National Library of Medicine. (US); 2000. Available from: https://clinicaltrials.gov/ct2/show/NCT03356782

[149] ClinicalTrials.gov. Identifier NCT04483778, B7H3 CAR T Cell Immunotherapy for Recurrent/Refractory Solid Tumors in Children and Young Adults; 2020 Jul 14. [Internet]. Bethesda (MD): National Library of Medicine. (US); 2000 Available from: https://clinicaltrials.gov/ct2/show/NCT04483778

[150] Chen N, Li X, Chintala NK, Tano ZE, Adusumilli PS. Driving CARs on the uneven road of antigen heterogeneity in solid tumors. Current Opinion in Immunology. 2018;**51**:103-110. DOI: 10.1016/j.coi.2018.03.002

[151] Kirtane K, Elmariah H, Chung CH, Abate-Daga D. Adoptive cellular therapy in solid tumor malignancies: Review of the literature and challenges ahead. Journal for Immunotherapy of Cancer. 2021;**9**(7)

[152] Patel AP, Tirosh I, Trombetta JJ, Shalek AK, Gillespie SM, Wakimoto H, et al. Single-cell RNA-seq highlights intratumoral heterogeneity in

primary glioblastoma. Science. 2014;**344**(6190):1396-1401. DOI: 10.1126/science.125425

[153] Ricketts CJ, Linehan WM. Intratumoral heterogeneity in kidney cancer. Nature Genetics. 2014;**46**(3):214-215. DOI: 10.1038/ng.2904

[154] Zhang J, Fujimoto J, Zhang J, Wedge DC, Song X, Zhang J, et al. Intratumor heterogeneity in localized lung adenocarcinomas delineated by multiregion sequencing. Science. 2014;**346**(6206):256-259. DOI: 10.1126/science.1256930

[155] Hutchinson L. Adaptive resistance to CARs in glioma. Nature Reviews Clinical Oncology. Oct 2017;**14**(10):586

[156] Srivastava S, Riddell SR. Chimeric antigen receptor T cell therapy: Challenges to bench-to-bedside efficacy. Journal of Immunology. 2018;**200**(2):459-468. DOI: 10.4049/jimmunol.1701155

[157] Craddock JA, Lu A, Bear A, Pule M, Brenner MK, Rooney CM, et al. Enhanced tumor trafficking of GD2 chimeric antigen receptor T cells by expression of the chemokine receptor CCR2b. Journal of Immunotherapy. 2010;**33**(8):780-788. DOI: 10.1097/CJI.0b013e3181ee6675

[158] Moon EK, Carpenito C, Sun J, Wang LC, Kapoor V, Predina J, et al. Expression of a functional CCR2 receptor enhances tumor localization and tumor eradication by retargeted human T cells expressing a mesothelin-specific chimeric antibody receptor. Clinical Cancer Research. 2011;**17**(14):4719-4730. DOI: 10.1158/1078-0432.CCR-11-0351

[159] Adusumilli PS, Cherkassky L, Villena-Vargas J, Colovos C, Servais E, Plotkin J, et al. Regional delivery of mesothelin-targeted CAR T cell therapy generates potent and long-lasting CD4-dependent tumor immunity. Science Translational Medicine. 2014;**6**(261):261ra151. DOI: 10.1126/scitranslmed.3010162

[160] Newick K, O'Brien S, Sun J, Kapoor V, Maceyko S, Lo A, et al. Augmentation of CAR T-cell trafficking and antitumor efficacy by blocking protein kinase a localization. Cancer Immunology Research. 2016;**4**(6):541-551. DOI: 10.1158/2326-6066.CIR-15-0263

[161] Brown CE, Alizadeh D, Starr R, Weng L, Wagner JR, Naranjo A, et al. Regression of glioblastoma after chimeric antigen receptor T-cell therapy. The New England Journal of Medicine. 2016;**375**(26):2561-2569. DOI: 10.1056/NEJMoa1610497

[162] Liu Z, Zhou Z, Dang Q, Xu H, Lv J, Li H, et al. Immunosuppression in tumor immune microenvironment and its optimization from CAR-T cell therapy. Theranostics. 2022;**12**(14):6273-6290. DOI: 10.7150/thno.76854

[163] Bai R, Cui J. Development of immunotherapy strategies targeting tumor microenvironment is fiercely ongoing. Frontiers in Immunology. 2022;**13**:890166. DOI: 10.3389/fimmu.2022.890166

[164] Lv B, Wang Y, Ma D, Cheng W, Liu J, Yong T, et al. Immunotherapy: Reshape the tumor immune microenvironment. Frontiers in Immunology. 2022;**13**:844142. DOI: 10.3389/fimmu.2022.844142

[165] Gun SY, Lee SWL, Sieow JL, Wong SC. Targeting immune cells for cancer therapy. Redox Biology. 2019;**25**:101174. DOI: 10.1016/j.redox.2019.101174

Chapter 2

Allogeneic Stem Cell Transplantation in Developing Countries

Francisco Jaramillo, Alejandro Toro-Pedroza, Pedro Caraballo, Álvaro Mondragón and Mildrey Mosquera

Abstract

Allogeneic hematopoietic stem cell transplantation (allo-HSCT) is crucial for treating high-risk acute leukemias in developing countries but is costly and complex. Our cost-effective protocol achieves promising results, with 12-month disease-free survival rates of 52% for acute lymphoblastic leukemia and 65% for acute myeloid leukemia, close to those reported by JACIE-accredited units. Despite the lower transplant rates and limited access compared to developed nations, advancements in donor selection, monitoring, and infection prophylaxis have improved outcomes. Continued research and cost-effective strategies are essential to enhance accessibility and success rates in emerging economies.

Keywords: stem cell transplantation, developing countries, allogeneic, leukemia, lymphoma

1. Introduction

Currently, allogeneic hematopoietic stem cell transplantation (allo-HSCT) is the only curative therapy available for high-risk acute leukemias in developing countries. This therapy is technically complex and costly, posing a clinical and social challenge to balance the limited resources of national public health systems while achieving clinical outcomes similar to those reported by transplant units in developed countries.

The cumulative incidence rate of hematopoietic neoplasms requiring allogeneic transplantation in emerging economies is comparable to that of developed countries. Despite the increasing number of bone marrow transplants in recent years, coverage and access remains significantly lower compared to developed regions. Most hospitals with allogeneic transplant units in developing countries are private and centralized, with very few public hospitals capable of performing allogeneic transplants, further limiting access for the entire population. Similarly, access to cytogenetic and molecular monitoring is not universal, making morphology and flow cytometry our primary diagnostic and follow-up tools, while NGS, qPCR, and chimerism analysis are almost exclusively available to patients in private health systems.

Here, we present our allogeneic transplant protocol in a cost-effective manner for public and private health regimes in emerging economies. After many years of work and analysis, we have achieved procedure-related mortality rates (TRM <15%) and 12-month disease-free survival rates (DFS of 52% in ALL and 65% in AML), slightly lower than those reported by units accredited under the Joint Accreditation Committee ISCT-Europe & EBMT (JACIE).

2. Epidemiology of bone marrow transplant in developing countries

The history of bone marrow transplantation in Latin America began in 1980 at the Federal University Hospital in Curitiba, Brazil [1]. Since then, there has been an increase in the number of programs and certified hospitals for these procedures. According to the Latin America Bone Marrow Transplant group (LABMT), in 2018, 5445 transplants were performed (2196 – 40% allogeneic and 3249–60% autologous), reported by 127 transplant units in 14 Latin American countries [2].

Despite this, the transplant rate (number of transplants per 10 million inhabitants per year) in Colombia and Latin America is 8 to 10 times lower than those reported in the United States and Europe. This is evidenced in the latest publication of the Worldwide Network for Blood and Marrow Transplantation (WBMT), which reports 2016 transplant rates of 560.8 for North America, 438.5 in Europe, 76.7 in Latin America, 53.6 in Southeast Asia/Pacific West, and 27.8 in Africa/Mediterranean East [3]. Regarding transplant indications, they vary according to the type of transplant. For allogeneic transplants, the primary indication is acute leukemia, while for autologous transplants, the main indications are multiple myeloma and relapsed high-grade lymphomas [4].

In recent decades, there has been a shift in the proportion of haploidentical transplants worldwide. Particularly in emerging economies, there is an increase in the relative frequency of haploidentical transplants. The Southeast Asia/Pacific West region accounts for 32%, and 26% of haploidentical transplants correspond to Latin America [5]. In contrast, in Europe and the United States, there is an increase in the number of unrelated identical allogeneic transplants compared to emerging economies, explained by the need to import unrelated hematopoietic cells, which is economically unfeasible, limiting this type of transplant for the minority of patients [6]. In Colombia, the bone marrow transplant rate has shown a progressive increase, rising from 75 transplants per 10 million inhabitants in 2017 to 108 in the following year, as reported by the LABMT [5].

3. Advancements in bone marrow transplantation in developing countries

In developing countries, significant improvements have been made in survival rates and hematologic outcomes for patients undergoing related allo-HSCT. Key advancements include better donor selection, enhanced monitoring and stratification of hematologic diseases, improved prophylactic tools for infections and graft-versus-host disease (GVHD), and the formation of dedicated multidisciplinary teams for bone marrow transplantation. These advancements follow the guidelines and management protocols of JACIE-accredited bone marrow units [7] (see **Table 1**).

Currently, the transplant-related mortality (TRM), relapse rate (R), and disease-free survival (DFS) for related allogeneic transplants in developing countries are comparable to those in non-emerging countries (TRM + R = 1-DFS).

DOI: http://dx.doi.org/10.5772/intechopen.1008566

Characteristic	1997	2024
HLA Typing	Serology	SSO (luminometry)
CD34 Progenitor Count	Nucleated cells (blood count)	Flow cytometry
Isolation	Open	High-pressure system HEPA filters
Advanced support	General ICU	Oncologic ICU
Infectology	Serology Limited antibiotics	PCR viral loads Specific antibiotics and antiretrovirals
Indication	Hematologist	Multidisciplinary board
Monitoring	Microscope	Chimerism Quantitative PCR

Table 1.
Main transplant-related advancements (1997–2024).

4. Exploration of donor-related variables in developing countries

GVHD is undoubtedly the most significant complication currently associated with allo-HSCT in our region. The best strategy to mitigate GVHD remains the careful selection of the donor prior to transplantation. However, to date, it is necessary to employ multiple additional strategies to attenuate the allo-reactivity of the donor's T lymphocyte populations, thereby ensuring that GVHD does not become severe [8].

In developing countries, various pharmacological options are available for both prophylaxis and treatment of GVHD, though their effectiveness can vary depending on the stage of the disease, sometimes proving minimally effective. Occasionally, it becomes necessary to use costly medications over extended periods, which poses a financial challenge for our healthcare system.

Reported outcomes for severe acute GVHD in haploidentical transplantation and for histocompatible transplants in benign conditions such as bone marrow aplasia are significantly better when direct marrow collection from the donor is employed compared to peripheral blood stem cell apheresis. This cell collection method is currently used more frequently in children than in adults in developing countries, especially for benign transplant indications [9]. In adult haploidentical transplants and non-neoplastic conditions, direct marrow collection also presents a viable alternative in our region for controlling acute GVHD, and it offers a lower long-term cost for allogeneic bone marrow transplantation (see **Table 2**).

Hierarchy for donor selection:

1. HLA (9/10–10/10) identical donor, <55 years of age.

2. Haploidentical HLA (5/10–8/10), <55 years of age

    a. Youngest male relative

    b. Females generally not eligible (discussed in medical boards for selected cases)

3. Antibodies anti-HLA absent in haploidentical transplants

Cell collection method	Peripheral	Medullary
CD34/kg	3.670.786	2.662.000
CD3/μml	150.296	509
Apheresis	Yes (Mahurkar/Amikus)	No
Cell freezing/DMSO	Yes	No
Total volume (mL)	300	650

Table 2.
A typical example of peripheral blood stem cell apheresis versus direct marrow collection from the donor [10].

4. ABO compatibility

5. CMV serological status

6. KIR ligand compatibility

Nowadays, in emergent countries, allogenic transplants from unrelated donors are almost nonexistent.

5. Receptor, disease, and indications for transplantation in developing countries

Evaluation for potential bone marrow transplant recipients in developing countries includes beyond HLA compatibility studies, assessing the viability of allo-HSCT using internationally validated risk scores. These scores help identify patients who may have a higher risk of transplant-related mortality [11].

Initially, we use the Hematopoietic Cell Transplantation-specific Comorbidity Index (HCT-CI) published by Sorror et al. [12], which attempts to objectively measure comorbidities and their severity. The comorbidities evaluated by the score include pulmonary function and pathological history (mild, moderate, or severe); cardiac; hepatic (mild, moderate, or severe); renal; infectious; psychiatric conditions; obesity; and a history of previous solid malignant tumors.

Additionally, we use the European Group for Blood and Marrow Transplantation (EBMT) risk score, which provides a score from 0 to 7 to evaluate outcomes and transplant-related mortality. Five factors are assessed: the patient's age; disease stage; time from diagnosis; donor type; and the correlation between donor and recipient gender. This risk score can be used regardless of the type of transplant, the type of conditioning, and the technology applied to the transplant [13].

Using these risk scores and many other variables related to unfavorable transplant outcomes helps us determine transplant viability and identify high-risk patients who require strict monitoring.

6. Conditioning strategies before allo-HSCT in developing countries

The EBMT and EWALL recommends MAC TBI-based regimens for ALL and selected AML patients, a stance also supported by the American Society for

Transplantation and Cellular Therapy [14, 15]. These recommendations are based on large retrospective studies showing that TBI-based regimens significantly improve PFS and OS compared to chemotherapy-only conditioning [16–22].

We conducted a retrospective study at a single center that analyzed 100 adult patients with ALL who underwent allo-HSCT, using two different TBI-based MAC regimens. No significant differences were found in OS (p = 0.305), DFS (p = 0.23), nor cumulative incidence of relapse (p = 0.32) between the SD-TBI and LD-TBI groups. The risk of aGVHD was lower in patients with SD-TBI, specifically in patients with HLA-Identical donors (RR 0.26 95% CI 0.08–0.82). This suggests that SD-TBI MAC regimens have comparable outcomes to LD-TBI and are associated with a lower incidence of aGVHD.

The EBMT reported significantly reduced NRM at 2 years and improved leukemia-free survival (LFS) with TBI-based conditioning in the setting of haploidentical HSCT with post-transplant cyclophosphamide, although the overall survival benefit was not significant [23]. Other studies have found that fludarabine-TBI reduced NRM compared to thiotepa, busulfan, and fludarabine (TBF), but increased relapse risk without affecting OS or GVHD. This suggest that fludarabine could be a reasonable alternative to cyclophosphamide and/or etoposide in TBI-based MAC [24, 25]. However, some controversy exists regarding the survival advantage of TBI-based conditioning regimens [26]. A randomized study of ALL patients in CR1 receiving either Bu-Cy or TBI-Cy found Bu-Cy to be noninferior, with similar rates of relapse at 2 years and NRM. There were no significant differences in toxicity, GVHD, or late effects between the groups [27]. Thus, multiple conditioning schemes are used today in clinical practice and the ideal regimen prior to allo-HCST in ALL remains controversial. Our results show that high dose Flu-TBI 12 Gy does not significantly differ in outcomes and relapse incidence compared to lower intensity TBI (4 Gy) with Bu-Flu myeloablative chemotherapy regimens.

Globally, the number of allogeneic transplants is increasing, particularly in treating conditions like ALL, where they offer a curative possibility. However, the

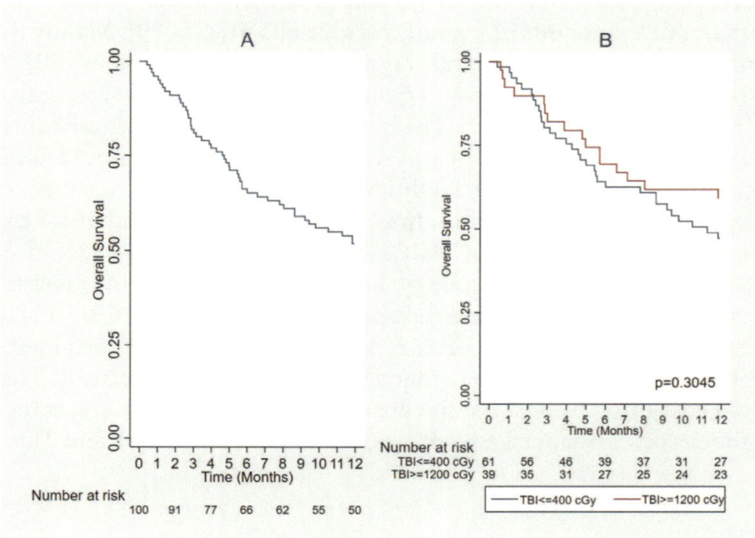

Figure 1.
Overall post-transplant survival in a cohort of 100 patients who underwent allo-HSCT for adult ALL (A). Overall survival in the cohort according to the use of TBI ≥ 1200 cGy and TBI ≤ 400 cGy conditioning regimens (B) [32].

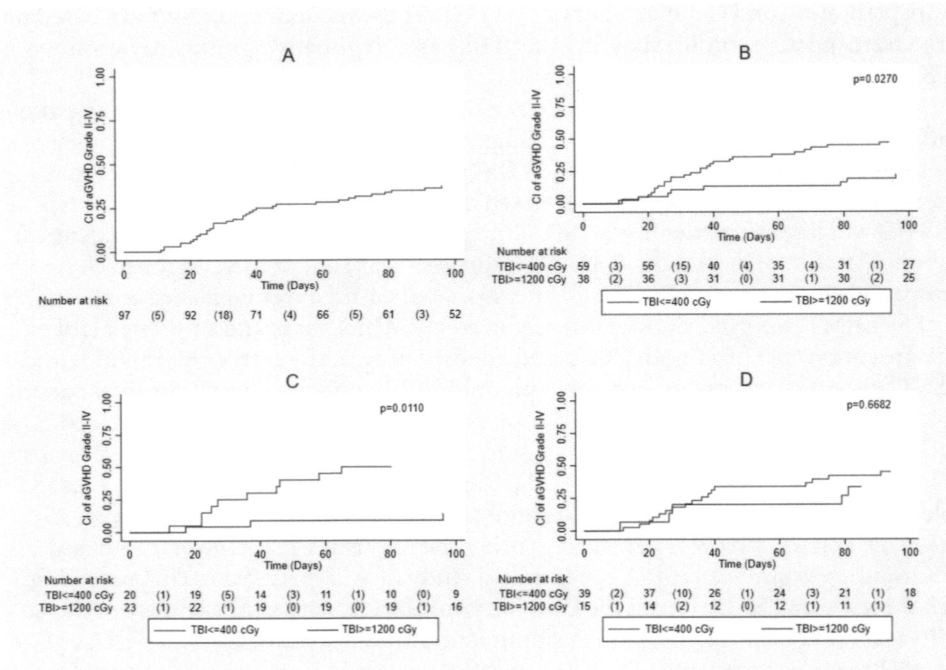

Figure 2.
Cumulative incidence (AI) of aGVHD (A). Cumulative incidence of aGVHD according to the use of TBI ≥ 1200 and TBI ≤ 400 (B). Cumulative incidence of aGVHD according to the use of TBI ≥ 1200 and TBI ≤ 400 in HLA-identical recipients (C). Cumulative incidence of GVHD according to the use of TBI ≥ 1200 and TBI ≤ 400 in HLA haploidentical recipients (D) [32].

cost of these procedures, especially in countries with emerging economies, poses a significant challenge for public health systems [28]. MAC has also been reported as one of the major determinants of overall costs for allo-HSCT [29]. Mainly due to its higher complexity, prolonged hospital stay and related complications [30]. Thus, there is critical need for conditioning regimens that are effective, safe, cost-efficient, and feasible to implement in LMICs. TBI-based protocols can be significantly more cost-effective compared to chemotherapy-containing regimens, provided the center is well-equipped with radiotherapy facilities. Additionally, the scarcity of crucial chemotherapy drugs such as busulfan, thiotepa, and treosulfan enhances the appeal of TBI-based regimens for HSCT in LMICs [31]. At our institution, costs are individualized by drug and radiation therapy. Based on unpublished intrainstitutional data, we have observed that costs are reduced when conditioning TBI >12 Gy is used. As high-cost treatments become accessible to more patients, it is crucial not only to evaluate clinical outcomes but also to manage financial costs effectively. Therefore, in LMIC, future cost-effectiveness studies are essential to determine if certain conditioning regimens offer financial advantages over others (**Figures 1** and **2**).

7. Conclusions

Globally, the number of allo-HSCT procedures is increasing, and although it is considered the only curative therapy for high-risk acute leukemias, its price and technical complexity poses significant challenges for developing countries. Here, we present our allogeneic transplant protocol in a cost-effective manner for public and private health regimes in emerging economies, with outcomes comparable to those reported by units accredited under the Joint Accreditation Committee ISCT-Europe & EBMT (JACIE). There is a critical need for development and future research for cost-effective strategies in these regions.

Author contribution

All authors declare to have contributed to the conception and design of the work, the acquisition, analysis, and interpretation of data for the work; participation in drafting and revising the work; approval of the manuscript version to be published; agreement to be accountable for all aspects of the work in ensuring that questions related to the accuracy or integrity of any part of the work are appropriately investigated and resolved.

Author details

Francisco Jaramillo[1*], Alejandro Toro-Pedroza[2], Pedro Caraballo[3], Álvaro Mondragón[3] and Mildrey Mosquera[4]

1 Bone Marrow Transplant Unit, Department of Hematology, Fundación Valle del Lili, Cali, Colombia

2 Faculty of Health Sciences, Universidad ICESI, Cali, Colombia

3 Hematology and Clinical Oncology, Universidad ICESI, Cali, Colombia

4 Universidad del Valle, Cali, Colombia

*Address all correspondence to: fcojaramillo77@hotmail.com

References

[1] Ferreira E, Dulley FL, Morsoletto F, Neto JZ, Pasquini R. Bone marrow transplantation in Brazil. Human Immunology. 1985;**14**(3):324-332

[2] Correa C, Gonzalez-Ramella O, Baldomero H, Basquiera AL, Baena R, Arcuri L, et al. Increasing access to hematopoietic cell transplantation in Latin America: Results of the 2018 LABMT activity survey and trends since 2012. Bone Marrow Transplantation. 2022;**57**(6):881-888

[3] Niederwieser D, Baldomero H, Bazuaye N, Bupp C, Chaudhri N, Corbacioglu S, et al. One and a half million hematopoietic stem cell transplants: Continuous and differential improvement in worldwide access with the use of non-identical family donors. Haematologica. 2022;**107**(5):1045-1053

[4] Jaimovich G, Martinez Rolon J, Baldomero H, Rivas M, Hanesman I, Bouzas L, et al. Latin America: The next region for haematopoietic transplant progress. Bone Marrow Transplantation. 2017;**52**(5):671-677

[5] Jaimovich G, Gale RP, Hanesman I, Vazquez A, Hammerschlak N, Simoes BP, et al. The paradox of haematopoietic cell transplant in Latin America. Bone Marrow Transplantation. 2021;**56**(10):2382-2388

[6] Rivera-Franco MM, Leon-Rodriguez E, Castro-Saldaña HL. Costs of hematopoietic stem cell transplantation in a developing country. International Journal of Hematology. 2017;**106**(4):573-580

[7] Aljurf M, Weisdorf D, Hashmi SK, Nassar A, Gluckman E, Mohty M, et al. Worldwide network for blood and marrow transplantation (WBMT) recommendations for establishing a hematopoietic stem cell transplantation program in countries with limited resources (part II): Clinical, technical and socio-economic considerations. Hematology/Oncology and Stem Cell Therapy. 2020;**13**(1):7-16

[8] Zeiser R, Blazar BR. Acute graft-versus-host disease - biologic process, prevention, and therapy. The New England Journal of Medicine. 2017;**377**(22):2167-2179

[9] Yu X, Liu L, Xie Z, Dong C, Zhao L, Zhang J, et al. Bone marrow versus peripheral blood as a graft source for haploidentical donor transplantation in adults using post-transplant cyclophosphamide-a systematic review and meta-analysis. Critical Reviews in Oncology/Hematology. 2019;**133**:120-128

[10] FJJ E, Chantre A, Lugo LP, Carvajal S. Trasplante alogénico de médula ósea con depleción de células T. Recolección directa por punción múltiple en un centro hospitalario de Cali, Colombia. Reporte de caso. Revista Colombiana de Hematología y Oncología [Internet]. 2022;**9**(2):133-137. Available from: https://revista.acho.info/index.php/acho/article/view/395 [Accessed: August 5, 2024]

[11] Snowden JA, Sánchez-Ortega I, Corbacioglu S, Basak GW, Chabannon C, de la Camara R, et al. Indications for haematopoietic cell transplantation for haematological diseases, solid tumours and immune disorders: Current practice in Europe, 2022. Bone Marrow Transplantation. 2022;**57**(8):1217-1239

[12] Sorror ML, Maris MB, Storb R, Baron F, Sandmaier BM, Maloney DG,

et al. Hematopoietic cell transplantation (HCT)-specific comorbidity index: A new tool for risk assessment before allogeneic HCT. Blood. 2005;**106**(8):2912-2919

[13] Gratwohl A. The EBMT risk score. Bone Marrow Transplantation. 2012;**47**(6):749-756

[14] DeFilipp Z, Advani AS, Bachanova V, Cassaday RD, Deangelo DJ, Kebriaei P, et al. Hematopoietic cell transplantation in the treatment of adult acute lymphoblastic leukemia: Updated 2019 evidence-based review from the American Society for Transplantation and Cellular Therapy. Biology of Blood and Marrow Transplantation. 2019;**25**(11):2113-2123

[15] Giebel S, Marks DI, Boissel N, Baron F, Chiaretti S, Ciceri F, et al. Hematopoietic stem cell transplantation for adults with Philadelphia chromosome-negative acute lymphoblastic leukemia in first remission: A position statement of the European working Group for Adult Acute Lymphoblastic Leukemia (EWALL) and the acute leukemia working Party of the European Society for blood and marrow transplantation (EBMT). Bone Marrow Transplantation. 2019;**54**(6):798-809

[16] Bunin N, Aplenc R, Kamani N, Shaw K, Cnaan A, Simms S. Randomized trial of busulfan vs total body irradiation containing conditioning regimens for children with acute lymphoblastic leukemia: A pediatric blood and marrow transplant consortium study. Bone Marrow Transplantation. 2003;**32**(6):543-548

[17] Davies SM, Ramsay NK, Klein JP, Weisdorf DJ, Bolwell B, Cahn JY, et al. Comparison of preparative regimens in transplants for children with acute lymphoblastic leukemia. Journal of Clinical Oncology. 2000;**18**(2):340-347

[18] Granados E, de La Cámara R, Madero L, Díaz MA, Martín-Regueira P, Steegmann JL, et al. Hematopoietic cell transplantation in acute lymphoblastic leukemia: Better long term event-free survival with conditioning regimens containing total body irradiation. Haematologica. 2000;**85**(10):1060-1067

[19] Kiehl MG, Kraut L, Schwerdtfeger R, Hertenstein B, Remberger M, Kroeger N, et al. Outcome of allogeneic hematopoietic stem-cell transplantation in adult patients with acute lymphoblastic leukemia: no difference in related compared with unrelated transplant in first complete remission. Journal of Clinical Oncology. 2004;**22**(14):2816-2825

[20] Marks DI, Forman SJ, Blume KG, Pérez WS, Weisdorf DJ, Keating A, et al. A comparison of cyclophosphamide and total body irradiation with etoposide and total body irradiation as conditioning regimens for patients undergoing sibling allografting for acute lymphoblastic leukemia in first or second complete remission. Biology of Blood and Marrow Transplantation. 2006;**12**(4):438-453

[21] Giebel S, Labopin M, Socié G, Beelen D, Browne P, Volin L, et al. Improving results of allogeneic hematopoietic cell transplantation for adults with acute lymphoblastic leukemia in first complete remission: An analysis from the acute leukemia working Party of the European Society for blood and marrow transplantation. Haematologica. 2017;**102**(1):139-149

[22] Cahu X, Labopin M, Giebel S, Aljurf M, Kyrcz-Krzemien S, Socié G, et al. Impact of conditioning with TBI in adult patients with T-cell ALL who receive a myeloablative allogeneic stem cell transplantation: A report from the acute leukemia working party of EBMT. Bone Marrow Transplantation. 2016;**51**(3):351-357

[23] Dholaria B, Labopin M, Angelucci E, Tischer J, Arat M, Ciceri F, et al. Outcomes of Total body irradiation- versus chemotherapy-based myeloablative conditioning regimen in haploidentical hematopoietic cell transplantation with post-transplant cyclophosphamide for acute lymphoblastic leukemia: ALWP of the EBMT study. Blood. 2019;**134**:320

[24] Swoboda R, Labopin M, Giebel S, Angelucci E, Arat M, Aljurf M, et al. Total body irradiation plus fludarabine versus thiotepa, busulfan plus fludarabine as a myeloablative conditioning for adults with acute lymphoblastic leukemia treated with haploidentical hematopoietic cell transplantation. A study by the acute leukemia working party of the EBMT. Bone Marrow Transplantation. 2022;**57**(3):399-406

[25] Solomon SR, Solh M, Zhang X, Morris LE, Holland HK, Bashey A. Fludarabine and total-body irradiation conditioning before ablative haploidentical transplantation: Long-term safety and efficacy. Biology of Blood and Marrow Transplantation. 2019;**25**(11):2211-2216

[26] Dogliotti I, Levis M, Martin A, Bartoncini S, Felicetti F, Cavallin C, et al. Maintain efficacy and spare toxicity: Traditional and new radiation-based conditioning regimens in hematopoietic stem cell transplantation. Cancers. 2024;**16**(5):865

[27] Zhang H, Fan Z, Huang F, Han L, Xu Y, Xu N, et al. Busulfan plus cyclophosphamide versus total body irradiation plus cyclophosphamide for adults acute B lymphoblastic leukemia: An open-label, multicenter, phase III trial. Journal of Clinical Oncology. 2023;**41**(2):343-353

[28] Khera N, Zeliadt SB, Lee SJ. Economics of hematopoietic cell transplantation. Blood. 2012;**120**(8):1545-1551

[29] Jaime-Pérez JC, Heredia-Salazar AC, Cantú-Rodríguez OG, Gutiérrez-Aguirre H, Villarreal-Villarreal CD, Mancías-Guerra C, et al. Cost structure and clinical outcome of a stem cell transplantation program in a developing country: The experience in Northeast Mexico. The Oncologist. 2015;**20**(4):386-392

[30] Broder MS, Quock TP, Chang E, Reddy SR, Agarwal-Hashmi R, Arai S, et al. The cost of hematopoietic stem-cell transplantation in the United States. American Health & Drug Benefits. 2017;**10**(7):366-374

[31] Pandit S, Sapkota S, Adhikari A, Karki P, Shrestha R, Jha DS, et al. Breaking barriers: Supporting hematopoietic stem cell transplant program through collaborative radiation therapy service from a physically distant center. Journal of the Egyptian National Cancer Institute. 2024;**36**(1):17

[32] Jaramillo FJ, Coronel A, Zapata DM, Toro-Pedroza A, Manzi Tarapues E, Useche E, et al. Outcomes of Standard-Dose Versus Low-Dose Total Body Irradiation as Myeloablative Conditioning for Allogeneic Stem Cell Transplantation in Adults with Acute Lymphoblastic Leukemia. Fundación Valle del Lili; 2024

Chapter 3

Congenital Neutropenia Associated with Bone Marrow Failure

Baris Malbora

Abstract

Neutropenia encompasses a diverse array of acquired or congenital conditions, ranging from benign to premalignant, which can manifest at any age. These conditions pose a risk of severe infections and may predispose individuals to myelodysplastic neoplasms or acute myeloid leukemia. Recent genomic advancements have unveiled novel genes and pathways implicated in the etiology and progression of these disorders, offering fresh avenues for personalized therapeutic approaches. Despite strides in diagnostics, real-world data underscore the predominant reliance on physicians' experience and local protocols in diagnosing and managing neutropenic patients. This chapter aims to consolidate evidence and consensus-based insights concerning the definition, classification, diagnosis, and monitoring of bone marrow failures associated with neutropenia, excluding severe congenital cases. Additionally, it underscores the importance of evaluating clinical manifestations alongside conventional and innovative laboratory tests, as well as advanced genetic analyses, to better classify risks and track disease progression across the spectrum of affected individuals.

Keywords: congenital neutropenia, bone marrow failure, diagnosis, treatment, stem cell transplantation

1. Introduction

Neutropenia refers to an absolute neutrophil count (ANC) below the normal range depending on the age and ethnic background of the affected individual. This disorder can manifest as acute/transient or chronic/persistent; acquired, likely acquired, or congenital; primary or secondary depending on a variety of disease conditions; and benign or pre-malignant, predisposing to myelodysplastic syndromes (MDS) or acute myeloid leukemia (AML) [1]. The most common causes of neutropenia are acquired factors such as sequestration, viral infections, chemotherapy, and drug reactions [2]. Although inherited causes are much rarer, they are generally more severe [3]. Congenital neutropenias (CN) are rare diseases belonging to myelopoiesis disorders and are typically characterized by maturation arrest at the promyelocyte stage [4]. Congenital neutropenia presents with a persistent decrease in neutrophil counts and is often seen in infancy, while cyclic neutropenia is characterized by regular fluctuations in blood neutrophil counts, leading to periodic neutropenia with a 21-day turnover frequency [5].

In recent years, advances in the genomic field have identified new genes playing a role in the pathogenesis and/or development of neutropenia, unraveled fundamental pathogenic mechanisms, and paved the way for new personalized treatment approaches [6, 7]. Comprehensive evaluations have been conducted by examining individual patient cases and their families, aiming to guide clinicians in accurate diagnosis, monitoring, and treatment, particularly for those with chronic conditions. However, a survey conducted within the scope of collaboration among physicians, especially those dealing with the diagnosis of chronic neutropenias (EuNet-INNOCHRON; https://eunet-innochron.eu/), involving physicians from many countries including our own, has shown that real-world data predominantly rely on local experiences and practices [8]. Therefore, the diagnosis and management of neutropenic patients vary and are fraught with challenges. Additionally, the seamless transition of these patients from pediatric to adult care poses difficulties. Treating physicians must be knowledgeable about complications associated with neutropenia, the risk of progression to MDS/AML, and special circumstances such as pregnancy and genetic counseling.

Fever may cause the count of neutrophils in the bloodstream to drop below 0.2 x 10^9/L during febrile neutropenic days [9]. This can lead to negative outcomes on the overall quality of life for these patients with recurrent fevers and their parents. It may adversely affect school attendance or daily activities, proving to be quite distressing for both the patient and the family [10]. For severe CN, the risk of MDS and AML significantly increases. However, treatment options are limited, and reliable tools predicting such progression are lacking. Moreover, the time taken for a definitive diagnosis can span several months or even years, during which children often undergo unnecessary diagnostic procedures. Therefore, an updated perspective is needed that emphasizes considering CN and accurately identifying it.

2. Definition and classification of neutropenia

2.1 Definition of neutropenia

The definition of neutropenia varies according to the ethnic background and age of the patient (**Table 1**). For newborns and infants up to 1 year of age of Caucasian descent, the widely accepted lower limit of ANC for defining neutropenia is 1.0×10^9/L. However, the threshold ANC for defining neutropenia is 2.5×10^9/L within 72–240 hours after birth and 1.0×10^9/L for premature newborns. From 1 year of age to adulthood, the threshold for defining neutropenia is 1.5×10^9/L. For white adults, the World Health Organization and expert groups on MDS/AML have adopted an ANC threshold of 1.8×10^9/L for defining neutropenia.

Age	ANC*
14 days-1 year	< 1.0×10^9/L
Child >1 year –adulthood	< 1.5×10^9/L
adults*	< 1.8×10^9/L

**ANC <1.5×10^9/L is considered normal in individuals of African and Middle Eastern descent.*

Table 1.
Age-specific threshold levels for defining neutropenia based on ANC.

It is important to mention that certain individuals of African and Middle Eastern descent may exhibit a normal Absolute Neutrophil Count (ANC) ranging from 0.5 to 1.5×10^9/L and occasionally even lower values [11]. This variability is inherited as an autosomal recessive trait linked to a polymorphism (rs2814778, −46 T > C) in the GATA box of the promoter region of the atypical chemokine receptor-1 (*ACKR1*) gene, also known as the Duffy antigen receptor for chemokines (DARC). In the homozygous (C/C) state, this polymorphism particularly results in the absence of the Duffy antigen on erythrocytes, referred to as the Duffy-null phenotype. Instead of using the term 'ethnic neutropenia,' it is recommended to utilize the term 'neutropenia associated with *ACKR1/DARC*' (NAD) to underscore its genetic rather than ethnic origin [12].

2.2 Classification of neutropenia

The risk and consequence of bacterial infections originating from neutropenia have long been known. This condition relies not only on the ANC but also on individuals' capacity to recruit and transport neutrophils to tissues [16]. However, valid methods for estimating total neutrophil count are not clinically available. Therefore, still measuring the quantitative relationship between ANC and infections in chemotherapy recipients, neutropenia is classified as 'mild' when ANC is in the range of 1.0 to 1.5 (or 1.8 for adults) $\times 10^9$/L, 'moderate' when ANC is in the range of 0.5 to 1.0×10^9/L, and 'severe' when ANC is $<0.5 \times 10^9$/L [15]. In severe neutropenias, the term 'profound neutropenia' is used when ANC is $<0.2 \times 10^9$/L. These patients are generally at risk of serious, life-threatening infections. Neutropenia can also be classified as acute or chronic depending on whether its duration is <3 months or > 3 months, respectively [17].

In addition to the conventional categorizations described above, an expanded classification based on pathogenesis has gained acceptance for neutropenias. This expanded classification includes congenital (see **Table 2**), acquired, and potentially acquired (see **Table 3**) categories. Congenital neutropenia refers to a range of genetic disorders characterized by abnormalities in neutrophil production, differentiation, and survival in the bone marrow (BM), increased susceptibility to infections, and a tendency for progression to MDS/AML [7, 18, 19]. It can be further divided into disorders where neutropenia is the primary abnormality and those where it occurs alongside extra-hematologic manifestations, immune deficiency/immune dysregulation, metabolic disorders, nutritional deficiencies, or forms part of more complex BM failure syndromes. This classification also includes the specific genes implicated in each subtype of congenital neutropenia.

Acquired neutropenia is typically classified as primary or idiopathic, associated with anti-neutrophil antibodies or other unidentified mechanisms, and secondary to infections, autoimmune diseases, drug exposure, nutritional deficiencies, hypersplenism, or hematologic disorders [20–22]. Potential acquired neutropenia is often referred to as idiopathic or autoimmune neutropenia (AIN) in childhood. This condition usually follows a benign and uncomplicated course. However, it includes cases of persistent (long-term) or late-onset (delayed-onset) neutropenia, whether mediated by antibodies or not, occurring beyond 24–36 months [23–25]. Some of these patients may also have lymphopenia linked to low B-cell and natural killer cell subtypes, and genetic analysis of genes associated with immune deficiency/immune dysregulation disorders has identified pathogenic variants in certain cases [25]. Consensus has been reached that pediatric or adolescent patients in this group, provisionally classified as idiopathic acquired should undergo a comprehensive investigation primarily to rule out an underlying genetic etiology.

	Congenital Neutropenias		
	Gene	**Inheritance**	**Main features/notes**
Isolated			
Severe CN	*ELANE*	OD	Ceased neutrophil maturation in BM
	CSF3R	OD	Unresponsive to G-CSF in BM
	CXCR2	OR	Maturation arrest and myelokathexis absent
	WAS, gain of function	X	(WASp-XLN) Delayed maturation, monocytopenia
Cyclic Neutropenia	*ELANE*	OD	Intermittent/cyclical disturbances in differentiation
Associated with extra-hematological symptoms			
Barth Syndrome (3-methylglutaconic aciduria type II)	*TAZ*	X	No maturation arrest, hypertrophic cardiomyopathy, and myopathic syndrome
Charcot-Marie-Tooth neuronopathy type B	*DNM2*	OD	Weakness and atrophy of distal extremity muscles due to peripheral neuropathy
DiGeorge syndrome	*22q11* deletion	OD	Variable cognitive delay, immunodeficiency, hypoparathyroidism, congenital heart disease, palatal, gastrointestinal, and renal abnormalities
Cohen syndrome	*VPS13B*	OR	No maturation arrest, psychomotor retardation, microcephaly, facial features, hypotonia, joint laxity, progressive retinochoroidal dystrophy, and myopia
G6PC3 mutation	*G6PC3*	OR	Skin hyperelasticity and prominent superficial venous network, congenital heart disease, arrhythmias, uropathy, cryptorchidism, and exocrine pancreas dysfunction
GFI1 mutation	*GFI1*	OD	Sometimes maturation arrest, lymphopenia, peripheral blood, and inner ear defects
HYOU1 deficiency	*HYOU1*	OR	Hypoglycemia and inflammatory complications
JAGN1 mutation	*JAGN1*	OR	Sometimes maturation arrest, bone and dental abnormalities, and exocrine pancreas dysfunction
Kostmann disease	*HAX1*	OR	Maturation failure, mental retardation, seizures, and susceptibility to MDS/AML
P14/LAMTOR2 mutation	*LAMTOR2*	OR	Chronic neutropenia, hypogammaglobulinemia, failure of maturation, oculocutaneous albinism, and growth retardation, hypopigmentation, coarse face, and frequent bronchopulmonary infections

	Congenital Neutropenias		
	Gene	**Inheritance**	**Main features/notes**
Pearson syndrome	Mitochondrial DNA deletions	Mitochondrial	Resistant sideroblastic anemia, vacuolization of BM precursors, and exocrine pancreas dysfunction
Schimke immuno-osseous dysplasia	*SMARCAL1*	OR	Spondyloepiphyseal dysplasia, slowly progressive immune defect, and immune complex nephritis
SEC61A1 mutation	*SEC61A1*	OD	Maturation arrest and tubulointerstitial kidney disease
SMARCD2 mutation	*SMARCD2*	OR	Dysplastic syndrome, absence of granules in neutrophils, chronic diarrhea, bone abnormalities, and low-set ears
Specific granule deficiency	*CEBPE*	OR	Neutrophils with bilobed nuclei
TCIRG1 neutropenia	*TCIRG1*	OD	Variable/failure of maturation and cutaneous angiomatosis
VPS45 mutation	*VPS45*	OR	Myeloid hyperplasia, myelofibrosis, nephromegaly, HSM, mental retardation, epilepsy, and osteosclerosis
Wolcott-Rallison syndrome	*EIF2AK*	OR	Maturation arrest, insulin-dependent neonatal diabetes, epiphyseal dysplasia, growth retardation, liver and kidney dysfunction, developmental delay, and exocrine pancreatic insufficiency
Associated with immunodeficiency/immune dysregulation			
Adenosine deaminase 2 deficiency	*ADA2*	OR	Severe combined immunodeficiency, vasculitis, cerebrovascular disease, pure red cell aplasia, and BM failure
ALPS	*FAS, FASLG, CASP10*	OR	Lymphoproliferation and autoimmune cytopenia
CD40L/hyper-IgM syndrome type I	*CD40L*	X	
Chédiak-Higashi syndrome	*LYST*	OD	Susceptibility to severe infections, autoimmune disease, and cancer
CLPB syndrome	*CLPB*	OR	Decreased pigmentation of hair and eyes, peroxidase-positive inclusion bodies in myeloblasts and promyelocytes in BM, specific malignant lymphoma
Perforin deficiency	*FHL2*	OR	Cataracts and neurological involvement
UNC13D deficiency	*FHL3*	OR	Fever, hepatosplenomegaly, and cytopenia
GATA2 syndrome	*GATA2*	OD	Fever, HSM, and cytopenias
Griscelli syndrome, type II	*RAB27A*	OR	Monocytopenia, hearing loss, and HPV infections

	Congenital Neutropenias		
	Gene	**Inheritance**	**Main features/notes**
Hermansky-Pudlak syndrome, type 2	*AP3B1*	OR	Hypopigmentation and neurological disorder
Reticular dysplasia (adenylate kinase-2 deficiency)	*AK2*	OR	Albinism
STK4 mutation	*STK4*	OR	Frequent and severe immunodeficiency and sensorineural hearing loss
WHIM syndrome	*CXCR4*	OD	Intermittent neutropenia, monocytopenia, T and B lymphopenia, atrial defect, and HPV infections
Wiskott-Aldrich syndrome	*WAS*, loss of function	X	Eczema, thrombocytopenia, severe infections, and bloody diarrhea
CVID	Various genes, *TNFSRF13, BAFFR, CTL4, LRBA, PI3K*	OD, OR	Recurrent infections, hypogammaglobulinemia, and autoimmune cytopenia including neutropenia
Associated with metabolic disorders and nutritional deficiencies			
Gaucher disease type I	*GBA*	OR	Hepatosplenomegaly, thrombocytopenia, and osteolytic lesions
Glycogen storage disease type Ib	*SLC37A4/G6PT1*	OR	Hepatomegaly, inflammatory bowel disease, and fasting hypoglycemia
Isovaleric acidemia	*IVD*	OR	Neonatal ketoacidosis, growth retardation, weakness, and refusal to feed
Methylmalonic acidemia	*MMUT*	OR	Weakness, growth retardation, recurrent vomiting, hypotonia, hepatomegaly, and growth retardation
Propionic acidemia	*PCCB, PCCA*	OR	Weakness, cardiomyopathy, feeding difficulties, and acute encephalopathy
Transcobalamin II deficiency	*TCN2*	OR	Developmental delay, diarrhea, vomiting, weakness, and mucosal ulceration
Associated with bone marrow failure			
Fanconi anemia	*FANC* complementation group	OR-X (*FANCB*)	Congenital malformations and predisposition to cancer
Ribosomopathies			
Diamond-Blackfan anemia	*RPS7,10,15,17,19, 24,26,27,27a,28,29, RPL5,9,11,15,18,26,27,31, 35a*	OD	Erythroid hypoplasia, congenital malformations, growth retardation, osteosarcoma, MDS, and AML
	GATA1	X	Early-onset anemia, thrombocytopenia, and BM erythroid hypoplasia
	EPO	OR	Erythroid hypoplasia
	TSR2, HEATR3	X-OR	Erythroid hypoplasia, craniofacial defects, short stature, facial and acromelic deformities, and intellectual disability

	Congenital Neutropenias		
	Gene	**Inheritance**	**Main features/notes**
Cartilage-Hair hypoplasia	*RMRP*	OR	Short stature (dwarfism) along with other skeletal abnormalities, metaphyseal chondrodysplasia, joint laxity, thin and sparse hair (hypotrichosis), and abnormal immune system function (immunodeficiency), recurrent infections
Shwachman-Diamond syndrome	*SBDS, EFL1, DNAJC21*	OR	Mild neutropenia, dysgranulopoiesis, mild dysmegakaryopoiesis, dyserythropoiesis, exocrine pancreatic insufficiency, metaphyseal dysplasia, cognitive impairment, cardiomyopathy, growth retardation, and hair/skin/tooth abnormalities
SAMD9/SAMD9L syndromes	*SAMD9/SAMD9L*	OD	Adrenal insufficiency, congenital malformations, cerebellar ataxia, severe invasive infections, and susceptibility to MDS
SRP54 mutation	*SRP54*	OD	Maturation arrest, severe neurodevelopmental delay, and exocrine pancreatic insufficiency
Telomere diseases	*DKC1*	X	Mucocutaneous features, liver fibrosis, idiopathic pulmonary fibrosis, and predisposition to cancer
	TERT, hTR, TINF2, DKC1, ACD	OD	
	TERT, NHP2, NOP10, WRAP53, NOLA3, TCB1, RTEL1, CTC1, PARN	OR	
U6 Small Nuclear RNA Biogenesis	*USB1* (Clericuzio syndrome, retinopathy with neutropenia)	OR	Mucocutaneous features, liver fibrosis, idiopathic pulmonary fibrosis, and predisposition to cancer

Table 2.
Classification of congenital neutropenia.

2.3 Severe congenital neutropenia: Nosological considerations

The categorization of congenital neutropenia holds significant importance, and numerous authorities have weighed in on this subject [13, 14]. As per the Severe Chronic Neutropenia International Registry, SCN refers to 'a group of rare inherited hematologic disorders that result in persistent neutropenia spanning many months or years.' Conversely, cyclic neutropenia is typified by recurring bouts of neutropenia, typically lasting 3–5 days every 21 days. The utilization of the same two-letter abbreviation (CN) for chronic neutropenia, congenital neutropenia, and cyclic neutropenia could potentially engender ambiguity. Severe CN (SCN) represents a genetically diverse array of hematopoietic disorders that may be inherited through autosomal dominant, autosomal recessive, or X-linked recessive means. Moreover, contextualizing SCN amidst other immunodeficiency disorders, also known as primary immunodeficiencies, might prove beneficial [15].

Online Mendelian Inheritance in Man (OMIM), an extensive repository of human genes and genetic disorders, delineates that SCN can be subdivided into various

Acquired	
Primary or idiopathic: typically presents as isolated neutropenia	Antibody mediated • Primary autoimmune • Primary alloimmune
	Non-antibody mediated • Idiopathic infant neutropenia • Uncertain cause cytopenia-neutropenia
Secondary: associated with or caused by neutropenia	Hypersplenism (due to congestive, infiltrative, phagocytic, and reactive splenomegaly)
	Infections • Viral (HIV, HCV, HBV, CMV, EBV, HIV, influenza, parvovirus B19, measles, and Sars-Cov-2) • Bacterial (*Salmonella*, *Brucella*, *Rickettsia*, Mycobacterium, Mycoplasma, and *Helicobacter pylori*) • Parasitic (*Plasmodium* species, visceral leishmaniasis) • Fungal (Histoplasmosis)
	Autoimmune diseases • Organ-specific (thyroid diseases, inflammatory bowel disease, and primary biliary cirrhosis) • Systemic (systemic lupus erythematosus, including Felty syndrome and rheumatoid arthritis, Sjögren's syndrome, systemic sclerosis, and graft-versus-host disease)
	Nutritional deficiencies • Deficiencies in B12, folic acid, iron, copper, and calories
	Immune regulatory disorders • Common variable immunodeficiency, ALPS, ALPS-like disorders, HLH, and macrophage activation syndrome
	Hematological diseases • Primary benign (aplastic anemia) • Clonal (including myeloid malignancies/lymphoid malignancies such as LGL)
	Drug-induced • Chemotherapy • Non-chemotherapeutic medications: pain relievers and NSAIDs, antibacterial agents (beta-lactams, cephalosporins, trimethoprim-sulfamethoxazole, sulfasalazine, vancomycin, rifampicin, fluconazole, ketoconazole), diuretics (furosemide, spironolactone), antiretroviral therapy for HIV, antithyroid drugs (thiamazole, methimazole), clozapine (olanzapine), deferiprone, metamizole (dipyrone), phenothiazines (alimemazine), quinine/quinidine, intravenous immunoglobulin (IVIG), monoclonal antibodies (Rituximab), and biological therapies (infliximab, etanercept).
Likely acquired	Children or adolescents experiencing neutropenia, whether with or without specific antibodies against neutrophils, or enduring beyond the age of 5, might warrant a diagnosis of late-onset and prolonged neutropenia. Recent literature characterizes this uncommon neutropenia as a consequence of immune dysregulation, marked by distinctive biochemical/immunological traits and the presence of genetic variants affecting immune regulation.

Table 3.
Classification of acquired neutropenia.

subcategories based on the underlying genetic anomaly. Consequently, individuals with *ELANE* mutations are classified as *SCN1*, those with *GFI1* deficiency as *SCN2*, *HAX1* deficiency as SCN3, *G6PC3* deficiency as SCN4, *VPS45* deficiency as SCN5, *JAGN1* deficiency as SCN6, *CSF3R* deficiency as SCN7, *SRP54* deficiency as SCN8, and individuals with *WAS* deficiency as *SCNX* (given its X-linked nature; whereas SCN1, SCN2, and SCN8 are autosomal dominant conditions, SCN3-SCN7 are all categorized as autosomal recessive conditions). Recognizing the significance of molecular disease classification can potentially enhance clinical management for patients and their families [15]. However, the primary focus should always remain on clinical symptoms.

3. Diagnosis

3.1 What information is important to gather from a patient or family history?

When investigating neutropenia in patients, it is crucial to start by obtaining a thorough medical and family history. The age at which neutropenia was first diagnosed is significant; congenital neutropenia (CN) is rarely identified in adulthood but is commonly diagnosed in early childhood. Previous blood count records can provide insights into the duration of neutropenia, helping to distinguish between acute and chronic forms. It is important to understand whether neutropenia is incidental or linked to an acute ailment.

Digging into the clinical history involves exploring the frequency, type, and severity of infections, as well as instances of hospitalization. Notable clinical events include fever, oral ulcers, sore throat/odynophagia, gingivitis, sinusitis, otitis, skin ulcers and cellulitis, deep tissue infections, pneumonia, gastrointestinal symptoms, and perianal infections. The cyclical recurrence of infections may suggest cyclic neutropenia. It is also crucial to assess whether prior febrile or infection episodes coincided with a spontaneous increase in absolute neutrophil count (ANC) and how they responded to treatment.

3.2 Family history

A thorough account of familial background, encompassing parental consanguinity and ethnic origins, holds significance. Investigating any family members exhibiting similar symptoms or laboratory results should span across generations. Moreover, it is crucial to ascertain occurrences of MDS/AML-related infections or fatalities within the family. Instances of unexplained infant mortality or miscarriages might also correlate with CN. Negativity in family history regarding neutropenia could stem from oversight in previous blood analyses, especially in cases of mild or moderate neutropenia. Furthermore, neutropenia severity can fluctuate even within the same family, underscoring the importance of examining blood counts across relatives for diagnosing familial neutropenia.

3.3 Additional medical history

When congenital neutropenia is suspected, the medical history should include whether extra-hematologic areas are affected in any way. A detailed history of symptoms indicating underlying autoimmune or other diseases should be obtained.

Many adult AIN patients may present with various autoimmune symptoms (e.g., sicca symptoms, joint pains, and Raynaud phenomenon) without meeting the exact diagnostic criteria for autoimmune disease [26]. Many patients also complain of chronic fatigue and weakness. Chronic viral infections such as viral hepatitis or HIV should be addressed in the medical history. Additionally, a careful inquiry about drug administrations should be made, including over-the-counter medications, natural supplements, and recreational drugs. Neutropenia may be associated not only with drugs recently started by the patient but also with drugs recently discontinued.

4. Detailed physical examination

Depending on neutropenia severity and its underlying pathogenesis, physical assessments may uncover manifestations such as aphthous ulcers, dental and gingival anomalies due to chronic periodontitis, or signs of acute or chronic sinusitis, pharyngitis, or otitis. Dermatological abnormalities such as hyperpigmentation, partial albinism, rashes, ulcers or abscesses, and nail dystrophy could be evident. Abdominal scrutiny, especially in cases of pain and fever, may raise suspicion of neutropenic colitis or abdominal infections. Thorough examination of the perianal and perirectal regions is essential to rule out necrotizing cellulitis caused by *Clostridium* species or other anaerobic organisms, as infections might present only mild symptoms due to reduced inflammatory response. Presence of lymphadenopathy, splenomegaly, or hepatomegaly may suggest underlying infections, myeloid/lymphoid malignancies, or metabolic disorders.

Besides assessing growth and development, it is crucial to examine mucous membranes, gums, teeth, skin, nails, and tympanic membranes. *ELANE* mutations commonly correlate with periodontitis development in CN patients, while 30% of those with *HAX1* mutations may experience neurological issues [24]. CN syndromes may manifest as short stature, chondrodysplasia, skeletal or cardiac abnormalities, increased superficial vein visibility, myopathy, nail/hair/skin abnormalities, bronchiectasis symptoms due to recurrent chest infections, and hepatosplenomegaly. Additionally, photophobia, nystagmus, and oculocutaneous albinism might be linked with specific CN syndromes.

Adolescents warrant evaluation for potential autoimmune or other neutropenia-associated diseases, indicated by signs such as skin rashes, arthritis, vitiligo, hepatosplenomegaly, dry eyes or mouth, and lymphadenopathy. It is important to note that the absence of obvious clinical signs does not exclude the presence of a congenital disorder.

5. Which tests should be included in primary and further investigations?

Patients with acute neutropenia, especially those with signs of infection, should be promptly investigated and, if necessary, hospitalized depending on the severity of neutropenia and symptoms. For patients with chronic, isolated neutropenia and no phenotype suggesting any underlying CN syndrome, a basic evaluation scheme is recommended [18, 28, 29].

If the initial evaluation does not suggest autoimmune neutropenia, post-infection, or drug-induced neutropenias, the first-line investigation should be followed as adjusted for the appropriateness of the recommended tests, as shown in **Box 1**. If the initial evaluation is inconclusive, the second-line investigation should also be followed as shown in **Box 1**.

First-line investigations include:

- Biochemical tests including complete blood count, peripheral smear, liver and kidney function tests, immunoglobulin levels, CRP, vitamin B12, and folate.
- Flow cytometric analysis of peripheral blood lymphocyte subsets, virological antibody screenings (HBV, HCV, HIV, EBV, CMV, and Parvovirus), indirect antineutrophil antibodies (GIFT, GAT, and others), thyroid hormones (sT3, sT4, TSH), and antithyroid antibodies (anti-TG and anti-TPO).
- Flow cytometric analysis of TCR-α/β-positive double-negative (CD4- and CD8-) CD3 peripheral blood lymphocytes.
- Bone marrow examination.
- RF, ANA, ENA, and ds-DNA.

Second-line investigations include:

- Serial complete blood counts twice weekly for 6 weeks in family members to rule out cyclic neutropenia; copper; serum ceruloplasmin, anti-tTG-IgA, anti-gliadin antibodies IgA/IgG, and pancreatic isoamylase.

Genetic investigations.

Box 1.
What are the tests required for the evaluation of chronic neutropenia?

Genetic testing is recommended when the second-line investigation fails to rule out congenital neutropenia. Additionally, genetic testing should be expedited in young children with a family history of severe neutropenia or those with typical congenital anomalies or recurrent serious infections, and it should be performed after the first-line investigation.

6. The role of bone marrow examination and recommendations for type, timing, and frequency of assessments

The role of bone marrow examination is crucial in providing information regarding cellular composition, erythroid and myeloid precursor cells, their quantity, maturation, presence of megakaryocytes, and any dysplastic features. Additionally, it can provide insights into the presence of non-myeloid cells and contribute to differential diagnosis, particularly in adults.

In addition to morphology, cytogenetic evaluation can offer further information about acquired chromosomal abnormalities typically associated with MDS or AML. Bone marrow samples may also facilitate genetic testing for acquired somatic gene mutations (e.g., *RUNX-1, CSF3R, TP53*) contributing to leukemogenesis or indicate the risk of transformation in acquired neutropenias (e.g., *SRSF2* and *IDH1*) [7, 30–32].

Bone marrow examination in patients with severe chronic neutropenia can assist in differentiating MDS/AML and ruling out other conditions. If not performed during diagnostic workup, bone marrow examination should be conducted before initiating granulocyte colony-stimulating factor (G-CSF) therapy to exclude MDS/AML and monitor the risk of malignant transformation.

In many cases of chronic neutropenia, diagnostic bone marrow reveals early-stage maturation arrest of neutrophil precursors (promyelocyte/myelocyte) and contains few neutrophilic series cells beyond the promyelocyte stage. Eosinophilia is frequently observed, while cellularity is usually normal or mildly decreased [32]. Megakaryocytes are typically normal in number and morphology. When feasible, *in vitro* growth of granulocyte colonies in granulocyte-macrophage colony-forming unit assays may also demonstrate a maturation arrest mimicking the disease.

Combining bone marrow aspiration and biopsy with cytogenetic and flow cytometric analyses should be considered in patients with unexplained neutropenia to rule out MDS or other hematological disorders (e.g., leukemia, lymphoma, myeloma, and myelofibrosis) or even non-hematological malignancy infiltrating the bone marrow [33]. In patients with prolonged, isolated, mild neutropenia, if the condition remains stable and no underlying disease is evident during initial evaluation or there is no significant change in blood counts (if G-CSF therapy is not required), bone marrow examination can be skipped or does not need repetition [1, 29].

For patients with SCN and other BMF syndromes (independent of MDS and G-CSF treatment), annual BM examination (aspiration, biopsy, and cytogenetic evaluation) is recommended. It can also be considered for non-congenital idiopathic neutropenia requiring G-CSF therapy (**Box 2**).

Diagnostic bone marrow examination should be conducted concurrently with next-generation sequencing (NGS) of genes linked to morphology, cytogenetics, and myeloid malignancies in the following scenarios:

1. Patients exhibiting severe and moderate chronic neutropenia, excluding drug-induced neutropenias and primary autoimmune neutropenia (AIN) characterized by positive anti-granulocyte antibodies.

2. Patients suspected of AIN experiencing recurrent infections despite negative granulocyte antibody testing.

3. All patients presenting with unexplained chronic neutropenia, except those with long-standing mild isolated neutropenia that remains stable over time.

Annual bone marrow and cytogenetic monitoring should be carried out for the following patients:

1. Patients with congenital bone marrow failure (BMF) syndromes irrespective of absolute neutrophil count (ANC) and G-CSF treatment.

2. Patients with severe chronic neutropenia (CN) not diagnosed after thorough investigation, where G-CSF therapy might be considered.

Patients with a decreasing ANC or changes in other blood cell counts, such as anemia, thrombocytopenia, or altered erythrocyte indices, should undergo repeated BM follow-up.

Box 2.
What is the role of bone marrow examination? When and how often should it be performed?

7. Diagnosis of autoimmune neutropenia and the role of anti-neutrophils antibodies detection

Detection of antibodies in children typically confirms the diagnosis of autoimmune neutropenia (AIN), characterized by early onset, few infections, spontaneous resolution within a few years, and a low risk of leukemic development [34]. The specificity of autoantibodies in most patients is against FcyRIIIb (CD16b). The most commonly used method for detecting antineutrophil antibodies is the indirect granulocyte immunofluorescence test (GIFT). This test involves incubating test granulocytes with patient serum and then detecting bound antibodies with a fluorescently labelled anti-human immunoglobulin reagent.

Antibody positivity, particularly in the appropriate age group of children with a typical medical history, suggests the diagnosis of AIN. While considering alternative diagnoses may be less emphasized in such cases, the possibility of false-positive and false-negative results should always be considered. However, positive antibodies have also been detected in genetically proven cases of congenital neutropenia (CN). Therefore, this result should not be used to exclude this diagnosis [35]. Several studies have shown that small children with a typical medical history but no detectable antibodies have a disease course very similar to AIN patients [24]. Immunodeficiency neutropenia (IN) may be diagnosed in these patients after detailed examinations to exclude other diseases. IN and AIN have overlapping clinical features, but IN tends to have a longer duration of illness and presents at older ages with a different immunological profile [24, 25]. Repeat antibody testing with negative results is recommended in suspected AIN cases at reference laboratories after thorough evaluations to exclude other diseases.

7.1 Genetic testing: Which genes, methods, and cell sources play a role? What is their place in the diagnostic algorithm?

Genetic testing plays a crucial role in confirming diagnosis, predicting the relative risk of late complications such as MDS/AML, and providing genetic counseling to affected patients and family members in patients with CN. Due to the variability of clinical presentations, unaffected relatives who are potential donors for hematopoietic stem cell transplantation should always undergo genetic screening.

Following negative outcomes from initial assessments (**Box 1**), genetic analyses should be conducted on all individuals exhibiting AIN alongside recurrent infections, and/or those with a familial background of severe neutropenia and characteristic anomalies. This should be done utilizing either single-gene Sanger sequencing or multi-gene next-generation sequencing (NGS) techniques. Initially, Sanger sequencing of *ELANE* (found in around 45% of AIN cases) is advised for genetic diagnosis in typical scenarios [7]. Nevertheless, familial history or clinical indicators might indicate the necessity for sequencing another specific gene associated with neutropenia (refer to **Table 2**). For instance, sequencing of *TAZ* (Barth syndrome) could be diagnostic in cases featuring cardiomyopathy, whereas sequencing of *G6PC3* could be diagnostic in instances of cardiac and genitourinary malformations. Signs such as growth retardation, malabsorption, fatty stools, and bone malformations may indicate mutations in *SBDS*, which are prevalent in most Shwachman-Diamond syndrome (SDS) patients [36]. Overall, Sanger sequencing of suspected genes offers a relatively straightforward and cost-effective approach. Subsequent to the identification of responsible gene(s), Sanger sequencing for mutation screening is also recommended for affected family members [37].

Multi-gene next-generation sequencing (NGS) or whole exome sequencing (WES), preferably incorporating DNA samples from both the patient and parents (trio analysis), are indispensable. Targeted NGS panels encompassing all known mutated genes relevant to the clinical entity represent a logical initial step, ensuring homogeneous sequencing coverage for all pertinent genes and facilitating simpler bioinformatics analysis. The panel's gene selection should encompass not only those directly implicated in cyclic neutropenia (CN) upon mutation but also diseases featuring as secondary manifestations of CN (e.g., immune deficiencies/disorders, metabolic and nutritional deficiencies, and other hematopoietic failure syndromes). When mutations evade detection in targeted genes, WES becomes invaluable, broadening the analysis to encompass all coding regions of the genome. Given the intricate nature of bioinformatic interpretation regarding genetic variants' pathogenicity across all NGS applications, adherence to consensus guidelines is paramount. Assessing the pathogenicity of variants of unknown significance should heavily lean on family segregation studies and, if applicable, functional assays. In instances where suspicion of CN persists despite negative findings in both NGS panels and WES analysis, consideration should be given to whole-genome sequencing and RNA sequencing. Typically, DNA for potential germline mutation detection is sourced from peripheral blood. However, if blasts are present in peripheral blood, non-leukocyte cells such as skin fibroblasts, hair follicle keratinocytes, or oral mucosa cells are preferably utilized, albeit the latter carries a risk of blood contamination. Specialized NGS somatic panels can detect somatic mutations associated with leukemia acquired over time, playing a pivotal role in leukemic transformation. With the exception of severe CN/monocytopenia cases, peripheral blood and bone marrow serve as equivalent sources for somatic mutation detection in myeloid-specific genes. For RNA sequencing, RNA ideally derives from peripheral blood or bone marrow myeloid cells, preferably promyelocytes from bone marrow or neutrophils from peripheral blood [38].

Genetic tests can also be used to confirm ADAN in individuals of Middle Eastern, African, West Indian, Yemenite, and Ethiopian Jewish descent who exhibit mild to moderate (less frequently, MNS $<0.5 \times 10^9$/L) neutropenia without a tendency for infections [11, 40]. A precise genetic diagnosis can eliminate the need for extensive, unnecessary investigations. Research may identify the rs2814778 polymorphism in the *ACKR1/DARC-associated neutropenia (ADAN)* gene. The same polymorphism has also been detected in a cohort analysis of European patients (mostly Greeks), who constitute most of the chronic idiopathic neutropenia (CIN) cases [39]. Therefore, investigating the presence of the rs2814778 *ACKRI/DARC* polymorphism in all patients with chronic and mild neutropenia is sensible. All the above recommendations are summarized in **Box 3**.

- Genetic diagnosis holds paramount importance for confirming cyclic neutropenia (CN) diagnosis, predicting the risk of myelodysplastic syndrome (MDS) or acute myeloid leukemia (AML), aiding in stem cell donor selection for patients, and offering crucial family counseling.
- In cases where the clinical presentation, inheritance pattern, or bone marrow (BM) features, such as blocks in the promyelocytic stage, suggest a specific gene mutation, single gene analysis utilizing the Sanger sequencing technique can be employed.
- When the clinical picture fails to indicate a specific genetic cause, employing NGS techniques such as multi-gene panels for CN or targeted whole exome sequencing (WES) is advisable.

- For patients for whom a genetic cause remains elusive despite previous methods, whole genome sequencing (WGS) and RNA sequencing serve as potent diagnostic tools.
- In cases of chronic unexplained neutropenia, conducting NGS analysis of BM or peripheral blood to identify acquired somatic variants is recommended.
- Screening family members for known mutations is advised to ascertain genetic predispositions and enable appropriate monitoring or intervention.
- Confirming germline mutations is imperative when leukemic cells are detected in peripheral blood, particularly utilizing fibroblasts or hair follicle keratinocytes for sampling (while oral mucosa cells are less recommended due to potential blood contamination).

Box 3.
The role of genetic tests: Which genes, methods, and cell sources? Where do they fit into the diagnostic algorithm?

7.2 The role of flow cytometry in the diagnosis of chronic neutropenia

Flow cytometry analysis of peripheral blood and/or BM cell populations serves as a supportive tool in the diagnosis of chronic neutropenia types. It can evaluate the quantity of cell populations from different origins, as well as cell surface, intracellular, and intranuclear proteins, along with assessing the biological functions and immune properties of neutrophils and lymphocytes. Flow cytometry is beneficial in diseases associated with neutropenia such as myelodysplastic neoplasms, primary immunodeficiencies, and telomere biology disorders.

In cases of unexplained neutropenia, flow cytometry can provide high specificity (up to 95%) and sensitivity (up to 75%) in diagnosing myeloid precursor lesions or MDS cases [33]. Evaluating the BM myeloid series with flow cytometry using specialized panels recommended by the International Society for Advancement of Cytometry can aid in excluding MDS [41]. Myeloid series abnormalities indicative of MDS include quantitative and qualitative changes in CD34+ cells at the blast gate, abnormal distribution of mature and immature granulocyte cells, abnormal expression patterns of CD11b, CD13, CD33, CD16, CD10, CD5, CD7, and CD56 on granulocyte cells, and low granulocyte cell granularity based on the side scatter properties of granulocyte cells [33, 41]. Flow cytometry is also shown to be useful in identifying paroxysmal nocturnal hemoglobinuria clones [42].

While the definitive diagnosis of primary immunodeficiencies is typically confirmed through genetic analysis, flow cytometry particularly aids in the diagnosis of the following conditions: (a) autoimmune lymphoproliferative syndrome, identified by an elevated frequency of TCR-α/β positive double-negative (CD4- and CD8-) T cells. (b) Common variable immunodeficiency, characterized by a reduced frequency of CD19+ B cells, stemming from a diminished number of transitional memory (CD27 + IgD − IgM−) B cell subset. Additionally, it is distinguished by decreased expression of BAFFR in B cells among patients with BAFFR mutations and reduced expression of ICOS in activated T cells among patients with ICOS deficiency. (c) Hyper-IgM syndrome (HIGM), predominantly X-linked in individuals with deficiency of CD40 ligand (CD40L; CD154) expression on activated CD4+ T cells, and marked by the absence of CD40 expression on B cells in patients with autosomal recessive HIGM [43]. For the exclusion of telomere biology disorders, if feasible, Flow-FISH (or other clinically validated telomere length measurement techniques) is recommended. Given that this group of diseases cannot be entirely ruled out by genetic testing, telomere length measurement may help mitigate diagnostic gaps.

8. Natural history: Follow-up

8.1 How frequently should these disease groups be monitored (complete blood count, BM, cytogenetic, and somatic gene NGS analysis)?

Follow-up recommendations are also condensed in **Box 4**. Annual BM evaluation is advised to detect cytogenetic abnormalities, high-risk somatic mutations, or preceding leukemia development in children with CN. Morphological examination of BM smears, cytogenetic analysis, and somatic NGS-based sequencing for leukemia gene panels, or at least deep sequencing of *CSF3R*, *RUNX1*, and *TP53* should be conducted. Although *CSF3R* mutation sensitivity has been demonstrated to be high using NGS-based deep sequencing, BM morphology and cytogenetics remain requisite for MDS or leukemia diagnosis. The sensitivity of detecting mutations in the BM versus peripheral blood for other somatic leukemia-associated mutations is still under assessment. Annual NGS analysis permits the detection of new hematopoietic stem cell clones bearing somatic mutations and evaluation of associated variant allele frequencies (VAF). It also facilitates monitoring the behavior of hematopoietic stem cell (HSC) clones harboring acquired mutations over time. Notably, some CN patients with a notably high *CSF3R* HSC clone frequency do not develop leukemia for years. Furthermore, emerging evidence indicates that CN patients with diverse genetic backgrounds attributable to inherited CN may accumulate mutations associated with leukemia at a heightened frequency compared to age-matched healthy individuals. The same study

- All patients with congenital BM failure syndromes should undergo annual BM examination and cytogenetic analysis, along with NGS analysis of associated genes, regardless of MNS and G-CSF treatment, for early diagnosis of MDS or leukemia and evaluation of clonal hematopoiesis.
- For patients with chronic neutropenia, a complete blood count, peripheral smear, and morphological evaluation are recommended every 3–4 months.
- As adolescence approaches, referral of CN patients to adult hematologists is recommended.
- NGS analysis of acquired somatic variants is recommended in patients with unexplained chronic neutropenia, from both BM and peripheral blood.
- Patients with one or more genetic mutations associated with MDS/leukemia, especially those with high VAF (>10%), should be monitored more frequently with complete blood counts, peripheral smears, and morphological examinations (>4 times a year). BM examination, aspiration, biopsy, karyotyping, and NGS analysis are required if cytopenias worsen, macrocytosis occurs, and/or morphological abnormalities are observed.

Box 4.
Which patient groups should be monitored at what frequency (complete blood count, bone marrow, cytogenetics, somatic gene NGS)?

also suggests that the acquisition of multiple genetic lesions linked to *RUNX1, SETBP1, ASXL1, TP53, PTPN11*, particularly in the neutropenia stage in CN patients, could serve as a pre-leukemic marker and necessitate closer monitoring [44].

In a small subset of CN/AML patients without *CSF3R* or *RUNX1* mutations, other leukemogenic events do not develop. In SDS patients, the frequency of clonal hematopoiesis increases, particularly from a young age, especially in the presence of *TP53* and *EIF6* mutations. Patients who develop leukemia acquire biallelic alterations in the TP53 locus [45]. Therefore, annual NGS analysis of leukemia predisposition myeloid panels and SNP arrays should be performed in the BM or, if BM is unavailable, in peripheral blood myeloid cells.

The primary determinants of MDS or leukemia transformation in CN patients encompass typical dysplastic features evident in both peripheral blood and bone marrow, including pseudo-Pelger-Huët anomaly, hypogranulation, hypersegmentation, ring-shaped, and round nuclei, along with defective granulation, myelocyte stage maturation arrest, and an increase in monocyte-like forms [46]. Additionally, cytogenetic abnormalities and an elevated frequency of somatic mutations in genes associated with leukemia contribute significantly. Trisomy 21 and monosomy 7 are the most prevalent chromosomal defects observed in the MDS stage among CN patients. A distinct correlation exists between the high prevalence of somatic *CSF3R* and *RUNX1* mutations and the onset of leukemia in CN patients. Specifically, truncation mutations within the intracellular segment of the *CSF3R* receptor, resulting in the loss of one, two, or three tyrosine phosphorylation regions, are characteristic of leukemia-associated *CSF3R* mutations in CN patients. *RUNX1* missense or truncation mutations typify *de novo* AML.

Clonal hematopoiesis involving splicingosome gene mutations and mutation patterns implicating epigenetic regulators such as *TET2, DNMT3A, ASXL1*, or *IDH1/2* in CN patients carry a positive predictive value for MDS/leukemia transformation [47].

9. Special conditions

9.1 Pregnancy

Long-term G-CSF treatment due to severe neutropenia has been shown to be safe and effective in extending patients' lifespan. Desire for parenthood has become an increasingly important issue for adult patients. Two questions are frequently asked in this regard.

9.2 Is it permissible to continue using G-CSF as a treatment for neutropenia during pregnancy?

Data regarding the use of G-CSF during pregnancy are limited. However, publications have indicated that using G-CSF during pregnancy is safe, well-tolerated, and does not result in significant side effects [48].

9.3 What is the risk of having an affected child?

Given the rising number of adult patients with genetic neutropenia subtypes, preconception genetic counseling and supportive care for mothers during pregnancy are crucial. Patients should be informed that the likelihood of having affected children is reasonable considering the high quality of life achieved through G-CSF treatment.

9.4 Management of congenital neutropenia patients

In recent years, a consensus among various hematologists has emerged regarding the significance of a multidisciplinary approach, involving immunologists, radiologists, and dentists, to enhance patient monitoring. The management of severe chronic neutropenia should prioritize infection prevention, organ dysfunction management, and leukemia transformation prevention [49]. While all febrile episodes and infections necessitate prompt evaluation and treatment, peritonitis should be considered in cases presenting with abdominal pain. Severe infections are uncommon in cyclic neutropenia, with most infections typically responsive to antibiotic therapy. Long-term follow-up studies of cyclic neutropenia patients have indicated rare occurrences of significant complications such as spontaneous peritonitis, segmental intestinal necrosis, and sepsis [50].

Children afflicted with severe neutropenia necessitate the long-term administration of recombinant human G-CSF administered subcutaneously either daily or on consecutive days. This treatment approach remains similar for cyclic neutropenia patients. G-CSF treatment, whether filgrastim or pegfilgrastim, is deemed safe and well-tolerated. It effectively elevates neutrophil counts, ameliorates disease symptoms, and mitigates most infection-related complications in both congenital and cyclic neutropenia cases [51].

In cyclic neutropenia, G-CSF serves to abbreviate the duration of neutropenia and the length of neutropenic cycles. Following normalization of the absolute neutrophil count, resistance to infections tends to increase. Common side effects associated with G-CSF therapy encompass bone pain, headache, splenomegaly, and osteoporosis. Nevertheless, adverse effects on growth and development in children have not been documented. Consequently, G-CSF therapy empowers patients to engage in school and activities without apprehension. The optimal dosage for filgrastim in cyclic neutropenia typically ranges from 2 to 3 μg/kg/day, a lower dose compared to that employed for CN (5–10 μg/kg/day) or chemotherapy-related neutropenia, potentially necessitating adjustments contingent upon treatment response [52]. Treatment assumes critical importance in cases of complicated severe infections, often proving life-saving. Such infections may be instigated by a spectrum of complex anaerobic/aerobic pathogens. Synchronizing filgrastim treatment with the cyclic neutropenia period presents a challenge; however, G-CSF administrations frequently modulate the periodicity of neutropenia. The efficacy of the drug typically manifests beyond 6 months and may necessitate lifelong continuation as deemed necessary. The primary treatment objective revolves around maintaining the absolute neutrophil count above the lower threshold of 0.5×10^9/L, markedly diminishing the infection risk and enhancing oral mucosal health, thereby ameliorating both chewing ability and comfortable eating [53]. Pegfilgrastim, a long-acting pegylated G-CSF formulation, also proves effective and practical; however, dosing in children poses challenges and may elicit more pronounced bone pain as a side effect [54]. Although granulocyte-macrophage colony-stimulating factor has been utilized for severe chronic neutropenia, it demonstrates lesser efficacy compared to G-CSF and is accompanied by a higher incidence of side effects.

In granulocytes harboring *JAGN1* mutations, the response to G-CSF may be attenuated due to inadequate production of the *JAGN1* protein located in the endoplasmic reticulum. This deficiency is characterized by ultrastructural abnormalities, granule deficiencies, abnormal glycosylation of multiple proteins, and heightened apoptosis [55].

As children mature into adulthood, G-CSF dosage adjustments should be based on symptoms and blood counts rather than body weight. Given the fluctuating neutrophil levels in cyclic neutropenia, neutrophil counts should be periodically monitored for several weeks after commencing G-CSF therapy. In some instances, the neutrophil cycle may transition into mild chronic neutropenia by the third decade of life. Previous research has suggested that a combination therapy involving G-CSF and high-dose immunoglobulin might effectively abolish neutrophil oscillations; however, this treatment regimen has yet to receive approval [56].

To reduce the risk and severity of infections, supportive care is also crucial, including oral hygiene and dental care. Regular health assessments should be conducted at six-month intervals, and mouth rinses with chlorhexidine should be performed twice daily after brushing teeth [57]. This helps in maintaining oral hygiene. Regular monitoring of bone density should be performed during long-term treatment to reduce the risk of osteoporosis, which could be a potential side effect of G-CSF. Monitoring of vitamin D levels is also important in these patients [58]. Additionally, maintaining an age-appropriate vaccination schedule is important. However, avoiding public places where people gather due to local organisms in the intestinal microbiota is not necessary for most infections. Common viral infections can complicate bacterial infections. Prophylactic antibiotics are not recommended due to the potential emergence of resistant organisms [59].

9.5 Hematopoietic stem cell transplantation

Thanks to the effectiveness of G-CSF therapy, the need for HSCT to save the lives of patients with severe congenital neutropenia has been eliminated. However, there are two situations where HSCT remains the only viable treatment option. The first is a lack of or poor response to G-CSF treatment. The other indication is the development of AML or MDS or clonal hematopoiesis marked by somatic mutations. In the first case, successful transplantation results in normalized peripheral blood counts, eliminating the need for further G-CSF therapy [60–62]. Determining the optimal timing for HSCT in G-CSF non-responders is challenging due to potential toxicity concerns. For patients developing MDS or AML, survival prospects are significantly diminished without HSCT. Patients declining transplantation necessitate ongoing long-term monitoring to discern any malignant disease transformation. Historically, patients treated with standard chemotherapy for AML without HSCT had poor outcomes, with most succumbing to transplant-related complications [60]. However, since 2001, milder chemotherapy followed by HSCT conditioning has substantially improved survival rates, particularly by anticipating HSCT when feasible and avoiding aggressive chemotherapy prior to transplantation [63].

Regardless of the indication, the 5-year survival rate for HSCT recipients in the European SCNIR cohort is around 80% [63]. Nonetheless, HSCT-related morbidity, including GvHD and infections, as well as the risk of secondary malignancies, remains significant. AML relapse rates post-transplant are low, with a 71% three-year event-free survival reported in one analysis. Importantly, achieving full donor chimerism is crucial in preventing the re-emergence of predisposed clones, particularly in congenital neutropenia [60]. Preclinical data on gene-modified cell autologous transplantation for CN treatment are available for patients unresponsive to G-CSF therapy and who fail HSCT [64].

In the conditioning regimen for HSCT in *ELANE*-associated SCN, busulfan and cyclophosphamide are frequently used. However, these double alkylating agents carry significant risks [65]. MSD-BMT in patients with SCN using a myeloablative but reduced-toxicity conditioning regimen provides excellent outcomes with minimal toxicity. Additionally, other reduced-toxicity regimens, such as treosulfan/fludarabine, may be considered. Treosulfan has shown similar efficacy and a more favorable toxicity profile compared to busulfan, representing an opportunity to further optimize outcomes [66, 67].

For patients planned to undergo HSCT, although the most suitable donor is a fully matched sibling, it should be remembered that these donor candidates may be silent carriers of the disease. Therefore, it is necessary to investigate the sibling donor candidates before transplantation [68]. Recent reports have described no differences in terms of survival between sibling and MUD transplants, although GVHD was higher in MUD transplants. By contrast, data supporting HCT from mis-matched donors as standard procedure are scarce.

Patients with neutropenia other than severe congenital neutropenia often undergo HSCT using busulfan-based conditioning regimens. However, in special circumstances such as the patient's pre-transplant performance, organ insufficiencies, and other comorbidities, reduced-intensity or reduced-toxicity regimens may be preferred. Furthermore, parents of patients with this disease should be informed that congenital malformations seen in these patients may persist after HSCT, necessitating continued close monitoring by relevant clinical units.

10. Conclusion

Bone marrow failures with neutropenia other than severe congenital neutropenia are rare and clinically heterogeneous diseases. Various types of immune deficiencies, congenital malformations, and additional comorbidities can be observed. In isolated neutropenias, G-CSF therapy may be sufficient. However, in cases of G-CSF unresponsiveness, malignant transformation, or accompanying severe immune deficiency, HSCT is a curative treatment method. Reduced-intensity or reduced-toxicity regimens may be chosen in HSCT to avoid toxicity. Patients with congenital malformations should be closely monitored by relevant clinical units in the post-transplant period.

Conflict of interest

The authors declare no conflict of interest.

Author details

Baris Malbora
Gaziosmanpasa Hospital, Istanbul Yeni Yuzyil University, Paediatric Haematology and Oncology, Istanbul, Turkey

*Address all correspondence to: barismalbora@gmail.com

References

[1] Dale DC. How I diagnose and treat neutropenia. Current Opinion in Hematology. 2016;**23**:1-4. DOI: 10.1097/MOH.0000000000000208

[2] Badolato R, Fontana S, Notarangelo LD, Gianfranco S. Congenital neutropenia: Advances in diagnosis and treatment. Current Opinion in Allergy and Clinical Immunology. 2004;**4**:513-521. DOI: 10.1097/00130832-200412000-00007

[3] Segel GB, Halterman JS. Neutropenia in pediatric practice. Pediatrics in Review. 2008;**29**:12-23. DOI: 10.1542/pir.29-1-12

[4] Dale DC, Person RE, Bolyard AA, Aprikyan AG, Bos C, Bonilla MA, et al. Mutations in the gene encoding neutrophil elastase in congenital and cyclic neutropenia. Blood. 2000;**96**: 2317-2322. DOI: 10.1182/blood.V96.7.2317

[5] Dale DC, Hammond WP. Cyclic neutropenia: a clinical review. Blood Reviews. 1988;**2**:178-185. DOI: 10.1016/0268-960x(88)90023-9

[6] Furutani E, Newburger PE, Shimamura A. Neutropenia in the age of genetic testing: Advances and challenges. American Journal of Hematology. 2019; **94**:384-393. DOI: 10.1002/ajh.25374

[7] Skokowa J, Dale DC, Touw IP, Zeidler C, Welte K. Severe congenital neutropoenias. Nature Reviews. Disease Primers. 2017;**3**:17032. DOI: 10.1038/nrdp.2017.32

[8] Papadaki HA, Mavroudi I, Almeida A, Bux J, Cichy J, Dale DC, et al. Congenital and acquired chronic neutropoenias: Challenges, perspectives and implementation of the EuNet-INNOCHRON action. Hema. 2020;**4**: e406. DOI: 10.1097/HS9.0000000000000406

[9] Horwitz M, Benson KF, Person RE, Aprikyan AG, Dale DC. Mutations in ELA2, encoding neutrophil elastase, define a 21-day biological clock in cyclic haematopoiesis. Nature Genetics. 1999; **23**:433-436. DOI: 10.1038/70544

[10] Rigante D. When, how, and why do fevers hold children hostage? Journal of Evidence-Based Medicine. 2020;**13**: 85-88. DOI: 10.1111/jebm.12377

[11] Palmblad J, Höglund P. Ethnic benign neutropenia: A phenomenon finds an explanation. Pediatric Blood & Cancer. 2018;**65**:e27361. DOI: 10.1002/pbc.27361

[12] Merz LE, Achebe M. When non-whiteness becomes a condition. Blood. 2021;**137**:13-15. DOI: 10.1182/blood.2020008600

[13] Boxer LA, Newburger PE. A molecular classification of congenital neutropenia syndromes. Pediatric Blood & Cancer. 2007;**49**(5):609-614. DOI: 10.1002/pbc.21282

[14] Oyarbide U, Corey SJ. SRP54 and a need for a new neutropenia nosology. Blood. 2018;**132**(12):1220-1222. DOI: 10.1182/blood-2018-07-859959

[15] Thobakgale CF, Ndung'u T. Neutrophil counts in persons of African origin. Current Opinion in Hematology. 2014;**21**:50-57. DOI: 10.1097/MOH.0000000000000007

[16] Dale DC, Liles C. How many neutrophils are enough? Lancet. 1998;

351:1752-1753. DOI: 10.1016/S0140-6736(05)78742-0

[17] Newburger PE, Dale DC. Evaluation and management of patients with isolated neutropenia. Seminars in Hematology. 2013;**50**:198-206. DOI: 10.1053/j.seminhematol.2013.06.010

[18] Fioredda F, Onofrillo D, Farruggia P, Barone A, Veltroni M, Notarangelo LD, et al. Diagnosis and management of neutropenia in children: The approach of the study group on neutropenia and marrow failure syndromes of the paediatric Italian Hemato-oncology association (Associazione Italiana Emato-Oncologia Pediatrica - AIEOP). Pediatric Blood & Cancer. 2022;**69**: e29599. DOI: 10.1002/pbc.29599

[19] Tangye SG, Al-Herz W, Bousfiha A, Cunningham-Rundles C, Franco JL, Holland SM, et al. Human inborn errors of immunity: 2022 update on the classification from the International Union of Immunological Societies Expert Committee. Journal of Clinical Immunology. 2022;**42**:1473-1507. DOI: 10.1007/s10875-022-01289-3

[20] Newburger PE. Autoimmune and other acquired neutropoenias. Hematology. American Society of Hematology. Education Program. 2016; **2016**:38-42. DOI: 10.1182/asheducation-2016.1.38

[21] Fioredda F, Dufour C, Höglund P, Papadaki HA, Palmblad J. Autoimmune neutropoenias: Update on clinical and biological features in children and adults. Hema. 2022;**7**:e814. DOI: 10.1097/HS9.0000000000000814

[22] Villalba NL, Alonso-Ortiz MB, Maouche Y, Zulfiqar AA, Andrès E. Idiosyncratic drug-induced neutropenia and agranulocytosis in elderly patients. Journal of Clinical Medicine. 2020;**9**: 1808. DOI: 10.3390/jcm9061808

[23] Dale DC. How I manage children with neutropenia. British Journal of Haematology. 2017;**178**:351-363. DOI: 10.1111/bjh.14677

[24] Farruggia P, Fioredda F, Puccio G, Onofrillo D, Russo G, Barone A, et al. Idiopathic neutropenia of infancy: Data from the Italian neutropenia registry. American Journal of Hematology. 2019; **94**:216-222. DOI: 10.1002/ajh.25353

[25] Fioredda F, Rotulo GA, Farruggia P, Dagliano F, Pillon M, Trizzino A, et al. Late-onset and long-lasting autoimmune neutropenia: An analysis from the Italian neutropenia registry. Blood Advances. 2020;**4**:5644-5649. DOI: 10.1182/bloodadvances.2020002793

[26] Palmblad J, Nilsson CC, Höglund P, Papadaki HA. How we diagnose and treat neutropenia in adults. Expert Review of Hematology. 2016;**9**:479-487. DOI: 10.1586/17474086.2016.1142867

[27] Roques G, Munzer M, Barthez MA, Beaufils S, Beaupain B, Flood F, et al. Neurological findings and genetic alterations in patients with Kostmann syndrome and HAX1 mutations. Pediatric Blood & Cancer. 2014;**61**: 1041-1048. DOI: 10.1002/pbc.24964

[28] Palmblad J, Dufour C, Papadaki HA. How we diagnose neutropenia in the adult and elderly patient. Haematologica. 2014;**99**:1130-1133. DOI: 10.3324/haematol.2014.110288

[29] Gibson C, Berliner N. How we evaluate and treat neutropenia in adults. Blood. 2014;**124**:1251-1258. DOI: 10.1182/blood-2014-02-482612

[30] Olofsen PA, Touw IP. RUNX1 mutations in the leukemic progression of

severe congenital neutropenia. Molecules and Cells. 2020;**43**:139-144. DOI: 10.14348/molcells.2020.0010

[31] Klimiankou M, Kandabarau S, Zeidler C, Spanoudakis M, Farruggia P, Almeida A, et al. Accumulation of specific somatic leukaemia-associated mutations in congenital neutropenia precedes malignant transformation - new preconditions for treatment decisions. Blood. 2022;**140**(Supplement 1):994-995. DOI: 10.1182/blood-2022-158914

[32] Welte K, Boxer LA. Severe chronic neutropenia: Pathophysiology and therapy. Seminars in Hematology. 1997; **34**:267-278

[33] Valent P, Orazi A, Steensma DP, Ebert BL, Haase D, Malcovati L, et al. Proposed minimal diagnostic criteria for myelodysplastic syndromes (MDS) and potential pre-MDS conditions. Oncotarget. 2017;**8**:73483-73500. DOI: 10.18632/oncotarget.19008

[34] Porretti L, Farruggia P, Colombo FS, Cattaneo A, Ghilardi R, Mirra R, et al. Diagnostic value of cell bound and circulating neutrophil antibody detection in pediatric neutropenia. Pediatric Blood & Cancer. 2018;**65**: e26904. DOI: 10.1002/pbc.26904

[35] Dobrewa W, Madzio J, Babol-Pokora K, Lopacz P, Gierszon A, Guz K, et al. A high prevalence of neutrophil-specific antibodies in ELANE-mutated severe congenital neutropenia. Pediatric Blood & Cancer. 2023;**70**:e30247. DOI: 10.1002/pbc.30247

[36] Woloszynek JR, Rothbaum RJ, Rawls AS, Minx PJ, Wilson RK, Mason PJ, et al. Mutations of the SBDS gene are present in most patients with Shwachman-diamond syndrome. Blood. 2004;**104**:3588-3590. DOI: 10.1182/blood-2004-04-1516

[37] Lucas F, Michaels PD, Wang D, Kim AS. Mutational analysis of hematologic neoplasms in 164 paired peripheral blood and bone marrow samples by next-generation sequencing. Blood Advances. 2020;**4**:4362-4365. DOI: 10.1182/bloodadvances.2020002306

[38] Fioredda F, Skokowa J, Tamary H, Spanoudakis M, Farruggia P, Almeida A, et al. The European guidelines on diagnosis and Management of Neutropenia in adults and children: A consensus between the European Hematology Association and the EuNet-INNOCHRON COST action. HemaSphere. 2023;**7**(4):e872. DOI: 10.1097/HS9.0000000000000872

[39] Atallah-Yunes SA, Ready A, Newburger PE. Benign ethnic neutropenia. Blood Reviews. 2019;**37**: 100586. DOI: 10.1016/j.blre.2019.06.003

[40] Fragiadaki I, Papadakis S, Sevastaki G, et al. Increased frequency of the single nucleotide polymorphism of the DARC/ACKR1 gene associated with ethnic neutropenia in a cohort of European patients with chronic idiopathic neutropenia. American Journal of Hematology. 2020;**95**:E163-E166. DOI: 10.1002/ajh.25813

[41] van Dongen JJM, Lhermitte L, Böttcher S, Almeida J, van der Velden VHJ, Flores-Montero J, et al. EuroFlow consortium (EU-FP6, LSHB-CT-2006-018708). EuroFlow antibody panels for standardized n-dimensional flow cytometric immunophenotyping of normal, reactive and malignant leukocytes. Leukemia. 2012;**26**: 1908-1975. DOI: 10.1038/leu.2012.120

[42] Westers TM, Alhan C, Visser-Wisselaar HA, Chitu DA, van de Loosdrecht AA. Dysplasia and PNH-type cells in bone marrow aspirates of

myelodysplastic syndromes. Cytometry Part B: Clinical Cytometry. 2023;**104**(2): 162-172. DOI: 10.1002/cyto.b.22038

[43] Hirokazu Kanegane H, Hoshino A, Okano T, Yasumi T, Wada T, Takada H, et al. Flow cytometry-based diagnosis of primary immunodeficiency diseases. Allergology International. 2018;**67**:43-54. DOI: 10.1016/j.alit.2017.06.003

[44] Olofsen PA, Fatrai S, van Strien PMH, Obenauer JC, de Looper HWJ, Hoogenboezem RM, et al. Malignant transformation involving CXXC4 mutations identified in a leukemic progression model of severe congenital neutropenia. Cell Reports Medicine. 2020;**1**:100074. DOI: 10.1016/j.xcrm.2020.100074

[45] Kennedy AL, Myers KC, Bowman J, Gibson CJ, Camarda ND, Furutani E, et al. Distinct genetic pathways define pre-malignant versus compensatory clonal hematopoiesis in Shwachman-diamond syndrome. Nature Communications. 2021;**12**:1334. DOI: 10.1038/s41467-021-21588-4

[46] List AF, Sandberg AA, Doll DC. Myelodysplastic syndromes. In: Greer JP, editor. Wintrobe's Clinical Hematology. 12th ed. Baltimore: Lippincott Williams & Wilkins; 2008

[47] Malcovati L, Gallì A, Travaglino E, Ambaglio I, Rizzo E, Molteni E, et al. Clinical significance of somatic mutation in unexplained blood cytopenia. Blood. 2017;**129**:3371-3378. DOI: 10.1182/blood-2017-01-763425

[48] Berends C, Maggen C, Lok CAR, van Gerwen M, Boere IA, Wolters VERA, et al. Maternal and neonatal outcome after the use of G-CSF for cancer treatment during pregnancy. Cancers (Basel). 2021;**13**:1214. DOI: 10.3390/cancers13061214

[49] Pernu HE, Pajari UH, Lanning M. The importance of regular dental treatment in patients with cyclic neutropenia. Follow-up of 2 cases. Journal of Periodontology. 1996;**67**: 454-459. DOI: 10.1902/jop.1996.67.4.454

[50] Geelhoed GW, Kane MA, Dale DC, Wells SA. Colon ulceration and perforation in cyclic neutropenia. Journal of Pediatric Surgery. 1973;**8**: 379-382. DOI: 10.1016/0022-3468(73) 90105-x

[51] Horwitz MS, Duan Z, Korkmaz B, Lee HH, Mealiffe ME, Salipante SJ. Neutrophil elastase in cyclic and severe congenital neutropenia. Blood. 2007;**109**: 1817. DOI: 10.1182/blood-2006-08-019166

[52] Dale DC, Cottle TE, Fier CJ, Bolyard AA, Bonilla MA, Boxer LA, et al. Severe chronic neutropenia: Treatment and follow-up of patients in the severe chronic neutropenia international registry. American Journal of Hematology. 2003;**72**:82-93. DOI: 10.1002/ajh.10255

[53] Hammond WP, Price TH, Souza LM, Dale DC. Treatment of cyclic neutropenia with granulocyte colony-stimulating factor. The New England Journal of Medicine. 1989;**320**:1306-1311. DOI: 10.1056/NEJM198905183202003

[54] Matarasso S, Daniele V, Iorio Siciliano V, Mignogna MD, Andreuccetti G, Cafiero C. The effect of recombinant granulocyte colony-stimulating factor on oral and periodontal manifestations in a patient with cyclic neutropenia: A case report. International Journal of Dentistry. 2009; **2009**:654239. DOI: 10.1155/2009/654239

[55] Boztug K, Järvinen PM, Salzer E, Racek T, Mönch S, Garncarz W, et al. JAGN1 deficiency causes aberrant

myeloid cell homeostasis and congenital neutropenia. Nature Genetics. 2014;**46**:1021-1027. DOI: 10.1038/ng.3069

[56] Ishida Y, Higaki A, Tauchi H, Yokota Y, Matsuda H. Disappearance of neutrophil fluctuations in a child with cyclic neutropenia by combination therapy of granulocyte colony-stimulating factor and high-dose immunoglobulin. Acta Paediatrica Japonica. 1995;**37**:388-390. DOI: 10.1111/j.1442-200x.1995.tb03338.x

[57] Okada M, Kobayashi M, Hino T, Kurihara H, Miura K. Clinical periodontal findings and microflora profiles in children with chronic neutropenia under supervised oral hygiene. Journal of Periodontology. 2001;**72**:945-952. DOI: 10.1902/jop.2001.72.7.945

[58] Zeidler C, Reiter A, Yakisan E, Koci B, Riehm H, Welte K. Long-term treatment with recombinant human granulocyte colony stimulating factor in patients with severe congenital neutropenia. Klinische Pädiatrie. 1993;**205**:264-271. DOI: 10.1055/s-2007-1025236

[59] Shiohara M, Shigemura T, Saito S, Tanaka M, Yanagisawa R, Sakashita K, et al. ELA2 mutations and clinical manifestations in familial congenital neutropenia. Journal of Pediatric Hematology/Oncology. 2009;**31**:319-324. DOI: 10.1097/MPH.0b013e3181984dbe

[60] Zeidler C, Welte K, Barak Y, Barriga F, Bolyard AA, Boxer L, et al. Stem cell transplantation in patients with severe congenital neutropenia without evidence of leukemic transformation. Blood. 2000;**95**:1195-1198. DOI: 10.1182/blood.V95.4.1195.004k36_1195_1198

[61] Fioredda F, Iacobelli S, van Biezen A, Gaspar B, Ancliff P, Donadieu J, et al. Stem cell transplantation in severe congenital neutropenia: An analysis from the European Society for Blood and Marrow Transplantation. Blood. 2015;**126**:1885-1892. quiz 1970. DOI: 10.1182/blood-2015-02-628859

[62] Oshima K, Hanada R, Kobayaşi R, Kato K, Nagatoshi Y, Tabuchi K, et al. Hematopoietic stem cell transplantation in patients with severe congenital neutropenia: An analysis of 18 Japanese cases. Pediatric Transplantation. 2010;**14**:657-663. DOI: 10.1111/j.1399-3046.2010.01319.x

[63] Zeidler C, Nickel A, Sykora KW, Welte K. Improved outcome of stem cell transplantation for severe chronic neutropenia with or without secondary leukaemia: A long-term analysis of European data for more than 25 years by the SCNIR. Blood. 2013;**122**:3347. DOI: 10.1182/blood.V122.21.3347.3347

[64] Rao S, Yao Y, Soares de Brito J, Yao Q, Shen AH, Watkinson RE, et al. Dissecting ELANE neutropenia pathogenicity by human HSC gene editing. Cell Stem Cell. 2021;**28**:833-845. DOI: 10.1016/j.stem.2020.12.015

[65] Connelly JA, Choi SW, Levine JE. Hematopoietic stem cell transplantation for severe congenital neutropenia. Current Opinion in Hematology. 2012;**19**:44-51. DOI: 10.1097/MOH.0b013e32834da96e

[66] van der Stoep M, Bertaina A, Moes D, Algeri M, Bredius RGM, Smiers FJW. Impact of treosulfan exposure on early and long-term clinical outcomes in pediatric allogeneic hematopoietic stem cell transplantation recipients: A prospective multicenter study. Transplantation and Cellular Therapy. 2022;**28**(99):e1-e7. DOI: 10.1016/j.jtct.2021.09.018

[67] Suh JK, Im HJ, Kang SH, Kim H, Choi ES, Koh KN. Treosulfan-based conditioning regimen for allogeneic hematopoietic stem cell transplantation in children with non-malignant diseases. Bone Marrow Transplantation. 2022;**57**: 681-684. DOI: 10.1038/s41409-021-01556-8

[68] Donadieu J, Beaupain B, Fenneteau O, et al. Congenital neutropenia in the era of genomics: Classification, diagnosis, and natural history. British Journal of Haematology. 2017;**79**:557-754. DOI: 10.1111/bjh.14887

Chapter 4

Current Advances in Stem Cell-Based Therapies: Adult Stem Cell Applications

Neslihan Mandacı Şanlı and Aysu Timuroğlu

Abstract

Recent progress in stem cell technology offers promise for patients afflicted with various diseases and injuries. Stem cell-based therapy, encompassing multipotent mesenchymal stem cells (MSCs) and hematopoietic stem cells (HSCs) has emerged as a pivotal component in regenerative medicine. The aim is to repair or substitute diseased or damaged organs or tissues with healthy new cells through stem cell transplantation. Both MSCs and HSCs are multipotent precursors endowed with the ability to self-renew. Over 5000 registered clinical trial phases spanning diverse areas have utilized MSCs or HSCs for therapeutic interventions. This overview presents a comprehensive update on recent clinical applications employing either HSCs or MSCs sourced from allogeneic origins for addressing human ailments such as neurological disorders, pulmonary dysfunctions, leukemia, cardiovascular conditions, and skin injuries.

Keywords: mesenchymal stem cells, hematopoietic stem cells, stem cell therapy, regenerative therapy, advances

1. Introduction

In the last 10 years, the use of stem cells as a therapeutic method for treating diverse disorders has steadily increased. While stem cell therapy has traditionally been employed for hematopoietic disorders, it has now become recognized for treating non-hematologic conditions as well [1, 2].

Recent advancements in stem cell technology have spurred clinical investigations into stem cell-based treatment methods. Consequently, regenerative medicine is swiftly emerging as a critical domain in healthcare, aiming to improve body repair mechanisms and rejuvenate the cell reservoir for tissue homeostasis and regeneration. This is achieved through the stimulation, modulation, and regulation of the endogenous stem cell population [3]. Stem cell-based therapies fall within the realm of regenerative medicine and encompass any approach aimed at treating diseases or medical conditions by leveraging stem cells to regenerate damaged cells and tissues within the human body. This involves replacing compromised cells with fresh, healthy, and fully functional ones through the introduction of exogenous cells into a patient [4]. Stem cells possess distinctive characteristics such as self-renewal and differentiation [3]. Moreover, there's a growing understanding that beyond direct cell replacement, stem/progenitor cells can exert diverse effects like paracrine actions, cell recruitment, and modulation of the microenvironment to facilitate organ repair [5, 6]. Consequently, stem cell-based therapy holds promise for patients grappling with incurable diseases, particularly those where current treatment strategies primarily manage symptoms rather than targeting the root cause.

Two types of stem cell transplantation are employed in cell-based therapy: autologous transplantation, sourcing stem cells from the patients themselves, and allogeneic transplantation, utilizing stem cells from a healthy donor [3]. Regenerative medicine focuses on tissue regeneration and cellular replacement, employing various stem cell types such as human pluripotent stem cells (hPSCs) and adult stem cells to achieve these objectives [7, 8]. Given the intricate nature of stem cell-based treatments, researchers are exploring readily available, safe, and multipotent stem cell reservoirs. Consequently, meticulous selection of the appropriate stem cell type for clinical use is paramount [9, 10]. This review elucidates advancements in stem cell-based therapies, particularly those utilizing adult stem cells, which possess multipotent properties and are sourced allogenically, emphasizing their application in regenerative treatment modalities.

2. Stem cell-based therapies

2.1 Hierarchy of stem cells

There are primarily three types of stem cells, with the most rudimentary and least specialized being totipotent stem cells. These cells possess the remarkable ability to mature into a complete embryo. Totipotent cells, found in the zygote and early blastomeres resulting from the fusion of sperm and egg (1–3 days post-oocyte fertilization), can generate all cell types. They belong to the embryonic stem cell category, located within the inner cell mass of a blastocyst formed 4–6 days after oocyte fertilization. Consequently, they can give rise to all three germ layers (ectoderm, mesoderm, endoderm), displaying pluripotent characteristics capable of producing various cell types [8]. Following successive divisions, stem cells lose their pluripotent property and transition to a multipotent state, restricting their differentiation capacity to specific lineages associated with their tissue of origin. Stem cells at this stage are termed adult stem cells [11]. Many researchers consider adult stem cells as the benchmark in stem cell-based therapies [12, 13], particularly in the realm of hematopoietic transplants, where they have shown promising clinical outcomes [14, 15]. Presently, over 5700 trials are registered in the WHO International Clinical Trials Registry involving the utilization of adult stem cells. Among these, HSCs and MSCs are the most frequently employed [8].

2.2 Adult MSCs in regenerative medicine

MSCs are presently among the most thoroughly researched therapeutic cellular products for clinical use [16]. Among progenitor cells, MSCs have garnered substantial attention and have demonstrated promise in regenerative medicine. A primary objective of regenerative medicine is to restore failing organs by replacing dysfunctional cells. Progress in techniques for isolating, culturing, and expanding MSCs has facilitated their widespread therapeutic implementation. This advancement, coupled with rapid enhancements in transplantation methodologies, has bolstered the utilization of MSCs in regenerative medicine [16, 17]. MSCs possess the capacity for self-renewal and restricted differentiation. Initially characterized as stromal cells originating from the bone marrow, they exhibit properties akin to hematopoietic stem cells but lack the ability to differentiate into hematopoietic cells [18].

MSCs are found in various tissues and organs, including adipose tissue, bone marrow, skin, fallopian tubes, cord blood, liver, and lungs [19]. To establish uniformity and define MSCs, the International Society for Cell and Gene Therapy (ISCT) has established minimal identification criteria applicable to MSCs derived from multiple tissue sources [20]. Among these sources, MSCs obtained from adipose tissue (AT), umbilical cord (UC), and bone marrow (BM) are the most extensively investigated in human clinical trials [21], representing the three primary tissue origins of MSCs [3].

Over the past few decades, MSCs have exhibited diverse biological functions, including multilineage differentiation, immunomodulation, angiogenesis, anti-apoptotic and anti-fibrotic activities, chemo-attraction, and tissue repair enhancement [22–25]. Moreover, MSCs have the capability to stimulate the proliferation and differentiation of other stem cells through the secretion of various molecules. Leveraging these properties, regenerative medicine has advanced treatments for a wide array of diseases, such as lung, liver, and neurological disorders [26, 27]. Given these remarkable attributes, research efforts have been predominantly directed toward MSCs [25], underscoring their pivotal therapeutic potential in clinical trials.

2.2.1 Therapeutic implementations of MSCs

The trophic and immunomodulatory characteristics of MSCs have positioned these cellular products as highly promising and vigorously pursued therapies.

2.2.1.1 Tissue trophic effect of MSCs

Numerous studies have demonstrated the capacity of MSCs to migrate, engraft, and effectively impact the repair process at the site of injury and damage [28, 29]. Post-injury, their anti-inflammatory activities play a crucial role in mitigating damage, clearing dead tissue, facilitating the migration and proliferation of reparative cell types, and enhancing vascularization and nutrient delivery [30]. In the presence of MSCs, the healing process accelerates, and the inflammatory response diminishes [31]. Paracrine signaling and differentiation, as suggested by Kim et al. [32], have both been implicated in the wound healing process [16].

MSCs actively contribute to the tissue repair process through their immunomodulatory, trophic antibacterial, antifibrotic, and proangiogenic capacities [33]. Additionally, during wound healing, MSCs assume a central role by orchestrating interactions among local cells/progenitors, cytokines, chemokines, and extracellular matrix proteins [34, 35]. Numerous in vitro and in vivo studies have affirmed the immunomodulatory prowess of MSCs, demonstrating their ability to suppress the proliferation and functions of T cells, B cells, dendritic cells, and natural killer cells [36–39]. As a result of these attributes, investigations are underway to assess the efficacy of mesenchymal stem cells in treating graft versus host disease, a complication that arises post-stem cell transplantation [40].

2.2.1.2 Immune-modulating effect of MSCs

MSCs are recognized for their immunomodulatory effect, endowing them with the potential to facilitate wound repair and alleviate inflammation [41]. Their immunological properties render MSCs pivotal during the tissue repair process [41, 42]. MSCs exert robust immunomodulatory effects through the secretion of various molecules, including transforming growth factor β (TGF-β), prostaglandin E2 (PGE2), interleukin-10 (IL-10), human leukocyte antigen class I molecule (HLA)-G5, inducible nitric oxide synthase (NOS2), CD39, and CD73 molecules. These factors impede the proliferation of numerous immune cells and the secretion of cytokines [IL-1, IL-6, IL-8, IL-12, tumor necrosis factor α (TNF-α), interferon-γ (IFN-γ), TNF-α], as well as chemokines (CCL2, CCL5) [43, 44]. MSCs also suppress the activation and proliferation of CD4+ and CD8+ T cells and reduce immunoglobulin production by B cells, rendering them suitable for allogeneic transplantation [45]. Moreover, it has been demonstrated that MSCs suppress the allogeneic T lymphocyte response, thereby facilitating the persistence of skin grafts [45]. In their study, Han and colleagues illustrated that MSCs lack significant immune-stimulating antigens, making them viable for biomedical applications and tissue engineering without the risk of graft rejection post-transplantation [17]. MSCs possess the ability to modulate the function of lymphocytes and macrophages by secreting PGE2 and IL-10 [31, 46]. On one hand, PGE2 plays a crucial role in regulating the transition of TH1 cells to TH2 cells, thereby reducing the activation and proliferation of proinflammatory lymphocytes within the injured tissue [47]. Conversely, IL-10 aids in inhibiting scar formation by averting the buildup of collagen I and III and the release of reactive oxygen species (ROS) in the wound vicinity [48]. The heightened presence of regulatory T (Treg) cells, inhibition of differentiation into antigen-presenting cells (APCs), and enhancement of natural killer (NK) cell cytotoxicity collectively amplify the anti-inflammatory effect. By modulating both innate and adaptive immune responses, MSCs effectively curb the progression of fibrosis [49–51].

2.2.2 Clinical indications and considerations of MSCs

Although MSCs can be sourced from various tissues, they are primarily categorized into two types based on their origins: adult-type sources and perinatal sources. Adult sources of MSCs are obtained from tissues such as dental pulp, BM, peripheral blood, AT, lungs, hair, or heart [52–59]. BM was the first adult tissue from which MSCs were isolated and has since been extensively studied. BM-derived MSCs (BM-MSCs) are extracted and cultured in vitro from BM aspirates, while AT-derived MSCs (AT-MSCs) are obtained from stromal vascular fractions of enzymatically digested adipose tissue acquired through liposuction [56], lipoplasty, or lipectomy procedures [60]. These tissue collection procedures can be invasive and painful for the patient. Perinatal sources of MSCs originate from components such as UC, Wharton's jelly, UC blood, as well as placental structures like the placental membrane, amnion, chorion membrane, and amniotic fluid [53]. Harvesting perinatal MSCs, such as UC-MSCs, is noninvasive since the placenta, UC, UC blood, and amnion are typically considered waste products after birth [61].

Despite the similar phenotypes of MSCs, they exhibit heterogeneous biological and functional characteristics. This diversity stems from variations in their growth and proliferation rates, multilineage differentiation potential, immunomodulatory capacities, and proangiogenic properties [17]. For instance, in terms of differentiation ability, donor-matched BM-MSCs demonstrate a greater propensity to differentiate into chondrogenic and osteogenic cell types compared to AT-MSCs, whereas AT-MSCs exhibit a stronger inclination toward the adipogenic lineage [62]. Additionally, MSCs obtained from fetal tissues exhibit higher proliferation rates and lower immunogenicity when compared to BM-MSCs and AT-MSCs [16, 47].

As of now, there are over 1400 registered clinical trial phases encompassing various trials utilizing MSCs for therapeutic purposes. Through numerous preclinical and clinical investigations, MSCs have demonstrated efficacy in treating a range of conditions including nervous system and brain disorders, lung diseases, cardiovascular conditions, wound healing, and more [3, 63, 64]. While existing data suggest diverse mechanisms underlying the beneficial effects of MSCs, such as their differentiation potential, secretion of trophic factors like cytokines, chemokines, and growth factors, as well as their immunomodulatory capabilities, it remains unclear which type of MSC is most effective for specific diseases [3, 65].

2.2.2.1 MSCs for neurodegenerative diseases

Due to the successful generation of neural cells from stem cells in laboratory settings, clinical studies employing stem cell-based therapies for neurodegenerative diseases have commenced [66, 67]. These treatments seek not only to halt the progression of irreversible conditions like Parkinson's, Alzheimer's, amyotrophic lateral sclerosis (ALS), and multiple sclerosis (MS) but also aim to achieve complete remission of such disorders [8]. MSCs are being explored for the treatment of degenerative disorders and diseases due to their demonstrated antioxidative, anti-apoptotic, and immunomodulatory effects [16, 25]. MS is characterized as an inflammatory and neurodegenerative autoimmune ailment affecting the central nervous system, with involvement from both the peripheral nervous system and tissue protection mechanisms in its etiology. Numerous trial studies utilizing various types of MSCs have demonstrated moderate to beneficial outcomes in cell applications. For instance, a phase 1b trial employing human placental MSCs revealed the safety of intravenous MSC treatment in 16 patients, with observed stability or improvement in patients following the procedure [68].

In a separate investigation led by Fernández, oil-based MSC treatments were administered to 19 MS patients at varying doses, resulting in safe procedures and the observation of beneficial clinical and radiological effects [69]. Similarly, Riordan and colleagues noted comparable clinical enhancements in 20 patients who received UC-MSC interventions [70].

A recent study revealed that administration of UC-MSCs significantly enhanced both gross motor function and cognitive skills in patients diagnosed with cerebral palsy. The observed improvements reached their peak at 6 months post-administration and persisted for 12 months following the initial transplantation [71]. In another open-label, phase I study, UC-MSCs were utilized as treatment cells for 12 patients with autism spectrum disorders via intravenous infusion. Results showed that only 50% of the patients exhibited improvement on at least two autism-specific measures [72]. Additionally, a study involving 22 patients with spinal cord injury (SCI) demonstrated potential therapeutic effects in 13 patients following UC-MSC infusion [73]. AT-MSCs have also been employed in SCI treatment, with a single case report indicating neurological and motor function improvement in a domestic ferret patient [74]. However, results from another phase I trial utilizing AT-MSCs showed only slight improvements in neurological function in a small cohort of patients [75]. Recently, de Celis-Ruiz and colleagues conducted a phase II, randomized, double-blind, placebo-controlled, single-center study, demonstrating that intravenous administration of adipose MSCs within the first 2 weeks of ischemic stroke onset is safe at 24 months of follow-up [76].

In summary, these findings suggest that MSCs may offer a safe and efficacious treatment option for neurological diseases, leading to notable enhancements in patients' quality of life and clinical symptoms. Consequently, MSCs hold the potential to emerge as a novel therapeutic avenue for the management of neurological disorders.

2.2.2.2 MSCs for respiratory diseases

The surge in respiratory diseases attributed to factors like air pollution, tobacco use, aging populations, and the emergence of respiratory viruses such as coronavirus (COVID-19) has placed substantial strains on global public health and healthcare systems [77]. Among the most prevalent respiratory ailments in both children and adults are bronchopulmonary dysplasia, Chronic Obstructive Pulmonary Disease (COPD), and acute respiratory distress syndrome (ARDS). The progression of these conditions typically involves inflammatory cell infiltration, deterioration of alveolar structure, reduced ability to clear alveolar fluid, release of cytokines leading to cytokine storms, airway remodeling, and the onset of pulmonary fibrosis.

MSCs are renowned for their immunomodulatory prowess, holding promise for mitigating injuries and facilitating lung recovery following damage. They represent an alternative therapeutic approach for respiratory diseases, particularly MSCs sourced from UC [3]. COPD, affecting over 65 million individuals globally, ranked as the third-leading cause of death in 2020. Characterized by chronic inflammation and progressive lung function decline, COPD poses significant health challenges. A pilot clinical study showcased the safety and potential efficacy of allogeneic administration of UC-MSCs in COPD treatment [78].

In another study involving 20 patients with a history of smoking, who underwent cell-based therapy, those treated with UC-MSC therapy exhibited notable improvements. Six months post-application, there were significant reductions in Modified Medical Research Council scores, COPD assessment test scores, and the frequency of pulmonary exacerbations observed. Additionally, 3 months after treatment, there were increases in FEV1/FVC ratios along with enhancements in respiratory system-specific measurements [79]. The objective of MSC therapy for COPD treatment is to foster the regeneration of parenchymal cells and alveolar structure, ultimately restoring lung function. While comprehensive assessments of UC-MSC effectiveness are still in early stages, the mounting number of studies utilizing UC-MSCs for COPD treatment signifies a promising trend, offering substantial data to support their future applications [80].

In a separate investigation, Averyanov et al. [81] conducted a randomized, placebo-controlled phase I/IIa study involving 20 patients diagnosed with mild to moderate idiopathic pulmonary fibrosis (IPF). Patients in the treatment cohort received intravenous allogeneic MSCs every 3 months, while those in the control group were administered a placebo. Over the course of 1 year, follow-up assessments revealed that patients treated with MSCs experienced a consistent increase in their 6-minute walk test distance (6MWD) from the 13th week, a improvement sustained until the 52nd week. Moreover, pulmonary function indicators exhibited significant improvement pre- and post-treatment in the MSC-treated group, whereas no significant changes were observed in the placebo group.

Similarly, several preclinical studies have highlighted the protective effects of MSCs in murine models of sepsis and septic shock [82]. A single-center prospective randomized clinical trial conducted in Russia evaluated the use of BM-MSCs in neutropenic patients with septic shock, revealing hemodynamic stabilization, withdrawal of vasopressors, mitigation of respiratory failure, and shortened duration of neutropenia [83]. Additionally, a preliminary analysis of a phase 1 and 2 study utilizing a good manufacturing practice product of allogeneic BM-MSCs in ARDS demonstrated improvements in oxygenation, reduction in lung edema, and decreased levels of proinflammatory cytokines [84].

While MSCs have emerged as promising candidates for the treatment of lung diseases, further randomized controlled trials are warranted to validate these findings.

2.2.2.3 MSCs for COVID-19

COVID-19 is caused by infection with severe acute respiratory syndrome coronavirus-2 (SARS-CoV-2) [85]. The COVID-19 pandemic has emerged as a significant global health crisis, particularly affecting lung health and leading to a high mortality rate worldwide. Numerous studies, reviews, and systematic analyses have been conducted to enhance our understanding of the virus. One of the most critical effects of the virus on the immune system is the initiation of hyperreactions within the adaptive immune response, resulting in the formation of cytokine storms [86, 87]. These cytokine storms are triggered by increased expression of IL-6, IL-10, or TNF-α from immune cells and propagate throughout the body via circulation once initiated [88].

Due to their immunosuppressive effects, MSCs have been increasingly utilized in COVID-19 treatment studies. Among 1047 patients enrolled in these studies, UC-MSCs have been the most commonly used MSCs, with 15 studies employing them, encompassing approximately 600 patients [88]. Eight of these studies administered allogeneic UC-MSC transplantation to critically ill patients [87]. Evaluation of case reports utilizing UC-MSCs has demonstrated their safety and tolerability in 14 patients with COVID-19. Key outcomes include increased percentages of T cells, improved respiratory and renal capacity, decreased levels of inflammatory biomarkers, and enhanced PaO_2/FiO_2 ratio [89–91].

A phase 2 trial demonstrated the beneficial effects of UC-MSCs on lung damage in severe COVID-19 patients. The trial involved 101 severe COVID-19 patients with lung damage, who were randomly assigned to two groups. The non-placebo group received intravenous administrations of UC-MSCs on days 0, 3, and 6, whereas 35 patients in the placebo group received placebo infusions. Compared to the placebo group, UC-MSCs treatment led to a significant improvement in whole lung lesion volume, with a median difference of 13.31%. Additionally, patients in the treatment group exhibited a longer 6-minute walking distance compared to those in the placebo group. No adverse effects were observed during the study [92].

While the efficacy of MSC therapy in treating COVID-19 has been demonstrated, phase 3 trials are necessary to assess its impact on reducing mortality and preventing long-term pulmonary disability.

2.2.2.4 MSCs for liver regeneration

Current studies indicate that stem cell therapy has demonstrated significant improvements in liver function by exerting anti-apoptotic and immune regulatory effects. MSCs possess the ability to differentiate into hepatocytes, playing a crucial therapeutic role in treating liver fibrosis. In vivo experiments have shown that MSCs treatment leads to an increase in anti-apoptotic factors such as hepatocyte growth factor (HGF) and insulin-like growth factor (IGF-1), as well as angiogenic and mitogenic factors [93]. In vitro experiments have revealed that MSCs can promote apoptosis of hepatic stellate cells and inhibit collagen synthesis [94, 95]. Clinical studies have demonstrated the effectiveness of BM-MSC treatment in patients with liver disease [96]. Wang and colleagues have reported that intravenous injection of UC-MSCs is feasible and well-tolerated in patients with primary biliary cirrhosis (PBC) [97]. They found that MSCs significantly reduced the levels of alkaline phosphatase (ALP) and gamma-glutamyl transferase (GGT), although there were no significant changes in serum aspartate aminotransferase (AST), alanine aminotransferase (ALT), total bilirubin, albumin, and prothrombin time activity. In another study, Zhang et al. demonstrated that MSC therapy reduced serum ALP, GGT, and total bilirubin levels in patients with ischemic-type biliary lesions following liver transplantation [98].

In a single-center phase I–II study, nine patients diagnosed with acute-on-chronic liver failure (ACLF) grades 2 and 3 were enrolled [99]. The experimental group (n = 4) received BM-MSCs for a duration of 3 weeks. Remarkable improvements in ACLF severity (from grade 3 to 0) and the model of end-stage liver disease (MELD) score were observed. Lin et al. conducted a study involving 110 patients with HBV-related ACLF, with half of the patients undergoing allogeneic BM-MSC transplantation and being monitored for 6 months [100]. No serious adverse effects related to transplantation were reported, and improvements were noted in parameters such as serum total bilirubin levels and MELD scores.

The group receiving MSCs treatment also showed a reduction in rates of organ failure and infections.

In conclusion, MSCs can significantly improve the clinical symptoms of these patients, diminish liver fibrosis, and impede disease progression.

2.2.2.5 MSCs for bone regeneration

Research on bone regeneration is currently a focal point in clinical studies due to its significance in various conditions such as bone fracture, defect, osteoarthritis, and osteoporosis, all of which necessitate resolution [101–103]. With advancements in understanding bone tissue biology and recent innovations in tissue regeneration, the use of MSCs, either allogeneic or autologous, has emerged as an appealing approach to enhance bone tissue formation [104, 105]. Shim and colleagues [106] reported beneficial outcomes from the intramedullary (4×10^7 cells) and intravenous (2×10^8 cells) infusion of Wharton's Jelly-derived MSCs combined with teriparatide in individuals with osteoporotic vertebral compression fractures, demonstrating stimulation of bone architecture. Similarly, a study by Dilogo et al. [107] found that UC-MSCs (10×10^6 cells) significantly reduced the osteoarthritis index, suggesting their potential as a novel regenerative treatment for patients with knee osteoarthritis.

Overall, these findings indicate that MSCs hold promise as a safe and effective alternative for bone regeneration, leading to significant improvements in the clinical outcomes of patients.

2.2.2.6 MSCs for heart regeneration

Stem cell transplantation has been a focus of research and clinical application in patients with ischemic heart disease (IHD) for over 20 years, with extensive worldwide utilization [108, 109]. Numerous clinical studies have been undertaken to explore this avenue [109–111]. Given the limited regenerative capacity of the myocardium, stem cell transplantation has emerged as a promising approach for accelerating myocardial regeneration.

Heart disease stands as the foremost and most frequently diagnosed ailment, representing the leading cause of disease-related deaths [112]. When cardiomyocytes undergo damage due to ischemic and other factors, they are replaced by non-contractile fibrous tissue. Transplantation of MSCs into failing hearts has shown promise in providing functional benefits, as reported in the literature [113–115].

The RIMECARD trial, a pilot study employing UC-MSCs in 30 patients with heart failure, demonstrated improved ejection fraction, left ventricular function, functional status, and quality of life following UC-MSC administration [116].

In a study by Ulus and colleagues involving patients with chronic ischemic cardiomyopathy, 54 patients were randomized to receive either UC-MSCs or bone marrow mononuclear cells (BM-MNCs) in conjunction with coronary artery bypass-grafting surgery (CABG), while control patients underwent only CABG intervention without cell therapy. Both cell-treated groups exhibited a decline in NT-proBNP levels, with an increase in left ventricular ejection fraction and stroke volume observed only in the UC-MSC group. Furthermore, reductions in necrotic myocardium were more pronounced in both the BM-MNC and UC-MSC groups compared to the control group [117].

In another study conducted by He et al., patients with chronic ischemic heart disease were administered intramyocardial injections of UC-MSCs embedded in collagen hydrogel in the collagen/cell group, while the cell group received UC-MSCs alone. The control group received no cell therapy and was monitored accordingly, with a notable decrease in infarct area observed in the cell therapy groups [111].

In conclusion, MSC therapy holds promise as an effective and safe approach that significantly improves cardiac function and enhances patients' quality of life.

2.2.2.7 MSCs for posttransplant graft versus host disease

Allogeneic hematopoietic stem cell transplantation (allo-HSCT) represents a potentially curative treatment option for various disease conditions, particularly hematological malignancies. However, its therapeutic benefits are often hindered by severe complications, such as graft versus host disease (GVHD), triggered by donor T-lymphocytes attacking host tissues. GVHD manifestations are typically classified as acute or chronic, historically based on the time of occurrence (<100 or >100 days post-transplantation) [118]. Nevertheless, acute GVHD can persist beyond 100 days, leading to overlapping features between acute and chronic syndromes [119].

GVHD contributes to the mortality of 8–16% of adult allogeneic HSCT recipients [120], likely impacting posttransplant mortality rates through predisposition to other common causes such as organ failure, infection, and bleeding. The management of GVHD is intricate and varies widely among medical centers globally [121]. Corticosteroids remain the primary treatment for both acute and chronic GVHD, yet approximately 50% of acute GVHD cases are refractory to high-dose steroids, resulting in a dismal prognosis with 2-year overall survival rates below 20% [122]. Despite investigations into diverse second-line agents for GVHD treatment post-steroid failure, no optimal therapy has been established for either acute or chronic GVHD, necessitating the exploration of new therapies with improved safety and efficacy profiles. Over the past two decades, there has been substantial interest in the therapeutic potential of MSCs across various clinical contexts, including the treatment of GVHD [16, 18, 119–123].

MSCs play a crucial role in tissue repair and regeneration, engaging various mechanisms, including anti-apoptotic, anti-inflammatory, immunomodulatory, anti-fibrotic, and angiogenic agents [16, 18]. Additionally, MSCs promote the generation of regulatory T cells, which exert immunosuppressive effects [16, 18]. As our comprehension of MSCs expands, they have emerged as an exciting tool for GVHD in the HSCT setting [40].

The pioneering use of MSCs to address GVHD in human subjects was documented by Le Blanc et al. in 2004 [124]. Since this seminal case report, over 40 clinical trials have been conducted to explore MSCs as a therapeutic avenue for GVHD, with registrations on clinicaltrials.gov. The predominant focus of these investigations has been on allogeneic MSCs, primarily derived from bone marrow. However, a minority of trials have utilized MSCs sourced from cord blood or adipose tissue. Notably, clinical studies investigating MSCs for steroid-refractory acute GVHD (SR-aGvHD) have demonstrated encouraging efficacy outcomes, albeit with overall response rates varying within the range of 50–80% [40, 125–129].

2.2.3 Application of MSCs in tissue engineering

In the realm of regenerative medicine, MSCs offer diverse properties, making them valuable assets for tissue engineering applications [18]. The advancement of bioprinting technology has provided a robust means of constructing tissue and organ structures within the field of tissue engineering [130]. Tissue engineering comprises three fundamental components: cells, scaffolds, and a variety of biomolecules [130]. By employing various bioprinting techniques, cells are integrated with scaffolds, while the presence of biomolecules such as growth factors facilitates the induction and sustenance of viability within the newly formed structure [130, 131]. Scaffolds play a pivotal role in tissue engineering, furnishing a conducive environment for cellular growth and development.

In the realm of regenerative therapy, the challenges of low engraftment and diminished stem cell viability present significant hurdles. Local application is particularly problematic in the context of damaged tissue environments. To address these issues, stem cells are often combined with biomaterial scaffolds to enhance cell viability and bolster survival rates. Among the most utilized biomaterials for scaffolds is hyaluronic acid (HA), a constituent of the extracellular matrix [132–134]. The integration of stem cells with HA has found widespread application in cartilage repair in animal models such as rats and rabbits [135, 136]. Moreover, studies have demonstrated favorable outcomes in wound healing in mice when MSCs are combined with HA [137].

The advent of bioprinting technology has opened avenues for the production of heart valves, myocardial tissue, vessels, and tracheal structures [138–144]. Notably, in a study by Gaebel et al. [145], a heart patch comprising human MSCs and endothelial cells seeded on polyester urethane urea was implanted into damaged rat hearts, showcasing enhanced vessel formation. Similarly, Duan et al. [146] observed robust viability and remodeling at day 7 with the construction of a trileaflet heart valve conduit comprised of hyaluronic acid, gelatin, and human aortic valve interstitial cells. These findings underscore the potential of bioprinting technology to generate cellularized tissue valves or vessels.

2.3 Hematopoietic stem cells

HSCs are versatile precursors with the capacity for self-renewal and the ability to regenerate all cell types within the blood-forming system [5, 147]. These cells are harbored in various organs, including peripheral blood (PB), BM, and UC blood. Understanding the molecular mechanisms governing the self-renewal and fate

DOI: http://dx.doi.org/10.5772/intechopen.1007380

determination of HSCs/progenitor cells is pivotal for tailoring clinical applications to specific disease types and severities [5, 148].

HSCT stands as the cornerstone of cellular immunotherapy, serving as a fundamental component of consolidation therapy in cancer treatments and offering curative or palliative measures for numerous hematologic and genetic disorders [149, 150]. This procedure entails administering HSCs to replace the recipient's compromised hematopoietic system, often ravaged by disease or high doses of chemotherapy and/or radiation therapy.

With remarkable strides in the field, hematopoietic stem cell transplantation has evolved into a well-established approach for addressing a myriad of hematologic malignancies, immunological disorders, and hereditary conditions, presenting promising prospects for cure [148].

2.3.1 Hematopoietic stem cell-based therapies

In vitro, cultivation of stem cells, which is capable of differentiating, brought hope that cells for transplantation may be grown in the laboratory, and as a result, many new treatment opportunities will become possible. As of now, the www.clinicaltrials.gov portal lists over 3800 registered clinical studies focused on hematopoietic stem cell therapies. These studies encompass the utilization of HSCs in the treatment of a wide array of conditions, including cancers, leukemia, lymphoma, cardiac failure, neural disorders, immunodeficiency, and metabolic or genetic diseases [151].

2.3.1.1 Hematopoietic stem cell transplant

HSCT has become a commonplace procedure for numerous congenital or acquired disorders affecting the hematopoietic system. Globally, approximately 45,000–50,000 marrow transplant procedures and transplants of CD34+ HSCs from peripheral blood or UC blood are conducted annually. These procedures predominantly fall into two categories based on stem cell donors: allogeneic and autologous transplants [148].

2.3.1.1.1 Allogeneic transplantation

Allo-HSCT involves transplanting stem cells from a donor other than the patient, such as a relative or an unrelated individual. With advancements in transplantation procedures and supportive care, the safety of SCT has improved, leading to enhanced patient survival rates. Consequently, allo-HSCT remains a crucial therapeutic option for a broad spectrum of hematologic malignancies and non-hematologic disorders [148, 149, 151].

The transplantation process entails procuring HSCs from either a related or an unrelated donor, categorized as follows: (1) Match-Related Donor (MRD): stem cells are sourced from an HLA-identical sibling of the patient; (2) Match Unrelated Donor (MUD): stem cells are obtained from an unrelated donor with identical HLA, identified through thorough molecular matching in international registries of voluntary donors; (3) Haploidentical: this involves transplanting a substantial quantity of T-cell depleted stem cells from a donor who is either a sibling or a parent, possessing at least a 50% HLA match [148].

According to the 2017 EBMT survey, acute myeloid leukemia (AML) ranks first among diseases treated with allo-HSCT [152]. Allo-HSCT holds a pivotal role in treating AML patients due to the cytoreduction resulting from conditioning and the immunological graft-versus-leukemia (GVL) effect it elicits [152]. AML predominantly affects the elderly population, with an average age of approximately 67 [153]. Elderly patients often present with fitness challenges and comorbidities, rendering them more susceptible to unfavorable outcomes [153]. Despite differences in disease trajectory, chemotherapy resistance and poor prognosis are commonly observed in elderly AML patients [153].

Wetzler et al. demonstrated that the 5-year overall survival (OS) rates were less than 25% for patients aged 60–65 years and less than 10% for those over 70 years, in contrast to approximately 50% for patients younger than 50 years [154]. Reduced intensity conditioning (RIC) regimens, offering a less toxic alternative, are well-tolerated in elderly patients [153, 155]. Post-induction chemotherapy, consolidation with allo-HSCT is increasingly advocated for this older age group [156]. Recent studies have underscored the effectiveness of haploidentical (HAPLO) transplants, which are increasingly preferred in the absence of an HLA-matched donor, yielding promising outcomes, particularly in elderly patients. Ciurea et al. reported a 2-year progression-free survival (PFS) rate of 63% and a relapse rate of 14% in patients undergoing HAPLO transplants [157].

The prompt identification of suitable donors has emerged as a significant concern in HSCT [148, 158]. Recent advancements have broadened donor availability for patients lacking compatible donors. Progress in controlling alloreactive responses in allo-HSCT has narrowed the outcome disparities between MRD and alternative donors over time [159]. Initially, haploidentical (HAPLO) transplantation using T-cell-depleted (TCD) grafts yielded unacceptable transplant-associated mortality rates due to delayed immune reconstitution and severe infections [160]. However, the introduction of post-transplantation cyclophosphamide (PTCy) has represented a substantial leap forward in managing the alloreactivity of haploidentical grafts. This innovation has markedly enhanced HAPLO transplant outcomes, significantly reducing the incidence of GVHD, lowering nonrelapse mortality (NRM), and improving OS [160].

A recent analysis by the Center for International Blood and Marrow Transplant Research (CIBMTR) concerning AML demonstrated comparable outcomes between HAPLO and MUD transplants, including OS [161]. In conclusion, in accordance with existing studies, these findings affirm HAPLO transplantation as a favorable option in the absence of MRD, particularly in scenarios where the search for an MUD may be prolonged or when urgent transplantation is warranted, as it does not compromise transplant outcomes [160, 162–165].

2.3.1.1.2 Allogeneic cord blood transplantation

UC blood serves as an alternative source for allo-HSCT, containing a substantial quantity of stem cells capable of long-term engraftment in both children and adults. Cord blood transplants are employed in scenarios where patients lack a sibling donor and an unrelated donor cannot be located in international registries. Notably, due to their immunologic immaturity, cord blood cells mitigate the risk of graft-versus-host reaction, a primary barrier to the efficacy of allo-HSCT [148].

2.3.1.2 Other hematopoietic stem cell-based therapies

Stem cell therapy for cardiac repair has emerged as a promising approach for treating heart diseases such as heart failure and myocardial infarction. Over the past decade, more than 1000 patients with ischemic heart disease have undergone cell transplantation in numerous clinical trials. Each study has yielded valuable insights into the use of diverse cell types for therapy, varying cell doses, and routes of administration. Encouragingly, these trials have reported enhancements in myocardial perfusion, ejection fraction, and contractility, alongside improvements in physical tolerance, coupled with reductions in heart failure symptoms and pain [166]. However, despite the apparent safety of some cell therapies, further investigations are warranted to ascertain the efficacy of transplantation.

2.3.1.2.1 Neurological applications

HSCs offer a promising avenue for addressing neurological disorders and neurodegenerative diseases. Several clinical trials have explored the therapeutic potential of umbilical cord blood cells in treating conditions such as ischemia, spinal cord injury, Alzheimer's disease, and amyotrophic lateral sclerosis. The mechanism underlying graft-mediated neural repair involves neurotrophic effects and the release of various growth factors that foster cell survival, angiogenesis, and anti-inflammatory responses [167, 168]. Additionally, there are reports supporting the apparent generation of neurons in vitro [169, 170]. However, to date, there is no evidence demonstrating the transdifferentiation of stem cells into functional neurons or glial cells in vivo.

2.3.1.2.2 Immunological applications

Systemic sclerosis, MS, systemic lupus erythematosus (SLE), rheumatoid arthritis (RA), and Crohn's disease represent primary targets for therapies involving HSCT in immune-mediated and inflammatory conditions [171]. Proposed strategies center on depleting active autoimmune T and B-cells and resetting the immune system through autologous HSCT [172].

Daikeler conducted a retrospective analysis of 900 patients who underwent HSCT for severe autoimmune diseases [173]. The study reported a 5-year OS of 85% and a PFS of 43%. Transplant-related mortality ranged from 1% for RA to 11% for SLE and juvenile idiopathic arthritis. Approximately 30% of patients achieved a complete response, often with lasting effects.

3. Conclusion

Rapid advancements characterize stem cell research, with particular focus on progenitor cells such as MSCs and HSCs in numerous clinical trials. The findings outlined in this review offer fresh insights into the utilization of MSCs and HSCs in clinical applications, presenting a novel perspective. The potential implications suggest that therapies utilizing MSCs and HSCs could exert a substantial influence on human health within the realm of regenerative medicine in the foreseeable future.

Acknowledgements

The authors acknowledge the use of Grammarly premium with its updated new version including AI, checking both the similarity and grammar and syntax errors of the article for language polishing of the manuscript.

Conflict of interest

The authors declare no conflict of interest.

Author details

Neslihan Mandacı Şanlı[1*] and Aysu Timuroğlu[2]

1 Department of Hematology and Bone Marrow Transplant Center, Erciyes University Faculty of Medicine, Kayseri, Turkey

2 Department of Hematology, Erzurum Training and Research Hospital, Erzurum, Turkey

*Address all correspondence to: ortoforia@hotmail.com

References

[1] Fugger L, Jensen LT, Rossjohn J. Challenges, progress, and prospects of developing therapies to treat autoimmune diseases. Cell. 2020;**181**(1):63-80

[2] Swart JF, Delemarre EM, Wijk FV, et al. Haematopoietic stem cell transplantation for autoimmune diseases. Nature Reviews Rheumatology. 2017;**13**(4):244-256

[3] Hoang DM, Pham PT, et al. Stem cell-based therapy for human diseases. Signal Transduction and Targeted Therapy. 2022;**7**:272

[4] Biehl JK, Russell B. Introduction to stem cell therapy. The Journal of Cardiovascular Nursing. 2009;**24**:98-103

[5] Lee JY, Hong SH. Hematopoietic stem cells and their roles in tissue regeneration. International Journal of Stem Cells. 2020;**13**:1

[6] Zaret KS, Grompe M. Generation and regeneration of cells of the liver and pancreas. Science. 2008;**322**:1490-1494

[7] Mousaei Ghasroldasht M, Seok J, Park HS, et al. Stem cell therapy: From idea to clinical practice. International Journal of Molecular Sciences. 2022;**23**:2850

[8] Aly RM. Current state of stem cell-based therapies: An overview. Stem Cell Investigation. 2020;**7**:8

[9] De Luca M, Aiuti A, Cossu G, et al. Advances in stem cell research and therapeutic development. Nature Cell Biology. 2019;**21**:801-811

[10] Bobba S, Di Girolamo N, Munsie M, et al. The current state of stem cell therapy for ocular disease. Experimental Eye Research. 2018;**177**:65-75

[11] Poulos J. The limited application of stem cells in medicine: A review. Stem Cell Research & Therapy. 2018;**9**:1

[12] Prentice DA. Adult stem cells. Circulation Research. 2019;**124**:837-839

[13] Lukomska B, Stanaszek L, Zuba-Surma E, et al. Challenges and controversies in human mesenchymal stem cell therapy. Stem Cells International. 2019;**2019**:9628536

[14] Morrison SJ, Scadden DT. The bone marrow niche for haematopoietic stem cells. Nature. 2014;**505**:327-334

[15] Sharpe M, Leoni G, Hyllner J. Stem cells. In: McQueen CA, editor. Comprehensive Toxicology. 3rd ed. Elsevier; 2017. pp. 23-59. ISBN: 9780081006122

[16] Merimi M, El-Majzoub R, Lagneaux L, et al. The therapeutic potential of mesenchymal stromal cells for regenerative medicine: Current knowledge and future understandings. Frontiers in Cell and Developmental Biology. 2021;**9**:661532

[17] Han Y, Li X, Zhang Y, et al. Mesenchymal stem cells for regenerative medicine. Cells. 2019;**8**:886

[18] Vasanthan J, Gurusamy N, Rajasingh S, et al. Role of human mesenchymal stem cells in regenerative therapy. Cells. 2021;**10**:54

[19] Mohammadian M, Shamsasenjan K, Lotfi Nezhad P, et al. Mesenchymal stem cells: New aspect in cell-based regenerative therapy. Advanced Pharmaceutical Bulletin. 2013;**3**:433-437

[20] Dominici M, Le Blanc K, Mueller I, et al. Minimal criteria for

defining multipotent mesenchymal stromal cells. The International Society for Cellular Therapy position statement. Cytotherapy. 2006;**8**:315-317

[21] Zhou T, Yuan Z, Weng J, et al. Challenges and advances in clinical applications of mesenchymal stromal cells. Journal of Hematology & Oncology. 2021;**14**:24

[22] Iang XX, Zhang Y, Liu B, et al. Human mesenchymal stem cells inhibit differentiation and function of monocyte-derived dendritic cells. Blood. 2005;**105**(10):4120-4126

[23] Marinescu C-I, Preda MB, Burlacu A. A procedure for in vitro evaluation of the immunosuppressive effect of mouse mesenchymal stem cells on activated T cell proliferation. Stem Cell Research & Therapy. 2021;**12**(1):319

[24] Malekpour K, Hazrati A, Zahar M, et al. The potential use of mesenchymal stem cells and their derived exosomes for orthopedic diseases treatment. Stem Cell Reviews and Reports. 2022;**18**(3):933-951

[25] Jovic D, Yu Y, Wang D, et al. A brief overview of global trends in MSC-based cell therapy. Stem Cell Reviews and Reports. 2022;**18**:1525-1545

[26] Saeedi P, Halabian R, Imani Fooladi AA. A revealing review of mesenchymal stem cells therapy, clinical perspectives and modification strategies. Stem Cell Investigation. 2019;**6**:34

[27] Patel DM, Shah J, Srivastava AS. Therapeutic potential of mesenchymal stem cells in regenerative medicine. Stem Cells International. 2013;**2013**:496218

[28] Wang Y, Zhang D, Shen B, et al. Stem/progenitor cells and biodegradable scaffolds in the treatment of retinal degenerative diseases. Current Stem Cell Research & Therapy. 2018;**13**:160-173

[29] Shojaei F, Rahmati S, Banitalebi DM. A review on different methods to increase the efficiency of mesenchymal stem cell-based wound therapy. Wound Repair and Regeneration. 2019;**27**:661-671

[30] Toh WS, Zhang B, Lai RC, Lim SK. Immune regulatory targets of mesenchymal stromal cell exosomes/small extracellular vesicles in tissue regeneration. Cytotherapy. 2018;**20**:1419-1426

[31] Li H, Wang C, He T, et al. Mitochondrial transfer from bone marrow mesenchymal stem cells to motor neurons in spinal cord injury rats via gap junction. Theranostics. 2019;**9**:2017-2035

[32] Kim HK, Lee SG, Lee SW, et al. A subset of paracrine factors as efficient biomarkers for predicting vascular regenerative efficacy of mesenchymal stromal/stem cells. Stem Cells. 2019;**37**:77-88

[33] Huayllani MT, Sarabia-Estrada R, Restrepo DJ, et al. Adipose-derived stem cells in wound healing of full-thickness skin defects: A review of the literature. Journal of Plastic Surgery and Hand Surgery. 2020;**54**:263-279

[34] Zahorec P, Koller J, Danisovic L, Bohac M. Mesenchymal stem cells for chronic wounds therapy. Cell and Tissue Banking. 2015;**16**:19-26

[35] Merimi M, Buyl K, Daassi D, et al. Transcriptional profile of cytokines, regulatory mediators and TLR in mesenchymal stromal cells after inflammatory signaling and cell-passaging. International Journal of Molecular Sciences. 2021;**22**:7309

[36] Puissant B, Barreau C, Bourin P, et al. Immunomodulatory effect of human adipose tissue-derived adult stem cells: Comparison with bone marrow mesenchymal stem cells. British Journal of Haematology. 2005;**129**(1):118-129

[37] Corcione A, Benvenuto F, Ferretti E, et al. Human mesenchymal stem cells modulate B-cell functions. Blood. 2006;**107**(1):367-372

[38] Ramasamy R, Fazekasova H, Lam EW, et al. Mesenchymal stem cells inhibit dendritic cell differentiation and function by preventing entry into the cell cycle. Transplantation. 2007;**83**(1):71-76

[39] Spaggiari GM, Capobianco A, Becchetti S, et al. Mesenchymal stem cell-natural killer cell interactions: Evidence that activated NK cells are capable of killing MSCs, whereas MSCs can inhibit IL-2-induced NK-cell proliferation. Blood. 2006;**107**(4):1484-1490

[40] Kelly K, Rasko JEJ. Mesenchymal stromal cells for the treatment of graft versus host disease. Frontiers in Immunology. 2021;**12**:761616

[41] Praveen Kumar L, Kandoi S, Misra R, et al. The mesenchymal stem cell secretome: A new paradigm towards cell-free therapeutic mode in regenerative medicine. Cytokine & Growth Factor Reviews. 2019;**46**:1-9

[42] Fu X, Liu G, Halim A, et al. Mesenchymal stem cell migration and tissue repair. Cells. 2019;**8**:784

[43] Vladimirovna IL, Sosunova E, Nikolaev A, Nenasheva T. Mesenchymal stem cells and myeloid derived suppressor cells: Common traits in immune regulation. Journal of Immunology Research. 2016;**2016**:7121580

[44] Jiang W, Xu J. Immune modulation by mesenchymal stem cells. Cell Proliferation. 2020;**53**:e12712

[45] Fan XL, Zhang Y, Li X, Fu QL. Mechanisms underlying the protective effects of mesenchymal stem cell-based therapy. Cellular and Molecular Life Sciences. 2020;**77**:2771-2794

[46] Hu C, Li L. Preconditioning influences mesenchymal stem cell properties in vitro and in vivo. Journal of Cellular and Molecular Medicine. 2018;**22**:1428-1442

[47] Du WJ, Chi Y, Yang ZX, et al. Heterogeneity of proangiogenic features in mesenchymal stem cells derived from bone marrow, adipose tissue, umbilical cord, and placenta. Stem Cell Research & Therapy. 2016;**7**:163

[48] Honarpardaz A, Irani S, Pezeshki-Modaress M, et al. Enhanced chondrogenic differentiation of bone marrow mesenchymal stem cells on gelatin/glycosaminoglycan electrospun nanofibers with different amount of glycosaminoglycan. Journal of Biomedical Materials Research. Part A. 2019;**107**:38-48

[49] Ti D, Hao H, Fu X, Han W. Mesenchymal stem cells-derived exosomal microRNAs contribute to wound inflammation. Science China. Life Sciences. 2016;**59**:1305-1312

[50] Julier Z, Park AJ, Briquez PS, Martino MM. Promoting tissue regeneration by modulating the immune system. Acta Biomaterialia. 2017;**53**:13-28

[51] Najar M, Lombard CA, Fayyad-Kazan H, et al. Th17 immune response to adipose tissue-derived mesenchymal stromal cells. Journal of Cellular Physiology. 2019;**234**:21145-21152

[52] Tuan RS, Boland G, Tuli R. Adult mesenchymal stem cells and cell-based tissue engineering. Arthritis Research & Therapy. 2003;**5**:32-45

[53] Witkowska-Zimny M, Wrobel E. Perinatal sources of mesenchymal stem cells: Wharton's jelly, amnion and chorion. Cellular & Molecular Biology Letters. 2011;**16**:493-514

[54] Alkhalil M, Smajilagic A, Redzic A. Human dental pulp mesenchymal stem cells isolation and osteoblast differentiation. Medicinski Glasnik (Zenica). 2015;**12**:27-32

[55] Ouryazdanpanah N, Dabiri S, Derakhshani A, Vahidi R, Farsinejad A. Peripheral blood-derived mesenchymal stem cells: Growth factor-free isolation, molecular characterization and differentiation. Iranian Journal of Pathology. 2018;**13**:461-466

[56] Francis MP, Sachs PC, Elmore LW, Holt SE. Isolating adipose-derived mesenchymal stem cells from lipoaspirate blood and saline fraction. Organogenesis. 2010;**6**:11-14

[57] Gong X, Sun Z, Cui D, et al. Isolation and characterization of lung resident mesenchymal stem cells capable of differentiating into alveolar epithelial type II cells. Cell Biology International. 2014;**38**:405-411

[58] Wang B, Liu XM, Liu ZN, et al. Human hair follicle-derived mesenchymal stem cells: Isolation, expansion, and differentiation. World Journal of Stem Cells. 2020;**12**:462-470

[59] Pilato CA, Stadiotti I, Maione AS, et al. Isolation and characterization of cardiac mesenchymal stromal cells from endomyocardial bioptic samples of arrhythmogenic cardiomyopathy patients. Journal of Visualized Experiments. 2018;**132**:e57263

[60] Schneider S, Unger M, van Griensven M, Balmayor ER. Adipose-derived mesenchymal stem cells from liposuction and resected fat are feasible sources for regenerative medicine. European Journal of Medical Research. 2017;**22**:17

[61] Torre P, Flores AI. Current status and future prospects of perinatal stem cells. Genes. 2020;**12**:6

[62] Mohamed-Ahmed S et al. Adipose-derived and bone marrow mesenchymal stem cells: A donor-matched comparison. Stem Cell Research & Therapy. 2018;**9**:168

[63] Chen L, Qu J, Kalyani FS, et al. Mesenchymal stem cell-based treatments for COVID-19: Status and future perspectives for clinical applications. Cellular and Molecular Life Sciences. 2022;**79**:142

[64] Borow KM, Yaroshinsky A, Greenberg B, Perin EC. Phase 3 DREAM-HF trial of mesenchymal precursor cells in chronic heart failure. Circulation Research. 2019;**125**:265-281

[65] Wei X, Yang X, Han ZP, et al. Mesenchymal stem cells: A new trend for cell therapy. Acta Pharmaceutica Sinica. 2013;**34**:747-754

[66] McGinley LM, Kashlan ON, Bruno ES, et al. Human neural stem cell transplantation improves cognition in a murine model of Alzheimer's disease. Scientific Reports. 2018;**8**:14776

[67] Fang Y, Gao T, Zhang B, et al. Recent advances: Decoding Alzheimer's disease with stem cells. Frontiers in Aging Neuroscience. 2018;**10**:77

[68] Lublin FD, Bowen JD, Huddlestone J, et al. Human placenta-derived cells (PDA-001) for the treatment of adults with multiple sclerosis: A randomized, placebo-controlled, multiple-dose study. Multiple Sclerosis and Related Disorders. 2014;**3**:696-704

[69] Fernández O, Izquierdo G, Fernández V, et al. Adipose-derived mesenchymal stem cells (AdMSC) for the treatment of secondary-progressive multiple sclerosis: A triple blinded, placebo controlled, randomized phase I/II safety and feasibility study. PLoS ONE. 2018;**13**:e0195891

[70] Riordan NH, Morales I, Fernández G, et al. Clinical feasibility of umbilical cord tissue-derived mesenchymal stem cells in the treatment of multiple sclerosis. Journal of Translational Medicine. 2018;**16**:57

[71] Gu J, Huang L, Zhang C, et al. Therapeutic evidence of umbilical cord-derived mesenchymal stem cell transplantation for cerebral palsy: A randomized, controlled trial. Stem Cell Research & Therapy. 2020;**11**:43

[72] Sun JM, Dawson G, Franz L, et al. Infusion of human umbilical cord tissue mesenchymal stromal cells in children with autism spectrum disorder. Stem Cells Translational Medicine. 2020;**91**:137-1146

[73] Liu J, Han D, Wang Z, et al. Clinical analysis of the treatment of spinal cord injury with umbilical cord mesenchymal stem cells. Cytotherapy. 2013;**15**:185-191

[74] Przekora A, Juszkiewicz L. The effect of autologous adipose tissue-derived mesenchymal stem cells' therapy in the treatment of chronic posttraumatic spinal cord injury in a domestic ferret patient. Cell Transplantation. 2020;**29**:963689720928982

[75] Hur JW, Cho TH, Park DH, Lee JB, Park JY, Chung YG. Intrathecal transplantation of autologous adipose-derived mesenchymal stem cells for treating spinal cord injury: A human trial. The Journal of Spinal Cord Medicine. 2016;**39**:655-664

[76] de Celis-Ruiz E, Fuentes B, Alonso de Leciñana M, et al. Final results of allogeneic adipose tissue-derived mesenchymal stem cells in acute ischemic stroke (AMASCIS): A phase II, randomized, double-blind, placebo-controlled, single-center, pilot clinical trial. Cell Transplantation. 2022;**31**:9636897221083863

[77] Murray CJL. COVID-19 will continue but the end of the pandemic is near. Lancet. 2022;**399**:417-419

[78] Le Thi BP, Nguyen Thi H, Dang Ngo Chau H, et al. Allogeneic umbilical cord-derived mesenchymal stem cell transplantation for treating chronic obstructive pulmonary disease: A pilot clinical study. Stem Cell Research & Therapy. 2020;**60**:11

[79] Karaoz E, Kalemci S, Ece F. Improving effects of mesenchymal stem cells on symptoms of chronic obstructive pulmonary disease. Bratislavské Lekárske Listy. 2020;**121**:188-191

[80] Hoang DM, Nguyen KT, Nguyen AH, et al. Allogeneic human umbilical cord-derived mesenchymal stem/stromal cells for chronic obstructive pulmonary disease (COPD): Study protocol for a matched case-control, phase I/II trial. BMJ. 2021;**11**:e045788

[81] Averyanov A, Koroleva I, Konoplyannikov M, et al. First-in-human high-cumulative-dose stem cell therapy in idiopathic pulmonary fibrosis with rapid lung function decline. Stem Cells Translational Medicine. 2020;**9**:6-16

[82] Laroye C, Gibot S, Reppel L, Bensoussan D. Concise review: Mesenchymal stromal/stem cells: A new treatment for sepsis and septic shock? Stem Cells. 2017;**35**:2331-2339

[83] Galstyan GM, Parovichnikova P, Kuzmina E, et al. The results of the

single center pilot randomized Russian clinical trial of mesenchymal stromal cells in severe neutropenic patients with septic shock (RUMCESS). Inflammation Research. 2018;**5**:1-8

[84] Bellingan GJ, Bannard-Smith F, Brealey J, et al. Primary Analysis of a Phase 1/2 Study to Assess MultiStemR cell therapy, a regenerative advanced therapy medicinal product (ATMP), in acute respiratory distress syndrome (MUST-ARDS). American Journal of Respiratory and Critical Care Medicine. 2019;**199**:1

[85] Umakanthan S, Sahu P, Ranade AV. Origin, transmission, diagnosis and management of coronavirus disease 2019 (COVID-19). Postgraduate Medical Journal. 2020;**96**:753-758

[86] Xu R, Feng Z, Wang FS. Mesenchymal stem cell treatment for COVID-19. eBioMedicine. 2022;**77**:103920

[87] Primorac D, Čemerin M, Matišić V, et al. Mesenchymal stromal cells: Potential option for COVID-19 treatment. Pharmaceutic. 2021;**13**:1481

[88] Zhang Y, Ding J, Ren S, et al. Intravenous infusion of human umbilical cord Wharton's jellyderived mesenchymal stem cells as a potential treatment for patients with COVID-19 pneumonia. Stem Cell Research & Therapy. 2020;**11**:207

[89] Shu L, Niu C, Li R, et al. Treatment of severe COVID-19 with human umbilical cord mesenchymal stem cells. Stem Cell Research & Therapy. 2020;**11**:361

[90] Tao J, Nie Y, Wu H, et al. Umbilical cord blood-derived mesenchymal stem cells in treating a critically ill COVID-19 patient. Journal of Infection in Developing Countries. 2020;**14**:1138-1145

[91] Saleh M, Vaezi AA, Aliannejad R, et al. Cell therapy in patients with COVID-19 using Wharton's jelly mesenchymal stem cells: A phase 1 clinical trial. Stem Cell Research & Therapy. 2021;**12**:410

[92] Shi L, Huang H, Lu X, et al. Effect of human umbilical cord-derived mesenchymal stem cells on lung damage in severe COVID-19 patients: A randomized, double-blind, placebo-controlled phase 2 trial. Signal Transduction and Targeted Therapy. 2021;**6**:58

[93] MohammadiGorji S, Joseph A, Mihandoust S, et al. Anesthesia workspaces for safe medication practices: Design guidelines. Cell. 2012;**14**(2):142-151

[94] Liang TZ, Zhu L, Gao WL, et al. Coculture of endothelial progenitor cells and mesenchymal stem cells enhanced their proliferation and angiogenesis through PDGF and Notch signaling. FEBS Open Bio. 2017;**7**:1722-1736

[95] Chen LJ, Zhang CF, Chen L, et al. Human menstrual blood-derived stem cells ameliorate liver fibrosis in mice by targeting hepatic stellate cells via paracrine mediators. Stem Cells Translational Medicine. 2017;**6**:272-284

[96] Hu C, Zhao L, Zhang L, Bao Q, Li L. An-free strategies: Safe and effective treatments for liver injury. Stem Cell Research & Therapy. 2020;**11**(1):377

[97] Wang L, Li J, Liu H, et al. Pilot study of umbilical cord-derived mesenchymal stem cell transfusion in patients with primary biliary cirrhosis. Journal of Gastroenterology and Hepatology. 2013;**28**(Suppl 1):85-92

[98] Zhang YC, Liu W, Fu BS, et al. Therapeutic potentials of umbilical cord-derived mesenchymal stromal cells for ischemic-type biliary lesions following

liver transplantation. Cytotherapy. 2017;**19**(2):194-199

[99] Schacher FC, Annelise Martins Pezzi da Silva A, Silla LMDR, Álvares-da-Silva MR. Bone marrow mesenchymal stem cells in acute-onchronic liver failure grades 2 and 3: A phase I-II randomized clinical trial. Canadian Journal of Gastroenterology and Hepatology. 2021;**2021**:3662776

[100] Lin BL, Chen JF, Qiu WH, et al. Allogeneic bone marrow-derived mesenchymal stromal cells for hepatitis B virus-related acute-on-chronic liver failure: A randomized controlled trial. Hepatology. 2017;**66**(1):209-219

[101] Oryan A, Alidadi S. Reconstruction of radial bone defect in rat by calcium silicate biomaterials. Life Sciences. 2018;**201**:45-53

[102] Qaseem A, Forciea MA, McLean RM, et al. Treatment of low bone density or osteoporosis to prevent fractures in men and women: A clinical practice guideline update from the American College of Physicians. Annals of Internal Medicine. 2017;**166**(11):818-839

[103] Čamernik K, Mihelič A, Mihalič R, Haring G, et al. Comprehensive analysis of skeletal muscle- and bone-derived mesenchymal stem/stromal cells in patients with osteoarthritis and femoral neck fracture. Stem Cell Research & Therapy. 2020;**11**(1):146

[104] Fu J, Wang Y, Jiang Y, Du J, Xu J, Liu Y. Systemic therapy of MSCs in bone regeneration: A systematic review and meta-analysis. Stem Cell Research & Therapy. 2021;**12**(1):377

[105] Gjerde C, Mustafa K, Hellem S, et al. Cell therapy induced regeneration of severely atrophied mandibular bone in a clinical trial. Stem Cell Research & Therapy. 2018;**9**(1):213

[106] Shim J, Kim KT, Kim KG, et al. Safety and efficacy of Wharton's jelly-derived mesenchymal stem cells with teriparatide for osteoporotic vertebral fractures: A phase I/IIa study. Stem Cells Translational Medicine. 2021;**10**(4):554-567

[107] Dilogo IH, Canintika AF, Hanitya AL, Pawitan JA, Liem IK, Pandelaki J. Umbilical cord-derived mesenchymal stem cells for treating osteoarthritis of the knee: A single-arm, open-label study. European Journal of Orthopaedic Surgery and Traumatology. 2020;**30**(5):799-807

[108] Yerebakan C, Kaminski A, Liebold A, Steinhoff G. Safety of intramyocardial stem cell therapy for the ischemic myocardium: Results of the Rostock trial after 5-year follow-up. Cell Transplantation. 2007;**16**:935-940

[109] Song J, He K, Hou J. Autologous bone marrow stem cell transplantation for patients undergoing coronary artery bypass grafting: A meta-analysis of 22 randomized controlled trials. Journal of Cardiothoracic Surgery. 2022;**17**:167

[110] Soetisna TW, Sukmawan R, Setianto B, et al. Combined transepicardial and transseptal implantation of autologous CD133+ bone marrow cells during bypass grafting improves cardiac function in patients with low ejection fraction. Journal of Cardiac Surgery. 2020;**35**:740-746

[111] He X, Wang Q, Zhao Y, et al. Effect of intramyocardial grafting collagen scaffold with mesenchymal stromal cells in patients with chronic ischemic heart disease. JAMA Network Open. 2020;**3**(9):e2016236

[112] Virani SS, Alonso A, Aparicio HJ, et al. Heart disease and stroke statistics—2021 update. Circulation. 2021;**143**(8):e254-e743

[113] Chien KR, Frisén J, Fritsche-Danielson R, et al. Regenerating the field of cardiovascular cell therapy. Nature Biotechnology. 2019;**37**(3):232-237

[114] Murry CE, MacLellan WR. Stem cells and the heart-the road ahead. Science. 2020;**367**(6480):854-855

[115] Gao LR, Chen Y, Zhang NK, et al. Intracoronary infusion of Wharton's jelly-derived mesenchymal stem cells in acute myocardial infarction: Double-blind, randomized controlled trial. BMC Medicine. 2015;**13**:162

[116] Bartolucci J, Verdugo FJ, González PL, et al. Safety and efficacy of the intravenous infusion of umbilical cord mesenchymal stem cells in patients with heart failure: A phase 1/2 randomized controlled trial (RIMECARD Trial [randomized clinical trial of intravenousinfusion umbilical cord mesenchymal stem cells on cardiopathy]). Circulation Research. 2017;**121**:1192-1204

[117] Ulus AT, Mungan C, Kurtoglu M, et al. Intramyocardial transplantation of umbilical cord mesenchymal stromal cells in chronic ischemic cardiomyopathy: A controlled, randomized clinical trial (HUC-HEART trial). International Journal of Stem Cells. 2020;**13**(3):364-376

[118] Kanakry CG, Fuchs EJ, Luznik L. Modern approaches to HLA-haploidentical blood or marrow transplantation. Nature Reviews. Clinical Oncology. 2016;**13**(2):132

[119] Hashmi S, Ahmed M, Murad MH, et al. Survival after mesenchymal stromal cell therapy in steroid-refractory acute graft-versus-host disease: Systematic review and meta-analysis. The Lancet Haematology. 2016;**3**(1):e45-e52

[120] Ferrara JL, Levine JE, Reddy P, Holler E. Graft-versus-host disease. Lancet. 2009;**373**(9674):1550-1561

[121] Gao L, Zhang Y, Hu B, et al. Phase II multicenter, randomized, double-blind controlled study of efficacy and safety of umbilical cord-derived mesenchymal stromal cells in the prophylaxis of chronic graft-versus-host disease after HLA-haploidentical stem-cell transplantation. Journal of Clinical Oncology. 2016;**34**(24):2843-2850

[122] Cahn JY, Klein JP, Lee SJ, et al. Prospective evaluation of 2 acute graftversus- host (GVHD) grading systems: A joint Societe Francaise de Greffe de Moelle et Therapie Cellulaire (SFGM-TC), Dana Farber Cancer Institute (DFCI), and International Bone Marrow Transplant Registry (IBMTR) prospective study. Blood. 2005;**106**(4):1495-1500

[123] Dazzi F, Ramasamy R, Glennie S, Jones SP, Roberts I. The role of mesenchymal stem cells in haemopoiesis. Blood Reviews. 2006;**20**(3):161-171

[124] Le Blanc K, Rasmusson I, Sundberg B, et al. Treatment of severe acute graft-versus-host disease with third party haploidentical mesenchymal stem cells. Lancet. 2004;**363**:1439-1441

[125] Soder RP, Dawn B, Weiss ML, et al. A phase I study to evaluate two doses of Wharton's jelly-derived mesenchymal stromal cells for the treatment of de novo high-risk or steroid refractory acute graft versus host disease. Stem Cell Reviews and Reports. 2020;**16**:979-991

[126] Von Bonin M, Stölzel F, Goedecke A, et al. Treatment of refractory acute GVHD with third-party MSC expanded in platelet lysate-containing medium. Bone Marrow Transplantation. 2009;**43**:245-251

[127] Lucchini G, Introna M, Dander E, et al. Platelet-lysate-expanded mesenchymal stromal cells as a salvage

therapy for severe resistant graft-versus-host disease in a pediatric population. Biology of Blood and Marrow Transplantation. 2010;**16**:1293-1301

[128] Muroi K, Miyamura K, Ohashi K, et al. Unrelated allogeneic bone marrow-derived mesenchymal stem cells for steroid-refractory acute graft-versus-host disease: A phase I/II study. International Journal of Hematology. 2013;**98**:206-213

[129] Introna M, Lucchini G, Dander E, et al. Treatment of graft versus host disease with mesenchymal stromal cells: A phase I study on 40 adult and pediatric patients. Biology of Blood and Marrow Transplantation. 2014;**20**:375-378

[130] Seol YJ, Kang HW, Lee SJ, Atala A, Yoo JJ. Bioprinting technology and its applications. European Journal of Cardio-Thoracic Surgery. 2014;**46**:342-348

[131] Gao G, Schilling AF, Yonezawa T, et al. Bioactive nanoparticles stimulate bone tissue formation in bioprinted three-dimensional scaffold and human mesenchymal stem cells. Biotechnology Journal. 2014;**9**:1304-1311

[132] Fraser JR, Laurent TC, Laurent UB. Hyaluronan: Its nature, distribution, functions and turnover. Journal of Internal Medicine. 1997;**242**:27-33

[133] Boeckel DG, Sesterheim P, Peres TR, Augustin AH, Wartchow KM, Machado DC, et al. Adipogenic mesenchymal stem cells and hyaluronic acid as a cellular compound for bone tissue engineering. The Journal of Craniofacial Surgery. 2019;**30**:777-783

[134] Kangari P, Talaei-Khozani T, Razeghian-Jahromi I, Razmkhah M. Mesenchymal stem cells: Amazing remedies for bone and cartilage defects. Stem Cell Research & Therapy. 2020;**11**(1):492

[135] Chung JY, Song M, Ha CW, Kim JA, Lee CH, Park YB. Comparison of articular cartilage repair with different hydrogel-human umbilical cord blood-derived mesenchymal stem cell composites in a rat model. Stem Cell Research & Therapy. 2014;**5**:39

[136] Park YB, Ha CW, Kim JA, Han WJ, Rhim JH, Lee HJ, et al. Single-stage cell-based cartilage repair in a rabbit model: Cell tracking and in vivo chondrogenesis of human umbilical cord blood-derived mesenchymal stem cells and hyaluronic acid hydrogel composite. Osteoarthritis and Cartilage. 2017;**25**:570-580

[137] Su N, Hao Y, Wang F, et al. Mesenchymal stromal exosome-functionalized scaffolds induce innate and adaptive immunomodulatory responses toward tissue repair. Science Advances. 2021;**7**:eabf7207

[138] Giraud MN, Guex AG, Tevaearai HT. Cell therapies for heart function recovery: Focus on myocardial tissue engineering and nanotechnologies. Cardiology Research and Practice. 2012;**2012**:971614

[139] Akhyari P, Kamiya H, Haverich A, Karck M, Lichtenberg A. Myocardial tissue engineering: The extracellular matrix. European Journal of Cardio-Thoracic Surgery. 2008;**34**:229-241

[140] Steinhoff G, Stock U, Karim N, et al. Tissue engineering of pulmonary heart valves on allogenic acellular matrix conduits: In vivo restoration of valve tissue. Circulation. 2000;**102**:III50-III55

[141] Ott HC, Matthiesen TS, Goh SK, et al. Perfusion-decellularized matrix: Using nature's platform to engineer a bioartificial heart. Nature Medicine. 2008;**14**:213-221

[142] Siepe M, Giraud MN, Liljensten E, et al. Construction of

skeletal myoblast-based polyurethane scaffolds for myocardial repair. Artificial Organs. 2007;**31**:425-433

[143] Hamilton N, Bullock AJ, Macneil S, Janes SM, Birchall M. Tissue engineering airway mucosa: A systematic review. Laryngoscope. 2013;**124**:961-968

[144] Delaere P, Vranckx J, Verleden G, De Leyn P, Van Raemdonck D, Leuven Tracheal Transplant Group. Tracheal allotransplantation after withdrawal of immunosuppressive therapy. The New England Journal of Medicine. 2010;**362**:138-145

[145] Gaebel R, Ma N, Liu J, Guan J, Koch L, Klopsch C, et al. Patterning human stem cells and endothelial cells with laser printing for cardiac regeneration. Biomaterials. 2011;**32**:9218-9230

[146] Duan B, Kapetanovic E, Hockaday LA, Butcher JT. Three-dimensional printed trileaflet valve conduits using biological hydrogels and human valve interstitial cells. Acta Biomaterialia. 2014;**10**:1836-1846

[147] Till JE, Mcculloch EA. A direct measurement of the radiation sensitivity of normal mouse bone marrow cells. Radiation Research. 1961;**14**:213-222

[148] Chivu-Economescu M, Rubach M. Hematopoietic stem cells therapies. Current Stem Cell Research & Therapy. 2017;**12**:124-133

[149] Balassa K, Danby R, Rocha V. Haematopoietic stem cell transplants: Principles and indications. British Journal of Hospital Medicine. 2019;**80**:1

[150] Sureda A, Bader P, Cesaro S, et al. Indications for Allo- and auto-SCT for haematological diseases, solid tumours and immune disorders: Current practice in Europe. Bone Marrow Transplantation. 2015;**50**(8):1037-1056

[151] Shenoy S, Boelens JJ. Advances in unrelated and alternative donor hematopoietic cell transplantation for nonmalignant disorders. Current Opinion in Pediatrics. 2015;**27**(1):9-17

[152] Dohner H, Weisdorf DJ, Bloomfield CD. Acute myeloid leukemia. The New England Journal of Medicine. 2015;**373**:1136-1152

[153] Prassek VV, Rothenberg-Thurley M, Sauerland MC, et al. Genetics of acute myeloid leukemia in the elderly: Mutation spectrum and clinical impact in intensively treated patients aged 75 years or older. Haematologica. 2018;**103**(11):1853-1861

[154] Wetzler M, Mrozek K, Kohlschmidt J, et al. Intensive induction is effective in selected octogenarian acute myeloid leukemia patients: Prognostic significance of karyotype and selected molecular markers used in the European Leukemia Net classification. Haematologica. 2014;**99**(2):308-303

[155] Jethava Y, Sica S, Savani B, et al. Conditioning regimens for allogeneic hematopoietic stem cell transplants in acute myeloid leukemia. Bone Marrow Transplantation. 2017;**52**:1504-1511

[156] Estey E, de Lima M, Tibes R, et al. Prospective feasibility analysis of reduced-intensity conditioning (RIC) regimens for hematopoietic stem cell transplantation (HSCT) in elderly patients with acute myeloid leukemia (AML) and high-risk myelodysplastic syndrome (MDS). Blood. 2007;**109**(4):1395-1400

[157] Ciurea SO, Shah MV, Saliba RM, et al. Biology of blood and marrow transplantation haploidentical

transplantation for older patients with acute myeloid leukemia and myelodysplastic syndrome. Biology of Blood and Marrow Transplantation. 2018;**24**(6):1232-1236

[158] Ciurea SO, Bittencourt MCB, Milton DR, et al. Is a matched unrelated donor search needed for all allogeneic transplant candidates? Blood Advances. 2018;**2**(17):2254-2261

[159] Oran B, Kongtim P, Popat U, et al. Cytogenetics, donor type, and use of hypomethylating agents in myelodysplastic syndrome with allogeneic stem cell transplantation. Biology of Blood and Marrow Transplantation. 2014;**20**(10):1618-1625

[160] Ciurea SO, Mulanovich V, Saliba RM, et al. Improved early outcomes using a T cell replete graft compared with T cell depleted haploidentica hematopoietic stem cell transplantation. Biology of Blood and Marrow Transplantation. 2012;**18**(12):1835-1844

[161] Ciurea SO, Zhang M, Bacigalupo AA, et al. Haploidentical transplant with posttransplant cyclophosphamide vs matched unrelated donor transplant for acute myeloid leukemia. Blood. 2015;**126**(8):1033-1040

[162] Shouval R, Fein JA, Labopin M, et al. Outcomes of allogeneic haematopoietic stem cell transplantation from HLA-matched and alternative donors: A European Society for Blood and Marrow Transplantation registry retrospective analysis. The Lancet Haematology. 2019;**6**(11):e573-e584

[163] Rashidi A, Slade M, DiPersio JF, Westervelt P, Vij R, Romee R. Post-transplant high-dose cyclophosphamide after HLA-matched haploidentical hematopoietic cell transplantation for AML. Bone Marrow Transplantation. 2016;**51**(12):1561-1564

[164] Di Stasi A, Milton DR, Poon LM, et al. Similar transplantation outcomes for acute myeloid leukemia and myelodysplastic syndrome patients with haploidentical versus 10/10 human leukocyte antigen-matched unrelated and related donors. Biology of Blood and Marrow Transplantation. 2014;**20**(12):1975-1981

[165] How J, Slade M, Vu K, et al. T cell-replete peripheral blood haploidentical hematopoietic cell transplantation with post-transplantation cyclophosphamide results in outcomes similar to transplantation from traditionally matched donors in active disease acute myeloid leukemia. Biology of Blood and Marrow Transplantation. 2017;**23**(4):648-653

[166] Sanchez A, Garcia-Sancho J. Cardiac repair by stem cells. Cell Death and Differentiation. 2007;**14**:1258-1261

[167] Park DH, Lee JH, Borlongan CV, et al. Transplantation of umbilical cord blood stem cells for treating spinal cord injury. Stem Cell Reviews. 2011;**7**:181-194

[168] Sun T, Ma Q-h. Repairing neural injuries using human umbilical cord blood. Molecular Neurobiology. 2013;**47**:938-945

[169] Wang T, Choi E, Monaco MC, et al. Direct induction of human neural stem cells from peripheral blood hematopoietic progenitor cells. Journal of Visualized Experiments. 2015;**95**:52298

[170] Ali H, Forraz N, McGuckin CP, et al. Stem Cell Reviews. 2012;**8**:210-223

[171] Trounson A, Thakar RG, Lomax G, et al. Clinical trials for stem cell therapies. BMC Medicine. 2011;**9**:52

[172] Szodoray P, Varoczy L, Szegedi G, et al. Autologous stem cell transplantation in autoimmune and rheumatic diseases: From the molecular background to clinical applications. Scandinavian Journal of Rheumatology. 2010;**39**:1-11

[173] Daikeler T, Tyndall A. Stem cell treatment of autoimmune disease. Deutsche Medizinische Wochenschrift. 2011;**136**:1684-1686

Chapter 5

Effective Techniques in Improving Stem Cell Function and Success in Stem Cell Transplantation

Parisa Fayyazpour, Amir Mehdizadeh, Khadijeh Abbasi, Ali Fayyazpour and Mortaza Raeisi

Abstract

Stem cell-based therapies have been considered as an effective treatment approach during the last few decades. Stem cells are used as a promising tool in the treatment of various diseases due to their unique properties. However, despite the advantages and benefits of stem cells, there are a number of obstacles that lead to poor performance and even death of transplanted stem cells in the host tissue. Therefore, by considering the effective factors in stem cell transplantation, including the system of transferring stem cells and the desired conditions and environment in the host tissue, which are effective in increasing the survival and function of these cells, we can lead to success in stem cell transplantation. Therefore, the design and engineering of scaffolds as a new technology leads to an increase in the efficiency of stem cells. Autologous serum condition (ASC) and human platelet lysate (HPL) provide the necessary requirements for the success of stem cell transplantation. They have been noticed because of their many advantages. Current research is focusing on developing biomaterials and blood product injections to protect cells during delivery and to stimulate endogenous regeneration through interactions of transplantation cells and the host tissue.

Keywords: stem cell transplantation, biomaterials, autologous conditioned serum (ACS), human platelet lysate (HPL), cell therapy, regenerative medicine, injectable hydrogels, scaffolds, stem cells, cell survival

1. Introduction

Stem cells are defined as a special population of undifferentiating cells of the human body with unique characteristics including the capacity of self-renewal (the ability to extensively proliferate) and clonality (usually arising from a single cell) and the ability to differentiate into a wide range of specialized tissues, which make them an ideal source for cellular therapy and transplantation in regenerative medicine [1, 2]. In 1868 for the first time, the term "stem cell" was used by German biologist Ernst Haeckel to describe the properties of the fertilized egg to give rise to all cells of the organism [3]. Stem cells from various origins have varying abilities of proliferation, migration, and differentiation, which influence their application in therapy [4].

Cell types	advantage	disadvantage
Embryonic Stem Cells (ESCs)	Can make virtually any tissue	Allogenic only
	Amenable to gene transfer	Teratoma formation
	Some tissue essays to generate	Ethical issue
		Most have limited self-renewal
		Some tissues difficult to generate
		Differentiation conditions to be established
		Differentiation outside lineage
		Some tissues difficult to generate
Adult Stem Cells (ASCs)	Autologous	
	Many types and sources	
	Not tumorigenic	
	Not ethical issue	
	Some types have extensive self-renewal	
	Default differentiation	
	Amenable to gene transfer	
	Potentially delivery methods attractive	

Table 1.
Advantages and disadvantages of adult and embryonic stem cells.

Stem cells originate from two main sources: adult body tissue and embryos. Adult human stem cells (ASCs) are found in mature tissue and can give rise to many specialized cell types of the tissue and organ. Hematopoietic and mesenchymal are two types of adult stem cells often used in stem cell transplantation [4]. Embryonic stem cells qre obtained from the inner cell mass of the embryonic blastocyst [5]. The use of embryonic stem cells, despite its high potential in differentiating into different tissues, is limited due to immunogenicity, rejection, ethical considerations, teratogenic character, and other things that have been mentioned [4, 6]. In contrast, adult stem cells overcome limitations imposed by ESCs; these cells do not have ethical problems and are available for use in cell therapy. The use of each of these sources has advantages and disadvantages that are stated in **Table 1**.

2. The history and adult stem cells

As mentioned, adult stem cells are widely used in stem cell transplantation and regenerative medicine due to their advantages.

Hematopoietic stem cells (HSCs) are one of the types of adult stem cells that are widely used in stem cell transplantation. The primary source of these cells is the bone marrow [7].

The hematopoietic system, as the most "prolific" system in a body, produces half a trillion new cells every day. HSCs are defined by the capabilities of multi-lineage differentiation and long-term self-renewal. These cells can produce all mature blood and immune cells of the body, which have many applications in the clinic and for treating both hematopoietic and non-hematopoietic disorders [6, 8].

In 1939, the first case report described the transplantation of human bone marrow for a patient diagnosed with aplastic anemia. Twenty years later, in 1958, the first stem cell transplantation was performed by French oncologist George Mathe to treat six nuclear researchers who were accidentally exposed to radioactive substances using bone marrow transplantation [9]. Another study by George Mathe in 1963 shed light on the scientific community, as he successfully conducted hematopoietic stem cell transplantation (HSCT), sometimes referred to as bone marrow transplant, in a patient with leukemia. The first allogeneic HSCT was pioneered by Dr. E. Donnall Thomas in 1957 [10, 11].

Another type of adult stem cells that is widely used in stem cell transplantation is mesenchymal stem cells (MSCs).

Fibroblast-like and plastic-adherent cells were first identified in mouse bone marrow. These cells were later named MSCs by Arnold L. Caplan [12].

MSCs are a heterogeneous population of pluripotent cells that have the ability to differentiate into a wide range of non-hematopoietic cell types and play an important role in tissue repair [13]. These cells are obtained from different sources in the following order: bone marrow (BM) (51%), umbilical/placental tissues (23%), adipose tissue (13%), derived from ESCs or iPSCs (Induced pluripotent stem cells) (8%), and others (5%) [13]. In 1990, for the first time, MSCs derived from bone marrow were used for the treatment of hematopoietic malignancies [14]. Bone marrow as a traditional source of MSCs is widely used in the treatment of many disorders and malignancies, but acquiring bone marrow (BM)-derived MSCs (BMMSCs) involves an invasive and painful procedure, and the number of stem cells has an inverse relationship with donor ages, so that the number of stem cells significantly decreases with age. Therefore, in the discussion of stem cell transplantation, choosing the right source for extracting stem cells is considered as a challenging issue. Fetal-derived MSCs have been considered as an ideal alternative source. The process of obtaining fetal-type MSCs is painless and harmless to the donor [15–17]. Compared with BMMSCs, it is shown that fetal-type MSCs can proliferate faster [18]. Therefore, fetal-type MSCs, such as umbilical cord-derived MSCs (UCMSCs), have been considered as an attractive source in therapeutic processes.

The therapeutic potential of MSCs is due to the following features: easy isolation, *in vitro* expansion, non-recognition by the immune system, and the mechanism of escape from the immune system [6, 19]. MSCs are not immunogenic due to lack of expression of class 2 antigen or the T-cell costimulatory molecule B7 and may actually inhibit both primary and secondary mixed lymphocyte reactions (MLR). Due to its immunosuppressive property, it inhibits Graft-versus-host disease (GVHD) reactions and increases the success of stem cell transplantation. Based on this, studies have shown that the simultaneous injection of MSCs along with HSCs accelerate lymphocyte recovery, reduce the risk of graft failure, and repair tissue damage in patients receiving HSCT [20, 21]. By creating a suitable microenvironment for HSCs, MSCs increase the function of these cells at the transplant site. In fact, MSC stromal cells seem to create a suitable scaffold and a complex network of cytokines, adhesion molecules, and extracellular matrix proteins, reducing the possibility of transplant rejection and increasing transplant success [20].

Another interesting feature of MSCs in stem cell transplantation is genetic modification. Studies have shown that by introducing genes for secreted proteins into MSCs

Potential clinical applications of Mesenchymal Stem Cells (MSCs) in stem cells transplantation
1- Promotion of Hematopoietic Engraftment
Autologous (lymphoma, myeloma, leukemia)
Allogenic (cord blood, unrelated, haploidentical)
2- Immune modulation
Prevention and Treatment of Graft Failure
Tolerance induction for solid organ allografting
Treatment of GVHD
3- Repairing Tissue Damage
4- Delivery of corrective genes
Fabry disease
Hemophilia

Table 2.
Clinical Application of Mesenchymal Stem Cells (MSCs) in Hematopoietic Stem Cells Transplantation (HSCT).

and by injecting them systemically, they home into the bone marrow and secrete the therapeutic protein [22]. The mentioned features have led to the development of MSC-based treatment methods as an effective, safe, and possible method [23, 24]. Clinical application of MSCs in HSCT is summarized in **Table 2**.

3. Stem cell transplantation

Stem cell transplantation is a sub-type of regenerative medicine [11]. Stem cell transplantation involves administering healthy stem cells to replace damaged cells or to regenerate organs in patients with various ailments [1]. Stem cells for cell-based therapy can be of (1) autologous, also known as self-to-self therapy, an approach using the patient's own cells, and of (2) allogeneic sources, which use cells from a healthy donor for treatment. Stem cell transplantation as a novel treatment will transform medical practice [11].

Healthy stem cells can be taken from the bone marrow, peripheral blood, or umbilical cord blood [25]. Although stem cells in the bone marrow have been used as the primary source of stem cells for a long time in the process of stem cell transplantation, later, *in vitro* studies and animal models have shown that peripheral blood stem cells are also capable of fully repopulating hematopoietic cells in circulation, considering that the use of buffy coat cells from the bone marrow donor reduces the high rejection rate of bone marrow transplants, but the use of this source was not considered for two reasons:

1. It was thought that the number of stem cells in the blood circulation is low, and therefore the autologous and allogeneic collection of these cells would be difficult.

2. The number of contaminating T cells is high for a safe allogeneic transplant.

But further studies in solid tumors and other disorders showed that by choosing the right time for sampling, usually as a rebound phenomenon after intensive chemotherapy, the number of stem cells in the blood increases significantly, so there is no problem for autologous stem cell transplantation. Due to the benefits of using peripheral blood stem cells in the transplant process, the source of stem cells was changed from bone marrow to peripheral blood [26].

This procedure was first used to treat congenital immune deficiencies, bone marrow failure, and hematologic malignant lesions and is now used routinely for some intractable diseases and nonmalignant conditions such as thalassemia, stroke, liver cirrhosis, neurodegenerative diseases, myocardial infarction, and bone defect repair that cannot be treated with conventional treatments [27]. The use of various treatment methods such as hormone therapy, chemotherapy, and radiotherapy to treat diseases has some side effects [28]. Nevertheless, stem cell transplantation is the most effective method in this regard. Basic research and clinical studies have proven that stem cell transplantation as a medical intervention can enhance tissue perfusion, contribute to angiogenesis, and preserve or regenerate damaged tissue [29].

4. Limitation of stem cells for stem cell transplantation

Despite the therapeutic benefits of stem cell transplantation, studies have shown that the use of stem cells and cell therapy has major hurdles including GVHD, low cell survival, and significant mortality of donor cells in the first hours and days of transplantation; cell injury from shear stress during injection, lack of growth factor, absence of sufficient cell adhesive ligands, lack of nutrients and oxygen, uncontrolled and/or low differentiation into desired phenotypes, induction of the host immune response, oxidative stress, inflammation, cell aggregation, lack of cell integration into the host tissue, the risk of teratoma formation and poor retention at the wound site have been glaring challenges for scientists that they need to overcome before stem cells can become a viable source of cells for regenerative medicine [30–34].

In general, success in stem cell transplantation is influenced by three factors: the stem cells' fate, delivery system, and integration of transplanted cells into host.

Studies have shown that the fate of transplanted stem cells resides in the host tissue, influenced by factors and signals from the extracellular matrix (ECM) and secretion of neighboring cells [35]. Thus, mimicking the complex *in vivo* milieu with essential biophysiochemical signals during delivery to host tissue and in host tissue by overcoming the limitations mentioned in the previous section affects many characteristics of transplanted cells including retention of stemness, self-renewal, differentiation outcomes, promotion of reprogramming, manipulation of genomic and epigenomic traits, or selection of functional phenotypes [5]. As mentioned, immune responses and GVHD are one of the limiting factors in the success of stem cell transplantation. Although the use of immunosuppressive therapies (IST) typically can increase the survival of transplanted cells, it faces limitations due to resistance to treatment and limitations in increasing the drug dose due to drug toxicity [36]. Fortunately, in the absence of next-generation drug and transplant therapies, in order to achieve this goal, the use of various biomaterials has been proposed to create conditions similar to the host tissue and delivery vehicles for transplanted cells [5]. The presence of growth factors and nutrients is another factor that causes success in stem cell transplantation. The presence of growth factors helps the proliferation, expansion, and differentiation of transplanted stem cells. Autologous conditioned serum (ACS) and human platelet lysate (HPL) are two blood products that play a key role in the success of stem cell transplantation due to the various compounds they release.

In this section, we review the novel methods used for stem cell transplantation and the benefits of each and discuss their potential to overcome some of the current limitations in stem cells therapy. By using biomaterials, ACS and HPL, these obstacles can be removed to a large extent.

4.1 Biomaterials for controlling stem cell fate

In recent years, special attention has been paid to biomaterials because it leads to more bio-functional capabilities of stem cells and increases the probability of success of stem cell transplantation. Biomaterials as novel cell culture matrices play an integral role in regenerative medicine [37]. Biomaterials have been gaining in popularity for stem cell culture and expansion because they serve as temporary scaffolding to guide new tissue growth and organization [38]. Scaffolds are a broad term for three dimensional (3D) biocompatible structures produced from various biomaterials and are better for cell culturing and expansion than their two-dimensional counterparts (i.e., polystyrene plates). These structures are extensively used both *in vitro* and *in vivo* due to their biomechanical and physical properties [39]. The size of pores and surfaces of biomaterials, which affect the growth and migration of transplanted stem cells, can be easily manipulated and changed into 3D structures [40]. Other changeable mechanical properties of biomaterials can be referred to as stiffness and elasticity, which affect the differentiation and expansion of stem cells [41].

An ideal scaffold should provide chemical stability or degradability and physical properties matching the surrounding tissue to provide cytocompatibility and support adhesion, proliferation, stability, and mechanical strength [42]. Polymer materials that are commonly used to produce scaffolds are classified into two groups: Synthetic and natural. Synthetic polymers used in scaffolds include polylactic acid (PLA or PDLLA), polyglycolic acid (PGA) and their co-polymers (PGLA), polydioxanone, and so forth. Natural polymer scaffolds, in turn, include different types of collagens, elastin, fibrins, glycosaminoglycans (GAGs), gelatins, silk fibroins, alginates, and chitins [39]. In modern medicine, natural and synthetic biomaterials play key role in regenerative medicine because they can mimic the structure of the niche that regulates the behavior of stem cells [43]. HSCs grow in a niche, a specific microenvironment that regulates their growth and proliferation. Transplantation of stem cells along with biomaterials reduces problems related to transplantation. Biomaterials based on the common principle of dynamic reciprocity and tissue-specific tensional homeostasis interact with stem cells. Biomaterials serve as a biomimetic platform and, by creating a suitable environment for stem cells, lead to success in stem cell transplantation. Biomaterials can function as intracellular delivery vehicles for stem cells to target areas and induce local tissue regeneration [44]. Biomaterials have a protective role and protect the transplanted stem cells against stress, hypoxia, and immune attack, thus facilitating long-term viability and maintenance. Another advantage of using biomaterial is the ability to modulate the host tissue niche. After delivery of the transplanted stem cells to the target region, in addition to recapitulating the microenvironment for stem cells and modulating the immune system, biomaterials stimulate host cell recruitment, homing, and differentiation and play a key role in the success of the transplant by reducing contact-dependent blood-mediated reactions, fibrosis, and adverse immune reactions, leading to increased survival of transplanted stem cells [45–47].

Hydrogels are a promising class of biomaterials that have been noticed due to their unique characteristics, including dynamic and diffusive nature [48]. Hydrogel scaffolds are frequently used in tissue engineering [49]. Self-healing injectable hydrogels play a key role in regenerative medicine and are considered as an ideal method. By providing mechanical tissue support, the hydrogels can be structurally and mechanically adjusted to resemble different tissues and promote regeneration [50, 51]. These hydrogels can be applied with high spatial control and mold into patient-specific tissue defects without unintended off-target leakage because of their quick self-healing [52–54]. The architecture of engineered biomaterials is such that it is possible to manipulate the environment and load various types of cytokines, growth factors, mechanical stimuli, vascularization, and immune modulators to favorably stimulate the host tissue niche. Therefore, the combination of biomaterials and stem cells and the simultaneous injection of this combination into the damaged area play an important role in the expansion and differentiation of transplanted stem cells [2, 55]. The mechanism of action of biomaterials in stem cell transplantation is shown in **Figure 1**.

4.1.1 Injectable biomaterial-based stem cell transplantation

In line with the development of cellular therapies and regenerative medicine, researchers are trying to develop injectable hydrogel systems as a means of stem cell transfer. Injectable biomaterials are one of the grooves of biomaterials that are used in the stem cell transplantation process.

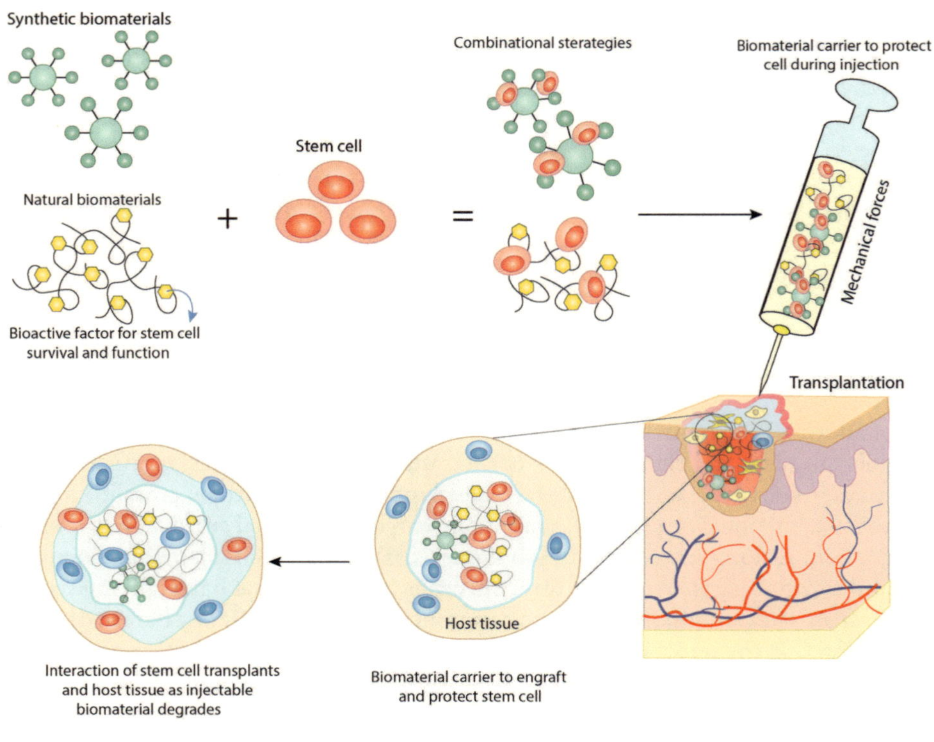

Figure 1.
Schematic illustration of injectable biomaterial for improved stem cell-derived therapeutics. Combinatorial regenerative medicine strategies often include encapsulation of stem cell-derived transplants in biomaterial precursors which are subsequently transplanted into the patient by injection using a needle. Biomaterials provide the initial structural support and retain cells in the defective area for cell growth, metabolism and new ECM synthesis. The biomaterial is readily degradable when the cells secrete ECM. This strategy enables the clinician to transplant the stem cells and biomaterial combination in a minimally invasive manner.

Hydrogels due to their potential to recapitulate the microenvironment are mainly used in injectable biomaterial-based stem cell transplantation. Hydrogels are three-dimensional polymeric networks that are insoluble in water, and due to the presence of hydrophilic groups in the main chain, they have the ability to absorb a large amount of water and biological fluids, and they can absorb the maximum amount of water and swell, and this feature has made hydrogels have a high exchange of ions and nutrients and metabolism with the surrounding fluids, which causes the preservation of stem cells. Due to the high water content, these polymer networks have the potential to mimic the extracellular matrix (ECM).

Hydrogels are connected to each other through chemical and physical cross-links, and due to their mechanical properties and creation of protein gradients or regions, they play a key role in determining the fate of stem cells. In a way, many characteristics of transplanted stem cells, including angiogenesis, local cell adhesion, morphology, migration, and differentiation by affecting the expression of genes, cause success in stem cell transplantation [56].

In addition, hydrogel scaffolds as a carrier system cause the transfer of stem cells to the host tissue. They play a key role in the homing of transplanted stem cells in the host tissue and facilitating the interaction of stem cells with the host tissue. The success of stem cell transplantation using hydrogels has a number of conditions, including allowing the formation of the hydrogel *in situ*, which improves the interaction of stem cells with the host tissue [57]. Hydrogel stiffness plays an important role in the regulation of stem cell fate such as adhesion, migration, and differentiation. Also, based on Ref. [58–60], the similarity of the elasticity of hydrogels with the damaged tissue plays an important role in the differentiation of stem cells transplanted to that tissue, so that softer or stiffer hydrogels do not find a significant difference. It is useful to divide an injectable hydrogel therapy into the following phases: pre-injection, injection, acute post-injection, and long-term survival and function, in order to identify the distinct challenges and potential solutions to maintaining cell survival.

4.1.2 Pre-injection phase

How cells interact with biomaterials before injection and during injection is one of the effective factors in the success of stem cell transplantation. So that even the choice of crosslinking mechanism potentially affects the survival of cells. Hydrogels are biomaterials that are injected with stem cells. The crosslinking mechanism of hydrogels, physical and chemical, due to their unique characteristics, is effective on the characteristics of stem cells and the transplant process. Hydrogel crosslinking mechanisms, chemical or physical, potentially can influence cell survival in the syringe pre-injection. As long-term exposure to these crosslinking mechanisms can decrease cell viability, one way to overcome this concern is the use of dual barrel syringes to isolate cells from the potentially cytotoxic component until immediately before delivery. Considering that the injection site has its own local microenvironment, controlling crosslinks can be difficult to reproducibly control because it affects crosslinking kinetics. To solve this problem, using a new strategy where cells are pre-encapsulated in injectable hydrogels that are already in the gel phase *ex vivo* is suggested [61].

4.1.3 Injection phase

Studies have shown that when stem cells are exposed to syringe needle flow, shear and extensional forces cause damage to the cell membrane. Plug flow, where shear-thinning at the walls forms a lubricating layer that enables the rest of the gel and encapsulated cells to slip through the needle relatively unreformed in hydrogels, is a factor that protects stem cells from membrane damage. Another strategy to prevent cell membrane damage is encapsulation of individual cells (or small clusters of cells) into microbeads [61].

4.1.4 Acute post-injection phase

Rapid gelation of the pre-polymer solution or rapid self-healing of an injectable gel is needed to achieve the maximum number of cells at the site of injury and the effectiveness of the treatment. In the acute post-injection phase, transplanted cells can be confronted with a host of survival challenges including hypoxia, lack of nutrients, and lack of a three-dimensional supporting matrix.

Depending on the specific clinical application, the relative importance of these different challenges can vary dramatically. Thus, biomaterials designed for different potential therapies often focus on addressing different challenges.

The immune and inflammatory response after tissue injury results in a harsh environment that transplanted cells must overcome to survive. In order to achieve this, materials that were once thought to be bioinert can be used to modulate the immune system [61].

4.1.5 Long-term survival and function phase

Transplanted cells show their therapeutic effects through different mechanisms. Sometimes the stem cells directly participate in the formation of new tissue, and sometimes they affect the host tissue indirectly through the secretion of pro-regenerative factors [61].

4.2 Autologous conditioned serum (ACS)

Inflammatory processes play an important role in the pathogenesis of many disorders and injuries. Damaged tissues release various inflammatory cytokines such as interleukin-1ß (IL-1ß) and tumor necrosis factor alpha (TNF-α). The release of inflammatory cytokines causes the release of reactive oxygen species (ROS), cytokines such as Interleukin 6 (IL-6) and Interleukin 8 (IL-8), cyclooxygenase 2 (COX-2), metalloproteinase (MMP 1, 3, and 13), and prostaglandin 2 (PGE2). As a natural protein, interleukin-1 receptor antagonist (IL-Ra) is a competitive antagonist of IL-1α and IL-1β, both of which are proinflammatory via activation of nuclear factor kappa-light-chain-enhancer of activated B cells (NF-κB) [62].

IL-Ra binds to IL-1β and down-regulates the effect of this inflammatory cytokine. The increase of endogenous IL-Ra is considered as a molecular approach because, as mentioned, IL-1β plays an important role in the pathogenesis of diseases. Studies show that in diseases, the concentration of IL-Ra decreases, so it cannot inhibit the effect of IL-1β. Therefore, recent biochemical modulatory therapies with the aim of increasing IL-R and reducing pro-inflammatory cytokines have been considered [63]. Since the discovery of IL-1Ra in 1986, several methods have been identified to increase the concentration of this protein in hemoderivatives because the endogenous increase of IL-1Ra as a therapeutic approach plays an important role in healing and treating inflammatory diseases and progressive muscular, skeletal, and bone degeneration [63].

Meijer et al. [63] devised a biologic therapeutic preparation known as Autologous Conditioned Serum (ACS). ACS is one of the methods in which IL-Ra particles are increased from whole blood flow. ACS is a simple-to-produce, safe, well-tolerated treatment and easily accessible treatment method with limited side effects. ACS (ACS; synonyms irap®, Orthokine®, Orthogen, Düsseldorf, Germany) was first introduced in the mid-1990s for the treatment of osteoarthritis and injected as a therapeutic agent [64]. Since 1998, ACS has been marketed under the brand name Orthokinew and is widely used in orthopedic patients [65].

ACS is used as one of the therapeutic approaches, alone or together with stem cells. The simultaneous use of ACS and transplanted stem cells at the site of injury and disease leads to an increase in the effectiveness of stem cells and success in stem cell transplantation. As mentioned, host immune response, oxidative stress, inflammation, and lack of growth factors are among the limitations of stem cell transplantation, and ACS, as a blood product enrichment of anti-inflammatory factors and cytokines, can overcome these obstacles. The study has shown that rat Achilles tendons exposed to ACS causes enhanced expression of the Collagen, type I, alpha 1 (Col1A1) gene and type1 collagen secretion and increases the speed of repair and recovery at the site of injury [66].

Actually, ACS is a cell-free blood product. The process of preparing ACS starts with the incubation of whole blood with sterile medical glass beads coated with chromium sulfates (CrSO4) at 37 for 24 hours. During the incubation, the platelets are degranulated, and the peripheral blood mononuclear cells (PBMC) stimulate the production of various cytokines and growth factors such as IL-4, TNF-α, IL-10, IL-6, basic fibroblast growth factor (FGFb), and IL-1Ra [67]. After incubation, the blood is recovered, and blood clots and serum are centrifuged, collected, and filtrated before administration [68]. This process stimulates a vigorous and rapid increase in production of IL-1RA so that the concentration increased 140-fold, and other anti-inflammatory cytokines (IL-4, IL-10) and GFs (insulin-like growth factor (IGF-I), Platelet-derived growth factor (PDGF) and Transforming growth factor beta (TGF-b)) that are slightly induced by stimulated by the interaction between the glass surface and the blood cells. The researchers observed that during the incubation, the concentration of IL-1Ra increased continuously, which is probably due to the de novo synthesis and caused by the interaction between the surface of glass beads and blood cells, especially monocytes [65]. The process of preparing ACS is shown in **Figure 2**.

The use of ACS is recommended for several reasons:

1. This product is rich in all kinds of growth factors, including IGF-1, bFGF, and TGF-β, which have the potential to attract resident precursor cells to injuries site, which leads to increased proliferation of stem cells and causes success in regenerative medicine [69, 70].

2. This product is rich in IL-1Ra protein, which leads to the reduction of prostaglandin E-2 production [70].

4.3 Human platelet lysate (HPL)

Another product that increases the efficiency of transplanted stem cells and increases effectiveness in regenerative medicine is human platelet lysate (HPL). HPL is a bioproduct derived from platelets. Platelet, in addition to participating in

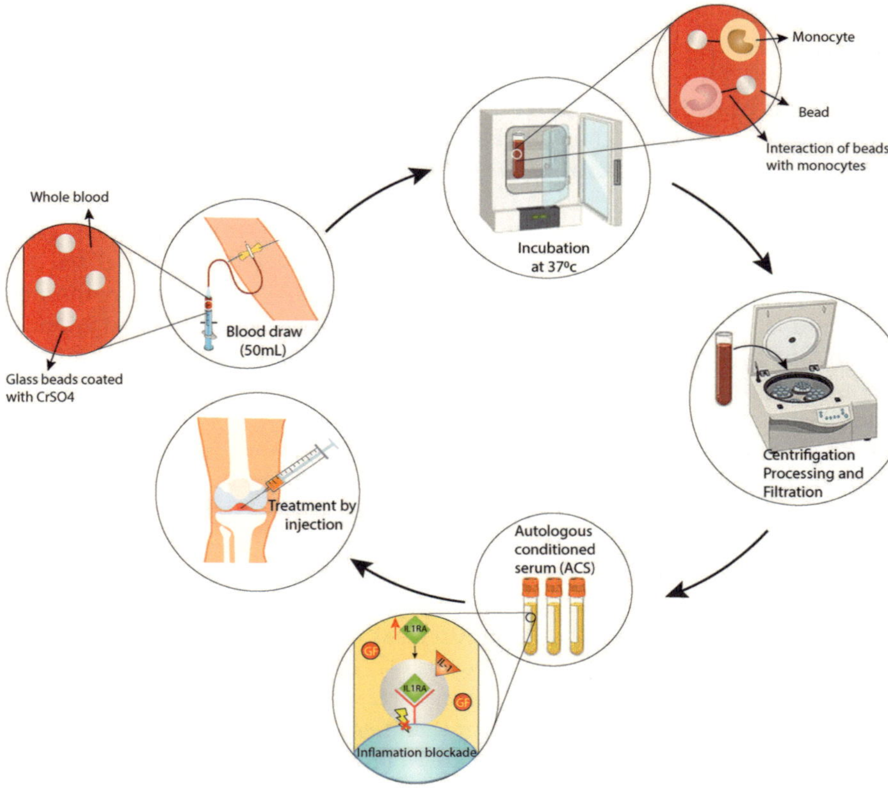

Figure 2.
A diagram for the preparation and administration of autologous conditioned serum (ACS). Whole blood is drawn into syringes containing glass beads, Incubation at 37 for 24 h, Centrifugation at 3500 rpm for 10 min, serum corresponding to 20% of origin blood volume is collected as ACS, ACS is injected into the injury site for treatment.

hemostasis and coagulation processes, as a secretory agent can also participate in wound healing and healing of damaged tissues [71]. Platelets have granules that are rich in growth factors, proteins, chemokines, and cytokines, which are used in the treatment of various diseases and disorders such as oral mucositis, radicular pain, osteoarthritis, infertility, androgenetic alopecia, wound healing, nerve repair and regenerative tissue, such as bone regeneration.

Platelet cargo is a term that refers to the types of growth factors, cytokines, and chemokines found in alpha granules, dense granules, lysosomes, and micro particles [72]. Alpha granules in platelets contain various growth factors, diverse regulatory proteins, and angiogenic factors, which they play an important role in regenerative medicine and success in stem cell transplantation [72]. Various studies have shown that HPL, due to the various factors and proteins mentioned below, can overcome the obstacles and limitations of stem cell transplantation and lead to increased proliferation and *ex vivo* expansion of stem cells, preservation and viability of transplanted stem cells at the site of injury, and angiogenesis and better differentiation of transplanted stem cells [72, 73]. It is also a good substitute to fetal bovine serum (FBS) [74]. The use of blood products has a number of advantages: due to the human origin of these products, it is devoid of immunological risks specifically associated with the use of xenoprotein-based products. Second, the infrastructure for blood collection and quality control and safety testing in industrialized countries is ideally established using advanced blood collection systems and good national regulatory supervision [75].

The contents of HPL include different types of key growth factors including mostly insulin-like growth factor 1 (IGF1); platelet-derived growth factor AA (PDGF-AA), -BB, -AB; transforming growth factor beta 1 (TGF-b1); brain-derived neurotrophic factor (BDNF); epidermal growth factor (EGF); vascular endothelial growth factor (VEGF); hepatocyte growth factor (HGF); and basic fibroblast growth factor (bFGF). Some reports indicate that TGF-b1, PDGF-AB, IGF1, and BDNF are present in higher quantities than other growth factors [74]. Other plasma proteins include albumin, protease inhibitors such as alpha 1-antitrypsin and alpha 2-macroglobulin, and immunoglobulins G, A, and M. Minerals like sodium, potassium, magnesium, and iron as well as vitamins and glucose are also present in HPL [72].

In addition to the mentioned compounds, PL contains chemo-attractants and mitogens, which play an important role in the activation and proliferation of stem cells and angiogenesis in the damaged area, and as a result, by increasing the success of stem cell transplantation, they reduce damage [76].

In recent years, clinical use of this product has attracted attention since its preparations are acellular, thereby reducing concerns over immunogenicity [74].

There are different strategies for preparing platelet lysate; repeated freezing and thawing processes, sonication, and chemical treatments lead to the preparation of a cocktail that is rich in various growth factors [74]. The use of single growth factors such as PDGF and TGF-β is one of the therapeutic methods according to Food and Drug Administration (FDA) approval, but the use of a growth factor compared to HPL, which is a combination of various growth factors and cytokines, shows limited success and heterogeneous results [77, 78]. PL, as a novel and multifactorial material, has a high concentration of various growth factors compared to other blood-derived products, and it can also be stored for a long time in lower temperatures in a freezer [79]. Several studies have shown that growth factors released from platelets interact with tyrosine kinase receptors, promote cell proliferation, and inhibit apoptosis of MSCs [80–82]. According to the study of Usman Rashid and colleagues [83], HPL increases the proliferation potential of MSCs, as it was observed in the scratch test that it filled the scratched area within 24 hours. In fact, it seems that the presence of two factors, PDGF and FGF-2, has the ability to initiate cellular DNA replication in the G0 or early G1 phases of the cell cycle. Also, this study shows that the differentiation capacity of MSCs increases in the presence of HPL [83]. Based on the study of Yael Hayon et al. [76] in the *in vivo* model, it was observed that the use of PL in rats whose middle cerebral artery was blocked compared to the control group that only received FGF2 caused a significant improvement in neurological severity scores. Also, immunohistochemistry showed increased endogenous neural stem cells in the peri-lesion cortex and increased angiogenesis in the sub ventricular zone (which is rich in stem cell niche and blood vessels), and infarct volumes were significantly reduced in rats treated with PL, which reveals a new potential of PL in reducing behavioral deficits after brain ischemia. So, in general, HPL has a positive effect on the activity and function of transplanted stem cells, which makes the transplant process successful.

The advantages of using this strategy and their effect on stem cell transplantation are listed in **Table 3**.

The stages of PL preparation and the mechanism of action of released factors in stem cell transplantation are shown in **Figure 3**.

Platelet Lysate (PL) advantage
1- Autologous setting
2- Application in different cell types
3- Proliferation of Mesenchymal Stem Cells (MSCs)
4- Enriched growth factors
5- No xenogeneic immune reactions
6- Used in clinical trial with no side effect

Table 3.
Platelet Lysate (PL) as a promising medium in stem cells transplantation.

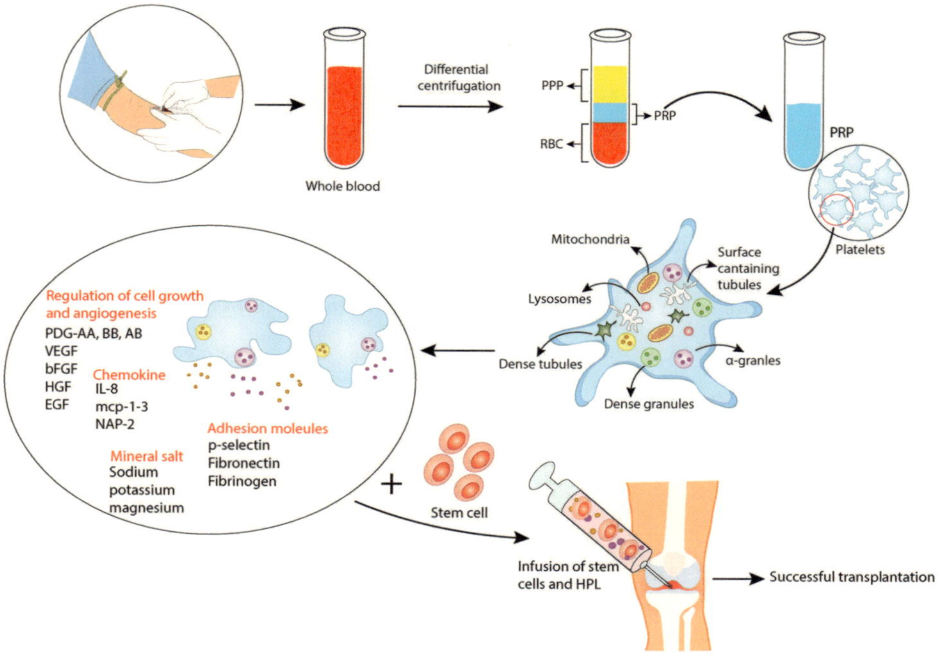

Figure 3.
A diagram for the preparation and use of human platelet lysate (HPL) in regenerative medicine. Blood is drawn from healthy donors and it is differential centrifuged at 2000 × g for 15 min to obtain platelet-rich plasma (PRP). The lower 1/3rd is PRP and upper 2/3rd is platelet-poor plasma (PPP). Remove PPP and PRP transfer to a new tube. Platelet lysate obtention by CaCl2 addition. A fraction of this platelet lysate is transferred to a new tube and use for treatment.

5. Conclusions

The ability of stem cells to congregate in damaged areas and produce new cells and differentiate into damaged tissue has led to the widespread use of these cells in the process of stem cell transplantation. Stem cells and stem cell transplantation have great potential in regenerative medicine. One of the challenges faced in the transplantation of transplanted cells is the survival of stem cells and the process of transferring transplanted cells to the target tissue, which is considered as an obstacle in the function of stem cell transplantation. Biomaterials are designed to improve transplanted cell survival, engraftment, and integration with host tissue. Hydrogels as a class of biomaterials have received attention in the past few decades. The loading of stem cells in designed injectable hydrogels can stop physical and chemical stress damage and protect and increase the survival of stem cells and has a significant effect on increasing therapeutic efficiency. Injectable hydrogels act as a potentiator of stem cells. Hydrogels create biomimetic stem cell niches with suitable mechanical environments and as a delivery system can improve many characteristics of stem cells, such as stemness, self-renewal, lineage commitment, cell physiology, and metabolism. Therefore, biomaterials play an effective role in the success of stem cell transplantation. The use of ACS and PL provides a suitable environment for the growth and proliferation of transplanted stem cells and plays an important role in the success of the transplantation process. Both products, ACS and PL, provide more physiological alternative to conventional treatments. These products are taken from the patient's own blood, so it is safe and does not create an immune response. It is also rich in various types of growth factors and non-inflammatory cytokines, which leads to increased efficiency in the processes of cell transplantation and regenerative medicine.

Acknowledgements

The authors would like to acknowledge the Department of Biochemistry and Clinical Laboratories at Hamadan University of Medical Science and Hematology and Oncology Research Center at Tabriz University of Medical Sciences for its great help.

Conflict of interest

The authors declare no conflict of interest.

Author details

Parisa Fayyazpour[1,2]*, Amir Mehdizadeh[3], Khadijeh Abbasi[4,5], Ali Fayyazpour[6] and Mortaza Raeisi[3]

1 Faculty of Medicine, Department of Clinical Biochemistry, Hamadan University of Medical Sciences, Hamadan, Iran

2 Student Research Committee, Hamadan University of Medical Sciences, Hamadan, Iran

3 Hematology and Oncology Research Center, Tabriz University of Medical Sciences, Tabriz, Iran

4 Student Research Committee, Tabriz University of Medical Sciences, Tabriz, Iran

5 Faculty of Medicine, Department of Biochemistry and Clinical Laboratories, Tabriz University of Medical Sciences, Tabriz, Iran

6 School of Medicine, Iran University of Medical Sciences, Tehran, Iran

*Address all correspondence to: fayyazpourparisa@gmail.com

References

[1] Kolios G, Moodley Y. Introduction to stem cells and regenerative medicine. Respiration; International Review of Thoracic Diseases. 2013;**85**(1):3-10. DOI: 10.1159/000345615

[2] Kharbikar BN, Mohindra P, Desai TA. Biomaterials to enhance stem cell transplantation. Cell Stem Cell. 2022;**29**(5):692-721. DOI: 10.1016/j.stem.2022.04.002

[3] Ramalho-Santos M, Willenbring H. On the origin of the term "stem cell". Cell Stem Cell. 2007;**1**(1):35-38

[4] Chu DT, Nguyen TT, Tien NLB, Tran DK, Jeong JH, Anh PG, et al. Recent Progress of stem cell therapy in cancer treatment: Molecular mechanisms and potential applications. Cells. 2020;**9**(3):1-19. DOI: 10.3390/cells9030563

[5] Ishihara M, Kishimoto S, Nakamura S, Fukuda K, Sato Y, Hattori H. Biomaterials as cell carriers for augmentation of adipose tissue-derived stromal cell transplantation. Bio-Medical Materials and Engineering. 2018;**29**(5):567-585. DOI: 10.3233/bme-181009

[6] Gurudutta GU, Satija NK, Singh VK, Verma YK, Gupta P, Tripathi RP. Stem cell therapy: A novel & futuristic treatment modality for disaster injuries. The Indian Journal of Medical Research. 2012;**135**(1):15-25. DOI: 10.4103/0971-5916.93419

[7] Rossi L, Challen GA, Sirin O, Lin KK, Goodell MA. Hematopoietic stem cell characterization and isolation. Methods in Molecular Biology (Clifton, NJ). 2011;**750**:47-59. DOI: 10.1007/978-1-61779-145-1_3

[8] Ellis MJ, Patel UD. Hematopoietic stem-cell transplantation. The New England Journal of Medicine. 2006;**355**(10):1070; author reply 1070-1071. DOI: 10.1056/NEJMc061443

[9] Jansen J. The first successful allogeneic bone-marrow transplant: Georges Mathe. Transfusion Medicine Reviews. 2005;**3**(19):246-248

[10] Blume KG, Weissman IL. E. Donnall Thomas (1920-2012). Proceedings of the National Academy of Sciences. 2012;**109**(51):20777-20778

[11] Khaddour K, Hana CK, Mewawalla P. Hematopoietic Stem Cell Transplantation. Treasure Island (FL) ineligible companies. Disclosure: Caroline Hana declares no relevant financial relationships with ineligible companies. Disclosure: Prerna Mewawalla declares no relevant financial relationships with ineligible companies.: StatPearls Publishing Copyright © 2024, StatPearls Publishing LLC.: StatPearls; 2024

[12] Caplan AI. Mesenchymal stem cells. Journal of Orthopaedic Research. 1991;**9**(5):641-650

[13] Musiał-Wysocka A, Kot M, Majka M. The pros and cons of mesenchymal stem cell-based therapies. Cell Transplantation. 2019;**28**(7):801-812. DOI: 10.1177/0963689719837897

[14] Torrente Y, Polli E. Mesenchymal stem cell transplantation for neurodegenerative diseases. Cell Transplantation. 2008;**17**(10-11):1103-1113. DOI: 10.3727/096368908787236576

[15] Caplan AI. Adult mesenchymal stem cells for tissue engineering versus

regenerative medicine. Journal of Cellular Physiology. 2007;**213**(2):341-347

[16] Le Blanc K, Frassoni F, Ball L, Locatelli F, Roelofs H, Lewis I, et al. Mesenchymal stem cells for treatment of steroid-resistant, severe, acute graft-versus-host disease: A phase II study. The Lancet. 2008;**371**(9624):1579-1586

[17] Lazarus HM, Koc ON, Devine SM, Curtin P, Maziarz RT, Holland HK, et al. Cotransplantation of HLA-identical sibling culture-expanded mesenchymal stem cells and hematopoietic stem cells in hematologic malignancy patients. Biology of Blood and Marrow Transplantation: Journal of the American Society for Blood and Marrow Transplantation. 2005;**11**(5):389-398

[18] Chan C-K, Wu K-H, Lee Y-S, Hwang S-M, Lee M-S, Liao S-K, et al. The comparison of interleukin 6–associated immunosuppressive effects of human ESCs, fetal-type MSCs, and adult-type MSCs. Transplantation. 2012;**94**(2):132-138

[19] Ankrum JA, Ong JF, Karp JM. Mesenchymal stem cells: immune evasive, not immune privileged. Nature Biotechnology. 2014;**32**(3):252-260

[20] Wu KH, Wu HP, Chan CK, Hwang SM, Peng CT, Chao YH. The role of mesenchymal stem cells in hematopoietic stem cell transplantation: From bench to bedsides. Cell Transplantation. 2013;**22**(4):723-729. DOI: 10.3727/096368912x655217

[21] Ball LM, Bernardo ME, Roelofs H, Lankester A, Cometa A, Egeler RM, et al. Cotransplantation of ex vivo–expanded mesenchymal stem cells accelerates lymphocyte recovery and may reduce the risk of graft failure in haploidentical hematopoietic stem-cell transplantation. Blood. 2007;**110**(7):2764-2767

[22] Devine SM, Hoffman R. Role of mesenchymal stem cells in hematopoietic stem cell transplantation. Current Opinion in Hematology. 2000;**7**(6):358-363

[23] Huang Y-Z, Gou M, Da L-C, Zhang W-Q, Xie H-Q. Mesenchymal stem cells for chronic wound healing: Current status of preclinical and clinical studies. Tissue Engineering Part B: Reviews. 2020;**26**(6):555-570

[24] Gentile P, Garcovich S. Systematic review: Adipose-derived mesenchymal stem cells, platelet-rich plasma and biomaterials as new regenerative strategies in chronic skin wounds and soft tissue defects. International Journal of Molecular Sciences. 2021;**22**(4):1538

[25] Msheik A, Assi F, Hamed F, Jibbawi A, Nakhl AM, Khoury A, et al. Stem cell transplantation for multiple sclerosis: A 2023 review of published studies. Cureus. 2023;**15**(10):e47972. DOI: 10.7759/cureus.47972

[26] Gratwohl A, Baldomero H, Schmid O, Horisberger B, Bargetzi M, Urbano-Ispizua A. Change in stem cell source for hematopoietic stem cell transplantation (HSCT) in Europe: A report of the EBMT activity survey 2003. Bone Marrow Transplantation. 2005;**36**(7):575-590. DOI: 10.1038/sj.bmt.1705104

[27] Chrostek MR, Fellows EG, Crane AT, Grande AW, Low WC. Efficacy of stem cell-based therapies for stroke. Brain Research. 2019;**1722**:146362. DOI: 10.1016/j.brainres.2019.146362

[28] Shareghi-Oskoue O, Aghebati-Maleki L, Yousefi M. Transplantation of human umbilical cord mesenchymal stem cells to treat premature ovarian

failure. Stem cell Research & Therapy. 2021;**12**(1):454. DOI: 10.1186/s13287-021-02529-w

[29] Karantalis V, DiFede DL, Gerstenblith G, Pham S, Symes J, Zambrano JP, et al. Autologous mesenchymal stem cells produce concordant improvements in regional function, tissue perfusion, and fibrotic burden when administered to patients undergoing coronary artery bypass grafting: The prospective randomized study of mesenchymal stem cell therapy in patients undergoing cardiac surgery (PROMETHEUS) trial. Circulation Research. 2014;**114**(8):1302-1310

[30] Li J, Ding Z, Zheng X, Lu G, Lu Q, Kaplan DL. Injectable silk nanofiber hydrogels as stem cell carriers to accelerate wound healing. Journal of Materials Chemistry B. 2021;**9**(37):7771-7781. DOI: 10.1039/d1tb01320c

[31] Mitrousis N, Fokina A, Shoichet MS. Biomaterials for cell transplantation. Nature Reviews Materials. 2018;**3**(11):441-456. DOI: 10.1038/s41578-018-0057-0

[32] Hayward M-K, Muncie JM, Weaver VM. Tissue mechanics in stem cell fate, development, and cancer. Developmental Cell. 2021;**56**(13):1833-1847

[33] Stokes RA, Cheng K, Lalwani A, Swarbrick MM, Thomas HE, Loudovaris T, et al. Transplantation sites for human and murine islets. Diabetologia. 2017;**60**:1961-1971

[34] Zhao X, Cui K, Li Z. The role of biomaterials in stem cell-based regenerative medicine. Future Medicinal Chemistry. 2019;**11**(14):1777-1790

[35] Lutolf MP, Gilbert PM, Blau HM. Designing materials to direct stem-cell fate. Nature. 2009;**462**(7272):433-441. DOI: 10.1038/nature08602

[36] Koch KC, Jadon N, Thesmar I, Tew GN, Minter LM. Combating bone marrow failure with polymer materials. Frontiers in Immunology. 2024;**15**:1396486. DOI: 10.3389/fimmu.2024.1396486

[37] Bai T, Li J, Sinclair A, Imren S, Merriam F, Sun F, et al. Expansion of primitive human hematopoietic stem cells by culture in a zwitterionic hydrogel. Nature Medicine. 2019;**25**(10):1566-1575. DOI: 10.1038/s41591-019-0601-5

[38] Kim B-S, Baez CE, Atala A. Biomaterials for tissue engineering. World Journal of Urology. 2000;**18**:2-9

[39] Afflerbach AK, Kiri MD, Detinis T, Maoz BM. Mesenchymal stem cells as a promising cell source for integration in novel In vitro models. Biomolecules. 2020;**10**(9):1-30. DOI: 10.3390/biom10091306

[40] Raic A, Rödling L, Kalbacher H, Lee-Thedieck C. Biomimetic macroporous PEG hydrogels as 3D scaffolds for the multiplication of human hematopoietic stem and progenitor cells. Biomaterials. 2014;**35**(3):929-940. DOI: 10.1016/j.biomaterials.2013.10.038

[41] Han SB, Kim JK, Lee G, Kim DH. Mechanical properties of materials for stem cell differentiation. Advanced Biosystems. 2020;**4**(11):e2000247. DOI: 10.1002/adbi.202000247

[42] Neuss S, Apel C, Buttler P, Denecke B, Dhanasingh A, Ding X, et al. Assessment of stem cell/biomaterial combinations for stem cell-based tissue engineering. Biomaterials. 2008;**29**(3):302-313. DOI: 10.1016/j.biomaterials.2007.09.022

[43] Facklam AL, Volpatti LR, Anderson DG. Biomaterials for personalized cell therapy. Advanced Materials (Deerfield Beach, Fla). 2020;**32**(13):e1902005. DOI: 10.1002/adma.201902005

[44] Kearney CJ, Mooney DJ. Macroscale delivery systems for molecular and cellular payloads. Nature Materials. 2013;**12**(11):1004-1017

[45] Eichinger JF, Haeusel LJ, Paukner D, Aydin RC, Humphrey JD, Cyron CJ, et al. Mechanical homeostasis in tissue equivalents: A review. Biomechanics and Modeling in Mechanobiology. 2021;**20**(3):833-850

[46] Kimura S, Tsuchiya A, Ogawa M, Ono M, Suda N, Sekimoto K, et al. Tissue-scale tensional homeostasis in skin regulates structure and physiological function. Communications Biology. 2020;**3**(1):637

[47] Stamenović D, Smith ML. Tensional homeostasis at different length scales. Soft Matter. 2020;**16**(30):6946-6963

[48] Bertsch P, Diba M, Mooney DJ, Leeuwenburgh SCG. Self-healing injectable hydrogels for tissue regeneration. Chemical Reviews. 2023;**123**(2):834-873. DOI: 10.1021/acs.chemrev.2c00179

[49] Lee KY, Mooney DJ. Hydrogels for tissue engineering. Chemical Reviews. 2001;**101**(7):1869-1879. DOI: 10.1021/cr000108x

[50] Nih LR, Carmichael ST, Segura T. Hydrogels for brain repair after stroke: An emerging treatment option. Current Opinion in Biotechnology. 2016;**40**:155-163. DOI: 10.1016/j.copbio.2016.04.021

[51] Tous E, Purcell B, Ifkovits JL, Burdick JA. Injectable acellular hydrogels for cardiac repair. Journal of Cardiovascular Translational Research. 2011;**4**(5):528-542. DOI: 10.1007/s12265-011-9291-1

[52] Massensini AR, Ghuman H, Saldin LT, Medberry CJ, Keane TJ, Nicholls FJ, et al. Concentration-dependent rheological properties of ECM hydrogel for intracerebral delivery to a stroke cavity. Acta Biomaterialia. 2015;**27**:116-130. DOI: 10.1016/j.actbio.2015.08.040

[53] Wang H, Rodell CB, Zhang X, Dusaj NN, Gorman JH, Pilla JJ, et al. Effects of hydrogel injection on borderzone contractility post-myocardial infarction. Biomechanics and Modeling in Mechanobiology. 2018;**17**(5):1533-1542. DOI: 10.1007/s10237-018-1039-2

[54] Schotman MJ, Dankers PY. Factors influencing retention of injected biomaterials to treat myocardial infarction. Advanced Materials Interfaces. 2022;**9**(7):2100942

[55] Adu-Berchie K, Mooney DJ. Biomaterials as local niches for immunomodulation. Accounts of Chemical Research. 2020;**53**(9):1749-1760. DOI: 10.1021/acs.accounts.0c00341

[56] Yan C, Altunbas A, Yucel T, Nagarkar RP, Schneider JP, Pochan DJ. Injectable solid hydrogel: Mechanism of shear-thinning and immediate recovery of injectable β-hairpin peptide hydrogels. Soft Matter. 2010;**6**(20):5143-5156. DOI: 10.1039/C0SM00642D

[57] Alvarado-Velez M, Pai SB, Bellamkonda RV. Hydrogels as carriers for stem cell transplantation. IEEE Transactions on Bio-Medical Engineering. 2014;**61**(5):1474-1481. DOI: 10.1109/tbme.2014.2305753

[58] Engler AJ, Sen S, Sweeney HL, Discher DE. Matrix elasticity directs

stem cell lineage specification. Cell. 2006;**126**(4):677-689. DOI: 10.1016/j.cell.2006.06.044

[59] Banerjee A, Arha M, Choudhary S, Ashton RS, Bhatia SR, Schaffer DV, et al. The influence of hydrogel modulus on the proliferation and differentiation of encapsulated neural stem cells. Biomaterials. 2009;**30**(27):4695-4699. DOI: 10.1016/j.biomaterials.2009.05.050

[60] Engler AJ, Griffin MA, Sen S, Bönnemann CG, Sweeney HL, Discher DE. Myotubes differentiate optimally on substrates with tissue-like stiffness : Pathological implications for soft or stiff microenvironments. Journal of Cell Biology. 2004;**166**(6):877-887. DOI: 10.1083/jcb.200405004

[61] Marquardt LM, Heilshorn SC. Design of injectable materials to improve stem cell transplantation. Current Stem Cell Reports. 2016;**2**:207-220

[62] Kapoor M, Martel-Pelletier J, Lajeunesse D, Pelletier J-P, Fahmi H. Role of proinflammatory cytokines in the pathophysiology of osteoarthritis. Nature Reviews Rheumatology. 2011;**7**(1):33-42

[63] Camargo Garbin L, Morris MJ. A comparative review of autologous conditioned serum and autologous protein solution for treatment of osteoarthritis in horses. Frontiers in Veterinary Science. 2021;**8**:602978. DOI: 10.3389/fvets.2021.602978

[64] Wehling P, Moser C, Frisbie D, Wayne McIlwraith C, Kawcak CE, Krauspe R, et al. Autologous conditioned serum in the treatment of orthopedic diseases: The Orthokine® therapy. Bio Drugs: Clinical Immunotherapeutics, Biopharmaceuticals and Gene Therapy. 2007;**21**:323-332

[65] Frizziero A, Giannotti E, Oliva F, Masiero S, Maffulli N. Autologous conditioned serum for the treatment of osteoarthritis and other possible applications in musculoskeletal disorders. British Medical Bulletin. 2013;**105**(1):169-184

[66] Majewski M, Ochsner PE, Liu F, Flückiger R, Evans CH. Accelerated healing of the rat Achilles tendon in response to autologous conditioned serum. The American Journal of Sports Medicine. 2009;**37**(11):2117-2125

[67] Shakouri SK, Dolati S, Santhakumar J, Thakor AS, Yarani R. Autologous conditioned serum for degenerative diseases and prospects. Growth Factors. 2021;**39**(1-6):59-70. DOI: 10.1080/08977194.2021.2012467

[68] Frisbie DD. Autologous-conditioned serum: Evidence for use in the knee. The Journal of Knee Surgery. 2015;**28**(01):063-066

[69] Dahlgren LA, van der Meulen MC, Bertram JE, Starrak GS, Nixon AJ. Insulin-like growth factor-I improves cellular and molecular aspects of healing in a collagenase-induced model of flexor tendinitis. Journal of Orthopaedic Research. 2002;**20**(5):910-919

[70] Geburek F, Lietzau M, Beineke A, Rohn K, Stadler PM. Effect of a single injection of autologous conditioned serum (ACS) on tendon healing in equine naturally occurring tendinopathies. Stem Cell Research & Therapy. 2015;**6**(1):126. DOI: 10.1186/s13287-015-0115-0

[71] Burnouf T, Strunk D, Koh MB, Schallmoser K. Human platelet lysate: Replacing fetal bovine serum as a gold standard for human cell propagation? Biomaterials. 2016;**76**:371-387

[72] Henschler R, Gabriel C, Schallmoser K, Burnouf T, Koh MB. Human platelet lysate current standards

and future developments. Transfusion. 2019;**59**(4):1407-1413

[73] Barsotti MC, Losi P, Briganti E, Sanguinetti E, Magera A, Al Kayal T, et al. Effect of platelet lysate on human cells involved in different phases of wound healing. PLoS One. 2013;**8**(12):e84753. DOI: 10.1371/journal.pone.0084753

[74] da Fonseca L, Santos GS, Huber SC, Setti TM, Setti T, Lana JF. Human platelet lysate – A potent (and overlooked) orthobiologic. Journal of Clinical Orthopaedics and Trauma. 2021;**21**:101534. DOI: 10.1016/j.jcot.2021.101534

[75] Safari FA, Safari FS. WHO guidelines on good manufacturing practices for blood establishments. Official Journal of Iranian Association of Clinical Laboratory Doctors Supplement issue for IQC 14 Tehran-Iran (The 9th International & 14th National Congress on Quality Improvement in Clinical Laboratories April,19-22, 2016). 2016

[76] Hayon Y, Dashevsky O, Shai E, Varon D, Leker RR. Platelet lysates stimulate angiogenesis, neurogenesis and neuroprotection after stroke. Thrombosis and Haemostasis. 2013;**110**(2):323-330. DOI: 10.1160/th12-11-0875

[77] Papanas N, Eleftheriadou I, Tentolouris N, Maltezos E. Advances in the topical treatment of diabetic foot ulcers. Current Diabetes Reviews. 2012;**8**(3):209-218

[78] Occleston NL, O'Kane S, Laverty HG, Cooper M, Fairlamb D, Mason T, et al. Discovery and development of avotermin (recombinant human transforming growth factor beta 3): A new class of prophylactic therapeutic for the improvement of scarring. Wound Repair and Regeneration: Official publication of the Wound Healing Society, the European Tissue Repair Society. 2011;**19**:s38-s48

[79] Rauch C, Feifel E, Amann E-M, Spötl HP, Schennach H, Pfaller W, et al. Alternatives to the use of fetal bovine serum: Human platelet lysates as a serum substitute in cell culture media. Altex. 2011;**28**(4):305-316

[80] Doucet C, Ernou I, Zhang Y, Llense JR, Begot L, Holy X, et al. Platelet lysates promote mesenchymal stem cell expansion: A safety substitute for animal serum in cell-based therapy applications. Journal of Cellular Physiology. 2005;**205**(2):228-236

[81] Warnke PH, Humpe A, Strunk D, Stephens S, Warnke F, Wiltfang J, et al. A clinically-feasible protocol for using human platelet lysate and mesenchymal stem cells in regenerative therapies. Journal of Cranio-Maxillofacial Surgery. 2013;**41**(2):153-161. DOI: 10.1016/j.jcms.2012.07.003

[82] Bari E, Perteghella S, Faragò S, Torre ML. Association of silk sericin and platelet lysate: Premises for the formulation of wound healing active medications. International Journal of Biological Macromolecules. 2018;**119**:37-47. DOI: 10.1016/j.ijbiomac.2018.07.142

[83] Rashid U, Saba E, Yousaf A, Tareen WA, Sarfraz A, Rhee MH, et al. Autologous platelet lysate is an alternative to Fetal bovine serum for canine adipose-derived mesenchymal stem cell culture and differentiation. Animals (Basel): An Open Access Journal from MDPI. 2023;**13**(16):2655

Chapter 6

Ethical Consideration of Stem Cell Transplantation

Luis Gerardo Jiménez Arias

Abstract

The utilization of stem cells holds considerable promise, prompting calls for heightened scrutiny in ethical considerations, particularly concerning experimental therapeutic applications and associated fees. We shall delve into various ethical facets of therapeutic stem cell usage, encompassing stem cell sourcing. The use of embryonic stem cells has found its limitations in the definition of a person and also in purely religious issues associated with divinity—political divergences, efficacy, and safety assessments, as well as considerations of treatment accessibility and cost. Patients, often grappling with insecurity and uncertainty regarding treatment success, place their hopes in stem cells. In this context, reevaluating the precautionary principle and acknowledging patient vulnerability is imperative.

Keywords: bioethics, ethical issues, precautionary principle, vulnerability, patient's rights

1. Introduction

Although the use of stem cells is quite promising, some want to be more strict about the ethical aspects, especially when it comes to experimental therapeutic use and charging for the treatment.

Cellular senescence, first characterized in 1961, denotes the aging of cells attributable to various factors. Notably, mesenchymal stem cells (MSCs) are susceptible to this aging process.

Regenerative medicine is at the forefront of exploration yet faces constraints like MSCs' senescence, oncogenic risks, and functional decline. However, advancements in regenerative medicine, such as induced pluripotent stem cells (iPSCs) and molecular genetics, particularly through novel gene editing techniques and clustered regularly interspaced short palindromic repeats (CRISPR), offer promise in overcoming these limitations. Yang et al. [1], for instance, demonstrated the modification of a single gene as a potential avenue for addressing these challenges.

Creating clones for reproductive purposes or generating embryonic cells from adult cells represents a significant ethical boundary. Utilizing iPSCs in biomedical research is morally acceptable, and the intent is therapeutic. Conversely, cloning or human reproduction deviates from this therapeutic purpose, potentially leading to dangerous outcomes such as embryo creation and destruction and raising risks of eugenic or reproductive misuse.

 IntechOpen

Next, we will analyze the various ethical issues concerning the use of stem cells and emerging genetic editing technologies, particularly concerning human dignity and human rights.

It is crucial to explore the potential applications of these technologies in human medicine, particularly their impact on treating genetic and infectious diseases. This entails examining the concerns and ethical implications surrounding the genetic manipulation of humans, especially in terms of altering the human species and its evolution. Our work also evaluates arguments for and against a moratorium on experiments involving genome editing of human embryos and germ cell lines. Furthermore, it is essential to analyze the roles of ethicists, the scientific community, public opinion, and policymakers in navigating the ethical challenges posed by genome editing technologies and iPSCs.

Considerations also need to be made regarding the ethical implications of stopping clinical trials on nonviable or leftover embryos from in vitro fertilization (IVF) compared to creating uniform embryos with iPSCs and genetically altering them. This analysis assesses the potential risks and benefits of using genetically edited human embryos and germ cell lines, focusing on the methods and the ultimate objectives of these experiments. Ensuring that the research adheres to ethical principles and minimizes risks associated with creating and destroying human lives is crucial.

1.1 Stem cells' potential and ethical limits

While stem cells have attracted attention from the scientific community, media, and the general public for their potential applications, they have also intrigued philosophers, bioethicists, and authors of science fiction literature. In a comprehensive review of the scientific realities of stem cells, Volarevic et al. [2] thoroughly examined the ethics and safety of cell therapies in 2018. They argue that while the creation and clinical use of human embryonic stem cells (hESCs) have been the primary focus of ethical discussions, attention must also be directed toward the moral implications of iPSCs due to their unlimited differentiation potential.

The authors delve into the ethical challenges hESCs pose, the research, therapeutic potential, and clinical application of iPSCs, and the safety concerns associated with therapies based on MSCs. They warn that iPSCs could be utilized in human cloning, posing risks for the creation of cloned humans and human-animal chimeras. Additionally, iPSCs may exhibit unwanted differentiation and potential transformation into malignant cells, raising safety concerns.

Volarevic et al. emphasize the importance of optimizing protocols approved for iPSC research to ensure the purity of cell populations derived from iPSCs before their clinical use. They also stress the need for close monitoring of MSCs, which are widely used in human treatment but may promote tumor growth and metastasis. Long-term continuous monitoring studies of MSCs in human and animal models are essential to ensure their safe and effective application.

Crohn's disease (CD), especially perianal fistulas, can be effectively treated with MSCs. Due to its complex treatment, CD poses a significant challenge for medical professionals. A study by Panés et al. [3] explores the potential use of stem cells expanded from adipose tissue. A randomized double-blind trial involving adult patients over 18 years old concluded that adipose tissue–derived stem cells were effective and safe for treating fistulas in patients unresponsive to conventional biological treatments or a combination thereof.

The Japanese government has taken a distinctive approach to therapeutic stem cell treatment by establishing over 100 private ethics committees to review treatment protocols involving MSCs. This novel strategy is designed to showcase the safety and effectiveness of stem cell treatments, bypassing international standards on biosafety and efficacy and positioning Japan's healthcare system as one of the most progressive globally. Nonetheless, concerns have been raised about the ethical implications of Japan's approach and whether these protocols may include patients who could benefit from traditional medicine.

The ultimate objective of the Japanese government is to attain coverage by health insurers [4]. Ethical considerations revolve around the lack of transparency regarding observed side effects and the true effectiveness of these treatments, as the primary goal appears to be achieving medical coverage.

In 2021, a National Institutes of Health (NIH) report highlighted the use of iPSCs and cells derived from embryonic cells for treating various diseases. Japan is a noteworthy example of how public-private coordination can yield tangible short- and medium-term achievements. Collaborative efforts between private enterprises and appropriate government policies can facilitate the development of more innovative cellular lines while not adhering to the precautionary principle.

Furthermore, ethical concerns arise regarding whether fundamental principles of biomedical research, such as informed consent, autonomy, beneficence, and nonmaleficence, are being upheld.

Currently, iPSCs can be obtained through cell reprogramming from skin or other tissues, alleviating ethical concerns associated with embryonic cells. Reprogrammed adult cells have shown improved safety and efficacy since 2006, leading to a surge in their utilization.

The iPSCs exhibit two distinct properties: indefinite division, yielding an identical lineage, and asymmetric cell division, generating two daughter cells with differing genetic expressions—one maintains parental identity, while the other undergoes genetic modifications.

Additionally, iPSCs display limited proliferation potential and developmental capacity compared to their progenitors yet retain the ability to differentiate into various cell types such as muscle or neuronal cells. Moreover, advancements in techniques such as cellular exploration, formerly known as therapeutic cloning, have emerged over time, contributing to the expanding repertoire of stem cell research methodologies.

For over two decades, embryonic stem cells have been a subject of discussion in regenerative medicine, particularly regarding the utilization of cell lines derived from the destruction of preimplantation blastocysts.

This ethical dilemma has long posed a significant challenge to embryonic stem cell research. During the early stages of stem cell development, certain authors emphasized the primary concerns surrounding the evaluation of utilizing blastocysts versus the potential therapeutic benefits for treating various diseases. This debate primarily centered on the moral implications related to two key aspects: the destruction of embryos and the potential risk of tumor formation. Meanwhile, the political discourse revolved around regulating the usage of embryos or embryonic materials, which in turn impacted the allocation of federal funds from the United States for basic research endeavors.

The debate over the destruction of blastocysts, which consist of around 100–200 cells, sparked extensive discussions, even reaching the United States Congress. The primary focus of the discourse was on the pluripotency of these cells, leading to profound ethical and moral considerations about creating life through IVF only to subsequently destroy it for the benefit of other human lives.

Derived from fetal or embryonic tissue, the potential use of embryonic stem cells for Parkinson's disease therapy has been under scrutiny for over a decade. The main focus of treatment is a group of dopaminergic neurons in the midbrain that are susceptible to degeneration in Parkinson's patients. Despite its promise, ethical concerns about this approach have been raised by numerous experts. These concerns revolve around potential risks associated with the implantation process, including the possibility of improper migration of the implanted cells and the formation of tumors.

It is essential to prioritize providing viable alternatives to participants before their involvement in clinical trials using embryonic cells. Ethical dilemmas may arise concerning the acquisition of fetal cells from induced abortions compared to spontaneous or uninduced abortions. Regrettably, this matter has not been adequately discussed by the aforementioned authors, particularly in terms of emphasizing the risk-benefit analysis.

The argument that hESCs are morally superior because they could potentially replace aborted fetal tissue in cell replacement therapies is flawed. This is because the process of obtaining embryonic stem cells involves the destruction of embryos, which raises ethical concerns, while aborted fetuses may not always result from intentional terminations. Furthermore, hES tissue typically lacks the maturity found in fully developed tissues, making it less therapeutically advantageous compared to fetal tissue.

Considerable advancements have been made in the utilization of diverse cell sources, including iPSCs, to address brain disorders. This encompasses patient enrollment, securing informed consent, and managing the translation of these breakthroughs into intricate clinical applications. Ethical considerations emerge when it comes to charging for experimental treatments, particularly when they represent the sole hope for patients selected within the clinical trial framework, given the absence of assured success.

The debate surrounding embryonic cells versus adult stem cells centers on their differentiation potential and ethical implications. Embryonic cells, being pluripotent, can differentiate into a broader array of tissues compared to adult stem cells, which have a more restricted differentiation capacity. However, adult stem cells are linked to fewer ethical controversies. This is not only due to concerns about embryo destruction and the use of voluntarily aborted fetuses but also because of their lower propensity for tumor formation.

1.2 The juridic status of the human embryo

The legal framework concerning the therapeutic use of stem cells varies widely among different countries, with a strong emphasis on the effectiveness and safety of MSCs. Costa Rica provides an interesting example, as the College of Physicians does not formally acknowledge stem cell treatment as a specialized practice. Nonetheless, these treatments are available at substantial prices, sometimes as high as $15,000, despite the lack of definitive evidence supporting their efficacy.

In 2010, Costa Rica's Supreme Constitutional Court banned the use of therapeutic stem cells. However, 6 years later, Executive Decree 39986-S reinstated their use, albeit with some ambiguity. According to this decree, the use of stem cells must be autologous, and any manipulation beyond the minimum requires a demonstration of efficacy and safety through clinical trials, especially regarding the expansion of MSCs.

Costa Rica is the first instance in Latin America where the country's Constitutional Chamber has hindered access to healthcare and disrupted doctor-patient relationships. The use of Executive Decrees to bypass constitutional rulings raises concerns about legality, but no legal challenges have emerged so far.

Moreover, Costa Rican law does not explicitly forbid the extraction of stem cells from embryos for research. Instead, a protracted legal dispute over the past 20 years between the Chamber and the Inter-American Court of Human Rights, especially concerning IVF, has left this matter unsettled.

Exploring the use of embryonic cells necessitates understanding their legal status, particularly recognizing them as human embryos. The extraction of embryonic cells has been controversial, grounded in the notion of protecting the inherent dignity of human life, a principle enshrined in Costa Rica's Constitution.

Research involving hESCs utilizes in vitro fertilized eggs. Moral and bioethical concerns arise because the use of these hESCs requires the destruction of human embryos, a practice that is viewed as ethically problematic by certain lawmakers and bioethicists due to the belief that human embryos should be considered human persons.

But are embryos protected by the international law system? Are they considered humans? Are they subjects of human rights? What is the status of the human embryos?

The core of human identity cannot be achieved gradually, like physical traits or specifics. Assigning value to customs, redefining terms broadly, and then shaping them within an interpretive framework amount to a type of exploitation. The right to life is not a matter of gradual acquisition; it is either inherent or absent and therefore should not depend on age or developmental stage [5].

As per the Inter-American Court of Human Rights [the Court], the human embryo is not deemed a human person in its early stages. The court establishes the commencement of human life as the moment of embryo implantation, leaving the initial stages ambiguous. Legally, the court does not acknowledge the embryo as human life during these early phases, and its interpretation of the right to life implies a gradual and incremental understanding [6].

Adriasola [7] argues that according to the legal framework of the Inter-American System of Human Rights, the embryo is not recognized as a person. It lacks legal subjectivity and does not have the right to life. Therefore, it does not possess the inherent dignity afforded to human beings. Consequently, any dignity it holds is granted externally, posing challenges in establishing a universal moral standard of protection as its dignity is contingent rather than intrinsic.

On the other hand, Barker [6] argues that Costa Rica, purportedly a staunch defender of human rights, paradoxically infringes upon these rights by neglecting to uphold the most fundamental and basic right—the right to life.

The concept of the right to life becomes complex when determining the beginning of life and its related rights are deliberately altered, placing it under the authority of secondary rights. Bohórguez and Aguirre [8] argue that international agreements on human dignity demonstrate different levels of significance when applied to "particular cases" in the Court's decisions or when they are no longer employed.

Andruet [5] argues that the right to life should not depend on the individual's age under protection. Similarly, Spaemann and Odonovan [9] assert that while human beings possess specific attributes defining personhood, the individuals who possess these attributes, not the attributes themselves, constitute persons. Personhood is closely linked to its inviolability, suggesting that it is not conferred by society but inherently granted by nature itself.

In essence, the Court has depersonalized the human embryo, and its rulings hold authority over all member nations.

1.3 The dignity of the human person

In his analysis of the clarification and origins of the concept of human dignity, Illie [10] underscores the inherent challenge in defining it, pointing out the perceived ambiguity of its essence. He mentions that one approach to defining human dignity has been to outline what is unworthy of a human being.

Defining human dignity has always been a challenging task due to its intricate and multifaceted nature. Throughout history, attempts have been made to define concepts by comparing them with their opposites. However, this challenge is compounded by the fact that the concept of dignity exists across various cultures and historical periods.

In debating the elucidation and origin of the concept of human dignity, the author highlights its inherent complexity, which stems from the perceived ambiguity of its essence. One approach to defining human dignity has been to delineate what is unworthy of a human being, a method often employed in philosophy to establish definitions through oppositions.

Throughout history, attempts have been made to define concepts by contrasting them with their opposites. However, the theory of definitions, rooted in logic, can encounter challenges when describing abstract concepts, particularly those spanning cultures and epochs.

Augustine of Hippo, for instance, conceptualized evil as the absence of virtue, illustrating how definitions are often contingent on context and relative to prevailing moral frameworks.

Applying this logical framework to the concept of dignity, defining it in terms of what is not worthy disregards the evolving nature of dignity. Both worthiness and unworthiness are relative concepts, subject to the morality of each society and the passage of time.

Contemporary understandings of dignity often emphasize the distinctive attributes of *Homo sapiens* that set them apart from other living beings, particularly the capacity for rational thought. However, this distinction does not imply superiority over other creatures, as such claims can lead to political exploitation.

The author argues for establishing principles that uphold individuals' dignity while acknowledging philosophical inquiries into whether human dignity is advancing or facing unprecedented threats. Recognizing and specifying these threats are essential if we are to address them effectively.

The author ultimately concludes that human dignity is absolute and not contingent upon its violations. Attempts to define it in concrete terms are futile, as the concept evolves over time. The Western concept of human dignity, rooted in Judeo-Christian beliefs and developing into a notion inherent to intelligent species, underscores the teleological question of human purpose. According to Kant, every individual is an end in themselves, not a means to an end, encapsulating the intrinsic value of human dignity.

As Pope Francis [11] has expressed, every person has an inherent and infinite dignity that is an essential part of their humanity. This dignity remains constant and transcends any circumstance, condition, or state that an individual may face. These principles are in harmony with those set forth in the Universal Declaration of Human Rights.

According to the Pope, human dignity must be upheld irrespective of one's interpretation, particularly in contexts involving the sourcing of stem cells and their therapeutic utilization. This encompasses both human embryos—despite lacking protection under the inter-American legal system—and the inherent risks associated with human research.

The application of clinical practices must strongly uphold the core principle of human dignity, which is rooted in Judeo-Christian ethics and upheld in international agreements overseeing human subject research. These agreements include the Declaration of Helsinki, adopted by the 1964 World Medical Association General Assembly (WMA), and the Nuremberg Code, which emphasizes informed consent and delineates the four fundamental principles governing medical practice [12].

1.4 Ethics of reprogramming and emerging technologies

Some people raise important ethical concerns about stem cell reprogramming, specifically regarding potential issues such as abnormal reprogramming resulting in iPSCs, the possibility of clone creation, and the generation of embryonic cells, while Scesa [13] contends that ethical dilemmas are not associated with the origin of the cells but rather with technical concerns such as tumorigenicity and rejection.

For Kass [14], the zygote and early embryonic stages clearly demonstrate signs of life, such as metabolism, respiration, and responsiveness to environmental stimuli. Additionally, they undergo division and growth. While eggs and sperm are living cells, they do not develop into living beings like zygotes do. As a result, a distinct living entity emerges with its own unique genetic composition.

On the other hand, for Lucas [15], The body acts as the medium for interacting with the physical world. The author distinguishes between the body and corporeality: the body relates to the external, factual reality, while corporeality delves into the subjective domain. While the body embodies a purely material aspect, corporeality embodies the inner essence of humanity.

The ethical and philosophical discourse surrounding the utilization of embryonic stem cells has escalated to international legal arenas. In our specific case, the Inter-American Court of Human Rights has stripped the human embryo of its dignity, reducing it to a mere object and placing it in a legal void. According to this court, human life commences not at fertilization but upon implantation, implying that the embryo lacks personhood and intrinsic dignity until this stage. Consequently, it is considered devoid of absolute rights during the initial days postfertilization, opening the door to potential manipulation, including the creation and destruction of embryonic stem cells, which are deemed objects lacking inherent rights.

It's noteworthy that a judicial body, rather than scientific consensus, has defined this biological fact, thereby curtailing what could be considered the most fundamental of rights, thus potentially paving the way for the curtailment of other rights.

This situation prompts questions about the ideological underpinnings of human rights. Do human rights exist primarily to safeguard life and health? Is every life accorded equal value, or are some lives deemed less significant than others? Furthermore, should scientific truths dictate legal frameworks, or should the law adapt to the realities of biotechnological progress?

The differing viewpoints emphasize a significant ethical divide, separating technical challenges from concerns arising from the nature and use of iPSCs, which require careful bioethical examination. Zheng is worried about the potential abnormal outcomes of reprogramming, while Scesa views these as technical obstacles rather than ethical dilemmas, similar to tumor formation. As a result, Scesa considers the risks associated with therapeutic misuse to be technical rather than moral issues.

The use of clones for reproduction or embryonic cells from adult cells raises ethical concerns. While using iPSCs in biomedical research without involving human subjects is considered morally acceptable for therapeutic purposes, pursuing cloning or human reproduction could lead to problematic outcomes such as embryo creation and destruction, as well as potential eugenic or reproductive exploitation.

Ultimately, the distinction lies between the origin of the cells, which are somatic, and the potential transformations or abuses they undergo. Biomedical research is considered morally licit when aimed at alleviating human suffering, as long as the means employed align with moral principles.

In advocating for protection, the central argument posits that "we all possess the entitlement to exhaust all technically feasible avenues to regain our health," thereby restoring "our quality of life". This right, perceived as inherent to individuals of legal age, is deemed unjust for any political authority to curtail. Subsequently, they substantiate their legal plea for protection by citing Article 2 of the American Convention on Human Rights [the Convention], which mandates that all authorities undertake measures, including legislative actions, to ensure the effective realization of the freedoms and rights enshrined within the Convention.

While the Convention does not explicitly mention the right to health, it inherently intertwines with the right to life. This is because access to health services facilitates the practical exercise of the right to life.

Global studies analyzing approved therapies have increased from 903 to 1120 as of 2023. Despite this increase, the tally of approved treatments worldwide remains at 12, with 9 originating in Asia. Among these, 5 have emerged in the Republic of Korea, 2 in Japan, and 1 each in India and Iran, covering various diseases such as cardiac issues and Crohn's disease [16].

A noticeable gap exists between clinical trials and approved therapies, with roughly one therapy gaining approval for every 100 phase 3 clinical trials. However, it's worth noting that significant strides have been made in therapies involving the use of MSCs, particularly in rare diseases or where traditional medicine has shown limited efficacy. Therapies incorporating MSCs exhibit remarkable effectiveness, offering genuine hope to patients for whom conventional treatments have proven insufficient. This underscores the importance of prioritizing the doctor-patient relationship, as Riordan et al. [17] emphasized.

Fukuyama [18] discusses the regulatory dynamics of scientific-biological advancements, highlighting the relationship between the NIH and the industry. Despite the widespread establishment of the precautionary principle (PP) [19] in international legislation [20], which dictates that precautionary measures should be taken in activities posing potential threats to human health or the environment, regardless of established cause-effect relationships, there remains a tension with the Hippocratic Oath's principle of "do no harm" [21].

The tension arises from the resistance of many individuals to treatments that lack scientific certainty [22]. However, this resistance fails to acknowledge the inherent uncertainty in scientific advancement. Science inherently operates within probabilities, so expecting any therapy to be 100% effective is unrealistic. The effectiveness of a therapy is influenced by numerous factors [23], including the patient's clinical profile, age, disease progression, and timing of treatment. Furthermore, in the case of therapies involving MSCs, additional variables such as nutrition, exercise, and environmental conditions also play significant roles.

While the precautionary principle has traditionally been applied in environmental contexts like regulating genetically modified crops (GMOs) [24] to mitigate potential adverse effects, some argue that its application in public health might be inappropriate [25]. This is because PP primarily aims to minimize or prevent undesirable technological outcomes, which may not fully align with the complexities of medical practice.

Furthermore, while the medical principle of "nonmaleficence" underscores the importance of avoiding harm, it's crucial to recognize that virtually all medical treatments carry some risk of unintended side effects.

Liang et al. [26] reported the first study using CRISPR-Cas9 on human embryos in 1975, marking the beginning of human genome editing research.

New technologies, such as genome editing with CRISPR/Cas9 iPSCs, have significantly advanced stem cell research and their potential applications in human medicine [27]. One of the immediate frontiers is achieving more precise and efficient reprogramming to avoid the risks associated with poor programming.

There is widespread concern that genetic manipulation of humans could lead to alterations in the human species and its evolution [28]. This concern extends beyond the scientific community to the general population. Various conferences on this topic have called for a moratorium on all experiments that could potentially alter the human species. This caution arises from the fear that new technologies involving embryonic stem cells or iPSCs could fundamentally change the human species. Despite this, the rapid advancement of new technologies holds promise for treating previously incurable genetic diseases and infectious diseases such as HIV.

Particularly according to these authors, using CRISPR-Cas9 in embryonic material carries significant risks, such as altering human evolution and the potential for eugenic practices. Based on the precautionary principle, they advocate for a moratorium on such experiments until there is scientific certainty that they will not be used for human reproduction.

However, the dilemma remains that some of these experiments could advance scientific knowledge. Should we halt clinical trials on nonviable or leftover embryos from IVF, or should we create homogeneous embryos with iPSCs and genetically edit them?

2. Conclusions

It is not ideal to stop scientific research due to potential abuses and jeopardize the advancement of knowledge. Ethicists, in collaboration with the scientific community, public opinion, and policymakers, are responsible for navigating this challenge within the human rights framework. The goal should be to minimize risks, particularly those involving the creation and destruction of human lives and their modification for reproductive purposes. Such actions would undermine the international community's efforts to align scientific development with ethical principles.

Using genetically edited human embryos and germ cell lines represents the ethical frontier. The focus should be on the laboratory tests and the ultimate goals and means of these tests. For a human action—one that is willful and knowledge-based—to be considered good, both the means and the end must be inherently good.

Acknowledgements

We express our sincere appreciation to the Bioethical Foundation for their invaluable support. Their generous contributions and unwavering commitment have played a vital role in advancing our research and promoting discussions on crucial bioethical issues. This support has allowed us to delve into complex ethical dilemmas, champion human dignity, and uphold human rights in the realm of emerging genetic technologies. We are grateful for partnering with us in this essential endeavor.

Conflict of interest

The authors declare no conflict of interest.

Notes/thanks/other declarations

The author would like to express gratitude to Shirley Schley for her invaluable technical support.

Author details

Luis Gerardo Jiménez Arias
Fundación Bioética, San Jose, Costa Rica

*Address all correspondence to: luisgjime@yahoo.com

References

[1] Yang J et al. Genetic enhancement in cultured human adult stem cells conferred by a single nucleotide recoding. Cell Research. 2017;**27**(9):1178-1181

[2] Volarevic V et al. Ethical and safety issues of stem cell-based therapy. International Journal of Medical Sciences. 2018;**15**(1):36-45

[3] Panés J et al. Expanded allogeneic adipose-derived mesenchymal stem cells (Cx601) for complex perianal fistulas in Crohn's disease: A phase 3 randomised, double-blind controlled trial. The Lancet. 2016;**388**(10051):1281-1290

[4] Enosawa S. Clinical trials of stem cell therapy in Japan: The decade of progress under the National Program. Journal of Clinical Medicine. 2022;**11**(23):7030

[5] Andruet A. La eticidad en las ciencias médicas en general y en la fecundación in vitro en particular. Revista Jurídica El Derecho. 1988;**127**(127):807

[6] Barker RS. Inverting human rights: The inter-American court versus Costa Rica. University of Miami Inter-American Law Review. 2015;**47**:1

[7] Adriasola G. El fallo de la Corte Interamericana de Derechos Humanos en el caso Artavia Murillo y otros vs. Costa Rica. Removiendo conceptos sobre el estatus jurídico del embrión. Revista Médica del Uruguay. 2013;**29**(3):181-186

[8] Monsalve VB, Aguirre Román J. As tensões da dignidade humana: conceituação e aplicação no direito internacional dos direitos humanos, Sur. Revista Internacional de Direitos Humanos. 2009;**6**(11):40-63

[9] Spaemann, R. Persons: The Difference between 'Someone' and 'Something'. USA: Oxford University Press; 2006

[10] Illie AP. Una aproximación al concepto de dignidad humana. Universitas: Revista de filosofía, derecho y política. 2004;**1**:9-13

[11] Dicastery for the Doctrine of the Faith. Dignitas Infinita. Vatican City: Holy See; 2024. Available from: https://press.vatican.va/content/salastampa/en/bollettino/

[12] Van Walraven SM et al. Addressing ethical and procedural principles for unrelated allogeneic hematopoietic progenitor cell donation in a changing medical environment. Biology of Blood and Marrow Transplantation. 2018;**24**(5):887-894

[13] Scesa G, Adami R, Bottai D. iPSC preparation and epigenetic memory: Does the tissue origin matter? Cells. 2021;**10**(6):1470

[14] Kass LR. Toward a More Natural Science. New York: Free Press; 2014

[15] Lucas Lucas R. Antropología y problemas bioéticos. Madrid: Biblioteca de Autores Cristianos; 2017

[16] Fernández-Garza LE, Barrera-Barrera SA, Barrera-Saldaña HA. Mesenchymal stem cell therapies approved by regulatory agencies around the world. Pharmaceuticals. 2023;**16**(9):1334

[17] Riordan NH, Jiménez Arias LG, Coronado R. Ethics of international stem cell treatments and the risk-benefit of helping patients. In: Possibilities and Limitations in Current Translational

Stem Cell Research. London, UK: IntechOpen; 2022. pp. 183-194

[18] Fukuyama F. Our Posthuman Future: Consequences of the Biotechnology Revolution. London, UK: Farrar, Straus and Giroux; 2003

[19] Goklany IM. The Precautionary Principle: A Critical Appraisal of Environmental Risk Assessment. Washington DC, US: Cato Institute; 2001

[20] Mcintyre O, Mosedale T. The precautionary principle as a norm of customary international law. Journal of Environmental Law. 1997;**9**:221

[21] Goodman JC. Applying the "do no harm" principle to health policy. Journal of Legal Medicine. 2007;**28**(1):37-52

[22] Freudenburg WR, Gramling R, Davidson DJ. Scientific certainty argumentation methods (SCAMs): Science and the politics of doubt. Sociological Inquiry. 2008;**78**(1):2-38

[23] Mastri M, Lin H, Lee T. Enhancing the efficacy of mesenchymal stem cell therapy. World Journal of Stem Cells. 2014;**6**(2):82

[24] Jimenez-Arias LG. Bioethics and the Environment. A Brief Review of the Ethical Aspects of the Precautionary Principle and Genetic Modified Crops. Buenos Aires, Argentina: LibrosEnRed; 2007

[25] Weed DL. Is the precautionary principle a principle? IEEE Technology and Society Magazine. 2002;**21**(4):45-48

[26] Liang P et al. CRISPR/Cas9-mediated gene editing in human tripronuclear zygotes. Protein and Cell. 2015;**6**(5):363-372

[27] Waddington SN et al. A broad overview and review of CRISPR-Cas technology and stem cells. Current Stem Cell Reports. 2016;**2**(1):9-20

[28] Morange M. Genetic modification of the human germ line: The reasons why this project has no future. Comptes Rendus. Biologies. 2015;**338**(8-9):554-558

Chapter 7

Fundamentals of Stem Cells and Application in Complex Disorders

Gaurav Verma and Srividhya Ravichandran

Abstract

Stem cells possess an outstanding ability to differentiate into various cell types, making them instrumental in human physiology and regenerative medicine. This chapter explores the fundamentals of stem cells and their applications in complex disorders like neurodegeneration and cardiovascular disorders. Stem cells are undifferentiated cells capable of self-renewal and differentiation into specialised cell types, including neurons, muscle cells and blood cells. They are further categorised into embryonic stem cells and adult stem cells. Understanding stem cell biology is crucial for harnessing their potential in treating diabetes, neurodegenerative diseases, cardiovascular disorders and autoimmune conditions. Stem cell-based therapy offers promising avenues for reversing damaged tissues and organs, as well as providing models for studying disease mechanisms and drug discovery. Recent advancements in induced pluripotent stem cells (iPSCs) have enabled the generation of patient-specific stem cells, allowing for personalised therapies for personalised medicine and disease modelling. However, extreme challenges remain, including the availability of donors, characterisation of stem cells, efficacy, ethical concerns, tumourigenicity and immune rejection. In summary, stem cells hold immense promise for next-generation medicine by offering novel therapeutic strategies for treating neurodegenerative diseases, albeit further investigation is essential to address existing challenges and realise the unprecedented potential of stem cells in reversing complex disorders.

Keywords: stem cells, pluripotency, differentiation, regenerative medicine, complex disorder

1. Introduction

Stem cells could be defined as "the cells which possess an inherent capacity to renew themselves and perpetuate so as to yield mature cells of a particular lineage after multiple rounds of differentiation and cell division" [1]. From the above definition, it is clear that stem cells must possess important properties like self-renewing capacity and a differentiation capacity under controlled conditions [2]. During the development of an embryo, the originally imbibed property of the embryo slowly evades and starts to acquire new properties [3]. In adulthood, only a few stem cells could be seen and most of them are dispersed across the tissues. Thus, a multicellular organism has its own design in the context of its origin from the stem cells of the embryo.

The two important terminologies that require attention in this context are totipotency and pluripotency. Totipotency is the capacity to regenerate the whole organism from an implanted single cell [3, 4], whereas pluripotency is the tendency of a cell to develop into any type of cell seen in an organism except for the placental trophoblast. Totipotency starts from the zygote, whereas pluripotency starts from the early embryonic stages [4]. A zygote has a unique property of totipotency which other cells including stem cells are devoid of.

A pluripotent cell is always compared to an embryonic stem cell, where the cells derived from the blastocyst stage of an embryo give rise to different types of cells [5] that fall under the three layers of the neural crest, and a totipotent stem cell is the actual zygote which has the potency to give rise to the whole organism upon cell division. The developmental stage of an embryo is very crucial for understanding the concepts of totipotency and pluripotency. From the zygote, a blastocyst is formed which gives rise to trophoblasts to become the placenta and the inner mass of this blastocyst forms the embryo after repeated cell divisions [6]. The inner mass of the blastocyst can be extracted and cultured *in vitro* under restricted conditions, to give rise to stem cells which are pluripotent in nature and differentiate to give rise to cells of various lineages. To understand the concepts of totipotency and multipotency, the readers are requested to refer to (**Figure 1**). The figure gives a description of the developmental stages of an embryo and the isolation of embryonic stem cells from the embryo.

The classification of stem cells employs an important concept of "developmental potency" [7], which means that the zygote which has the potency to give rise to the whole organism can be used as a source of a stem cell. In spite of the inherent property of the zygote to develop into blastomeres of equal developmental potency, the limitation lies in the self-replenishing potential [7]. The efficiency of the embryo to develop into a stem cell lineage, which means that the cells isolated from the embryo develop into the inner mass of blastocysts in culture, leads to the development of "embryonic stem cells" (ESCs) [8]. The property of stem cells depends upon their origin, which is also governed by their microenvironment [9].

ESCs were the first of their kind to be isolated and studied for their highly unusual pluripotency. They were first isolated and characterised from the mouse embryos [10] followed by the human ESCs [11]. During their isolation, these ESCs were seen to retain their potency of developing into germ layers under specific conditions *in vitro*. In comparison with the ESCs, the non-embryonic stem cells are multipotent or unipotent. Multipotency is defined as the potential of a cell to develop into multiple types of cells within a particular tissue [12]. For example, haematopoietic stem cells are isolated from the bone marrow. Unipotency is defined as the potential to develop into only one type of a cell within a particular tissue. A well-studied example of a unipotent cell is the satellite cell of skeletal muscle, which has the potential to differentiate into only one cell type. The developmental potency of ESCs is huge when compared to unipotent cells. During their developmental stages, the plant kingdom also exhibits their stem cell properties. As we describe the ESCs, in plants, they are described as meristematic tissues [13]. The "Primary meristems" are those that arise at an early stage of embryogenesis, whereas the cells that have evolved post-embryonic development are called "Secondary meristems" [13]. A picture of several meristematic tissues present in a plant cell is shown in **Figure 2**.

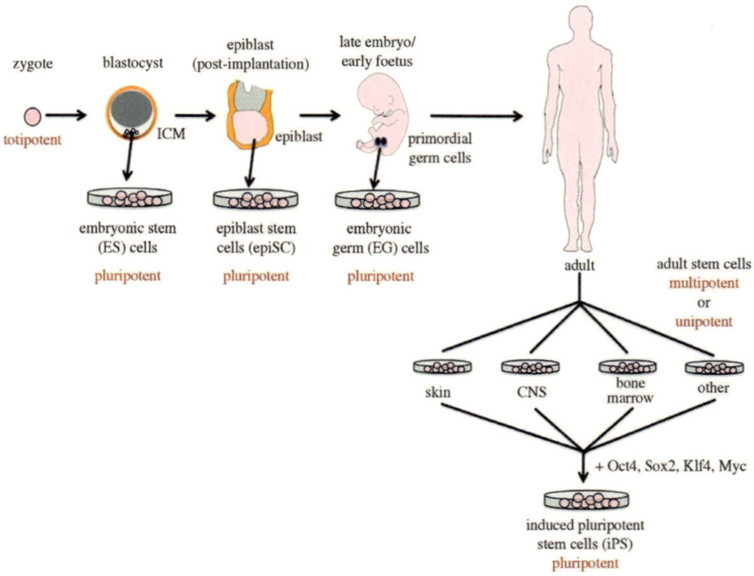

Figure 1.
The generation of an amniotic functional cell from a pluripotent stem cell. Also, this give a clear idea of totipotency and pluripotency. The pluripotent stem cell (embryonic stem cell) would give rise to a progenitor cell, which is mitotic and non-functional. From the progenitor cell of a lineage, an actual functional cell with a limited lifespan is formed. The figure also shows how the unipotent/multipotent adult stem cells give rise to induced pluripotent stem cells in vitro. *Image source: Watt Fiona M. and Driskell Ryan R. The therapeutic potential of stem cells 365 Phil. Trans. R. Soc. B.*

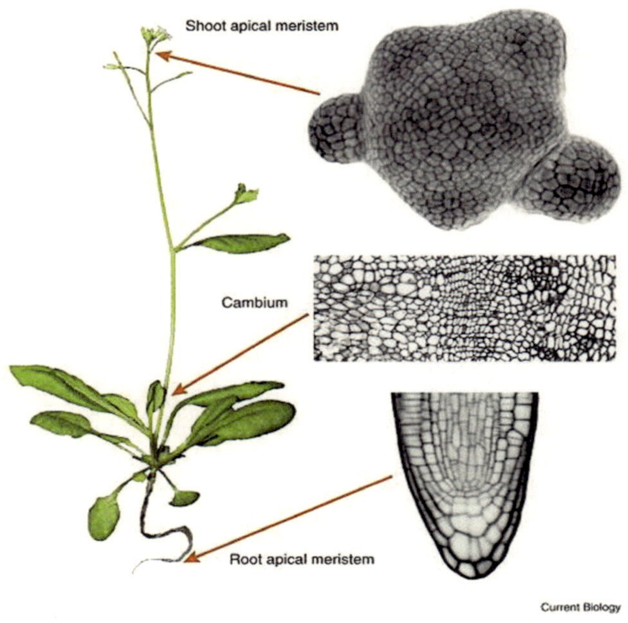

Figure 2.
Showing the meristematic tissues of a plant as denoted by the shoot and root apical meristems and cambium. Meristems being the stem cells of the plant are also totipotent and capable of giving rise to the whole plant. Source: Modified figure from 2. Gaillochet, C., and Lohmann, J.U. (2015). The neverending story: from pluripotency to plant developmental plasticity. Development 142, 2237-2249 as indexed in: Plant Stem Cells Thomas Greb and Jan U. Lohmann Plant Stem Cells. Greb T et al. Curr Biol. (2016).

Some somatic cells like neurons and skeletal muscle cells are postmitotic and hence, could not be replenished [14] and myocardial cells. However, the stem cells have the capacity for continuous regeneration and the capacity to replenish the somatic cells during the time of tissue damage/injury. This property of stem cells to replace damaged tissues is very important for tissue homeostasis in the body. The concept of adult stem cells (ASCs) was a ground-breaking discovery in stem cell research. With reference to their origin, these non-embryonic stem cells isolated from the other non-embryonic tissues had an important role to play in the physiological system with the pretext of their functional capacity, and their pluripotent nature was exploited for their huge therapeutic potential. These cells can be isolated from amniotic fluid, umbilical cord tissue, bone marrow, heart, brain, etc. [15]. The therapeutic potential of stem cells will be useful in treating complex disorders like heart diseases and neurodegenerative disorders, which will be discussed later in the chapter.

2. Stem cells

2.1 Why are stem cells important?

Stem cells have become an indispensable part of human life. As discussed in the previous section, stem cells have the capacity to form a whole organism. Their potential for differentiation is attributed to their ability to develop into numerous cells of the body, including neurons. The self-perpetuating potential of stem cells in therapeutic applications has not been explored much in the past with reference to the ethical considerations of usage. The discovery of ASCs reframed the entire system of stem cell biology and paved the way for the implication of stem cells in clinical settings in a better format.

With the huge value of their differentiation potential, the embryonic stem cells were seen as a magic medicine to most of the therapeutic cell/tissue engineering applications, but the immune response mounted by these cells hampered their usage. Therapeutic cloning (somatic cell nuclear transfer) procedure handled the problem of immunogenicity effectively [16]. The somatic cell nuclear transfer solved some of the ethical problems encountered with the ESCs, thanks to yet another tool called "induced pluripotency" that solved the rest of the issues. Induced pluripotency is a state where the normal body cells are induced to attain a stage of pluripotency similar to that of the ESCs [17]. The rejuvenating potential of stem cells has proved to be useful in cell-based therapies for many diseases like heart diseases, neurodegenerative diseases [18], and genetic disorders. Thanks to the work of Yamanaka, the process of reorienting or dedifferentiating the somatic cells by many cancer genes (oncogenes), a new group of cancer stem cells are formed which would aid the processes of trans-differentiation along with dedifferentiation [19]. The properties of cancer cells and stem cells are completely different. Though the proliferation capacity of the stem cells is very high as a cancer cell, some properties like the destruction of the immune system and metastasis clearly differentiate the stem cells from the cancer cells [20].

In the present scenario, the role of stem cells in yielding their therapeutic potential to replace damaged tissues and tissue repair processes has been overwhelming in all fields of applications including immune and gene therapy. The most common clinical application of stem cells is in haematopoietic stem cell transplantations for haematological disorders such as haemophilia, leukaemia and thrombocytopenia [21]. Stem cell banking, which collects self-tissue from the bone marrow or umbilical cord, led to several applications [22, 23].

The potency of stem cells in revolutionising health and medical research is so tremendous that there would not be any medical research in the future without stem cells [24]. Identification of stem cells from various tissue sources might provide appropriate targets for gene therapy applications and generating new drug targets/ establishing new disease models [25]. Research on stem cells to fight several diseases like autoimmune beta cell dysfunction, chromosomal disorders, congenital heart diseases and neurodegenerative disorders has opened new vistas in the field of regenerative medium [18, 26, 27].

2.2 Properties of stem cells

The properties of stem cells hugely depend upon their nature of cell division, function, developmental pattern and diversity. A stem cell progeny also imbibes the inherent potential of self-replenishment and differentiation; hence, it is sincerely committed to any of the lineage which involves a lot of cell division processes to yield a functional cell, tissue or an organ [28]. Depending on the developmental source (source tissue), they could be named according to their descendants. For example, the umbilical cord blood consists of cells that could develop into blood cells and mesenchymal cells and hence, they are named as the haematopoietic stem cells [29] and mesenchymal stem cells [30], respectively. For lineage differentiation, an interesting pictorial representation can be referred to in **Figure 3**.

The lineages exhibit several properties, the most important of them being stemness, often defined as the ability of a single cell to differentiate into various lineages

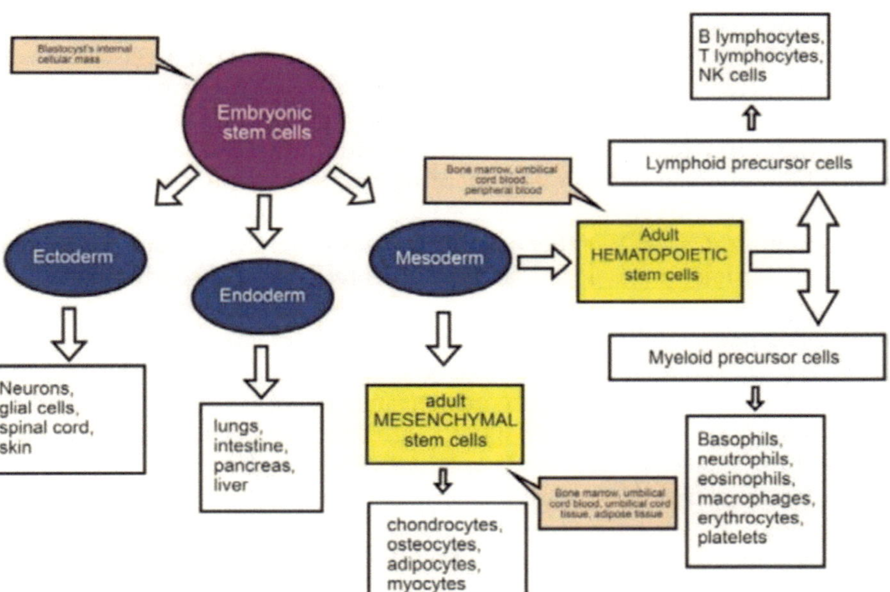

Figure 3.
The figure depicts the pluripotency of embryonic stem cells and the multipotency of the adult stem cells. The blastocysts of the embryo becomes the source of the embryonic stem cells which then develops into three germ layers by gastrulation. From the pluripotent embryonic stem cells, several organs and organ systems are formed from the germ layers. Adult stem cells are obtained from these three germ layers and they are multipotent or unipotent.
Photo source: *An Bras Dermatol. 2014;**89**(2):286-292.*

that are normally found in that tissue [31]. The regulatory mechanisms governing the behaviour of the cells might vary across these groups. Another important property is plasticity which refers to the ability of ASCs to develop into a new phenotype [32]. The molecular and cellular changes associated with plasticity are often governed by a related complex phenomenon called trans-differentiation, wherein a differentiated cell type can be destined to develop into another [33].

Cells are often not seen separately. They associate themselves with a tissue and exhibit their functions as an organ or an organ system. In this preface, stem cells are frequently located in a distinguished microenvironment called the "niche" which helps them nourish and exhibit their stemness [34]. A niche is often described as the local microenvironment of these stem cells with a lower tension of oxygen during the early embryonic developmental stages [35]. Several transcription factors like Oct-4 are involved in the cytoprotection and adhesion of the cells to the niche [36]. Also, the anaerobic environment helps the stem cells and plays a major role as a signalling molecule to regulate the stemness of these cells [35].

Cell proliferation and growth of a stem cell are very important and the factors and mechanisms regulating the same are also very essential in the first place. During the development and differentiation of stem cells, a large number of signalling pathways such as Wnt signalling, PI3K, MAPK and NO and calcium signalling pathways are involved [37–40]. It is noteworthy that the factors and mechanisms responsible for the maintenance of the stem cell population without specialisation are crucial during the lifespan of the stem cell. Most of the somatic cells like myocytes, blood cells or neurons do not replicate for many times during their entire lifespan. But the stem cells have the extraordinary property of undergoing multiple rounds of cell division during their lifetime and proliferating for numerous cycles [5]. The stem cells possess an inherent property of differentiating into a number of cell types [24] but they do not exhibit the features pertaining to a tissue. To make it precise, stem cells are not committed to any repertoire of cells pertaining to a specific physiological milieu. For example, a β cell could be specialised to secrete insulin or a hepatocyte to possess a specific receptor on its cell surface. This property of a stem cell makes it hard to attain coherence with its neighbouring cells to perform specific functions, such as a heart muscle cell that pumps the blood throughout the body or a nerve cell, which communicates among other cells using electrochemical signals. But whenever there is any physiological trigger or an experimentally induced condition, they could perform a specific function of a tissue or an organ [41]. This property of stem cells is exploited in many applications of stem cells. The differentiation process involves the development of the stem cells to undergo different stages. Molecular and cellular signals within the stem cell niche (microenvironment) result in the culmination of the stem cells towards differentiation [42], with internal signals being controlled by the DNA of the stem cells itself, while external signals might include physical and chemical factors secreted by neighbouring cells present in the microenvironment, to be specific [43].

Certain stem cells have a property of teratogenicity that means that the stem cell eventually develops into a benign carcinoma [10]. Both embryonic and adult cells give rise to such types of carcinomas. In ovaries, these carcinomas arise from the oocytes and are benign tumours that undergo a parthenogenetic activation leading to the formation of a relatively disorganised embryonic tissue (teratoma) [44]. We could see such tumours in the testis, but these solid tumours found in the testis are malignant in nature and are often called teratocarcinomas which are classical stem cell tumours [44]. Readers could appreciate the difference between the tumourigenicity of the tumour cells and stem cells and the terms "teratoma" and "teratocarcinoma" in this section.

2.2.1 Section summary

Stem cells have the unique ability to self-renew and differentiate into specialised tissues, influenced by their environment and signalling pathways. They reside in niches that support regeneration and growth. While highly valuable for therapy, some stem cells can become teratogenic, forming benign or malignant tumours like teratomas and teratocarcinomas.

2.3 Sources of embryonic stem cells

2.3.1 In vitro *fertilisation*

The year 2010 was regarded as a golden year for the ESCs. The development of IVF by Robert Edwards was a key discovery in the field of stem cell biology [45]. This IVF laid a foundation stone for scientists who worked on early embryos for the isolation and identification of human embryonic stem cells. Fresh or frozen cleavage-stage human embryos, produced by IVF for clinical purposes, were obtained from individuals with prior consent and are regarded as the best source of human ESCs [46]. These embryos would be cultured up to the stage of a blastocyst (14-day cell mass) and processed carefully to obtain various embryonic stem cell lines after a number of passages.

2.3.2 Nuclear transfer

As an alternative treatment to IVF, any adult cells of the body could be reprogrammed to a pluripotent (embryonic) state by transfer of the adult nucleus into the cytoplasm of an oocyte or by fusion with a pluripotent cell [1]. Somatic cell nuclear transfer techniques succeed in rendering a full developmental potency of the nuclear genome of mammalian cells in many animals. The famous "Dolly" was rendered by the transfer of a somatic nucleus into an oocyte [47]. This process, though not very successful in generating human ESCs, was helpful in the preservation of somatic nuclei which renders hope for the regeneration or restoration of endangered animals. When compared to IVF, nuclear transfer has lesser ethical regard when it comes to the destruction of early embryos for research purposes [48].

2.4 Growing of embryonic stem cells in laboratory

In order to understand the embryonic stem cells *in vitro*, it is mandatory to refresh the morphology of the embryonic tissue in a higher detail. First of all, the zygote, which is totipotent by nature, undergoes a defined set of cleavage divisions, resulting in the formation of embryos [4]. The embryo consists of two cells initially, which grows into four cells and eight cells after cell division. The 16–32 cell embryo is called the morula, which then gives rise to the early, mid and late blastocysts. The germ layers are seen after the late blastocyst stage [4]. The correct stage of isolation of ESCs from the embryo is during the early blastocyst stage. The inner mass of the blastocyst is isolated carefully and grown in the laboratory using aseptic methods [49]. The stages of the embryo and the stage of identification and isolation of ESCs from the embryo are listed in **Table 1**.

2–4 cell stage Day 2 post-IVF	The different blastomeres are easily visible and their size and shape may be regular or not. Extracellular fragmentations may be present. Utilisation of the stores of maternal mRNA for protein synthesis.
4–8 cell stage Day 3 after IVF	Activation of the embryonic genome: utilisation of embryonic mRNA for protein synthesis.
8-cell stage	Compaction: Blastomeres form a compact sphere, tight junctions develop between the outside blastomeres to seal them and isolate the inside cells; gap junctions form between the internal cell to allow small molecules and ions to pass between the cells.
Morula	Compacted embryo, 16-32 cell External cells will give rise to the trophoblast cells; internal cells will become the inner cell mass.
Blastocyst Day 5–6	Cavitation: The trophoblast cells secrete a fluid to create a blastocoel cavity. The inner cell mass, which will give rise to the embryo, is clearly distinct from the trophoblast cells which are required for implantation.

Table 1.
Stages of human embryo highlighting the correct stage for embryonic stem cell extraction.

The process of establishing mouse ESC lines is laborious. If the source of ESCs is IVF, post-IVF procedures involve isolating ESCs from the inner mass of a single blastocyst and culturing them on a mitotically inactivated embryonic fibroblast feeder layer with mitomycin C in an enriched ESC medium [11, 50] and continuously culturing them for more than 40 passages. If the source of ESCs is by nuclear transfer, the procedure involves the transfer of nuclei obtained from cumulus cells into oocytes of mice and the establishment of cell lines following the previous protocol. If the cells are to be used for transplantation into mice, then they must be cultured on gelatin-coated dishes at least for two or more passages to remove the feeder layer [51]. The cells must be subjected to cytogenetic analysis for evaluation and screening [52]. The harvested cells must be screened for any chromosomal anomalies and also karyotyped using conventional procedures like Giemsa staining [52]. The resultant mouse ESCs can be maintained and propagated as immortal cell lines on a layer of mitotically inactivated mouse embryonic fibroblasts or LIF in culture since the mouse ESCs are highly dependent on LIF for their survival [53, 54].

The culture of human embryonic stem cells is also a laborious process. The protocol involves the transfer of cells carefully from a preimplantation-stage embryo into a sterile laboratory culture dish containing a culture medium [11]. The culture medium must be a nutritive medium required for the growth of the cells. Usually, physiological temperature of 37°C and pH of 7.4 are used along with 5% CO_2 [55]. Along with the pH, humidity and temperature, these cells need a medium (a sticky surface) to attach themselves and grow. Hence, a coating of cells is required for them to attach themselves. This coating layer of cells is called a feeder layer which would be often composed of mouse embryonic fibroblast cells or undefined basement membrane extracts like Matrigel and Geltrex containing laminin, collagen, proteoglycans and other growth factors necessary for the cell survival [56]. This would allow only the embryonic stem cells to proliferate indefinitely under controlled conditions.

Scientists have now developed methods to grow these stem cells in the absence of feeder layers to avoid the problems of contamination and undesirable mutations [57]. When the plated cells grow and become confluent (adhere to the dish and spread over), they are removed gently and plated into several fresh culture dishes (subculture) as several passages to form an established cell line [58]. The human ESCs must be in a complete pluripotent state for at least 6 months to be regarded as a cell line. Readers could recollect that pluripotency means the state of a single cell that is capable of differentiating into all tissues of an organism. Batches of these cells can be frozen and shipped to other laboratories for further culture and experimentation. A figure depicting the growth of hESCs in a laboratory setting is shown in **Figure 4**.

2.5 Laboratory test to identify embryonic stem cells

The establishment of ESC lines follows a process of validating these cell lines for their fundamental properties, a process called characterisation. The characterisation of the ESCs would be done by any of the following methods.

- The cells must be grown for many months under aseptic conditions in a laboratory. Each time the cells are passaged, their morphology is observed through a microscope, and the karyotype is determined using standard techniques [59, 60]. Even after a number of passages, it is essential that the cells stay healthy and remain undifferentiated.

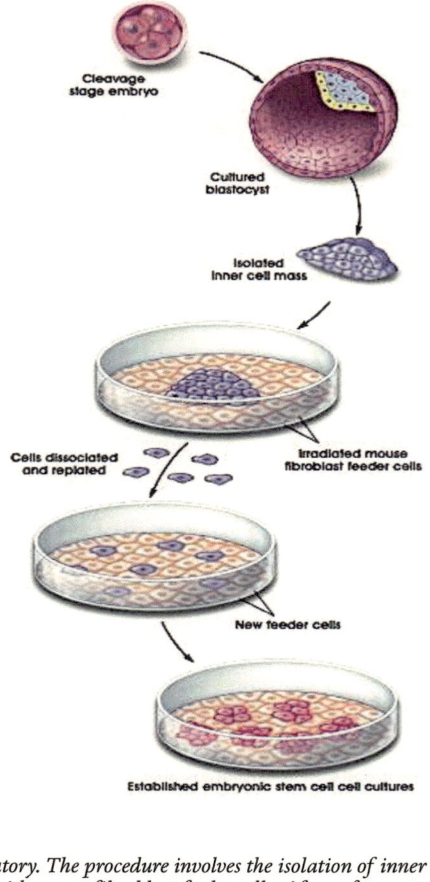

Figure 4.
Growing of hESCs in the laboratory. The procedure involves the isolation of inner mass cells from a blastocyst aseptically and cultured along with mouse fibroblast feeder cells. After a few passages, an new embryonic stem cell culture is established. Image source: Appendix C: Human Embryonic Stem Cells and Human Embryonic Germ Cells https://stemcells.nih.gov/info/2001report/appendixC.htm.

- Using gene expression studies, the cells are analysed for the expression of transcription factors like Sox2, Nanog and Oct4 which are specific for undifferentiation. Since these transcription factors are involved in the processes of cell differentiation and embryonic development, they give detailed information about the state of the cells in culture [61]. It is essential to know that Oct4, Sox2 and Nanog are some of the transcription factors expressed by the ESCs which are essential for the cells to maintain their pluripotent state [61].

- Karyotyping of chromosomes using microscopy and examining them for the presence of particular cell surface markers that are typical for stem cells. The procedure also helps in determining the next steps like re-growth, subculture etc. [62, 63].

After they remain differentiated for a number of passages, the potential of these cells for specialisation could also be monitored using the following methods:

1. By growing the cells so that they grow and differentiate by themselves in culture.

2. By changing the medium and allowing them to differentiate and undergo gastrulation to form the three germ layers of the embryo [1].

3. By introducing certain growth factors and supplementing them with an appropriate medium and analysing them for the markers of the differentiated cell lineage [64].

4. By analysing the cells by introducing them into an immunocompromised mouse that has the tendency to form a benign tumour called teratoma (a mass of cells). Since the experiment employs a mouse with a suppressed immune system, there will not be any condition of graft rejection by the recipient. These teratomas contain a mixed population of many differentiated cells and exhibit the pluripotent nature of the stem cells [65].

5. This could be useful for creating a model system for any rare disease without animal models and also for drug screening experiments in various cell lineages.

Apart from the conventional assays, there are several novel biomarker-based approaches and proteomic approaches to identify the induced pluripotent stem cells. A mass spectrometry assay that identifies the stem cell markers in a proteomics-based approach is a cost-effective and reliable tool instead of genomic analysis, which is time-consuming and laborious [66]. A novel proteomic-based screening approach using mass spectrometry for the analysis of embryonic stem cells is depicted in **Figure 5**.

2.6 Differentiation of embryonic stem cells

The differentiation of stem cells involves certain characteristics. The common phenomenon of the ESCs is that they could be maintained as a pluripotent population for a longer time in cell culture until a trigger for differentiation is given to them. The overall process of differentiation is divided into three important stages: competence, a state when a cell is capable of responding to a particular signal; specification, the stage when the signal could be removed without changing the fate of the cell and third the determination, when the cell is independent of the signal and the signal

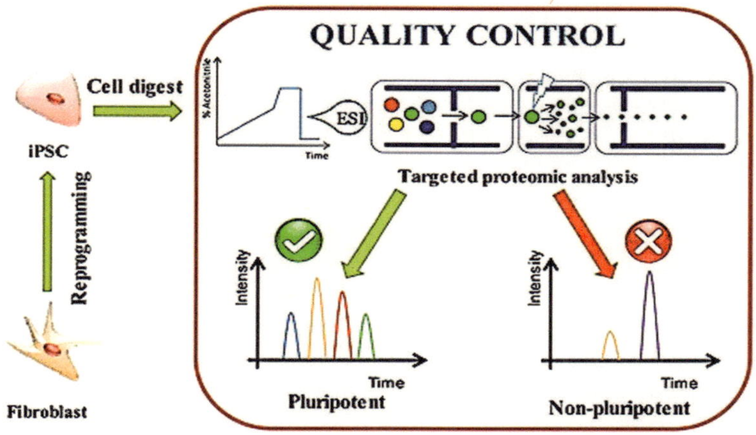

Figure 5.
*A proteomics based approached described for screening the induced pluripotent stem cells. One of the novel ways to identify and screen the induced pluripotent stem cells is subjecting them to proteome analysis using mass spectrometry. The proteome which is obtained after mass spectrometry is screened for any bio-marker specific to that stem cell lineage. Image source: Multiplex High-Throughput Targeted Proteomic Assay To Identify Induced Pluripotent Stem Cells. Baud et al., 2017. Anal. Chem. 2017;**894**:2440-2448.*

could not influence the cell anymore [67, 68]. The signals could be of physical or chemical environmental factors that surround them, that is the signals are provided by the surrounding cells.

The pluripotent nature of the ESCs is clearly illustrated by their ability to contribute to adult tissues, including the germline, following their injection into host blastocysts. Unlike transformed tumour cell lines, the ESCs can retain normal karyotypes following extensive passaging in culture [69]. The normal karyotype is retained until the ESCs grow in culture under appropriate conditions and remain undifferentiated (unspecialised). The pluripotency itself is divided into two stages during their culture *in vitro*. The earliest stage of it is termed "naive", which represents the ground state or unprimed state of the stem cell [70]. The naive state cell resembles the embryonic epiblast in the preimplantation blastocyst stage [71], whereas the "primed" state refers to the state of the early stage of post-implantation epiblast prior to gastrulation [72]. Generally, the human ESC lines are in the primed state. It is a common phenomenon that these cells, when allowed, try to clump together to form embryoid bodies and start to differentiate spontaneously [73]. In order to generate pure cultures of specific types of differentiated cells like heart muscle cells, blood cells or nerve cells, it is mandatory to regulate the process of the differentiation of these ESCs [74]. Any changes in the culture medium, the matrix composition present in the medium or any genetic modification induced in the cells might interfere with the actual lineage to which these cells are to be committed, and it is indispensable that the growth media should be supported with LIF and foetal calf serum for the mouse ESC growth [75]. The human ESCs do not depend on LIF or STAT3 for their survival or differentiation [76]. The cells could be grown either on a support feeder layer or independently using Matrigel or laminin-coated plates along with transcription factors like Oct4 or Nanog [77].

Several protocols have been developed to date for the maintenance of survival factors and the composition of the nutritive medium determines the fate of the ESCs towards future development. They could either form structures like embryoid bodies or proper germ layers. Such embryoid bodies are made to differentiate on stromal cells or on extracellular matrix proteins in order to nurture the cells by providing a three-dimensional structure and enhancing cell-cell interactions [78]. The disadvantage of this procedure is that the embryoid bodies secrete several cytokines that could interfere with the experimental protocols. Growth of ESCs on the stromal cells generally supports the growth of the stem cells in a good manner [79] but a disadvantage of this practice is that the stromal cells tend to secrete some undesirable factors which could meddle with the lineage commitment of the ESCs to an undesirable one.

2.6.1 Section summary

ESCs can be generated by *in vitro* methods like IVF and nuclear transfer. Each method has its own merits and demerits. The gene expression pattern of these cells can be used as a tool to identify them during growth. Research is being carried out for the growth of these stem cells under aseptic conditions and to make them suitable for experiments in the long run.

2.7 Adult stem cells

Cell growth and proliferation are the common pathways in the metabolism of a cell during its lifetime. During cell growth and survival, certain cells enter into a state of quiescence (calmness), which is a reversible state of growth arrest. A good example of quiescence is the striated muscle cell called the satellite cell. When there is any signal from the external environment, this quiescence would stall, and the cells might resume their proliferation [80]. As discussed in a previous section, the microenvironment in which they live is termed a "niche", which plays a major role in the maintenance of quiescence and growth states of the stem cells thereof.

The occurrence of adult stem cells came to the limelight more than 60 years ago when the bone marrow was found to contain at least two different types of cells *viz.*, haematopoietic stem cells (HSCs) and stromal stem cells/mesenchymal stem cells. The sources of these MSCs are tissues like adipocytes, chondrocytes and osteoblasts [81]. These cells possess characteristics similar to mesenchymal lineage and thus give rise to MSCs. MSCs are seen in circulation as peripheral blood cells, and also seen in the primary and secondary lymphoid organs like the bone marrow, spleen, thymus and lymph nodes [82]. In a broader sense, the ASCs could be grouped into two: the haematopoietic and the non-haematopoietic stem cells.

ASCs are seen in many tissues and organs of the body. Even in post-mitotic organs like the brain, these ASCs are seen who can renew themselves and become specialised to yield all the cell types of the tissue from where they originated [83]. The special feature of ASCs is that they are capable of making identical copies of themselves throughout the lifetime of the organism. During the process of differentiation, they give rise to specific progenitor cells which are non-functional but committed to a lineage, which then differentiate or develop into mature cell types with characteristic shapes and specialised functions [83]. The continuous proliferation of ASCs helps to maintain homeostasis of the cells/tissue where they are homed. For instance, a blood-forming ASC in the bone marrow normally gives rise to red blood cells, white blood cells and platelets. However, after several years of research, it was postulated that the stem cell from one tissue may be able to give rise to cell types of a completely different tissue, a phenomenon known as plasticity or trans-differentiation [84]. For instance, blood cells could trans-differentiate into neurons, bone marrow stem cells could trans-differentiate into hepatocytes and so on.

This property of trans-differentiation helps them to give rise to mature cell types with characteristic morphologies and specialised functions. Though they are relatively rare, they have an important role to play in tissue homeostasis; that is, they have functions of tissue/cell replacement during an injury or a disease [85]. They are dispersed in tissues throughout the body, yet the characteristics of these cells are not uniform, and they mostly depend on the local environment in which they reside [86]. For instance, the HSCs of the bone marrow have the capacity to replenish all types of blood cells, and they cannot give rise to a cell with a different function, such as a neuron or a myocyte, depending upon the survival factors and the supplemented medium *per se*.

The HSCs are extensively studied for their potential to develop into numerous types of blood cells. The importance of the non-haematopoietic stem cells is also on par with the HSCs, as these cells give rise to support cells like cartilages and fibrous connective tissue. In addition to their role in supporting the organs as connective tissues, the stromal cells adapt themselves to the changing metabolic requirements of the surrounding tissue. It would be interesting to note that among the post-mitotic (cannot regenerate during the entire lifetime) neurons also, there are some adult stem cells like oligodendrocytes and astrocytes in the brain. These ASCs are isolated from the adult tissues and studied extensively for their clinical significance in cell therapies. Like the ESCs, the ASCs are also isolated using aseptic methods under controlled conditions in cell culture. A simple protocol for isolation of adult stem cells from a source tissue is given in **Figure 6**.

2.8 Cellular location of adult stem cells

Cellular location of stem cells and their identification is essential for studying their properties and research in elucidating the pathways of regulation in maintaining their basic properties of proliferation and differentiation has opened a new vista in the

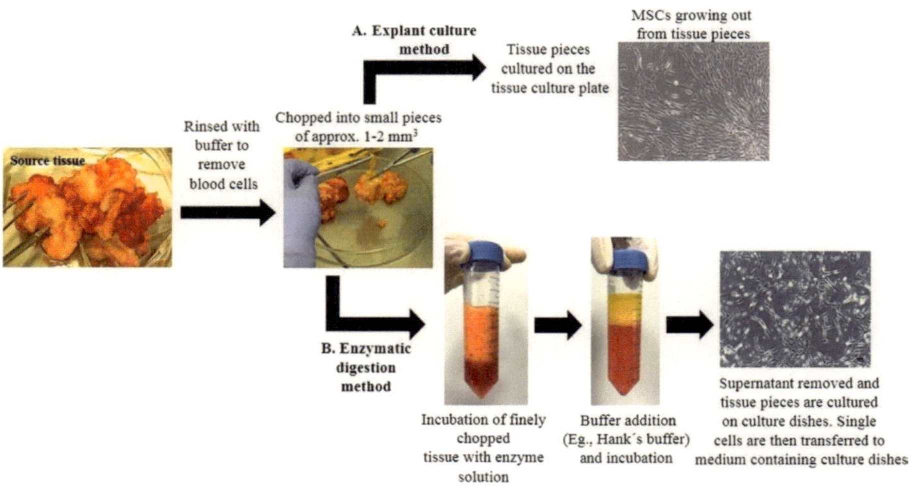

Figure 6.
The aseptic methods of growing the adult (mesenchymal)stem cells in a laboratory setting. Adult stem cells are aseptically removed from the source tissue and rinsed with appropriate buffer and minced into small pieces or subjected to mild homogenisation prior to digestion. Mild enzymatic digestion of the tissue pieces in suitable buffer followed by the culture of these tissues pieces on culture dishes in appropriate culture medium or plates.

field of therapeutic perspective in regenerative medicine. As discussed in a previous section, ASCs are often found in a niche that plays an important role in the regulatory processes, which retain their location, adhesion, mobilisation, quiescence or activation, cell division, differentiation and several other properties of them [87]. These niches are identified using protein markers which are expressed in the stem cells and their niches or using labelling techniques using tritiated thymidine or bromodeoxyuridine [88]. The stem cells would retain their labels owing to the property of slow (cell) cycling time [88]. Another property of stem cells also contributes to this label retention. The stem cells have a strategical phenomenon called "Selective DNA segregation" (schematically explained in **Figure 7**), which makes them retain their stemness for a longer time [89]. Cairns and his coworkers [90] explained the process as the "Immortal strand hypothesis," wherein the stem cells always designate one strand of DNA (in each chromosome) as a template strand. During replication, only the template strand and its copy go to the daughter cell that remains as a stem cell.

Among the cells of the niche, there are some cells called "Transit amplifying cells" or committed progenitor cells as well [91]. The latter, with more rapid though limited proliferation, low self-renewal and restricted differentiation, increase the number of differentiated cells produced by one ASC division. This could be possible only if these niches express specific markers like c-kit, Sca-1, integrins, CD-34, Oct-4, Nanog, nestin, Bmi-1, bcl-2 and CD 133 [92]. Several signalling pathways like Wnt, notch and hedgehog pathways are seen in these adult stem cells [93]. ASCs have been identified and isolated from the bone marrow, brain, liver, peripheral blood, gut, skeletal muscles, blood vessels, skin, teeth, heart, ovary, testis [94]. Like ESCs, the inherent property of stem cells to remain pluripotent or unspecialised is conserved even in an ASC.

As a good example of ASCs, the MSCs, which were first reported in the bone marrow, were also isolated from other tissues of the body, such as the umbilical cord blood, amniotic fluid, placenta, muscle cell or adipose tissue [95]. Some of the biological features are common to all ASCs as seen in MSCs. To name a few, the adhering capacity of these cells is very high, and they express several cell surface markers

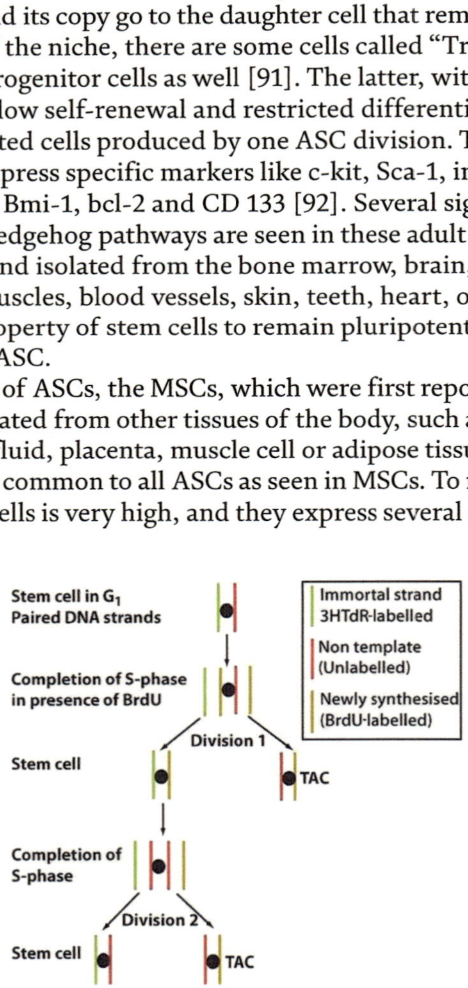

Figure 7.
The DNA selective strand segregation by the stem cells. During cell division, the stem cells always preserve the template strand for the daughter cells which also continue to inherit the immortality of the parent stem cells. This selective segregation of DNA is a plausible explanation for the stem cells to have a slow cycling time and retention of labelling during their identification in their niches. Figure source: The Journal of Pathology Volume 217, Issue Attributes of adult stem cells MR Alison S Islam.

through which they could be identified easily. MSCs express various cell surface markers like CD44, CD29, CD73, CD90, CD105, CD106, CD166 and ICAM-1 and lack the expression of HSC markers like CD45, CD34, CD14, CD79a, CD11b and HLA-DR [96, 97]. MSCs also express mild levels of HLA and MHC Class I antigens and lack MHC Class II antigens on their surface [98].

2.9 Laboratory test to identify adult stem cells

The understanding of the nature of the ASCs would help us to better identify, isolate and culture them *in vitro*. The role of ASCs is very important in the context of the regenerative capacity in the absence of the actual cells performing their function. Adult stem cells have both the properties of unipotency and multipotency, the majority of them residing in the bone marrow. It becomes indispensable to follow a meticulous protocol to isolate them from bone marrow or any other source since they are very low in number [99]. When comparing an MSC population, they are almost 10-fold lesser in number compared to the HSCs in the bone marrow. The proliferation capacity of the HSCs is so high that these cells, which are said to be multipotent, residing in the bone marrow, have the capacity to regenerate the whole haematopoietic cell population [100]. The rest of the cells that reside in the bone marrow are the MSCs, also called the stromal cells. These cells could regenerate/differentiate into bone, chondrocytes and adipose cells [101]. After their isolation, it becomes essential that they are propagated as a pure population. Hence, their identification using advanced molecular methods becomes essential. With the help of several molecular markers, these cells could be labelled and after they differentiate and be committed to a particular lineage or become specialised rather, their tissue of origin could be easily spotted out. For this purpose, several methods like labelling with fluorescence tags and extracting the fluorescence information using flow cytometry are widely used techniques to date [102, 103]. In turn, the primary cells could be isolated from the animal, and they could be labelled while they are in culture. Once these cells adhere and grow, they could be transplanted back into the animal and their potential to re-proliferate into the specified tissue can be followed.

The enrichment of stem cells after their isolation helps them differentiate, specialise and be committed to a particular lineage. For instance, the enrichment of HSCs would be done by exploiting a natural cytoprotective mechanism used by the stem cells for the defence against xenobiotics and toxins. This includes the ATP-binding cassette superfamily of proteins which are membrane transporters functioning as drug efflux mechanisms for dyes like Hoechst 33342 and Rhodamine-123 [104]. These cells are able to efflux these drugs and appear at the bottom left corner of the FACS panel as often called the side population, a special feature of stem cells [104]. Thus, the HSCs could be isolated and sorted using such a dye exclusion method (Hoechst 33342). The quiescent cells exist in a side population that could be seen on FACS (Fluorescence-activated cell sorting) machine. For a detailed pictorial representation, kind refer to (**Figure 8**).

The protocol involves careful removal of ASC (tissue) from the animal using sterile conditions. After washing with sterile saline or ice-cold PBS, the cells are carefully and gently homogenised using the same medium and also treated with enzymes like collagenase to remove any fibrous tissue around them [105]. The adherent cells are filtered through a series of sterile filters like 100 and 40 microns. After gentle centrifugation, live/dead cell discrimination is done using a Fluorescence-activated cell sorter. After sorting, the live cells could be plated at a volume of 50,000 to 100,000

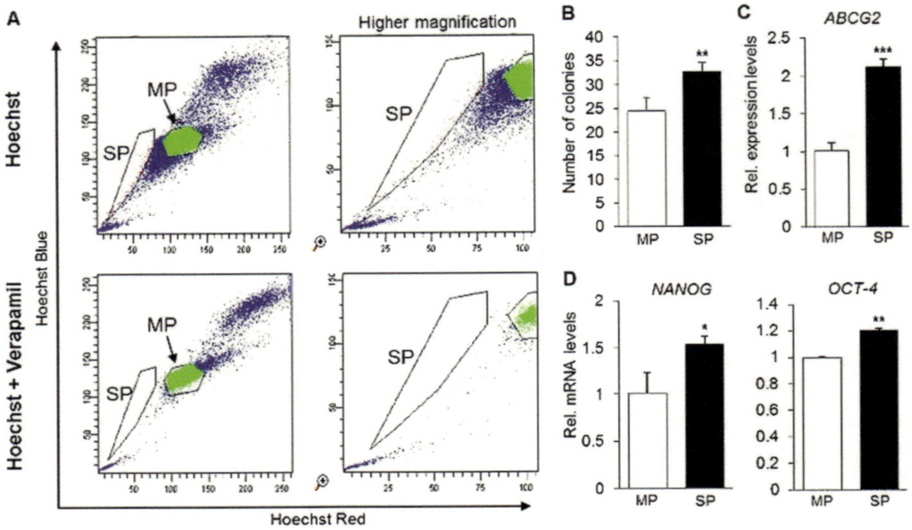

Figure 8.
The identification of adult stem cells using the dye Hoechst-33342 and verapamil. Dye-exclusion method and side-population analysis have been a versatile technique to identify a stem cell population. The inherent nature of the ABC membrane transport proteins which function as a drug efflux mechanism for dyes like Hoechst render a side-population at the left side end of the panel in a FACS machine.

cells/well in a six-well culture dish with a growth medium supported with serum and antibiotics and grown in a CO_2 chamber [105, 106]. The adherent cells were expanded as primary cell lines after a few generations. The characterisation of adult stem cells involves a meticulous procedure. The cells should be almost 80–90% confluent after three passages. The cells would be stained for expression of CD markers for cell surface proteins (receptors) expressed on stem cells and sorted using FACS. A complete understanding of the cell surface antigens is necessary for their characterisation. The following table (**Table 2**) gives a complete tabulation of stem cell markers to identify stem cells. The staining methods are important and involve an aseptic/sterile handling of the cells. Several methods, which involve the insertion of a reporter gene that expresses green fluorescence protein (or red fluorescence protein) into the stem cells, could also be used. GFP expression would be driven by the promoter of a specific

Marker name	Cell type
CD4, CD8, CD45 and CD38	WBCs and WBC lineage cells
CD34	HSC
CD34 + Sca1	MSC
C-kit, Sca-1, Thy-1	HSC and MSC
Colony-forming unit (CFU)	HSC, MSC progenitor
CD-133	NSC, HSC
Stro-1 antigen	MSC (stromal)

Table 2.
Cell surface makers for identification of basic stem cell types. WBC—white blood cells; HSC—haematopoietic stem cell; MSC—mesenchymal stem cell; NSC—neural stem cell.

marker expressed in that particular cell. The GFP expressing cells, once activated, express GFP and could be employed to identify and sort them using FACS [107, 108].

2.10 Differentiation of adult stem cells

The process of developing into cells of multiple lineages, as dictated well by their inherent property of multipotency, is called differentiation [109]. The process of differentiation is very elaborate and tends to follow the normal molecular pathways of a cell. The differentiation property of these ASCs is biologically relevant and physiologically significant if they could be used for tissue repair/therapy. In case of a physiological trigger such as tissue repair/damage, the ASCs start to differentiate into a mature cell, which has the shape and structure characteristic to that of the particular cell [110]. The ASCs from the bone marrow have sources of both haematopoietic and stromal stem cells. The HSCs could give rise to both lymphoid and myeloid progenitor (precursor) cells. The lymphoid progenitor cells would give rise to T and B lymphocytes and natural killer cells, whereas the myeloid progenitor cells would give rise to red blood cells, white blood cells and platelets [111]. In a similar way, the stromal (mesenchymal) stem cells would give rise to pre-osteoblasts, pre-osteoclasts, which give rise to osteoblasts and osteoclasts, which are major cells of the bone, or also to other cells like adipocytes and chondrocytes [112].

The differentiation could be induced using several techniques using biological/pharmacological reagents (e.g. Hormones). Their differentiation could also be regulated by mechanical cues, like changing the cellular morphology and matrix elasticity [113]. Environmental factors like oxygen tension would also solicit stem cell differentiation. Also, external electrical and mechanical forces like low-intensity ultrasound could be introduced to regulate their differentiation [114, 115]. It appears that several physicochemical differentiation factors govern the trans-differentiation of adult stem cells into specific cell types and the protocols have to be optimised in order to facilitate the application of these stem cells in tissue engineering or regenerative medicine.

The external factors usually evoke a signalling cascade in these cells that triggers them to get lineage committed. For instance, the vascular endothelial growth factor (VEGF) and shear stress induce a synergistic differentiation of endothelial progenitor cells into endothelial cells [116], while hydraulic pressure induces a synergistic chondrogenic differentiation of MSCs [117]. A combination of fluid shear stress and the bone-like extracellular matrix also induced an osteo-differentiation of MSC in the presence/absence of dexamethasone, a substrate analogue [118]. Several molecular pathways, such as MAP kinase signalling, calcium signalling pathways, GPCR pathways and others, are involved in the differentiation of stem cells into specific cell types.

2.11 Similarities and dissimilarities between embryonic and adult stem cells

Though the origin of ESCs and the ASCs is completely different from each other, they have certain properties in common. Both ESCs and ASCs have the developmental potency that defines the stem cell property (discussed elaborately in the previous sections). Both the population of cells could be isolated and cultured *in vitro* for a number of generations as a pure population without specialisation. The isolation protocols include a meticulous preparation of both the media as well as the reagents. Both ESCs and the ASCs remain in a state of quiescence for a longer duration. Once a trigger is given, they would exhibit their property of differentiation or specialisation and get committed to a particular lineage (lineage-dependent). These two types of cells have a tumourigenic nature and could form teratomas though it is more prevalent in the ESCs. Both of them are well-identified for their value in cell-based therapy.

2.11.1 Section summary

The source of ASCs is extensive and the clinical potential of these cells in stem cell therapies is exemplary. Several useful molecular tools like flow cytometry have been indispensable in the characterisation of ASCs. Like the ESCs, the ASCs are also isolated using aseptic methods under controlled conditions in cell culture.

3. Applications of human stem cells in complex disorders

3.1 Potential use of human stem cells in CVD

Cardiovascular disorders remain the major cause of morbidity and mortality throughout the world. As we all know, the cardiomyocytes are post-mitotic, and their regeneration capacity is very minimal or sometimes negligible. Hence, the human heart is unable to counteract the severe loss of heart muscle tissue during myocardial infarction or other myocardial disorders. Heart transplantation is the standard therapy to replace the injured heart, which is highly expensive, and the success rate is also not appreciable. Several experimental studies and recent clinical trials assumed that stem cell transplantation might be indicated as an alternative and promising therapeutic strategy for heart failure.

In the present scenario, about four classes of stem cells have been identified for cardiac regeneration. They include: (1) Embryonic stem cells, (2) Bone Marrow-derived stem cells, (3) Skeletal myoblasts and (4) Cardiac stem cells and cardiac progenitor cells [119]. As elaborately discussed in the previous sections, the ESCs could not be preferred owing to their inherent tumourigenicity and ability to form teratomas, their inherent potential for immunologic cellular rejection and also their low efficiency in differentiating into desired cardiomyocytes. In a situation like this, the only resort is the bone marrow-derived stem cells but those have not yet been proven to independently differentiate into functional cardiomyocytes. Also, several scientists have experimented with skeletal myoblasts, which have come a long way and created in vivo myotubes. These in vivo myotubes often pose a challenge of electrical integration into the existing myocardium, rendering them functionally deprived [120]. The only cell types which are promising candidates for cellular therapy for cardiac failure are the cardiac stem cells and cardiac progenitor cells. The versatility of stem cells in cardiac cell replacement therapy is pictorially represented in **Figure 9**.

Multiple candidate stem cell types, including mesenchymal stem cells and pluripotent stem cells, have also been evaluated in animal models and in clinical human studies for their ability to regenerate cardiac damage and reconstitute cardiomyocyte loss [121]. Mesenchymal stem cells could differentiate into various cardiac cells like cardiomyocytes, endothelial cells or smooth muscle cells [122]. Both allogeneic (from different individuals) and autologous (self) MSCs were safe and effective for treating ischemic heart disease in clinical trials. Also, allogeneic MSCs did not stimulate any donor-specific immune reaction [123]. The behaviour and fate of the stem cells also depend upon the route of injection of these cells. There are various routes for administering stem cells to the recipient [124]. They are intravenous injection, intramyocardial injection and intracoronary injection. Of these, intramyocardial injection is the most successful route followed [125]. The immunomodulatory and differentiation

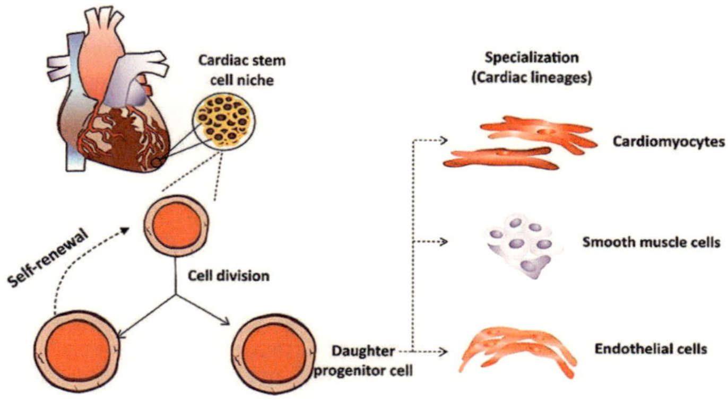

Figure 9.
The role of cardiac stem cells in generating several cardiac cells for cell therapy. Cardiac stem cells have been promising candidates in the treatment of heart diseases as they could develop into different types of heart cells like cardiomyocytes, smooth muscle cells and endothelial cells.

potential of these cells are very high, which is helpful in the expression of various cytokines and receptors required for the cardiomyocytes to counteract the injury and initiate a tissue repair mechanism [126]. Transplantation of cells derived from stem cells has proven to be efficient in some animal models of cardiac disease but failed to be effective for in the clinical trials [127]. The embryonic stem cells, though effective, could not meet the high demands of the ethical and social concerns *per se*.

The cardiac stem cells are those which are isolated from the heart itself. Though the heart is a post-mitotic organ that stops replenishing after its birth, the discovery of cardiac stem cells is really a boon for the cardiac cell therapy procedures followed [119]. These cells were isolated from the heart and grown under *in vitro* conditions before their cardiac cell replacement *in vivo*. These cells were identified using side population analysis or several stem cell markers like cardiac c-kit or Sca-1 [92]. These cardiac stem cells grow and differentiate into smooth muscle cells, cardiomyocytes and endothelial cells under laboratory conditions and some of the trials showed good replenishment and functional capacity of these cardiac stem cells [128].

The pluripotency of embryonic stem cells would give rise to a variety of cell types like cardiomyocytes, endothelial and smooth muscle cells. These cells would be useful in the regeneration of a damaged myocardium. The heart contains a small population of endogenous stem cells that could facilitate minor repair and turnover-mediated cell replacement [129]. These cells have been isolated from rats, mice and humans. These cells can be harvested in a limited number from human endomyocardial biopsy specimens and can be injected into the site of infarction to promote cardiomyocyte formation and improvements in systolic function [130].

The first transplantation reports of human ESC-derived cardiomyocytes into pigs and guinea pigs have shown their potential to function as biological pacemakers in electrophysiologically silenced or atrioventricular-blocked hearts [121]. Successful engraftment and damage repair have also been solicited in primates. Cardiomyocytes have been derived from human ESCs (71–95% purity) using a BMP-4/Activin-A system. Their transplantation into infarcted cardiac tissue offers promising, non-invasive alternatives to the placement of pacemakers [131]. In the clinical setting, a transplantation venture to humans has several crucial limitations like tumourigenic nature of the stem cells, genetic instability, ethical issues and so on [48]. The purity of the isolated cardiomyocytes is still a question when it comes to a therapeutic application.

Robust protocols are being developed to address the issue of obtaining physical as well as functional maturation of the stem cell-derived cardiomyocytes to be used for clinical applications and several milestones would be crossed to successfully authenticate the regenerative cardiac medicine using stem cell-derived cells.

Apart from the role of stem cells in cardiac disorder, it has a promising effect on other complex disorders like neurodegenerative disorders. Stem cells are known for their pluripotent nature and, therefore, play a crucial role in brain regeneration and recovery after injury by modulating the local tissue environment and interacting with surrounding cells. As a result, therapies involving stem cell transplantation have been explored and evaluated. Over the past decade, more than 15 clinical trials across various countries have investigated the effectiveness of autologous stem cells derived from bone marrow, umbilical cord blood, peripheral blood and adipose tissue in treating brain injuries and neurodegenerative disorders [132]. Among these, mesenchymal stem cell (MSC) therapy has shown promise in reducing excessive inflammation and mitigating brain damage. MSCs and MSC-derived exosomes have also demonstrated potential benefits for brain health, particularly in premature infants, as shown in a small-scale clinical trial [133]. While stem cell therapies hold great promise, establishing a standardised, universally accepted protocol remains a significant challenge. Critical factors such as cell type, delivery method, dosing regimen and patient tolerance need to be carefully considered. Continued research, at the both preclinical and clinical levels, is essential to fully establish stem cell therapies as effective treatments for other complex disorders.

3.2 Potential use of human stem cells in NDD

Neurodegeneration is a potential area of interest in stem cell therapy. The utility of ESCs in treating some of the physiological conditions of the disease has been explored in numerous studies. Researchers have explored the possibility of replacing the damaged neurons in the affected areas of the brain. The potential of ESCs to differentiate into specific neuronal subpopulations is suitable for treating dopaminergic and cholinergic irregularities in PD and AD, respectively [134]. Stem cell therapies can be useful in clinical treatments of neurodegenerative disorders (NDDs) owing to their efficiency in targeting multiple pathways and no blood-brain barrier limitation in contrast to conventional drugs. MSCs have been proven essential in treating PD, ALS and AD at the preclinical level. The plausible mechanisms would be the promotion of neuronal survival, microglial activation and release of neurotrophic factor [135]. iPSCs have been a great tool in the disease modelling of neurodegenerative disorders. Genetic modifications by CRISPR/Cas9 to produce isogenic grafts or organoids have been a continuous line of research in the area of NDD [136]. MSC transplantation successfully improved motor neuron function and reduced ubiquitin in ALS patients [137]. iPSCs have been highly used in stem cell therapies of NDD but the clinical efficacy is still a question. Given the advantage of autologous transplantation with minimal immune rejection and ethical concerns, iPSCs have always been the wisest choice in NDD [138]. Neural stem cell transplantation has been a powerful tool for stroke, mental illness and cancer [139]. Graft rejections in allogenic transplants in RPE cells and dopaminergic neurons have been witnessed in several clinical trials [140]. Neural stem cells have been studied for their regeneration potential by involving in potential mechanisms such as immune regulation, neuroprotection, release of neurotrophic factors and decrease of neuroinflammation [141].

Combined transplantation strategies have been tried in the past which include mesenchymal stem cells and NSCs with controlled culture conditions that would lead to their differentiation into dopaminergic neurons and improve the spatial learning ability in PD [142]. Clinical evaluation of stem cell transplanted patients who received intraspinal injections of NSCs showed a decrease in the disease progression, thereby proving the efficiency of these cells in slowing down the process of neurodegeneration. In another clinical trial, improvement in the clinical symptoms of PD with no significant side effects was seen.

3.2.1 Section summary

Stem cell therapies have been a promising tool in fighting the detrimental impacts of CVD. Research in stem cell therapies is very active and working towards the goal of creating robust methods to mitigate cardiomyopathies and associated disorders.

4. Useful scientific concepts in stem cell biology

4.1 Scientific concept: New embryonic stem cell experiment: Robert Lanza experiment

The IVF methods for the isolation of human ESCs have a prerequisite of embryo destruction *in vitro*. There are several ethical issues addressing this concern of destroying ex-utero embryos for human ESC research [48]. Animal studies have proven and identified ways to generate ESCs using a single cell from biopsy samples. These cells are obtained during preimplantation genetic diagnosis and these samples would not hamper the developmental potential of the embryo in any form. The single blastomere, which was grown overnight, resulted in cells sufficient for both genetic screening and stem cell isolation in the current clinical setting, also addressing the ethical concerns of society.

Dr. Robert Lanza was the one to report a series of experiments to demonstrate that human ESCs can be derived from single blastomeres [143]. The experiment involved the extraction of biopsies from each embryo multiple times using micromanipulation methods, and the embryos were not allowed to grow *in vitro*. The embryos were obtained from fertility clinics that practised IVF regularly, following the consent of the patients and a completed ethical procedure. Approximately 16 pronuclear and multi-cell stage embryos were obtained and continued for *in vitro* culture up to 8–10 cell stages in an appropriate medium. The outer layer was disrupted and the individual blastomeres were carefully separated from the denuded embryos with a micropipette. The separated blastomeres were cultured together in the same medium in a tissue culture plate. Almost 58% of the isolated blastomeres were divided at least once, and approximately half of them formed vesicles/clumps that produced outgrowths within 2 days. Co-culture of ESCs with the blastomeres was done in the notion that the human blastomeres do not tend to aggregate much with the ESCs. These cells were seeded with GFP-tagged human ESCs for their easy screening. After they formed outgrowths, they were mechanically passed into fresh feeder layers containing human ESC medium. Many common outcomes were seen in those grown cultures. Trophoderm morphology, differentiating cells and undifferentiated proliferation were some among them.

The cultures were regularly subcultured in the appropriate medium in fresh feeder layers and continuously screened for their morphology and pluripotency markers. When these human ESC cultures were allowed to overgrow, they started forming embryoid bodies and readily differentiated into three germ layers of the embryo. After successful passages, two embryonic stem cell lines with female karyotypes were formed. These cell lines were maintained without differentiating for more than 8 months, and also examined for normal karyotype and pluripotency markers like Oct-4, TRA-1-60/81, Nanog and Alkaline phosphatase. The experiment was successful, and the new stem cell lines retained their pluripotency to develop into three germ layers both in vitro and in teratomas. A pictorial representation from the original paper is shown in **Figure 10** for the ease of understanding. Cell therapy has been tried as a tool for the treatment of many diseases; hence, much research has been in the limelight, which involved experimentation of stem cells to grow into desirable cell types [24]. Ethical concerns literally hamper the ESC research since destroying human embryos was considered a social stigma. Scientists are experimenting with stem cells derived from bone marrow or umbilical cord blood, but such cells are limited in the types of cells they can grow into. It was proved from Robert Lanza's experiment that these human ESCs readily grow into many vital cell types including neural progenitors under given conditions. Another key finding was that the 8–16 cell embryo stage was successfully utilised owing to its pluripotency, whereas the two-cell and the four-cell embryo stages have shown totipotency towards the trophectoderm lineage and hence not be used for the generation of ESC lines.

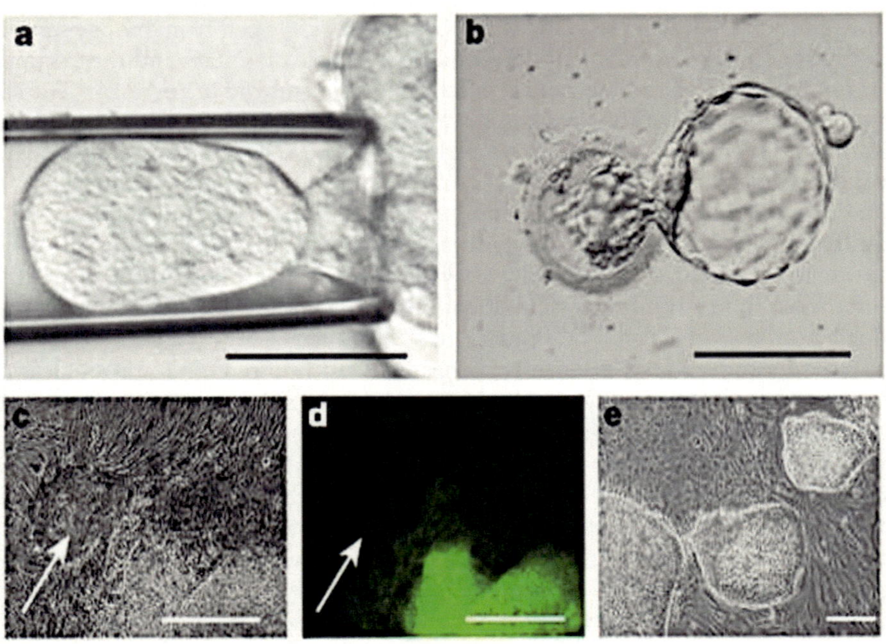

Figure 10.
The picture represents a sequential steps in Robert Lanza's experiment. a. Excision of a blastomere using biopsy technique from embryo. b. Development of the biopsied embryo. c. Embryonic outgrowth from the blastocyst (arrow represents a colony of human ESCs). d. Fluorescence image showing GFP selection of positive colonies of human ESCs. e. Human ESC line after growth. Image source: Klimanskaya I, Chung Y, Becker S, Lu SJ, Lanza R. Human embryonic stem cell lines derived from single blastomeres. Nature. 2006;444(7118):481-5.

Robert Lanza was the first one to derive human ESCs from blastomeres [144]. Previous experimental groups could not succeed in the development of pluripotent stem cell lines from human blastomeres. The blastomeres could proliferate from the cleavage stage of the embryo but they generated only trophectoderm cells. The isolation of a single blastomere and growing it successfully into an established human ES cell line was a landmark achievement in stem cell biology. It is noteworthy that the blastomeres in these experiments were from embryos that came for preimplantation genetic diagnosis. These human ESCs could have a greater potential for biomedical research and also these blastomere-derived cells would be useful for therapeutic applications.

4.2 Scientific concept: Integrated experimental and theoretical studies of stem cells: Hanna L. Sladitschek experiment

The very established properties of stem cells are their proliferative and their differentiating capacity into numerous cell types. The fate of the stem cells is highly controlled, and homeostasis is generally maintained in a given mass of stem cells. The underlying factors for the stemness are really extensive and it becomes a laborious process to identify and study them all over a period of time. The identification of molecular markers and tools for genetic analysis of the progenies of the same is also very tedious. Hanna L. Sladitschek devised a strategy and developed an experimental approach to study the two main properties of stem cells [145]. Hanna's experiments revolve around the mouse ESCS. It is well known that these cells are derived from the inner cell mass of mouse blastocysts before implantation, and they possess dual properties of stem cells as mentioned above.

These stem cells highly depend on the LIF pathway wherein LIF binds to its receptor and forms a complex with gp130 and induces STAT3 and PI3K/AKT pathways [146]. For a detailed review of these pathways, readers are requested to go through a standard book on molecular signalling. These two pathways determine the two fates of a stem cell, namely, cell renewal and cell survival. Also, these two pathways are involved in the differentiation of stem cells into various other types. In order to prove the heterogeneity of LIF signalling in mouse ESC fate, Hanna's group used single-cell microRNA (miRNA) reporter studies. They found out that miRNA miR-142 is bimodally expressed in the LIF-dependent pluripotency of these cells [147]. They found out that this miRNA is differentially expressed in two subpopulations of cells, with those undergoing differentiation expressing low miR-142 and those not differentiating expressing high miR-142 levels [147]. A simple model was developed to follow the reaction kinetics to follow the reconstitution rates of these two subpopulations. Cell sorting experiments were also evident in proving this concept. The reaction kinetics model was helpful in finding out the interconversion rates and switch-over frequencies of the two subpopulations which occurred every 12 cell divisions on average and the proliferation rates remained the same in both states. The state switching was thought to be a stochastic process, so the model was helpful in determining this as well. After the cells were sorted as single cells, they were propagated in cell culture. The clonogenicity of the populations was identified by the high/low miR-142 levels. The bimodal expression of miR-142 could also theoretically satisfy the negative feedback loop between the miR-142 and Kras/Erk activation, where a miR-142 could repress Kras translation, while Erk activity represses miR-142 transcription. The heterogeneous expression of LIF genes was proved using experimental and mathematical models by Hanna's group.

The self-organising and differentiating capacity of the stem cells was also a question to address. ESCs have been proven to be pluripotent and could differentiate into three germ layers of the embryo under specific conditions. These cells would also produce embryoid bodies which could develop eventually into organoids, organs and organ systems as explained by Dr. Robert Lanza's experiment. Hanna's group tried to prove the heterogeneity of the LIF expression in the self-renewal and differentiation capacities of ESCs. The combination of quantitative experiments using single-cell sorting and single-cell RNA sequencing and reporter gene constructs were used as experimental tools for their experiments along with the mathematical modelling to prove their experimental points [140]. These modelling experiments could offer a theoretical explanation for the mechanisms put forth for the fate of the stem cell *in vitro* and also offer a plausible explanation of strategies employed by the stem cells for organogenesis and tissue homeostasis *in vivo*. These prediction models should be applicable to the biological system being studied and the experimental approach could be helpful in unravelling the molecular mechanisms involved in the determination of cellular/molecular fates.

5. Chapter summary

- The stem cells are characterised by their distinct property of proliferation and differentiation into various cell types. The zygote is said to be totipotent, whereas the embryo is pluripotent which develops into three germ layers. The developmental potency of an embryonic stem cell explains its pluripotent nature. The embryonic stem cells are isolated from the inner cell mass of the embryo blastocysts.

- Embryonic stem cells are isolated from the embryo and are pluripotent, whereas the non-embryonic stem cells obtained from adult tissues like the bone marrow are multipotent/unipotent. A multipotent cell has the capacity to develop into various types of cells, whereas a unipotent cell can develop into only one type of cell. The haematopoietic stem cells and the mesenchymal stem cells of the bone marrow are well-studied examples of adult stem cells.

- Owing to their capacity of pluripotency, embryonic stem cells are capable of regenerating almost all types of cells. Sources of embryonic stem cells include IVF and somatic cell nuclear transfer. The distinct properties of stem cells include their capacity for self-renewal, differentiation and trans-differentiation. Stem cells are also encountered in the plant kingdom and are referred to as meristems.

- The properties of a stem cell highly depend upon the microenvironment or niche of a stem cell. The niche helps to maintain the stemness and nurture them to grow and differentiate. Initially, stem cells live in a phase of quiescence without differentiation or specialisation but once a trigger is given, they start differentiating and get committed to a particular lineage.

- There are specific markers for the identification and isolation of embryonic and adult stem cells. After their isolation, the lineage-specific markers help researchers to trace the lineage identity. Several molecular paradigms identify these populations by techniques like flow cytometry.

- The adult stem cells are helpful in attaining specific types of cells, but their rare occurrence does not permit the ease of their isolation and differentiation. Several researchers have employed the use of stem cells in various disciplines of regenerative medicine *viz*., heart diseases, spinal cord injury, leukaemia.

- Stem cell Biology seems to be an attractive and promising area of therapeutic medicine. Researchers were successful in obtaining embryonic stem cells from the adult stem cells which minimises the ethical issues concerned with the extraction of embryos and related procedures.

- There are some ethical issues when we consider the embryonic stem cells for regenerative medicine, but they are easy to grow in a laboratory setting. Thus, stem cell biology is an emerging field of medicine that could bring a cure for many genetic disorders, and more and more research has to flow in to substantiate the procedures to utilise stem cells in a therapeutic arena.

Author details

Gaurav Verma[1*] and Srividhya Ravichandran[2]

1 Institute of Neuroscience and Physiology, University of Gothenburg, Sweden

2 Department of Biotechnology, Indian Institute of Technology, Chennai, India

*Address all correspondence to: gaurav.verma@gu.se

References

[1] Zakrzewski W, Dobrzyński M, Szymonowicz M, et al. Stem cells: Past, present, and future. Stem Cell Research & Therapy. 2019;**10**:68. DOI: 10.1186/s13287-019-1165-5

[2] Zhang H, Wang ZZ. Mechanisms that mediate stem cell self-renewal and differentiation. Journal of Cellular Biochemistry. 2008;**103**:709-718. DOI: 10.1002/jcb.21460

[3] Zhu M, Zernicka-Goetz M. Principles of self-organization of the mammalian embryo. Cell. 2020;**183**:1467-1478. DOI: 10.1016/j.cell.2020.11.003

[4] Mitalipov S, Wolf D. Totipotency, pluripotency and nuclear reprogramming. Advances in Biochemical Engineering/Biotechnology. 2009;**114**:185-199. DOI: 10.1007/10_2008_45

[5] Romito A, Cobellis G. Pluripotent stem cells: Current understanding and future directions. Stem Cells International. 2016;**2016**:9451492. DOI: 10.1155/2016/9451492

[6] Pfeffer PL. Building principles for constructing a mammalian blastocyst embryo. Biology. 2018;7:41. DOI: 10.3390/biology7030041

[7] El-Kadiry AE, Rafei M, Shammaa R. Cell therapy: Types, regulation, and clinical benefits. Frontiers in Medicine (Lausanne). 2021;**8**:756029. DOI: 10.3389/fmed.2021.756029

[8] Nichols J, Smith A. Pluripotency in the embryo and in culture. Cold Spring Harbor Perspectives in Biology. 2012;**4**(8):a008128. DOI: 10.1101/cshperspect.a008128

[9] Zhang P, Zhang C, Li J, et al. The physical microenvironment of hematopoietic stem cells and its emerging roles in engineering applications. Stem Cell Research & Therapy. 2019;**10**:327. DOI: 10.1186/s13287-019-1422-7

[10] Evans MJ, Kaufman MH. Establishment in culture of pluripotential cells from mouse embryos. Nature. 1981;**292**:154-156. DOI: 10.1038/292154a0

[11] Thomson JA, Itskovitz-Eldor J, Shapiro SS, Waknitz MA, Swiergiel JJ, Marshall VS, et al. Embryonic stem cell lines derived from human blastocysts. Science. 1998;**282**(5391):1145-1147. DOI: 10.1126/science.282.5391.1145. Erratum in: Science 1998; 282(5395):1827

[12] Khanlarkhani N, Baazm M, Mohammadzadeh F, Najafi A, Mehdinejadiani S, Sobhani A. Multipotent stem cell and reproduction. Journal of Stem Cells. 2016;**11**:219-229

[13] Gaillochet C, Lohmann JU. The never-ending story: From pluripotency to plant developmental plasticity. Development. 2015;**142**(13):2237-2249. DOI: 10.1242/dev.117614

[14] von Zglinicki T, Wan T, Miwa S. Senescence in post-mitotic cells: A driver of aging? Antioxidants & Redox Signaling. 2021;**34**:308-323. DOI: 10.1089/ars.2020.8048

[15] Abdulrazzak H, Moschidou D, Jones G, Guillot PV. Biological characteristics of stem cells from foetal, cord blood and extraembryonic tissues. Journal of the Royal Society Interface. 2010;7(Suppl. 6):S689-S706. DOI: 10.1098/rsif.2010.0347

[16] Kfoury C. Therapeutic cloning: Promises and issues. McGill Journal of Medicine. 2007;**10**:112-120

[17] Ye L, Swingen C, Zhang J. Induced pluripotent stem cells and their potential for basic and clinical sciences. Current Cardiology Reviews. 2013;**9**:63-72. DOI: 10.2174/157340313805076278

[18] Cecerska-Heryć E, Pękała M, Serwin N, Gliźniewicz M, Grygorcewicz B, Michalczyk A, et al. The use of stem cells as a potential treatment method for selected neurodegenerative diseases: Review. Cellular and Molecular Neurobiology. 2023;**43**:2643-2673. DOI: 10.1007/s10571-023-01344-6

[19] Yamanaka S. Induced pluripotent stem cells: Past, present, and future. Cell Stem Cell. 2012;**10**:678-684. DOI: 10.1016/j.stem.2012.05.005. PMID: 22704507

[20] Papaccio F, Paino F, Regad T, Papaccio G, Desiderio V, Tirino V. Concise review: Cancer cells, cancer stem cells, and mesenchymal stem cells: Influence in cancer development. Stem Cells Translational Medicine. 2017;**6**:2115-2125. DOI: 10.1002/sctm.17-0138

[21] Zakas PM, Spencer HT, Doering CB. Engineered hematopoietic stem cells as therapeutics for hemophilia A. Journal of Genetic Syndromes & Gene Therapy. 2011;**1**:2410. DOI: 10.4172/2157-7412.S1-003

[22] Harris DT. Stem cell banking for regenerative and personalized medicine. Biomedicine. 2014;**2**:50-79. DOI: 10.3390/biomedicines2010050

[23] Weiss ML, Troyer DL. Stem cells in the umbilical cord. Stem Cell Reviews. 2006;**2**:155-162. DOI: 10.1007/s12015-006-0022-y

[24] Hoang DM, Pham PT, Bach TQ, Ngo ATL, Nguyen QT, Phan TTK, et al. Stem cell-based therapy for human diseases. Signal Transduction and Targeted Therapy. 2022;**7**:272. DOI: 10.1038/s41392-022-01134-4

[25] Lin X, Tang J, Lou Y-R. Human pluripotent stem-cell-derived models as a missing link in drug discovery and development. Pharmaceuticals. 2021;**14**:525. DOI: 10.3390/ph14060525

[26] Julian K, Garg N, Hibino N, Jain R. Stem cells and congenital heart disease: The future potential clinical therapy beyond current treatment. Current Cardiology Reviews. 2023;**19**(2):1-4. DOI: 10.2174/1573403X18666220531093326. ISSN: 1573-403X/1875-6557

[27] Nelson TJ, Cantero Peral S. Stem cell therapy and congenital heart disease. Journal of Cardiovascular Development and Disease. 2016;**3**:24. DOI: 10.3390/jcdd3030024

[28] Fuchs E, Chen T. A matter of life and death: Self-renewal in stem cells. EMBO Reports. 2013;**14**:39-48. DOI: 10.1038/embor.2012.197

[29] Bernitz JM, Moore KA. Haematopoietic stem cells: Uncovering the origins of a niche. eLife. 2014;**3**:e05041

[30] Bianco P, Robey PG, Simmons PJ. Mesenchymal stem cells: Revisiting history, concepts, and assays. Cell Stem Cell. 2008;**2**:313-319. DOI: 10.1016/j.stem.2008.03.002

[31] Wu J, Izpisua Belmonte JC. Stem cells: A renaissance in human biology research. Cell. 2016;**165**:1572-1585. DOI: 10.1016/j.cell.2016.05.043

[32] Chatterjee T, Sarkar RS, Dhot PS, Kumar S, Kumar VK. Adult stem cell

plasticity: Dream or reality? Medical Journal, Armed Forces India. 2010;**66**:56-60. DOI: 10.1016/S0377-1237(10)80095-4

[33] Alvarado AS, Yamanaka S. Rethinking differentiation: Stem cells, regeneration, and plasticity. Cell. 2014;**157**(1):110-119. DOI: 10.1016/j.cell.2014.02.041. ISSN: 0092-8674

[34] Li Z, Li L. Understanding hematopoietic stem-cell microenvironments. Trends in Biochemical Sciences. 2006;**31**:589-595. DOI: 10.1016/j.tibs.2006.08.001

[35] Mohyeldin A, Garzón-Muvdi T, Quiñones-Hinojosa A. Oxygen in stem cell biology: A critical component of the stem cell niche. Cell Stem Cell. 2010;7(2):150-161. DOI: 10.1016/j.stem.2010.07.007. ISSN: 1934-5909

[36] Wagner RT, Zwaka TP. Development: Sketch for theory of Oct4. Current Biology. 2013;**23**(22):R1014-R1016. DOI: 10.1016/j.cub.2013.10.003. ISSN: 0960-9822

[37] Tonelli FM, Santos AK, Gomes DA, da Silva SL, Gomes KN, Ladeira LO, et al. Stem cells and calcium signaling. Advances in Experimental Medicine and Biology. 2012;**740**:891-916. DOI: 10.1007/978-94-007-2888-2_40

[38] Haston S, Pozzi S, Carreno G, Manshaei S, Panousopoulos L, Gonzalez-Meljem JM, et al. MAPK pathway control of stem cell proliferation and differentiation in the embryonic pituitary provides insights into the pathogenesis of papillary craniopharyngioma. Development. 2017;**144**(12):2141-2152. DOI: 10.1242/dev.150490

[39] Caballano-Infantes E, Cahuana GM, Bedoya FJ, Salguero-Aranda C, Tejedo JR. The role of nitric oxide in stem cell biology. Antioxidants (Basel). 2022;**11**(3):497. DOI: 10.3390/antiox11030497

[40] Snoeck HW. Calcium regulation of stem cells. EMBO Reports. 2020;**21**(6):e50028. DOI: 10.15252/embr.202050028

[41] Yin X, Mead BE, Safaee H, Langer R, Karp JM, Levy O. Engineering stem cell organoids. Cell Stem Cell. 2016;**18**(1):25-38. DOI: 10.1016/j.stem.2015.12.005. ISSN: 1934-5909

[42] Velikic G, Maric DM, Maric DL, Supic G, Puletic M, Dulic O, et al. Harnessing the stem cell niche in regenerative medicine: Innovative avenue to combat neurodegenerative diseases. International Journal of Molecular Sciences. 2024;**25**:993. DOI: 10.3390/ijms25020993

[43] Scodellaro C, Pina RR, Ferreira FC, Sanjuan-Alberte P, Fernandes TG. Unlocking the potential of stem cell microenvironments in vitro. Bioengineering. 2024;**11**:289. DOI: 10.3390/bioengineering11030289

[44] Stevens LC. Teratocarcinogenesis and spontaneous parthenogenesis in mice. Results and Problems in Cell Differentiation. 1980;**11**:265-274. DOI: 10.1007/978-3-540-38267-6_34

[45] Nobel. 2010. Available from: http://nobelprize.org/nobel_prizes/medicine/laureates/2010/announcement.html

[46] Mehta RH. Sourcing human embryos for embryonic stem cell lines: Problems & perspectives. The Indian Journal of Medical Research. 2014;**140**(Suppl. 1):S106-S111

[47] Loi P, Czernik M, Zacchini F, Iuso D, Scapolo PA, Ptak G. Sheep: The first large animal model in nuclear transfer research. Cellular Reprogramming.

2013;**15**(5):367-373. DOI: 10.1089/cell.2013.0032

[48] Lo B, Parham L. Ethical issues in stem cell research. Endocrine Reviews. 2009;**30**(3):204-213. DOI: 10.1210/er.2008-0031

[49] Cowan CA, Klimanskaya I, McMahon J, Atienza J, Witmyer J, Zucker JP, et al. Derivation of embryonic stem-cell lines from human blastocysts. The New England Journal of Medicine. 2004;**350**:1353-1356. DOI: 10.1056/NEJMsr040330

[50] Jiang G, Wan X, Wang M, Zhou J, Pan J, Wang B. A reliable and economical method for gaining mouse embryonic fibroblasts capable of preparing feeder layers. Cytotechnology. 2016;**68**:1603-1614. DOI: 10.1007/s10616-014-9815-z

[51] Choi NY, Park YS, Ryu JS, Lee HJ, Araúzo-Bravo MJ, Ko K, et al. A novel feeder-free culture system for expansion of mouse spermatogonial stem cells. Molecules and Cells. 2014;**37**:473-479. DOI: 10.14348/molcells.2014.0080

[52] Meisner LF, Johnson JA. Protocols for cytogenetic studies of human embryonic stem cells. Methods. 2008;**45**:133-141. DOI: 10.1016/j.ymeth.2008.03.005

[53] Fan Y, Wu J, Ashok P, Hsiung M, Tzanakakis ES. Production of human pluripotent stem cell therapeutics under defined xeno-free conditions: Progress and challenges. Stem Cell Reviews. 2015;**11**:96-109. DOI: 10.1007/s12015-014-9544-x

[54] Williams RL, Hilton DJ, Pease S, et al. Myeloid leukaemia inhibitory factor maintains the developmental potential of embryonic stem cells. Nature. 1988;**336**(6200):684-687. DOI: 10.1038/336684a0

[55] Malik P, Mukherjee TK. Large-scale culture of mammalian cells for various industrial purposes. In: Mukherjee TK, Malik P, Mukherjee S, editors. Practical Approach to Mammalian Cell and Organ Culture. Singapore: Springer; 2023. DOI: 10.1007/978-981-19-1731-8_15-2

[56] Hagbard L, Cameron K, August P, Penton C, Parmar M, Hay DC, et al. Developing defined substrates for stem cell culture and differentiation. Philosophical Transactions of the Royal Society of London. Series B, Biological Sciences. 2018;**373**:20170230. DOI: 10.1098/rstb.2017.0230

[57] Herberts CA, Kwa MS, Hermsen HP. Risk factors in the development of stem cell therapy. Journal of Translational Medicine. 2011;**9**:29. DOI: 10.1186/1479-5876-9-29

[58] Verma A, Verma M, Singh A. Animal tissue culture principles and applications. Animal Biotechnology. 2020. pp. 269-293. DOI: 10.1016/B978-0-12-811710-1.00012-4

[59] Heng BC, Ye CP, Liu H, Toh WS, Rufaihah AJ, Yang Z, et al. Loss of viability during freeze-thaw of intact and adherent human embryonic stem cells with conventional slow-cooling protocols is predominantly due to apoptosis rather than cellular necrosis. Journal of Biomedical Science. 2006;**13**:433-445. DOI: 10.1007/s11373-005-9051-9

[60] Schwartz PH, Brick DJ, Nethercott HE, Stover AE. Traditional human embryonic stem cell culture. Methods in Molecular Biology. 2011;**767**:107-123. DOI: 10.1007/978-1-61779-201-4_8

[61] Carlin R, Davis D, Weiss M, Schultz B, Troyer D. Expression of early transcription factors Oct-4, Sox-2 and Nanog by porcine umbilical cord

(PUC) matrix cells. Reproductive Biology and Endocrinology. 2006;**4**:8. DOI: 10.1186/1477-7827-4-8

[62] Geraghty R, Capes-Davis A, Davis J, et al. Guidelines for the use of cell lines in biomedical research. British Journal of Cancer. 2014;**111**:1021-1046. DOI: 10.1038/bjc.2014.166

[63] Lai D, Wang Y, Sun J, et al. Derivation and characterization of human embryonic stem cells on human amnion epithelial cells. Scientific Reports. 2015;**5**:10014. DOI: 10.1038/srep10014

[64] Gharibi B, Hughes FJ. Effects of medium supplements on proliferation, differentiation potential, and in vitro expansion of mesenchymal stem cells. Stem Cells Translational Medicine. 2012;**1**:771-782. DOI: 10.5966/sctm.2010-0031

[65] Wesselschmidt RL. The teratoma assay: An in vivo assessment of pluripotency. Methods in Molecular Biology. 2011;**767**:231-241. DOI: 10.1007/978-1-61779-201-4_17

[66] Lindoso RS, Kasai-Brunswick TH, Monnerat Cahli G, Collino F, Bastos Carvalho A, Campos de Carvalho AC, et al. Proteomics in the world of induced pluripotent stem cells. Cells. 2019;**8**:703. DOI: 10.3390/cells8070703

[67] Wilson R. Competent steps in determination of cell fate. BioEssays. 1999;**21**(6):455-458. DOI: 10.1002/(SICI)1521-1878(199906)21:6<455::AID BIES2>3.0.CO; 2-4

[68] Davidson EH, Andrew Cameron R, Ransick A. Specification of cell fate in the sea urchin embryo: Summary and some proposed mechanisms. Development. 1998;**125**(17):3269-3290. DOI: 10.1242/dev.125.17.3269

[69] Keller G. Embryonic stem cell differentiation: Emergence of a new era in biology and medicine. Genes & Development. 2005;**19**(10):1129-1155. DOI: 10.1101/gad.1303605

[70] Takahashi S, Kobayashi S, Hiratani I. Epigenetic differences between naïve and primed pluripotent stem cells. Cellular and Molecular Life Sciences. 2018;**75**:1191-1203. DOI: 10.1007/s00018-017-2703-x

[71] Guo G, Stirparo GG, Strawbridge SE, Spindlow D, Yang J, Clarke J, et al. Human naive epiblast cells possess unrestricted lineage potential. Cell Stem Cell. 2021;**28**(6):1040-1056.e6. DOI: 10.1016/j.stem.2021.02.025

[72] Nichols J, Smith A. Naive and primed pluripotent states. Cell Stem Cell. 2009;**4**:487-492. DOI: 10.1016/j.stem.2009.05.015

[73] Martin GR, Evans MJ. Differentiation of clonal lines of teratocarcinoma cells: Formation of embryoid bodies in vitro. Proceedings of the National Academy of Sciences of the United States of America. 1975;**72**:1441-1445

[74] Murry CE, Keller G. Differentiation of embryonic stem cells to clinically relevant populations: Lessons from embryonic development. Cell. 2008;**132**(4):661-680. DOI: 10.1016/j.cell.2008.02.008. ISSN: 0092-8674

[75] Nikolits I, Nebel S, Egger D, Kreß S, Kasper C. Towards physiologic culture approaches to improve standard cultivation of mesenchymal stem cells. Cells. 2021;**10**(4):886. DOI: 10.3390/cells10040886

[76] Ludwig TE, Levenstein ME, Jones JM, Berggren WT, Mitchen ER, Frane JL, et al. Derivation of human embryonic stem cells in defined

conditions. Nature Biotechnology. 2006;**24**(2):185-187

[77] Hua Y, Yoshimochi K, Li J, et al. Development and evaluation of a novel xeno-free culture medium for human-induced pluripotent stem cells. Stem Cell Research & Therapy. 2022;**13**:223. DOI: 10.1186/s13287-022-02879-z

[78] Rungarunlert S, Techakumphu M, Pirity MK, Dinnyes A. Embryoid body formation from embryonic and induced pluripotent stem cells: Benefits of bioreactors. World Journal of Stem Cells. 2009;**1**(1):11-21. DOI: 10.4252/wjsc.v1.i1.11

[79] Kaufman DS, Thomson JA. Human ES cells—Haematopoiesis and transplantation strategies. Journal of Anatomy. 2002;**200**(Pt 3):243-248. DOI: 10.1046/j.1469-7580.2002.00028.x

[80] Rocheteau P, Vinet M, Chretien F. Dormancy and quiescence of skeletal muscle stem cells. Results and Problems in Cell Differentiation. 2015;**56**:215-235. DOI: 10.1007/978-3-662-44608-9_10

[81] Marion NW, Mao JJ. Mesenchymal stem cells and tissue engineering. Methods in Enzymology. 2006;**420**:339-361. DOI: 10.1016/S0076-6879(06)20016-8

[82] Chan AML, Sampasivam Y, Lokanathan Y. Biodistribution of mesenchymal stem cells (MSCs) in animal models and implied role of exosomes following systemic delivery of MSCs: A systematic review. American Journal of Translational Research. 2022;**14**(4):2147-2161

[83] Hombach-Klonisch S, Panigrahi S, Rashedi I, Seifert A, Alberti E, Pocar P, et al. Adult stem cells and their trans-differentiation potential—Perspectives and therapeutic applications. Journal of Molecular Medicine (Berlin, Germany). 2008;**86**(12):1301-1314. DOI: 10.1007/s00109-008-0383-6

[84] Eguizabal C, Montserrat N, Veiga A, Izpisua Belmonte JC. Dedifferentiation, transdifferentiation, and reprogramming: Future directions in regenerative medicine. Seminars in Reproductive Medicine. 2013;**31**(1):82-94. DOI: 10.1055/s-0032-1331802

[85] Zhang W, Lou YR, Zhou J, He Z. Editorial: Stem cells in tissue homeostasis and disease. Frontiers in Cell and Development Biology. 2022;**10**:876060. DOI: 10.3389/fcell.2022.876060

[86] Drummond-Barbosa D. Stem cells, their niches and the systemic environment: An aging network. Genetics. 2008;**180**(4):1787-1797. DOI: 10.1534/genetics.108.098244

[87] Rojas-Ríos P, González-Reyes A. Concise review: The plasticity of stem cell niches: A general property behind tissue homeostasis and repair. Stem Cells. 2014;**32**(4):852-859. DOI: 10.1002/stem.1621

[88] Salic A, Mitchison TJ. A chemical method for fast and sensitive detection of DNA synthesis in vivo. Proceedings of the National Academy of Sciences of the United States of America. 2008;**105**:2415-2420

[89] Bussard KM, Boulanger CA, Kittrell FS, Behbod F, Medina D, Smith GH. Immortalized, pre-malignant epithelial cell populations contain long-lived, label-retaining cells that asymmetrically divide and retain their template DNA. Breast Cancer Research. 2010;**12**(5):R86. DOI: 10.1186/bcr2754

[90] Cairns J. Somatic stem cells and the kinetics of mutagenesis and carcinogenesis. Proceedings of

the National Academy of Sciences of the United States of America. 2002;**99**(16):10567-10570. DOI: 10.1073/pnas.162369899. Epub 2002 Jul 29

[91] Hsu YC, Li L, Fuchs E. Transit-amplifying cells orchestrate stem cell activity and tissue regeneration. Cell. 2014;**157**(4):935-949. DOI: 10.1016/j.cell.2014.02.057

[92] Tu SM, Lin SH. Prostate cancer stem cells. Clinical Genitourinary Cancer. 2012;**10**(2):69-76. DOI: 10.1016/j.clgc.2012.01.002

[93] Blank U, Karlsson G, Karlsson S. Signaling pathways governing stem-cell fate. Blood. 2008;**111**(2):492-503. DOI: 10.1182/blood-2007-07-075168

[94] Pisciotta A, Carnevale G, Meloni S, Riccio M, De Biasi S, Gibellini L, et al. Human dental pulp stem cells (hDPSCs): Isolation, enrichment and comparative differentiation of two sub-populations. BMC Developmental Biology. 2015;**15**:14. DOI: 10.1186/s12861-015-0065-x

[95] Kulus M, Sibiak R, Stefańska K, Zdun M, Wieczorkiewicz M, Piotrowska-Kempisty H, et al. Mesenchymal stem/stromal cells derived from human and animal perinatal tissues-origins, characteristics, signaling pathways, and clinical trials. Cells. 2021;**10**(12):3278. DOI: 10.3390/cells10123278

[96] Fonseca LN, Bolívar-Moná S, Agudelo T, Beltrán LD, Camargo D, Correa N, et al. Cell surface markers for mesenchymal stem cells related to the skeletal system: A scoping review. Heliyon. 2023;**9**(2):e13464. DOI: 10.1016/j.heliyon.2023.e13464

[97] Sousa BR, Parreira RC, Fonseca EA, Amaya MJ, Tonelli FM, Lacerda SM, et al. Human adult stem cells from diverse origins: An overview from multiparametric immunophenotyping to clinical applications. Cytometry. Part A. 2014;**85**(1):43-77. DOI: 10.1002/cyto.a.22402

[98] Machado Cde V, Telles PD, Nascimento IL. Immunological characteristics of mesenchymal stem cells. Revista Brasileira de Hematologia e Hemoterapia. 2013;**35**(1):62-67. DOI: 10.5581/1516-8484.20130017

[99] Bozdağ SC, Yüksel MK, Demirer T. Adult stem cells and medicine. Advances in Experimental Medicine and Biology. 2018;**1079**:17-36. DOI: 10.1007/5584_2018_184

[100] Lee JY, Hong SH. Hematopoietic stem cells and their roles in tissue regeneration. International Journal of Stem Cells. 2020;**13**(1):1-12. DOI: 10.15283/ijsc19127

[101] Galipeau J, Sensébé L. Mesenchymal stromal cells: Clinical challenges and therapeutic opportunities. Cell Stem Cell. 2018;**22**:824-833

[102] Li Z, Leung M, Hopper R, Ellenbogen R, Zhang M. Feeder-free self-renewal of human embryonic stem cells in 3D porous natural polymer scaffolds. Biomaterials. 2010;**31**:404-412

[103] Lin CS, Xin ZC, Dai J, Lue TF. Commonly used mesenchymal stem cell markers and tracking labels: Limitations and challenges. Histology & Histopathology 2013;**28**(9):1109-1116. DOI: 10.14670/HH-28.1109

[104] Bertoncello I, Williams B. Hematopoietic stem cell characterization by Hoechst 33342 and rhodamine 123 staining. Methods in Molecular Biology. 2004;**263**:181-200. DOI: 10.1385/1-59259-773-4:181

[105] Beane OS, Darling EM. Isolation, characterization, and differentiation

of stem cells for cartilage regeneration. Annals of Biomedical Engineering. 2012;**40**(10):2079-2097. DOI: 10.1007/s10439-012-0639-8

[106] Liao X, Makris M, Luo XM. Fluorescence-activated cell sorting for purification of plasmacytoid dendritic cells from the mouse bone marrow. Journal of Visualized Experiments. 2016;**117**:54641. DOI: 10.3791/54641

[107] Soboleski MR, Oaks J, Halford WP. Green fluorescent protein is a quantitative reporter of gene expression in individual eukaryotic cells. The FASEB Journal. 2005;**19**(3):440-442. DOI: 10.1096/fj.04-3180fje

[108] Den Hartogh SC, Passier R. Concise review: Fluorescent reporters in human pluripotent stem cells: Contributions to cardiac differentiation and their applications in cardiac disease and toxicity. Stem Cells. 2016;**34**(1):13-26. DOI: 10.1002/stem.2196. Epub 2015 Oct 8

[109] Zhou ZY, Yang M, Jiang YH. Adult stem cells and possible mechanisms of its differentiation—Editorial. Zhongguo Shi Yan Xue Ye Xue Za Zhi. 2005;**13**(3):353-357. Chinese

[110] Zhang X, Zhang S, Wang T. How the mechanical microenvironment of stem cell growth affects their differentiation: A review. Stem Cell Research & Therapy. 2022;**13**:415. DOI: 10.1186/s13287-022-03070-0

[111] Lai AY, Kondo M. T and B lymphocyte differentiation from hematopoietic stem cell. Seminars in Immunology. 2008;**20**(4):207-212. DOI: 10.1016/j.smim.2008.05.002. Epub 2008 Jun 25

[112] Muruganandan S, Ionescu AM, Sinal CJ. At the crossroads of the adipocyte and osteoclast differentiation programs: Future therapeutic perspectives. International Journal of Molecular Sciences. 2020;**21**(7):2277. DOI: 10.3390/ijms21072277

[113] Pittenger MF, Mackay AM, Beck SC, Jaiswal RK, Douglas R, Mosca JD, et al. Multilineage potential of adult human mesenchymal stem cells. Science. 1999;**284**(5411):143-147. DOI: 10.1126/science.284.5411.143

[114] Kaitsuka T, Hakim F. Response of pluripotent stem cells to environmental stress and its application for directed differentiation. Biology (Basel). 2021;**10**(2):84. DOI: 10.3390/biology10020084

[115] Tan Y, Guo Y, Reed-Maldonado AB, Li Z, Lin G, Xia SJ, et al. Low-intensity pulsed ultrasound stimulates proliferation of stem/progenitor cells: What we need to know to translate basic science research into clinical applications. Asian Journal of Andrology. 2021;**23**(6):602-610. DOI: 10.4103/aja.aja_25_21

[116] dela Paz NG, Walshe TE, Leach LL, Saint-Geniez M, D'Amore PA. Role of shear-stress-induced VEGF expression in endothelial cell survival. Journal of Cell Science. 2012;**125**(Pt 4):831-843. DOI: 10.1242/jcs.084301

[117] Yang X, Tian S, Fan L, Niu R, Yan M, Chen S, et al. Integrated regulation of chondrogenic differentiation in mesenchymal stem cells and differentiation of cancer cells. Cancer Cell International. 2022;**22**(1):169. DOI: 10.1186/s12935-022-02598-8

[118] Datta N, Pham QP, Sharma U, Sikavitsas VI, Jansen JA, Mikos AG. In vitro generated extracellular matrix and fluid shear stress synergistically

enhance 3D osteoblastic differentiation. Proceedings of the National Academy of Sciences of the United States of America. 2006;**103**(8):2488-2493. DOI: 10.1073/pnas.0505661103

[119] Kocher AA, Schuster MD, Szabolcs MJ, Takuma S, Burkhoff D, Wang J, et al. Neovascularization of ischemic myocardium by human bone-marrow-derived angioblasts prevents cardiomyocyte apoptosis, reduces remodeling and improves cardiac function. Nature Medicine. 2001;**7**:430-436

[120] Durrani S, Konoplyannikov M, Ashraf M, Haider KH. Skeletal myoblasts for cardiac repair. Regenerative Medicine. 2010;**5**(6):919-932. DOI: 10.2217/rme.10.65

[121] Duelen R, Sampaolesi M. Stem cell technology in cardiac regeneration: A pluripotent stem cell promise. eBioMedicine. 2017;**16**:30-40. DOI: 10.1016/j.ebiom.2017.01.029. Epub 2017 Jan 27

[122] Karantalis V, Hare JM. Use of mesenchymal stem cells for therapy of cardiac disease. Circulation Research. 2015;**116**(8):1413-1430. DOI: 10.1161/CIRCRESAHA.116.303614

[123] Hare JM, Fishman JE, Gerstenblith G, DiFede Velazquez DL, Zambrano JP, Suncion VY, et al. Comparison of allogeneic vs autologous bone marrow–derived mesenchymal stem cells delivered by transendocardial injection in patients with ischemic cardiomyopathy: The POSEIDON randomized trial. JAMA. 2012;**308**:2369-2379

[124] Hong Y, Kai L, Zhu J, Wang J'a. Stem cell therapy for ischemic heart diseases. British Medical Bulletin. 2017;**121**(1):135-154

[125] Makkar RR et al. Intramyocardial injection of allogenic bone marrow-derived mesenchymal stem cells without immunosuppression preserves cardiac function in a porcine model of myocardial infarction. Journal of Cardiovascular Pharmacology and Therapeutics. 2005;**10**(4):225-233

[126] Wang Y, Qi Z, Yan Z, Ji N, Yang X, Gao D, et al. Mesenchymal stem cell immunomodulation: A novel intervention mechanism in cardiovascular disease. Frontiers in Cell and Development Biology. 2022;**9**:742088. DOI: 10.3389/fcell.2021.742088

[127] Chowdhury MA, Zhang JJ, Rizk R, Chen WCW. Stem cell therapy for heart failure in the clinics: new perspectives in the era of precision medicine and artificial intelligence. Frontiers in Physiology. 2024;**14**:1344885. DOI: 10.3389/fphys.2023.1344885

[128] Goichberg P, Chang J, Liao R, Leri A. Cardiac stem cells: Biology and clinical applications. Antioxidants & Redox Signaling. 2014;**21**(14):2002-2017. DOI: 10.1089/ars.2014.5875

[129] Boyle AJ, Schulman SP, Hare JM, Oettgen P. Is stem cell therapy ready for patients? Stem cell therapy for cardiac repair. Ready for the next step. Circulation. 2006;**114**:339-352. DOI: 10.1161CIRCULATION AHA.105.590653

[130] Becher UM, Tiyerili V, Skowasch D, Nickenig G, Werner N. Personalized cardiac regeneration by stem cells-hype or hope? The EPMA Journal. 2011;**2**(1):119-130. DOI: 10.1007/s13167-011-0068-z

[131] Oikonomopoulos A, Kitani T, Wu JC. Pluripotent stem cell-derived cardiomyocytes as a platform for cell

therapy applications: Progress and hurdles for clinical translation. Molecular Therapy. 2018;**26**(7):1624-1634. DOI: 10.1016/j.ymthe.2018.02.026

[132] Eugenia DN, Jenni VLN, Vladimir NC, Vadim VT, Helgi BS. Stem cells in clinical trials on neurological disorders: Trends in stem cells origins, indications, and status of the clinical trials. International Journal of Molecular Sciences. 2022;**23**(19):11453. DOI: 10.3390/ijms231911453

[133] Fei T, Xuran L, Zhao W, Jiaojiao L, Khawar S, Jialin Z. Clinical applications of stem cell-derived exosomes. Signal Transduction and Targeted Therapy. 2024;**9**:17. DOI: 10.1038/s41392-023-01704-0

[134] Bonaventura G, Munafò A, Bellanca CM, La Cognata V, Iemmolo R, Attaguile GA, et al. Stem cells: Innovative therapeutic options for neurodegenerative diseases? Cells. 2021;**10**:1992. DOI: 10.3390/cells10081992

[135] Chen KS, Koubek EJ, Sakowski SA, Feldman EL. Stem cell therapeutics and gene therapy for neurologic disorders. Neurotherapeutics. 2024;**21**(4):e00427. DOI: 10.1016/j.neurot.2024.e00427

[136] Li C, Fleck JS, Martins-Costa C, Burkard TR, Themann J, Stuempflen M, et al. Single-cell brain organoid screening identifies developmental defects in autism. Nature. 2023;**621**:373-380. DOI: 10.1038/s41586-023-06473-y

[137] Blanquer M, Moraleda JM, Iniesta F, Gómez-Espuch J, Meca-Lallana J, Villaverde R, et al. Neurotrophic bone marrow cellular nests prevent spinal motoneuron degeneration in amyotrophic lateral sclerosis patients: A pilot safety study. Stem Cells. 2012;**30**:1277-1285. DOI: 10.1002/stem.1080

[138] Sivandzade F, Cucullo L. Regenerative stem cell therapy for neurodegenerative diseases: An overview. International Journal of Molecular Sciences. 2021;**22**(4):2153. DOI: 10.3390/ijms22042153

[139] Tuazon JP, Castelli V, Lee JY, Desideri GB, Stuppia L, Cimini AM. Borlongan CV neural stem cells. Advances in Experimental Medicine and Biology. 2019;**1201**:79-91. DOI: 10.1007/978-3-030-31206-0_4

[140] Kashani AH, Lebkowski JS, Hinton DR, Zhu D, Faynus MA, Chen S, et al. Survival of an HLA-mismatched, bioengineered RPE implant in dry age-related macular degeneration. Stem Cell Reports. 2022;**17**:448-458. DOI: 10.1016/j.stemcr.2022.01.001

[141] De Gioia R, Biella F, Citterio G, Rizzo F, Abati E, Nizzardo M, et al. Neural stem cell transplantation for neurodegenerative diseases. International Journal of Molecular Sciences. 2020;**21**(9):3103. DOI: 10.3390/ijms21093103

[142] Yao Y, Huang C, Gu P, Wen T. Combined MSC-secreted factors and neural stem cell transplantation promote functional recovery of PD rats. Cell Transplantation;**25**(6):1101-1113. DOI: 10.3727/096368915x689938

[143] Klimanskaya I, Chung Y, Becker S, Lu SJ, Lanza R. Human embryonic stem cell lines derived from single blastomeres. Nature. 2006;**444**(7118):481-485. DOI: 10.1038/nature05142

[144] Klimanskaya I, Chung Y, Becker S, Lu SJ, Lanza R. Derivation of human embryonic stem cells from single blastomeres. Nature Protocols. 2007;**2**(8):1963-1972. DOI: 10.1038/nprot.2007.274

[145] Sladitschek HL, Neveu PA. Integrated experimental and theoretical studies of stem cells. Current Stem Cell Reports. 2017;**3**:248-252. DOI: 10.1007/s40778-017-0096-2

[146] Nicola NA, Babon JJ. Leukemia inhibitory factor (LIF). Cytokine & Growth Factor Reviews. 2015;**26**(5):533-544. DOI: 10.1016/j.cytogfr.2015.07.001

[147] Sladitschek HL, Neveu PA. The bimodally expressed microRNA miR-142 gates exit from pluripotency. Molecular Systems Biology. 2015;**11**(12):850. DOI: 10.15252/msb.20156525

Chapter 8

Home Care and Pharmacy in Hematopoietic Transplantation

Sonia Ruiz Boy and Francesc Fernández Avilés

Abstract

Home hospitalization, and other models of outpatient care models, represents an alternative to traditional hospitalization, providing comparable clinical safety and efficacy for hematological patients. At-home therapies can range from the delivery of intravenous antibiotics to more complex scenarios such as the care during the early period after hematopoietic stem cell transplantation. Early discharge from conventional hospitalization is feasible, contributing to a reduction in hospital resource utilization and waiting lists, as well as to a better quality of life and level of patient satisfaction. The coordinated efforts of multidisciplinary teams, including hematologists, nurses, and pharmacists, ensure patient safety and continuity of care. The active involvement of a clinical pharmacist in home hospitalization programs is essential, not only for the proper logistical management of the medication but also to ensure its appropriateness, optimize treatment, address queries from the team and patients, and promote adherence.

Keywords: home care, outpatient, multidisciplinary team, quality of life, saving costs, hematopoietic transplantation

1. Introduction

Hematopoietic Stem Cell Transplantation (HSCT) practice is a highly specialized and resource-intense medical procedure and it is associated with a high occurrence of adverse events. In recent years, there has been a substantial increase in the number of HSCTs performed annually [1].

In order to alleviate the workload in hematology units, as well as to improve patient comfort, among other advantages, outpatient HSCT care models have been developed. These initiatives have occurred in autologous and allogeneic HSCT. The outpatient HSCT model can be complete, with the patient receiving the conditioning regimen, transplantation, and management of the aplastic on an ambulatory basis, or mixed, with only one or more of the above steps being performed at the outpatient level [1].

On the other hand, ambulatory management can be carried out in two settings: in the outpatient clinic or at home.

A multidisciplinary team is necessary for the successful implementation of these ambulatory models. These will include hematologists, nurses, pharmacists, and other health specialists. Pharmacists are an integral part of this team, contributing to the safe and successful administration of medications, adequate composition of complex solutions, and adherence to standards of care [2].

In this chapter, we will describe the different care models of outpatient HSCT. We will observe the main features of each model, as well as published outcomes, and the different roles and tasks of each professional and caregiver, highlighting the essential role of an ambulatory pharmacist.

2. Classification of ambulatory HSCT care models

Table 1 shows the classification of the different outpatient models. The different models can be classified according to where the patient is treated and what processes he or she receives on an ambulatory basis.

2.1 Depending on where the patient is treated

- *At-home:* the patient receives part or all of the treatment/monitoring at home. Normally, the patient is visited daily (once or twice) by a nurse or advanced practice provider. In some cases, patients can also be continuously monitored via an advanced technological kit located in a centralized command center. Via telephone or videoconference, there is usually also daily contact between the patient and the hematologist, as well as daily meetings between nurse and doctor to discuss the patient's condition and progress. In addition to the face-to-face visits, the patient usually receives one or two calls per day from the nurse. Daily follow-up in the home would continue until the patient is discharged from the acute HSCT phase (normally until engraftment is achieved) to continue their

Complete ambulatory program	
Conditioning regimen, Hematopoietic Progenitor Cells (HPC) infusion, and management of the aplastic phase	Ambulatory clinic
Delayed admission	
Conditioning regimen, HPC infusion, and management of the aplastic phase	Inpatient Early discharged (+1) and readmission (+5)
Mixed inpatient-outpatient	
Conditioning regimen and management of the aplastic phase	Ambulatory clinic ± at home
HPC infusion	Inpatient
Early discharge outpatient	
Conditioning regimen and HPC infusion	Inpatient
Management of the aplastic phase	Ambulatory clinic
Early discharge at home	
Conditioning regimen and HPC infusion	Inpatient
Management of the aplastic phase	At home

Table 1.
Ambulatory HSCT models.

care in the regular return clinic. In case of emergency, the patient can usually call the doctor or nurse at any time of the day.

- *At ambulatory clinic*: the clinic is usually located in close proximity to the hospital's emergency room and inpatient and laboratory facilities. The patient usually goes to the clinic on a daily basis. Transport to the clinic may be provided by the center or by the patient. In some cases, the clinic has lodging facilities close to the unit so that the patient can stay there during treatment/follow-up. Depending on the center, the clinic is only open on weekdays from 8 am to 5 pm, or it is open non-stop. In any case, the patient is usually always provided with a medical contact number in case of emergency.

  In some cases, the patient receives most part of the medical care at home but in some situations, the patient has to be referred to the ambulatory clinic. The most common situations would be the process of transplantation/conditioning administration, the need for receiving transfusions of packed red blood cells or platelets, face-to-face visits with the doctor due to adverse effects or to evaluate fever.

- *Mixed inpatient-outpatient model*: the patient receives HPC infusion at the hospital, while the conditioning regimen and management of the aplastic phase are on an ambulatory basis.

In any of the models, the hospital usually reserves a bed in case the outpatient has to be admitted due to complications.

2.2 Depending on which processes the patient receives outside the hospital

- *Complete ambulatory program:* the patient receives an out-of-hospital conditioning regimen, HPC infusion, and management of the aplastic phase. Hospitals that implement this model use the ambulatory clinic. There is no article describing that the transplantation or conditioning chemotherapy processes take place at the patient's home.

- *Early discharge*: a model in which the patient would receive conditioning chemotherapy followed by HPC infusion, either in the hospital or at a day clinic. The patients would then be discharged home on post-HSCT days 0–1 (if use of cyclophosphamide post in allogeneic HSCT: day +5), unless other medical conditions (e.g., sepsis, hypotension) required them to stay in the hospital. In this model, the management of the aplastic phase of the outpatient can be carried out both at the patient's home and in the ambulatory clinic.

- *Delayed admission:* The patient is discharged on the day of transplantation for ambulatory management and readmitted on day +5 of transplantation for management of aplasia.

2.3 Depending on the type of transplant

Depending on the type of transplant, ambulatory models for allogeneic (including haploidentical) and autologous HSCT have been described. In addition, some studies have included outpatients who had received HSCT for a non-malignant hematological

Patient location	Processes carried out as outpatient	Type of HSCT	First author (year of publication, country), n° patients
At-home	Early discharge	Autologous	Fernández-Avilés [3] (2006, Spain), 50 Sung [4] (2022, USA), 17
		Allogeneic	Gutiérrez [5] (2020, Spain), 41 Ringdén [6] (2018, Sweden), 259 Svahn [7] (2002, Sweeden), 36
At ambulatory unit	Complete ambulatory model	Autologous	Holbro* [8] (2012, Canada), 91 Jagannath* [9] (1997, USA), 91 Kassar* [10] (2007, USA), 90 Landau [11] (2022, USA), 15 Marini* [12] (2022, USA), 50 McDiarmid** [13] (2010, USA), 508
		Allogeneic	Cantú** [14] (2007, Mexico), 103 Colunga [15] (2021, Mexico), 60 McDiarmid** [13] (2010, USA), 178 Singhal [16] (2023, USA), 856
	Early discharge	Autologous	Kharfan* [17] (2023, USA), 4 Martino* [18] (2014, Italy), 528 Paul* [19] (2015, USA), 301
		Allogeneic	Nicolau
	Delayed admission	Autologous	Anastasia** [20] (2009, Italy), 119
Mixed inpatient-outpatient model		Autologous	Clemmons* [21] (2017, USA), 18 Morabito* [22](2002, Italy), 29
		Allogeneic	Solomon [23] (2010, USA), 100

**multiple myeloma patients only.*
***malignant and non-malignant hematological diseases.*

Table 2.
Classification and description of outpatient model papers in adult patients.

disease, such as anemia or autoimmune disorders. The type of HSCT in which most ambulatory models have been described is autologous HSCT for patients with multiple myeloma.

Table 2 shows the types of transplantation performed according to the paper published.

3. Services provided to outpatients

Table 3 shows a summary of the different services provided in both home care and the ambulatory clinic.

Services provided in general in the patient's home:

- Full physical exam (nurse and/or hematologist). In one model [17], patients are continuously monitored via an advanced technological kit located in a centralized command center.

- Sample collection (blood, stool, and/or urine).

Home	• Full physical exam • Sample collection • Administration of IV or oral electrolytes, IV fluids ± other IV drugs, like G-CSF. • Infectious and GVHD prophylaxis. • Control of pain. • Optional services: • Red blood cell or platelet transfusion • Administration of parenteral nutrition • IV antibiotic therapy in febrile neutropenia
Ambulatory clinic	• Full physical exam • Sample collection • Administration of IV or oral electrolytes, IV fluids ± other IV drugs, like G-CSF. • Infectious and GVHD prophylaxis. • Control of pain. • Red blood cell or platelet transfusion • IV antibiotic therapy in febrile neutropenia • Optional services: o Administration of chemotherapy and radiotherapy/total body irradiation. o Stem cell mobilization o Graft infusion o Placement, removal, or replacement of venous catheters. o Evaluation of fever/febrile neutropenia in hemodynamically stable patients

Table 3.
Services provided to outpatients.

- Intravenous (IV) or oral electrolyte supplementation or/and IV fluids, and administration of another IV medication, like antiemetics or antibiotics. Depending on the home model, none [20], some [4, 11, 17, 19], or many types [3, 5–7] of IV medication are administered. In case the patient needs a type of IV medication that the home care model does not provide, he/she will have to go to the ambulatory clinic or be admitted.

- Infectious prophylaxis is usually the same or similar to what an inpatient would receive, varying according to the type of transplant, the patient's serology, his history of bacterial colonization and/or active infections, as well as the hospital's internal protocol. Antibiotics are usually administered orally but can be administered intravenously if there is significant mucositis [3, 5–7, 11].

- Granulocyte colony-stimulating factor (G-CSF) is either self-administered or given by a caregiver or a nurse. Posology is according to patient characteristics, type of transplant, and internal hospital protocol. Some groups do not use it during the aplasia period after transplant [24, 25].

- GVHD prophylaxis: different drugs will be administered depending on the characteristics of the conditioning, the transplant, and the patient. In most models, patients have to return to the day clinic/be admitted for the administration of methotrexate or post-transplant cyclophosphamide.

- Control of pain (generally by oral mucositis): early with topical treatments (nystatin, viscous lidocaine) and oral paracetamol or non-steroidal anti-inflammatory medications. If this is not sufficient, IV analgesics or oral [3]/sublingual [5]/transdermic [5]/IV narcotics [7] are used, although in some models that is an admission criterion for the outpatient [11]. In one model [7], continuous IV morphine is given, using a home pump.

- Optional services (if not offered, the patient will have to go to the ambulatory clinic or to be admitted to hospital):

  - Red blood cell or platelet transfusion [4, 6, 7].

  - Administration of parenteral nutrition, if the patient's fluid intake was less than 2 liters in 24 hours and weight had decreased by more than 2 kg and/or if the patient could not feed himself or herself at all [6, 7].

  - Administration of IV antibiotic therapy in febrile neutropenia: after the assessment of fever at the ambulatory clinic, if the patient is stable and feels well, IV antibiotic treatment can be continued at home through a portable intermittent infusion pump. Empiric therapy will differ according to center protocols, but general guidelines are as follows: if the patient has not received quinolone prophylaxis, he/she will receive levofloxacin (oral or IV), moxifloxacin (oral), ciprofloxacin (oral) ± amoxicillin/clavulanate or linezolid (oral). If the patient has received quinolone prophylaxis, ceftriaxone, piperacillin/tazobactam, or meropenem (IV) are commonly used. The stability time of the pumps, as well as their storage (e.g., meropenem should be kept in a refrigerator), should be taken into account in order to find out which treatments are feasible in the home model and also to know how often the nurse will have to visit the patient to replace the pump. Patient and/or caregiver training on the use and care of these pumps, as well as their active collaboration, will be of key importance at this point. In case of cytomegalovirus infection or disease, ganciclovir and foscarnet have been administered at home [26].

Services provided in general in ambulatory clinic:

- All of the above services, except for the administration of parenteral nutrition.

- Optional services (if not offered, the patient will have to be admitted to hospital):

  - Administration of chemotherapy and radiotherapy/total body irradiation. Both typical myeloablative and reduced-intensity myeloablative protocols can be administered in ambulatory clinic [4, 8–13, 15, 16, 23, 27].

  - Stem cell mobilization. Drugs most used are cyclophosphamide and G-SCF [4, 8–13, 15, 16, 27].

- Graft infusion [4, 8–13, 15, 16, 27].

- Placement, removal, or replacement of venous catheters [4, 8, 13].

- Evaluation of fever/febrile neutropenia in hemodynamically stable patients [3, 6, 7, 9, 13]: if the patient develops fever at home, he/she should go to the ambulatory clinic for a check-up. There, blood cultures and a chest x-ray are taken and IV antibiotics are started. If the patient feels well and in the absence of hemodynamic instability, pneumonia, or cardiac or respiratory distress, he can go home even with a fever, continuing IV antibiotics there or in the ambulatory clinic according to the model. In one ambulatory clinic care model [9], for example, outpatients are provided with an infuser each of vancomycin 1 g and ceftazidime 2 g (stable for 7 days) and these are self-administered during non-clinic hours, every 12 h. In another [13], patients with uncomplicated febrile neutropenia receive immediate monotherapy treatment with IV ceftriaxone 1 g daily at the clinic.

4. Patient inclusion criteria

The following are the most common inclusion criteria, according to published articles. The overview is shown in **Table 4**.

- It is essential that the patient agrees to this ambulatory monitoring. It is also often required that the patient speaks the language of the geographical area where the hospital is located.

- In all cases, the presence of a caregiver is necessary throughout the ambulatory process, with a 24/7 availability.

- Some hospitals require the patient to have insurance to cover ambulatory medical expenses and/or adequate Wi-Fi connection [9, 11].

- The patient's residence should be close to the ambulatory clinic or hospital. There is usually a driving time limit from home to the center of between 30 and 120 minutes.

- A specific age limit has only been imposed on some models, with 70–80 years being the usual maximum age of patients [4, 8, 9, 20]. The ability to maintain self-care activities, as well as the absence of significant comorbidities, is taken more into account. The most common inclusion criteria in this regard are creatinine <2–3 mg/dl, liver function abnormalities less than twice the upper range of normal, cardiac left ventricular ejection fraction >50%, carbon monoxide diffusion capacity >50%, Karnofsky score of at least 70–80%, and preserved oral route.

- In the case of at-home models, some hospitals carry out a home inspection beforehand in order to confirm cleanliness and suitability [4, 6, 7, 11]. The most common things they usually check are: the absence of black mold, functional heating and cooling, clean water source, and water temperature must be more than 50°C.

Patient agreed.
Presence of caregiver.
Patient's residence close to the ambulatory clinic/hospital (30–120 min driving).
Ability to maintain self-care activities.
Absence of significant comorbidities.
At-home model: clean house.

Table 4.
Most common inclusion criteria.

5. Figure of the caregiver and patient responsibilities

Caregivers are generally relatives or friends, who are required to undergo educational training and subsequently be approved by the transplant team before the patient proceeds forward with the therapy. Their responsibilities are listed in **Table 5**.

Both the patient and the caregiver are instructed with the following recommendations or rules to follow during their ambulatory period (listed in **Table 6**).

6. Readmission criteria

The criteria for hospital admission vary between models, depending on the services offered to outpatients, such as control of fever without hemodynamic instability responsive to first-line antibiotics, or whether or not parenteral nutrition, IV antibiotics, or fluids are administered. The most frequent criteria for hospital admission are shown in **Table 7**.

To be company for the patient.
To make food if needed and assist in personal hygiene.
To be alert to any complications that require medical attention (signs alert). If necessary or any doubts, to give a call to the hospital. The most common alert signs are: • Development of fever (temperature > 37.6–38.5°C), rigor or chills • Recurrent nausea or emesis • Significant diarrhea or rash • Neurologic alterations • Dyspnoea • Jaundice, bleeding
Provide updated information on patient compliance with oral medication, diet, and physical activity, as well as feedback on symptoms and emotional status.
Control the scheduling and dosage of the prescribed medicine (only oral).
Accompany patient to ambulatory clinic/hospital. In case of using the car, drive himself/herself.

Table 5.
Caregiver responsibilities and the most common signs of alert.

Isolation measures	Sleeping in a single bedroom, wearing a mask (surgical or FFP3) outdoors, avoiding persons with symptoms of or having contagious diseases, not going near construction areas (due to the risk of aspergillosis), visiting no one and not visiting indoor public spaces such as shopping malls, restaurants, or theaters.
State of residence	Removing flowers in pots (because the earth may contain Aspergillus), avoiding pet animals at home, changing the sheets 3 times a week, and cleaning the house once a week.
Physical activity	Each patient is expected to stay physically active and to walk up to one mile daily or the equivalent exercise activity, unless unable due to a medical condition.
Other recommendations	Not driving and taking a neutropenic diet [9, 13, 23].

Table 6.
Recommendations to patient/caregiver during the ambulatory period.

Febrile neutropenia/severe infection/temperature > 37.6–38.5°C (threshold varies according to the model)/ requirement of greater than twice-daily IV antibiotic therapy.
Uncontrolled nausea and/or vomiting or significant diarrhea requiring continuous IV fluids/parenteral alimentation.
Severe mucositis necessitating IV narcotic analgesics/parenteral nutrition.
GVHD grade 3 or 4
Central venous catheter-related complications.
Significant organ dysfunction like renal, cardiac, or respiratory distress.
Insufficient pain control.
Hemodynamic instability or deterioration of patient condition.
Grade III or IV hemorrhagic cystitis.
Patient request.
Loss of caregiver support.
Judgment of the transplantation physician based on the patient's clinical course and performance status.

Table 7.
Criteria for hospital admission.

The median readmission rate during the first 100 days after HSCT of outpatients varies according to the published articles [3, 5, 7–13, 15–19, 23], ranging from 0 to 88% (median: 60%). Hospital admission typically occurs at the nadir of neutropenia, being readmitted from the median on day +7 (range: 3–69). The median length of hospitalization in outpatients is 8 days (range: 0–100), which is significantly shorter than that of inpatients (21.5; 3–115 days). Admission to the intensive care unit (ICU) is not common (median 0; range 0–7%). The outpatient, if admitted to the hospital, is mostly admitted only once. It has not yet been possible to establish clinical factors that might affect the hospitalization rate/duration in hematological outpatients, varying these depending on the article [3, 8–10, 15, 18].

7. Efficacy and safety

There are currently not a large number of articles that have compared the safety and efficacy of ambulatory post-transplant patients vs. a matched cohort of inpatient patients. And, at the time of writing, no published article has been found that compares the two types of ambulatory models (home care vs. ambulatory unit care).

A systematic review [1] concluded that mortality rates without disease recurrence at 1 year were similar between inpatient and ambulatory models of both allogeneic and autologous HSCT. Nor have any differences been found in progression-free survival. In addition, some articles have noted longer long-term survival (2–5 years) in the ambulatory model for both types of HSCT. In any case, no article has observed a shorter survival in the ambulatory model [1]. Regarding the median engraftment time, most articles have not found significant differences [3–5, 7, 9], although one of them [12] observed that it was shorter if patients had received conditioning and HSCT as outpatients ($p < 0.001$).

Concerning security, a systematic review observed that the frequency of neutropenic fever was usually lower in the ambulatory model, especially in allogeneic HSCT [1]. It also noted that the frequency of mucositis was similar in autologous HSCT and six times lower in allogeneic HSCT in the outpatient vs. inpatient model [1]. In the case of allogeneic HSCT, a lower frequency of acute incidence of transplant rejection or GVHD was also observed [1]. Also, individual studies have noted that the outpatient cohort spent fewer days on parenteral nutrition [13], had minor admission to an ICU [9, 19], had fewer erythrocyte transfusions [9, 13], had fewer days on IV antibiotics and IV analgesics [13], and had a shorter duration of fever [3] than the inpatients. On the other hand, it was a concern for some hematologists whether the risk of aspergillus infections would increase in the home care patients, who are not isolated. However, it can be seen from the different articles that the number of cases of invasive aspergillosis was very low (between 0 and 4%), and no different from the inpatients cohort [3, 5–7, 15, 23]. This may be due to the cold climate [7], as well as the isolation recommendations provided [7] and/or the administration of prophylactic antibiotic therapy [3, 5, 15, 23]. It appears that voriconazole is more effective than inhaled amphotericin B for the prevention of Aspergillus [5].

8. Cost

Regarding costs, several studies, especially of autologous HSCT, have compared the direct costs of an inpatient vs. an outpatient cohort. In all cases, the ambulatory model was less costly than the inpatient care model. The percentage of direct cost reduction varied between 10 and 60% [1, 7, 28], depending on the study. The median cost reduction was 32% for autologous HSCT and 27% for allogeneic HSCT. The main cause is the reduction or absence of the cost of hospital stay (room, meal, nurse staff...). According to a systematic review, the number of days of hospital stay in the ambulatory model was 55 and 19% less than the hospital-based model, in the autologous and allogeneic cases, respectively [1]. It should be noted that in the case of an ambulatory model where the clinic pays for the patient's hotel accommodation and meals, the savings are still significant [9]. Minor financial savings are also found in pharmacy, pathology/laboratory charges, and physician fees. In contrast, some studies have noted that there was a trend for a higher cost in at-home groups for nurse fees [3]. The study of "real" costs in this regard is still to be carried out. On the other hand, from the few studies available, there do not appear to be significant differences in cost savings between the two models of ambulatory care.

9. Quality of life and patient/caregiver satisfaction

The results are generally positive, although there are some inconsistencies between articles.

Through satisfaction questionnaires given at the last follow-up, all articles agreed that the majority of patients and caregivers were satisfied with the ambulatory program [13, 23]. In general, in home-model care, no one treated at home regretted their decision [7], feeling safe there [3]. The primary advantages of it mentioned were the quiet and increased home comfort, the familiar environment, the free choice of activity and food, and increased privacy [3, 11]. On the other hand, some patients reported some disadvantages, like anxiety and fatigue, and the cost of a local hotel or apartment [3]. And, with regard to carers, some reported that caregiving had a moderate impact on their physical strengths, finances, and schedules, but they also endorsed deriving esteem from providing care [3, 11].

A few articles have conducted quality of life scales on their patients, the vast majority in the field of autologous HSCT and the ambulatory clinic model. One of them [4] showed that home HSCT recipients had preserved the quality of life throughout the transplant, without a significant peri-transplant change in the mean score survey, but being this at Day +100 significantly greater than their initial score at Day 0 and also significantly greater than that of the inpatient cohort at day 100. These findings are in line with another paper's results [29], whose conclusions were that the ambulatory clinic model neither improved nor impaired global patient quality of life in the first 30 days after HSCT. Another author [30] also reported significantly higher scores for emotional well-being and global quality of life in outpatients vs. inpatients. On the other hand, one article's conclusions [31] were that the quality of life decreased immediately post-treatment but then increased to above pretreatment levels by 6 months. Also, additional works exposed that the outpatient group referred to a better quality of life in terms of health status, daily work activity, and social/family well-being, respectively, but no statistical differences were found when compared to their respective inpatient group [27, 29]. Lastly, a good clinical outcome following HSCT was associated with better quality of life and greater satisfaction with care [28].

10. Pediatric ambulatory care experience

There is much less experience in pediatrics in this area than in adults. This approach has been undertaken with less enthusiasm by pediatric centers, mainly because of concerns about infection, safety, limited isolation, patient and family compliance, and the lack of adequate resources, both in personnel and facilities. Also, performing ambulatory transplantation in young children presents unique challenges, including an inability to articulate their clinical status and their well-recognized clinical deterioration in the setting of sepsis relative to adolescents and adults. The few published articles have used an ambulatory clinic care model, in patients aged 1–18 years. There is more experience with autologous HSCT than with allogeneic HSCT. The home care model does not seem to have been used in pediatric patients so far. These published articles show results very similar to those of adults, concluding that the ambulatory model in children is feasible, safe, and economical, with no apparent increase in morbidity and mortality. **Table 8** shows the characteristics of the articles published on this topic.

Patient location	Processes carried out as outpatient	Type of HSCT	First author (year of publication, country), n° patients
At ambulatory unit	Complete ambulatory model	Autologous	González-Ryan [32] (2001, USA), 48 Koo [33] (2019, USA), 49 Jaime-Pérez [34] (2021, Mexico), 4
		Allogeneic	González-Ryan [32] (2001, USA), 7 Jaime-Pérez [34] (2021, Mexico), 11

Table 8.
Classification and description of outpatient model papers in pediatric patients.

11. Pharmaceutic role

The Clinical Pharmacist (CP) has been acknowledged as an essential member of the multidisciplinary team in several settings, including the pharmaceutical care of HSCT inpatients. Some of the pharmacist's core competencies include medication management and counseling, drug-drug or drug-food interactions, symptom management, therapeutic drug monitoring, fluids and drugs/electrolytes compatibilities, discharge planning and transitions of care, and patient and caregiver education, and evaluation of medication adherence. All of these are fully applicable to ambulatory programs [35, 36].

Just as the role of the CP is essential within the multidisciplinary group that manages the post-hematological HSCT inpatient, it is also essential at the ambulatory level [37]. CP ensures the safety and appropriateness of anticancer therapy for patients because of their training and expertise in oncology/hematology diseases and management [38]. However, the specific role and core duties of CP can vary depending upon an institution's or healthcare system's pharmacy practice model, size, and setting [38, 39].

In one model [23], for example, a CP was assigned to the allogeneic ambulatory HSCT program, and along with the clinical team verified with the patient and caregiver on each visit that the patient was taking their medications correctly, monitored medication drug levels, potential drug interactions, cytomegalovirus surveillance testing, and adverse events. At the time of hospital discharge, a written medication calendar, discharge instructions, and treatment plan were prepared by the CP. He then provided this information in writing as well as orally to the patient and caregiver, including a home medicine administration record outlining the medication name, indication, drug concentration, and administration instructions. Also, each week, an independent audit was performed by the clinical pharmacist team to verify that chemotherapy, immunosuppressive, and antimicrobial agents had been administered as per institutional guidelines in the allogeneic HSCT patients.

Another task for CP is overseeing the sterile compounding of anticancer, antimicrobial, analgesic, or nutritional treatments and ensuring that the individualized regimen is appropriate and safe for outpatients. It is essential for the CP to ensure that these sterile medications prepared in the pharmacy remain physically, microbiologically, and chemically stable from preparation to administration to the outpatient [38].

Also, an adequate pharmacy infrastructure is recommended to provide continuity of care for these patients 7 days a week [9].

Integrating a CP into an ambulatory/home team has been shown to reduce drug-related problems, improve adherence, and save costs [38]. The CP needs to be able to effectively clarify any doubts raised by physicians, nurses, patients, and/or caregivers.

Medication management and monitoring
Provide thorough medication review.
Participate in multidisciplinary rounding.
Assistance with chemotherapy planning and review or preparation of it.
Therapeutic drug monitoring: immunosuppressants, anti-infectives, anti-seizure medications, anticoagulants, chemotherapeutic agents...
Provide medication therapy management: diabetes, hypertension, dyslipidemia...
Manage anti-infective therapies/prophylaxis and promote stewardship.
Ensure correct preparation of sterile preparations and correct stability ± compatibility at the time of administration.
Preparation or review of magistral formulas for topical conditions that the patient may require.
Patient care
Assist in symptom management of supportive care and pain/palliative care.
Optimize GVHD management.
Facilitate post-transplant vaccine administration.
Adherence evaluation.
Transition Planning
Assist with transitions of care: provide medication reconciliation and collaborate on discharge planning and management.
Facilitate transitions to hospice/palliative care where appropriate.
Facilitate access to medications available through patient assistance programs.
Education
Patients and caregivers (in person or online, in case of questions or for education).
Healthcare providers and trainees.
Research and Quality Improvement
Contribute to institutional and collaborative research and scholarly activities.
Assist with policy/guideline development.
Serve as patient & professional advocate.
Monitor, evaluate, and report transplant-related outcomes to assist in improvements to clinical practice.
Implementing/participating in mHealth programs specifically designed for HSCT outpatient follow-up.

Table 9.
Roles of clinical pharmacist in ambulatory HSCT care.

For that, the use of eHealth is perceived by some authors as an opportunity to improve care in the ambulatory/home setting [36].

The following table lists the different activities of the CP in ambulatory HSCT care (**Table 9**) [35, 36].

12. Conclusions

Over the last decades, several studies have shown that the ambulatory/home model improves the quality perceived by the patient and their families, reduces the duration of hospitalization and the waiting list for autologous and allogeneic HSCT, and minimizes exposure to nosocomial germs. In short, it optimizes healthcare resources and clearly represents economic savings in healthcare, with no difference in mortality rates compared to the hospital model. Integrating a Clinical Pharmacist into multidisciplinary ambulatory/home teams has the potential to improve patient monitoring, medication management, and patient and caregiver education, optimizing the outcomes of ambulatory care in HSCT recipients.

Acknowledgements

To my colleagues of the Pharmacy Service of Hospital Clínic of Barcelona for their support of my scientific and teaching activity.

Conflict of interest

The authors declare no conflict of interest.

Author details

Sonia Ruiz Boy[1] and Francesc Fernández Avilés[2]*

1 Pharmacy Service, Division of Medicine, Hospital Clinic of Barcelona, University of Barcelona, Barcelona, Spain

2 Bone Marrow Transplantation Unit, Hospital Clinic of Barcelona, University of Barcelona, Barcelona, Spain

*Address all correspondence to: ffernand@clinic.cat

References

[1] González MJ, Urizar E, Urtaran-Laresgoiti M, Nuño-Solinís R, Lázaro-Pérez E, Vázquez L, et al. Hospital and outpatient models for hematopoietic stem cell transplantation: A systematic review of comparative studies for health outcomes, experience of care and costs. PLoS One. 2021;**16**(8):e0254135. DOI: 10.1371/journal.pone.0254135

[2] The Pharmaceutical Journal. Haematopoietic stem cell transplant in children. Pharmaceutical Journal. 2021;**306**(7945):1-27. DOI: 10.1211/PJ.2020.20208648

[3] Fernández-Avilés F, Carreras E, Urbano-Ispizua A, Rovira M, Martínez C, Gaya A, et al. Case-control comparison of at-home to total hospital care for autologous stem-cell transplantation for hematologic malignancies. Journal of Clinical Oncology. 2006;**24**(30):4855-4861. DOI: 10.1200/JCO.2006.06.4238

[4] Sung AD, Giri VK, Tang H, Nichols KR, Lew MV, Bohannon L, et al. Home-based hematopoietic cell transplantation in the United States. Transplantation and Cellular Therapy. 2022;**28**(4):207.e1-207.e8. DOI: 10.1016/j.jtct.2022.01.015

[5] Gutiérrez-García G, Rovira M, Arab N, Gallego C, Sánchez J, Ángeles Álvarez M, et al. A reproducible and safe at-home allogeneic haematopoietic cell transplant program: First experience in central and southern Europe. Bone Marrow Transplantation. 2020;**55**(5):965-973. DOI: 10.1038/s41409-019-0768-x

[6] Ringdén O, Sadeghi B, Moretti G, Finnbogadottir S, Eriksson B, Mattsson J, et al. Long-term outcome in patients treated at home during the pancytopenic phase after allogeneic haematopoietic stem cell transplantation. International Journal of Hematology. 2018;**107**(4):478-485. DOI: 10.1007/s12185-017-2363-5

[7] Svahn BM, Remberger M, Myrbäck KE, Holmberg K, Eriksson B, Hentschke P, et al. Home care during the pancytopenic phase after allogeneic hematopoietic stem cell transplantation is advantageous compared with hospital care. Blood. 2002;**100**(13):4317-4324. DOI: 10.1182/blood-2002-03-0801

[8] Holbro A, Ahmad I, Cohen S, Roy J, Lachance S, Chagnon M, et al. Safety and cost-effectiveness of outpatient autologous stem cell transplantation in patients with multiple myeloma. Biology of Blood and Marrow Transplantation. 2013;**19**(4):547-551. DOI: 10.1016/j.bbmt.2012.12.006

[9] Jagannath S, Vesole DH, Zhang M, Desikan KR, Copeland N, Jagannath M, et al. Feasibility and cost-effectiveness of outpatient autotransplants in multiple myeloma. Bone Marrow Transplantation. 1997;**20**(6):445-450. DOI: 10.1038/sj.bmt.1700900

[10] Kassar M, Medoff E, Seropian S, Cooper DL. Outpatient high-dose melphalan in multiple myeloma patients. Transfusion. 2007;**47**:115-119. DOI: 10.1111/j.1537-2995.2007.01073.x

[11] Landau HJ, Orlando E, Rodriguez ES, Applebaum A, Mitchell HR, Peled JU, et al. Pilot trial of homebound hematopoietic cell transplantation. Transplantation and Cellular Therapy. 2022;**28**(12):832.e1-832.e7. DOI: 10.1016/j.jtct.2022.09.014

[12] Marini J, Maldonado A, Weeda E, Neppalli A, Hashmi H,

Edwards K. Effectiveness, safety, and cost implications of outpatient autologous hematopoietic stem cell transplant for multiple myeloma. Hematology/Oncology and Stem Cell Therapy. 2023;**16**(4):351-357. DOI: 10.56875/2589-0646.1038

[13] Mcdiarmid S, Hutton B, Atkins H, Bence-Bruckler I, Bredeson C, Sabri E, et al. Performing allogeneic and autologous hematopoietic SCT in the outpatient setting: Effects on infectious complications and early transplant outcomes. Bone Marrow Transplantation. 2010;**45**(7):1220-1226. DOI: 10.1038/bmt.2009.330

[14] Cantú-Rodríguez OG, Sánchez-Cárdenas M, Treviño-Montemayor OR, Gutiérrez-Aguirre CH, Tarín-Arzaga L, Jaime-Pérez JC, et al. Impact of outpatient non-myeloablative haematopoietic stem cell transplantation in quality of life vs. conventional therapy. Psychology, Health & Medicine. 2016;**21**(1):10-19. DOI: 10.1080/13548506.2015.1054843

[15] Colunga-Pedraza PR, Gómez-De León A, Rodríguez-Roque CS, Morcos-Sandino M, Colunga-Pedraza JE, Cantú-Rodriguez OG, et al. Outpatient Haploidentical stem cell transplantation using post-transplant cyclophosphamide is safe and feasible. Transplantation and Cellular Therapy. 2021;**27**(3):259.e1-259.e6. DOI: 10.1016/j.jtct.2020.12.006

[16] Singhal S, Saadeh SS, Durani U, Kansagra A, Alkhateeb HB, Shah MV, et al. Allogeneic hematopoietic stem cell transplantation in the outpatient setting: The Mayo clinic experience. Transplantation and Cellular Therapy. 2023;**29**(3):183.e1-183.e6. DOI: 10.1016/j.jtct.2022.12.016

[17] Kharfan-Dabaja MA, Roy V, Murthy H, Fischer D, Mohty R, Greathouse A, et al. Post-autologous hematopoietic cell transplant care in the "home sweet home" setting: A treatment paradigm shift. Hematology/Oncology and Stem Cell Therapy. 2023;**16**(4):407-411. DOI: 10.56875/2589-0646.1088

[18] Martino M, Montanari M, Ferrara F, Ciceri F, Scortechini I, Palmieri S, et al. Very low rate of readmission after an early discharge outpatient model for autografting in multiple myeloma patients: An Italian Multicenter retrospective study. Biology of Blood and Marrow Transplantation. 2014;**20**(7):1026-1032. DOI: 10.1016/j.bbmt.2014.03.027

[19] Paul TM, Liu SV, Chong EA, Luger SM, Porter DL, Schuster SJ, et al. Outpatient autologous stem cell transplantation for patients with myeloma. Clinical Lymphoma, Myeloma & Leukemia. 2015;**15**(9):536-540. DOI: 10.1016/j.clml.2015.05.006

[20] Anastasia A, Giglio F, Mazza R, Sarina B, Todisco E, Bramanti S, et al. Early discharge after high-dose melphalan and peripheral blood stem cell reinfusion in patients with hematological and non-hematological disease. Leukemia & Lymphoma. 2009;**50**(1):80-84. DOI: 10.1080/10428190802535098

[21] Clemmons AB, Anderegg S. Mixed outpatient-inpatient autologous stem cell transplant for multiple myeloma: A cost-saving initiative in a resource constrained environment. Journal of Oncology Pharmacy Practice. 2017;**23**(5):384-388. DOI: 10.1177/1078155216639753

[22] Morabito F, Martino M, Stelitano C, Oliva E, Kropp M, Irrera G, et al. Feasibility of a mixed inpatient-outpatient model of peripheral blood stem cell transplantation for multiple myeloma. Haematologica. 2002;**87**(11):1192-1199

[23] Solomon SR, Matthews RH, Barreras AM, Bashey A, Manion KL, McNatt K, et al. Outpatient myeloablative Allo-SCT: A comprehensive approach yields decreased hospital utilization and low TRM. Bone Marrow Transplantation. 2010;**45**(3):468-475. DOI: 10.1038/bmt.2009.234

[24] Martínez-Cibrian N, Magnano L, Gutiérrez-García G, Andrade X, Correa JG, Suárez-Lledó M, et al. At-home autologous stem cell transplantation in multiple myeloma with and without G-CSF administration: A comparative study. Bone Marrow Transplantation. 2016;**51**(4):593-595. DOI: 10.1038/bmt.2015.287

[25] Rodríguez-Lobato LG, Martínez-Roca A, Castaño-Díez S, Palomino-Mosquera A, Gutiérrez-García G, Pedraza A, et al. The avoidance of G-CSF and the addition of prophylactic corticosteroids after autologous stem cell transplantation for multiple myeloma patients appeal for the at-home setting to reduce readmission for neutropenic fever. PLoS One. 2020;**15**(11):e0241778. DOI: 10.1371/journal.pone.0241778

[26] Ruiz-Boy S, Pedraza A, Prat M, Salas MQ, Carcelero E, Riu-Viladoms G, et al. At-home Foscarnet Administration in Patients with cytomegalovirus infection post-allogeneic stem cell transplantation: A Unicentric, safe, and feasible program. Pharmaceuticals (Basel). 2023;**16**(12):1741. DOI: 10.3390/ph16121741

[27] Cantú-Rodríguez OG, Jaime-Pérez JC, Gutiérrez-Aguirre CH, González-Llano O, Mancías-Guerra C, Tarín-Arzaga LC, et al. Outpatient allografting using non-myeloablative conditioning: The Mexican experience. Bone Marrow Transplantation. 2007;**40**(2):119-123. DOI: 10.1038/sj.bmt.1705700

[28] Fernandez-Avilés F, Gutiérrez-García G. At-home HSCT. In: Carreras E, Dufour C, Mohty M, Kröger N, editors. The EBMT Handbook: Hematopoietic Stem Cell Transplantation and Cellular Therapies. 7th ed. Berlin: Springer; 2018. pp. 467-471. DOI: 10.1007/978-3-030-02278-5_63

[29] Martino M, Ciavarella S, De Summa S, Russo L, Meliambro N, Imbalzano L, et al. A comparative assessment of quality of life in patients with multiple myeloma undergoing autologous stem cell transplantation through an outpatient and inpatient model. Biology of Blood and Marrow Transplantation. 2018;**24**(3):608-613. DOI: 10.1016/j.bbmt.2017.09.021

[30] Summers N, Dawe U, Stewart DA. A comparison of inpatient and outpatient ASCT. Bone Marrow Transplantation. 2000;**26**(4):389-395. DOI: 10.1038/sj.bmt.1702534

[31] Schulmeister L, Quiett K, Mayer K. Quality of life, quality of care, and patient satisfaction: Perceptions of patients undergoing outpatient autologous stem cell transplantation. Oncology Nursing Forum. 2005;**32**(1):57-67. DOI: 10.1188/05.ONF.57-67

[32] Gonzalez-Ryan L, Haut PR, Coyne K, Van SK, Duerst R, Haro D, et al. Developing a pediatric outpatient transplantation program. The children's memorial hospital experience. Frontiers in Bioscience. 2001;**6**:G1-G5. DOI: 10.2741/ryan

[33] Koo J, Silverman S, Nuechterlein B, Keating AK, Verneris MR, Foreman NK, et al. Safety and feasibility of outpatient autologous stem cell transplantation in pediatric patients with primary central nervous system tumors. Bone Marrow Transplantation. 2019;**54**(10):1605-1613. DOI: 10.1038/s41409-019-0479-3

[34] Jaime-Pérez JC, González-Treviño M, Meléndez-Flores JD, Ramos-Dávila EM, Colunga-Pedraza JE, Colunga-Pedraza PR, et al. Safety and feasibility of outpatient hematopoietic cell transplantation in pediatric patients during the COVID-19 pandemic: A single-center experience. Pediatric Blood & Cancer. 2021;**68**(12):e29252. DOI: 10.1002/pbc.29252

[35] Clemmons AB, Alexander M, DeGregory K, Kennedy LA. The hematopoietic cell transplant pharmacist: Roles, responsibilities, and recommendations from the ASBMT pharmacy special interest group. Biology of Blood and Marrow Transplantation. 2018;**24**(5):914-922. DOI: 10.1016/j.bbmt.2017.12.803

[36] Moreno-Martínez ME, Riba M, García-Cadenas I, Esquirol A, Yusta M, Redondo S, et al. Optimization of a home hospitalization program for hematopoietic stem cell transplantation with ehealth integration and clinical pharmacist involvement. Frontiers in Immunology. 2024;**15**:1397115. DOI: 10.3389/fimmu.2024.1397115

[37] Holle LM, Segal EM, Jeffers KD. The expanding role of the oncology pharmacist. Pharmacy (Basel). 2020;**8**(3):130. DOI: 10.3390/pharmacy8030130

[38] Segal EM, Bates J, Fleszar SL, Holle LM, Kennerly-Shah J, Rockey M, et al. Demonstrating the value of the oncology pharmacist within the healthcare team. Journal of Oncology Pharmacy Practice. 2019;**25**(8):1945-1967. DOI: 10.1177/1078155219859424

[39] Peinovich M, Darracott R, Dow J. Developing pharmacy services in a home hospital program: The Mayo Clinic experience. American Journal of Health-System Pharmacy. 2022;**79**(21):1925-1928. DOI: 10.1093/ajhp/zxac200

Chapter 9

The Role of Long Noncoding RNAs in Gene Expression and Epigenetic Regulation in Spermatogonial Stem Cells and Therapeutic Implications

Danial Hashemi Karoii and Zahra Hasani Mahforoozmahalleh

Abstract

Spermatogonial stem cells (SSCs) play a pivotal role in male germ cell development and are essential for the continuous production of spermatozoa throughout a male's reproductive life. The regulation of gene expression in SSCs is a highly intricate process crucial for maintaining spermatogenesis. Long noncoding RNAs (lncRNAs) have emerged as significant players in the epigenetic and transcriptional regulation of gene expression. This abstract provides an overview of the current state of knowledge regarding the involvement of lncRNAs in the regulation of gene expression in spermatogonial stem cells and explores the potential therapeutic applications associated with targeting lncRNAs in the context of male infertility and reproductive health. This review delves into the mechanisms by which lncRNAs influence gene expression in SSCs, including their involvement in chromatin remodeling, transcriptional regulation, and post-transcriptional modifications. The intricate network of interactions between lncRNAs and protein-coding genes is highlighted, showcasing their impact on the delicate balance required for SSC self-renewal and differentiation. Furthermore, the potential therapeutic interventions targeting lncRNAs for the treatment of male infertility and reproductive disorders are discussed, offering insight into the future directions of research in this field.

Keywords: spermatogonial stem cells, long noncoding RNAs, differentiation, infertility, spermatogenesis

1. Introduction

The regulation of gene expression in spermatogonial stem cells (SSCs) is a complex and essential process that is vital for the continuous production of sperm throughout a male's reproductive life [1]. Spermatogenesis, the process of sperm cell development, relies on the precise control of when and how genes are turned on or off within these specialized cells [2, 3]. The regulation of gene expression in spermatogonial stem cells is a multifaceted process that is essential for maintaining spermatogenesis. Long noncoding RNAs have emerged as significant players in this process, exerting their influence through epigenetic modifications and transcriptional control [4]. This intricate web of interactions underscores the importance of lncRNAs in ensuring the proper development and maintenance of sperm, with far-reaching implications for male fertility and reproductive health. The study of lncRNAs in SSCs represents an exciting avenue of research that could potentially revolutionize our understanding of male reproductive biology and pave the way for innovative therapeutic interventions [5].

Long noncoding RNAs (lncRNAs) are a type of genetic material that, unlike protein-coding genes, do not provide instructions for making proteins. Instead, lncRNAs were initially thought to have no functional role, but research has shown that they play a significant role in controlling gene expression [6, 7]. In SSCs, these lncRNAs are key players in the regulation of gene expression, particularly in two major ways.

2. Epigenetic regulation

LncRNAs are involved in epigenetic modifications, which are changes that occur in the DNA or the proteins associated with DNA (histones) that can influence whether a gene is active or silent [8]. Epigenetic changes can be inherited from one cell generation to the next and are crucial for the long-term maintenance of gene activity. LncRNAs can interact with proteins and guide them to specific genomic locations, where they can modify the epigenetic marks, thereby influencing whether nearby genes are turned on or off [9].

2.1 Transcriptional regulation

LncRNAs also participate in controlling the transcription of genes. Transcription is the process by which genetic information is used to produce RNA molecules (messenger RNAs) that are later translated into proteins. LncRNAs can interact with the machinery involved in transcription, either promoting or inhibiting the production of RNA from specific genes [10]. In this way, lncRNAs can directly impact, which genes are actively producing proteins within SSCs.

Promotion of Transcription: Some lncRNAs act as facilitators, encouraging the transcription machinery to bind to specific gene promoters, which then leads to increased mRNA production. This can activate genes necessary for SSC self-renewal or differentiation, depending on the context.

2.2 Inhibition of transcription

Conversely, some lncRNAs act as repressors. They can prevent the transcription machinery from binding to gene promoters, thereby decreasing the production of specific mRNAs [11]. This repression can silence genes that should not be expressed in SSCs under certain conditions.

2.3 Implications for male fertility and reproductive health

Understanding the intricate interplay between lncRNAs and gene expression regulation in SSCs has profound implications for male fertility and reproductive health [12]. Any disruption in this delicate balance can result in infertility or other reproductive issues. Investigating how lncRNAs influence SSC behavior and sperm production can lead to potential therapeutic strategies for male infertility. Targeting specific lncRNAs or the molecular pathways, the influence could be a novel approach to enhance fertility or address reproductive disorders [5].

3. Regulation of gene expression in spermatogonial stem cells

SSCs serve as the foundation for male germ cell development and sperm production. The regulation of gene expression in these cells is central to maintaining the delicate balance between SSC self-renewal and differentiation into sperm [13–17]. LncRNAs have gained prominence as regulators in this process, contributing to the epigenetic and transcriptional control of gene expression.

3.1 Epigenetic regulation

LncRNAs are involved in the epigenetic control of gene expression through their ability to interact with chromatin-modifying enzymes, such as histone modifiers and DNA methyltransferases [6]. By guiding these enzymes to specific genomic loci, lncRNAs influence the addition or removal of epigenetic marks on DNA and histones. This dynamic interplay leads to the activation or repression of nearby genes and plays a pivotal role in maintaining the SSC population and directing differentiation [18].

3.2 Transcriptional regulation

LncRNAs are also involved in the transcriptional regulation of genes in SSCs. Some lncRNAs facilitate transcription by promoting the binding of the transcription machinery to gene promoters, while others inhibit transcription by preventing such binding. This intricate orchestration of gene expression ensures the proper function of SSCs.

4. Potential therapeutic applications in male infertility and reproductive health

The comprehension of lncRNA-mediated gene regulation in SSCs carries substantial therapeutic potential, particularly in the context of male infertility and reproductive health. The study of lncRNAs provides insights into new strategies to address fertility-related issues [19].

4.1 Therapeutic targeting of LncRNAs

By understanding the roles of specific lncRNAs in SSC gene regulation, researchers can potentially design therapeutic interventions aimed at modulating the expression or activity of these lncRNAs [6]. This could be pivotal in restoring or enhancing male fertility in cases where dysregulation of gene expression within SSCs is implicated in infertility.

4.2 Reproductive health advancements

Insights into lncRNA involvement in SSC function may lead to advancements in reproductive health beyond addressing infertility. The potential to influence SSC behavior through lncRNA-targeted strategies opens up possibilities for managing various reproductive disorders and optimizing sperm production [20].

5. Mechanisms by which lncRNAs influence gene expression in SSCs, including their involvement in chromatin remodeling, transcriptional regulation, and post-transcriptional modifications

In an academic context, elucidating the mechanisms through which lncRNAs exert their influence on gene expression in SSCs is paramount to comprehending the intricate regulatory processes at play. This encompasses their participation in chromatin remodeling, transcriptional regulation, and post-transcriptional modifications.

5.1 Chromatin remodeling

Chromatin, a complex of DNA and associated proteins, can be in a condensed or open state, influencing the accessibility of genes for transcription. LncRNAs play a significant role in modulating chromatin structure in SSCs, thus impacting gene expression.

Guiding Epigenetic Modifiers: LncRNAs interact with proteins responsible for epigenetic modifications, such as DNA methyltransferases and histone modifiers. They guide these proteins to specific genomic loci, where they introduce or remove epigenetic marks on DNA and histones. These modifications can either promote or inhibit gene expression by dictating whether the associated genes are in a transcriptionally permissive or repressive state [21].

Establishing Enhancer-Promoter Interactions: Some lncRNAs facilitate long-range interactions between enhancers (regulatory regions) and gene promoters by tethering them together. This proximity enhances the likelihood of gene activation, as enhancers are brought in closer proximity to the genes they regulate. Thus, lncRNAs can influence the expression of genes that are important for SSC self-renewal and differentiation [22].

5.2 Transcriptional regulation

Transcription is the process by which genetic information in DNA is transcribed into RNA, which can subsequently be translated into proteins. LncRNAs are involved in transcriptional regulation in SSCs through various mechanisms.

Enhancing Transcription: Some lncRNAs act as transcriptional activators by promoting the binding of RNA polymerase and transcription factors to gene promoters. This enhances the initiation and progression of transcription, leading to increased mRNA production from specific genes. In SSCs, these lncRNAs can facilitate the expression of genes crucial for SSC functions [23].

Repressing Transcription: Conversely, some lncRNAs function as transcriptional repressors by preventing the binding of RNA polymerase and transcription factors to gene promoters. This inhibition decreases the transcription of specific genes. In SSCs, these lncRNAs can silence genes that should remain inactive under certain conditions [24].

5.3 Post-transcriptional modifications

Post-transcriptional modifications encompass processes that occur after RNA is transcribed but before it is translated into a protein. LncRNAs can participate in these modifications, further influencing gene expression in SSCs.

RNA Splicing: LncRNAs can interact with pre-messenger RNAs (pre-mRNAs) and influence their splicing patterns. Alternative splicing can generate diverse mRNA isoforms from a single gene, which can lead to the production of different protein variants. LncRNA-mediated splicing regulation can thus affect the repertoire of proteins expressed in SSCs.

RNA Stability and Translation: LncRNAs can also interact with mRNAs to regulate their stability and translation efficiency. By influencing the degradation rate or translational potential of specific mRNAs, lncRNAs can fine-tune the expression of critical genes in SSCs. lncRNAs, in SSCs, play a multifaceted role in the regulation of gene expression by participating in chromatin remodeling, transcriptional control, and post-transcriptional modifications [25]. Their ability to influence these key regulatory processes underscores their significance in orchestrating the precise gene expression patterns necessary for SSC self-renewal, differentiation, and spermatogenesis. A thorough understanding of these mechanisms is pivotal for unraveling the intricacies of male reproductive biology and exploring potential therapeutic interventions in male fertility and reproductive health.

6. The intricate network of interactions between lncRNAs and protein-coding genes is highlighted, showcasing their impact on the delicate balance required for SSC self-renewal and differentiation

The intricate network of interactions between lncRNAs and protein-coding genes is a focal point of study, emphasizing their influence on the delicate equilibrium essential for spermatogonial stem cell (SSC) self-renewal and differentiation.

6.1 Regulatory interplay between LncRNAs and protein-coding genes

Binding Partners: LncRNAs are molecules that can interact with various cellular components, including protein-coding genes. These interactions may involve direct physical associations or more complex regulatory mechanisms mediated through other cellular factors. These interconnections are dynamic and context-dependent, with specific lncRNAs engaging with particular protein-coding genes in response to developmental or environmental cues [26].

Cis- and Trans-Regulation: LncRNAs can exert their influence on protein-coding genes in a manner that is either “cis” or “trans.” In cis-regulation, a lncRNA acts on neighboring genes on the same chromosome, modulating their expression. In trans-regulation, lncRNAs can affect genes located on different chromosomes or at distant genomic loci [10].

6.2 Impact on SSC self-renewal and differentiation

Self-Renewal: SSCs possess the remarkable ability to self-renew, producing identical copies of themselves. This process requires a balanced gene expression profile to maintain the stem cell population. The intricate network of interactions between lncRNAs and protein-coding genes contributes to this balance. LncRNAs may promote the expression of genes associated with self-renewal while repressing those linked to differentiation. This ensures the continual existence of the SSC pool [27].

Differentiation: In contrast, differentiation involves the transformation of SSCs into sperm cells, necessitating the activation of specific genes responsible for sperm production. The same network of interactions between lncRNAs and protein-coding genes is also involved in controlling this process. Here, lncRNAs may facilitate the expression of differentiation-related genes while suppressing self-renewal-associated genes, guiding SSCs toward a differentiated state [28].

6.3 Delicate balance

Maintenance of Pluripotency: The dual roles played by lncRNAs in both promoting self-renewal and facilitating differentiation underscore their role in preserving SSC pluripotency, the capacity to give rise to different cell types. It is this balance that allows SSCs to persist and fulfill their critical roles in spermatogenesis.

Dynamic Regulation: The interactions between lncRNAs and protein-coding genes are highly dynamic and context-dependent. They respond to signals from the SSC microenvironment, hormonal cues, and other external factors, allowing SSCs to adapt and respond to changing physiological demands. The intricate network of interactions between lncRNAs and protein-coding genes is a focal point of investigation, with a particular focus on their impact on the delicate equilibrium essential for spermatogonial stem cell self-renewal and differentiation [29]. The precise coordination of these interactions is crucial for maintaining the pool of SSCs while enabling their differentiation into sperm, a process that is fundamental to male fertility and reproductive health. A comprehensive understanding of these regulatory mechanisms has the potential to shed light on male reproductive biology and the development of targeted interventions for male infertility and reproductive disorders.

7. Potential therapeutic interventions targeting lncRNAs for the treatment of male infertility and reproductive disorders

In an academic context, the exploration of potential therapeutic interventions targeting lncRNAs for the treatment of male infertility and reproductive disorders represents a promising area of research. This academic explanation highlights the rationale, strategies, and implications of such therapeutic approaches.

7.1 Rationale for targeting LncRNAs in male infertility and reproductive disorders

LncRNAs as Regulatory Hubs: LncRNAs have emerged as pivotal regulators of gene expression in the male reproductive system. Their multifaceted roles in spermatogenesis and fertility-related processes make them attractive candidates for therapeutic intervention.

Genetic and Epigenetic Dysregulation: Aberrant expression or function of specific lncRNAs has been associated with male infertility and reproductive disorders. These lncRNAs may be linked to disrupted spermatogenesis, impaired sperm production, or other fertility-related issues, highlighting their potential as therapeutic targets [29].

Specificity and Precision: Targeting lncRNAs offers a level of specificity not always achievable with traditional therapeutic approaches. By focusing on the molecular players directly involved in gene regulation, it is possible to modulate gene expression in a highly targeted manner [30].

7.2 Strategies for targeting LncRNAs

Antisense Oligonucleotides (ASOs): ASOs are synthetic nucleic acid sequences designed to complement and bind to specific lncRNAs, thereby preventing their interaction with target genes or other regulatory elements. ASOs can be used to inhibit or degrade lncRNAs with aberrant or detrimental functions [6].

Small Molecules: Small molecule compounds that can disrupt the structure or function of lncRNAs are under investigation. These molecules may interfere with lncRNA-protein interactions, thereby modulating gene expression.

7.3 Genome editing technologies

Emerging genome editing tools, such as CRISPR-Cas systems, can be adapted to target lncRNAs for manipulation. This approach allows for the precise modification or deletion of lncRNA-encoding genomic regions.

7.4 RNA interference (RNAi)

RNAi technology employs small RNA molecules to trigger the degradation of target RNAs, including lncRNAs. It can be used to downregulate specific lncRNAs that contribute to reproductive disorders.

8. Investigation of lncRNA-mediated mechanisms in SSCs represents an exciting and promising area of research with the potential to revolutionize our understanding of male fertility

The investigation of lncRNA-mediated mechanisms in SSCs is an exciting and promising area of research with the potential to revolutionize our comprehension of male fertility. This academic explanation highlights the significance, potential implications, and avenues of impact associated with this research endeavor [12].

8.1 Significance of investigating LncRNA-mediated mechanisms in SSCs

Critical Role of SSCs in Male Fertility: SSCs are the foundation of male germ cell development and sperm production. Their ability to self-renew and differentiate is fundamental for the continuous production of sperm throughout a male's reproductive life. Any disruptions or dysregulations in SSC behavior can lead to male infertility.

Emerging Role of LncRNAs: LncRNAs have recently emerged as key regulators of gene expression, including that of genes governing SSC self-renewal and differentiation. They play essential roles in epigenetic, transcriptional, and post-transcriptional gene regulation, all of which are critical for SSC functions [27].

8.2 Potential implications of investigating LncRNA-mediated mechanisms

Precision and Targeted Therapies: Understanding the mechanisms through which lncRNAs influence SSCs could pave the way for the development of precise and targeted therapeutic interventions. By modulating specific lncRNAs, it may be possible to influence SSC behavior and restore or enhance male fertility in cases where gene expression dysregulation is implicated in infertility.

Deeper Understanding of Male Reproductive Biology: Investigating lncRNA-mediated mechanisms in SSCs provides a more comprehensive picture of the intricate regulatory processes underpinning male reproductive biology. This knowledge can inform not only potential therapies but also our fundamental understanding of the biological mechanisms governing male fertility.

Uncovering Novel Biomarkers: The identification of specific lncRNAs associated with healthy SSC function or infertility could yield novel biomarkers for assessing male fertility and reproductive health. These biomarkers could aid in the diagnosis and prognosis of fertility-related conditions [12].

8.3 Potential for a paradigm shift in male reproductive health

As researchers delve deeper into the roles of lncRNAs in SSCs, new research avenues and questions arise. These investigations expand our knowledge of male reproductive health, shedding light on previously unexplored areas. The outcomes of this research have the potential to revolutionize male fertility research and, by extension, clinical practice. Innovative approaches, grounded in an understanding of lncRNA-mediated mechanisms, could transform the treatment of male infertility and related disorders.

9. Hub lncRNAs involved in SSCs

9.1 LncRNA H19

LncRNA H19 is a well-characterized and extensively studied long noncoding RNA (lncRNA) that plays a crucial role in various biological processes, including those related to SSCs. H19 is a maternally expressed gene that was initially identified as an imprinted gene in mammals and is typically highly expressed during embryonic development but downregulated in most adult tissues [31]. While its exact role in SSCs is still the subject of ongoing research, several aspects of its functions and potential implications have been explored:

Regulation of SSC Self-Renewal and Differentiation: H19 has been associated with the control of SSC self-renewal and differentiation. The balance between these two processes is crucial for the continuous production of sperm throughout a male's reproductive life. H19 has been found to interact with various protein-coding genes and regulatory elements to influence this balance. It may promote self-renewal by suppressing differentiation-related genes or vice versa, depending on the cellular context and environmental cues.

Epigenetic Regulation: H19 is known to have epigenetic regulatory functions. It can interact with chromatin-modifying proteins and guide them to specific genomic loci, impacting the addition or removal of epigenetic marks on DNA and histones. These epigenetic modifications can affect the accessibility of genes for transcription, influencing the overall gene expression profile in SSCs.

Interaction with Imprinted Genes: H19 is part of an imprinted gene cluster that also includes the insulin-like growth factor 2 (IGF2) gene. This cluster is involved in growth and development. H19 and IGF2 are reciprocally imprinted, meaning only one of them is actively expressed from the allele inherited from a specific parent, while the other is silenced. This imprinted expression pattern can influence growth and development, including the development of germ cells [32]. Beyond its role in SSCs, H19 has been associated with various cancer types, where it often acts as a tumor suppressor or oncogene, depending on the context. Its functions in SSCs may overlap with its roles in cancer biology, as it regulates cell proliferation and differentiation [33].

Potential Therapeutic Target: Given its involvement in SSC regulation, H19 has been explored as a potential therapeutic target in the context of male infertility and reproductive health. By understanding how H19 influences SSC behavior, researchers aim to develop strategies to modulate its expression or function, with the goal of ameliorating male reproductive issues. LncRNA H19 is a multifaceted lncRNA that plays a significant role in the regulation of gene expression and epigenetic control, with implications for SSC self-renewal and differentiation [34]. Its intricate involvement in the balance of these processes makes it a key player in male fertility and spermatogenesis, and further research is expected to unveil additional details about its precise mechanisms and potential therapeutic applications.

10. Signaling pathway, biological process, and molecular functions in LncRNA H19

Signaling pathway shows that LncRNA H19 can contribute to "EGFR tyrosine kinase inhibitor resistance," "Endocrine resistance," "MicroRNAs in cancer," and "HIF-1 signaling pathway" (**Figure 1B**). Enrich tool analysis revealed that LncRNA H19 is enriched in 3 gene ontology (GO) terms. Functional enrichment analysis demonstrated that the biological process (BP) terms "GO:0045596 negative reg. of cell differentiation," "GO:0008284 positive reg. of cell population proliferation," "GO:0051093 negative reg. of developmental proc," and "GO:0035239 tube morphogenesis" were significantly involved in spermatogonia stem cell (**Figure 1C**). Molecular function (MF) investigation of LncRNA H19 highlighted "GO:0001221 transcription coregulator binding," "GO:0140297 DNA-binding transcription factor binding," "GO:0008134 transcription factor binding," and "GO:0004712 protein serine/threonine/tyrosine kinase activity" (**Figure 1D**).

10.1 LncRNA TDRG1

Long noncoding RNA TDRG1, denoting Testis development-related gene 1, has emerged as a pivotal molecular component in the intricate landscape of SSCs and male reproductive biology. As research in this field progresses, an in-depth examination of TDRG1 reveals its known functions and potential implications [5]. A central role of TDRG1 is its participation in the regulation of SSC self-renewal and differentiation, two fundamental facets of male fertility. SSCs are responsible for maintaining the population of germ cells and ensuring the continuous production of sperm throughout a male's lifespan. Maintaining the delicate balance between self-renewal and differentiation is a finely tuned process, and TDRG1 plays a role in orchestrating this equilibrium [13]. TDRG1 has been shown to interact with specific protein-coding genes and signaling pathways within SSCs. Depending on the specific context, it can promote either self-renewal or differentiation by modulating gene expression. In doing so, TDRG1 contributes to the regulation of SSC fate, ensuring that a subset of cells continues to self-renew to sustain the stem cell pool, while others embark on the path of differentiation to produce sperm. The high degree of tissue-specific expression exhibited by TDRG1 places it predominantly within the testes. This testis-specific expression implies a significant role in testicular development and spermatogenesis, the process of sperm cell production. It is reasonable to infer that TDRG1 plays a pivotal role during key developmental stages and in adult testicular tissues. Throughout testicular development, TDRG1 likely contributes to the formation and maintenance

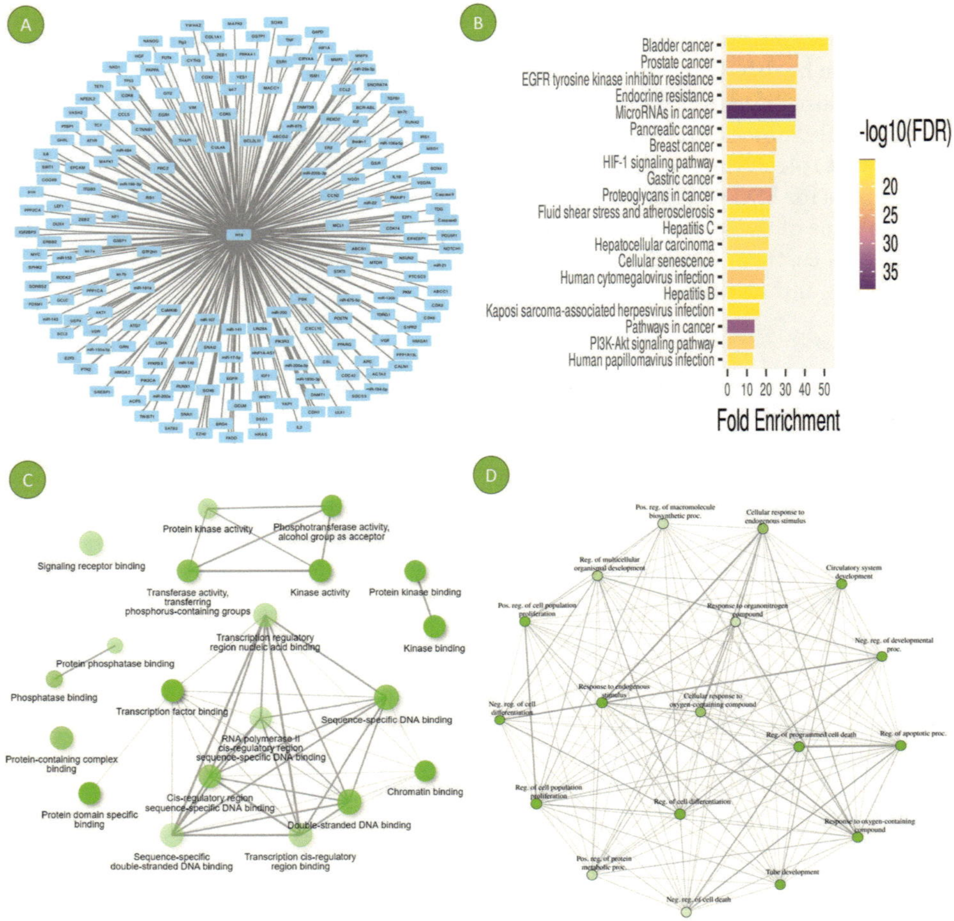

Figure 1.
LncRNA-mRNA-miRNA interaction, signaling pathway, biological process, and molecular functions in LncRNA H19 involved in spermatogonia stem cell.

of germ cells, which are paramount for fertility. Its role may encompass the regulation of genes crucial for proper testicular function, establishing it as a prominent actor in the intricate orchestration of male reproductive biology.

11. Epigenetic regulation and chromatin remodeling

Epigenetic modifications, including DNA methylation and histone modifications, wield a profound influence on gene regulation within SSCs. TDRG1 is believed to engage with these epigenetic processes, contributing to chromatin remodeling and influencing the accessibility of specific genes. Alterations in chromatin structure can have a substantial impact on the transcriptional activity of genes integral to SSC fate determination [35]. Through its interaction with the epigenetic landscape in SSCs, TDRG1 aids in directing the expression of genes vital for self-renewal, differentiation, and the overall process of spermatogenesis. The exclusive expression of TDRG1 within the testes, combined with its role in SSC regulation, positions it as a potential diagnostic marker for male infertility and reproductive disorders. Analyzing TDRG1 expression levels in clinical contexts offers valuable insights into the health and functionality of SSCs.

12. Therapeutic implications for male infertility

A comprehensive grasp of TDRG1's precise mechanisms and its role in SSC biology could potentially lead to novel therapeutic strategies in the realm of male infertility and reproductive health. Anomalous TDRG1 expression or function may contribute to infertility, and the exploration of TDRG1 as a therapeutic target could offer a promising avenue for restoring or enhancing male fertility, particularly in cases where SSC dysfunction is identified as a contributing factor. TDRG1 stands as a compelling subject of research that has the potential to reshape our comprehension of male reproductive biology and introduce novel approaches for the diagnosis and treatment of male infertility and reproductive disorders. Its multifaceted functions within SSCs, combined with its tissue-specific expression and roles in epigenetic regulation, make it a captivating focus of study in the sphere of male fertility and reproductive health [36].

13. Signaling pathway, biological process, and molecular functions in TDRG1

Signaling pathway shows that TDRG1 can contribute to "MicroRNAs in cancer," "Chemical carcinogenesis-receptor activation," and "Focal adhesion" (**Figure 2B**). Enrich tool analysis revealed that TDRG1 is enriched in 4 GO terms. Functional enrichment analysis demonstrated that the BP term "GO:0045944 positive reg. of transcription by RNA polymerase II," "GO:0045595 reg. of cell differentiation," "GO:0045893 positive reg. of transcription DNA-templated" and "GO:1903508 positive reg. of nucleic acid-templated transcription" was significantly involved in spermatogonia stem cell (**Figure 2C**). MF investigation of TDRG1 highlighted "GO:0043426 MRF binding," "GO:0008083 growth factor activity," "GO:0008201 heparin binding," "GO:0001228 DNA-binding transcription activator activity RNA polymerase II-specif," "GO:0001216 DNA-binding transcription activator activity," and "GO:0048018 receptor ligand activity" (**Figure 2D**).

13.1 LncRNA Sirt1-AS

Long noncoding RNA Sirt1-AS, denoting Sirtuin 1 antisense RNA, is a compelling molecular component that has garnered attention in the context of gene regulation and cellular processes, including its implications within SSCs and male reproductive biology [37]. This detailed academic exposition delves into the intricacies of Sirt1-AS, its known functions, and the potential consequences in the framework of SSC biology.

Sirt1-AS serves as a pivotal regulator of gene expression within SSCs, where the delicate balance between self-renewal and differentiation underpins the continuous production of spermatozoa in males. SSCs, being the foundation of spermatogenesis, require precise gene expression regulation to fulfill their critical functions. Sirt1-AS is instrumental in orchestrating this regulation. Sirt1-AS interfaces with specific protein-coding genes and signaling pathways in SSCs. Depending on the specific context and cellular signals, it can either promote self-renewal by repressing genes associated with differentiation or enhance differentiation by inhibiting the expression of genes linked to self-renewal [38]. Epigenetic modifications, encompassing DNA methylation and

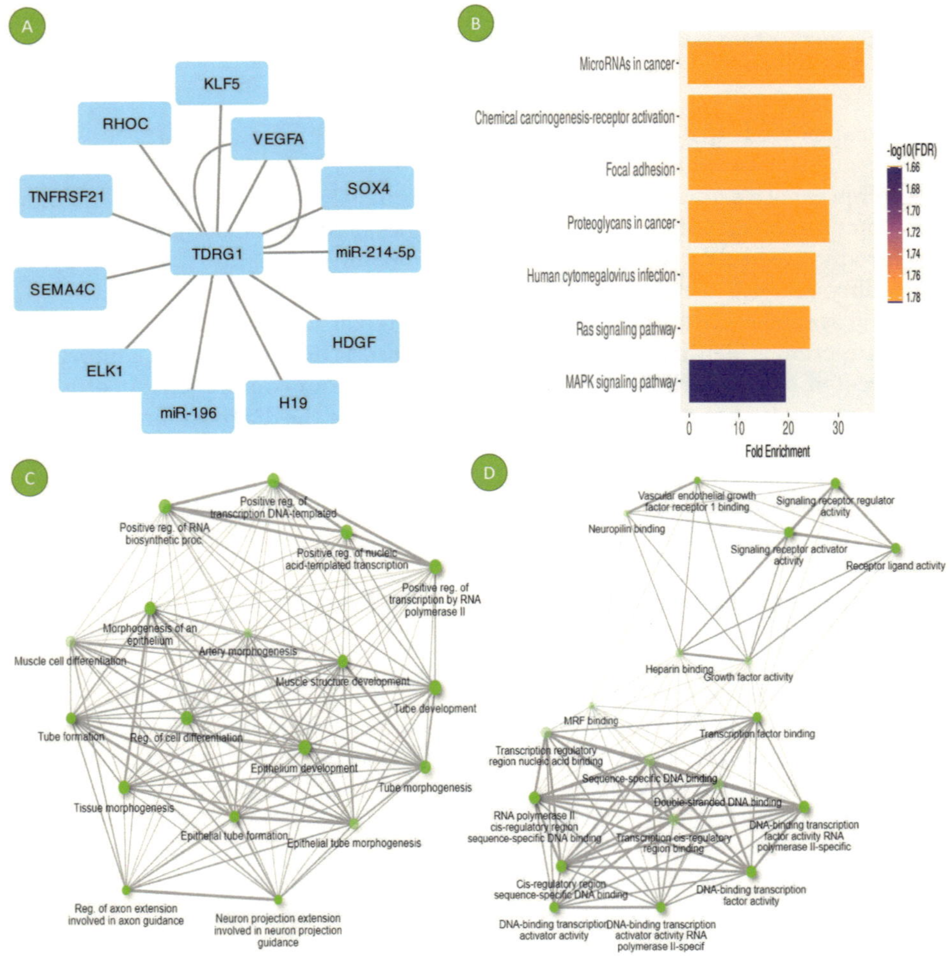

Figure 2.
LncRNA-mRNA-miRNA interaction, signaling pathway, biological process, and molecular functions in TDRG1 involved in spermatogonia stem cell.

histone modifications, exert a profound impact on SSC gene regulation. Sirt1-AS is intricately involved in these epigenetic processes, thus contributing to chromatin remodeling and influencing the accessibility of specific genes. This interaction with the epigenetic landscape profoundly shapes the transcriptional activity of genes central to SSC fate determination. By mediating changes in the chromatin structure, Sirt1-AS plays a pivotal role in guiding the expression of genes essential for self-renewal, differentiation, and the overarching process of spermatogenesis. Sirt1-AS is closely associated with the regulation of the Sirtuin 1 (Sirt1) gene, which is a vital determinant of SSC self-renewal and differentiation. Sirt1 functions as an NAD-dependent deacetylase and is implicated in various cellular processes, including those relevant to SSCs.

Sirt1-AS modulates SSC behavior by controlling the expression of Sirt1. The level of Sirt1 expression influences the balance between self-renewal and differentiation in SSCs, and Sirt1-AS serves as a key mediator in this regulatory process. A comprehensive understanding of Sirt1-AS's intricate mechanisms and its role in SSC biology holds promise for therapeutic interventions targeting male infertility and reproductive health. Dysregulation of Sirt1-AS or Sirt1 expression may contribute to male infertility, making these molecules potential therapeutic targets [39].

Exploring strategies to modulate Sirt1-AS expression or function, or targeting Sirt1 directly, may offer avenues for the restoration or enhancement of male fertility, particularly in cases where SSC dysfunction is a recognized contributing factor. Sirt1-AS represents a pivotal and multifaceted lncRNA in the realm of SSC biology, male fertility, and reproductive health. Its roles in gene expression modulation, epigenetic regulation, and interaction with the Sirt1 gene underscore its significance in orchestrating the fine balance between SSC self-renewal and differentiation. As research progresses, the insights gained into Sirt1-AS's regulatory mechanisms hold promise for innovative diagnostic and therapeutic approaches within the domain of male infertility and reproductive disorders [40].

13.2 LncRNA HOTAIR

Long noncoding RNA HOTAIR, denoting HOX transcript antisense intergenic RNA, is a prominent molecular player with profound implications in gene regulation and cellular processes. In this comprehensive academic elucidation, we delve into the intricacies of HOTAIR, its known functions, and the potential ramifications in various biological contexts:

HOTAIR is chiefly recognized for its role as an epigenetic regulator, particularly in the context of chromatin modification. It exerts its influence by interacting with chromatin-modifying complexes and guiding them to specific genomic loci. Through these interactions, it plays a pivotal role in altering the epigenetic landscape, particularly in relation to histone modifications and DNA methylation.

The epigenetic modifications orchestrated by HOTAIR are influential in shaping the chromatin structure and regulating the accessibility of specific genes. These changes profoundly impact gene expression and play a vital role in various cellular processes, including those related to spermatogonial stem cells (SSCs) and male reproductive biology. Within the complex regulatory network governing SSCs, HOTAIR is a critical contributor. SSCs have the remarkable ability to self-renew and differentiate, processes that are crucial for sustaining spermatogenesis and male fertility. HOTAIR's multifaceted role in epigenetic regulation is instrumental in governing this delicate balance [35].

HOTAIR is known to influence gene expression profiles in SSCs, favoring either self-renewal or differentiation, depending on the specific molecular and cellular context. It can promote self-renewal by repressing genes associated with differentiation and vice versa. In this manner, HOTAIR shapes the fate of SSCs, ensuring the equilibrium between self-renewal and differentiation is maintained. HOTAIR is intricately linked to the regulation of HOX genes, a family of genes that play a fundamental role in development, patterning, and differentiation. Specifically, HOTAIR is associated with the silencing of genes located within the HOXC cluster. The interaction between HOTAIR and HOX genes underscores its significance in modulating developmental processes, including those that pertain to germ cell development and spermatogenesis [41].

14. Signaling pathway, biological process, and molecular functions in HOTAIR

Signaling pathway shows that HOTAIR can contribute to "Endocrine resistance," "Cellular senescence," and "PI3K-Akt signaling pathway" (**Figure 3B**). Enrich tool

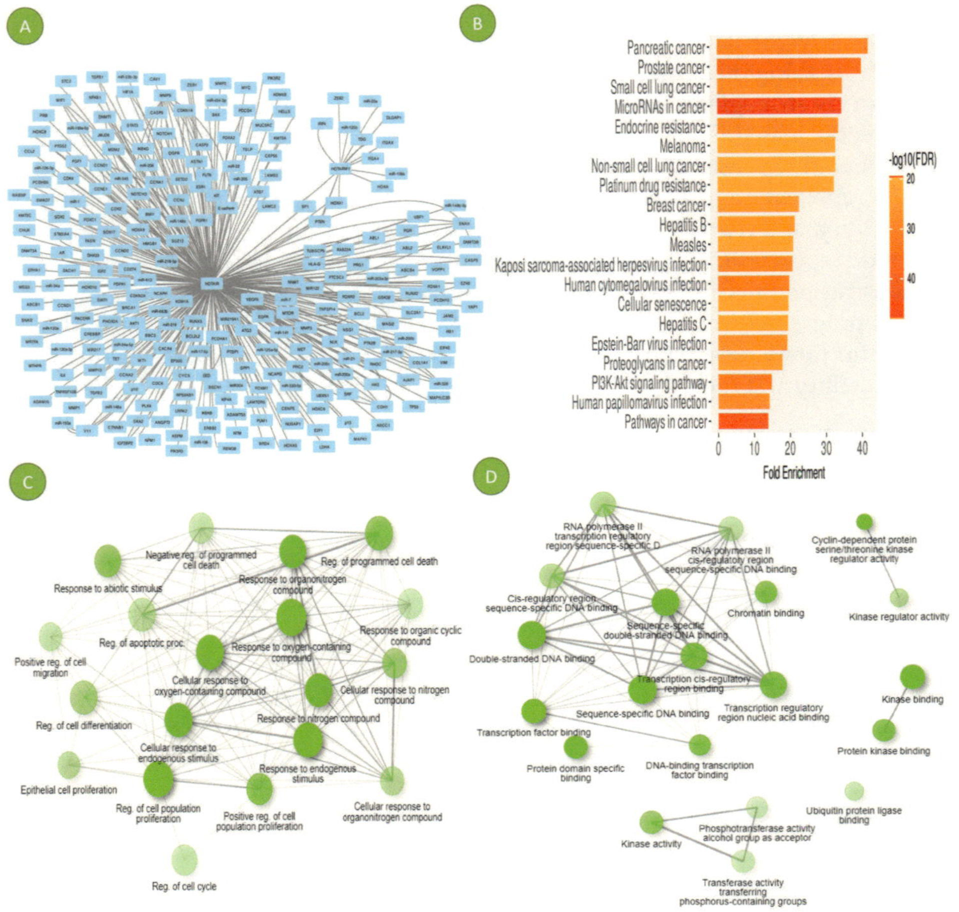

Figure 3.
LncRNA-mRNA-miRNA interaction, signaling pathway, biological process, and molecular functions in HOTAIR involved in spermatogonia stem cell.

analysis revealed that HOTAIR is enriched in 3 GO terms. Functional enrichment analysis demonstrated that the BP term "GO:0050673 epithelial cell proliferation," "GO:0030335 positive reg. of cell migration," "GO:0071417 cellular response to organonitrogen compound "and" GO:1901699 cellular response to nitrogen compound" was significantly involved in spermatogonia stem cell (**Figure 3C**). MF investigation of HOTAIR highlighted "GO:0016538 cyclin-dependent protein," "serine/threonine kinase regulator activity," "GO:0019207 kinase regulator activity," "GO:0031625 ubiquitin protein ligase binding," "GO:0008134 transcription factor binding," "GO:0140297 DNA-binding transcription factor binding," and "GO:0003682 chromatin binding" (**Figure 3D**).

14.1 LncRNA Gm2044

Long noncoding RNA Gm2044, denoted as "Gene model 2044," is a molecular entity that has drawn scientific interest due to its potential involvement in gene regulation and cellular processes, particularly in the context of spermatogonial stem cells (SSCs) and male reproductive biology. In this academic elucidation, we explore the intricacies of lncRNA Gm2044, its established functions, and the implications it holds in various biological contexts:

Gm2044 assumes a pivotal role in the finely tuned balance between SSC self-renewal and differentiation. SSCs are fundamental in supporting continuous spermatogenesis and male fertility. Ensuring the precise regulation of these processes is paramount, and Gm2044 is a crucial contributor to this equilibrium [42].

Gm2044 interacts with specific protein-coding genes and signaling pathways within SSCs. Depending on the context and cellular cues, it can either promote self-renewal by repressing genes associated with differentiation or encourage differentiation by inhibiting genes linked to self-renewal. This regulatory duality enables Gm2044 to dictate the fate of SSCs, ensuring the necessary equilibrium is maintained. Epigenetic modifications, including DNA methylation and histone modifications, have profound implications for gene regulation in SSCs. Gm2044 is believed to influence these epigenetic processes, contributing to the remodeling of chromatin and modifying the accessibility of specific genes. This interaction with the epigenetic landscape plays a pivotal role in determining SSC fate.

By mediating changes in chromatin structure, Gm2044 plays a vital role in directing the expression of genes that are essential for self-renewal, differentiation, and the overall process of spermatogenesis. These are just a few examples of lncRNAs that have been implicated in the regulation of gene expression and SSC behavior. Research in this field is ongoing, and new lncRNAs and mechanisms are continually being discovered, further expanding our understanding of how lncRNAs mediate processes in SSCs [5].

14.2 LncRNA PlncRNA-1 (promoter-associated long noncoding RNA 1; CBR3-AS1)

PlncRNA-1 has been shown to be expressed in differentiating SSCs. It regulates SSC differentiation by interacting with transcription factors and promoting the expression of genes associated with differentiation.

15. Signaling pathway, biological process, and molecular functions in PlncRNA-1

Signaling pathway shows that PlncRNA-1 can contribute to "Adherens junction," "AGE-RAGE signaling pathway in diabetic complications," "Hippo signaling pathway," "TGF-beta signaling pathway," "Cell cycle," and "Tight junction" (**Figure 4B**). Enrich tool analysis revealed that PlncRNA-1 is enriched in 3 GO terms. Functional enrichment analysis demonstrated that the BP term "GO:0010718 positive reg. of epithelial to mesenchymal transition," "GO:0072132 mesenchyme morphogenesis," "GO:0001837 epithelial to mesenchymal transition," and "GO:0045216 cell-cell junction organization" was significantly involved in spermatogonia stem cell (**Figure 4C**). MF investigation of PlncRNA-1 highlighted "GO:0070411 I-SMAD binding," "GO:0005160 transforming growth factor beta receptor binding," "GO:0070412 R-SMAD binding," "GO:0043425 bHLH transcription factor binding," "GO:0070888 E-box binding," "GO:0008013 beta-catenin binding," and "GO:0003682 chromatin binding" (**Figure 4D**).

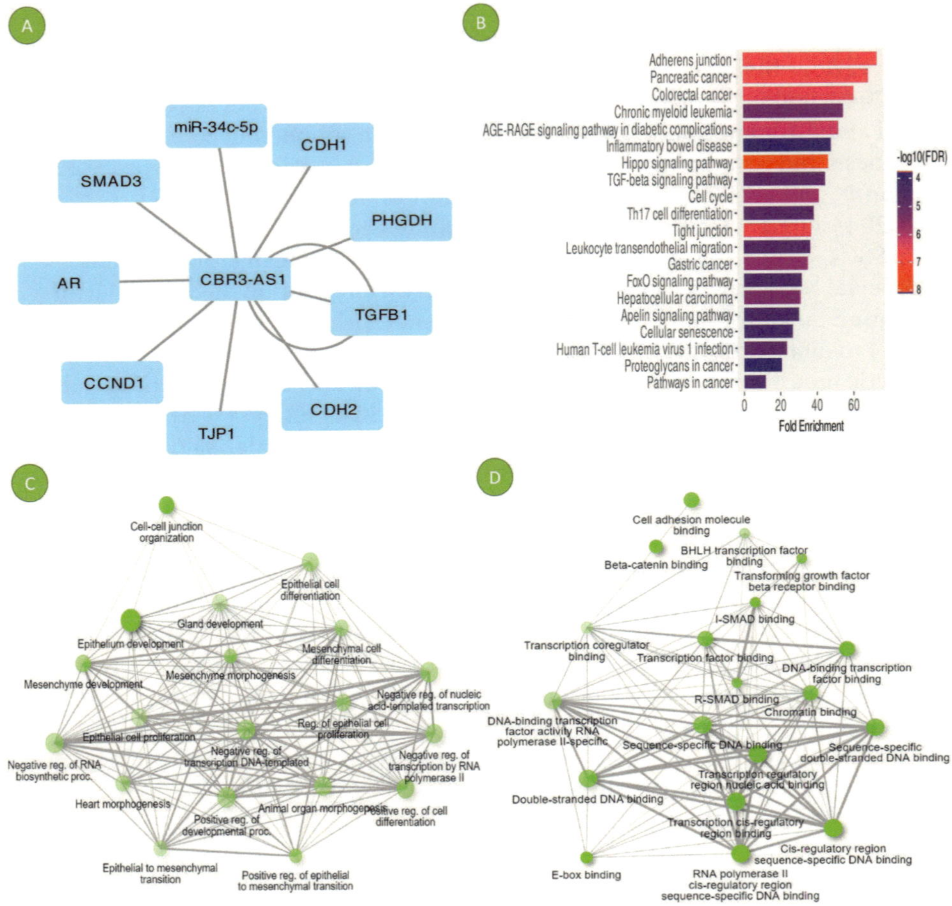

Figure 4.
LncRNA-mRNA-miRNA interaction, signaling pathway, biological process, and molecular functions in PlncRNA-1 involved in spermatogonia stem cell.

15.1 LncRNA MIAT (myocardial infarction-associated transcript)

While primarily known for its role in cardiac biology, MIAT has been found to play a role in the regulation of SSC differentiation. It modulates the expression of genes important for SSC differentiation, contributing to the process [43].

16. Signaling pathway, biological process, and molecular functions in MIAT

The signaling pathways analysis indicates that Myocardial Infarction Associated Transcript (MIAT) plays a contributory role in several pathways, including the AGE-RAGE signaling pathway, Relaxin signaling pathway, FoxO signaling pathway, and Cellular senescence, as illustrated in **Figure 5B**. Further, an Enrich tool analysis reveals that MIAT is significantly enriched in three GO terms. The functional enrichment analysis elaborates that MIAT is notably involved in the BP terms such as "GO:0050679 positive regulation of epithelial cell proliferation," "GO:0050673 epithelial cell proliferation," "GO:0030335 positive regulation of cell migration," "GO:2000147 positive regulation of cell motility," and "GO:0051272 positive regulation of cellular component movement." These processes are crucial in the context of spermatogonia stem cell function, as depicted in **Figure 5C**. Additionally, MF analysis of MIAT highlights its significant involvement in various functions, including "GO:0004407 histone deacetylase activity," "GO:0005160 transforming growth factor beta receptor binding," "GO:0033558 protein lysine deacetylase activity," "GO:1990841 promoter-specific chromatin binding," "GO:0030295 protein kinase activator activity," and "GO:0031490 chromatin DNA binding," as demonstrated in **Figure 5D**.

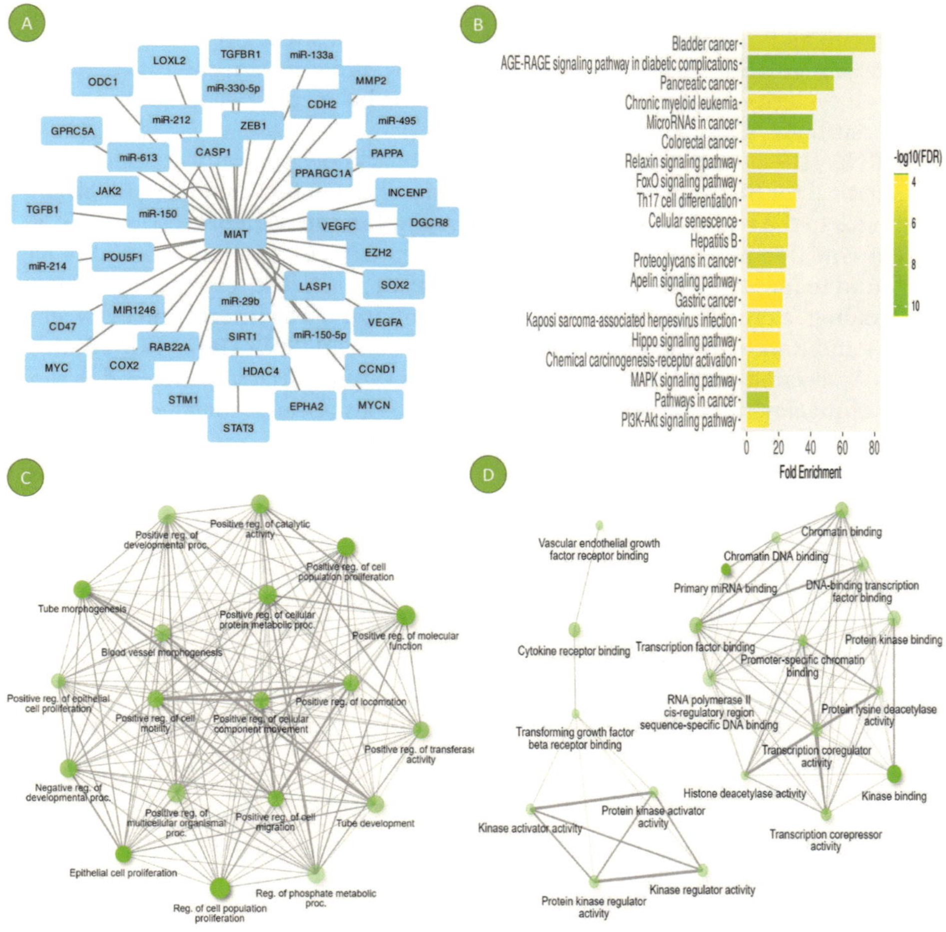

Figure 5.
LncRNA-mRNA-miRNA interaction, signaling pathway, biological process, and molecular functions in MIAT involved in spermatogonia stem cell.

16.1 LncRNA HOTAIRM1 (HOX antisense intergenic RNA myeloid 1)

Although initially identified in the context of myeloid differentiation, HOTAIRM1 has also been linked to SSC differentiation. It interacts with chromatin modifiers and contributes to the regulation of genes involved in SSC differentiation [44].

16.2 LncRNA Neat1 (nuclear enriched abundant transcript 1)

Neat1 is another lncRNA associated with SSC differentiation. It interacts with chromatin-modifying proteins and has been found to influence the expression of genes involved in the differentiation of SSCs into more mature germ cell types.

17. mRNAs and LncRNAs and their interactions in SSCs

LncRNA TDRG1 and mRNA Cyclin D2 (CCND2): TDRG1 has been shown to interact with and modulate the expression of CCND2 in SSCs. Cyclin D2 is involved in cell cycle regulation and has implications for SSC proliferation and differentiation. TDRG1's interaction with CCND2 can influence the balance between self-renewal and differentiation in SSCs [45].

LncRNA H19 and mRNA Sirtuin 1 (Sirt1): H19 is known to interact with Sirt1, a gene critical for SSC fate determination. The interaction between H19 and Sirt1 contributes to the regulation of self-renewal and differentiation processes in SSCs. LncRNA Gm2044 and mRNA Retinoic Acid Receptor Gamma (Rarg): Gm2044 has been found to interact with Rarg, a gene that plays a role in the response to retinoic acid signaling, a crucial pathway for SSC differentiation. The interaction between Gm2044 and Rarg may modulate SSC differentiation. LncRNA HOTAIRM1 and mRNAs Associated with Myeloid Differentiation: While initially discovered in the context of myeloid differentiation, HOTAIRM1 has also been associated with SSCs. It interacts with mRNAs encoding proteins involved in differentiation processes [46].

LncRNA MIAT and mRNAs Associated with SSC Differentiation: MIAT has been found to interact with mRNAs encoding proteins critical for SSC differentiation. This interaction contributes to the regulation of the differentiation process in SSCs. LncRNA HOTAIR and mRNA HOX Genes: HOTAIR interacts with genes from the HOX cluster, such as HOXA9 and HOXA10. The interaction between HOTAIR and these HOX genes has implications for the regulation of differentiation processes in SSCs. HOTAIR plays a role in the fine-tuned control of gene expression associated with SSC self-renewal and differentiation.

LncRNA Gm2044 and mRNA BMP4 (Bone Morphogenetic Protein 4): Gm2044 has been shown to interact with BMP4, a gene involved in various cellular processes, including SSC differentiation. This interaction modulates the expression of BMP4 and, consequently, influences SSC differentiation and self-renewal. LncRNA MIAT and mRNAs Associated with Mitochondrial Function: MIAT has been associated with SSC differentiation and is known to interact with mRNAs encoding proteins involved in mitochondrial function. This interaction is significant in the regulation of SSC differentiation and energy metabolism, which is crucial for their differentiation into more mature germ cell types [47].

LncRNA Neat1 and mRNAs Encoding Epigenetic Regulators: Neat1 interacts with mRNAs that encode epigenetic regulatory proteins, including histone modifiers. These interactions are critical for shaping the epigenetic landscape within SSCs, which in turn influences their differentiation and self-renewal. LncRNA H19 and mRNA IGF2 (Insulin-Like Growth Factor 2): H19, part of the H19-IGF2 imprinted gene cluster, interacts with IGF2. This interaction is relevant to the regulation of growth and development within the testes, and it may have implications for SSC differentiation processes [48].

18. Signaling pathway, biological process, and molecular functions in Neat1

The signaling pathway analysis indicates that Neat1 may contribute to the development of endocrine resistance and the activation of the PI3K-Akt signaling pathway, as depicted in **Figure 6B**. Furthermore, an analysis utilizing the Enrich tool has identified that Neat1 is associated with two GO terms. The functional enrichment

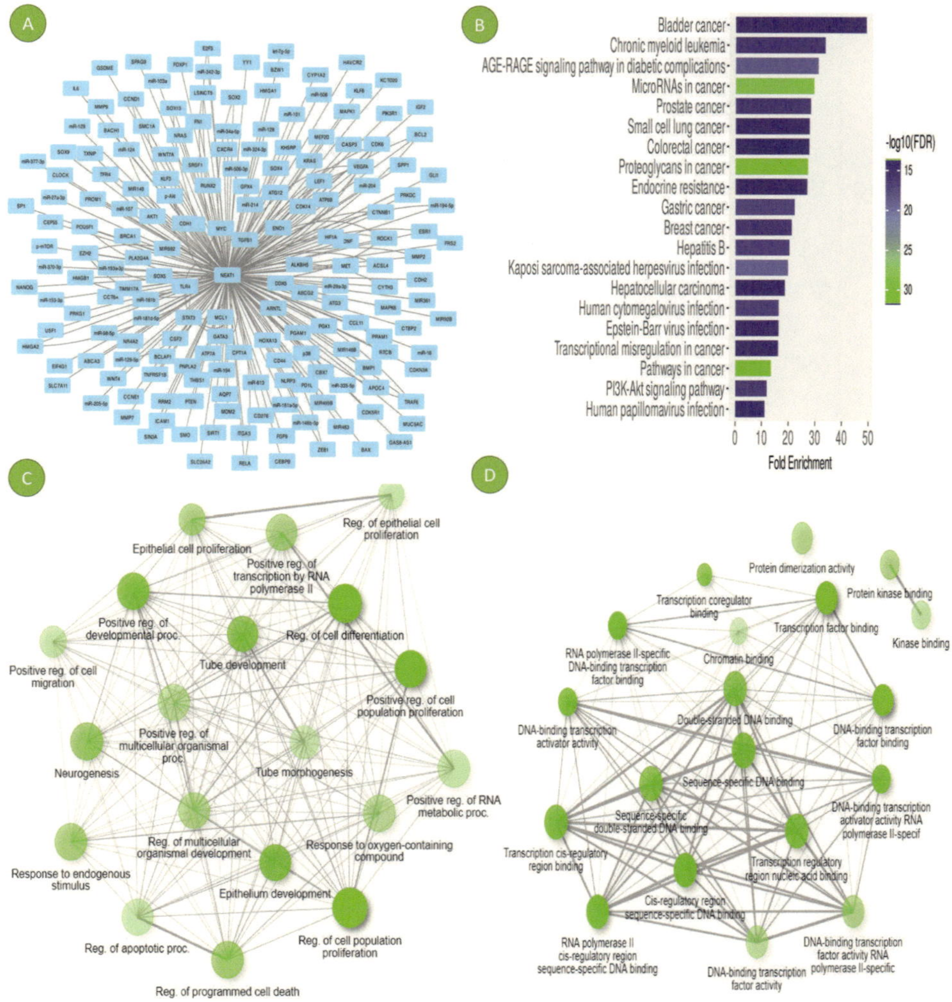

Figure 6.
LncRNA-mRNA-miRNA interaction, signaling pathway, biological process, and molecular functions in Neat1 involved in spermatogonia stem cell.

analysis further elucidates this association, showing significant involvement of Neat1 in three BP terms: epithelial cell proliferation (GO:0050673), positive regulation of cell migration (GO:0030335), and positive regulation of cell population proliferation (GO:0008284). These processes are particularly relevant in the context of spermatogonia stem cells, as illustrated in **Figure 6C**. Additionally, the MF analysis of Neat1 underscores its role in several key areas: transcription coregulator binding (GO:0001221), RNA polymerase II-specific DNA-binding transcription factor

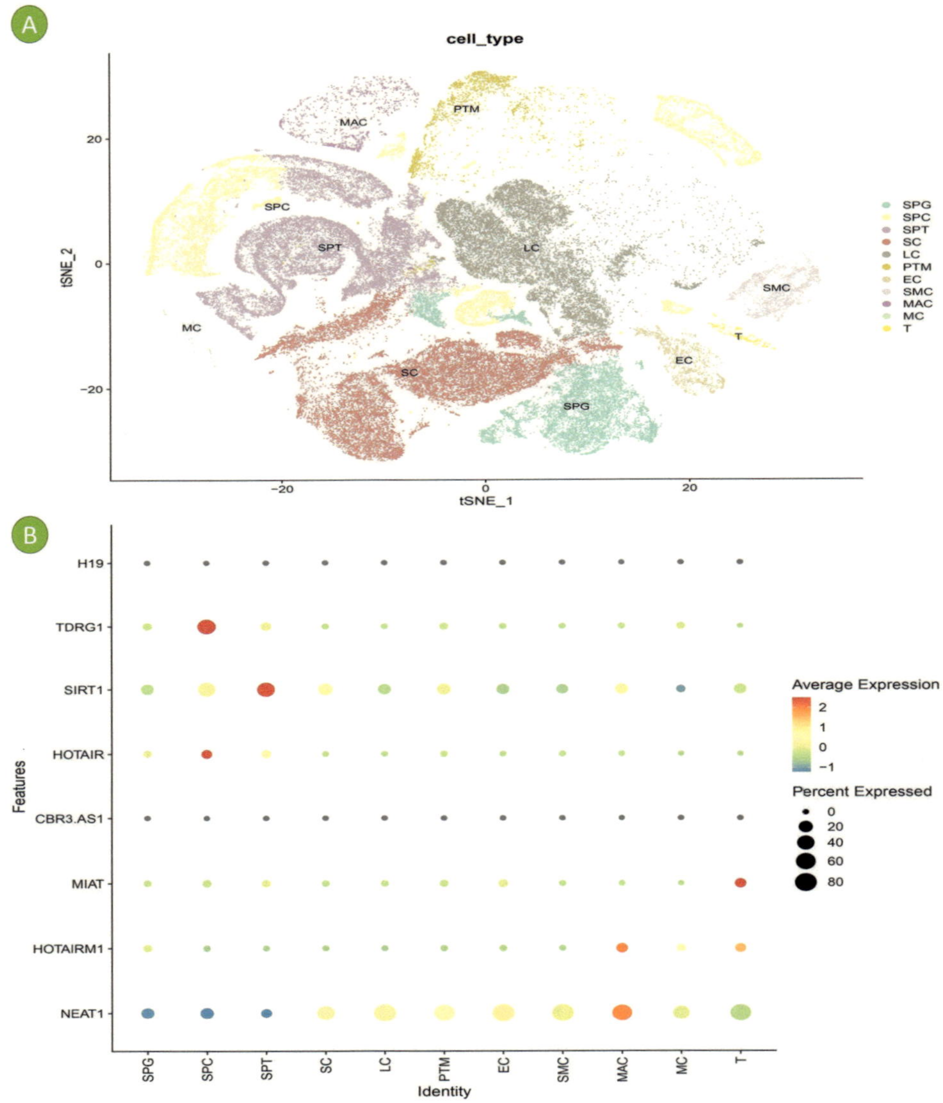

Figure 7.
Identification of a testis-specific hub gene in single cell. (A) The 10 samples were split into 11 clusters. These clusters included spermatogonia (SPG), spermatids/sperm (SPT), spermatocytes (SPS), Sertoli cells (SC), Leydig cells (LC), peritubular myoid cells (PTM), endothelial cells (EC), vascular smooth muscle cells (SMC), macrophages (MAC), mast cells (MC), and T-cells (T). Single-cell transcriptomic analysis identified nine hub genes—H19, TDRG1, SIRT1, HOTAIR, CBR3-AS1, MIAT, HOTAIRM1, and NEAT1—as significantly associated with spermatogonia stem cells.

binding (GO:0061629), DNA-binding transcription factor binding (GO:0140297), transcription factor binding (GO:0008134), DNA-binding transcription activator activity RNA polymerase II-specific (GO:0001228), and DNA-binding transcription activator activity (GO:0001216), as shown in **Figure 6D**.

19. Identification of a hub gene from single-cell RNA sequencing (scRNA-seq)

To investigate the heterogeneity of spermatogonia stem cells, we employed unsupervised clustering on a dataset comprising 10 samples (sourced from 10x Genomics Datasets: GSE109033 and GSE163027). This methodology enabled the identification of four discrete clusters, each corresponding to cell types integral to the process of spermatogenesis. These cell types include spermatogonia (SPG), spermatids/sperm (SPT), spermatocytes (SPS), Sertoli cells (SC), Leydig cells (LC), peritubular myoid cells (PTM), endothelial cells (EC), vascular smooth muscle cells (SMC), macrophages (MAC), mast cells (MC), and T-cells (T). Further validation of these clusters was achieved through Principal Component Analysis (PCA), the results of which are depicted in **Figure 7**. Moreover, our single-cell transcriptomic analysis identified eight central genes—H19, TDRG1, SIRT1, HOTAIR, CBR3-AS1, MIAT, HOTAIRM1, and NEAT1—which exhibited significant associations, particularly in spermatogonia stem cells within the testicular environment (**Table 1**).

Signaling pathway	Genes	Enrichment FDR
hsa05219 Bladder cancer	E2F1 E2F3 EGFR ERBB2 MMP2 MMP9 MYC MAPK1 MAPK3 RB1 TP53 VEGFA	4.62E-17
hsa05215 Prostate cancer	CDK2 CTNNB1 E2F1 E2F3 EGFR ERBB2 AKT1 MTOR GSTP1 IGF1 MMP9 LEF1 PIK3CA MAPK1 MAPK3 RB1 BCL2 ZEB1 TP53 PIK3R3	1.75E-24
hsa01521 EGFR tyrosine kinase inhibitor resistance	BCL2L11 EGFR EIF4EBP1 ERBB2 AKT1 MTOR IGF1 IL6 NF1 PIK3CA MAPK1 MAPK3 BCL2 STAT3 VEGFA PIK3R3	1.01E-19
hsa01522 Endocrine resistance	E2F1 E2F3 EGFR ERBB2 AKT1 ESR1 MTOR IGF1 MMP2 MMP9 NOTCH1 PIK3CA MAPK1 MAPK3 PTK2 RB1 BCL2 TP53 PIK3R3	4.92E-23
hsa05206 MicroRNAs in cancer	BCL2L11 CDK6 DNMT1 DNMT3B E2F1 E2F3 EGFR ERBB2 EZH2 SIRT1 MTOR APC IRS1 ITGB3 MCL1 MMP9 MYC NOTCH1 ABCB1 PIK3CA MAPK1 MAPK3 BCL2 SOX4 STAT3 ZEB1 TP53 VEGFA VIM HMGA2 PIK3R3 ZEB2	7.30E-39
hsa05212 Pancreatic cancer	CDK6 E2F1 E2F3 EGFR ERBB2 AKT1 MTOR PIK3CA MAPK1 MAPK3 RB1 STAT3 TP53 VEGFA PIK3R3	2.16E-18
hsa05224 Breast cancer	CDK6 CTNNB1 E2F1 E2F3 EGFR ERBB2 AKT1 ESR1 MTOR APC IGF1 MYC NOTCH1 LEF1 PIK3CA MAPK1 MAPK3 RB1 TP53 WNT1 PIK3R3	1.98E-22
hsa04066 HIF-1 signaling pathway	EGFR EIF4EBP1 ERBB2 AKT1 MTOR IGF1 IL6 PFKFB3 PIK3CA MAPK1 MAPK3 BCL2 STAT3 VEGFA PIK3R3	4.42E-16
hsa05226 Gastric cancer	CDK2 CTNNB1 E2F1 E2F3 EGFR ERBB2 AKT1 MTOR APC MYC LEF1 ABCB1 PIK3CA MAPK1 MAPK3 RB1 BCL2 TP53 WNT1 PIK3R3	6.20E-21

Signaling pathway	Genes	Enrichment FDR
hsa05205 Proteoglycans in cancer	COL1A1 CTNNB1 EGFR ERBB2 AKT1 ESR1 MTOR IGF1 ITGB3 MMP2 MMP9 MYC PIK3CA PPP1CA MAPK1 MAPK3 PTK2 STAT3 TP53 TWIST1 VEGFA WNT1 NANOG PIK3R3 CBL ROCK2	2.09E-26
hsa05418 Fluid shear stress and atherosclerosis	CTNNB1 NQO1 AKT1 GSTP1 IL1B ITGB3 MMP2 MMP9 NFE2L2 PIK3CA PRKAA1 PTK2 BCL2 CCL2 TP53 VEGFA PIK3R3	2.90E-17
hsa05160 Hepatitis C	CDK2 CDK6 CTNNB1 E2F1 E2F3 EGFR AKT1 CXCL10 MYC PIK3CA PPP2CA MAPK1 MAPK3 RB1 STAT3 TP53 YWHAZ PIK3R3 FADD	4.80E-19
hsa05225 Hepatocellular carcinoma	CDK6 CTNNB1 NQO1 E2F1 E2F3 EGFR AKT1 MTOR GSTP1 APC MYC NFE2L2 LEF1 PIK3CA MAPK1 MAPK3 RB1 TP53 WNT1 PIK3R3	6.73E-20
has 04218 Cellular senescence	CDK2 CDK6 E2F1 E2F3 EIF4EBP1 AKT1 FOXM1 SIRT1 MTOR IL6 MYC PIK3CA PPP1CA MAPK1 MAPK3 RB1 TP53 PIK3R3	9.70E-18
hsa05163 Human cytomegalovirus infection	CDK6 CTNNB1 E2F1 E2F3 EGFR EIF4EBP1 AKT1 MTOR IL1B IL6 ITGB3 MYC PIK3CA MAPK1 MAPK3 PTK2 RB1 CCL2 STAT3 TP53 VEGFA PIK3R3 FADD ROCK2	1.32E-22
hsa05161 Hepatitis B	CDK2 E2F1 E2F3 AKT1 IL6 MMP9 MYC PIK3CA MAPK1 MAPK3 RB1 BCL2 STAT3 TP53 YWHAZ PIK3R3 FADD	4.01E-16
hsa05167 Kaposi sarcoma-associated herpesvirus infection	CDK6 CTNNB1 E2F1 E2F3 AKT1 MTOR IL6 MYC LEF1 PIK3CA MAPK1 MAPK3 RB1 STAT3 TP53 VEGFA PIK3R3 FADD	4.01E-16
hsa04151 PI3K-Akt signaling pathway	BCL2L11 CDK2 CDK6 COL1A1 EGFR EIF4EBP1 ERBB2 AKT1 MTOR IGF1 IL2 IL6 IRS1 ITGB3 MCL1 MYC PIK3CA PPP2CA PRKAA1 MAPK1 MAPK3 PTK2 BCL2 TP53 VEGFA YWHAZ PIK3R3	1.01E-21
hsa05200 Pathways in cancer	BCL2L11 CDK2 CDK6 CTNNB1 NQO1 E2F1 E2F3 EGFR ERBB2 AKT1 ESR1 MTOR GSTP1 APC IGF1 IL2 IL6 MMP2 MMP9 MYC NFE2L2 NOTCH1 LEF1 PIK3CA PMAIP1 PPARG MAPK1 MAPK3 PTK2 RB1 BCL2 STAT3 TP53 VEGFA WNT1 PIK3R3 RUNX1 CBL FADD ROCK2	2.81E-32
hsa05165 Human papillomavirus infection	CDK2 CDK6 COL1A1 CTNNB1 E2F1 EGFR EIF4EBP1 AKT1 MTOR APC ITGB3 NOTCH1 PIK3CA PKM PPP2CA MAPK1 MAPK3 PTK2 RB1 TP53 VEGFA WNT1 PIK3R3 FADD	6.83E-19

Table 1.
Signaling pathway in spermatogenesis related to hub gene from single-cell RNA sequencing.

20. Conclusions

In conclusion, this review has shed light on the pivotal role of long noncoding RNAs (lncRNAs) in the intricate gene expression regulation within spermatogonial stem cells (SSCs) and has outlined the promising therapeutic implications associated with their manipulation. Through a comprehensive analysis of the current literature, it becomes evident that lncRNAs play a multifaceted role in the intricate orchestration of SSC development, maintenance, and differentiation. LncRNAs have been demonstrated to exert a profound influence on SSC self-renewal and differentiation. They participate in the fine-tuning of the SSC transcriptome, facilitating the maintenance of the stem cell pool while guiding the transition toward meiosis. Furthermore, lncRNAs have been implicated in the modulation of SSC responses to external cues, thereby playing a key role in the adaptation to various microenvironmental conditions. The therapeutic potential of lncRNAs in the context of male infertility and other reproductive disorders is a topic of great interest. The ability to manipulate lncRNA expression, either through their upregulation or silencing, presents a promising avenue for addressing these conditions. Potential therapeutic strategies include the use of lncRNA-based molecular tools and the development of novel pharmacological agents targeting specific lncRNAs. Moreover, the prospect of utilizing lncRNAs as biomarkers for diagnosing reproductive disorders holds great promise, as it may enable early detection and intervention.

In summary, the study of long noncoding RNAs in the context of gene expression regulation in spermatogonial stem cells has provided valuable insights into the intricacies of male reproduction and has laid the foundation for potential therapeutic interventions. Continued research in this field will undoubtedly yield a deeper understanding of the mechanisms at play and, ultimately, contribute to the development of novel strategies for addressing male infertility and related disorders. The potential to harness lncRNAs as diagnostic markers and therapeutic targets offers an exciting avenue for future research and clinical applications in the field of reproductive biology and medicine.

Acknowledgements

This study was funded by Tehran university. We used the AI for the purpose of language polishing and grammar check.

Conflict of interest

The authors declare no conflict of interest.

Abbreviations

SSCs	spermatogonia stem cells
GO	gene ontology
ASOs	antisense oligonucleotides
RNAi	RNA interference
MIAT	myocardial infarction-associated transcript
HOTAIRM1	HOX antisense intergenic RNA myeloid 1
SPG	spermatogonia
SPT	spermatids/sperm
SPS	spermatocytes
SC	sertoli cells
LC	leydig cells

Author details

Danial Hashemi Karoii[1,2,3]* and Zahra Hasani Mahforoozmahalleh[4]

1 Department of Cell and Molecular Biology, School of Biology, College of Science, University of Tehran, Tehran, Iran

2 Department of Genetics, Reproductive Biomedicine Research Centre, Royan Institute for Reproductive Biomedicine, ACECR, Tehran, Iran

3 Department of Biotechnology, Amol University of Special Modern Technologies, Amol, Iran

4 Department of Microbial Biotechnology, Amol University of Special Modern Technologies, Amol, Iran

*Address all correspondence to: d.hashemi.karoii@ut.ac.ir; danialhashemiac@gmail.com

References

[1] Singh SR et al. Spermatogonial stem cells, infertility and testicular cancer. Journal of Cellular and Molecular Medicine. 2011;**15**(3):468-483

[2] Jan SZ et al. Unraveling transcriptome dynamics in human spermatogenesis. Development. 2017;**144**(20):3659-3673

[3] Danial Hashemi K, Hossein A. Undifferentiated and differentiated spermatogonial stem cells. In: Leisheng Z, editor. Advances in Pluripotent Stem Cells. London, UK, Rijeka: IntechOpen; 2023. p. Ch. 10

[4] Phillips BT, Gassei K, Orwig KE. Spermatogonial stem cell regulation and spermatogenesis. Philosophical Transactions of the Royal Society of London. Series B, Biological Sciences. 2010;**365**(1546):1663-1678

[5] Joshi M, Rajender S. Long non-coding RNAs (lncRNAs) in spermatogenesis and male infertility. Reproductive Biology and Endocrinology. 2020;**18**(1):103

[6] Statello L et al. Gene regulation by long non-coding RNAs and its biological functions. Nature Reviews Molecular Cell Biology. 2021;**22**(2):96-118

[7] Azizi H, Hashemi Karoii D, Skutella T. Whole exome sequencing and in silico analysis of human sertoli in patients with non-obstructive azoospermia. International Journal of Molecular Sciences. 2022;**23**(20):12570

[8] Morlando M, Fatica A. Alteration of epigenetic regulation by long noncoding RNAs in cancer. International Journal of Molecular Sciences. 2018;**19**(2):570

[9] Frías-Lasserre D, Villagra CA. The importance of ncRNAs as epigenetic mechanisms in phenotypic variation and organic evolution. Frontiers in Microbiology. 2017;**8**:2483

[10] Wang KC, Chang HY. Molecular mechanisms of long noncoding RNAs. Molecular Cell. 2011;**43**(6):904-914

[11] Sebastian-delaCruz M, Gonzalez-Moro I, Olazagoitia-Garmendia A, Castellanos-Rubio A, Santin I. The role of lncRNAs in gene expression regulation through mRNA stabilization. Non-coding RNA. 2021;**7**(1):3

[12] Zhao S, Heng N, Sahlu BW, Wang H, Zhu H. Long noncoding RNAs: Recent insights into their role in male infertility and their potential as biomarkers and therapeutic targets. International Journal of Molecular Sciences. 2021;**22**(24):13579

[13] Oatley JM, Brinster RL. Regulation of spermatogonial stem cell self-renewal in mammals. Annual Review of Cell and Developmental Biology. 2008;**24**:263-286

[14] Hashemi Karoii D, Azizi H. OCT4 protein and gene expression analysis in the differentiation of spermatogonia stem cells into neurons by immunohistochemistry, immunocytochemistry, and bioinformatics analysis. Stem Cell Reviews and Reports. 2023;**19**(6):1828-1844

[15] Hashemi Karoii D, Azizi H. Functions and mechanism of noncoding RNA in regulation and differentiation of male mammalian reproduction. Cell Biochemistry and Function. 2023;**41**(7):767-778

[16] Niazi Tabar A et al. Testicular localization and potential function of vimentin positive cells during

spermatogonial differentiation stages. Animals. 2022;**12**(3):268

[17] Hashemi Karoii D, Azizi H, Skutella T. Microarray and in silico analysis of DNA repair genes between human testis of patients with nonobstructive azoospermia and normal cells. Cell Biochemistry and Function. 2022;**40**(8):865-879

[18] Kumari P, Khan S, Wani IA, Gupta R, Verma S, Alam P, et al. Unravelling the role of epigenetic modifications in development and reproduction of angiosperms: A critical appraisal. Frontiers in Genetics. 2022;**13**:819941

[19] He C, Wang K, Gao Y, Wang C, Li L, Liao Y, et al. Roles of noncoding RNA in reproduction. Frontiers in Genetics. 2021;**12**:777510

[20] Hasani Mahforoozmahalleh Z, Azizi H. An overview of novel transcription factors involved in spermatogonial stem cells [Internet]. Obstetrics and Gynecology (IntechOpen). 2024. DOI: 10.5772/intechopen.1004178

[21] Hanly DJ, Esteller M, Berdasco M. Interplay between long non-coding RNAs and epigenetic machinery: Emerging targets in cancer? Philosophical Transactions of the Royal Society B: Biological Sciences. 2018;**373**(1748):20170074

[22] Lewis MW, Li S, Franco HL. Transcriptional control by enhancers and enhancer RNAs. Transcription. 2019;**10**(4-5):171-186

[23] Kim TK, Hemberg M, Gray JM. Enhancer RNAs: A class of long noncoding RNAs synthesized at enhancers. Cold Spring Harbor Perspectives in Biology. 2015;**7**(1):a018622

[24] Long Y et al. How do lncRNAs regulate transcription? Science Advances. 2017;**3**(9):eaao2110

[25] Zhang X, Wang W, Zhu W, Dong J, Cheng Y, Yin Z, et al. Mechanisms and functions of long non-coding RNAs at multiple regulatory levels. International Journal of Molecular Sciences. 2019;**20**(22):5573

[26] McDonel P, Guttman M. Approaches for understanding the mechanisms of long noncoding RNA regulation of gene expression. Cold Spring Harbor Perspectives in Biology. 2019;**11**(12):a032151

[27] Luo M et al. Long non-coding RNAs control hematopoietic stem cell function. Cell Stem Cell. 2015;**16**(4):426-438

[28] Ben Maamar M, Nilsson EE, Skinner MK. Epigenetic transgenerational inheritance, gametogenesis and germline development†. Biology of Reproduction. 2021;**105**(3):570-592

[29] Xiao Y et al. Predicting the functions of long noncoding RNAs using RNA-seq based on Bayesian network. BioMed Research International. 2015;**2015**:839590

[30] Patil VS, Zhou R, Rana TM. Gene regulation by non-coding RNAs. Critical Reviews in Biochemistry and Molecular Biology. 2014;**49**(1):16-32

[31] Liao J et al. Long noncoding RNA (lncRNA) H19: An essential developmental regulator with expanding roles in cancer, stem cell differentiation, and metabolic diseases. Genes and Diseases. 2023;**10**(4):1351-1366

[32] Nordin M et al. Epigenetic regulation of the Igf2/H19 gene cluster. Cell Proliferation. 2014;**47**(3):189-199

[33] Liu K-S et al. Advances of long noncoding RNAs-mediated regulation in reproduction. Chinese Medical Journal. 2018;**131**(2):226-234

[34] Li XP et al. H19 gene methylation status is associated with male infertility. Experimental and Therapeutic Medicine. 2016;**12**(1):451-456

[35] Handy DE, Castro R, Loscalzo J. Epigenetic modifications: Basic mechanisms and role in cardiovascular disease. Circulation. 2011;**123**(19):2145-2156

[36] Assidi M. Infertility in men: Advances towards a comprehensive and integrative strategy for precision theranostics. Cells. 2022;**11**(10):1711

[37] Li B et al. Sirt1 antisense Long noncoding RNA promotes cardiomyocyte proliferation by enhancing the stability of Sirt1. Journal of the American Heart Association. 2018;**7**(21):e009700

[38] Diao L et al. Roles of Spermatogonial stem cells in spermatogenesis and fertility restoration. Frontiers in Endocrinology(Lausanne). 2022;**13**:895528

[39] Ma CY et al. SIRT1 suppresses self-renewal of adult hippocampal neural stem cells. Development. 2014;**141**(24):4697-4709

[40] Yu W et al. Modulation of SIRT1 expression improves erectile function in aged rats. Asian Journal of Andrology. 2022;**24**(6):666-670

[41] Gao S et al. Long noncoding RNA HOTAIR promotes the self-renewal of leukemia stem cells through epigenetic silencing of p15. Experimental Hematology. 2018;**67**:32-40.e3

[42] Liang M et al. Upregulated lncRNA Gm2044 inhibits male germ cell development by acting as miR-202 host gene. Animal Cells and Systems (Seoul). 2019;**23**(2):128-134

[43] Fioretto BS et al. The contribution of epigenetics to the pathogenesis and gender dimorphism of systemic sclerosis: A comprehensive overview. Therapeutic Advances in Musculoskeletal Disease. 2020;**12**:1759720x20918456

[44] Li F et al. lncRNA HotairM1 depletion promotes self-renewal of cancer stem cells through HOXA1-Nanog regulation loop. Molecular Therapy—Nucleic Acids. 2020;**22**:456-470

[45] Zhao M et al. Cyclin D2 overexpression enhances the efficacy of human induced pluripotent stem cell-derived cardiomyocytes for myocardial repair in a swine model of myocardial infarction. Circulation. 2021;**144**(3):210-228

[46] Su D, Gudas LJ. Gene expression profiling elucidates a specific role for RAR gamma in the retinoic acid-induced differentiation of F9 teratocarcinoma stem cells. Biochemical Pharmacology. 2008;**75**(5):1129-1160

[47] Setiawan AM, Kamarudin TA, Abd Ghafar N. The role of BMP4 in adipose-derived stem cell differentiation: A minireview. Frontiers in Cell and Developmental Biology. 2022;**10**:1045103

[48] Butler AA, Johnston DR, Kaur S, Lubin FD. Long noncoding RNA NEAT1 mediates neuronal histone methylation and age-related memory impairment. Science Signaling. 2019;**12**(588):eaaw9277

Chapter 10

The Role of NK Cell, HLA, and Peptide Interaction in Hematopoietic Stem Cell Transplantation

Daniela Cardozo, Jeane Eliete Laguila Visentainer and Carmino De Souza

Abstract

NK cells are cytotoxic lymphocytes, representing about 5–15% of circulating mononuclear lymphocytes, and are defined as CD3-CD56 + NKp46+. They play crucial roles in host defense by recognizing and eliminating virus-infected and tumor cells without prior sensitization. The function of NK cells is regulated by a repertoire of receptors that issue activating or inhibitory intracellular signals. The ligands for these receptors can be diverse. Recent studies show that interactions between NK cell receptors and their ligands can depend on the peptide presented by HLA at the time. This chapter explores key findings in the literature regarding the interaction between NK cell receptors, HLA, and peptides and aims to understand the immunological mechanisms involved in these interactions and how these interactions might impact hematopoietic stem cell transplantation.

Keywords: interactions, HSCT, NK cells, HLA, peptide

1. Introduction

The innate lymphoid cells (ILCs) are heterogeneous lymphocytes, considered distinct from B or T cell subsets, and, like these cells, they are involved in immune responses, inflammation, and tissue homeostasis. ILCs can be classified into five main subsets: natural killer (NK) cells, lymphoid tissue inducer cells (LTi), ILC type 1 (ILC1s), ILC type 2 (ILC2s), and ILC type 3 (ILC3s). ILC1s express T-Bet and produce interferon-gamma (IFN-γ); ILC2s are dependent on GATA-3 and secrete interleukin (IL)-12 and -5, while ILC3s express RORγt and produce IL-17 and IL-22 [1, 2].

Initially, natural killer (NK) cells were considered part of the ILC type 1 subset because both share many characteristics, such as the expression of T-Bet and the production of IFN-γ. However, several markers have been proposed to discriminate between these cells, such as the integrin α1 (CD49a), CD200R, interleukin-7 receptor

(CD127), CD69, and chemokine receptor CXCR6. Additionally, type 1 ILCs reside in tissues, whereas NK cells circulate in peripheral blood [3, 4].

Natural killer cells are characterized as cytotoxic lymphocytes of the innate immune response, constituting approximately 5–15% of circulating mononuclear lymphocytes and are phenotypically defined as CD3-CD56 + NKp46+ [5]. These cells play crucial roles in host defense due to their ability to recognize and eliminate virus-infected cells and transform tumor cells without the need for prior sensitization. NK cells develop in the bone marrow niche from hematopoietic stem cell progenitors, distributing throughout lymphoid organs and other tissues, undergoing a continuous maturation process that confers tissue-specific functions. For instance, this is evident in NK cells found in the uterus, liver, and tonsils [5, 6].

The function of NK cells is governed by a repertoire of receptors responsible for issuing intracellular activating or inhibitory signals. The balance of this signaling directs the cytotoxic activity of these cells, as well as cell proliferation and cytokine release. Typical activating receptors on these cells include the natural cytotoxicity receptors (NCRs) NKp44 (CD336), NKp46 (CD335), and NKp30 (CD337), as well as some of the killer cell immunoglobulin-like receptors (KIRs), which recognize human leukocyte antigen (HLA) class I antigens expressed on the surface of target cells [7]. The NCRs recognize a diverse set of ligands, such as heparan sulfate proteoglycans, cell surface proteins, and proteins that reach the surface after their intracellular cleavage. These ligands are not exclusively activating but can also have an inhibitory effect, depending on the splice variant of the receptor [7].

Some of the most well-studied receptors include the transcript associated with B7-H6 and HLA-B 3 (BAT3), which binds to NKp30, and the proliferating cell nuclear antigen (PCNA), which binds to NKp44. The mechanism of positive regulation of NK cell-activating ligands is not entirely elucidated, although increasing evidence suggests transcriptional and post-translational modifications occur as a result of the cellular response to stressful stimuli and DNA damage [8].

In NK cells, receptors such as CD224 (2B4), CD226 (DNAM-1), CD94:NKG2C, and NKG2D are also expressed. Several ligands for these receptors, considered activators, are upregulated after cellular stress, infection, or malignant transformation. In tumor formation, positively regulated ligands include MICA and MICB (MHC class I polypeptide-related sequence A and B), UL16 binding proteins (ULBPs), and the adhesion molecules PVR (poliovirus receptor, also known as CD155) and Nectin-2. MICA/B and ULBPs mediate activating signals by binding to NKG2D, while PVR and Nectin-2 bind to DNAM-1 [9].

In addition to activating KIRs, many of the KIR group receptors are known to propagate inhibitory intracellular signaling upon interaction with their ligands. Inhibitory signals are also mediated by the sialic acid-binding Ig-like lectin 7 (siglec-7) and 9 (siglec-9) receptors, which bind to sialic acid-containing carbohydrates (e.g., mucins) expressed aberrantly in tumor cells. Other inhibitory receptors include the CD94-NKG2A complex and the receptors CD161 and KLRG1, which bind to HLA-E, lectin-like transcript 1, and cadherin, respectively [8].

Many studies seek to explain the interactions between NK cell receptors and their ligands expressed on target cells as either low- or high-affinity interactions. According to these findings, it is not merely the binding of an activating/inhibitory receptor to its ligand that matters, but the affinity at which these interactions occur can dictate the intracellular signaling processes in NK cells, either activating or inhibiting them. Therefore, affinity represents a key factor in directing NK cell responses [10–12].

Recent investigations have shown that certain specific interactions like KIR/HLA go beyond mere affinity and describe that many of these interactions depend on the peptide being presented by the HLA at that moment. Analysis of the peptidome (HLA class I + bound peptide) has demonstrated that some NK cell receptors are peptide-dependent. Therefore, the need to understand these interactions by examining the entire complex of NK cell receptor + HLA class I + peptide is fundamental in various human diseases across different immunological contexts [13].

Given the above, the objective of this chapter is to explore the main findings in the literature regarding the interaction between NK cell receptors, HLA, and peptides and to understand the immunological mechanisms involved in these interactions.

2. Activation of natural killer cells and inhibitory and activating receptors

The cytolytic action of NK cells is determined by the balance of inhibitory and activating signals resulting from the interaction of receptors present in NK cells with their ligands on the target cell [14, 15]. Among the main inhibitory receptors are those that recognize human leukocyte antigen (HLA) class I molecules and include killer immunoglobulin-like receptors (KIR), which recognize classical class I HLA molecules, namely groups A, B, and C; the CD94-NKG2A receptor, which recognizes class I HLA of the E group; and the ILT2 receptor (immunoglobulin-like transcript 2 or LILRB1), which recognizes a relatively conserved region on the class I HLA molecule, providing broad specificity for class I HLA, including group G [15–18].

Additionally, there is the KLRG1 receptor, which binds to cadherin proteins on neural and epithelial cells, potentially contributing to the inhibition of NK cell activity [19–22].

Among the activating receptors, the natural cytotoxic receptors (NCR) NKp46 and NKp30, which are constitutively expressed on NK cells, and the NKp44 receptor, which is expressed after NK cell activation by interleukin 2 (IL-2), stand out [23].

These receptors are involved in the recognition of tumor cells, and virus-infected cells, and in the regulation of dendritic cell maturation. The density of NCR expression varies among individuals and correlates directly with the ability of NK cells to eliminate altered cells [24]. Another important activating receptor is NKG2D, which is also expressed on cytotoxic T lymphocytes (γ/δ T and α/β T-CD8). NKG2D recognizes as ligands HLA class I homologous molecules represented by transmembrane proteins such as MIC (MHC class I related-chain protein, MIC/A and MIC/B) and ULBP4 (UL16-binding protein) and proteins anchored to the cell surface by glycosylphosphatidylinositol (GPI) such as ULBPs 1, 2, and 3 [25–28].

In humans, the increased expression of MIC and ULBPs is associated with different forms of cellular stress, particularly viral infection and malignant transformation [29]. Indeed, MICA/B expression has been observed in most human epithelial tumors, including breast, ovarian, colon, kidney, and lung carcinomas [30, 31], contributing to the susceptibility of these tumors to the cytotoxic action of NK cells. DNAM-1 (DNAX accessory molecule-1) is another activating receptor present in NK cells that contributes to immune surveillance against tumors [32, 33].

The ligands for the DNAM-1 receptor are the CD155 (poliovirus receptor, PVR) and CD112 (nectin-2) molecules, and recent studies show that the CD155 molecule is overexpressed in various types of tumors, including ovarian carcinoma [15, 32]. Class I HLA is expressed by almost all nucleated cells in an individual, playing a fundamental role in enabling the immune system to recognize the body's cells as self. Under normal conditions, once engaged with "self" class I HLA, inhibitory KIR receptors signal dominantly over the signals from activating receptors, preventing the cytolytic action of NK cells [31, 34, 35]. Thus, NK cells spare autologous cells that express normal levels of class I HLA while eliminating altered cells, such as tumor cells and virus-infected cells that frequently lose class I HLA expression [36–38].

3. Influence of natural killer cells on hematopoietic stem cell transplants for the treatment of onco-hematological diseases

As previously mentioned, the beneficial effect of NK cells in hematopoietic stem cell transplantation (HSCT) would depend on the development of alloreactive NK lymphocytes from the graft and the functional integrity of these cells throughout the reconstitution of the patient's immune system. In this context, an alloreactive NK cell is understood as that NK cell subpopulation that develops from the graft and therefore expresses the donor's inhibitory KIR receptor, for which the specific class I HLA ligand is absent in the patient. Thus, it is proposed that determining the absence of the class I HLA ligand in the patient for the donor's inhibitory KIR receptor could be used as a prognostic factor for transplant outcomes between HLA-identical or partially identical donor-receptor pairs [39].

The clinical significance of the absence of the class I HLA ligand was first evidenced in haploidentical transplants for the treatment of acute myeloid leukemia (AML), where a reduced risk of relapse, improved survival, and reduced post-transplant complications, such as graft-versus-host disease (GVHD), were observed [40, 41]. However, the fact that KIR and class I HLA genes are located on different chromosomes (chromosomes 19q13.4 and 6p, respectively) and therefore segregate independently allows for asynchrony between an individual's KIR and class I HLA repertoires. Consequently, it is proposed that alloreactive NK cells could emerge even in HLA-compatible allogeneic HSCTs [41]. Since KIR receptors recognize groups of class I HLA alleles, it has been possible to establish the main KIR receptors involved in the allorecognition of cells deficient in the expression of one or more class I HLA ligands. Thus, KIR2DL1 recognizes as self the HLA-C group 2 (HLA-C2) alleles, which are characterized by the presence of a lysine residue at position 80 of the molecule. The inhibitory receptors KIR2DL2 and KIR2DL3 recognize HLA-C group 1 (HLA-C1) alleles, which are characterized by the presence of an asparagine residue at position 80 of the molecule.

Therefore, in an HSCT from a donor with the KIR2DL1 gene to a patient homozygous for HLA-C1 alleles (C1C1), although it is an HLA-compatible transplant, NK cells expressing the KIR2DL1 receptor could arise without a specific ligand. Similarly, in a transplant from a donor with the KIR2DL2/3 gene to a patient homozygous for HLA-C2 alleles (C2C2), NK cells expressing the KIR2DL2/3 receptor could emerge without a specific ligand. In addition to the HLA-C groups, HLA alleles of the Bw4 group are specific ligands for the inhibitory receptor KIR3DL1, and the absence of this HLA in a patient transplanted with hematopoietic progenitor cells from a donor with the KIR3DL1 gene could also give rise to NK cells without a specific ligand, that is, alloreactive NK cells [24].

Although possible, the question of the beneficial effect of alloreactive NK cells in HLA-compatible post-transplantation remains open due to controversial results.

Some studies have shown improved overall survival and event-free survival, such as reduced relapse rates [42–44]. However, other studies have not observed improved overall survival, event-free survival, or protection against GVHD, indicating that NK alloreactivity might not occur in HLA-compatible transplants since donor NK cells are transferred to a "genetically" identical environment, and thus tolerance mechanisms may render NK cells hyporesponsive [45]. A possible explanation for the differences in results is that during the post-transplant period, when the patient's immune system is reconstituting, tolerance to self may break down, allowing these cells to recognize and attack the patient's malignant cells, promoting graft-versus-leukemia [43, 46].

Yu et al. [43] used tumor cells transfected with different class I HLA genes and co-incubated with NK cells collected from the patient at consecutive times during post-transplantation to show that in the absence of specific HLA expression for the inhibitory KIR, NK cell degranulation occurs. Degranulation was not observed when tumor cells were transfected with the specific HLA ligand for the inhibitory KIR. Another factor that appears to influence the performance of alloreactive NK cells is related to the depletion of T lymphocytes from the graft. According to Sobecks et al. [44], alloreactivity is possible but only in T-cell-depleted transplants. In the presence of T lymphocytes, NK cell maturation is delayed. In the absence of T lymphocytes, NK cell maturation is more evident, favoring alloreactivity in the early post-transplant months [47].

4. Interaction of NK cell receptors with the HLA + peptide

In humans, class I HLA molecules are highly polymorphic and are expressed on the surface of all nucleated cells. Variations in the class I HLA molecule enable the display of thousands of distinct peptides on the cell surface. As these molecules become ligands for NK cell receptors, their assembly and expression are subject to intricate cellular quality control mechanisms [48].

The assembly pathway of class I HLA molecules begins in the endoplasmic reticulum (ER), migrates to the Golgi apparatus, and follows the secretory pathway to reach the cell surface. Assembly in the ER involves the peptide-loading complex (PLC), a large macromolecular complex comprising subunits such as TAP1, TAP2, tapasin, ERp57, and calreticulin, in addition to the heavy chain of HLA-I and β2m. Alongside the PLC proteins, variants like ERAP1, ERAP2, and TAPBPR are crucial in peptide-trimming and also in quality control mechanisms [49, 50].

Natural killer (NK) cells and CD8+ T lymphocytes play complementary roles in cellular immune responses. While CD8+ T cells recognize and eliminate cells presenting non-self-peptides bound to HLA class I, NK cells kill cells with deficient surface expression of HLA class I. Analyzing these distinct functions, T cells are highly sensitive to changes in the repertoire of bound peptides. On the other hand, killer immunoglobulin-like receptors (KIRs), representing the most polymorphic receptors on NK cells, are less sensitive to changes in the peptide repertoire, and capable of recognizing a broader range of peptide-HLA combinations [51].

Despite their broader specificity, pioneering studies in this field have suggested that patterns of NK cell-mediated cytotoxicity are influenced by the nature of the peptide bound to HLA class I. Although NK cell receptors are not as peptide-specific as T cell receptors (TCRs), there is considerable evidence that they exhibit sophisticated peptide selectivity [51–53].

Based on these findings, a growing number of studies have been exploring the potential peptide selectivity of NK cells, its extent, and functional consequences [54–59].

5. Interaction of KIR receptors with the HLA + peptide

Natural killer cells express, among other receptors, different combinations of inhibitory KIR receptors (KIR2DL1, KIR2DL2, KIR2DL3, KIR2DL4, KIR2DL5, KIR3DL1, KIR3DL2, and KIR3DL3) and activating ones (KIR2DS1, KIR2DS2, KIR2DS3, KIR2DS4, KIR2DS5, and KIR3DS1). KIR receptors exhibit a high degree of polymorphism, second only to molecules in the major histocompatibility complex, and specifically recognize epitopes present on HLA-A, HLA-B, and HLA-C molecules, divided into serological groups: C1, C2, Bw4, and A3/A11 [60, 61].

Five inhibitory KIRs and three activators recognize specific HLA class I ligands. Inhibitory KIR2DL1 recognizes HLA-C alleles from group 2, characterized by the presence of a lysine residue at position 80 in the peptide-binding groove of the HLA molecule. KIR2DL2 and KIR2DL3 recognize HLA-C alleles from group 1, which are characterized by the presence of an asparagine residue at position 80 in the peptide-binding groove. KIR3DL1 recognizes HLA-Bw4 alleles, and KIR3DL2 recognizes HLA-A3/−A11 alleles (**Table 1**) [60, 62].

Recent studies indicate that the classical recognition between KIR receptors and HLA molecules might have exceptions and alternative forms of interaction. In multiplex binding assays using KIR2DL1-Fc and class I HLA, it has been demonstrated that KIR2DL1 specifically and with high affinity binds to all HLA-C group 2 molecules but

Inhibitory receptors	Ligands	Activating receptors	Ligands	Coreceptors	Ligands
KIR2DL1	HLA-C2	KIR2DS1	HLA-C2	2B4	CD48
KIR2DL2	HLA-C1	KIR2DS2	HLA-C1	NTB-A	NTB-AL
KIR2DL3	HLA-C1	KIR2DS3	Unknown	CS1	CS1L
KIR2DL4	HLA-G	KIR2DS4	HLA-A11	NKp80	AICL
KIR2DL5	Unknown KIR2DS5		Unknown	TLR	TLRL
KIR3DL1	HLA-Bw4		HLA-F	DNAM-1 (CD226)	VR, Nectin-2
KIR3DL2	HLA-A3/A11	NKG2C	HLA-E	CD96	PVR
KIR3DL3	Unknown NKG2D		MICA, MICB, ULBP-4		
NKG2A LIR-1	HLA-E HLA class I	NKp30 NKp44 NKp46 CD16	B7-H6, BAT3, CMV pp65 Hemagglutinins viral Hemagglutinins viral IgG-1, 3, 4		

Table 1.
Classical NK cell receptors and their ligands.

DOI: http://dx.doi.org/10.5772/intechopen.1007711

not to HLA-C group 1 or any HLA-A or HLA-B alleles. However, there is an exception represented by KIR2DL1022*, a receptor that carries Lysine at position 44 instead of Methionine 44, making this receptor specific to the HLA-C*08:02 alleles (HLA-C1), which in turn presents a restricted number of peptides [10].

KIR2DL2 and KIR2DL3 are characterized by higher selectivity for peptides than KIR2DL1, and this capacity seems particularly relevant for the recognition of low-affinity HLA-C2 allotypes. Growing evidence suggests that both KIR2DL2/3 and KIR2DL1, apart from their ability to discriminate between C1 and C2 epitopes, bind to numerous combinations of peptide/HLA-C, maintaining a variable degree of peptide selectivity. We know that KIR2DL2/3 binds to HLA-C1 with high affinity; however, these receptors can also bind to HLA-B*46:01 and HLA-B*73:01 with high affinity, as well as to HLA-C2 with low affinity. These findings suggest that NK cells are capable of detecting not only the negative regulation of HLA class I but also alterations in the peptidome presented by HLA, which can occur during viral infections or malignant transformations [12].

Sim and colleagues recently evaluated over 3500 specific interactions to determine the specificity of five types of KIR receptors for peptides presented by four HLA-C ligands, groups 1 and 2. Among their findings, the inhibitory KIR2DL1 was largely nonspecific regarding peptide sequences and could bind to roughly 60% of the hundreds of tested HLA-peptide complexes. The inhibitory KIR2DL2, KIR2DL3, and activating KIR2DS1 and KIR2DS4 only bound to 10% and 1% of the tested HLA-peptide complexes, respectively. The binding of KIR2DS1 with HLA-C2, previously described as having low affinity, showed high-affinity binding to HLA-C with high specificity for peptide sequence. Hence, NK cell responses can be shaped by peptidomes (HLA class I + bound peptide) in healthy cells as well as in infectious, inflammatory, and tumor processes. However, due to the high degree of polymorphism in HLA class I molecules and the vast range of peptides they can present, coupled with the high polymorphism of KIR receptors, many questions regarding these interactions and their influences on various human diseases are still not fully elucidated [59].

Few studies have been published evaluating human diseases and the peptidomes associated with NK cell receptors. The vast majority of these studies assess these interactions in the context of HIV infection. However, there are limited reports explaining these specificities with other diseases, whether infectious, inflammatory, or even fewer within the tumor microenvironment [63–65].

6. Influence of the interaction between KIR receptors and HLA ligands in allogeneic hematopoietic stem cell transplantation

Understanding the interactions between NK cell receptors and their ligands requires identifying characteristics that differentiate one interaction from another. This is crucial since these interactions dictate the recognition and the type of response NK cells will mount against the target cell. For instance, certain KIR/HLA associations are linked to a spontaneous resolution against the Hepatitis C virus; however, other combinations between fetal HLA ligands and maternal NK cell KIR receptors increase the risk of preeclampsia. Some NK cell receptors are strongly associated with individuals' resistance to malaria infection, and numerous studies demonstrate associations of receptor/ligand pairs with either favorable or poor prognoses in the development of solid tumors and onco-hematologic conditions [66–71].

A class I HLA molecule is expressed by all nucleated cells in an individual, playing a crucial role in allowing the immune system to recognize the body's cells as self. Under normal conditions, when engaged with "self" class I HLA, inhibitory KIR receptors dominantly signal over activation receptors, preventing the cytolytic action of NK cells [34, 35, 72]. Thus, NK cells spare autologous cells expressing normal levels of class I HLA while eliminating altered cells such as tumor cells and virus-infected cells, which often lose or have reduced class I HLA expression [38, 73].

In the context of hematopoietic stem cell transplantation (HSCT), the beneficial effect of NK cells relies on the development of alloreactive NK cells from the graft and their functional integrity during the patient's immune system reconstitution. NK cells represent the first subset of lymphocytes to reconstitute following a transplant and thus play a crucial role in controlling early relapses and preventing infections. Additionally, they possess the unique ability to eliminate both recipient T cells and antigen-presenting cells (APCs), which is essential in preventing graft rejection and Graft-versus-Host Disease (GVHD) [74]. Due to their location on different chromosomes (19q13.4 for KIR genes and 6p-21.3 for HLA genes), there can be KIR/HLA incompatibility between donor and recipient. Typically, NK cells acquire self-tolerance and functional competence through an educational process, wherein inhibitory KIRs can be inhibited by self-HLA ligands and activated in a non-self HLA environment. Additionally, the reduced activation response capacity of KIRs in the presence of their cognate ligands also helps prevent autoimmunity [75]. An alloreactivity of NK cells is intriguing in the context of HSCT, as alloreactive NK cells have the capability to eliminate residual leukemic cells, reducing the relapse rates of the underlying disease and preventing graft-versus-host response (GVHD). There are three proposed models to enhance donor selection for HSCT based on KIR receptors, aiming for improved post-transplant outcomes. The first model considers that donor NK cells may be alloreactive against recipient cells if the HLA ligand for the inhibitory KIR receptor is present in the donor's cells but absent in the recipient's cells. Therefore, this model is focused on donor/recipient HLA incompatibility [41].

The second model posits that donor NK cells may become alloreactive if the HLA ligand in the patient's cells is incompatible with the inhibitory KIR on the donor's NK cells (receptor-ligand or missing ligand model). This model was initially proposed by Leung and colleagues based on the compatibilities between the recipient's HLA and the donor's inhibitory KIR. The results of this study suggested that the receptor-ligand model better predicted the risk of disease relapse, especially for lymphoid malignancies [75, 76].

Recently, our group demonstrated that the absence of the HLA ligand in the patient's cells for the donor NK cell's inhibitory KIR receptor in HLA-matched related allogeneic hematopoietic stem cell transplantation (allo-HSCT) without T-cell depletion may result in improved overall survival and event-free survival rates in patients with acute myeloid leukemia (AML) and myelodysplastic syndrome (MDS) [71].

The third model proposed by Cooley and colleagues focused on the presence of HLA ligands in the patient for donor's activating KIRs (KIR-B haplotype model). This study demonstrated that unrelated donors with KIR-B haplotypes (presence of up to five activating KIRs) conferred a significant benefit in relapse-free survival in patients with acute myeloid leukemia undergoing non-T-cell-depleted hematopoietic stem cell transplantation [77].

However, the results of the studies are still quite controversial, and several important questions remain regarding the biology of NK cells post-HSCT. For instance, what are the exact effects of NK cell alloreactivity in patients after HSCT? Why does NK cell alloreactivity work better for some onco-hematologic diseases than for others? Could the presentation of specific peptides by HLA molecules in the patient modulate the responses of NK cells?

Understanding how the peptidome might influence the recognition of donor NK cells through KIR receptors is of paramount importance in elucidating issues that are still unclear regarding NK cell alloreactivity post-allogeneic transplantation. Furthermore, comprehending the specificity of KIR receptors for specific HLA peptide-bound alleles could enhance the donor selection process and predict post-transplant response.

Therefore, understanding the complexity of interaction between NK cell receptors and the peptidome could enhance the effectiveness of these cells in recognizing altered and infected cells, thus improving therapy efficacy and reducing adverse events.

Many questions regarding the interaction between KIR receptors and the peptidome need to be elucidated, and new investigations will undoubtedly contribute to a better understanding of NK cell response post-transplant, particularly in the context of cytomegalovirus reactivation.

Author details

Daniela Cardozo[1], Jeane Eliete Laguila Visentainer[2] and Carmino De Souza[1*]

1 Medical Science Faculty, State University of Campinas, Campinas, São Paulo, Brazil

2 University of Maringá, Immunology Department, Maringá, Paraná, Brazil

*Address all correspondence to: carmino@unicamp.br

References

[1] Vivier E, Artis D, Colonna M, Diefenbach A, Di Santo JP, Eberl G, et al. Células linfóides inatas: 10 anos depois. Célula. 2018;**174**:1054-1066. DOI: 10.1016/j.cell.2018.07.017

[2] Lopes N, Galluso J, Escalière B, Carpentier S, Kerdiles YM, Vivier E. Tissue-specific transcriptional profiles and heterogeneity of natural killer cells and group 1 innate lymphoid cells. Cell Reports Medicine. 2022;**3**(11):100812. DOI: 10.1016/j.xcrm.2022.100812

[3] Gasteiger G, Fan X, Dikiy S, Lee SY, Rudensky AY. Residência tecidual de células linfóides inatas em órgãos linfóides e não linfóides. Ciência. 2015;**350**:981-985. DOI: 10.1126/science.aac9593

[4] Seillet C, Brossay L, Vivier E. Assassinos naturais ou ILC1s? Essa é a questão. Current Opinion in Immunology. 2021;**68**:48-53. DOI: 10.1016/j.coi.2020.08.009

[5] Bryceson YT, Março ME, Barbeiro DF, Ljunggren HG, Long EO. Polarização e degranulação citolítica de grânulos controlada por diferentes receptores em células NK em repouso. The Journal of Experimental Medicine. 2005;**202**(7):1001-1012. DOI: 10.1084/jem.20051143

[6] Bashirova AA1, Martin MP, McVicar DW, Carrington M. The killer immunoglobulin-like receptor gene cluster: Tuning the genome for defense. Annual Review of Genomics and Human Genetics. 2006;**7**:277-300

[7] Hsu KC, Dupont B. Natural killer cell receptors: Regulating innate immune responses to hematologic malignancy. Seminars in Hematology. 2005;**42**(2):91-103. DOI: 10.1053/j.seminhematol.2005.01.010

[8] Karvouni M, Vidal-Manrique M, Lundqvist A, Alici E. Engineered NK cells against cancer and their potential applications beyond. Frontiers in Immunology. 2022;**13**:825979. DOI: 10.3389/fimmu.2022.825979

[9] Bottino C, Castriconi R, Pende D, Rivera P, Nanni M, Carnemolla B, et al. Identification of PVR (CD155) and Nectin-2 (CD112) as cell surface ligands for the human DNAM-1 (CD226) activating molecule. The Journal of Experimental Medicine. 2003;**198**(4):557-567. DOI: 10.1084/jem.20030788. Epub 2003 Aug 11

[10] Hilton HG, Norman PJ, Nemat-Gorgani N, Goyos A, Hollenbach JA, Henn BM, et al. Loss and gain of natural killer cell receptor function in an African hunter-gatherer population. PLoS Genetics. 2015;**11**:e1005439. DOI: 10.1371/journal.pgen.1005439

[11] Hilton HG, Guethlein LA, Goyos A, Nemat-Gorgani N, Bushnell DA, Norman PJ, et al. Polymorphic HLA-C receptors balance the functional characteristics of KIR haplotypes. Journal of Immunology. 2015;**195**:3160-3170

[12] Pende D, Falco M, Vitale M, Cantoni C, Vitale C, Munari E, et al. Killer Ig-like receptors (KIRs): Their role in NK cell modulation and developments leading to their clinical exploitation. Frontiers in Immunology. 2019;**10**:1179. DOI: 10.3389/fimmu.2019.01179

[13] Sim MJW, Brennan P, Wahl KL, Lu J, Rajagopalan S, Sun PD, et al. Innate receptors with high specificity for HLA class I-peptide complexes. Science

Immunology. 2023;**8**(87):eadh1781. DOI: 10.1126/sciimmunol.adh1781. Epub 2023 Sep 8

[14] Bryceson YT, Fauriat C, Nunes JM, Wood SM, Björkström NK, Long EO, et al. Functional analysis of human NK cells by flow cytometry. Methods in Molecular Biology. 2010;**612**:335-352. DOI: 10.1007/978-1-60761-362-6_23

[15] Lanier LL. NK cell recognition. Annual Review of Immunology. 2005;**23**:225-274

[16] Malmberg KJ, Bryceson YT, Carlsten M, Anderson S, Bjorklund A, Bjorkstrom NK, et al. NK cells-mediated targeting of human cancer and possibilities for new means of immunotherapy. Cancer Immunology, Immunotherapy. 2008;**57**:1541-1552

[17] Moretta L, Moretta A. Unravelling natural killer cell function: Triggering and inhibitory human NK receptors. The EMBO Journal. 2004;**23**(2):255-259

[18] Alici E, Sutlu T, Bjorkstrand B, Gilljam M, Stellan B, Nahi H, et al. Autologous antitumor activity by NK cells expanded from myeloma patients using GMP-compliant components. Blood. 2008;**111**:3155-3162

[19] Butcher S, Arney KL, Cook GP. MAFA-L, an ITIM-containing receptor encoded by the human NK cell gene complex and espressed by basophils and NK cells. European Journal of Immunology. 1998;**28**(11):3755-3762

[20] Colonna M, Navarro F, Bellon T, Llano M, Garcia P, Samaridis J, et al. A common inhibitory receptor for major histocompatibility complex class I molecules on human lymphoid and myelomonocytic cells. The Journal of Experimental Medicine. 1997;**186**(11):1809-1818

[21] Ito M, Maruyama T, Saito N, Koganei S, Yamamoto K, Matsumoto N. Killer cell lectin-like receptor G1 binds three members of the classical cadherin family to inhibit NK cell cytotoxicity. The Journal of Experimental Medicine. 2006;**203**(2):289-295

[22] Parham P. MHC class I molecules and KIR in human history, health and survival. Nature Reviews Immunology. 2005;**5**:201-214

[23] Moretta A, Bottino C, Vitale M, Pende D, Cantoni C, Mingari MC, et al. Activating receptors and coreceptors involved in human natural killer cell-mediated cytolysis. Annual Review of Immunology. 2001;**19**:197-223

[24] Pende D, Marcenaro S, Falco M, et al. Anti-leukemia activity of alloreactive NK cells in KIR ligandmismatched haploidentical HSCT for pediatric patients: Evaluation of the functional role of activating KIR and redefinition of inhibitory KIR specificity. Blood. 2009;**113**(13):3119-3129

[25] Raulet DH. Roles of the NKG2D immunoreceptor and its ligands. Nature Reviews. Immunology. 2003;**3**(10):781-790

[26] Poggi A, Zocchi MR. Human natural killer lymphocytes through the engagement of natural cytotoxity receptors and NKG2D can togger self-agression. Autoimmunity Reviews. 2007;**6**(5):295-299

[27] Coudert JD, Held W. The role of the NKG2D receptor for tumor immunity. Seminars in Cancer Biology. 2006;**16**(5):333-343

[28] Guerra N, Tan YX, Joncker NT, Choy A, Gallardo F, Xiong N, et al. NKG2D-deficient mice are defective in tumor surveillance in models od

spontaneous malignancy. Immunity. 2008;**28**(4):571-580

[29] Cosman D, Mulberg J, Sutherland CL, Chin W, Armitage R, Fanslow W, et al. ULBPs, novel MHC class I-related molecules, bind to CMV glycoprotein UL16 and stimulated NK citotoxicity through the NKG2D receptor. Immunity. 2001;**14**(2):123-133

[30] Biassoni R, Cantoni C, Pende D, Sivori S, Parolini S, Vitale M, et al. Human natural killer cell receptors and co-receptors. Immunological Reviews. 2001;**181**:203-214

[31] Diefenbach A, Jensen ER, Jamieson AM, Raulet DH. Rael and H60 ligands of the NKG2D receptor stimulate tumour immunity. Nature. 2001;**98**(20):11521-11526

[32] Carlsen M. Molecular specificities of NK cell-mediated recognition of human tumor cells [Thesis]. Karolinska Institutet; 2010

[33] Iguchi-Manaka A, Kai H, Yamashita Y, Shibata K, Tahara-Hanaoka S, Honda S, et al. Accelerated tumor growth in mice deficient in DNAM-1 receptor. The Journal of Experimental Medicine. 2008;**205**(13):2959-2964

[34] Bryceson YT, March ME, Ljunggren HG, Long EO. Synergy among receptors on resting NK cells for the activation of natural cytotoxicity and cytokine secretion. Blood. 2006;**107**(1):159-166

[35] Caligiuri MA. Human natural killer cells. Blood. 2008;**112**(3):461-469

[36] Ljunggren HG, Karre K. In search of the missing self: MHC molecules and NK cell recognition. Immunology Today. 1990;**11**:237-244

[37] Garcia-Lora A, Martinez M, Algarra I, Gaforio JJ, Garrido F. MHC class I-deficient metastatic tumor variants immunoselected by T lymphocytes originate from the coordinated downregulation of APM components. International Journal of Cancer. 2003;**106**(4):521-527

[38] Schanoski AS, Cavalcanti TC, Campos CB, Viera-Matos AN, Rettori O, Guimarães F. Walker 256 tumor MHC class I expression during the shift from a variant to the immunogenic AR variant. Cancer Letters. 2004;**211**(1):119-127

[39] Yawata M, Yawata N, Draghi M, Partheniou F, Little A, Parham P. MHC class Ispecific inhibitory receptors and their ligands structure diverse human NK cell repertoires towards a balance of missing-self response. Blood. 2008;**112**:2369-2380

[40] Asai O, Longo DL, Tian Z, et al. Suppression of graft-versus-host disease and amplification of graft-versus-tumor effects by activated natural killer cells after allogeneic bone marrow transplantation. The Journal of Clinical Investigation. 1998;**101**:1835-1842

[41] Ruggeri L, Capanni M, Urbani E, et al. Effectiveness of donor natural killer cell alloreactivity in mismatched hematopoietic transplants. Science. 2002;**295**:2097-2100

[42] Miller J, Cooley S, Parham P, et al. Missing KIRligands is associated with less relapse and increased graft versus host disease (GVHD) following unrelated donor allogeneic HCT. Blood. 2007;**109**:5058-5061

[43] Yu J, Venstrom JM, Liu X, Pring J, Hasan RS, O'Reilly RJ, et al. Breaking tolerance to self, circulating natural killer cells expressing inhibitory KIR for non-self HLA exhibit effector

function after T cell–depleted allogeneic hematopoietic cell transplantation. Blood. 2009;**113**(16):3875-3884

[44] Sobecks RM, Ball EJ, Maciejewski JP, Rybicki LA, Brown S, Kalaycio M, et al. Survival of AML patients receiving HLA-matchedsibling donor allogeneic bone marrow transplantation correlates with HLA-Cw ligandgroups for killer immunoglobulin-like receptors. Bone Marrow Transplantation. 2007;**39**:417-424

[45] Björklund AT, Schaffer M, Fauriat C, Ringdén O, Remberger R, Hammarstedt C, et al. NK cells expressing inhibitory KIR for non-self-ligands remain tolerant in HLA-matched sibling stem cell transplantation. Blood. 2010;**115**:2686-2694

[46] Hsu KC, Taylor CAK, Wilton A, et al. Improved outcome in HLA-identical sibling hematopoietic stem-cell transplantation for acute myelogenous leukaemia predicted by KIR and HAL genotypes. Blood. 2005;**105**:4878-4884

[47] Andersson S, Fauriat C, Malmberg JA, Ljunggren HG, Malmberg KJ. KIR acquisition probabilities are independent of self-HLA class I ligands and increase with cellular KIR expression. Blood. 2 Jul 2009;**114**(1):95-104. DOI: 10.1182/blood-2008-10-184549. Epub 2009 Mar 20. PMID: 19304956

[48] Zaitoua AJ, Kaur A, Raghavan M. Variations in MHC class I antigen presentation and immunopeptidome selection pathways. F1000Res. 2020;**9**:F1000 Faculty Rev-1177. DOI: 10.12688/f1000research.26935.1

[49] Raghavan M, Geng J. HLA-B polymorphisms and intracellular assembly modes. Molecular Immunology. 2015;**68**(2 Pt A):89-93. DOI: 10.1016/j.molimm.2015.07.007

[50] Garcia KC, Adams EJ. How the T cell receptor sees antigen--A structural view. Cell. 2005;**122**(3):333-336

[51] Malnati MS et al. Peptide specificity in the recognition of MHC class I by natural killer cell clones. Science. 1995;**267**(5200):1016-1018

[52] Peruzzi M et al. Peptide sequence requirements for the recognition of HLA-B*2705 by specific natural killer cells. Journal of Immunology. 1996;**157**(8):3350-3356

[53] Storkus WJ et al. Peptide-induced modulation of target cell sensitivity to natural killing. Journal of Immunology. 1992;**149**(4):1185-1190

[54] Zappacosta F et al. Peptides isolated from HLA-Cw*0304 confer different degrees of protection from natural killer cell-mediated lysis. Proceedings of the National Academy of Sciences of the United States of America. 1997;**94**(12):6313-6318

[55] Rajagopalan S, Long EO. The direct binding of a p58 killer cell inhibitory receptor to human histocompatibility leukocyte antigen (HLA)-Cw4 exhibits peptide selectivity. The Journal of Experimental Medicine. 1997;**185**(8):1523-1528

[56] Boyington JC et al. Crystal structure of an NK cell immunoglobulin-like receptor in complex with its class I MHC ligand. Nature. 2000;**405**(6786):537-543

[57] Fan QR, Long EO, Wiley DC. Crystal structure of the human natural killer cell inhibitory receptor KIR2DL1-HLA-Cw4 complex. Nature Immunology. 2001;**2**(5):452-460

[58] Stewart CA et al. Recognition of peptide-MHC class I complexes by activating killer immunoglobulin-like

receptors. Proceedings of the National Academy of Sciences of the United States of America. 2005;**102**(37):13224-13229

[59] Sim MJ et al. Canonical and cross-reactive binding of NK cell inhibitory receptors to HLA-C allotypes is dictated by peptides bound to HLA-C. Frontiers in Immunology. 2017;**8**:193

[60] Khakoo SI, Carrington M. KIR and disease: A model system or system of models? Immunological Reviews. 2006;**214**:186-201

[61] Purdy AK, Campbell KS. Natural killer cells and cancer: Regulation by the killer cell Ig-like receptors (KIR). Cancer Biology and Therapy. 2009;**8**(23):2211-2220. Epub 2009 Dec 28

[62] Pende D, Marcenaro S, Falco M, Martini S, Bernardo ME, Montagna D, et al. Anti-leukemia activity of alloreactive NK cells in KIR ligand-mismatched haploidentical HSCT for pediatric patients: Evaluation of the functional role of activating KIR and redefinition of inhibitory KIR specificity. Blood. 2009;**113**(13):3119-3129. DOI: 10.1182/blood-2008-06-164103. Epub 2008 Oct 22

[63] Alter G et al. HIV-1 adaptation to NK-cell-mediated immune pressure. Nature. 2011;**476**(7358):96-100

[64] van Teijlingen NH et al. Sequence variations in HIV-1 p24 gag-derived epitopes can alter binding of KIR2DL2 to HLA-C*03:04 and modulate primary natural killer cell function. AIDS. 2014;**28**(10):1399-1408

[65] Holzemer A et al. Selection of an HLA-C*03:04-restricted HIV-1 p24 gag sequence variant is associated with viral escape from KIR2DL3+ natural killer cells: Data from an observational Cohort in South Africa. PLoS Medicine. 2015;**12**(11):e1001900. discussion e00

[66] Khakoo SI, Thio CL, Martin MP, Brooks CR, Gao X, Astemborski J, et al. HLA and NK cell inhibitory receptor gene in resolving hepatitis C virus infection. Science. 2004;**305**:872-874

[67] Hiby SE, Walker JJ, O'Shaughnessy KM, Redman CWG, Carrington M, Trowsdale J, et al. Combinations of maternal KIR and fetal HLA-C genes influence the risk of preeclampsia and reproductive success. The Journal of Experimental Medicine. 2004;**200**:957-965

[68] Ruggeri L, Capanni M, Urbani E, Perruccio K, Shlomchik WD, Tosti A, et al. Effectiveness of donor natural killer cell alloreactivity in mismatched hematopoietic transplants. Science. 2002;**295**:2097-2100

[69] Nelson GW, Martin MP, Gladman D, Wade J, Trowsdale J, Carrington M. Cutting edge: Heterozygote advantage in autoimmune disease: Hierarchy of protection/susceptibility conferred by HLA and killer Ig-like receptor combinations in psoriatic arthritis. Journal of Immunology. 2004;**173**:4273-4276

[70] Venstrom G, Pittari TA, Gooley JH, Chewning S, Spellman M, Haagenson MM, et al. Hsu, HLA-C-dependent prevention of leukemia relapse by donor activating KIR2DS1. The New England Journal of Medicine. 2012;**367**:805-816

[71] Cardozo DM, Marangon AV, da Silva RF, Aranha FJP, Visentainer JEL, Bonon SHA, et al. Synergistic effect of KIR ligands missing and cytomegalovirus reactivation in improving outcomes of haematopoietic stem cell transplantation from HLA-matched

sibling donor for treatment of myeloid malignancies. Human Immunology. 2016;**77**(10):861-868. DOI: 10.1016/j.humimm.2016.07.003. Epub 2016 Jul 6

[72] Diefenbach A, Raulet DH. Strategies for target cell recognition by natural killer cells. Immunological Reviews. 2001;**181**:170-184. DOI: 10.1034/j.1600-065x.2001.1810114.x

[73] García-Lora A, Martinez M, Pedrinaci S, Garrido F. Different regulation of PKC isoenzymes and MAPK by PSK and IL-2 in the proliferative and cytotoxic activities of the NKL human natural killer cell line. Cancer Immunology, Immunotherapy. 2003;**52**(1):59-64. DOI: 10.1007/s00262-002-0336-9. Epub 2002 Oct 17

[74] Gao F, Ye Y, Gao Y, Huang H, Zhao Y. Influence of KIR and NK cell reconstitution in the outcomes of hematopoietic stem cell transplantation. Frontiers in Immunology. 2020;**11**:2022. DOI: 10.3389/fimmu.2020.02022

[75] Leung W. Use of NK cell activity in cure by transplant. British Journal of Haematology. 2011;**155**:14-29. DOI: 10.1111/j.1365-2141.2011.08823.x

[76] Leung W, Iyengar R, Turner V, Lang P, Bader P, Conn P, et al. Determinants of antileukemia effects of allogeneic NK cells. Journal of Immunology. 2004;**172**(1):644-650. DOI: 10.4049/jimmunol.172.1.644

[77] Cooley S, Trachtenberg E, Bergemann TL, Saeteurn K, Klein J, Le CT, et al. Donors with group B KIR haplotypes improve relapse-free survival after unrelated hematopoietic cell transplantation for acute myelogenous leukemia. Blood. 2009;**113**:726-732. DOI: 10.1182/blood-2008-07-171926

Section 2

Mesechymal Stem Cells Applications in Regenerative Medicine

Chapter 11

Dissecting the Multifaceted Therapeutic Implications of Mesenchymal Stem Cell-Derived Extracellular Vesicles: An Intricate Exploration

Naina Soni and Rashmi Rameshwari

Abstract

Mesenchymal stem cells (MSCs) are prized for their versatility and therapeutic potential. Derived from various sources such as the umbilical cord, adipose tissue, or bone marrow, MSCs have garnered attention over the past 2 decades for their regenerative and immunomodulatory capabilities. What sets MSCs apart is their ability to facilitate repair through paracrine pathways, notably via extracellular vesicles (EVs). These EVs offer a novel therapeutic approach, eliminating the need for direct cell transfer or injection and mitigating risks like graft rejection and tumor formation. Recent studies highlight the therapeutic advantages of MSC EVs, loaded with mRNA, proteins, and miRNA from their parent cells. Operating primarily through the endosomal compartment, they influence disease processes through intricate signaling pathways, offering promise for diverse medical conditions. Our review delves into the multifaceted role of MSCs and their EVs in mediating therapeutic outcomes.

Keywords: paracrine signaling, clinical studies, therapeutics approach, inflammation, regeneration

1. Introduction

Cells that can self-renew and display multilineage differentiation are known as MSCs. MSCs can be isolated from various tissues, notably the umbilical cord, endometrium, menstrual blood, bone marrow, adipose tissue, and other ectodermal lineage cells (**Figure 1**) [1]. Friedenstein and his team found the first evidence of MSCs' multipotent behavior in 1966 [2]. Following an extended period, Caplan et al. introduced the term MSCs in 1991 [3]. Before this, scientists disagreed on the stemness and nomenclature of MSCs. Some articles favored mesenchymal "stromal" cells above stem cells. Due to their secretory function at the sites of illnesses, traumas, and inflammations, several researchers tried to rename MSCs as therapeutic signaling cells [4]. Later studies revealed that prostaglandin E2 (PGE2), which MSCs release, is

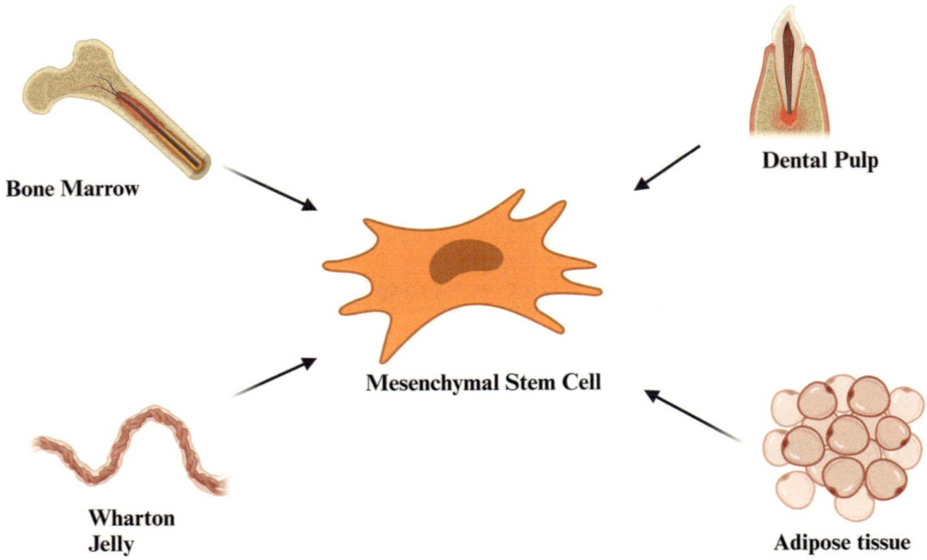

Figure 1.
Significant sources of mesenchymal stem cells.

also involved in the immunomodulation of MSCs and is responsible for preserving the ability of MSCs to self-renew and demonstrating the stemness of MSCs [5]. Thus, the term "mesenchymal stem cells" found suitable.

1.1 Identity and roles of the MSC

MSCs are frequently extracted from neonatal tissue, such as the umbilical cord, adult bone marrow, and adipose tissue [5]. MSCs are believed to have a variety of cell subsets in culture due to intrinsic and extrinsic factors and inherent differences linked to various sources and donors [6]. Multipotent cells and diverse but committed progenitors are included in MSCs in culture. MSCs from different sources may not all be comparable. However, MSC cultures may have characteristics in common that meet the International Society of Cell Therapy (ISCT) standards established in 2006 [7]. Notably, ISCT proposed three minimal criteria to identify MSCs:

1. adherent cells with a spindle-shaped morphology under standard culture conditions.

2. cell surface expression of cluster of differentiation (CD)105, CD73, and CD90, but not of CD45, CD34, CD14 or CD11b, CD79, or CD19.

3. *in vitro* differentiation of MSCs into osteoblasts, adipocytes, and chondroblasts in response to a specific stimulation [8].

Importantly, MSC culture systems must be optimized and standardized to meet the requirements of excellent manufacturing output and to fulfill the demands for the highest safety, quality, and quantity of MSCs [9]. Efforts to enhance MSC culture include promoting advanced, stringent aseptic techniques. These hypoxic conditions mimic the native microenvironment of MSCs, avoiding non-human products (such as xenogeneic-free medium) and limiting ambiguous medium composition (e.g., by encouraging the use of serum-free medium). Additionally, MSCs are subcultured using conventional cell culture systems or through substantial bioprocessing employing high-capacity bioreactors designed for high cell production yield to satisfy therapeutic demands [10].

Even though MSCs have a fibroblast-like shape in vitro, they are proliferative. The proliferation function of MSCs appears to be tightly controlled early in culture when wingless-type (Wnt)/−catenin signaling is not active. Additionally, oxygen controls the growth of MSCs by altering the transcription factor hypoxia-inducible factor 1, which allows the expression of genes that direct cell cycle development [11].

1.2 Multipotent MSC characteristics

A cell must be able to proliferate for self-renewal and cell expansion, which is connected to "stemness" [10]. Under specialized in vitro stimulation, MSCs differentiate into adipocytes, chondroblasts, and osteoblasts (**Figure 2**); the differentiation is seen morphologically and with particular biomarker expression. Depending on the biological circumstances, MSCs may undergo an overhaul of their intracellular signaling and transcriptional alterations in vivo or through in vitro treatments. Dexamethasone, insulin, isobutylmethylxanthine, and indomethacin frequently stimulate MSCs in vitro to induce adipogenesis. Lipid vacuoles that gradually coalesce into a single big vacuole in terminally differentiated adipocytes indicate MSCs developing into adipocytes. Adipogenesis can be measured by lipoprotein lipase enzyme,

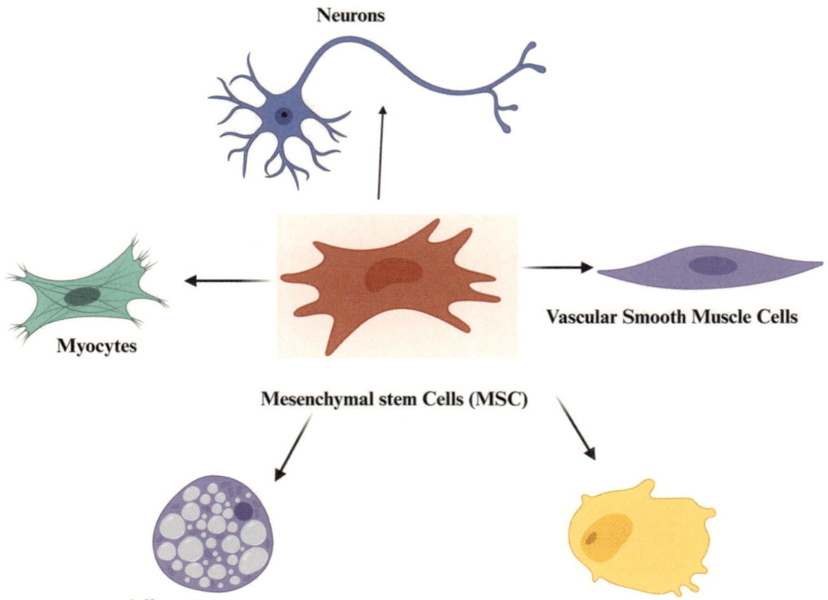

Figure 2.
Mesenchymal stem cells and their differentiations.

activity and the buildup of the fatty acid- binding protein adipocytes P2 in mature adipocytes [12].

MSCs are often cultured in aggregate or pellet cultures and treated with transforming growth factor (TGF), TGF-1, TGF-3, insulin-like growth factor-1 (IGF-1), fibroblast growth factor-2 (FGF-2), or bone morphogenic protein-2 (BMP-2) to produce chondrogenesis in vitro [13]. By activating MSCs with ascorbic acid, glycerophosphate, vitamin D3, and BMP-2, -4, -6, and -7, osteogenesis can be accomplished in vitro [14]. According to some research, MSCs may also develop into myoblasts, endothelial progenitors, and specialized cells other than mesoderm lineage, most notably neuroblasts [15, 16].

1.3 MSCs diverse functionality

MSCs are thought to control homeostasis in hematopoietic niches in vivo by assisting hematopoietic stem cell (HSC) growth, maintenance, and differentiation. Additionally, MSCs may assist other progenitors' in vivo homeostasis [17]. The MSCs may help promote tissue healing and regeneration in several organs and maintain the homeostasis of HSCs and bone tissue [18]. MSC functionality has been tested in a variety of degenerative experimental models, including the kidney (acute renal injury), the liver (liver cirrhosis), or several organs (diabetes complications) [19]. Improvements in organ health and function have been linked to MSC production of growth factors, specifically hepatocyte growth factor (HGF), IGF-1, and occasionally growth factors related to angiogenesis and neurogenesis such as vascular endothelial growth factor (VEGF), erythropoietin (EPO), and glial cell-derived neurotrophic factor (GDNF) [20, 21].

1.4 MSC homing

MSC homing is the halted movement of MSCs within the tissue's vasculature, followed by their transmigration through the endothelium. Given the lack of a precise MSC homing mechanism, in contrast to the well-known leukocyte adhesion cascade that defines leukocyte homing, such a non-mechanistic characterization is reasonable. It is challenging to identify whether the cells have halted trans endothelial migration (homing) or have arrested inside the vasculature (localization) because of MSC infusion due to a lack of information describing their precise location.

1.4.1 Are host MSCs capable of peripheral blood mobilization?

There is conflicting data on the likelihood of MSC trafficking under physiological steady-state circumstances. Several studies report being unable to find any circulating MSCs at all, whereas others describe just trace amounts of MSCs in the blood being present [22, 23].

1.5 MSCs and their mode of action

There have been several hypothesized mechanisms by which MSCs could exert their beneficial effects. Early research suggested that MSCs could travel to damaged areas before differentiating into functional cells. Or that they could combine with damaged cells to repair or regenerate tissues. More recent research has shown that extracellular vesicle secretion, mitochondrial transfer, and paracrine factors all play significant mediating roles in the effects of MSCs. Because of the paradigm change toward paracrine signaling as the fundamental mechanism of MSC therapeutic efficacy, the emphasis on MSC conditioned media's regenerative and immunomodulatory potential contains numerous growth factors, and other small molecular weight conveys cues. The bundle of the medicinal advantages offered by MSCs has been linked to these MSC-secreted components found in conditioned media, or the MSC "secretome" [24, 25]. The MSC secretome has been used in numerous investigations, showing therapeutic potential in widely used injury models [26, 27]. Depending on different cellular and preparation factors, the content and concentration of the secretome's constituents might vary considerably. An exosome-specific type of extracellular vesicles can mimic most of the characteristics of human stem cells (hMSCs) [28].

2. MSC-derived EVs

EV research's growth and more profound comprehension era began in the early 1980s. The coherence started here, even the significant burst of publications, theories, and discussions about nomenclature and EV-related organizations would not start for another 20 years or more. The Johnstone and Stahl groups published two vital and complementary publications that established beyond a shadow of a doubt that intraluminal vesicles are released from all types of cells and are known as exosomes [29, 30]. EVs are functional vesicles and have been widely studied in terms of their clinical aspects—diagnostics and therapeutics. If we talk about the importance of EVs in therapeutics, then MSC-derived EVs are significant; they mediate and preserve their origin's biological capabilities, highlighting their importance in regulating biological processes and immunoregulation [31]. The International Society for Extracellular Vesicles (ISEV) surveyed the utility of MSC-EV isolation methods, revealing that ultracentrifugation-based methods are the most used despite many drawbacks. Several alternatives are available for the isolation with varied purity and concentration [32]. EVs can be modified to enhance therapeutic potential, incorporating drugs, antibodies, proteins, RNA, and molecules for targeted delivery and traceability, generating novel approaches like bio-engineered EVs [33].

2.1 EVs' chemical makeup

EVs biogenesis is a meticulously controlled process that involves three main steps: the formation of endocytic vesicles through invagination of the plasma membrane, the construction of multivesicular bodies (MVBs) through inward budding of the endosomal membrane, and finally, shaping communication between produced MVBs and plasma membrane and release of the vesicular composites, known as exosomes [33]. The exosome comprises a lipid bilayer and a tiny cytoplasm devoid of cellular organelles. The tetraspanin proteins CD9, CD63, and CD81, typically present on exosome membranes, are distinctive exosome proteins that participate in membrane fusion, signaling, and protein trafficking. Additionally essential for exosome biogenesis are ALIX, flotillin, and TSG101. ESCRT mediates MVB formation and ILV engulfment and is a common exosome cargo protein with the related proteins flotillin, TSG101, ALIX, and hepatocyte growth factor-regulated tyrosine kinase substrate (Hrs). MVB binding to the plasma membrane and exosome secretion are mediated by the vesicle fusion and transport proteins rab27a and annexin. Additionally concentrated in exosomes are the heat shock proteins Hsp70 and Hsp90, as well as the cytoskeletal proteins actin and myosin. Together, these proteins act as indicators that help isolate exosomes from other EVs and can be utilized to identify exosomes [34–37].

2.2 EVs and their activities

In biological circuits, exosomes carry out several important tasks. Microvesicles, for instance, take a role in cell signaling, apoptotic bodies transmit the contents of dying cells to healthy cells, and so on.

2.3 EVs and their therapeutic application

Delivery of exosomal cargo may contribute to the development of illness or protection from the disease. It entirely depends on the source of exosomes. Exosome-based therapies and diagnostics must be developed, which requires an understanding of the involvement of exosomes in disease pathways. The roles of exosomes in cancer, cardiovascular illness, and neurodegenerative disease are discussed in this part, followed by descriptions of clinical studies using exosomes as diagnostic and therapeutic tools.

2.4 MSCs and EVs in various diseases

2.4.1 Skin repair and wound healing

Several studies have been conducted or are being conducted to examine the efficacy, viability, and safety of MSC-based therapies for wound healing and skin regeneration in human subjects. A recent study on 24 patients with aged skin who received ablative fractional laser (AFL) treatment found that (adipose tissue—AT) AT-MSCS-CM generated positive effects, most likely through reducing MMP-1 and MMP-2 expression coupled with improving collagen one expression [38]. Additionally, the injection of allogeneic (UC—umbilical cord) UC-MSCs comprising serum and cream sped up wound healing. It reduced post-treatment erythema, demonstrating their vital capability to assist recovery following AFL [39]. Additionally, several studies have shown that MSC-EVs can help with various chronic wounds. A study reported that BMMSC-EVs increased the ex vivo proliferation and migration of fibroblasts from healthy and chronic wound subjects in a dose-dependent manner [40]. These EVs also facilitated tube formation by endothelial cells by activating pathways (Akt, ERK, and STAT3) involved with wound healing [41]. Zhang J. et al. reported that—an in vitro study revealed that therapy with human-induced pluripotent mesenchymal stem cells (iPSCs)—EVs could improve cutaneous wound healing by boosting collagen production and angiogenesis [42]. Higher collagen and elastin synthesis was observed when human ADMSC-EVs were introduced to photodamaged human dermal fibroblasts in vitro. Another finding suggests that in vivo wound healing was increased by injecting human ADMSC-EVs into a mouse skin incision model. Collagen I and III distributions were specifically secreted. Fibroblasts were shown to be more abundant in the initial phases of wound healing, but collagen was increased in the late stages. In vivo, wound healing was increased by injecting human ADMSC-EVs into a mouse skin incision model [43]. Another researcher compared the effect of MSC-EVs from two different sources and found that when rabbit ADMSC-EVs and BMMSC-EVs were tested on rat cutaneous wound models, ADMSC-EVs demonstrated significantly superior healing [44].

2.4.2 Cardiovascular disease (CVD)

A vascular condition that affects diabetic people is called DR. The pathophysiology of DR is linked to inflammatory mechanisms that harm microcirculation. As a result, the blood-retina barrier (BRB) is disrupted, primarily due to pericyte loss, one of the initial signs of DR [45]. In actuality, retinal sensory dysfunction, mainly brought on by retinal ganglion cell (RGC) loss, is related to the visual loss that manifests in the early stages of DR [46]. In advanced RP patients, a phase I clinical research assessed the efficacy, feasibility, and safety of intravitreal injection of autologous BMSC. Despite minor side effects, BCVA significantly improved, visual field (VF) and central subfield thickness (CST) remained steady, and one profound side effect appeared 3 years after the procedure [47]. After administering UCMSC to 82 retinopathy (RP) patients, the phase III clinical trial identified no severe side effects or ocular problems. As a result, best-corrected visual acuity (BCVA), VF, and multifocal electroretinography recordings significantly improved [48]. In phase I/II clinical research, including 32 patients with advanced RP, UCMSCs were found to be safe and beneficial. Most patients improved BCVA, particularly in the first 3 months, suggesting that they may enhance their quality of life.

2.4.3 Neurodegenerative diseases

CVD is the primary cause of death worldwide. The most prevalent manifestations of CVD include hypertension, coronary artery disease, myocardial injury, and heart failure. Despite tremendous breakthroughs in the care of cardiac disease over the previous 2 decades, there is an unmet clinical need to treat this vast and growing patient population. Endogenous myocardial repair following injury is highly sluggish, relying on the restricted self-division of pre-existing cardiomyocytes and the recruitment and differentiation of local cardiac stem cells. Early research found that human embryonic-derived MSC-EVs may reduce infarct size in a mouse model of myocardial ischemia/reperfusion damage (MI) by activating the PI3K/Akt signaling pathway [49], which promoted myocardial viability and reduced unfavorable remodeling [50]. Another study reported that murine (iPSCs) EVs and human amniotic fluid-derived mesenchymal stem cells (hAFS) were reported to promote cardiac repair in MI murine models and activate heart regeneration via paracrine activation of endogenous processes [51, 52]. Additionally, MSC-EVs can be encased in hydrogels for regulated and targeted administration as an alternative to injecting them into the heart defect. When human UCMSC-EVs were loaded with functional peptides, a sustained release profile and enhanced heart regeneration were observed. The EV/hydrogel compound improved myocardial function by lowering inflammation, fibrosis, and apoptosis and increasing angiogenesis in the rat-infarcted border zone hearts [53].

2.4.4 Diabetic retinopathy (DR)

During the last few years, a considerable body of evidence has accumulated addressing the use of MSCs and their EVs in the treatment of neurodegenerative disorders, namely, amyotrophic lateral sclerosis (ALS), Alzheimer's, and Parkinson's disease (PD). ALS is a fatal paralytic neurodegenerative disease caused by motor neuron degeneration in the brain and spinal cord [54]. Animal studies show promising results in ALS treatment with MSCs, showing therapeutic potential and efficacy in preclinical studies using MSC-conditioned media and exosomes [55, 56]. MSC therapy in ALS has exhibited a satisfactory safety profile in early-phase human clinical trials. A phase I trial indicated that transient back/leg pain was safe. Other investigations using naive bone marrow-derived MSCs to increase neurotrophic factor secretion reported comparable safety characteristics [57–60]. Dementia is a neurodegenerative disorder causing cognitive decline, often causing Alzheimer's disease (AD), with memory decline appearing early in the disease's clinical phase. The pathologic hallmarks of Alzheimer's disease (plaques and neurofibrillary tangles) have long been known to be made up of beta-amyloid and tau protein aggregates. Mouse models of Alzheimer's disease (AD) have reported a decrease in amyloid deposits with bone marrow-derived MSCs treatment. These MSCs have been shown to reduce microglia cell counts but not amyloid plaques. The effect of MSCs on AD pathology and cognition may be mediated through modulatory effects on neuroinflammation (**Figure 3**), inhibition of apoptosis, and enhanced neurogenesis via the Wnt signaling pathway [61–63]. According to many research, the anti-inflammatory action of MSC-EVs occurs through the reduction/modulation of activated microglia, which is a significant mediator of inflammation when chronically active [64]. There was limited evidence that neurogenesis and neuroprotection resulted in the recovery of synaptic abnormalities or, more broadly, an overall decrease in neuronal degeneration. Many studies reported on how the extracellular environment influences such activities. For example, MSCs exposed to hypoxia produced EVs that performed better in repairing learning memory deficits than MSCs grown in normoxic settings [65]. Tremors, bradykinesia, and rigidity clinically define PD; several non-motor symptoms and findings commonly involve cognitive dysfunction, mood disorders, sleep disturbances, and autonomic dysfunction, which go beyond the classification of PD as a movement disorder. Lewy bodies—intracytoplasmic neuronal alpha-synuclein inclusions—and neuronal loss are pathologic hallmarks of PD in selected areas of the brain, including the substantia nigra, locus coeruleus, dorsal vagal nucleus, and cerebral cortex; however, Lewy pathology is also found in the spinal cord, sympathetic ganglia, as well

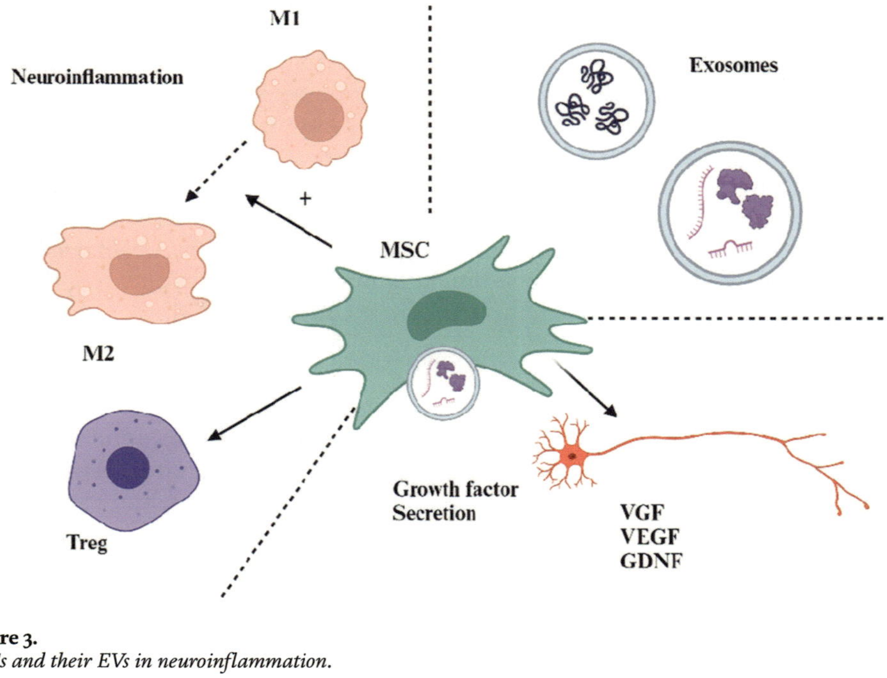

Figure 3.
MSCs and their EVs in neuroinflammation.

as the cardiac sympathetic denervation [66]. MSCs, which secrete neurotrophic factors, modulate inflammation, and potentially act as mitochondria donors, are being explored to treat PD. Animal studies show promising results, including improved motor function, nigrostriatal protection, and dopamine release [67]. Preliminary data on using bone marrow-derived MSCs in PD is available. In a 2010 study, Indian researchers gave 106 MSCs to 7 PD patients, resulting in long-term improvement on the unified PD rating scale. In a 2012 study, 2×106 MSCs were administered to 8 PD and 8 PD plus patients [68, 69]. The proteomic analysis demonstrates that MSC EVs contain important neuroregulatory molecules responsible for the protective effects. Additional evidence suggests that BM-MSC EVs can lessen aberrant α-synuclein accumulation, oxidative stress, and neuroinflammation [70]. There have been no human investigations on the efficacy of MSC-EVs. MSC-EVs from human placenta have been studied in a mouse model of MS and reported that MSC-EVs stimulated myelin regeneration in vitro by differentiating endogenous oligodendrocyte precursor cells into mature myelinating oligodendrocytes and boosted myelination in the spinal cord followed by enhanced motor function outcomes in treated mice [71]. Furthermore, EVs derived from human BMMSCs treated with interferon-gamma (IFN-) decreased neuroinflammation and demyelination while improved motor abilities in an MS mice model of autoimmune encephalomyelitis (EAE) model [72].

2.4.5 Acute liver failure (ALF)

ALF is a fatal liver fibrosis syndrome causing hepatocyte necrosis and inflammatory infiltration. It progresses rapidly and presents poorly for medical treatment. Cirrhosis, the end stage, is prone to complications [73]. Many research has been undertaken to confirm the role of MSCs in ALF animal model therapy. MSCs can reduce mortality, improve liver functions, suppress hepatocyte apoptosis, and stimulate proliferation [74–76]. Currently, the use of MSCs in ALF is restricted to basic research. However, MSCs can promote liver regeneration and decrease inflammation. After 1 week of therapy, BM-MSCs effectively lowered the levels of ALT and ALB compared to the control group. These investigations suggested that MSCs have a wide range of applications in the clinical treatment of ALF [77]. In a study examining the benefits of MSC-EVs in connection to liver illness, human embryonic MSC-EVs were found to stimulate hepatic regeneration in a carbon tetrachloride (CCl_4)-induced liver injury mice model by boosting hepatocyte proliferation and decreasing hepatocyte apoptosis [78]. Further, human IPSC-EVS were also examined and showed beneficial effects by promoting hepatic regeneration in rat models of hepatic ischemia-reperfusion injury by preventing hepatic cell death, reducing inflammatory responses, and attenuating oxidative stress. IPSC-EVs were reported to enhance hepatocyte proliferation. In vitro and in vivo, in a dose-dependent manner, are associated with sphingosine activation [79]. Human embryonic MSC-EVs were also encased in PEG hydrogels instead of traditional injections for sustained systemic distribution against hepatic failure to increase the advantages of EVs further and harness the beneficial effects [80].

3. The benefits and difficulties of MSC-EVs therapy

There are various advantages to adopting MSC-EV therapy as opposed to traditional MSC-based treatment. Crucial points highlighting the supremacy of MSC-EVs therapy are as follows. (I) They are highly stable and can be preserved for an extended period without adding cryopreservatives. (II) MSC-EVs can produce more significant signaling in intercellular communication. They accomplish this by delivering functional proteins and microRNAs directly to target cells. (II) They have a lesser chance of immunological response when used allogeneically (between people with diverse genetic backgrounds). Despite various advantages, several substantial hurdles must be overcome before MSC-EVs can be used in clinical settings. The most significant benefits include fewer unwanted effects and problems and improved efficacy compared to parent MSCs. A few things need to be addressed to overcome the shortcomings: (I) Standard method for EV isolation because different ways yield varying purity and concentration. (II) Marker identification and defined nomenclature for various EVs shed from the cells. (II) Thorough understanding of the EVs content at protein and transcriptomics level to get detailed information about the effector molecule present in EVs. By keeping all this in view, one more critical thing is that the production of EVs needs to be standardized before moving further. An essential visit to their detrimental effect, which is unknown, is necessarily required [33, 75, 81, 82].

4. Conclusion

In recent decades, stem cell-based techniques have been widely researched to produce therapeutic procedures to counter human illnesses when other treatments fail. Although various stem cells have impressive potential therapeutic applications, they are currently not used in many circumstances due to ethical and safety concerns. MSCs, which are readily available even for autologous injection, are an appropriate compromise. Many experimental data from in vitro studies and animal models show promising results, even if only a small percentage of injected cells engrafted and differentiated into functional cells following transplantation. Because of the widespread paracrine action of MSCs, cell-free therapies have been tested and found to be effective. Clinical trials have been conducted to assess the safety and efficacy of MSC transplantation in neurodegenerative diseases, with intravenous and intrathecal techniques being the most used. Despite promising results in animal models, MSC therapy has yet to be shown beneficial to patients. Regardless, no malignancies or deaths have been reported because of the transplant.

Prospects: There is currently a considerable amount of research from preclinical and early-phase clinical trials that elucidates the mechanism of action of MSC treatment and identifies potential problems. It is also widely acknowledged that one of the significant constraints to the long-term success of MSC therapy is the transplanted cells' limited retention and survival.

Author contribution

NS and RR wrote the chapter, and both authors contributed to editing the document. Both authors have read and agreed to the published version of the chapter.

Conflicts of interest

The authors declare no conflict of interest.

Diagrams were created with BioRender.com.

Abbreviations

MSCs	mesenchymal stem cells
EVs	extracellular vesicles
TGF	transforming growth factor
VEGF	vascular endothelial growth factor
EPO	erythropoietin
GDNF	glial cell-derived neurotrophic factor
ESCRT	endosomal sorting complex required for transport
ALIX	ALG-2 interacting protein X
TSG101	tumor susceptibility gene 101
CD	(any number): cluster of differentiation
HSC	hematopoietic stem cells
ILV	internal limiting membrane
HSP	(any number): heat shock protein
DR	diabetic retinopathy
RGC	retinal ganglion cells
RP	retinopathy
BCVA	best corrected visual activity
VF	visual field
UCMSCs	umbilical cord derived mesenchymal stem cells
CVD	cardiovascular diseases
PI3K/Akt	phosphoinositide 3 – kinase/protein kinase B
M1	macrophage type – I
ALS	amyotrophic lateral sclerosis
PD	Parkinson disease
ALF	acute liver failure
MS	multiple sclerosis
EAE	autoimmune encephalomyelitis

Author details

Naina Soni* and Rashmi Rameshwari
Manav Rachna International Institute of Research and Studies,
Faridabad, Haryana, India

*Address all correspondence to: nainaneuro@gmail.com

References

[1] Salehinejad P, Moshrefi M, Eslaminejad T. An overview on mesenchymal stem cells derived from extraembryonic tissues: Supplement sources and isolation methods. Stem Cells and Cloning: Advances and Applications. 2020;**13**:57-65. DOI: 10.2147/SCCAA.S248519

[2] Friedenstein AJ, Piatetzky-Shapiro II, Petrakova KV. Osteogenesis in transplants of bone marrow cells. Journal of Embryology and Experimental Morphology. 1966;**16**(3):381-390. DOI: 10.1242/dev.16.3.381

[3] Caplan AI. Caplan-1991.pdf. Journal of Orthopaedic Research. [Online]. 1991. pp. 641-650. doi: 10.1002/jor.1100090504

[4] Dewindt TS, Vonk LA, Saris DBF. Response to: Mesenchymal stem cells: Time to change the name! Stem Cells Translational Medicine. 2017;**6**(8):1747-1748. DOI: 10.1002/sctm.17-0120

[5] Lee BC et al. PGE 2 maintains self-renewal of human adult stem cells via EP2-mediated autocrine signaling and its production is regulated by cell-to-cell contact. Scientific Reports. 2016;**6**(April):1-12. DOI: 10.1038/srep26298

[6] Naji A, Eitoku M, Favier B, Deschaseaux F, Rouas-Freiss N, Suganuma N. Biological functions of mesenchymal stem cells and clinical implications. Cellular and Molecular Life Sciences. 2019;**76**(17):3323-3348. DOI: 10.1007/s00018-019-03125-1

[7] Phinney DG, Galipeau J, Krampera M, Martin I, Shi Y, Sensebe L. MSCs: Science and trials. Nature Medicine. 2013;**19**(7):812. DOI: 10.1038/nm.3220

[8] Soni N, Gupta S, Rawat S, Krishnakumar V, Mohanty S, Banerjee A. MicroRNA-enriched exosomes from different sources of mesenchymal stem cells can differentially modulate functions of immune cells and neurogenesis. Biomedicine. 2022;**10**(1): 1-20. DOI: 10.3390/biomedicines10010069

[9] Galipeau J, Sensébé L. Mesenchymal stromal cells: Clinical challenges and therapeutic opportunities. Cell Stem Cell. 2018;**22**(6):824-833. DOI: 10.1016/j.stem.2018.05.004

[10] Fotia C, Massa A, Boriani F, Baldini N, Granchi D. Hypoxia enhances proliferation and stemness of human adipose-derived mesenchymal stem cells. Cytotechnology. 2015;**67**(6):1073-1084. DOI: 10.1007/s10616-014-9731-2

[11] Fehrer C et al. Reduced oxygen tension attenuates differentiation capacity of human mesenchymal stem cells and prolongs their lifespan. Aging Cell. 2007;**6**(6):745-757. DOI: 10.1111/j.1474-9726.2007.00336.x

[12] Chen Q et al. Fate decision of mesenchymal stem cells: Adipocytes or osteoblasts? Cell Death and Differentiation. 2016;**23**(7):1128-1139. DOI: 10.1038/cdd.2015.168

[13] Tuli R et al. Transforming growth factor-β-mediated chondrogenesis of human mesenchymal progenitor cells involves N-cadherin and mitogen-activated protein kinase and Wnt signaling cross-talk. The Journal of Biological Chemistry. 2003;**278**(42): 41227-41236. DOI: 10.1074/jbc.M305312200

[14] Friedman MS, Lone MW, Hankenson KD. Osteogenic differentiation of human mesenchymal stem cells is regulated by bone

morphogenetic protein-6. Journal of Cellular Biochemistry. 2006;**98**(3):538-554. DOI: 10.1002/jcb.20719

[15] Ullah I, Subbarao RB, Rho GJ. Human mesenchymal stem cells—Current trends and future prospective. Bioscience Reports. 2015;**35**:1-18. DOI: 10.1042/BSR20150025

[16] Janeczek Portalska K et al. Endothelial differentiation of mesenchymal stromal cells. PLoS One. 2012;**7**(10):1-16. DOI: 10.1371/journal.pone.0046842

[17] Morrison SJ, Scadden DT. The bone marrow niche for haematopoietic stem cells. Nature. 2014;**505**(7483):327-334. DOI: 10.1038/nature12984

[18] Caplan AI. Why are MSCs therapeutic? New data: New insight. The Journal of Pathology. 2009;**217**(2):318-324. DOI: 10.1002/path.2469

[19] Lee RH et al. Multipotent stromal cells from human marrow home to and promote repair of pancreatic islets and renal glomeruli in diabetic NOD/scid mice. Proceedings of the National Academy of Sciences of the United States of America. 2006;**103**(46):17438-17443. DOI: 10.1073/pnas.0608249103

[20] Morigi M, Benigni A. Mesenchymal stem cells and kidney repair. Nephrology, Dialysis, Transplantation. 2013;**28**(4):788-793. DOI: 10.1093/ndt/gfs556

[21] Zhang Y, Li Y, Zhang L, Li J, Zhu C. Mesenchymal stem cells: Potential application for the treatment of hepatic cirrhosis. Stem Cell Research and Therapy. 2018;**9**(1):1-7. DOI: 10.1186/s13287-018-0814-4

[22] Kuznetsov SA, Mankani MH, Gronthos S, Satomura K, Bianco P, Robey PG. Circulating skeletal stem cells. The Journal of Cell Biology. 2001;**153**(5):1133-1139. DOI: 10.1083/jcb.153.5.1133

[23] He Q, Wan C, Li G. Concise review: Multipotent mesenchymal stromal cells in blood. Stem Cells. 2007;**25**(1):69-77. DOI: 10.1634/stemcells.2006-0335

[24] Pittenger MF, Discher DE, Péault BM, Phinney DG, Hare JM, Caplan AI. Mesenchymal stem cell perspective: Cell biology to clinical progress. NPJ Regenerative Medicine. 2019;**4**(1):1-15. DOI: 10.1038/s41536-019-0083-6

[25] Yin JQ, Zhu J, Ankrum JA. Manufacturing of primed mesenchymal stromal cells for therapy. Nature Biomedical Engineering. 2019;**3**(2):90-104. DOI: 10.1038/s41551-018-0325-8

[26] Iwasaki K et al. Pharmacological mobilization and recruitment of stem cells in rats stops abdominal adhesions after laparotomy. Scientific Reports. 2019;**9**(1):1-8. DOI: 10.1038/s41598-019-43734-1

[27] Lou G, Chen Z, Zheng M, Liu Y. Mesenchymal stem cell-derived exosomes as a new therapeutic strategy for liver diseases. Experimental and Molecular Medicine. 2017;**49**(6):1-9. DOI: 10.1038/emm.2017.63

[28] Yuca H. *Capsicum annuum L.* In: Novel Drug Targets with Traditional Herbal Medicines: Scientific and Clinical Evidence. 2022. pp. 95-108. DOI: 10.1007/978-3-031-07753-1_7

[29] Harding C, Heuser J, Stahl P. Receptor-mediated endocytosis of transferrin and recycling of the transferrin receptor in rat reticulocytes. The Journal of Cell Biology. 1983;**97**(2):329-339. DOI: 10.1083/jcb.97.2.329

[30] Pan BT, Johnstone RM. Fate of the transferrin receptor during maturation of sheep reticulocytes in vitro: Selective externalization of the receptor. Cell. 1983;**33**(3):967-978. DOI: 10.1016/0092-8674(83)90040-5

[31] Matheakakis A, Batsali A, Papadaki HA, Pontikoglou CG. Therapeutic implications of mesenchymal stromal cells and their extracellular vesicles in autoimmune diseases: From biology to clinical applications. International Journal of Molecular Sciences. 2021;**22**(18):1-26. DOI: 10.3390/ijms221810132

[32] Gardiner C et al. Techniques used for the isolation and characterization of extracellular vesicles: Results of a worldwide survey. Journal of Extracellular Vesicles. 2016;**5**(1):1-6. DOI: 10.3402/jev.v5.32945

[33] Hassanzadeh A et al. Mesenchymal stem/stromal cell-derived exosomes in regenerative medicine and cancer; overview of development, challenges, and opportunities. Stem Cell Research and Therapy. 2021;**12**(1):297. DOI: 10.1186/s13287-021-02378-7

[34] Andreu Z, Yáñez-Mó M. Tetraspanins in extracellular vesicle formation and function. Frontiers in Immunology. 2014;**5**:442. DOI: 10.3389/fimmu.2014.00442

[35] Hu Q et al. Clinical applications of exosome membrane proteins. Precision Clinical Medicine. 2020;**3**(1):54-66. DOI: 10.1093/pcmedi/pbaa007

[36] Popowski K, Lutz H, Hu S, George A, Dinh PU, Cheng K. Exosome therapeutics for lung regenerative medicine. Journal of Extracellular Vesicles. 2020;**9**(1):1-11. DOI: 10.1080/20013078.2020.1785161

[37] Valapala M, Vishwanatha JK. Lipid raft endocytosis and exosomal transport facilitate extracellular trafficking of annexin A2. The Journal of Biological Chemistry. 2011;**286**(35):30911-30925. DOI: 10.1074/jbc.M111.271155

[38] Lee YI, Kim S, Kim J, Kim J, Chung KB, Lee JH. Randomized controlled study for the anti-aging effect of human adipocyte-derived mesenchymal stem cell media combined with niacinamide after laser therapy. Journal of Cosmetic Dermatology. 2021;**20**(6):1774-1781. DOI: 10.1111/jocd.13767

[39] Kim J, Kim B, Kim S, Lee YI, Kim J, Lee JH. The effect of human umbilical cord blood-derived mesenchymal stem cell media containing serum on recovery after laser treatment: A double-blinded, randomized, split-face controlled study. Journal of Cosmetic Dermatology. 2020;**19**(3):651-656. DOI: 10.1111/jocd.13063

[40] Long Y et al. Effective wound healing enabled by discrete alternative electric fields from wearable nanogenerators. ACS Nano. 2018;**12**(12):12533-12540. DOI: 10.1021/acsnano.8b07038

[41] Shabbir A, Cox A, Rodriguez-Menocal L, Salgado M, Van Badiavas E. Mesenchymal stem cell exosomes induce proliferation and migration of normal and chronic wound fibroblasts, and enhance angiogenesis in vitro. Stem Cells and Development. 2015;**24**(14):1635-1647. DOI: 10.1089/scd.2014.0316

[42] Zhang J et al. Exosomes released from human induced pluripotent stem cells-derived MSCs facilitate cutaneous wound healing by promoting collagen synthesis and angiogenesis. Journal of Translational Medicine. 2015;**13**(1):1-14. DOI: 10.1186/s12967-015-0417-0

[43] Da Fonseca Ferreira A et al. Extracellular vesicles from

adipose-derived mesenchymal stem/stromal cells accelerate migration and activate AKT pathway in human keratinocytes and fibroblasts independently of mir-205 activity. Stem Cells International. 2017;**2017**:1-15. DOI: 10.1155/2017/9841035

[44] Pelizzo G et al. Extracellular vesicles derived from mesenchymal cells: Perspective treatment for cutaneous wound healing in pediatrics. Regenerative Medicine. 2018;**13**(4):385-394. DOI: 10.2217/rme-2018-0001

[45] Kim JM et al. Perivascular progenitor cells derived from human embryonic stem cells exhibit functional characteristics of pericytes and improve the retinal vasculature in a rodent model of diabetic retinopathy. Stem Cells Translational Medicine. 2016;**5**(9):1268-1276. DOI: 10.5966/sctm.2015-0342

[46] Beltramo E, Porta M. Pericyte loss in diabetic retinopathy: Mechanisms and consequences. Current Medicinal Chemistry. 2013;**20**(26):3218-3225. DOI: 10.2174/09298673113209990022

[47] Tuekprakhon A et al. Intravitreal autologous mesenchymal stem cell transplantation: A non-randomized phase I clinical trial in patients with retinitis pigmentosa. Stem Cell Research and Therapy. 2021;**12**(1):52. DOI: 10.1186/s13287-020-02122-7

[48] Kahraman NS, Oner A. Umbilical cord derived mesenchymal stem cell implantation in retinitis pigmentosa: A 6-month follow-up results of a phase 3 trial. International Journal of Ophthalmology. 2020;**13**(9):1423-1429. DOI: 10.18240/ijo.2020.09.14

[49] García-Prieto J et al. β3 adrenergic receptor selective stimulation during ischemia/reperfusion improves cardiac function in translational models through inhibition of mPTP opening in cardiomyocytes. Basic Research in Cardiology. 2014;**109**(4):1-15. DOI: 10.1007/s00395-014-0422-0

[50] Arslan F et al. Mesenchymal stem cell-derived exosomes increase ATP levels, decrease oxidative stress and activate PI3K/Akt pathway to enhance myocardial viability and prevent adverse remodeling after myocardial ischemia/reperfusion injury. Stem Cell Research. 2013;**10**(3):301-312. DOI: 10.1016/j.scr.2013.01.002

[51] Börschel CS, Schnabel RB. The imminent epidemic of atrial fibrillation and its concomitant diseases—Myocardial infarction and heart failure—A cause for concern. International Journal of Cardiology. 2019;**287**:162-173. DOI: 10.1016/j.ijcard.2018.11.123

[52] Adamiak M et al. Induced pluripotent stem cell (iPSC)-derived extracellular vesicles are safer and more effective for cardiac repair than iPSCs. Circulation Research. 2018;**122**(2):296-309. DOI: 10.1161/CIRCRESAHA.117.311769

[53] Han C et al. Human umbilical cord mesenchymal stem cell derived exosomes encapsulated in functional peptide hydrogels promote cardiac repair. Biomaterials Science. 2019;**7**(7):2920-2933. DOI: 10.1039/c9bm00101h

[54] Di Salvio M et al. Pur-alpha functionally interacts with FUS carrying ALS-associated mutations. Cell Death and Disease. 2015;**6**(10):1-11. DOI: 10.1038/cddis.2015.295

[55] Chen BK et al. A safety study on intrathecal delivery of autologous mesenchymal stromal cells in rabbits directly supporting phase I human trials. Transfusion. 2015;**55**(5):1013-1020. DOI: 10.1111/trf.12938

[56] Bonafede R et al. Exosome derived from murine adipose-derived stromal cells: Neuroprotective effect on in vitro model of amyotrophic lateral sclerosis. Experimental Cell Research. 2016;**340**(1):150-158. DOI: 10.1016/j.yexcr.2015.12.009

[57] Oh K-W et al. Phase I trial of repeated intrathecal autologous bone marrow-derived mesenchymal stromal cells in amyotrophic lateral sclerosis. Stem Cells Translational Medicine. 2015;**4**(6):590-597. DOI: 10.5966/sctm.2014-0212

[58] Mazzini L et al. Mesenchymal stromal cell transplantation in amyotrophic lateral sclerosis: A long-term safety study. Cytotherapy. 2012;**14**(1):56-60. DOI: 10.3109/14653249.2011.613929

[59] N. P. Staff et al. Safety of intrathecal autologous adipose-derived mesenchymal stromal cells in patients with ALS. Neurology. 2016;**87**(21): 2230-2234. DOI: 10.1212/WNL.0000000000003359

[60] Petrou P et al. Safety and clinical effects of mesenchymal stem cells secreting neurotrophic factor transplantation in patients with amyotrophic lateral sclerosis: Results of phase 1/2 and 2a clinical trials. JAMA Neurology. 2016;**73**(3):337-344. DOI: 10.1001/jamaneurol.2015.4321

[61] Lee JK, Jin HK, Bae J-S. Bone marrow-derived mesenchymal stem cells attenuate amyloid β-induced memory impairment and apoptosis by inhibiting neuronal cell death. Current Alzheimer Research. 2010;**7**(6):540-548. DOI: 10.2174/156720510792231739

[62] Duncan T, Valenzuela M. Alzheimer's disease, dementia, and stem cell therapy. Stem Cell Research and Therapy. 2017;**8**(1):111. DOI: 10.1186/s13287-017-0567-5

[63] Oh SH, Kim HN, Park H-J, Shin JY, Lee PH. Mesenchymal stem cells increase hippocampal neurogenesis and neuronal differentiation by enhancing the Wnt signaling pathway in an Alzheimer's disease model. Cell Transplantation. 2015;**24**(6):1097-1109. DOI: 10.3727/096368914X679237

[64] Ma X et al. ADSCs-derived extracellular vesicles alleviate neuronal damage, promote neurogenesis and rescue memory loss in mice with Alzheimer's disease. Journal of Controlled Release. 2020;**327** (September):688-702. DOI: 10.1016/j.jconrel.2020.09.019

[65] Cui GH et al. Exosomes derived from hypoxia-preconditioned mesenchymal stromal cells ameliorate cognitive decline by rescuing synaptic dysfunction and regulating inflammatory responses in APP/PS1 mice. The FASEB Journal. 2018;**32**(2):654-668. DOI: 10.1096/fj.201700600R

[66] Asahina M, Vichayanrat E, Low DA, Iodice V, Mathias CJ. Autonomic dysfunction in parkinsonian disorders: Assessment and pathophysiology. Journal of Neurology, Neurosurgery, and Psychiatry. 2013;**84**(6):674-680. DOI: 10.1136/jnnp-2012-303135

[67] Bouchez G et al. Partial recovery of dopaminergic pathway after graft of adult mesenchymal stem cells in a rat model of Parkinson's disease. Neurochemistry International. 2008;**52**(7):1332-1342. DOI: 10.1016/j.neuint.2008.02.003

[68] Venkataramana NK et al. Open-labeled study of unilateral autologous bone-marrow-derived mesenchymal stem cell transplantation in Parkinson's

disease. Translational Research. 2010;**155**(2):62-70. DOI: 10.1016/j.trsl.2009.07.006

[69] Venkataramana NK et al. Bilateral transplantation of allogenic adult human bone marrow-derived mesenchymal stem cells into the subventricular zone of Parkinson's disease: A pilot clinical study. Stem Cells International. 2012;**2012**:931902. DOI: 10.1155/2012/931902

[70] Cai Y, Zhang MM, Wang M, Jiang ZH, Tan ZG. Bone marrow-derived mesenchymal stem cell-derived exosomes containing gli1 alleviate microglial activation and neuronal apoptosis in vitro and in a mouse Parkinson disease model by direct inhibition of Sp1 signaling. Journal of Neuropathology and Experimental Neurology. 2022;**81**(7):522-534. DOI: 10.1093/jnen/nlac037

[71] Clark K et al. Placental mesenchymal stem cell-derived extracellular vesicles promote myelin regeneration in an animal model of multiple sclerosis. Cells. 2019;**8**(12);1-16. DOI: 10.3390/cells8121497

[72] Riazifar M et al. Stem cell-derived exosomes as nanotherapeutics for autoimmune and neurodegenerative disorders. ACS Nano. 2019;**13**(6):6670-6688. DOI: 10.1021/acsnano.9b01004

[73] Bernal W, Auzinger G, Dhawan A, Wendon J. Acute liver failure. Lancet (London, England). 2010;**376**(9736): 190-201. DOI: 10.1016/S0140-6736(10)60274-7

[74] Salomone F, Barbagallo I, Puzzo L, Piazza C, Li Volti G. Efficacy of adipose tissue-mesenchymal stem cell transplantation in rats with acetaminophen liver injury. Stem Cell Research. 2013;**11**(3):1037-1044. DOI: 10.1016/j.scr.2013.07.003

[75] Gupta S et al. Comparative evaluation of anti-fibrotic effect of tissue specific mesenchymal stem cells derived extracellular vesicles for the amelioration of CCl4 induced chronic liver injury. Stem Cell Reviews and Reports. 2022;**18**(3):318-324. DOI: 10.1007/s12015-021-10313-9

[76] Gupta S, Krishnakumar V, Soni N, Rao EP, Banerjee A, Mohanty S. Comparative proteomic profiling of small extracellular vesicles derived from iPSCs and tissue specific mesenchymal stem cells. Experimental Cell Research. 2022;**420**(2):1-12. DOI: 10.1016/j.yexcr.2022.113354

[77] Lin B-L et al. Allogeneic bone marrow-derived mesenchymal stromal cells for hepatitis B virus-related acute-on-chronic liver failure: A randomized controlled trial. Hepatology. 2017;**66**(1):209-219. DOI: 10.1002/hep.29189

[78] Tan CY, Lai RC, Wong W, Dan YY, Lim SK, Ho HK. Mesenchymal stem cell-derived exosomes promote hepatic regeneration in drug-induced liver injury models. Stem Cell Research and Therapy. 2014;**5**(3):1-14. DOI: 10.1186/scrt465

[79] Du Y et al. Exosomes from human-induced pluripotent stem cell-derived mesenchymal stromal cells (hiPSC-MSCs) protect liver against hepatic ischemia/reperfusion injury via activating sphingosine kinase and sphingosine-1-phosphate signaling pathway. Cellular Physiology and Biochemistry. 2017;**43**(2):611-625. DOI: 10.1159/000480533

[80] Mardpour S et al. Hydrogel-mediated sustained systemic delivery of mesenchymal stem cell-derived extracellular vesicles improves hepatic regeneration in chronic liver failure. ACS Applied Materials and Interfaces.

2019;**11**(41):37421-37433. DOI: 10.1021/acsami.9b10126

[81] Tieu A et al. Mesenchymal stromal cell extracellular vesicles as therapy for acute and chronic respiratory diseases: A meta-analysis. Journal of Extracellular Vesicles. 2021;**10**(12):1-25. DOI: 10.1002/jev2.12141

[82] Soni N et al. Bone marrow-derived extracellular vesicles modulate the abundance of infiltrating immune cells in the brain and exert an antiviral effect against the Japanese encephalitis virus. FASEB Bioadvances. 2022;**4**(12):798-815. DOI: 10.1096/fba.2022-00071

Chapter 12

Mesechymal Stem Cells Applications in Regenerative Medicine and in Biotherapies

Vasiliki E. Kalodimou

Abstract

This review illustrates the state of the art of regenerative medicine (RM) and mesenchymal stem cells (MSC) as applied to surgical diseases and demonstrates that this field has the potential to address some of the unmet needs in surgery. RM is a multidisciplinary field whose purpose is to regenerate *in vivo* or *ex vivo* human cells, tissues or organs in order to restore or establish normal function through exploitation of the potential to regenerate, which is intrinsic to human cells, tissues and organs. RM uses MSCs' and/or specially designed biomaterials to reach its goals and RM-based therapies are already in use in several clinical trials in most fields of surgery. The main challenges for investigators are threefold: Creation of an appropriate microenvironment *ex vivo* that is able to sustain MSC physiology and function in order to generate the desired MSC cells or body parts; identification and appropriate manipulation of MSCs that have the potential to generate parenchymal, stromal and vascular components on demand, both *in vivo* and *ex vivo*; and production of smart materials that are able to drive cell fate.

Keywords: regenerative medicine, tissue engineering, surgery, extracellular matrix, scaffold, mesenchymal stem cells

1. Introduction

The term regenerative medicine was first used by Leland Kaiser in a 1992-article on hospital administration. Regenerative medicine is an interdisciplinary approach whereby the normal function of the tissues is restored by repairing, replacing, reprogramming or engineering damaged or diseased tissues and organs [1, 2].

The term tissue engineering was first used by Y.C. Fung in 1985. This field applies the principles of engineering and life sciences in the production of biological substitutes that restore, maintain and improve cell function. Two such examples of engineered tissues are skin and cartilage which have been approved by FDA.

2. Different possibilities explored under regenerative medicine are

Previously, irreparable tissues and organs in the body can be engineered and healed by stimulating the body's own repair system either by cell injections or immunomodulation. If the body cannot heal, then it includes the possibility of growing the organs and tissues in the laboratory and then implanting them (tissue engineering). The cell on scaffold seeding technology (CSST) is being used where auto-or- heterologous cells are seeded on supporting scaffolds which are derived either from animal or human organ or bio-engineered, resembling the target organ [3–6].

Stem cells or progenitor cells (cell therapy) are also used. Biotherapies have been demonstrated in clinical trials across the globe to partially heal fractured bones and tendons, severe burns, heart injuries, nerve damage and a range of other conditions.

3. Stem cells and regenerative medicine

Stem cells are the miraculous cells of the human body that have the ability to differentiate and form other cell types with an ability to divide and change. Because of that ability, they have the potential to treat a wide range of human diseases. Embryonic stem cells are derived from blastocyst-stage embryos, and they can be pluripotent, giving rise to tissues from all three germ layers. Induced pluripotent stem cells are formed from differentiated somatic cells exposed to a suitable set of transcription factors that induce pluripotency.

Since the emergence of tissue engineering and regenerative medicine two decades ago, the Food and Drug Administration has approved some available commercial treatments. Products used in these therapies are differentiated cells which have retained their capacity for proliferation. Cells are harvested either autologously or allogenic, expanded in vivo and then injected at the site of injury.

For example, carticel contains autologous chondrocytes harvested from articular cartilage and used in the treatment of focal articular cartilage defects. Laviv contains autologous fibroblasts and injected to smoothen nasolabial fold wrinkles.

Apligraf, GINTUIT: it is a wound-healing graft produced allogenic from human foreskin. The advantage of allogeneic products with low immunogenicity is that they are ready for grafting whenever the need arises, while autologous products have to harvest from the patient and culture-expanded causing a delay in treatment [7, 8].

Dermagraft: used in the treatment of venous ulcer, a 3D polymer scaffold is used for fibroblasts.

Dermapure: decellularised donor tissues are used in this bone allograft.

Regranex: releases platelet-derived growth factor (PDGF) which promote healing and regeneration in biomaterial Infuse: delivers bone morphogenic proteins 2 and 7 which help in the healing of bone [9, 10].

4. Current scenario of regenerated artificial organs

Small arteries, cartilage, skin grafts, and supplementary bladders, even a complete trachea have been implanted in patients, but the procedures are still in the experimental stage and expensive. Large organs such as kidney, liver, and heart have been successfully created in a lab but still far away from being implanted in a patient, but they are used for researches especially drug trials reducing the experimental usage of animals. In addition to medical applications, these regenerated or bioengineered tissues can be used to detect biochemical threat agents and tissue chips which have the ability to test the toxicity in experimental drugs [11–14].

5. Therapies at preclinical stage and in clinical testing

They can be broadly categorised as follows: regenerating organ or tissue through scaffold, 3D bioprinting and self-assembly, graft integration by inducing vascularisation and innervation alteration of the host environment through immune-modulation and cell infusion to promote therapeutic response.

6. Regeneration of organ or tissue structure

Scaffold fabrication. Cells and tissues in the body are embedded in an extracellular matrix which not only supports them but also acts as a relay station for various signalling molecules. By understanding cellular responses to these signals, researchers have been able to manipulate the processes to repair damaged tissues or even create new ones [15].

To create an organ in the lab, first, a scaffold is built from sources ranging from proteins to plastics. Once a scaffold is created, cells with or without growth factors are added, and if conditions are optimum, a tissue develops.

Use of medical imaging technologies: with the aid of computed tomography (CT) and medical resonance imaging (MRI) 3D images of the replaced tissue can be created according to the patient's body which then acts as moulds for scaffold generation thus organs can be tailor made to suit the patient's needs. A synthetic trachea was created in this way and then seeded with patient's cells.

Decellularisation of organs and recellularising them before transplantation: scaffolds can also be made from a donor organ or human tissues discarded during surgery by chemically stripping cells from the organ (continuous perfusion of the organ with a detergent based solution- sodium doceyl sulphate) and using the remaining collagen scaffold to grow new tissue. This method holds the greater advantage as patient's cells can now be added to this scaffold to make customised organs which will not be rejected by the patient's immune system [16–21].

Decellularisation removes immunogenic cells, while the structure, mechanical properties, and material composition of the original extracellular matrix are retained theoretically. This technique has been used to repair large muscle defects. However, decellularisation may affect the mechanical properties of the tissue and remove certain ECM associated signalling molecules, and the regenerated tissue may degrade with time.

Currently, 20% of kidneys obtained for transplant are ultimately discarded as unsuitable for transplantation. These may become a valuable source for ECM scaffold. It is not too distant future when CSST reaches the bedside, and a patient with end-stage renal disease will undergo nephrectomy, and his own removed kidney will provide ECM scaffold which will be reseeded with his own cells. Thus, a new kidney will be created with his own biomaterials. This will solve the problem of organ shortage, produce organ on demand and immunosuppression free transplantation, ruling out the need of lifelong immunotherapy.

3D bioprinting: it creates structures that combine high-resolution control over material and cell placement within engineered constructs. Two techniques used in this are inkjet and microextrusion [22].

Cartilage has been engineered in this way. In microextrusion printing, a continuous stream of ink is dispensed on the stage. This technique has been used to fabricate aortic valve replacements.

Cell sheet technology: this technique has been utilised in the corneal reconstruction [23]. No scaffolds are produced, but it involves the retrieval of confluent sheets of cells, and thus all the signalling molecules, ECM molecules, as well as cell-cell adhesion, remains intact [24–27].

7. Graft integration by inducing vascularisation and innervation

For graft success, it is of primary importance that graft is adequately integrated with the host. Effective gaseous and nutrient exchange can take place within 100 μ of the capillary. For vascularisation of engineered tissues, the body's own angiogenic response in the form of growth factors can be utilised. A number of growth factors like vascular endothelial growth factor (VEGF), platelet-derived growth factor (PDGF), and basic fibroblast growth factor (bFGF) have been seen to play a role in angiogenesis as well as innervation (especially VEGF).

These growth factors can be applied from outside as well, and their sustained release has resulted in a fairly good angiogenic response. It can save ischemic limbs from necrosis. Another approach is to prevascularise the graft by first placing it on a large host vessel and then implanting it at its final location once it gets vascularised.

Nerve innervation and angiogenesis share certain signalling pathways and can be induced by growth factors like VEGF. Innervation is especially needed in tissues like skeletal muscle and skin where motor control and sensation is necessary [11–13, 28–30].

8. New cell sources

Obtaining ample therapeutic cells is often a challenge with the vast majority of approaches using adult tissue-derived cells. Embryonic stem cells (ES) and induced pluripotent stem cells (iPS) represent potentially infinite sources of cells for regeneration. iPS are holding a greater promise as they can be generated by patients own cells, thus eliminating ethical issues of ES and risk of rejection of the transplanted cells.

9. The future of regenerative medicine

Stem cells may transform the way doctors treat heart failure. It has been shown in animal models that stem cells migrate to the site of injury and improve vascular flow and overall heart function. Lost cells may be replaced in neurodegenerative disorders. Clinical trials are underway as to how children with type I diabetes can benefit by the infusion of cord blood stem cells.

For large organs like liver and kidneys to be regenerated and implanted, a way has to be found out to create a network of blood vessels. Soldier's wounds can be repaired by biological scaffolding material, by re-growing muscles and tendons. Neural stem cell transplants may 1 day help Parkinson's patients [31].

10. Aged stem cells rejuvenation

The ageing of niche cells and age-dependent alterations in the components of stem cell niche are able to cause a loss stem cell function in the adult body, happening gradually with age and as a result will cause organ and tissue failure. Many clinical trials globally are using 'young' blood in order to restore the aged stem cell population through regular infusions in order to reverse tissue ageing and ageing in general [32].

11. Will cell therapies increase-maintain the longevity in humans?

Cell therapies and their applications in regenerative medicine try to increase our lifespan. This research has shown positive results, and we can safely conclude that in the years to come, the doctors will have the ability to repair some the damage caused by ageing and at the same time will have the ability to reverse the stem cell function restoring stem cell maintenance. We are looking to a new medical era when the replacement of a damaged organ will be restored from the patient's own stem cells [33, 34].

12. National Institute of Health (NIH) funded researches in the area of regenerative medicine

Developing lattices of sugar solution which will act as a vascular network for engineered tissues pluripotent stem cells to be fully converted into mature bone cells which can be implanted. As of now, stem cells when injected, convert into primitive forms of cells which are only partially functional, implanting engineered human liver tissue in mice and looking for drug interactions and thus cutting the cost and time of monitoring drug reactions.

Pluripotent stem cells are grown in different types of growth solutions and studying the various types of gene networks that are triggered and their transformation into various cells.

Today, a number of diseases and lesions are being treated by regenerative medicine derived, and new FDA-approved therapies range from cell manipulation technologies for repairing tissues to the fabrication of scaffolds customised to suit the patient.

Strategies are being improvised to integrate the transplanted organ with host vasculature and nervous system with the augmentation of body's healing process by immunomodulation. New cell sources for transplantation are being developed to solve the problem of limited cell supply. 3D human culture models are being developed for testing different approaches to regenerative medicine instead of animal models currently used in preclinical trials [35].

However, for the advancement of regenerative medicine, few issues have to be addressed. First, strict monitoring of isolated stem cells and their microenvironment has to be conducted to increase their safety profile and efficacy. Second creation of technologies will enable to fully vascularise the graft at the time of implantation and anastomose it with the host vasculature to ensure graft survival. Third, a better understanding of the role of the immune system, age, disease and microenvironment of the patient is needed which affect the outcome of regeneration.

13. Regenerative medicine, cellular therapy and tissue engineering: current scenario

Is the process of engineering human cells, tissue or organs, to restore or establish normal function. It therefore essentially involves translational research, tissue engineering and/or molecular biology. Often, it stimulates the body's own repair mechanisms to facilitate the healing process. Use of epigenetics is therefore a critical component, wherein 'erasers' can actually reverse the disease process. It also includes artificial/semi artificial tissues and organs developed in the laboratory for human use. Imagining the value of patient's own cells capable of growing into any damaged organ or tissue, it would be disruptive technology solving the problem of organ donation and transplant rejection [36].

14. Tissue engineering

It refers to the use of cells, engineering and materials methods in various combinations and permutations under appropriate biochemical and physicochemical environment to facilitate, heal, restore, improve or replace biological functions in living creatures. It therefore combines scaffolds, cells and biologically active molecules into functional tissues. Artificial skin and cartilage are examples of engineered tissues that have been approved by US FDA.

In 2003, the first human genome was finally decoded at a cost of about $3 billion. This information is now exhibited in the 'Medicine Now' room, Wellcome Collection, London. It contains 3.4 billion units of DNA code into more than 100 volumes, each more than 1000 pages long with a font size that is barely legible (very small). Today, the same human genome analysis can be done in India in less than a week for $ 1000, and Google is offering to store this information on the cloud for $25 per year.

Today, 62% of human body parts can be artificial. In fact, 3D printing is able to provide such custom made scaffolds in real time inside the operation theatre itself—with more accuracy and at a fraction of the cost of conventional implants.

Artificial blood made from plastic at Sheffield University is a classic example. Others involve eye, ear, pancreas, spleen, kidney, heart, trachea, foot, ankle, hand and arm. The bionic arm wherein the amputated person can not only do motor movements but also feel sensations is a one of the wonders of this century. The École Polytechnique Fédérale de Lausanne—Project TIME—has been able to implant electrodes directly in key nerves—both motor AND sensory. ReWalk exoskeleton allowing paraplegics to walk on their own is another example [37–39].

The American Society for Gene Therapy was formed in 1996. The journal 'Gene Therapy' was commenced in 2004. GenBank (http://www.ncbi.nlm.nih.gov) has information on 39,000 species that is uploaded within 24 hours and receives more than 200,000 queries daily. The first human gene therapy was the approval of p53 in China for Head Neck Cancers in the year 2005. It dramatised the response rates in advanced cancers, giving a three times better outcome in studies involving 135 patients. Glybera became the first gene therapy approved by the Western world. It is for the treatment of lipoprotein lipase deficiency by replacing the mutant gene. It also has the dubious distinction of being the most expensive medicine in history, costing a whopping $ 1.4 million per patient.

15. Stem cell therapy

Recently, the European Commission approved Holoclar for use in cases of blindness caused by burning. This therapy has been developed by Graziella Pellegrini and colleagues from the University of Modena in a small Italian laboratory in Italy. Her story is a beacon of hope, having overcome several towering obstacles to take the therapy from bench to bedside.

The cornea requires constant renewal to keep it smooth and clear. This is possible because new corneal cells are constantly being generated from stem cells in the limbus. If the limbus is destroyed (commonly occurs due to burns), there is overgrowth of blood vessels, chronic pain and inflammation, ultimately leading to blindness. Holoclar treatment can reverse this within a year by adding new stem cells to seed the regrowth of a transparent cornea. Cells are harvested from 1 or 2 square mm of active limbus, cultured in the presence of modified human fibrin scaffold and transplanted. The results were first published way back in 1997, and currently (2013), they have an impressive 77% success rate [40].

16. Regulatory and ethical considerations

Any new treatment is always fraught with a regulatory hurdle. This is especially true for such disruptive technology which might have no precedence. We are living in times of desperate patients seeking unproven remedies and clutching at straws. Under the circumstances, there are a lot of ethical and regulatory considerations. At the same time, it is our collective responsibility to do the best for the patients under the limited knowledge. Hence, it is vital to remember that no evidence of benefit is not the same as evidence of no benefit. Whenever faced with such a question, the benefit of doubt should be given to the therapy and the patient.

It is clear that regenerative medicine, cellular therapy, biotherapies and tissue engineering is here to stay [41].

17. Mesenchymal stem cells application in clinical practice and regenerative medicine

Mesenchymal stem cells (MSCs) hold the ability to be used as therapeutic agents in cellular therapies because they can differentiate into many cell types and also have immunosuppressive properties.

It has been shown that the transplantation-infusion of MSCs could be the appropriate therapy for many diseases as well as syndromes. To date, hundreds of clinical trials using MSC's have been registered in the American database: http://www.clinicaltrials.gov, as well as in the rest of the world clinical trial platforms, with very promising outcomes.

The main goal is to isolate the MSCs from many different sources (such as placenta, umbilical cord tissue, bone marrow, peripheral blood and adipose tissue) and then expand the cells under *Good Manufacturing Practice* (*GMP*) conditions.

Most of the time, MSCs holds a very high expansion ability as well as cell viability in the long-term culture system. The source of MSCs plays a crucial role in their expansion potential, i.e., the adult bone marrow MSCs (BMMSCs) can grow identically in culture for 6–7 passages, whereas placenta umbilical cord MSCs can undergo 30–40 passages. Chromosomal abnormalities can be arising after prolonged culture of human embryonic stem cells (ESCs), while induced pluripotent stem cells (iPSC's) undergo deletions of tumour-suppressor genes during the process of reprogramming [42–44].

18. The mesenchymal stem cells

MSCs have the ability to be activated and mobilised if needed, but their efficiency is very low. MSCs have a large capacity for self-renewal while maintaining their multipotency.

Many successful clinical trials where resulting after intravenous infusion of the cells in systemic diseases and sepsis. However, diseases involving peripheral tissues may be better treated with methods that increase the local concentration of cells. Direct injection of MSC cells into a site need repair shown to be the preferred method of treatment. Many clinical trials in orthopaedics have been published and had been demonstrating the effectiveness from their use.

The important question rise about the use of MSCs to constitute a natural system for tissue repair are how MSCs are mobilised and how they get to the site of injury. Chemokines such as SDF1 and its cognate receptor CXCR4 may have an essential role in directing MSCs to sites of injury. More recently, it has been reported that the cytokine receptor CCR2 and its intracellular adaptor molecule FROUNT are necessary for homing of bone marrow-derived mesenchymal stem cells to sites of injury. Investigators around the globe had used and still used in clinical models MSCs as cellular vehicles for gene therapy [36, 40, 41, 45].

Adult human MSCs have been reported to express intermediate levels of major histocompatibility complex (MHC) class I proteins, but not MHC class II proteins and this feature that makes MSCs so extraordinary and unique. This gives the opportunity to doctors to use the cells and transplant them into an allogeneic host without any immunosuppression problems. MSCs have immunosuppressive properties by modulating specific T-cell functions in vitro, and some of these observations have even been translated into in vivo settings such as experimental models of graft versus host disease.

Finally, unlike their HSC counterparts, MSCs can be easily isolated and highly expanded in culture.

19. Umbilical cord tissue (UCT)

UCT represents a very unique, easily accessible and noncontroversial source of mesenchymal stem cells. Because of those unique characteristics, many studies compared the properties of UCT-derived mesenchymal stem cells with those derived from adult sources (such as adipose tissue), and their differences are summarised as follows:

- Cord tissue-derived mesenchymal stem cells lack some immune suppression properties compared with adult mesenchymal stem cells.

- Cord tissue-derived mesenchymal stem cells lack class II HLA, whereas adult mesenchymal stem cells express these antigens. This may be particularly important in facilitating the 'acceptance' of transplanted mesenchymal stem cells.

- Cord tissue-derived mesenchymal stem cells also express a range specific cytokines that are different from those expressed by adult mesenchymal stem cells.

- Thus, mesenchymal stem cells extracted from cord tissues have greater therapeutic potential than adult cells. Studies have also shown that cord tissue-derived mesenchymal stem cells appear to have greater in vitro capacity for expansion and shorter doubling times.

- Infusion of human umbilical cord blood (hUCB)-MSCs was found to be a possible treatment for brain injuries and degenerative disorders.

Therefore UCB derived MSCs can be an effective therapeutic applications of various diseases such as in degenerated neurons, cardiac cells, muscle cells and chondrocytes.

However, the potential application of these cells requires extensive characterisation as well as cell viability and standardisation of differentiation protocols with ultimate functional characterisation of the differentiated cells. Morphologically, the MSCs are adherent, fibroblast-like cells with multipotent differentiation potential and thus can be induced to differentiate into cells of multiple lineages such as adipocytes, osteocytes, chondrocytes, myocytes, hepatocytes, neurons and astrocytes.

Administration of human umbilical cord blood (hUCB)-MSCs was found to be feasible treatment for brain injuries from many researchers across the globe. Using spinal cord-injured animal models and infusing them with ThUCB-MSC has shown significant improvement in their neurological function. If we can use these results in human clinical trials, we could have a new era of possible regenerative medicine therapy treatments with a high potential for neuronal differentiation. In vitro treatments with ß-mercaptoethanol and retinoic acid resulted in a very drastic difference in cellular morphology of MSCs from fibroblastic to spindle-shaped with elongated processes resembling a neuronal phenotype. This multi-lineage differentiation capacity, the expression of neural properties and overlapping genetic programmes for haematopoiesis and neuropoiesis suggest that hUCB cells may have the ability to trans-differentiate in to neural cells [31, 35, 46–49].

20. Adipose mesenchymal stem cells (ADSC's)

Adipose-derived stem cells (ASCs) are mesenchymal stem cells (MSCs) that are obtained from abundant adipose tissue, mostly lipoaspiration, and they have the ability to adherent on plastic culture flasks, can be expanded in vitro, and have the capacity to differentiate into multiple cell lineages.

There is a big difference between the MSCs cell population and those of ADSCs derived cells. Specifically, the stem cell frequency in the bone marrow of adult humans is one stem cell per 200,000 total nucleated cells. As a result, cell culture is necessary to enrich these cells and expand the number available to a clinically adequate cell dose for a safe infusion. By contrast, the frequency of these cells in digested adipose tissue is one in 100 nucleated cells, 2000 times greater than bone marrow.

The specific characteristics of ADSCs make them an attractive tool in regenerative medicine and in biotherapies. They possess the capacity to engraft into various tissues and organs when infused systematically, and the engraftment has been shown to be stable and viable in the long term, i.e., bone defects. Moreover, when they are infused to the peripheral circulation, they have the ability to migrate to a specific site of injury. They have been also proposed to be an excellent potential tool for gene therapies.

The concept of cell assisted lipo-transfer by adding ADSCs has been successfully used in the clinic for soft tissue augmentation and in the treatment of wound healing.

Allergic diseases, autoimmune diseases, and graft versus host disease (GVHD) are another potential application of ADSCs which involves the harnessing of their immunomodulatory effects to treat those conditions. It has been reported that infused ADSCs can ameliorate steroid resistant severe GVHD after haematopoietic stem cell transplantation.

The infusion of ADSCs in cardiovascular diseases, the leading cause of death worldwide, is promising evidence which indicates the improved cardiac function via the differentiation into cardiomyocytes and vascular cells and through paracrine pathways. By ADSC infusion, paracrine factors secreted and enhance angiogenesis, reduce cell apoptosis rates and promote neuron sprouts in damaged myocardium [50]. The use of ADSCs could be design the future of cell therapy strategies for cardiac repair globally.

In the field of neurodegenerative diseases, ADSCs have been known to have a differentiation potential by the secretion of some nerve growth factors and are capable of promoting neuronal healing. They express, in a significantly high proportion, a marker for neural progenitor cells called nestin and can secrete vascular endothelial growth factor (VEGF) which is an angiogenic factor and insulin-like growth factor 1 which is a neuroprotective factor. At the moment, worldwide, there are many limitations in protocols in order to establish homogeneous populations of neural progenitor's stem cells for effective therapy applications in neurodegenerative diseases such as Parkinson's disease, multiple sclerosis and Alzheimer's.

The use of ADSCs in clinical trials has been increased the last years because of their ability to be harvest in large numbers with safety and efficacy. So far, very few clinical trials and in a very limited number of research areas have been conducted to assess the therapeutic potential of ADSCs compared with the large number of published preclinical studies globally.

Adipose stem cells are actively being researched in clinical trials for the treatment of a variety of diseases. Many studies have examined their characteristics, plasticity and inducibility.

Prior to ADSC use, there are many unclear areas that need to be taken under consideration:

1. The differentiation potential of ADSCs may depend on the anatomic location of the fat and the donor's gender and age.

2. The key molecular events and transcription factors that initially allocate the ADSCs to a specific lineage are still unknown, while their lineage-specific differentiation into mesodermal cell origin is well understood at their molecular level.

Despite these limitations, ADSCs have an extensive proliferative potential and are able to differentiate into various cell lineages and also have practical clinical advantages and as a result their use has become more realistic because of adipose tissue can be easy obtained. In comparison to BMSCs cells, the procedure is safe and minimally invasive, with minor to none donor morbidity. It has been well documented that the cells engraft successfully in patients and cause beneficial effects. Due to these important features, their use in clinical trials has been increased.

Lastly but most importantly, there is no risk of immune rejection with autologous cells in comparison with UCC or induced pluripotent stem cells (iPSCs). In conclusion, ADSCs are a useful tool in the field of regenerative medicine, and in gene therapy, in vitro and in vivo and will play a crucial role in the way we treat patents in the near future [40, 41, 45].

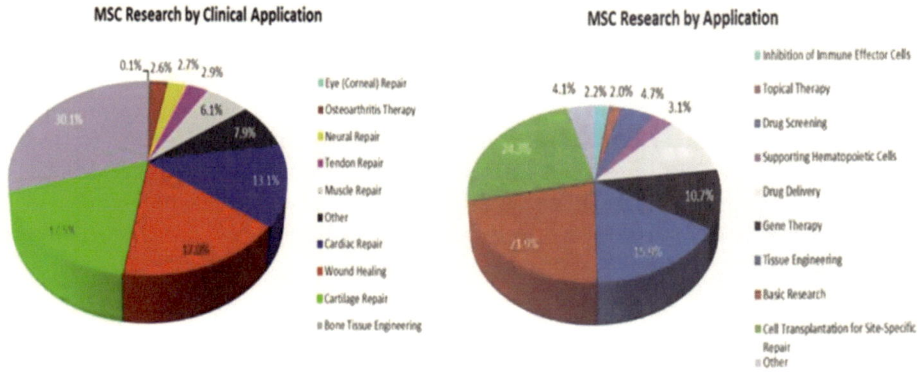

Figure 1.
Mesenchymal stem cells: clinical applications and research interest.

In general, mesenchymal stem cells give rise to different cell tissue types, and as a result, there are intense therapeutic interest, worldwide, because they have the potential to treat a wide range of acute and degenerative diseases as well as in biotherapies (**Figure 1**).

In addition, recent advances include the use of mesenchymal stem cells to:

- Regenerate damaged cartilage.
- Regeneration of the meniscus (knee).
- Regeneration of damaged tendons.
- Repair of bone fractures.

Regenerative medicine has the ability of regenerative cells to boost the natural repair mechanisms, and as a result, it holds an extremely promising avenue for the treatment of many diseases, injuries and conditions that are inadequately or cannot be addressed.

Adult mesenchymal stem cells have a clear advantage in terms of their ability to be used autologously, without side effects, eliminating this way, the issues associated with rejection, infection or teratoma formation.

The ability to use adipose tissue as a source of regenerative cells that can be delivered in large amounts without the need for cell culture creates a unique novel opportunity for bedside application/infusions using the patient's own cells. The promise of this medical approach is already being realised in numerous clinical settings globally.

21. Safety and efficacy

In tissue and organ engineering, animal studies and clinical trials using MSCs are accumulating, and some very good results were reported.

As a result, it is of great importance that those studies in adipose tissue engineering investigate the integration of the engineered tissue and organ as well as the short/long-term safety and efficacy of the whole procedure.

Safety issues address the safety of the procedure including the absence of any rejection. Long-term efficacy should be considered when an animal study or clinical trial is designing, and long-term follow-up is needed, as well as validation protocols, to monitor recurrence.

It is important to know if the treatment should be repeated after a certain time or whether the engineered tissue or organ will last for the whole life. Further, in order to be able to integrate with the surrounding tissue of the host to attain functional efficacy and survival, in vivo tissue or organ engineering or implanted ex vivo tissue or organ engineering product need to be addressed [36].

22. Strategies and future directions

MSCs retaining their multilineage differentiation potential makes them attractive candidates for biological cell-based tissue repair approaches while having extensive proliferative ability in culture in an uncommitted state.

The understanding of MSC will instruct the development of innovative therapeutic measures such as producing pharmacological substances that target endogenous MSCs and their niches in order to activate and guide intrinsic repair and to improve disease outcomes in the human population.

In order to fully understand the above assumptions, let us look as an example the different infusion method of MSCs as a therapy to acute myocardial infarction (MI):

1. The easiest, because it only requires peripheral venous access, and most practical method for delivery of MSC's is intravenous infusion; however, the cells must travel through the pulmonary circulation to reach the myocardium, where entrapment of cells is a concern.

2. Intracoronary infusion of MSCs is delivered with a standard over-the-wire balloon angioplasty catheter placed into the target coronary artery. After the angioplasty catheter is positioned, the balloon is inflated at a low pressure to block blood flow, while the cells are infused through the distal lumen. The concern lies in inducing ischemia during coronary artery occlusion, and the lack of vessels in chronically occluded areas of scar tissue may not allow effective cell delivery. However, this infusion may be able to effectively deliver cells to ischemic after an acute MI. The familiarity of angioplasty techniques to interventional cardiologists and the ability to deliver cells during percutaneous intervention for acute MI Transendocardial stem cell injection delivers MSC's directly into the myocardium with a catheter navigated in the LV by fluoroscopic guidance or electroanatomic mapping are the advantages of this technique.

3. A needle deployed from the tip of a catheter that engages and infuses cells into the myocardium called transendocardial stem cell injection. Using this method with transendocardial stem cell injection, we must monitor the perforation of myocardium, with the potential for cardiac tamponade, and induction of arrhythmias, but the minimally invasive ability to directly inject cells into scar and border zones makes it an attractive delivery technique.

4. Direct surgical intramyocardial injection of MSCs is the largest preclinical experience with allogeneic and autologous MSC therapy for cardiac disease used.

DOI: http://dx.doi.org/10.5772/intechopen.1005868

> The most invasive delivery technique is the surgical injection because it requires either a thoracotomy or sternotomy for cell delivery, but it allows direct visualisation of scarred myocardium and needle engagement. If perforation occurs, it can be controlled at the time of injection with sutures. Clinical studies using large animal have been conducted to study the direct surgical injection of MSCs with acute MI, subacute MI and chronic ischemic cardiomyopathy. The optimal time to deliver MSC's in the post-MI setting, whether acutely during the remodelling phase or in the chronic setting after a healed scar is developed, needs better and more rigorous definition in future studies. MSCs holds a great potential to treat damaged myocardium, but it is crucial to keep safety concerns in mind. Two are the key concerns: pro-arrhythmia and tumour formation.

MSCs and their therapeutic potential has evolved substantially over the past few decades in order to be able to understand their endogenous roles. MSCs are now appreciated as essential cells that govern tissue homeostasis by regulating niches, whereas, initially had been considered by standers of uncertain significance.

This property has greatly facilitated a role for these cells as a cell-based therapeutic agent, along with a capacity for multipotential differentiation. MSCs have been tested in preclinical models and clinical trials for their ability to both prevent and reverse ventricular remodelling, enhanced by ease of preparation and immunoprivilege.

Their remarkable safety profile as well as the early demonstrations of all the above effects have set the stage for pivotal testing of these cells as a therapeutic agent.

The MSC population is needed a systematic, phenotypic in vivo characterisation for numerous reasons:

1. Organogenesis and the role of MSCs.

2. Expand them in vivo with efficacy and appropriately.

3. Better understand the regeneration processes and modulate in vivo repair without the need for in vitro expansion

4. The development of a reproducible and faithful in vivo model is a necessary precondition for an adequate characterisation of the MSC population that would test the ability of MSCs to self-renew, proliferate and differentiate in vivo.

5. The identification of pharmacological tools that could be used to expand in vivo and/or in vitro the MSC pool is an important goal to be achieved along with the proper in vivo characterisation of the cells.

6. A detailed and systematic analysis of the complex network of signalling pathways and cells that regulate the ability of MSCs to self-renew, proliferate and eventually differentiate is what is been required for a successful identification of useful pharmacological tools.

Clear priority among scientific community is the definition of a possible niche for MSCs and its regulation. The importance lies in the identification of this network in order to reach a deep understanding of the rules that govern the size of the MSC pool in vivo, which would then ultimately allow for the appropriate pharmacological interventions.

23. Take home messages

Mesenchymal stem cells (MSCs) are pluripotent cells, and they have the ability to differentiate into a variety of cell types, including osteoblasts, myocytes, fat cells, chondrocytes and potentially neuronal cells.

They represent a population of cells with the potential to treat a wide range of acute and degenerative diseases, as state above, and as a result, they hold a strong therapeutic interest.

The superiority of MSCs in comparison with other stem cell types lies among two reasons:

1. We, as a scientific community, do not face any ethical issues/considerations upon their use as we do when using embryonic stem cells in research, and

2. MSC's are immune privileged and therefore are idyllic type of allogeneic cells, reducing the risk of rejection and complications upon infusion.

Recent clinical studies underlining the significant progress in the use of autologous mesenchymal stem cells for the regeneration of human tissues, i.e., cartilage, meniscus, tendon and bone fractures.

The uses of MSCs make clear that is a promising source to be used in tissue and gene therapy as well as in biotherapies and regenerative medicine because those cell can differentiate into many different cell types in vitro safely and with great efficacy and cell viability, they are relative ease to proliferate in culture, and they have immunological traits.

Although the current animal and clinical trials applied strategies have been increased and advanced in the last years, complete understanding of the mechanism of interactions between ADSCs, growth factors and biomaterials is needed in order to advance the end goal of developing regenerative engineering products.

This represents a remarkable innovate technique with major therapeutic potentials, but many studies/clinical trials are needed in order to reveal this potential.

Author details

Vasiliki E. Kalodimou[1,2]

1 EUC School of Medicine Frankfurt-Germany Branch, Germany

2 NCSR Demokritos, Greece

*Address all correspondence to: v.kalodimou@euc.ac.cy

References

[1] Hamilton NJ, Kanani M, Roebuck DJ, et al. Tissue engineered tracheal replacement in a child: A 4 year follow up study. American Journal of Transplantation. 2015;(10):2750-2757. DOI: 10.1111/ajt.13318

[2] Orlando G, Soker S, Stratta RJ. Organ bioengineering and regeneration as the new holy grail for organ transplantation. Annals of Surgery. 2013;**258**:221-232

[3] Keller G. Embryonic stem cell differentiation: Emergence of a new era in biology and medicine. Genes & Development. 2005;**19**(10):1129-1155

[4] Takahashi K, Yamanaka S. Induction of pluripotent stem cells from mouse embryonic and adult fibroblast cultures by defined factors. Cell. 2006;**126**(4):663-676

[5] Dewan AK, Gibson MA, Elisseeff JH, Trice ME. Evolution of autologous chondrocyte repair and comparison to other cartilage repair techniques. BioMed Research International. 2014;**2014**:272481

[6] Kim PJ, Dybowski KS, Steinberg JS. A closer look at bio engineered alternative tissues. Podiatry Today. 2006;**19**(7):1-9

[7] Okano T, Yamada N, Sakai H, Sakurai Y. A novel recovery system for cultured cells using plasma-treated polystyrene dishes grafted with poly(N-isopropyl acrylamide). Journal of Biomedical Materials Research. 1993;**27**(10):1243-1251

[8] Nakajima K et al. Intact microglia are cultured and non-invasively harvested without pathological activation using a novel cultured cell recovery method. Biomaterials. 2001;**22**(11):1213-1223

[9] Supp DM, Boyce ST. Engineered skin substitutes: Practices and potentials. Clinics in Dermatology. 2005;**23**(4):403-412

[10] O'Brien T, Barry FP. Stem cell therapy and regenerative medicine. Mayo Clinic Proceedings. 2009;**84**(10):859-861

[11] Sacco S et al. Renal extracellular matrix scaffolds from discarded kidneys maintain glomerular morphometry and vascular resilience and retains critical growth factors. Transplantation. 2015;**99**(9):1807-1816

[12] Tekin E, Smith PJ, Schubert US. Inkjet printing as a deposition and patterning tool for polymers and inorganic particles. Soft Matter. 2008;**4**(4):703

[13] Cui X, Boland T, D'Lima DD, Lotz MK. Thermal inkjet printing in tissue engineering and regenerative medicine. Recent Patents on Drug Delivery & Formulation. 2012;**6**(2):149-155

[14] Bhatia SN, Ingber DE. Microfluidic organs-on-chips. Nature Biotechnology. 2014;**32**(8):760-772

[15] Falanga V, Sabolinski M. A bilayered living skin construct (APLIGRAF) accelerates complete closure of hard-to-heal venous ulcers. Wound Repair and Regeneration. 1999;7(4):201-207

[16] Harding K, Sumner M, Cardinal M. A prospective, multicentre, randomised controlled study of human fibroblast-derived dermal substitute (Dermagraft) in patients with venous leg ulcers. International Wound Journal. 2013;**10**(2):132-137

[17] Song JJ, Ott HC. Organ engineering based on decellularized matrix scaffolds.

Trends in Molecular Medicine. 2011;**17**(8):424-432

[18] Walsh G. Biopharmaceutical benchmarks 2010. Nature Biotechnology. 2010;**28**(9):917-924

[19] Hynes RO. The extracellular matrix: Not just pretty fibrils. Science. 2009;**326**:1216-1219

[20] Ballyns JJ et al. Image-guided tissue engineering of anatomically shaped implants via MRI and micro-CT using injection molding. Tissue Engineering. Part A. 2008;**14**(7):1195-1202

[21] Appel AA, Anastasio MA, Larson JC, Brey EM. Imaging challenges in biomaterials and tissue engineering. Biomaterials. 2013;**34**(28):6615-6630

[22] Ajalloueian F et al. Biomechanical and biocompatibility characteristics of electrospun polymeric tracheal scaffolds. Biomaterials. 2014;**35**(20):5307-5315

[23] Nishida K et al. Corneal reconstruction with tissue-engineered cell sheets composed of autologous oral mucosal epithelium. The New England Journal of Medicine. 2004;**351**:1187-1196

[24] Crapo PM, Gilbert TW, Badylak SF. An overview of tissue and whole organ decellularization processes. Biomaterials. 2011;**32**(12):3233-3243

[25] Macchiarini P et al. Clinical transplantation of a tissue-engineered airway. Lancet. 2008;**372**(9655):2023-2030

[26] Mase VJ Jr et al. Clinical application of an acellular biologic scaffold for surgical repair of a large, traumatic quadriceps femoris muscle defect. Orthopedics. 2010;**33**(7):511

[27] Fishman JM et al. Airway tissue engineering: An update. Expert Opinion on Biological Therapy. 2014;**14**(10):1477-1491

[28] Badylak SF, Taylor D, Uygun K. Whole-organ tissue engineering: Decellularization and recellularization of three-dimensional matrix scaffolds. Annual Review of Biomedical Engineering. 2011;**13**:27-53

[29] Gifford S et al. Recycling organs–growing tailor-made replacement kidneys. Regenerative Medicine. 2015;**10**(8):913-915

[30] Derby B. Printing and prototyping of tissues and scaffolds. Science. 2012;**338**(6109):921-926

[31] Harris DT et al. The potential of cord blood stem cells for use in regenerative medicine. Expert Opinion on Biological Therapy. 2007;**7**(9):1311-1322

[32] Lovett M, Lee K, Edwards A, Kaplan DL. Vascularization strategies for tissue engineering. Tissue Engineering. Part B, Reviews. 2009;**15**(3):353-370

[33] Darland DC, D'Amore PA. Blood vessel maturation: Vascular development comes of age. The Journal of Clinical Investigation. 1999;**103**(2):157-158

[34] Conway EM, Collen D, Carmeliet P. Molecular mechanisms of blood vessel growth. Cardiovascular Research. 2001;**49**(3):507-521

[35] Warnke PH et al. Growth and transplantation of a custom vascularised bone graft in a man. Lancet. 2004;**364**(9436):766-770

[36] Trojahn Kølle SF, Oliveri RS, Glovinski PV, Elberg JJ, Fischer-Nielsen A, Drzewiecki KT.

Importance of mesenchymal stem cells in autologous fat grafting: A systematic review of existing studies. Journal of Plastic Surgery and Hand Surgery. 2012;**46**(2):59-68

[37] Borselli C et al. Functional muscle regeneration with combined delivery of angiogenesis and myogenesis factors. Proceedings of the National Academy of Sciences of the United States of America. 2010;**107**(8):3287-3292

[38] Shvartsman D et al. Sustained delivery of VEGF maintains innervation and promotes reperfusion in ischemic skeletal muscles via NGF/GDNF signalling. Molecular Therapy. 2014;**22**(7):1243-1253

[39] Lee K, Silva EA, Mooney DJ. Growth factor delivery-based tissue engineering: General approaches and a review of recent developments. Journal of the Royal Society Interface. 2011;**8**(55):153-170

[40] Colman A, Dreesen O. Pluripotent stem cells and disease modeling. Cell Stem Cell. 2009;**5**:244-247

[41] Casteilla L, Dani C. Adipose tissue-derived cells: From physiology to regenerative medicine. Diabetes & Metabolism. 2006;**32**:393-401

[42] Scales BS, Huffnagle GB. The microbiome in wound repair and tissue fibrosis. The Journal of Pathology. 2013;**229**(2):323-331

[43] Kalodimou VE. A Handbook to Mesenchymal Stem Cells in Regenerative Medicine. Speg Co, Speg Consulting; 2015

[44] Deans RJ, Moseley AB. Mesenchymal stem cells: Biology and potential clinical uses. Experimental Hematology. 2000;**28**:875-884

[45] Zuk PA, Zhu M, Mizuno H, et al. Multilineage cells from human adipose tissue: Implications for cell-based therapies. Tissue Engineering. 2001;**7**(2):211-228

[46] Silva EA, Mooney DJ. Spatiotemporal control of vascular endothelial growth factor delivery from injectable hydrogels enhances angiogenesis. Journal of Thrombosis and Haemostasis. 2007;**5**(3):590-598

[47] Hea L. The effect of the controlled release of basic fibroblast growth factor from ionic gelatin-based hydrogels on angiogenesis in a murine critical limb ischemia model. Biomaterials. 2007;**28**(16):8

[48] Dvir T et al. Prevascularization of cardiac patch on the omentum improves its therapeutic outcome. Proceedings of the National Academy of Sciences of the United States of America. 2009;**106**(35):14990-14995

[49] Hirschi KK, Li S, Roy K. Induced pluripotent stem cells for regenerative medicine. Annual Review of Biomedical Engineering. 2014;**16**:277-294

[50] Taguchi A et al. Administration of CD34+ cells after stroke enhances neurogenesis via angiogenesis in a mouse model. The Journal of Clinical Investigation. 2004;**114**(3):330-338

Chapter 13

Mesenchymal Stem Cells from Fat: From Differentiation Mechanisms to Biomedical Application in Patients

José Joaquín Merino, María Gabriela Villalba, Ricardo Martínez-Murillo and Ana I. Flores

Abstract

Adipose tissue mesenchymal stem cells (AD-MSC or ADSC) are multipotent cells that do not show immune rejection. In this work, we analyze the route of administration and its possible differentiation into specific lineages of adipogenic, chondrogenic, osteogenic, myogenic, or neurogenic phenotypes. Transplanted cells induced tissue repair by inducing angiogenic, anti-inflammatory, and immunomodulatory effects (IDO, PG-2, nitric oxide, and some cytokine signaling). The ADSC exert these tissue repair processes through the release of chemokines and growth factors in a paracrine manner. Other fat-derived stem cells such as perivascular adipose tissue cells (PVAT) and muse cells induced reparative effects. Cell-free therapy using stromal vascular fraction (SVF) or the use of exosomes releasing miRNAs and cytokines also confirmed their safety and efficacy *in vitro*. Several published preclinical and clinical trials with AD-MSC confirmed their beneficial effects to repair and prevent chronic-degenerative pathologies. In this chapter, we review AD-MSC-based therapies that have used preclinical rodent models of disease for cartilage repair, regeneration of the peripheral and central nervous system, dental bone, myocardium, and liver, and in the treatment of perianal fistula in Chron's disease, and in wound and skin fibrosis repair. In addition, this work also includes clinical studies with AD-MSC or other fat-derived stem cells in patients with various pathologies.

Keywords: regenerative medicine, stem cells, adipose tissue mesenchymal stem cells (AD-MSC), adipose-derived stem cells (ADSC), exosomes, stromal vascular fraction (SVF), clinical applications of adipose stem cells; neurorepair, skin repair, cartilage repair, myocardium repair, liver regeneration, perianal fistula, aesthetic medicine, stem cell therapy, signalling pathways, chemokines

1. Introduction

1.1 Mesenchymal stem cells from adipose tissue (AD-MSC): Characteristics and markers

Adipose tissue secretes adiponectin, leptin, and other adipokines involved in metabolic functions [1]. The incidence of obesity in the world is high, and adipose tissue aspiration (liposuction-lipoaspiration) is an esthetic surgical procedure that reduces the amount of unwanted fat and also allows the isolation of autologous adipose-derived stem cells (SC) (ADSC) [2]. Adipose-derived mesenchymal stromal cells (AD-MSC) play a role in maintaining adipocyte populations and promote tissue's regenerative effects. Adipose tissue is an abundant, reliable, safe and feasible source for mesenchymal stromal cell isolation. Adipose tissue contains a large number of multipotent stem cells called adipose mesenchymal stromal cells (AD-MSC) and lipoblasts, among other components, capable of differentiating into specific lineages, such as adipogenic, chondrogenic, osteogenic, myogenic or neurogenic [3]. However, AD-MSC obtained from subcutaneous fat deposits or abdominal fat is a heterogeneous cell population, and individual differences could affect autologous transplants while allogenic transplants have the risk of immune rejection [4]. On the other hand, the efficacy and functionality of transplanted MSC depend on number of transplanted cells, route of infusion, and frequency of administration [5–9]. However, the number of AD-MSC transplants that reach the damaged tissue is low, although MSC can differentiate into the specific cell type to replace it, given its multipotentiality.

It is known that multiple doses infusion of AD-MSC offer longer cell persistence and are more effective than a single dose [9, 10]. It is also worthy to mention that intravenous infusion of ADSC (or stem cells in general) lead to their accumulation into the lung [11, 12], which could reduce the clinical efficacy of these transplanted SC [13]. Inflammation produces chemokines (chemotactic cytokines) that recruit MSC to damaged tissues. Homing or recruitment of stem cells to damaged tissues depends on the route of administration [14–20].

One advantage of fat is its safety and easy procedure for SC isolation under local anesthesia, making it less invasive than the puncture technique [21]. In addition, fat from liposuction can be preserved in biobanks for long periods of time at −85°C with cryoprotective agents. This fat can be isolated throughout life, although there are differences regarding the origin of fat in the number of cells obtained [1, 2]. In addition, the proliferative capacity of isolated cells is greater in fatty deposits of young than in older patients [22]. However, the composition of fat-derived SCs (or their subpopulations) varies between laboratories and preparations, probably due to the lack of standardization of *in vitro* isolation protocols [23–26]. The number of MSCs is limited and its proliferation capacity decreases with successive passages *in vitro* [27]. ADSC can adhere to plastic *in vitro* and maintain a normal karyotype *in vitro* through several growth passages (teen), although its capacity of expansion progressively decreases *in vitro* [28]. Thus, the multipotential capacity of differentiation into a wide variety of cell types converts them into good candidates for biomedical applications. In fact, several studies have confirmed the long-term safety of AD-MSC in rodent models of disease and clinical trials without adverse effects or tumorigenesis [29–31]. Finally, adipose/fat tissue provides an abundant source of stromal vascular fraction (SVF) cells while exosomes (exo) are nanovesicles derived vesicles released by SC that contain many soluble factors involved in cell repair [32–38].

Mesenchymal stem cells (MSC) are a multipotential heterogeneous population of stromal cells that adhere to the plastic, including AD-MSC. MSC can be isolated from bone marrow, adipose tissue, peripheral blood, Wharton's jelly from the umbilical cord, venous walls, placenta, periosteum, trabecular bone, or teeth, including the periodontal ligament, among other tissues [39–44]. The main criteria that identify MSC are their ability to adhere to the plastic *in vitro*, their fibroblast-type morphology, and the capacity of differentiation *in vitro* into cells of mesodermal origin (chondrocytes, adipocytes, and osteoblasts) [44–48]. The first characterization of these cells, obtained from human lipoaspirates, was demonstrated by Patricia Zuk and coworkers [41, 42]. It was in 2004 when the International Federation for Adipose Therapeutic and Science (IFATS) decided to unify the different nomenclatures of SC derived from the vasculo-stromal fraction of adipose tissue with the acronym ADSC (Adipose-derived Stem cells), including the mesenchymal stem cells from adipose tissues (MS-ADSC) [33, 35]. ADSC are positive for the surface markers Stro-1 (stromal precursor antigen-1), CD73 CD29, and also are negative for the hematopoietic markers (CD34 and CD45) [36–45, 49]. **Table 1** compares the properties of bone marrow-derived MSC (BM-MSC) with ADSC (see table). In addition, ADSC also expresses CD44 (hyaluronic acid receptor), and adhesion molecules such as CD29 (integrin β1), CD90 (Thy-1: thymocyte antigen-1), surface enzymes such as CD13 (aminopeptidase), CD71, CD105 (endoglin), CD73 (ecto-5′-nucleotidase). They also express CD49d (VLA4), CD106 (VCAM-1), and CD54 (ICAM-1) markers [46–48]. However, tissue-resident MSC may also express the hematopoietic stem cell (HSC) marker CD34 [50]. Although freshly isolated ADSC can express CD34, long-term cultured ADSC do not, perhaps due to the artificial environment of the tissue culture plate [50].

On the other hand, perivascular stem cells are located in blood vessels, such as pericytes or vascular precursor SC [51, 52]. ADSC is predominantly associated with vascular structures in the adipose microvasculature having a CD34+/CD31 phenotype of capillary endothelial cells [50–53]. In fact, in fresh adipose tissues, CD34+ cells are located in the intima and adventitia layers of blood vessels. However, MSC even appear to be pericytes capable of stabilizing blood vessels and also contribute to tissue homeostasis, playing an active role in repairing against injury [53]. Following ADSC transplants, controversial results on their repair capabilities were observed in both preclinical models of the disease and in clinical trials [53, 54]. These discrepancies can

Parameter	BM-MSC (Bone marrow derived mesenchymal stem cells)	AD-MSC (Mesenchymal stem cells from fat)
Quantity of isolated cells from donnor	Less abundant	More abundant
Accessibility	Poorly accessible (located in the bone)	Highly accessible (located subcutaneously in fat)
Extraction procedure	It requires general anesthesia	It requires local anesthesia
Yield	Lower efficiency in the number of isolated stem cells	Higher efficiency in the number of isolated stem cells
Does it require the use of mobilizers?	Yes, the treatment with G-CSF (granulocyte-colony stimulating factor) enhances the number of isolated SC	Not required

BM-MSC: bone marrow mesenchymal stromal cells; MSC: mesenchymal stromal cells; AD-MSC: adipose-derived mesenchymal stromal cells.

Table 1.
Differences between bone mesenchymal stem cells (BM-MSC) and adipose stem cells (AD-MSC).

be explained by the infusion route used, differences in the cell type, and in the dose of MSC applied, which may affect their clinical efficacy [7, 55]. Even so, the therapeutic success of ADSC has been confirmed in multiple studies [56, 57], although there are particular cases where results differ, given the lack of standardization among multi-center clinical trials [58].

Finally, ADSC can be expanded *in vitro* as well as cryopreserved in biobanks under good manufacturing practices (GMP) [59]. In allogeneic transplants, the compatibility between donor and receptor depends on the histocompatibility antigens (HLA). Although allogeneic transplant often is well tolerated, the risk of immune rejection could occur after transplant [60, 61].

1.2 Differentiation capacity of adipose-derived stem cells (ADSC) into various cell types

ADSC are progenitor cells that have a high capacity to differentiate into mesenchymal lineages such as osteocytes, adipocytes, and chondrocytes [28], although they can also differentiate into non-mesenchymal lineages (hepatocytes, pancreatic β cells) [28, 62–65]. ADSC can differentiate into adipocytes or osteoblasts, as fat-inducing factors inhibit osteogenesis, and conversely, bone-inducing factors hinder adipogenesis. Several external signals regulate the balance of adipo-osteogenic differentiation, and dysregulations of these balances contribute to aging, obesity, osteoporosis, osteopenia, etc. The process of MSC differentiation into adipocytes or osteoblasts has attracted great attention because of its potential repair capabilities in treating certain pathologies [29, 65].

1.2.1 Differentiation of adipose stem cells (ADSC) into mesenchymal lineages

As mentioned above, ADSC can be differentiated into adipogenic, osteogenic, and chondrogenic lineages [65].

1.2.1.1 Adipogenic differentiation of ADSC

Given the origin of ADSC, undifferentiated cells express genes such as leptin, lipoprotein lipase, and PPARgmama-2 (peroxisome proliferator-activated receptor gamma-2). Their ability to differentiate into the adipogenic lineage is maintained even after the cells have been transplanted [65]. ADSC cells are grown in adipogenic medium with indomethacin, dexamethasone, 3-isobutyl-methylxanthine, penicillin/streptomycin, and rh-insulin, l-glutamine. Differentiated cells express the PPARγ2 marker (peroxisome proliferator-activated receptor gamma 2) or Glut4 [65].

1.2.1.2 Osteogenic differentiation of ADSC

Osteogenic induction of ADSC takes place for 21 days after adding osteogenic factors to the culture medium, such as TGF-β and bone morphogenetic proteins (BMP), dexamethasone, vitamin D3 and ascorbic acid/ascorbate, among others [65]. The capacity for osteogenic differentiation is confirmed by Von Kossa or Alizarin Red staining that identify mineralized matrix and calcium deposits at the histological level [66]. During differentiation of ADSC into osteoblasts, the expression of bone markers and mineralized extracellular matrix (ECM) increases, such as Runx-related transcription factor 2 (Runx2), osteonectin, osteocalcin, alkaline phosphatase (ALP) and collagen type I [65, 67, 68]. In fact, osteogenic differentiation is enhanced by transfection of ADSC with BMP-2 and Runx2 [69, 70].

1.2.1.3 Chondrogenic differentiation of ADSC

Chondrogenic differentiation is induced by the addition of several factors in the culture medium such as transforming growth factor family members (TGF-β beta1 and 3), bone morphogenic protein (BMP-4) or basic fibroblast growth factor (bFGF). The differentiation of ADSC takes place during 21–28 days, and it is confirmed by Alcian Blue staining, which binds strongly to sulfated glycosamyglycans (GAGs) and glycoproteins [65]. Chondrogenic differentiation is also confirmed by TGF-β and SOX-9 expression, which is essential for the expression of type II collagen in chondrocytes [67, 68]. In this way, SOX-9 overexpression enhances chondrogenic differentiation and inhibits the osteogenic differentiation of MSC [71]. In addition, proteoglycans in the cartilage-like matrix can be identified by toluidine blue or safranin O staining [72]. Interestingly, the combination of L-ascorbic acid and platelet-rich plasma (PRP) can increase the survival of ADSC and improve chondrogenic function under appropriate concentrations [3].

1.2.2 Differentiation of adipose stem cells (ADSC) into non-mesenchymal phenotypes (muscle, cardiac, neurogenic, or endothelial cells)

ADSC can also differentiate into non-mesenchymal phenotypes, such as myogenic, neuronal, or endothelial lineages [73, 75]. Myogenic capacity of ADSC was demonstrated after transplantation into damaged rabbit muscles. In rodent models of cardiac injury, ADSC transplantation promotes regeneration of damaged myocardial tissue [64] On the other hand, human or rat ADSC can be differentiated into neuronal lineages by treatment with beta-mercaptoethanol, which induces the expression of beta-enolase, a neuronal marker [74]. In another study, treatment of ADSC with endothelin-1 (a paracrine factor released by endothelial cells) promotes their differentiation into endothelial cells with angiogenic properties. Finally, ADSC may be useful in treating preeclampsia, a complication of hypertension in pregnant women that leads to endothelial dysfunction [75].

1.3 Stromal vascular fraction (SVF) and exosomes (exo-ADSC) are adipose tissue-derived products

1.3.1 Stromal vascular fraction (SVF)

Lipoaspirate is a medical waste that contains stromal precursor stem cells called stromal vascular fraction (SVF). These cells are reliably and viably isolated from fat by enzymatic dissociation with collagenase [41, 42, 76] or using commercial kits [77]. The application of the local anesthetic method for SVF isolation with the collagenase kit, which is subsequently inactivated, allows the SVF to be obtained from the lipoaspirate [77]. The multipotent capacity of adipose tissue-resident stromal cells was demonstrated in 2000, when it was discovered that adipose tissue stromal mesenchymal progenitor cells could differentiate into bone tissue *in vitro* under the right conditions [78]. In 2001, the presence of progenitor cells in subcutaneous adipose tissue with inherent properties of MSC was discovered, with potential ability to differentiate into chondrocytes, osteoblasts, and adipocytes. These cells exhibited good self-renewal capacity inherent to progenitor cells and typical mesenchymal-like markers on their surface, and at that time were termed processed lipoaspirate cells (PLA cells) [41, 76]. In SVF, ADSC are immunophenotypically characterized as CD45-/CD235a-/CD31-/CD31-/CD34+ markers, which represent approximately 20% of total SVF [33, 64]. In 2013, IFATS published a statement that included the minimum phenotypic criteria for characterizing uncultured SVF and also adherent ADSC isolated from adipose tissue, which typically account for up to 3%, approximately 2500 times more than bone marrow-derived cells (see **Figure 1**) [65].

Among the components of SVF, there are certain paracrine factors (VEGF, SDF1 alpha, BDNF, etc.) that protect against osteoarthritis in patients with joint injuries, especially of the knee [79]. The beneficial effects of SVF-associated therapy have been confirmed in preclinical models and also clinical trials, especially in patients with osteoarthrosis [80]. The clinical safety and efficacy of autologous SVF infusion performed in patients with grade 3 and 4 arthritis demonstrated 67% improvement in patients after stem cell therapy and with no adverse effects [81]. In this study, SVF-induced anti-inflammatory effects reduced osteoarticular injury [79] through the release of nitric oxide, TGF-β1, stromal-derived factor (SDF-1 alpha), and other chemokines [80, 82, 83]. The exact composition of the paracrine factors in this SVF fraction and the exact molecular mechanism(s) by which SVF induces tissue repair are questions to be clarified in further studies. In addition, the combination of SVF with platelet-rich plasma (PRP) could enhance anti-inflammatory responses through the release of platelet-derived growth factor (PDGF) in patients [79].

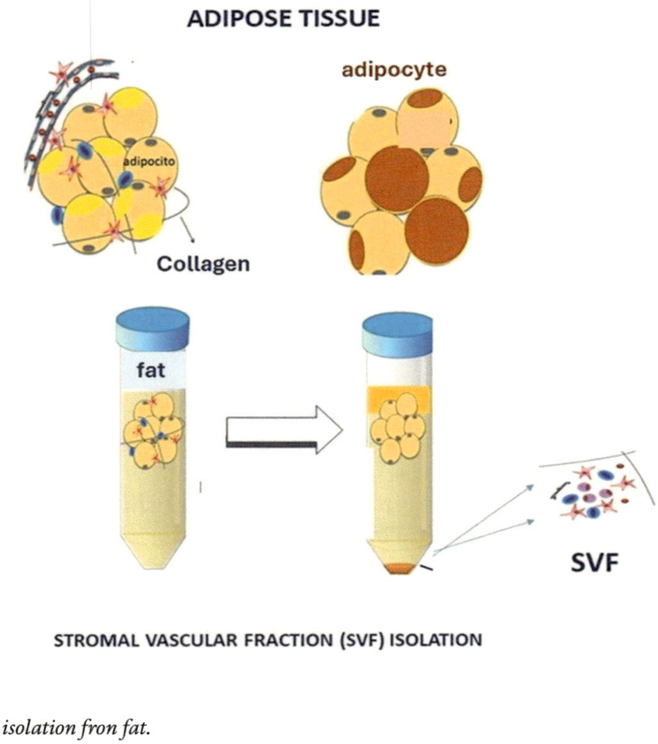

Figure 1.
Stromal vascular isolation fron fat.

1.3.2 Exosomes

Exosomes (exo) are small extracellular nanovesicles (size 40–150 nm) positive for CD63 and CD9 tetraspanin markers, and result from the fusion of vesicular bodies with the plasma membrane [84]. Exo contains various nucleic acids (miRNA, RNA), proteins, and lipids and are involved in the cellular trafficking processes [84]. Soluble factors released by exo-ADSC promote anti-inflammatory and angiogenic effects as exo contains a cholesterol-rich lipid bilayer and also carries miRNAs that contribute to cellular repair mechanisms (miR-23a, miR-26b, miR-125b, miR-130b, miR-140, miR-203a, miR-223, miR-224), but also several proteins involved in many physiological processes [57, 84]. **Figure 2** shows several molecules involved in cell adhesion, prosurvival, and migration pathways of MSC, including fat MSC (ADSC, see **Figure 2**).

1.4 Repair mechanisms of adipose stem cells (ADSC)

ADSC release a plethora of paracrine factors involved in their protective/reparative effects, as demonstrated in several disease models in rodents, such as in diabetes, liver regeneration, ankle pathologies or neurological diseases [85–89]. The angiogenic capacity of ADSC depends on their origin [90], and they are cells carrying non-coding miRNAs involved in angiogenesis (e.g., miR-126, miR-296, miR-378, and miR-210 [91]). Activation of metalloproteases MMP-2 and 9 induces extracellular matrix remodeling, whereas ADSC stem cell transplantation reduces their activity [92] (for the repair mechanisms of ADSC review, **Figures 2** and **3**).

The main protective and/or reparative mechanisms of transplanted ADSC are attributable to paracrine release of factors with angiogenic and/or immunomodulatory effects, making them ideal candidates for clinical transplants in patients [91, 93–96]. Indeed, MSC favors the conversion of proinflammatory M1 macrophage phenotypes to an anti-inflammatory M2 phenotype and exerts immunomodulatory effects [97, 98]. MSC may act as antigen-presenting cells (APCs) with immunosuppressive properties for allogeneic

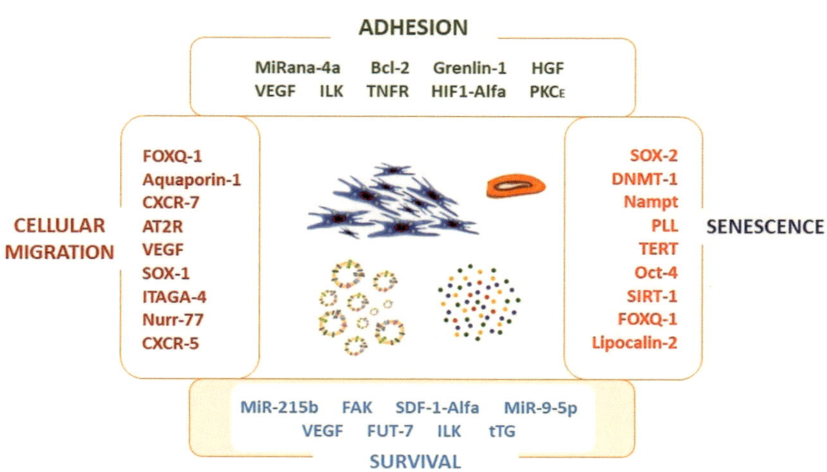

Figure 2.
Content of mesenchymal stem cell (MSC) exosomes (exo): miRNA and proteins involved in cell adhesion, migration, survival, and senescence.

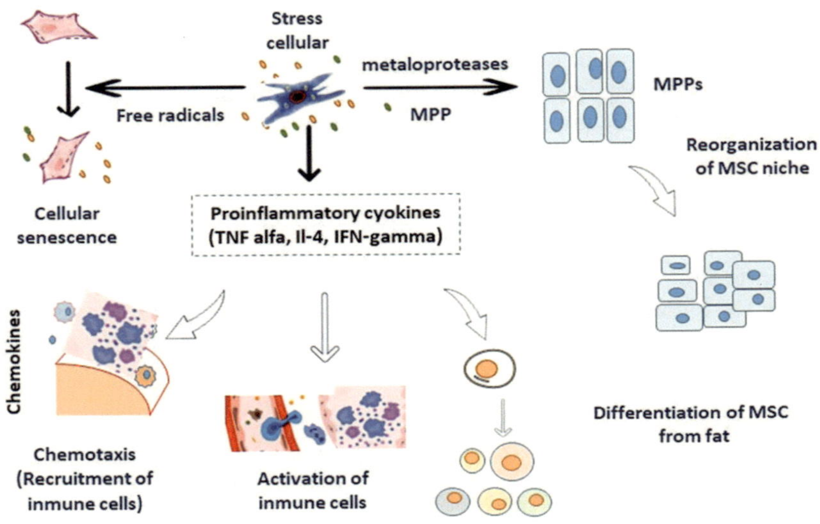

Figure 3.
Immunomodulatory properties of mesenchymal stem cells (MSC).

stem cell therapy [98, 99]. Prostaglandin E-2 (PGE-2) and IDO are released by MSC [15, 100, 101] and involved in their immunosuppressive activities [102–106]. In addition, TNF-stimulated gene 6 (TSG-6), another factor secreted by MSC, is stimulated by TNF-alpha or IL-1, resulting in anti-inflammatory effects [106] and prevention of vascular endothelial cell apoptosis ([107], see **Figures 2** and **3**). In a rodent infarction model,

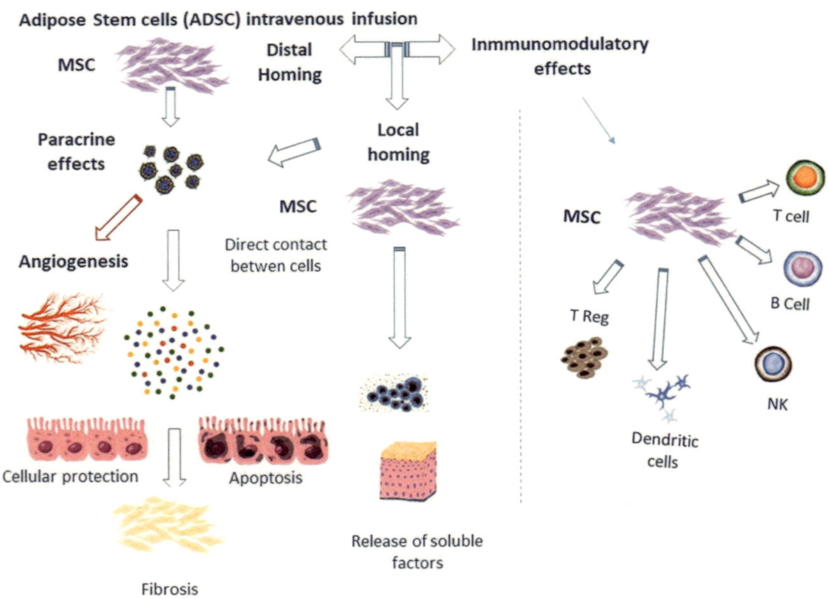

Figure 4.
Repair mechanisms of mesenchymal stromal cells, including ADSC, promote angiogenic, protective, anti-apoptotic, and immunomodulatory effects.

MSC transplantation reduced neutrophil recruitment by promoting anti-inflammatory effects dependent on TSG-6 protein levels concomitant with reduced MMP-6 activity in damaged areas ([108], and see **Figures 2** and **4**).

The inflammatory response is regulated by a delicate balance between proinflammatory signals released by M1-type macrophages as scavengers that remove cellular debris and M2-type macrophages that release cytokines that promote tissue repair. MSC can convert macrophages from an M1 proinflammatory phenotype toward an M2 anti-inflammatory effect. Activation of TNF-alpha-secreting macrophages by MSC could inhibit inflammatory responses [108]. The ability of MSC to regulate macrophage polarization through direct cell contact, although recent studies indicate that the immunomodulatory properties of MSC largely depend on paracrine mediators secreted by them [107].

2. Translational applications of adipose stem cells (ADSC)

Bone tissue engineering allows for improved therapeutic options of ADSC for the treatment of bone defects in situations such as reconstructive surgery, trauma, malformations, or other pathological conditions. The therapeutic role of bone tissue could also improve the quality of life of patients with osteoarticular problems. It is known that osteoprogenitor cells can grow *in vitro*, combined with an appropriate scaffold [65]. ADSC combined with various types of organic or inorganic scaffolds (polylactic acid, PLA, polyglycolic acid, PGA, glycolic acid, PLGA, fibrin, collagen, gelatin) have been shown to have improved their *in vitro* repair capabilities [107]. Ceramics are known to exhibit osteoconductive properties, support osteogenic potential, and affect growth direction *in vitro* [109]. ADSC growth, in combination with bioactive factors and scaffolds, exhibits strong osteoinductive properties (biocompatible and biodegradable) and increases osteogenic differentiation, enhancing bone trabeculae formation without forming toxic by products. Overall, ADSC treatments for bone regeneration are promising for the treatment of craniofacial defects [110, 111].

2.1 Cell-based therapies for cartilage repair

The infusion route (intravenous, subcutaneous, etc.), dose, and frequency/timing of administration affect the safety and clinical efficacy of autologous or allogeneic human stem cell transplants [7]. Osteoarthritis (OA) is a progressive degenerative joint disease characterized by deterioration of articular cartilage as well as pathological changes in the adjacent subchondral bone. Cartilage injury is a leading cause of disability worldwide, and current conventional treatments (physical therapy, chondroitin sulfate supplementation, and arthroscopic surgery) can improve joint function but without complete reversal of pain. Since cartilage injury is a leading cause of disability worldwide, autologous stem cell therapy treatment could promote cartilage regeneration [111]. However, stem cells injected directly into the pain site often have limited cell retention and a low survival rate, especially in large cartilage lesions, which might reduce their clinical efficacy [112, 113]. In a mouse model of OA injury, a single injection of AD-MSC into the knee during the early stage of OA inhibits cartilage destruction [114]. However, there was no effect when ADSC were infused at an advanced stage of the disease [115]. Given the limited efficacy of conventional treatments against OA, intra-articular infusion of autologous stem cells could be a potential approach for cartilage regeneration, including the application of exosomes or MSC-derived advanced therapy drugs (ATMPs) against osteoarticular lesions [115, 116].

MSC can differentiate into chondrocytes, osteoblasts, and myocytes, among other types of mesodermal cells (for review, see [117]). In one study, the authors demonstrated that clinical symptoms improved during a 6-month follow-up after intra-arterial injection of ADSC, in patients with osteoarticular problems [118, 119]. In another study, an MSC-conditioned culture medium protected against bone defects. However, the properties of MSC are affected by sex, age, and body mass index of donors [120], leading to inter-individual differences between patients. On the other hand, the use of scaffolds increases the viability of ADSC and their aggregation capacity *in vitro* [121]. The three-dimensional structure of ADSC loaded on scaffolds, as well as the pore size, the type of material, and the stiffness of the scaffolds, are important factors that could affect the osteoarticular regeneration processes. Indeed, certain biomaterials (scaffolds) can enhance cartilage formation *in vitro*, as ADSC can be differentiated into chondrocytes by the addition of certain recombinant proteins (IGF-1, TGF-β, or BMP) to the medium [117]. MSC-derived cartilage increases mineralization and promotes angiogenesis for bone formation in a SCID mouse model [116]. The use of biomimetic injectable hydrogels can induce the release of soluble factors capable of retaining stem cells at the target site. For example, natural materials favor cell adhesion, promote their biodegradability, and regulate their mechanical properties [122].

Scaffolds with smaller pore sizes (90–250 μm) preserved cell adhesion properties and favored proliferation, and also increased collagen type II expression [123]. In this sense, a combination of a 3D type I collagen scaffold with platelet-derived growth factor (PDGF) and insulin are biocompatible materials *in vitro* and increase the chondrogenic differentiation of ADSC [124]. Other studies have confirmed that hydrogel-based scaffold systems can create cartilage but with better mechanical properties [125]. ADSC cells in a PLGA-gelatin scaffold with immobilized TGF-β1 reduce cartilage damage and improve cartilage quality [118]. The efficacy of intra-articular injections of hydrogels (with and without ADSC) was evaluated in a collagenase-induced OA rat model 7 days after induction of cartilage degeneration. The results showed that both hydrogel and ADSC-hydrogel treated groups showed chondroprotective and anti-inflammatory effects, suggesting that the hydrogel induces cartilage tissue regeneration [126–128]. In this model, the OA lesion and inflammation were reduced by ADSC-hydrogel treatment. In conclusion, the use of biomaterials combined with stem cells could be a good therapeutic option for OA patients. However, it is important to establish a specific therapeutic option for each patient based on their degree of osteoarticular injury.

A recent systemic review of 4348 articles has evaluated the use of ADSC for ankle orthopedy treatment, demonstrating the beneficial effects of AD-MSC against osteoarticular lesions. For example, improvements in MRI outcomes after SVF treatment together with marrow stimulation were demonstrated as compared to marrow stimulation alone [129]. Other meta-analyses evaluated the effect of ADSC and SVF in 82 studies with a total of 3594 treated patients with osteoarticular diseases. 70% of these studies evaluated the effect of stem cells in osteoarthritis, and 26% dealt with expanded ADSC. This review confirmed the heterogeneity of cell types used for osteoarticular lesions from different sources of adipose tissue (for review consult [129]). Thus, the safety and efficacy of SVF treatment to treat osteoarticular lesions are associated with many protective mechanisms, among them the induction of immunomodulatory effects and graft survival, leading to cell repair [130]. However, differences in the methodology for SVF isolation, donor age, etc., are limitations that may reduce the efficacy of ADSC and adipose tissue-derived products. On the other hand, lack of standardization of clinical procedures, as well as hormonal status or comorbidities such as obesity may also affect the efficacy of ADSC transplantation.

Finally, differences in international regulations between countries may affect the products available on the market. In some cases, conducting double-blind randomized trials without a true control group is another limitation [131]. Despite these limitations, promising results highlight the clinical use of ADSC treatments for osteoarthritis pathologies (in general) by inducing anti-inflammatory and chondroprotective effects without serious complications [132]. Due to regulatory issues, most of the research is still in the experimental phase, and most are preclinical results; therefore, we are still far from reaching the generation of new cartilage due to the cellular and molecular complexity of this tissue (see, **Figure 5**, [133]).

2.2 Regeneration of the central and peripheral nervous system by fat-derived stem cells (ADSC)

Neuroprotective and promyelinating properties of conditioned culture medium have been demonstrated in human adipose mesenchymal stromal cells [134]. This study confirmed the repair capacity of isolated ADSC from intact adipose tissue of 10 subjects undergoing abdominal plastic surgery [134]. The therapeutic potential of ADSC in the central and peripheral nervous system has been associated with its capacity to differentiate into neurons, endothelial cells and Schwann cells, to the expression of neuronal markers and to a faster proliferation rate than other types of stem cells [135]. Neuronal differentiation of ADSC is mediated by several neurotrophins (NGF, BDNF, GDNF, FGF) and hormones (IGF-1), which regenerate damaged nerves and induce neuroprotective effects [89]. Neuronal degeneration and vascular damage can lead to inflammation, followed by loss of oligodendrocytes and neurons in the lesions. Therefore, ADSC administration could control the inflammatory response after injury and could also improve regenerative capacity, given the contribution of vascular factors in repair [136]. In fact, vascularization is another factor involved in cell survival

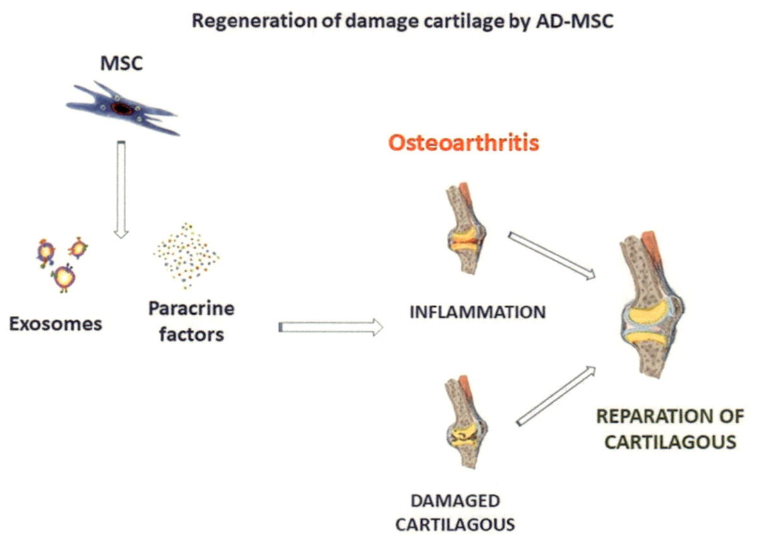

Figure 5.
Repairing effects of ADSC in osteoarthritis.

and proliferation, and released vascular factors released by damaged endothelium enhance ADSC-mediated regenerative effects (**Figure** 5, [137]).

ADSC exo-derivatives produce neuroprotective effects in rodent models of multiple sclerosis, cerebral ischemia, and traumatic brain injury (TBI) [138, 139]. ADSC infusion prevented TBI-induced neuroinflammation and also decreased secondary injury as a consequence of TSG-6 release, leading to reduced microglial overactivation [140].

ADSC exo-derivatives can decrease the cytotoxic effect of overactivated microglia by reducing the nuclear factor kappa-beta (NF-β) transactivation and regulating mitogen-activated protein kinase (MAPK) pathway [92]. Axonal regeneration and less reactive gliosis were evident after ADSC treatment in rats with spinal cord injury [141]. In a multidisciplinary clinical trial, intrathecal infusion of ADSC decreased the medullar lesion at L3–4 level of the spinal cord, and the evaluation of lesions with objective criteria following the International Standards for Neurological Classification of Spinal Cord Injury scores demonstrated improvements in these patients [142]. Preclinical findings confirmed that the combination of scaffolds and ADSC could prevent symptomatology in certain neurological disorders, such as amyotrophic lateral sclerosis, Alzheimer's (AD), Parkinson's (PD), or Huntington's disease [142, 143]. Some clinical trials evaluated the safety and efficacy of ADSC to treat pathological conditions in patients with Alzheimer's, Parkinson's or Huntington's, or amyotrophic disease (see ClinicalTrials.gov, www.clnical-trials.gov identification numbers NCT03117738 and NCT02184546) [144–149].

2.2.1 Regeneration of the peripheral nervous system by ADSC

ADSC infusion in rodent models promoted regenerative effects of the peripheral nervous system after sciatic nerve damage with varying degrees of severity. ADSC treatment enhanced Schwann cell regeneration by increasing the number of myelinated fibers in the damaged nerves. In fact, ADSC can be differentiated into Schwann cells [150], and the use of compatible biomaterials *in vitro*, such as PGA, PCL, and collagen, increase the repair capacity of ADSC [151, 152].

2.3 Regeneration of the dental pulp by ADSC

ADSC treatment also promotes the regeneration of bone tissue in alveolar bone defects, specifically in patients with periodontal disease [153]. The inclusion of ADSC in PLGA matrices enhances bone growth in animal models. The synergic combination of ADSC with fibrin also prevented inflammation and reduced bone resorption in the tooth, suggesting that MSC from the oral cavity contributes to repair [154]. The combination of autologous ADSC and Platelet-Rich Plasma (PRP) induces periodontal tissue regeneration. For example, CD31 positive adipose population contributes to promoting repair in dental pulp cells in dogs [153]. However, their applications in regenerative dentistry need further confirmation with clinical trials [154].

2.4 Regeneration of the myocardium by ADSC

Several routes of infusion have been evaluated in transplants with ADSC into the myocardium, such as intramyocardial, intravenous, or intracoronary injection. Transplantation with ADSC has promoted beneficial effects in cardiovascular pathology due to its capacity for differentiation to the required cell type [155] and the release of paracrine factors involved in angiogenic, anti-apoptotic, and anti-inflammatory effects [156, 157]. Intramuscular and intracoronary infusion of ADSC improved the left ventricular ejection fraction (LVEF) function [158], and inhibit the fibrosis process [159]. The efficacy of direct infusion of ADSC into infarcted myocardium in rodent models is limited, given the low survival rate of transplanted cells [158]. Intravenous administration of resistin-treated ADSC improved LVEF function, reduced fibrosis, and decreased cardiomyocyte apoptosis [160]. Intracoronary administration of ADSC also reduced apoptosis (programmed cell death) in the infarcted area of the myocardium without significantly affecting LVEF [161], clearly indicating the relevance of the infusion route for ADSC transplantation. This intracoronary administration significantly reduced infarct size and induced angiogenesis [161].

On the other hand, the combination of biomaterials with ADSC creates a favorable microenvironment for tissue repair in the infarcted area [162]. Some studies have addressed the factors involved in the reparative effect of ADSC. For example, the overexpression of the Stromal Cell Derived Factor (SDF-1 alpha) chemokine as a regenerative factor in stem cell therapy since it reduces apoptosis and enhances angiogenesis, leading to protective effects [162, 163] and its overexpression prevented detrimental consequences of myocardial infarction [164]. Overexpression of other trophic factors such as IGF-1, VEGF, hepatic growth factor (HGF), and FGF-2 also decreased apoptosis and enhanced angiogenesis [165, 166]. From this perspective, several clinical trials have reported that intracardiac ADSC transplantation improved the scales with improvement in cardiac function [164, 165].

2.5 Liver regeneration by ADSC

Acute liver failure or chronic liver diseases can be the consequence of infection with viruses, toxins or even genetic factors. These liver diseases are characterized by fibrosis and can provoke liver damage and inflammation with the consequent activation of hepatic stellate cells (HSC) [167]. Given that ADSC can differentiate into various types of liver cells and have anti-apoptotic and immunomodulatory effects, they could be useful for treating liver pathologies [168]. The administration of ADSC suppresses the expression of proinflammatory cytokines and decreases the proliferation of activated HSC cells [168, 169]. In a rodent model of induced liver fibrosis by CCl4 treatment, ADSC transplant or exosomes infusion markedly decreased liver fibrosis, leading to apoptosis of HSC cells [170, 171]. In another study, ADSC injection into liver parenchyma protected cells against liver injury by inducing Superoxide dismutase-1 (SOD-1) and produced anti-inflammatory effects by reducing IL-1 beta levels [172]. The culture of ADSC under hypoxia increase the expression Nrf2 transcription factor *in vitro* (a nuclear factor erythroid 2 with antioxidant effects) that protects against free radical-mediated toxicity in the liver [173]. Another factor released by ADSC, such as hepatic growth factor (HGF), induced anti-inflammatory effects by increasing IL-10, an anti-inflammatory cytokine [174, 175]. In addition, exosomes from ADSC infusion reduced the levels of liver enzymes such as aminotransferase (AST), alanine aminotransferase (ALT) activities and reduced lactate dehydrogenase (LDH) activity [168]. Intravenous infusion of ADSC in a murine model of liver injury prevented liver damage concomitant with decreased serum ALT and AST activities. In a carbon chloride liver injury model, ADSC-derived exosomes carrying the modified miRNA-181-5p reduced liver fibrosis [176, 177]. The combination of ADSC with scaffolds such as PLGA markedly improved liver function and reduced necrotic areas in the tissue [178]. Umbilical cord MSC (UC-MSC) infusion reduces fibrosis and increases caveolin-1 in hepatic stellate cells [179]. Finally, some clinical trials confirmed that ADSC infusion is safe and effective in treating liver cirrhosis in patients. In these clinical trials, ADSC transplantation improved liver function without any adverse effects [180, 181]. However, further clinical trials with a larger sample size should confirm the long-term safety of ADSC in patients.

2.6 ADSC infusion for the treatment of perianal fistula

Perianal fistulas are a complication suffered by 20 to 25% of Crohn's patients, and the infusion of autologous or allogenic ADSC can promote reparative effects through the release of paracrine, immunomodulatory, anti-apoptotic, and angiogenic factors. Several clinical trials have confirmed a complete cure rate of anoperineal fistulas in these patients. A phase III controlled trial confirmed the safety of treatment with allogenic ADSC (Alofisel®), which induced a radiological remission in 60% of cases [182, 183]. Another phase II multicenter randomized controlled trial has shown that infusion of expanded ADSC (20 to 60 million cells) in combination with fibrin glue is safe and achieves higher cure rates than fibrin glue treatment alone [184].

2.7 Skin repair by ADSC

Subcutaneous administration of AD-MSC is well-studied in esthetic medicine, especially in cases of facial rejuvenation [185, 186]. Topical administration of ADSC increases skin graft survival as well as wound healing capacity [187, 188]. Furthermore, subcutaneous injection of AD-MSC from different human donors did not form teratomas in immunodeficient SCID mice up to 17 months after infusion. Furthermore, increased cell survival, exclusively at the injection site, was observed for at least 17 months after infusion, as well as these infused ADSC were able to differentiate into subdermal tissue fibroblasts without migration to organs [188]. ADSC transplantation normalized blood flow at the wound site [189], facilitated fibroblast migration and proliferation, inhibited collagen deposition, and also decreased α-smooth muscle α-actin expression in scar fibroblasts [189, 190]. ADSC differentiate into skin stem cells and promote the accumulation of autologous skin SC. Several trophic factors are involved in skin repair by ADSC, including epithelial growth factor (EGF) receptor, while the increased release of GDF11 factor promotes skin anti-aging effects. In addition, severe burn injuries and intractable ulcers are reduced after ADSC transplantation during wound healing [191]. Reduced inflammation by ADSC infusion promotes the polarization of proinflammatory M1 macrophages into their anti-inflammatory M2 phenotype, contributing to wound healing repair in rodent models of skin injury [192]. Autologous fat grafts with ADSC induce antifibrotic and anti-inflammatory effects, but the exact mechanism of repair is not yet fully understood.

Infusion of ADSC promotes reparative effects associated with the release of cytokines (G-CSF, PDGF, HGF, IL-6, and IL-8), chemokines (Il-8, SDF-1 alpha), and vascular factors (VEGF). Since ADSC are also modulators of metalloprotease activity [193, 194], it is possible that they may promote anti-fibrotic effects by decreasing their activity. By simulating the interaction of ADSC with fibroblasts and endothelial cells (ECs) in the skin, the ADSC secretome could alter scleroderma vascular inflammatory foci (SSc) in the skin *in silico* [192]. In a study using ADSC expressing Bcl-2 embedded in collagen scaffolds to treat diabetic wounds, significant improvement in wound healing [194, 195], enhanced neovascularization, and decreased healing time compared to controls were observed [196]. ADSC exosomes stimulated wound healing by increasing fibroblast, keratinocyte, and endothelial cell proliferation [196, 197].

Several clinical trials have shown that ADSC transplantation has corrected facial skin defects in patients with radiation injury [196] and that ADSC are safe and effective in the treatment of this type of radiation injury [198]. In contrast, a single dose of autologous fat grafting was insufficient to improve burn scars, perhaps due to the small patient sample size. Overall, more clinical trials should confirm the clinical efficacy of ADSC to treat wound healing [199].

3. Preclinical trials developed with fat-derived stem cells (ADSC) in murine models

The following **Table 2** indicates the preclinical findings with ADSC infusion in murine models of disease, as well as the route of administration of the cells (for a review refer to Dong et al. [61]) [65, 200].

Author	Tissue	Rodent model and implantation of cells	Results
Arrigoni et al.	Bone/rabbit	Surgery implantation/ rabbit	ADSC-induced bone formation with increased bone density
Chen et al.	Bone/human	Surgery implantation/ rabbit	Overexpression of miR-375 enhances osteogenesis in vitro and in vivo Efficiently reduced inflammatory responses
Li et al.	Cartilagous	Injection or surgery implantation/rabbit	ADSC with scaffold promote long-term regeneration
Cho et al.	Cartilagous	Surgery implantation	The transplant of ADSC was able to improves the quality of cartilage
ter Huerne et al.	Cartilagous	injection	ADSC infusion improved synovial inflammation
Yin et al.	Cartilagous	Surgery implantation	ADSC combined with an immobilized TGF beta scaffold improves regeneration of defective cartilage
Hu et al.	Nerves	surgery implantation	Improves nerve regeneration by ADSC treatment
Khingam et al.	Nerves	surgery implantation	ADSC promotes axonal regeneration
Li et al.	Nerves	Injection	ADSC reduces neurodegeneration
Durco et al.	Nerves	Surgery implantation	The ADSC infusion increase the number of nerve fibers
Nagata et al.	miocardium	Transfusión	ADSC Transplant Promotes Better Cardiac Recovery
Bobi et al.	miocardium	Surgery implantation	ADSC infusion improves perfusion in the periischemic area of the myocardium
Mori et al.	miocardium	Surgery implantation	Increases vascular density in the peri-infarct area by ADSC infusion in the myocardial
Qiao et al.	miocardium	injection	The combination of scaffolds with ADSCs increase angiogenesis, decreases fibrosis, and reduces the infarct size.
Zhang et al.	liver	injection	ADSC combined with matrices increased angiogenesis, reduced fibrosis, and reduced infarct size

Table 2.
Preclinical studies with adipose stem cells (ADSC).

4. Clinical trials with ADSC in patients with various pathologies

ADSC treatment of corneal and retinal lesions or optic nerve degeneration is associated with corneal epithelial proliferation [61] and promotes anti-inflammatory and immunomodulatory effects [61, 201]. The feasibility of ADSC to stabilize retinal microvasculature has been conclusively demonstrated in a diabetic retinopathy model [88, 202]. Another study reported that ADSC-loaded collagen sponge promoted repair of tracheal defects and restored cilia motility function [203], as well as for the treatment of silicosis, asthma, or even inflammatory diseases [204, 205]. On the other hand, autologous transplantation of ADSC can improve the functional recovery of skeletal muscle and also favor its differentiation without direct involvement of new myofiber formation [206]. ADSC have been used to treat autoimmune diseases [207] such as systemic lupus erythematosus (SLE), Sjörgren's syndrome, or Crohn's syndrome [208] or even for the treatment of type I diabetes mellitus [172]. Moreover, the anti-inflammatory capacity of ADSC makes them ideal candidates for the treatment of tendon injuries [209]. Furthermore, ADSC-derived extracellular vesicles also prevent metabolic dysfunction due to steatosis liver disease [210]. The administration of fat from subcutaneous abdominal adipose tissue (lipoaspirate) was evaluated in patients with type 2 diabetes with insulin resistance compared with healthy donors. In this study, patients with insulin resistance had a lower capacity for proliferation of ADSC, which is usually the first stage of diabetes [211]. In addition, the expression of osteogenesis markers is higher in cells from patients with type 2 diabetes (T2D), leading to the conclusion that type 2 diabetes modifies stem cell activity and that insulin resistance reduces ADSC proliferation [212]. Stem cell therapy is a novel therapeutic strategy for erectile dysfunction in patients with bilateral cavernous nerve injury. A meta-analysis of 12 studies with 319 rats analyzed the efficacy of stem cell transplants. Further studies will investigate the role of nerve restoration and vascular cell recovery in urology [213].

Several trials have confirmed the safety and efficacy of autologous ADSC to treat different pathologies [200, 213]. For example, fat is ideal for treating vocal problems because it is a biocompatible material and is readily available. The main disadvantage of its use is that resorption can cause long-term failure after its transplantation. In one study, fat harvested from the liposuction procedure was centrifuged, and the fat cell layer was injected directly into the vocal muscle. Fourteen patients received ADSC treatment (18–74 years, mean age: 48 years) with respiratory dysphonia secondary to laryngeal hemiplegia (n = 7) or anatomical defects (n = 7). The results confirmed an improvement in voice quality after ADSC transplantation, and these patients remained stable for 3–26 months (mean 10.6) [214]. In addition, seven patients improved their paralytic dysphonia after stem cell therapy [214].

Immunophenotyping of SVF-enriched fresh fat grafts has been evaluated for enhancing reconstructed breasts. Although fat injection into and around the reconstructed breast may require repeated injections and fat necrosis may occur, it appears to be a safe technique that corrects significant contour deformities [215]. The performance of both grafts was evaluated through volume measurement by magnetic resonance imaging before and after 6 months of the operation in women for breast reconstruction. The finding of this study indicates that fat injection corrects contour deformities in the reconstructed breast [215].

A recent clinical trial has demonstrated the beneficial effects of allogenic ADSC treatment in acute stroke. In this single-center, randomized, double-blind, placebo-controlled, phase II pilot clinical trial, the safety and efficacy of allogeneic ADSC therapy has been confirmed [216]. The American Diabetes Association indicates that 85 percent of amputations result from complications arising from diabetic foot ulcers or peripheral arterial disease. In this context, cell therapy could prevent limb amputation by inducing the blood vessel formation and also reduce neuropathy relief in diabetic patients (**Table 3**) [214].

Clinical trials with ADSC	Treatment with ADSC	Diseases conditions or Therapeutic indication
Japanese UMIN clinical trial registry (UMIN000022601)	The 3.3×10^5/kg was administered via the hepatic artery using microcatheter IV. The follow-up period was 24 weeks after implantation.	The patients underwent liver biopsies prior to treatment. 24 weeks after treatment the evolution of nonalcoholic fatty liver disease (NAFLD) and fibrosis were measured [200]
NCT02904824	Injection using autologous fat enriched with ADSC vs. autologous fat for the functional reconstruction of the glottal gap provoked by unilateral vocal cord paralysis by laryngoplasty	Unilateral vocal cord paralysis (unilateral)
NCT02387723 without results yet (results will be published in 2025).	Phase I-II, multicenter, randomized, controlled clinical trial for the study of factibility of the cutaneous adult allogenic adipose-derived mesenchymal stem cells (AD-MSC) application expanded on fibrin hyaluronic biological matrix to treat venous ulcer of the lower limbs (fistula).	Adipose tissue stem cells (ADSC) on biological matrix for the treatment of venous ulcer of the lower limbs The incidence of adverse events (time frame through study completion, an average of 1 year) were evaluated.
NCT02287974 Andalusian network for design and translation of advanced Therapies n = 48 treated patients (Spain) Clinical Trial I/II opened, randomized and controlled for the study of the Use of stem Cells therapy in insulinized diabetic patients type 2 with critical ischemia in lower limbs (CLI): study of the needs of insulin	To study the effect of the stem cells autologous mononuclear treatment from the bone marrow (n = 12), autologous progenitor endothelial CD133 cells from the bone marrow (n = 12) and autologous mesenchymal stem cells from adipose tissue (n = 12) on proinflammatory cytokines, the resistance to insulin as well as the reduction of insulin treatment. Inclusion criteria: diabetic type 2 (treated with insulin at least 3 previous months). vascular disease infrapopliteal, atherosclerotic of severe degree	The critical ischemia of the foot is defined as a persistent/persistent pain that needs analgesia and/or not healing present sores>4 weeks, without evidence of improvement with conventional therapies patients were included by the criteria of impossibility of revascularization or fail in the surgery of revascularization (evaluated at least 30 days before, or entry in phase of critical ischemia).

Clinical trials with ADSC	Treatment with ADSC	Diseases conditions or Therapeutic indication
NCT01828723 Interventional (Clinical trial phase I, interventional) Estimated enrollment: phase one, open Label, single arm study single group assignment	Inclusion criteria: female or male: age 18 years or older that are scheduled for liposuction (BMI between and including 23 and 28) and facial fat grafting procedures for cosmetic purposes. Facial volume defects treated with a total graft volume of between 1 mL and 50 ml. Aim: to evaluate the safety of concentrated SVF-enriched fat graft by monitoring the number and types of adverse events, physical examinations, blood draws (CBC/LFT/BMP), and urinalysis (time Frame: 6 months) were also evaluated	Conditions of disease: lipoatrophy/aging/wrinkles Demonstrate the safety of antria cell preparation process during facial fat grafting assisted with autologous SVF treatment. Outcome: to demonstrate the efficacy of autologous SVF treatment via antria cell preparation process by observing graft survival time, volume, and quality of facial re-contouring (time frame: 6 months)
NCT01678534 (AMASCIS-01: phase II) Reparative effect of acute allogenic mesenchymal stem cells from adipose tissue against ischemic stroke; evaluation of safety assessment, in a randomized, double blind placebo controlled single center pilot clinical trial (n = 19) Masking: quadruple (Participant, care provider, Investigator, outcomes assessor) Intervention: ischemic Stroke and placebo group Quadruple (Participant, Care Provider, Investigator, Outcomes Assessor)	Aim: safety evaluation with allogeneic MSC-ADSC Study group: allogeneic stem cells from adipose tissue A single intravenous AD-MSC infusion, dose (1 million units/kg) infused within the first 2 weeks after the onset of stroke symptoms. Drug: allogenic mesenchymal stem cells from adipose in acute ischemic stroke patients (Follow-up 24 months) Inclusion criteria: Male or women with acute ischemia (aged 60–80 years with symptoms of acute cerebral infarction). • Patients should be treated within 2 weeks from the onset of stroke symptoms. • Patients with a focal neurological that must persist to the time of treatment without clinically improvement by computerized tomography (CT) and/or MRI. Patients must have a score on the NIH Stroke Scale 8–20, with at least 2 of these points and motor deficit at the time of inclusion.	Primary Outcome Measures: number of participants with Adverse events, complications. (Time Frame: 24 months) Adverse events (AES) were recorded during all the study period (24 months) Neurological and systemic complications (brain oedema, seizures, respiratory infections, deep venous thrombosis, development of tumors, etc) were measured. Secondary outcome measures: to assess the potential efficacy of allogeneic stem cells from adipose tissue in acute ischemic stroke patients by evaluating the outcome at 3 months (NIH Stroke Scale, total volume of stroke by performing MRI). Changes in biochemical markers for brain repair as VEGF, BDNF, MMP-9 and its relationship to neurological and functional outcomes were evaluated
NCT01649687 Phase 1 and II (n = 7)	Aim: Allogeneic SC treatment for the Treatment of cerebellar ataxia with fat-derived MSC (ADSC) Patients will receive intravenously one dose of 5–7 × 10^7 cells of allogeneic adipose-derived mesenchymal stem cells	Allogeneic adult adipose-derived mesenchymal stem cells for cerebellar ataxia treatment Biological: allogeneic adult adipose-derived mesenchymal stem cells for the treatment of diagnosed of spinocerebellar ataxia 3 (SCA3) or multiple system atrophy-cerebellar (MSA-C, ages between 20 ~ 70 years).

Clinical trials with ADSC	Treatment with ADSC	Diseases conditions or Therapeutic indication
NCT01585857 A Phase I, prospective, bi-centric, single-arm, open-label (n = 18)	Aim: to study the safety (primary aim) and efficacy (secondary aim) of a single injection of autologous ADSC on patients with moderate or severe osteoarthritis of the knee. Autologous ADSC for intra-articular infusion (three study groups with doses (2, 5, and 10 millions of ADSC intra-articular injection in 5 ml) Each patient will receive one single administration for 3 months. They will be follow-up during 1 year with routinely examinations for safety issues. The first patient will be followed during 12 weeks before inclusion of the second patient.	Osteoarthritis treatment with ADSC to evaluate safety of a single injection of autologous AD-MSC to treat severe osteoarthritis of the knee joint Primary outcome evaluation of adverse events (Time frame: during 1 year, following injection) Secondary outcome measures: functional status of the knee. Efficacy will be assessed by scales for evaluating the osteoarthritis index and range of motion of the target knee joint, pain-specific assessment and image analysis.
NCT01532076 Effectiveness of Adipose tissue derived mesenchymal stem cells as osteogenic. Composite Grafts Prospective randomized first in men proof of principle trial	Aim: to study the effectiveness of AD-MSC as osteogenic component in composite grafts vs. acellular bone graft substitutes for augmentation in the treatment of proximal humeral fractures as model for fractures of osteoporotic bone. Experimental: graft by lipoaspiration, SVF and embebed in fibrin gel, wrapping around hydroxyapatite granules The control acellular composite graft augmentation open reduction internal fixation (ORIF) of the fracture, augmentation with acellular bone graft substitute.	Primary outcome measurements: development of secondary dislocation within 12 months postoperative (Time frame: 12 months postoperative). Secondary outcome measures: functional outcome 6 weeks, 6 and 12 months after fixation (Time frame: 12 months postoperative). Pain at either surgical site will be recorded via the visual analogue scale. Finally, bone mineral density (Time frame: 12 months postoperative) in case of implant removal was analyzed with microcomputed tomography) for bone mineral density.
NCT0125776	Autologous hAD-MSC (human adipose tissue-derived MSC) to treat lower extremity ischemia in diabetes	Critical limb ischemia in diabetes
NCT01222039 Multicenter clinical trial phase I/II randomized, controlled, for the evaluation of safety and feasibility of two different doses of allogenic AD-MSC in patients with graft versus host disease GVHD (n = 19).	Aim: to assess the safety and feasibility of two-dose of allogeneic AD-MSC infusion expanded in vitro in patients undergoing hematopoietic stem cell transplantation, who have developed chronic and extensive GVHD. Conventional treatment: gradually descending dosage of prednisone and cyclosporin or tacrolimus for at least 46 weeks. Starting dose: 1 mg/Kg/24 h prednisone and 3 mg/Kg/12 h cyclosporin. Intravenous infusion of allogenic AD-MSC (Low dose: 1×10^6/Kg).	Conventional treatment plus intravenous infusion of allogenic mesenchymal stem cells from adipose tissue for the treatment of GVHD.

Clinical trials with ADSC	Treatment with ADSC	Diseases conditions or Therapeutic indication
NCT01157650 Interventional (clinical trial, n = 15)	Aim: evaluation of viability, security and tolerance of allogenic AD-MSC in fistulizing Chron's disease patients, collecting the reactions and adverse events (3 years of study) Experimental: Autologous mesenchymal stem cells Fistulizing Crohn's disease	Treatment of Fistulous Crohn's Disease by Implant of Autologous AD-MSC Tissue. Evaluating the ADSC therapeutic effect, in particular Fistulas healing efficiency.
NCT01056471 Multicenter clinical trial phase I/II randomized, placebo-controlled study.	Aim: to evaluate safety and feasibility of two different doses of autologous AD-MSC from patients with secondary progressive multiple sclerosis who do not respond to treatment Intravenous infusion of autologous AD-MSC (low 10^6 cells/Kg and high dose: $4*10^6$ cells/Kg), including placebo control.	Condition of diseases: autoimmune Diseases immune System Diseases demyelinating Diseases nervous System Diseases demyelinating autoimmune diseases, CNS Autoimmune diseases of the nervous system
NCT00442806 A randomized clinical trial of adipose-derived The APOLLO Trial (n = 14).	Subjects who have coronary artery disease and have suffered acute myocardial infarct Eligible subjects will undergo standard treatment after admission and will then undergo liposuction for further ADRC's or placebo treatment.	Stem cells for the treatment of myocardial Infarction. Primary outcome measures: to evaluate safety determined by major adverse cardiac and cerebral events (MACCE) (Time frame: 6 months) Secondary outcome measures: feasibility - assessment of cardiac function by MRI, SPECT, and Echocardiography (Time frame: 6 months)
NCT0177191324 Immunophenotyping of Fresh SVF from adipose tissue or enriched Fat grafts for refinements of reconstructed Breasts	Inclusion criteria for women: contour irregularities and volume insufficiency in reconstructed breasts. local flaps with conditions to receive fat grafts good health condition Exclusion criteria: breast cancer patients under chemotherapy bad health condition patients too thin patients that require secondary reconstruction	Study groups: centrifuged fat graft in women who underwent breast reconstruction. Study-2: ADSCs enriched centrifuged fat graft. ADSCs enriched from the abdominal subcutaneous tissue is isolated by lipectomy and SVF isolated and immediately added to the fat graft in order- to increase the volume insufficiency or correct contour irregularities for reconstructed breasts.
This is the first development safety update report FIBHGM-ECNC007–2010	The aim is to induce the overexpression and production of microvessels at local level. Route of administration: injection into the thickness of a parlay (Phase I/II). Clinical unicentric trial, randomized, controlled, two parallel-groups, to evaluate the safety of ADSC derived from fat in the glottal Gap zed vocal cord.	

Clinical trials with ADSC	Treatment with ADSC	Diseases conditions or Therapeutic indication
NCT02904824	The facial recontouring by fat was evaluated in the treatment of glottic incompetence. Study design and setting: fourteen patients (aged 18–74 years) with breathy dysphonia secondary to laryngeal hemiplegia or anatomical defects underwent vocal fold lipoinjection. Fat harvested by liposuction was centrifuged, and the fat cell layer injected into the vocalis muscle. The patients underwent pre- and postoperative video laryngostroboscopy, maximum phonation time (MPT) measurements, and voice handicap index (VHI) self-assessments	Unilateral vocal cord paralysis (unilateral)
NCT02387723	Allogeneic ADSC in patients with heart failure Patients with heart failure will be treated with culture expanded AD-MSC from healthy donors stored in nitrogen until use.	Heart failure The cells will be injected directly into the myocardium. The patients will be followed for 6 months for safety and efficacy registration.
Phase II multicenter, clinical trial, randomized controlled trial	Aim: to investigate the effectiveness and safety of ADSC for the treatment of complex perianal fistulas safety of stem cell-based therapy with expanded ADSC against Crohn's disease, and perianal cryptoglandular fistulas (n = 35; n = 14) Patients were randomly assigned to fibrin glue or 20 million ADSC cells plus plus fibrin glue treatment. Fistula healing was tested at 8 weeks and 1 year. If healing was not seen at 8 weeks, they received these treatments	Fistula healing was observed in 71 percent of ADSC-treated patients. The proportion of patients with healing was similar in Crohn's and non-Crohn's subgroups. However, ADSC were also more effective than fibrin glue alone in patients with a suprasphincteric fistulous tract. At 1 year follow-up, both treatments were well tolerated.

Table 3.
Clinical trials conducted with fat stem cells in patients (consult number of each trial for more detail in www.clinicaltrials.gov).

5. Future perspectives

ADSC are widely used to treat a wide range of clinical indications. Short follow-up periods in some studies and low power of findings in clinical trials are explained by low patient numbers. Furthermore, differences in regulatory practices between countries may hamper the application of stem cell therapies. Standardization of protocols and safety of long-term clinical trials need to be demonstrated before clinical application of AD-MSC and SVF against chronic diseases. Since complications in adipose tissue are strongly associated with obesity, cardiovascular disease and type 2 diabetes, a better understanding of the cellular and repair mechanisms associated with adipose organoids as promising tools together with stromal vascular fraction or pluripotent stem cells as stem cell sources in the future is also needed [213].

Future perspectives should focus on mechanistic pathways for a better understanding of SVF properties. Long-term clinical studies are needed to truly confirm safety in patients, as well as studying their optimal dosages for SVF-based therapies over time in chronic diseases. Furthermore, studying the signaling pathways associated with the repair effect of ADSC, including the specific function of each SVF cell type, may provide insight into the knowledge and clinical application of fat stem cell therapies [217]. Small sample sizes of some studies reduce the power to detect significant effects.

6. Conclusions

The application and development of ADSC are boosted by the field of biomaterials and tissue engineering, which contribute to the improvement of the field of regenerative medicine. Given its abundance and easy isolation procedure as a waste material, liposuction fat is a good source for the isolation of ADSC for biomedical applications. These ADSC infusion promotes anti-inflammatory and immunomodulatory effects and the release of autocrine and paracrine functions, including chemokines and growth factors contribute to repair processes. Finally, the differentiation capacity of ADSC into various cell types could into specific local cell types in damage tissue also prevent tissular damage. Several clinical trials with ADSC confirmed their protective effects against chronic-degenerative osteoarticular injuries. However, further studies are required to evaluate the long-term biosafety and efficacy of ADSC stem cell therapy. Standardization of protocols and the inclusion of a larger sample size are required to confirm their long-term safety. However, a systematic review confirmed the safety and efficacy of intra-articular application of autologous stem cells, fat-derived against knee osteoarthritis with good results [218].

Acknowledgements

We appreciate María Eugenia Cabaña-Muñoz's support.

Conflict of interest

The authors declare no conflict of interest.

Abbreviations

GVHD	chronic graft versus host disease.
AD-MSC	mesenchymal stem cells from adipose tissue

Author details

José Joaquín Merino[1*], María Gabriela Villalba[2], Ricardo Martínez-Murillo[3] and Ana I. Flores[4]

1 Facultad de Farmacia, Depatment of Farmacología, Farmacognosía y Botánica, Universidad Complutense de Madrid (UCM), Spain

2 Legislation and Regenerative Medicine, Celumed SL, Switzerland

3 Grupo de Investigación Neurovascular, Departamento de Neurobiología Traslacional, Cajal Institute, CSIC, Madrid, Spain

4 Grupo de Medicina Regenerativa, Instituto de Investigación Sanitaria del Hospital 12 de Octubre (imas12), Madrid, Spain

*Address all correspondence to: info@jjmerino.com

References

[1] Simonacci F, Bertozzi N, Grieco MP, Raposio E. From liposuction to adipose-derived stem cells: Indications and technique. Acta Bio-Medica. 2019;**90**(2):197-208

[2] Mikłosz A, Nikitiuk BE, Chabowski A. Using adipose-derived mesenchymal stem cells to fight the metabolic complications of obesity: Where do we stand? Obesity Reviews. 2022;**23**(5):e13413. DOI: 10.1111/obr.13413. Epub 2022 Jan 5

[3] Barlian A, Judawisastra H, Alfarafisa NM, Wibowo UA, Rosadi I. Chondrogenic differentiation of adipose-derived mesenchymal stem cells induced by L-ascorbic acid and platelet rich plasma on silk fibroin scaffold. Peer J. 2018;**9**(6):e5809

[4] Li C et al. Allogeneic vs. autologous mesenchymal stem/stromal cells in their medication practice. Cell & Bioscience. 2021;**11**(1):187

[5] Petrou P et al. Safety and clinical effects of mesenchymal stem cells secreting neurotrophic factor transplantation in patients with amyotrophic lateral sclerosis: Results of phase 1/2 and 2a clinical trials. JAMA Neurology. 2016;**73**(3):337-344

[6] Pinzur MS. Clinical use of stem cells. Foot & Ankle International. 2020;**41**(8):1027

[7] Kean TJ et al. MSCs: Delivery routes and engraftment, cell-targeting strategies, and immune modulation. Stem Cells International. 2013;**2013**:732742

[8] Lin CS, Xin ZC, Dai J, Lue TF. Commonly used mesenchymal stem cell markers and tracking labels: Limitations and challenges. Histology and Histopathology. 2013;**28**(9):1109-1116

[9] Tokita Y et al. Repeated administrations of cardiac progenitor cells are markedly more effective than a single administration: A new paradigm in cell therapy. Circulation Research. 2016;**119**(5):635-651

[10] Tang JN et al. Concise review: Is cardiac cell therapy dead? Embarrassing trial outcomes and new directions for the future. Stem Cells Translational Medicine. 2018;**7**(4):354-359

[11] Leichebaer J, Henschler R. Biodistribution, migration and homing of systemically applied mesenchymal stem/stromal cells. Stem Cell Research & Therapy. 2016;7:1

[12] Uwe M et al. Pulmonary passage is a major obstacle for intravenous stem cell delivery: The pulmonary first-pass effect. Stem Cells and Development. 2009;**18**(5):683-692

[13] Schrepfer S et al. Scell transplantation: The lung barrier. Transplantation Proceedings. 2007;**39**(2):573-576

[14] Sohni A, Verfaillie CM. Mesenchymal stem cells migration homing and tracking. Stem Cells International. 2013;**2013**:130763. 8 pages

[15] Spaeth E et al. Inflammation and tumor microenvironments: Defining the migratory itinerary of mesenchymal stem cells. Gene Therapy. 2008;**15**(10):730-738

[16] Imitola J et al. Directed migration of neural stem cells to sites of CNS injury by the stromal cell-derived factor 1alpha/

CXC chemokine receptor 4 pathway. Proceedings of the National Academy of Sciences of the United States of America. 2004;**101**(52):18117-18122

[17] Fehrer C, Lepperdinger G. Mesenchymal stem cell aging. Experimental Gerontology. 2005;**40**(12):926-930

[18] Prockop DJ. Concise review: Two negative feedback loops place mesenchymal stem/stromal cells at the center of early regulators of inflammation. Stem Cells (Dayton, Ohio). 2013;**31**(10):2042-2046

[19] Ren G et al. Mesenchymal stemcell-mediated immunosuppression occurs via concertedaction of chemokines and nitricoxide. Cell Stem Cell. 2008;**2**:141-150

[20] Vegh I, Grau M, Gracia M, Grande J, de la Torre P, Flores AI. Decidua mesenchymal stem cells migrated toward mammary tumors in vitro and in vivo affecting tumor growth and tumor development. Cancer Gene Therapy. 2013;**20**(1):8-16

[21] Ou MY, Zhang H, Tan PC, Zhou SB, Li QF. Adipose tissue aging: Mechanisms and therapeutic implications. Cell Death & Disease. 2022;**13**(4):300

[22] Burian E, Syväri J, Dieckmeyer M, Holzapfel C, Drabsch T, Sollmann N, et al. Age- and BMI-related variations of fat distribution in sacral and lumbar bone marrow and their association with local muscle fat content. Scientific Reports. 2020;**10**(1):9686

[23] Calabrese G, Giuffrida R, Fabbi C, Figallo E, Lo Furno D, Gulino R, et al. Collagen-hydroxyapatite scaffolds induce human adipose derived stem cells osteogenic differentiation in vitro. PLoS One. 2016;**11**:e0151181

[24] Calcat-I-Cervera S, et al. Harmonised culture procedures minimise but do not eliminate mesenchymal stromal cell donor and tissue variability in a decentralised multicentre manufacturing approach. Stem Cell Research & Therapy. 4 May 2023;**14**(1):120

[25] Rashid U, Yousaf A, Yaqoob M, Saba E, Moaeen-Ud-Din M, Waseem S, et al. Characterization and differentiation potential of mesenchymal stem cells isolated from multiple canine adipose tissue sources. BMC Veterinary Research. 2021;**17**(1):388

[26] Wang X, Zhang H, Nie L, Xu L, Chen M, Ding Z. Myogenic differentiation and reparative activity of stromal cells derived from pericardial adipose in comparison to subcutaneous origin. Stem Cell Research & Therapy. 2014;**5**(4):92

[27] Gu Y, Li T, Ding Y, Sun L, Tu T, Zhu W, et al. Changes in mesenchymal stem cells following long-term culture in vitro. Molecular Medicine Reports. 2016;**13**(6):5207-5215

[28] Chen Q, Shou P, Zheng C, Jiang M, Cao G, Yang Q, et al. Fate decision of mesenchymal stem cells: Adipocytes or osteoblasts? Cell Death and Differentiation. 2016;**23**(7):1128-1139

[29] Dai M et al. Identification and validation of hub genes and construction of miRNA-gene and transcription factor-gene networks in adipogenesis of mesenchymal stem cells. Stem Cells International. 2024;**29**(2024):5789593

[30] Kim KI, Lee MC, Lee JH, Moon YW, Lee WS, Lee HJ, et al. Clinical efficacy and safety of the intra-articular injection of autologous adipose-derived mesenchymal stem cells for knee osteoarthritis: A phase III, randomized, double-blind, placebo-controlled trial. The American Journal of Sports Medicine. 2023;**51**(9):2243-2253

[31] Ferreira MY, Carvalho Junior JDC, Ferreira LM. Evaluating the quality of studies reporting on clinical applications of stromal vascular fraction: A systematic review and proposed reporting guidelines (CLINIC-STRA-SVF). Regenerative Therapy. 2023;**26**(24):332-342

[32] Kim YS, Lee HJ, Choi YJ, et al. Does an injection of a stromal vascular fraction containing adipose-derived mesenchymal stem cells influence the outcomes of marrow stimulation in osteochondral lesions of the talus? A clinical and magnetic resonance imaging study. The American Journal of Sports Medicine. 2014;**42**:2424-2434

[33] Bourin P et al. Stromal cells from the adipose tissue-derived stromal vascular fraction and culture expanded adipose tissue-derived stromal/stem cells: A joint statement of the International Federation for Adipose Therapeutics and Science (IFATS) and the International Society for Cellular Therapy (ISCT). Cytotherapy. 2013;**15**(6):641-648

[34] Nahmgoong H, Jeon YG, Park ES, Choi YH, Han SM, Park J, et al. Distinct properties of adipose stem cell subpopulations determine fat depot-specific characteristics. Cell Metabolism. 2022;**34**(3):458-472.e6

[35] Dai Prè E, Busato A, Mannucci S, Vurro F, De Francesco F, Riccio V, et al. In vitro characterization of adipose stem cells non-enzymatically extracted from the thigh and abdomen. International Journal of Molecular Sciences. 2020;**21**(9):3081

[36] Almeria C, Kreß S, Weber V, Egger D, Kasper C. Heterogeneity of mesenchymal stem cell-derived extracellular vesicles is highly impacted by the tissue/cell source and culture conditions. Cell & Bioscience. 2022;**12**(1):51

[37] Caplan AI. Why are MSCs therapeutic? New data: New insight. The Journal of Pathology. 2009;**217**(2):318-324. DOI: 10.1002/path.2469

[38] Caplan AI. Adult mesenchymal stem cells for tissue engineering versus regenerative medicine. Journal of Cellular Physiology. 2017;**213**:341-347

[39] Caplan AI. Mesenchymal stem cells: Time to change the name! Stem Cells Translational Medicine. 2017;**6**(6):1445-1451

[40] Henry C et al. Mesenchymal stromal cell therapeutic delivery: Translational challenges to clinical application. Frontiers in Immunology. 2019;**10**:1645

[41] Zuk PA et al. Cells from human adipose tissue: Implications for cell-based therapies. Tissue Engineering. 2001;7:211-222

[42] Zuk PA et al. Human adipose tissue is a source of multipotent stem cells. Molecular Biology of the Cell. 2002;**13**:4279-4295

[43] Zampar AG, Farina Junior JA, Orellana MD, Caruso SR, Fernandes TR, Gomes R, et al. Analysis of adipose-derived stem cells from different donor areas and their influence on fibroblasts in vitro. Aesthetic Plastic Surgery. 2020;**44**(3):971-978

[44] Lin CS, Xin ZC, Deng CH, Ning H, Lin G, Lue TF. Defining adipose tissue-derived stem cells in tissue and in culture. Histology and Histopathology. 2010;**25**(6):807-815

[45] Ning H, Lin G, Lue TF, Lin CS. Mesenchymal stem cell marker Stro-1 is a 75 kd endothelial antigen. Biochemical and Biophysical Research Communications. 2011;**413**(2):353-357. Epub 2011 Aug 27

[46] Chung MT, Liu C, Hyun JS, Lo DD, Montoro DT, Hasegawa M, et al. CD90 (Thy-1)-positive selection enhances osteogenic capacity of human adipose-derived stromal cells. Tissue Engineering. Part A. 2013;**19**(7-8):989-997

[47] Lv Feng J, Tuan RS, Cheung KM, Leung VY. Concise review: The surface markers and identity of human mesenchymal stem cells. Stem Cells. 2014;**32**(6):1408-1419

[48] Ritter A, Friemel A, Fornoff F, Adjan M, Solbach C, Yuan J, et al. Characterization of adipose-derived stem cells from subcutaneous and visceral adipose tissues and their function in breast cancer cells. Oncotarget. 2015;**6**(33):34475-34493

[49] Sonoda Y. Human CD34-negative hematopoietic stem cells: The current understanding of their biological nature. Experimental Hematology. 2021;**96**:13-26

[50] Hatzmann FM, Ejaz A, Wiegers GJ, Mandl M, Brucker C, Lechner S, et al. Quiescence, stemness and adipogenic differentiation capacity in human DLK1-/CD34+/CD24+ adipose stem/progenitor cells. Cells. 2021;**10**(2):214

[51] da Silva ML, Bellagamba BC, Camassola M, Nardi NB. Mesenchymal stem cells and their relationship to pericytes. Frontiers in Biosciences (Landmark Ed). 2016;**21**(1):130-156

[52] da Silva ML, Caplan AI, Nardi NB. In search of the in vivo identity of mesenchymal stem cells. Stem Cells. 2008;**26**(9):2287-2299. Epub 2008 Jun 19

[53] Tseng SL et al. Adipose-derived stem cells in diabetic foot care: Bridging clinical trials and practical application. World Journal of Diabetes. 2024;**15**(6):1162-1177

[54] Wang G et al. Engineered three-dimensional bioactive scaffold for enhanced bone regeneration through modulating transplanted adipose derived mesenchymal stem cell and stimulating angiogenesis. Frontiers in Bioengineering and Biotechnology. 2024;**26**(12):1342590

[55] Kanelidis AJ et al. Route of delivery modulates the efficacy of mesenchymal stem cell therapy for myocardial infarction: A meta-analysis of preclinical studies and clinical trials. Circulation Research. 2017;**120**(7):1139-1150

[56] Zhai R et al. Comparative analysis of the therapeutic effects of mscs from umbilical cord, bone marrow, and adipose tissue and investigating the impact of oxidized rna on radiation-induced lung injury. Stem Cells International. 2024;**24**:7419270

[57] Yang S, Sun Y, Yan C. Recent advances in the use of extracellular vesicles from adipose-derived stem cells for regenerative medical therapeutics. Journal of Nanobiotechnology. 2024;**22**(1):316

[58] Tripepi G, Chesnaye NC, Dekker FW, Zoccali C, Jager KJ. Intention to treat and per protocol analysis in clinical trials. Nephrology (Carlton Vic.). 2020;**25**(7):513-517

[59] Agostini F, Rossi FM, Aldinucci D, Battiston M, Lombardi E, Zanolin S, et al. Improved GMP compliant approach to manipulate lipoaspirates, to cryopreserve stromal vascular fraction, and to expand adipose stem cells in xeno-free media. Stem Cell Research & Therapy. 2018;**9**(1):130

[60] Mohty B et al. Long-term complications and side effects after allogenic hematopoietic stem cell trasplantation: An update. Blood Cancer Journal. 2011;**1**:e16

[61] Dong L, Li X, Leng W, Guo Z, Cai T, Ji X, et al. Adipose stem cells in tissue regeneration and repair: From bench to bedside. Regenerative Therapy. 2023;**24**:547-e560

[62] Wakitani S et al. Mesenchymal cell-based repair of large, full-thickness defects of articular cartilage. The Journal of Bone and Joint Surgery America. 1994;**76**(4):579-592

[63] Wang L, Huang C, Li Q, Xu X, Liu L, Huang K, et al. Osteogenic differentiation potential of adipose-derived stem cells from ovariectomized mice. Cell Proliferation. 2017;**50**:e12328

[64] Feisst V, Meidinger S, Locke MB. From bench to bedside: Use of human adipose-derived stem cells. Stem Cells Cloning. 2015;**8**:149-162

[65] Lin CR et al. Effects of intra-articular stromal vascular fraction injection on clinical symptoms and cartilage health in osteoarthritic knees: A single-center pilot study. Life (Basel). 2024;**14**(11):1468

[66] Bogdanova-Jatniece A, Berzins U, Bruvere R, Eivazova R, Kozlovska T. Adipose-derived stem cells culture in autologous serum maintain the characteristics of mesenchymal stem cells. Proceedings of the Latvian Academy of Sciences Section B Natural Exact and Applied Sciences. 2011;**64**(3-4):106-113

[67] Lefebvre V, Huang W, Harley VR, Goodfellow PN, de Crombrugghe B. SOX9 is a potent activator of the chondrocyte-specific enhancer of the pro alpha1(II) collagen gene. Molecular and Cellular Biology. 1997;**17**(4):2336-2346

[68] Guasti L, Prasongchean W, Kleftouris G, Mukherjee S, Thrasher AJ, Bulstrode NW, et al. High plasticity of pediatric adipose tissue-derived stem cells: Too much for selective skeletogenic differentiation? Stem Cells Translational Medicine. 2012;**1**:384-395

[69] Lee SJ, Kang SW, Do HJ, Han I, Shin DA, Kim JH, et al. Enhancement of bone regeneration by gene delivery of BMP2/Runx2 bicistronic vector into adipose-derived stromal cells. Biomaterials. 2010;**31**:5652-5659

[70] Cha BH, Kim JH, Kang SW, Do HJ, Jang JW, Choi YR, et al. Cartilage tissue formation from dedifferentiated chondrocytes by codelivery of BMP-2 and SOX-9 genes encoding bicistronic vector. Cell Transplantation. 2013;**22**(9):1519-1528

[71] Liao J, Hu N, Zhou N, Lin L, Zhao C, Yi S, et al. Sox9 potentiates BMP2-induced chondrogenic differentiation and inhibits BMP2-induced osteogenic differentiation. PLoS One. 2014;**9**(2):e89025

[72] Taninaka A, Kabata T, Hayashi K, Kajino Y, Inoue D, Ohmori T, et al. Chondroprotective effects of chondrogenic differentiated adipose-derived mesenchymal stem cells sheet on degenerated articular cartilage in an experimental rabbit model. Bioengineering (Basel). 2023;**10**(5):574

[73] Bacou F et al. Transplantation of adipose tissue-derived stromal cells increases mass and functional capacity of damaged skeletal muscle. Cell Transplantation. 2004;**13**:103-111

[74] Safford KM et al. Neurogenic differentiation of murine and human adipose-derived stromal cells. Biochemical and Biophysical Research Communications. 2002;**294**:371-379

[75] Lee M-S et al. Endothelin-1 differentially directs lineage specification

of adipose- and bone marrow-derived mesenchymal stem cells. FASEB Journal. 2019;**33**(1):996-1007

[76] Uguten M, van der Sluis N, Vriend L, Coert JH, Harmsen MC, van der Lei B, et al. Comparing mechanical and enzymatic isolation procedures to isolate adipose-derived stromal vascular fraction: A systematic review. Wound Repair Regen. 2024;**32**(6):1008-1021

[77] Garber MG et al. Isolation of autologous stromal vascular fraction (SVF) from adipose tissue (human Lipoaspirates): A clinical protocol. International Journal of Science and Research (IJSR). 2018;**8**(4):1869-1875

[78] Lee H et al. Rapid-acting pain relief in knee osteoarthritis: Autologous-cultured adipose-derived mesenchymal stem cells outperform stromal vascular fraction: A systematic review and meta-analysis. Stem Cell Research & Therapy. 2024;**15**(1):446

[79] Boada-Pladellorens A, Avellanet M, Pages-Bolibar E, Veiga A. Stromal vascular fraction therapy for knee osteoarthritis: A systematic review. Therapeutic Advances in Musculoskeletal Disease. 2022;**14**:1759720X221117879

[80] Simunec D, Salari H, Meyer J. Treatment of grade 3 and 4 osteoarthritis with intraoperatively separated adipose tissue-derived stromal vascular fraction: A comparative case series. Cells. 2020;**9**(9):2096

[81] Riggle C et al. Complications of stem cell-based injections for knee osteoarthritis: A systematic review. HSS Journal. 2024;**16**:15563316241271058

[82] Cosenza S et al. Mesenchymal stem cells derived exosomes and microparticles protect cartilage and bone from degradation in osteoarthritis. Scientific Reports. 2017;**7**(1):16214

[83] Lee H et al. Rapid-acting pain relief in knee osteoarthritis: Autologous-cultured adipose-derived mesenchymal stem cells outperform stromal vascular fraction: A systematic review and meta-analysis. Stem Cell Research & Therapy. 2024;**15**(1):446

[84] de Nobrega M et al. The potential of cell-free and exosomal micro-RNAs as biomarkers in liquid biopsy in patients with prostate cancer. Journal of Cancer Research and Clinical Oncology. 2022;**148**:2893-2910. DOI: 10.1007/s00432-022-04213-9

[85] Kokai LE et al. Adipose stem cells: Biology and clinical applications for tissue repair and regeneration. Translational Research. 2014;**163**(4):399-408

[86] Lee SC et al. Hypoxic conditioned medium from human adipose-derived stem cells promotes mouse liver regeneration through JAK/STAT3 signaling. Stem Cells Translational Medicine. 2016;**5**(6):816-825

[87] Roura S et al. Potential of extracellular vesicle-associated TSG-6 from adipose mesenchymal stromal cells in traumatic brain injury. International Journal of Molecular Sciences. 2020;**21**(18):6761

[88] Cronk SM, Kelly-Goss MR, Ray HC, Mendel TA, Hoehn KL, Bruce AC, et al. Adipose-derived stem cells from diabetic mice show impaired vascular stabilization in a murine model of diabetic retinopathy. Stem Cells Translational Medicine. 2015;**4**(5):459-467

[89] Mazini L, Rochette L, Amine M, Malka G. Regenerative capacity of adipose derived stem cells (ADSCs),

comparison with mesenchymal stem cells (MSCs). International Journal of Molecular Sciences. 2019;**20**(10):2523

[90] Gurunathan S et al. Review of the isolation, characterization, biological function, and multifarious therapeutic approaches of exosomes. Cells. 2021;**8**:462

[91] Hamdan Y et al. Exosomes and Micro-RNAs in aging process. Biomedicines. 2021;**9**(8):968

[92] Zhang Y, Li Y, Wang Q, Zheng D, Feng X, Zhao W, et al. Attenuation of hepatic ischemia reperfusion injury by adipose stem cell derived exosome treatment via ERK1/2 and GSK 3β signaling pathways. International Journal of Molecular Medicine. 2022;**49**(2):13

[93] Almalki SG, Agrawal DK. Effects of matrix metalloproteinases on the fate of mesenchymal stem cells. Stem Cell Research & Therapy. 2016;**7**(1):129

[94] Bartholomew A et al. Mesenchymal stem cells suppress lymphocyte proliferation in vitro and prolong skin graft survival in vivo. Experimental Hematology. 2002;**30**(1):42-48

[95] De Miguel MP et al. Immunosuppressive properties of mesenchymal stem cells: Advances and applications. Current Molecular Medicine. 2012;**12**:574-591

[96] Murphy JM et al. Stem cell therapy in a caprine model of osteoarthritis. Arthritis and Rheumatism. 2003;**48**(12):3464-3474

[97] Rui L, Li D, Wang H, Chen K, Wang S, Jie X, et al. Exosomes from adipose-derived stem cells regulate M1/M2 macrophage phenotypic polarization to promote bone healing via miR-451a/MIF. Stem Cell Research Therapy. 2022;**13**:149

[98] Nauta AJ, Fibbe WE. Immunomodulatory properties of mesenchymal stromal cells. Blood. 2007;**110**(10):3499-3506

[99] Blanc L, Katarina, and Dimitrios Mougiakakos. Multipotent mesenchymal stromal cells and the innate immune system. Nature Reviews. Immunology. 2012;**12**(5):383-396

[100] Ling W. Mesenchymal stem cells employ IDO to regulate immunity in tumor microenvironment. Cancer Research. 2014;**74**(5):1576-1587

[101] Spaggiari GM et al. Mesenchymal stem cells inhibit natural killer-cell proliferation, cytotoxicity, and cytokine production: Role of indoleamine 2,3-dioxygenase and prostaglandin E2. Blood. 2008;**111**(3):1327-1333

[102] Spaggiari GM et al. MSCs inhibit monocyte-derived DC maturation and function by selectively interfering with the generation of immature DCs: Central role of MSC-derived prostaglandin E2. Blood. 2009;**113**(26):6576-6583

[103] English K. Mechanisms of mesenchymal stromal cell immunomodulation. Immunology and Cell Biology. 2013;**91**:1926

[104] English K et al. IFN-γ and TNF-α differentially regulate immunomodulation by murine mesenchymal stem cells. Immunology Letters. 2007;**110**:91-100

[105] Wisniewski HG, Vilcek J. TSG-6: An IL-1/TNF-inducible protein with anti-inflammatory activity. Cytokine & Growth Factor Reviews. 1997;**8**(2):143-156

[106] Song HB, Park SY, Ko JH, Park JW, Yoon CH, Kim DH, et al. Mesenchymal stromal cells inhibit inflammatory

lymphangiogenesis in the cornea by suppressing macrophage in a TSG-6-dependent manner. Molecular Therapy. 2018;**26**(1):162-172

[107] Kuppa SS, Kim HK, Kang JY, Lee SC, Seon JK. Role of mesenchymal stem cells and their paracrine mediators in macrophage polarization: An approach to reduce inflammation in osteoarthritis. International Journal of Molecular Sciences. 2022;**23**(21):13016. DOI: 10.3390/ijms232113016

[108] Choi H et al. Anti-inflammatory protein TSG-6 secreted by activated MSCs attenuates zymosan-induced mouse peritonitis by decreasing TLR2/NF-κB signaling in resident macrophages. Blood. 2011;**118**(2):330-338

[109] Elashry MI, Baulig N, Wagner AS, Klymiuk MC, Kruppke B, Hanke T, et al. Combined macromolecule biomaterials together with fluid shear stress promote the osteogenic differentiation capacity of equine adipose-derived mesenchymal stem cells. Stem Cell Research & Therapy. 2021;**12**(1):116

[110] Hatsushika D, Muneta T, Nakamura T, Horie M, Koga H, Nakagawa Y, et al. Repetitive allogeneic intraarticular injections of synovial mesenchymal stem cells promote meniscus regeneration in a porcine massive meniscus defect model. Osteoarthritis and Cartilage. 2014;**22**(7):941-950

[111] Kasir R, Vernekar VN, Laurencin CT. Regenerative engineering of cartilage using adipose-derived stem cells. Regenerative Engineering and Translational Medicine. 2015;**1**(1):42-49

[112] Makarczyk MJ. Cell therapy approaches for articular cartilage regeneration. Organogenesis. 2023;**19**(1):2278235

[113] Park JS, Piao J, Park G, Yoo KS, Hong HS. Osteoporotic conditions influence the activity of adipose-derived stem cells. Tissue Engineering and Regenerative Medicine. 2020;**17**(6):875-885

[114] Lin W, Yang Z, Shi L, Wang H, Pan Q, Zhang X, et al. Alleviation of osteoarthritis by intra-articular transplantation of circulating mesenchymal stem cells. Biochemical and Biophysical Research Communications. 2022;**636**(1):25-32

[115] Meng HY, Lu V, Khan W. Adipose tissue-derived mesenchymal stem cells as a potential restorative treatment for cartilage defects: A PRISMA review and meta-analysis. Pharmaceuticals (Basel). 2021;**14**(12):1280

[116] Zhao J et al. Hypoxia-treated adipose mesenchymal stem cell-derived exosomes attenuate lumbar facet joint osteoarthritis. Molecular Medicine. 2023;**29**(1):120

[117] Somoza RA, Welter JF, Correa D, Caplan AI. Chondrogenic differentiation of mesenchymal stem cells: Challenges and unfulfilled expectations. Tissue Engineering. Part B, Reviews. 2014;**20**(6):596-608. Epub 2014 May 27

[118] Schmitz C, Alt C, Pearce DA, Furia JP, Maffulli N, Alt EU. Methodological flaws in meta-analyses of clinical studies on the management of knee osteoarthritis with stem cells: A systematic review. Cells. 2022;**11**(6):965

[119] Vargel İ, Tuncel A, Baysal N, Hartuç-Çevik İ, Korkusuz F. Autologous adipose-derived tissue stromal vascular fraction (AD-tSVF) for knee osteoarthritis. International Journal of Molecular Sciences. 2022;**23**(21):13517

[120] Rady D, Abbass MMS, El-Rashidy AA, El Moshy S, Radwan IA,

Dorfer CE, et al. Mesenchymal stem/progenitor cells: The prospect of human clinical translation. Stem Cells International. 2020;**2020**:8837654

[121] Koh YG, Lee JA, Kim YS, Lee HY, Kim HJ, Kang KT. Optimal mechanical properties of a scaffold for cartilage regeneration using finite element analysis. Journal of Tissue Engineering. 2019;**28**(10):2041731419832133

[122] Gomez-Florit M et al. Natural-based hydrogels for tissue engineering applications. Molecules. 2020;**25**(24):5858

[123] Zhou J, Xiong S, Liu M, Yang H, Wei P, Yi F, et al. Study on the influence of scaffold morphology and structure on osteogenic performance. Frontiers in Bioengineering and Biotechnology. 2023;**27**(11):1127162

[124] Scioli MG, Bielli A, Gentile P, Cervelli V, Orlandi A. Combined treatment with platelet-rich plasma and insulin favours chondrogenic and osteogenic differentiation of human adipose-derived stem cells in three-dimensional collagen scaffolds. Journal of Tissue Engineering and Regenerative Medicine. 2017;**11**(8):2398-2410

[125] Wei W, Ma Y, Yao X, Zhou W, Wang X, Li C, et al. Advanced hydrogels for the repair of cartilage defects and regeneration. Bioactive Materials. 2020;**6**(4):998-1011

[126] Yin F, Cai J, Zen W, Wei Y, Zhou W, Yuan F, et al. Cartilage regeneration of adipose-derived stem cells in the TGF-β1-immobilized PLGA-gelatin scaffold. Stem Cell Reviews and Reports. 2015;**11**(3):453-459

[127] Szponder T, Latalski M, Danielewicz A, Krać K, Kozera A, Drzewiecka B, et al. Osteoarthritis: Pathogenesis, animal models, and new regenerative therapies. Journal of Clinical Medicine. 2022;**12**(1):5

[128] Bhattacharjee M, Escobar Ivirico JL, Kan HM, Shah S, Otsuka T, Bordett R, et al. Injectable amnion hydrogel-mediated delivery of adipose-derived stem cells for osteoarthritis treatment. Proceedings of the National Academy of Sciences of the United States of America. 2022;**119**(4):e2120968119

[129] Arceri A, Mazzotti A, Artioli E, Zielli SO, Barile F, Manzetti M, et al. Adipose-derived stem cells applied to ankle pathologies: A systematic review. Musculoskeletal Surgery. 2024;**108**(1):1-9. DOI: 10.1007/s12306-023-00798-7. Epub 2023 Nov 9. Erratum in: Musculoskelet Surg. 2024 108(2):239

[130] Ossendorff R, Menon A, Schildberg FA, Randelli PS, Scheidt S, Burger C, et al. A worldwide analysis of adipose-derived stem cells and stromal vascular fraction in orthopedics: Current evidence and applications. Journal of Clinical Medicine. 2023;**12**(14):4719

[131] Sterne JAC, Savović J, Page MJ, Elbers RG, Blencowe NS, Boutron I, et al. RoB 2: A revised tool for assessing risk of bias in randomised trials. BMJ. 2019;**366**:l4898

[132] Goncharov EN, Koval OA, Igorevich EI, Encarnacion Ramirez MJ, Nurmukhametov R, Valentinovich KK, et al. Analyzing the clinical potential of stromal vascular fraction: A comprehensive literature review. Medicina (Kaunas, Lithuania). 2024;**60**(2):221

[133] Ude CC, Shah S, Ogueri KS, Nair LS, Laurencin CT. Stromal vascular fraction for osteoarthritis of the knee regenerative engineering. Regenerative Engineering and Translational Medicine. 2022;**8**:210-224

[134] Piccinno MS, Petrachi T, Pignatti M, Murgia A, Grisendi G, Candini O, et al. Human adipose mesenchymal stromal/stem cells improve fat transplantation performance. Cells. 2022;**11**(18):2799

[135] Madani Neishaboori A, Eshraghi A, Tasouji Asl A, Shariatpanahi M, Yousefifard M, Gorji A. Adipose tissue-derived stem cells as a potential candidate in treatment of Alzheimer's disease: A systematic review on preclinical studies. Pharmacology Research & Perspectives. 2022;**10**(4):e00977

[136] Sochocka M, Diniz BS, Leszek J. Inflammatory response in the CNS: Friend or foe? Molecular Neurobiology. 2017;**54**(10):8071-8089. Epub 2016 Nov 26

[137] Saffari TM, Bedar M, Hundepool CA, Bishop AT, Shin AY. The role of vascularization in nerve regeneration of nerve graft. Neural Regeneration Research. 2020;**15**(9):1573-1579

[138] Sheykhhasan M, Amini R, Soleimani Asl S, Saidijam M, Hashemi SM, Najafi R. Neuroprotective effects of coenzyme Q10-loaded exosomes obtained from adipose-derived stem cells in a rat model of Alzheimer's disease. Biomedicine & Pharmacotherapy. 2022;**152**:113224

[139] Singh G, Mehra A, Arora S, Gugulothu D, Vora LK, Prasad R, et al. Exosome-mediated delivery and regulation in neurological disease progression. International Journal of Biological Macromolecules. 2024;**264**(Pt 2):130728

[140] Lu X, Lv C, Zhao Y, Wang Y, Li Y, Ji C, et al. TSG-6 released from adipose stem cells-derived small extracellular vesicle protects against spinal cord ischemia reperfusion injury by inhibiting endoplasmic reticulum stress. Stem Cell Research & Therapy. 2022;**13**(1):291

[141] Yin H, Jiang T, Deng X, Yu M, Xing H, Ren X. A cellular spinal cord scaffold seeded with rat adipose derived stem cells facilitates functional recovery via enhancing axon regeneration in spinal cord injured rats. Molecular Medicine Reports. 2018;**17**(2):2998-3004

[142] Bydon M, Qu W, Moinuddin FM, Hunt CL, Garlanger KL, Reeves RK, et al. Intrathecal delivery of adipose-derived mesenchymal stem cells in traumatic spinal cord injury: Phase I trial. Nature Communications. 2024;**15**(1):2201

[143] Jeon ES et al. Sphimgosylphosphorylcholine induced proliferation of humant tissue adipose-derived mesenchymal stem cells via activation of JNK. Journal of Lipid Research. 2006;**47**:653-664

[144] Wu H, Fan Y, Zhang M. Advanced Progress in the role of adipose-derived mesenchymal stromal/stem cells in the application of central nervous system disorders. Pharmaceutics. 2023;**15**(11):2637. DOI: 10.3390/pharmaceutics15112637

[145] Wu R et al. Concise review: Optimized strategies for stem cell-based therapy in myocardial repair: Clinical translatability and potential limitation. Stem Cells. 2008;**36**(4):482-500

[146] Soltani A, Moradi M, Nejad AR, Moradi S, Javandoost E, Nazari H, et al. Adipose-derived stem cells: Potentials, availability and market size in regenerative medicine. Current Stem Cell Research & Therapy. 2023;**18**(3):347-379

[147] Yeh DC, Chan TM, Harn HJ, Chiou TW, Chen HS, Lin ZS, et al. Adipose tissue-derived stem cells in

neural regenerative medicine. Cell Transplantation. 2015;**24**(3):487-492

[148] Sánchez-Castillo AI, Sepúlveda MR, Marín-Teva JL, Cuadros MA, Martín-Oliva D, González-Rey E, et al. Switching roles: Beneficial effects of adipose tissue-derived mesenchymal stem cells on microglia and their implication in neurodegenerative diseases. Biomolecules. 2022;**12**(2):219

[149] Duma C, Kopyov O, Kopyov A, Berman M, Lander E, Elam M, et al. Human intracerebroventricular (ICV) injection of autologous, non-engineered, adipose-derived stromal vascular fraction (ADSVF) for neurodegenerative disorders: Results of a 3-year phase 1 study of 113 injectioaurns in 31 patients. Molecular Biology Reports. 2019;**46**(5):5257-5272

[150] Cai A, Zheng ZM, Himmler M, Schubert DW, Fuchsluger TA, Weisbach V, et al. Schwann cells promote myogenic differentiation of myoblasts and adipogenic mesenchymal stromal cells on poly-ε-caprolactone-collagen I-nanofibers. Cells. 2022;**11**(9):1436

[151] Zhang J et al. Adipose-derived stem cells: Current applications and future directions in the regeneration of multiple tissues. Stem Cells International. 2020;**2020**:8810813

[152] Jiang W, Zhan Y, Zhang Y, Sun D, Zhang G, Wang Z, et al. Synergistic large segmental bone repair by 3D printed bionic scaffolds and engineered ADSC nanovesicles: Towards an optimized regenerative microenvironment. Biomaterials. 2024;**308**:122566

[153] Gaur S, Agnihotri R. Application of adipose tissue stem cells in regenerative dentistry: A systematic review. Journal of International Society of Preventive & Community Dentistry. 2021;**11**(3):266-271

[154] Cabaña-Muñoz ME, Pelaz Fernández MJ, Parmigiani-Cabaña JM, Parmigiani-Izquierdo JM, Merino JJ. Adult mesenchymal stem cells from oral cavity and surrounding areas: Types and biomedical applications. Pharmaceutics. 2023;**15**(8):2109

[155] Nagata H, Ii M, Kohbayashi E, Hoshiga M, Hanafusa T, Asahi M. Cardiac adipose-derived stem cells exhibit high differentiation potential to cardiovascular cells in C57BL/6 mice. Stem Cells Translational Medicine. 2016;**5**(2):141-151

[156] Yamada Y, Wang XD, Yokoyama S, Fukuda N, Takakura N. Cardiac progenitor cells in brown adipose tissue repaired damaged myocardium. Biochemical and Biophysical Research Communications. 2006;**342**(2):662-670

[157] Ahmad N et al. Autologous fat grafting-a panacea for scar tissue therapy? Cells. 2024;**13**(16):1384

[158] Jo HJ, Kim JH, Hong SJ. Adipose tissue-derived stem cells for myocardial regeneration. Korean. Circulation Journal. 2017;**47**(2):151-159

[159] Chai CY, Song J, Tan Z, Tai IC, Zhang C, Sun S. Adipose tissue-derived stem cells inhibit hypertrophic scar (HS) fibrosis via p38/MAPK pathway. Journal of Cellular Biochemistry. 2019;**120**(3):4057-4064

[160] He Y, Guo Y, Xia Y, Guo Y, Wang R, Zhang F, et al. Resistin promotes cardiac homing of mesenchymal stem cells and functional recovery after myocardial ischemia-reperfusion via the ERK1/2-MMP-9 pathway. American Journal of Physiology. Heart and Circulatory Physiology. 2019;**316**(1):H233-H244

[161] Bobi J, Solanes N, Fernández-Jiménez R, Galán-Arriola C,

Dantas AP, Fernández-Friera L, et al. Intracoronary administration of allogeneic adipose tissue-derived mesenchymal stem cells improves myocardial perfusion but not left ventricle function, in a translational model of acute myocardial infarction. Journal of the American Heart Association. 2017;**6**(5):e005771

[162] Omar AM, Meleis AE, Arfa SA, Zahran NM, Mehanna RA. Comparative study of the therapeutic potential of mesenchymal stem cells derived from adipose tissue and bone marrow on acute myocardial infarction model. Oman Medical Journal. 2019;**34**(6):534-543

[163] Zisa D, Shabbir A, Mastri M, Taylor T, Aleksic I, McDaniel M, et al. Intramuscular VEGF activates an SDF1-dependent progenitor cell cascade and an SDF1-independent muscle paracrine cascade for cardiac repair. American Journal of Physiology. Heart and Circulatory Physiology. 2011;**301**(6):H2422-H2432

[164] Mazo M, Planet-Bérnard V, Abizanda G, Pelacho B, Léobon B, Gávira JL, et al. Transplantation of adipose derived stromal cells is associated with functional improvement in a rat model of chronic myocardial infarction. European Journal of Heart Faillure. 2014;**10**(5):454-4612

[165] Pang JL, Shao H, Xu XG, Lin ZW, Chen XY, Chen JY, et al. Targeted drug delivery of engineered mesenchymal stem/stromal-cell-derived exosomes in cardiovascular disease: Recent trends and future perspectives. Frontiers in Bioengineering and Biotechnology. 2024;**15**(12):1363742

[166] Deng H, Zhang P, Gao X, Chen W, Li J, Wang F, et al. Emerging trophic activities of mesencephalic astrocyte-derived neurotrophic factor in tissue repair and regeneration. International Immunopharmacology. 2023;**114**: 109598

[167] Rani R, Gandhi CR. Stellate cell in hepatic inflammation and acute injury. Journal of Cellular Physiology. 2023;**238**(6):1226-1236

[168] Wu B, Feng J, Guo J, Wang J, Xiu G, Xu J, et al. ADSCs-derived exosomes ameliorate hepatic fibrosis by suppressing stellate cell activation and remodeling hepatocellular glutamine synthetase-mediated glutamine and ammonia homeostasis. Stem Cell Research & Therapy. 2022;**13**(1):494

[169] Gharbia S et al. Adipose-derived stem cells (ADSCs) supplemented with hepatocyte growth factor (HGF) attenuate hepatic stellate cell activation and liver fibrosis by inhibiting the TGF-β/smad signaling pathway in chemical-induced liver fibrosis associated with diabetes. Cells. 2022;**11**(21):3338

[170] Wu B, Feng J, Guo J, Wang J, Xiu G, Xu J, et al. ADSCs-derived exosomes ameliorate hepatic fibrosis by suppressing stellate cell activation and remodelling hepatocellular glutamine synthetase-mediated glutamine and ammonia homeostasis. Stem Cell Research & Therapy. 2022;**13**(1):494

[171] Wang X, Rao H, Liu F, Wei L, Li H, Wu C. Recent advances in adipose tissue dysfunction and its role in the pathogenesis of non-alcoholic fatty liver disease. Cells. 2021;**10**(12):3300

[172] Hou Y, Ding W, Wu P, Liu C, Ding L, Liu J, et al. Adipose-derived stem cells alleviate liver injury induced by type 1 diabetes mellitus by inhibiting mitochondrial stress and attenuating inflammation. Stem Cell Research & Therapy. 2022;**13**(1):132

[173] Chen X, Yan L, Guo Z, Chen Z, Chen Y, Li M, et al. Adipose-derived mesenchymal stem cells promote the survival of fat grafts via crosstalk between the Nrf2 and TLR4 pathways. Cell Death & Disease. 2016;**7**(9):e2369

[174] Park JS, Kim D, Hong HS. Priming with a combination of FGF2 and HGF restores the impaired osteogenic differentiation of adipose-derived stem cells. Cells. 2022;**11**(13):2042

[175] Qu Y, Zhang Q, Cai X, Li F, Ma Z, Xu M, et al. Exosomes derived from miR-181-5p-modified adipose-derived mesenchymal stem cells prevent liver fibrosis via autophagy activation. Journal of Cellular and Molecular Medicine. 2017;**21**(10):2491-2502

[176] Qu Y et al. Exosomes derived from miR-181-5p-modified adipose-derived mesenchymal stem cells prevent liver fibrosis via autophagy activation. Journal of Cellular and Molecular Medicine. 2017;**21**(10):2491-2502

[177] Liu P, Mao Y, Xie Y, Wei J, Yao J. Stem cells for treatment of liver fibrosis/cirrhosis: Clinical progress and therapeutic potential. Stem Cell Research & Therapy. 2022;**13**(1):356

[178] Chae YJ, Jun DW, Lee JS, Saeed WK, Kang HT, Jang K, et al. The use of Foxa2-overexpressing adipose tissue-derived stem cells in a scaffold system attenuates acute liver injury. Gut Liver. 2019;**13**(4):450-460

[179] Yao Y, Xia Z, Cheng F, Jang Q, He J, Pan C, et al. Human placental mesenchymal stem cells ameliorate liver fibrosis in mice by upregulation of Caveolin1 in hepatic stellate cells. Stem Cell Research & Therapy. 2021;**12**(1):294

[180] Tao YC, Chen EQ. Clinical application of stem cell in patients with end-stage liver disease: Progress and challenges. Annals of Translational Medicine. 2020;**8**(8):564

[181] Sakai Y et al. Regenerative therapy for liver cirrhosis based on intrahepatic arterial infusion of autologous subcutaneous adipose tissue-derived regenerative (stem) cells: protocol for a confirmatory multicenter uncontrolled clinical trial. JMIR Research Protocols. 2020;**9**(3):e17904

[182] Garcia-Olmo D, Herreros D, Pascual I, Pascual JA, Del-Valle E, Zorrilla J, et al. Expanded adipose-derived stem cells for the treatment of complex perianal fistula: A phase II clinical trial. Diseases of the Colon and Rectum. 2009;**52**(1):79-86

[183] Gilletta C, Buscail L. Adipose-derived stem cells in the treatment of perianal fistulas in Crohn's disease: Rationale, clinical results and perspectives. International Journal of Molecular Sciences. 2021;**22**(18):9967

[184] Buscail E et al. Adipose-derived stem cells in the treatment of perianal fistulas in crohn's disease: rationale, clinical results and perspectives. International Journal of Molecular Sciences. 2021;**22**(18):9967

[185] Surowiecka A et al. Stromal vascular fraction and emulsified fat as regenerative tolos in rejuvenation of the lower eyelid área. Dermatologic Therapy. 2021;**34**:e14937

[186] Hanson SE et al. Mesenchymal stem cell therapy for nonhealing cutaneous wounds. Plastic and Reconstructive Surgery. 2010;**125**(2):510-516

[187] Magne B et al. Mesenchymal stromal cell preconditioning: The next step toward a customized treatment for severe burn. Stem Cells and Development. 2018;**27**(20):1385-1405

[188] López-IglesiasP,Blázquez-MartínezA, Fernández-Delgado J, Regadera J, Nistal M, Miguel MP. Short and long term fate of human AMSC subcutaneously injected in mice. World. Journal of Stem Cells. 2011;**3**(6):53-62

[189] Zhou ZQ, Chen Y, Chai M, Tao R, Lei YH, Jia YQ, et al. Adipose extracellular matrix promotes skin wound healing by inducing the differentiation of adipose derived stem cells into fibroblasts. International Journal of Molecular Medicine. 2019;**43**(2):890-900

[190] Wang M, Zhao J, Li J, Meng M, Zhu M. Insights into the role of adipose-derived stem cells and secretome: Potential biology and clinical applications in hypertrophic scarring. Stem Cell Research & Therapy. 2024;**15**(1):137

[191] Hao Z, Qi W, Sun J, Zhou M, Guo N. Review: Research progress of adipose-derived stem cells in the treatment of chronic wounds. Frontiers in Chemistry. 2023;**13**(11):1094693

[192] Frommer ML, Langridge BJ, Awad L, Jasionowska S, Denton CP, Abraham DJ, et al. Single-cell analysis of ADSC interactions with fibroblasts and endothelial cells in scleroderma skin. Cells. 2023;**12**(13):1784

[193] Mazini L, Rochette L, Admou B, Amal S, Malka G. Hopes and limits of adipose-derived stem cells (ADSCs) and mesenchymal stem cells (MSCs) in wound healing. International Journal of Molecular Sciences. 2020;**21**(4):1306

[194] Ding S, Ye X, Xin Y, Yue L, Tan Q. Effect of collagen scaffold with Bcl-2 modified adipose-derived stem cells on diabetic mice wound healing. The International Journal of Lower Extremity Wounds. 2019;**19**(2):139-147

[195] Papadopoulos KS, Christina P, Penelope K. Clinical applications of adipose-derived stem cell (ADSC) exosomes in tissue regeneration. International Journal of Molecular Sciences. 2024;**25**(11):5916

[196] Lesmanawati FE, Windura CA, Saputro ID, Hariani L. Autologous fat grafting and adipose-derived stem cells therapy for acute burns and burn-related scar: A systematic review. Tzu Chi Medical Journal. 2024;**36**(2):203-211

[197] Wang K, Yang Z, Zhang B, Gong S, Wu Y. Adipose-derived stem cell exosomes facilitate diabetic wound healing: Mechanisms and potential applications. International Journal of Nanomedicine. 2024;**17**(19):6015-6033

[198] Berry CE, Abbas DB, Lintel HA, Churukian AA, Griffin M, Guo JL, et al. Adipose-derived stromal cell-based therapies for radiation-induced fibrosis. Advances in Wound Care (New Rochelle). 2024;**13**(5):235-252

[199] Gal S, Ramirez JI, Maguina P. Autologous fat grafting does not improve burn scar appearance: A prospective, randomized, double-blinded, placebo-controlled, pilot study. Burns. 2017;**43**(3):486-489

[200] Sakai Y, Fukunishi S, Takamura M, Kawaguchi K, Inoue O, Usui S, et al. Clinical trial of autologous adipose tissue-derived regenerative (stem) cells therapy for exploration of its safety and efficacy. Regenerative Therapy. 2021;**21**(18):97-101

[201] Finocchio L, Zeppieri M, Gabai A, Spadea L, Salati C. Recent advances of adipose-tissue-derived mesenchymal stem cell-based therapy for retinal diseases. Journal of Clinical Medicine. 2023;**12**(22):7015

[202] Nakamura R, Katsuno T, Tateya I, Omori K. Evaluation of cilia function in rat trachea reconstructed using collagen sponge scaffold seeded with adipose tissue-derived stem cells. The Anatomical Record (Hoboken). 2020;**303**(3):471-477

[203] Ji S, Peng Y, Liu J, Xu P, Tang S. Human adipose tissue-derived stem cell extracellular vesicles attenuate ocular hypertension-induced retinal ganglion cell damage by inhibiting microglia-TLR4/MAPK/NF-κB proinflammatory cascade signaling. Acta Neuropathologica Communications. 2024;**12**(1):44

[204] Chen S, Cui G, Peng C, Lavin MF, Sun X, Zhang E, et al. Transplantation of adipose-derived mesenchymal stem cells attenuates pulmonary fibrosis of silicosis via anti-inflammatory and anti-apoptosis effects in rats. Stem Cell Research & Therapy. 2018;**9**(1):110

[205] Trzil JE, Masseau I, Webb TL, Chang CH, Dodam JR, Liu H, et al. Intravenousadipose-derivedmesenchymal stem cell therapy for the treatment of feline asthma: A pilot study. Journal of Feline Medicine and Surgery. 2016;**18**(12):981-990

[206] Gorecka A et al. Autologous transplantation of adipose-derived stem cells improves functional recovery of skeletal muscle without direct participation in new myofiber formation. Stem Cell Research and Therapy. 2018;**9**(1):195

[207] Gorecka A, Salemi S, Haralampieva D, Moalli F, Stroka D, Candinas D, et al. Autologous transplantation of adipose-derived stem cells improves functional recovery of skeletal muscle without direct participation in new myofiber formation. Stem Cell Research & Therapy. 2018;**9**(1):195

[208] Sayegh S, El Atat O, Diallo K, Rauwel B, Degboé Y, Cavaignac E, et al. Rheumatoid synovial fluids regulate the immunomodulatory potential of adipose-derived mesenchymal stem cells through a TNF/NF-κB-dependent mechanism. Frontiers in Immunology. 2019;**28**(10):1482

[209] Senesi L, De Francesco F, Marchesini A, Pangrazi PP, Bertolini M, Riccio V, et al. Efficacy of adipose-derived mesenchymal stem cells and stromal vascular fraction alone and combined to biomaterials in tendinopathy or tendon injury: Systematic review of current concepts. Medicina (Kaunas, Lithuania). 2023;**59**(2):273

[210] Li W, Yu L. Role and therapeutic perspectives of extracellular vesicles derived from liver and adipose tissue in metabolic dysfunction-associated steatotic liver disease. Artificial Cells, Nanomedicine, and Biotechnology. 2024;**52**(1):355-369

[211] Yan D, Song Y, Zhang B, Cao G, Zhou H, Li H, et al. Progress and application of adipose-derived stem cells in the treatment of diabetes and its complications. Stem Cell Research & Therapy. 2024;**15**(1):3

[212] Skubis-Sikora A, Sikora B, Witkowska A, Mazurek U, Gola J. Osteogenesis of adipose-derived stem cells from patients with glucose metabolism disorders. Molecular Medicine (Cambridge, Mass.). 2020;**26**(1):67

[213] Fode M et al. Feasibility of minimally invasive, same-day injection of autologous adipose-derived stem cells in the treatment of erectile dysfunction. Scandinavian Journal of Urology. 2023;**57**(1-6):110-114

[214] Cantarella G et al. Vocal fold augmentation by autologous fat injection with lipostructure procedure. Otolaryngology and Head and Neck Surgery. 2005;**132**(2):239-243

[215] Spear SL, Wilson HB, Lockwood MD. Fat injection to correct contour deformities in the reconstructed breast. Plastic and Reconstructive Surgery. 2005;**116**(5):1300-1305

[216] Okumuş EB, Böke ÖB, Turhan SŞ, Doğan A. From development to future prospects: The adipose tissue & adipose tissue organoids. Life Sciences. 2024;**30**:122758

[217] Guess AJ, Daneault B, Wang R, Bradbury H, La Perle KMD, Fitch J, et al. Safety profile of good manufacturing practice manufactured interferon γ-primed mesenchymal stem/stromal cells for clinical trials. Stem Cells Translational Medicine. 2017;**6**(10):1868-1879

[218] Holzbauer M, Priglinger E, Kølle ST, Prantl L, Stadler C, Winkler PW, et al. Intra-articular application of autologous, fat-derived orthobiologics in the treatment of knee osteoarthritis: A systematic review. Cells. 2024;**13**(9):750

Chapter 14

Mesenchymal Stem Cell Applications in General Surgery

Vasiliki E. Kalodimou, Marios Kozakos and Maria Toumazou

Abstract

The present review illustrates the state of the art of regenerative medicine (RM) and mesenchymal stem cells (MSC) as applied to surgical diseases and demonstrates that this field has the potential to address some of the unmet needs in surgery. RM is a multidisciplinary field whose purpose is to regenerate *in vivo* or *ex vivo* human cells, tissues or organs in order to restore or establish normal function through exploitation of the potential to regenerate, which is intrinsic to human cells, tissues and organs. RM uses MSCs' and/or specially designed biomaterials to reach its goals and RM-based therapies are already in use in several clinical trials in most fields of surgery. The main challenges for investigators are threefold: creation of an appropriate microenvironment *ex vivo* that is able to sustain MSC physiology and function in order to generate the desired MSC cells or body parts; identification and appropriate manipulation of MSCs that have the potential to generate parenchymal, stromal and vascular components on demand, both *in vivo* and *ex vivo*; and production of smart materials that are able to drive cell fate.

Keywords: regenerative medicine, tissue engineering, surgery, extracellular matrix, scaffold, mesenchymal stem cells

1. Introduction

The present review illustrates the state of the art of regenerative medicine (RM) and mesenchymal stem cells (MSC) as applied to surgical diseases and demonstrates that this field has the potential to address some of the unmet needs in surgery. RM is a multidisciplinary field whose purpose is to regenerate *in vivo* or *ex vivo* human cells, tissues or organs in order to restore or establish normal function through exploitation of the potential to regenerate, which is intrinsic to human cells, tissues and organs. RM uses MSCs' and/or specially designed biomaterials to reach its goals and RM-based therapies are already in use in several clinical trials in most fields of surgery. The main challenges for investigators are threefold: creation of an appropriate microenvironment *ex vivo* that is able to sustain MSC physiology and function in order to generate the desired MSC cells or body parts; identification and appropriate manipulation of MSC's that have the potential to generate parenchymal, stromal and vascular components on demand, both *in vivo* and *ex vivo*; and production of smart materials that are able to drive cell fate.

In recent years, mesenchymal stem cell (MSC) therapy has emerged as a promising avenue in the realm of regenerative medicine, offering new hope for the treatment of various kidney diseases. This innovative approach harnesses the unique properties of MSCs to facilitate tissue repair, modulate inflammation and promote renal regeneration.

In 2006, the International Society for Cellular Therapy (ISCT) produced a position statement in which it suggested the minimum criteria required to define MSCs [1]. They stated that cells must:

1. Be plastic adherent

2. Express the cell surface antigens CD105, CD73 and CD90

3. Not express the cell surface antigens such as CD45, CD34, CD14, CD11b, CD79α, CD19 or HLA-DR

4. Differentiate into osteoblasts, adipocytes and chondro-blasts *in vitro*.

2. Skin

Primary mesenchymal stem cells (MSCs) can be obtained from various sources such as bone marrow, adipose tissue and umbilical cord blood. According to clinical studies, MSCs have now been recognized as therapeutic agents for the regeneration and rejuvenation of the skin.

The skin is the largest organ in the human body and can be damaged by cuts or breaks of the tissue and burns (**Figure 1**). Also, through the passage of time with a natural process, the aging of the skin is consequential and gradual. This gradual process, however, can be aggravated by extensive exposure to the sun, environmental pollution as well as some factors from some daily human habits such as smoking and alcohol.

MSCs have the ability to be applied to damaged and aging skin. They can drive regeneration as well as rejuvenation of the skin.

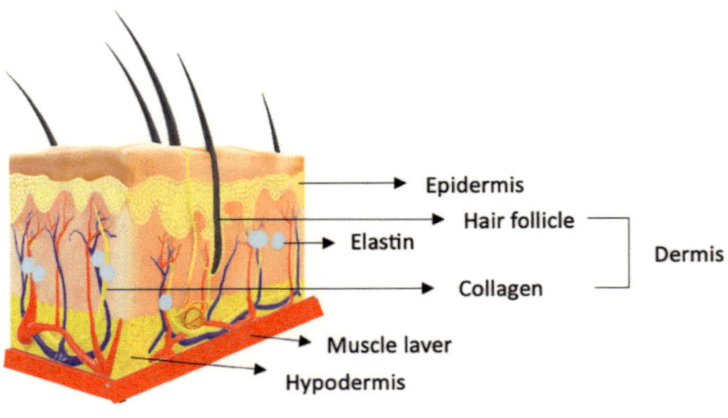

Figure 1.
The skin is compressed of three major layers: epidermis, dermis and hypodermis.

MSCs are suitable for cell therapy because:

a. They have stemness potency.

b. They are easy to isolate from original tissues.

c. They have less severe ethical issues as compared to embryonic stem cells (ESC).

d. Unlike induced pluripotent stem cells (iPSC), they carry a lower risk of teratoma-formation and

e. They are useful for a variety of therapeutic applications because of their ability to migrate to damaged tissue by chemoattraction.

Hence, it is possible to apply MSCs for the treatment of tissues of different origins.

Skin regeneration is an important area of research in the field of tissue engineering, especially for cases involving loss of massive areas of skin, where current treatments are not capable of inducing permanent satisfying replacements. An important area of research is skin regeneration specifically in the area of tissue engineering. This is very important especially for cases of loss of large areas of skin where current treatments are unable to heal the permanent condition of the damage [2–6].

The epidermis is subdivided into five distinct strata: stratum corneum, stratum lucidum, stratum granulosum, stratum spinosum and stratum basale. Keratinocyte stem cells are located in the basal layer and gradually differentiate across the layers until they become terminally differentiated in the stratum corneum, being gradually replaced by keratinocytes prevenient from the bottom layers.

The healing of wounds involving the skin tissue consists of four phases such as: hemostasis, inflammation, proliferation and maturation. MSCs migrate to the spot of skin injury, inhibit inflammation and elevate the proliferation and differentiation potential of fibroblasts, epidermal cells and endothelial cells.

One of the causes that causes skin degeneration is age where it is related to exogenous factors such as exposure to the sun. Through studies, treatments are emerging for the repair of aging skin using adipose-derived stem cells. Its main pathological causes are deterioration of the elastin matrix. That means in the deep dermis the radial degeneration of the elastin fibers loses their functional properties as well as in the subcutaneous area there is a loss of oxytalan and elaunin fibers.

As we have mentioned above, mesenchymal cells have been used in various studies and also in skin treatments. As we have mentioned above, MSCs have been used in various studies and also in skin treatments. MSCs have a broad paracrine effect on skin cells, stimulate angiogenesis and regulate inflammation, pain and immune tolerance. In various clinical studies, the use of stem cells derived from adipose tissue in the skin shows satisfactory results.

In general, through a general review of facial skin treatments using enriched adipose tissue grafts or adipose tissue-derived stem cells, it is concluded that sun-aged skin has significant modifications, such as skin regeneration and skin elastin regeneration. Where through this conclusion a promising treatment proposal for anti-aging action in the regeneration of photodamaged human skin can emerge.

3. Blood vessels

Mesenchymal cells (MSCs) can stabilize blood vessels and contribute to tissue and immune homeostasis under normal conditions and take a more active role in tissue repair in response to injury.

In cardiac regeneration, it was observed that pericyte treatment reduced myocardial fibrosis and a significant increase in host angiogenesis was observed. Pericytes supported microvascular structures *in vivo* and formed capillary-like networks with or without endothelial cells in 3D cultures. MSCs have a close association with pericytes and have the potential to produce various angiogenic growth factors and matrix molecules.

Angiogenesis involves the interaction of endothelial progenitor cells, pericytes, growth factors and cell matrix components. For there to be a proper formation of new blood vessels, there must be a harmonious coherence of some stages. In these stages, there is also a coordinating sequence of growth factors that have the main role [7, 8].

These steps are:

a. proliferation,

b. migration of endothelial progenitor cells (EPCs),

c. lumen formation,

d. basement membrane assembly,

e. pericyte/vascular smooth muscle cell recruitment,

f. vascular maturation and

g. blood flow.

Any absence of the above results in the production of a holey, immature and permeable blood vessel.

The combined use of MSCs and EPCs is recommended for the development of a mature and stable blood vessel. This combined use of MSCs and EPCs is a case of safe treatment since it promotes the creation of more stable and mature blood vessels, where they promote more improved clinical results. Specific studies conclude that, for example in patients with CMI, the combined use of a cell product containing EPC (BM-MNC), pericytes, angiogenic growth factor and extracellular matrix molecules (BM-MSC), during intracoronary infusion, creates the development of new mature and fixed capillaries.

4. Cardiac restoration

One of the main problems that need to be studied is the regeneration of the heart muscle, where after a heart attack its ability to regenerate is limited. The use of MSCs in angiogenesis has several important influences in the chapter of cardiac rehabilitation as well as in better blood flow.

As mentioned above, the combined use of EPC and MSC cell product contributes during intracoronary infusion to the development of new mature and stable blood

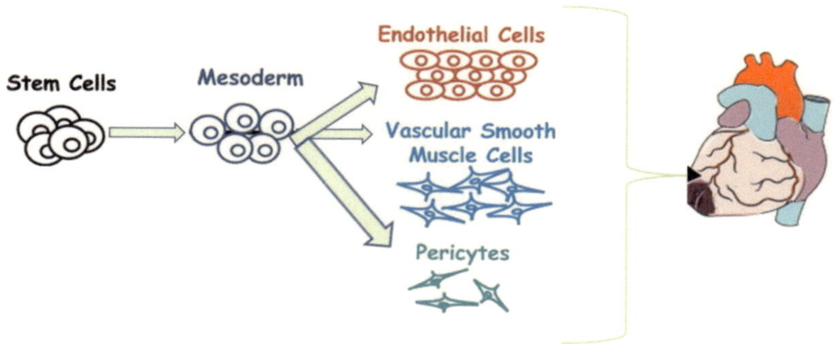

Figure 2.
Schematic representation of MSCs and their ability to form into cardiac muscle.

vessels. Specifically, in patients with CMI, the injection of bone marrow MNC or other cell products into the ischemic myocardium has led to the conclusion of a quite promising treatment for improving collateral flow as well as promoting angiogenesis [9, 10].

Randomized trials in which a single source of EPCs was infused into patients with CPI showed limited improvement in cardiac function. Specific combination of MSCs that have the ability to form into cardiac myocytes and endothelial progenitor EPCs that promote neoangiogenesis show to be able to facilitate tissue repair in myocardial scars (**Figure 2**).

Infusion of EPCs into humans with recent acute myocardial infarction (MI) has now been shown to improve contractility of the infarcted left ventricular (LV) region. Intramyocardial injections of Cardiology Stem Cell Centre Adipose tissue-derived mesenchymal Stromal Cell (CSCC-ASC) in patients with chronic heart failure with reduced ejection fraction (HFrEF) were safe but did not improve myocardial function or structure or clinical symptoms and need further investigations and clinical trials.

5. Respiratory tract

After a number of successful clinical studies conducted in recent years, treatments using mesenchymal cells (MSCs) are considered to be very promising for the treatment of pulmonary fibrosis. The ability of MSCs to produce a large number of biologically active substances with anti-inflammatory, immunosuppressive and angiogenic properties is what makes them relevant to their potential effect on pulmonary fibrosis [11]. The results of published clinical trials for the cell therapy of mild-to-moderate IPF have confirmed the safety of MSCs.

As for patients suffering from end-stage lung disease, the only curable and definitive option is a whole organ transplant. This option is limited since the lack of organs, the high cost, the high probability of complications and the low 10-year survival rate (<20%) are what characterize it.

The first successful human tracheobronchial reconstruction with a decellularized human airway construct repopulated with MSCs-derived epithelial cells and chondrocytes has been performed. This was used to restore lung function in a patient with end-stage airway disease. The patient is currently well active with normal lung function and does not require immunosuppressive drugs.

Several studies focused on the use of MSCs due to the Covid-19 pandemic [12]. Respiratory insufficiency/distress syndrome (ARDS) in Covid-19 has been a frequent phenomenon where it is associated with high mortality.

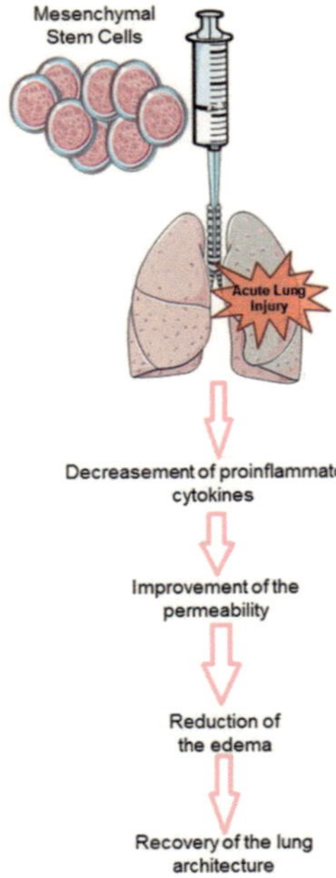

Figure 3.
Schematic representation of MSCs and their ability to treat pulmonary fibrosis.

MSCs are now known to exert immune and anti-inflammatory effects. According to studies done during the use of umbilical cord mesenchymal cells (US-MSC) in people with acute respiratory distress syndrome Covid-19, where they showed a significant improvement of the patients as well as their recovery time. US-MSC in Covid-19 with ARDS appears to be safe and associated with significant reduction of SAEs, mortality as well as recovery time (**Figure 3**).

6. Kidney

Kidney diseases represent a significant global health burden, affecting millions of people worldwide and contributing to substantial morbidity and mortality. Despite advances in medical management, many kidney disorders, including acute kidney injury (AKI), chronic kidney disease (CKD), diabetic nephropathy and glomerulonephritis, continue to pose formidable challenges to clinicians and patients alike. End-stage renal disease is treated with dialysis or transplantation. Dialysis can only partially replace the kidney function, but not the endocrine and homeostatic renal function. On the other hand, organ transplantations cannot be extended to a large patient population because of the shortage of donor organs. Traditional treatments often focus on managing symptoms rather than addressing the underlying pathology, highlighting the urgent need for alternative therapeutic strategies [13–18].

MSCs multipotent progenitor cells found in various tissues such as bone marrow, adipose tissue and umbilical cord possess remarkable immunomodulatory and regenerative capabilities. These cells exhibit the ability to migrate to sites of injury, where they participate in tissue repair processes through paracrine signaling, differentiation into specialized cell types and immunomodulation.

In the context of kidney diseases, MSCs exert their therapeutic effects through several mechanisms:

1. Immunomodulation: MSCs possess potent immunomodulatory properties, suppressing excessive inflammation and promoting tissue repair. They modulate the activity of immune cells, including T cells, B cells, dendritic cells and macrophages, thereby attenuating immune-mediated kidney injury.

2. Paracrine effects: MSCs secrete a variety of trophic factors, cytokines and growth factors that promote tissue repair and regeneration. These factors modulate cellular processes such as apoptosis, proliferation, angiogenesis and extracellular matrix remodeling, contributing to renal tissue repair.

3. Differentiation: While the extent of MSC differentiation into renal cell types *in vivo* is debated, some studies suggest that MSCs can differentiate into functional renal cells, such as podocytes, tubular epithelial cells and endothelial cells, contributing to tissue regeneration and functional recovery.

Preclinical studies utilizing animal models of kidney disease have demonstrated the potential of MSC therapy to ameliorate renal dysfunction and promote tissue regeneration. By releasing trophic factors and anti-inflammatory cytokines, MSCs can attenuate oxidative stress, inhibit fibrosis and promote the regeneration of damaged renal tissue. Furthermore, MSCs have been shown to modulate the immune response, suppressing aberrant immune activation and fostering an environment conducive to tissue repair.

In clinical trials, MSCs therapy has shown promising results in various kidney diseases, including acute kidney injury, diabetic nephropathy and chronic kidney disease. These studies have highlighted the safety and feasibility of MSC administration, with many patients experiencing improvements in renal function, reductions in proteinuria and amelioration of symptoms.

Despite these encouraging findings, challenges remain in optimizing the efficacy and safety of MSC therapy for kidney diseases. Important considerations include

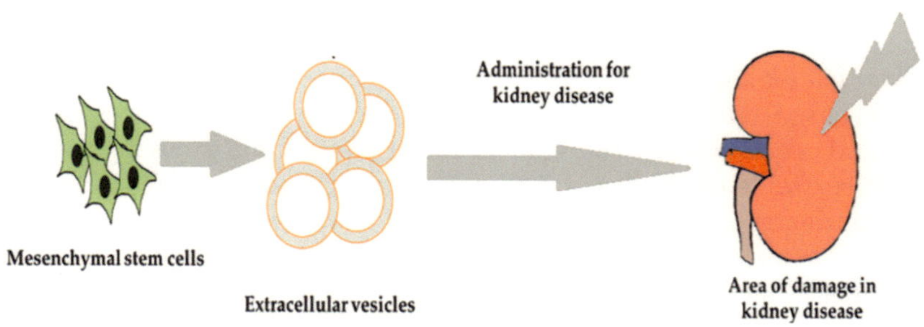

Figure 4.
Schematic representation of MSCs and their ability to kidney regeneration.

standardizing protocols for MSC isolation, expansion and administration, as well as elucidating the mechanisms underlying MSC-mediated renal repair (**Figure 4**).

7. Urinary tract

In the context of urinary tract disorders, such as urinary incontinence or urethral strictures, tissue engineering approaches aim to restore normal structure and function to the affected tissues.

One of the key strategies in regenerative medicine applications using scaffolds seeded with mesenchymal stem cells (MSCs) to facilitate tissue regeneration. Scaffolds serve as temporary frameworks or templates that support the growth and organization of cells into functional tissues. These scaffolds can be made from various biomaterials, such as synthetic polymers or natural extracellular matrix components, and are designed to mimic the native tissue architecture and provide mechanical support [19].

Scaffold-based strategies offer a promising solution by providing a supportive environment for cell growth and tissue formation, while MSCs contribute to tissue regeneration through their differentiation and paracrine effects.

While research in this field is still ongoing, early studies and preclinical trials have shown encouraging results, suggesting that regenerative medicine approaches incorporating scaffolds and MSCs could offer effective treatments for urinary tract disorders in the future.

8. Bladder

The necessity for developing tissue engineering technologies for bladder reconstruction is highlighted by the fact that de-tabularized bowel segments are still commonly used for this purpose. Besides the obvious physiological mismatch between the two tissue types, this technique has long been known to suffer from other limitations including infection, perforation, metabolic disturbances, urolithiasis and even malignancy [20].

The integration of surgical techniques with scaffold-based approaches and mesenchymal stem cells (MSCs) have emerged as a potential solution for bladder regeneration. Bladder reconstruction typically involves creating a functional bladder substitute that can store and expel urine effectively while minimizing complications and preserving quality of life. Tissue engineering strategies aim to achieve this by developing biomimetic scaffolds seeded with cells, including MSCs, to facilitate the regeneration of bladder tissue.

Scaffolds used in bladder reconstruction are designed to mimic the native bladder structure and properties, providing mechanical support and guiding tissue regeneration. These scaffolds act as a framework for cell attachment, proliferation and differentiation, ultimately leading to the formation of functional bladder tissue.

In bladder reconstruction procedures, scaffolds seeded with MSCs are often surgically implanted into the patient's body, where they gradually integrate with surrounding tissues and promote the formation of new bladder tissue.

Over time, the scaffold degrades, leaving behind regenerated bladder tissue that can functionally mimic the native bladder.

While challenges remain, such as optimizing scaffold design, enhancing cell viability and functionality, and ensuring long-term graft survival, bladder reconstruction

DOI: http://dx.doi.org/10.5772/intechopen.1007381

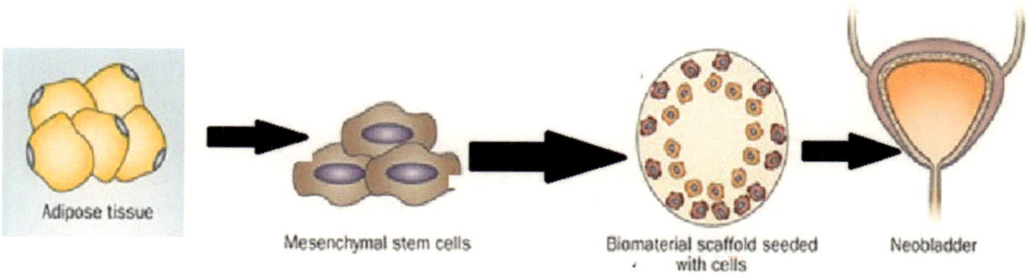

Figure 5.
Schematic representation of MSCs and their ability to bladder regeneration.

using regenerative medicine approaches holds great potential for improving outcomes and quality of life for patients with bladder dysfunction or damage (**Figure 5**).

9. Pancreas

Diabetes mellitus is a widespread disease in modern society. Insulin administration in patients with diabetes mellitus is unable to accurately replace normal pancreatic endocrine function and the resulting chronic carbohydrate dysmetabolism is associated with progressive diabetic complications such as diabetic microangiopathy and coronary artery disease, diabetic nephropathy and renal failure and diabetic foot syndrome.

These progressive diabetic complications increase the degree of difficulty in treatment, the costs of the health system, morbidity and mortality.

Regenerative medicine, particularly in the context of pancreatic tissue engineering, holds immense promise for revolutionizing the treatment of various diseases, including diabetes due to the mesenchymal stem cells (MSCs) multipotent nature [1].

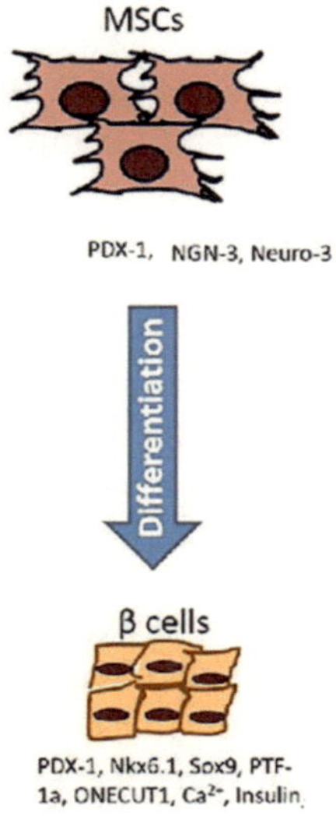

Figure 6.
Schematic representation of MSCs and their ability to pancreas regeneration.

In the realm of pancreatic tissue engineering, researchers are exploring ways to utilize MSCs to regenerate damaged pancreatic tissue or even engineer entirely new pancreatic tissue. This approach offers potential solutions for conditions such as type 1 diabetes, where the body's immune system attacks and destroys insulin-producing beta cells in the pancreas.

Another avenue of research involves using MSCs to stimulate the regeneration of pancreatic tissue by promoting the growth and differentiation of endogenous stem cells or by directly differentiating into insulin-producing beta cells. Significant difficulties remain the development of stable culture techniques for the production of large numbers of functional beta cells and the prevention of xenogeneic, allogeneic and autoimmune reactions of the immune system.

MSCs hold a great promise for advancing the treatment of pancreatic diseases through tissue engineering approaches, offering hope for improved outcomes and quality of life for patients (**Figure 6**).

10. Liver

Most untreated chronic liver diseases eventually lead to irreversible cirrhosis of the liver, in which liver tissue is destroyed and fibrin is transformed. Two basic mechanisms determine the clinical course and prognosis:

a. The increasing lack of function of the parenchyma leads to limited synthesis and metabolic function of the liver. The consequences are multiple; for example, decreased coagulation function leads to an increased tendency to bleed, while an increase in toxic metabolites can cause hepatic encephalopathy.

b. Progressive loss of functional and properly vascularized liver tissue (increasing fibrosis) causes portal hypertension with complications of ascites, portocaval bypass circuits (e.g., esophageal varicose veins) and splenomegaly.

Liver transplantation stands as the ultimate solution for end-stage liver failure, fulminant liver failure and certain types of inborn errors of metabolism.

Hepatocyte transplantation was the first attempt to replace liver transplantation, for the treatment of acute or chronic liver disease. The first patient treatment was performed in 1992 in a French-Canadian woman with familial hypercholesterolemia. After *ex vivo*, transduction with a retrovirus encoding the human LDL receptor, the patient's hepatocytes were injected *via* the inferior mesenteric vein into the liver. LDL and HDL levels improved over the next 18 months and transgene expression was detected in liver biopsy. After this promising success, other cases followed. However, not all patients clearly benefited from the procedure. Low efficacy and lack of long-term therapeutic effect were common in all procedures.

In various clinical studies was investigated the feasibility, safety, and efficacy of using autologous mesenchymal stem cells (MSCs), autologous bone marrow-derived stem cell and hematopoietic stem cell (HSC) as a treatment for end-stage liver disease. In those studies, patients with chronic end-stage hepatitis C-induced liver disease, end-stage autoimmune liver disease and severe alcoholic liver cirrhosis (ALC) were included. The stem cells were isolated, proliferated and injected into either the peripheral or portal vein or *via* the hepatic artery [21]. The results indicated that the treatment was well tolerated by almost all patients, with improvements observed in

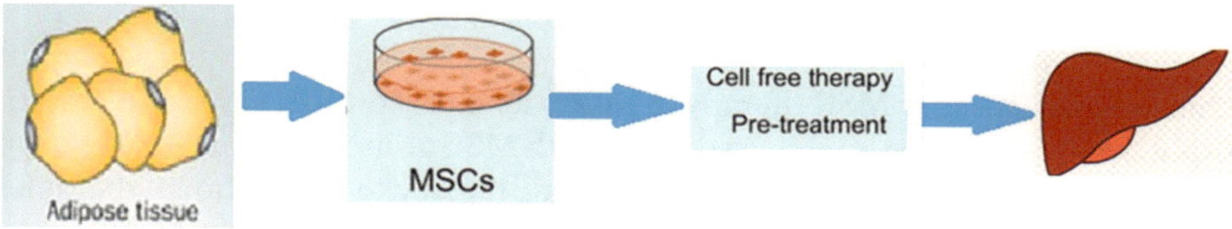

Figure 7.
Schematic representation of MSCs and their ability to liver regeneration.

liver function parameters such as Model for End-Stage Liver disease score, prothrombin complex, serum creatinine levels and significant improvements in ascites levels for all patients.

The findings suggest that MSC injection, autologous CD34$^{(+)}$ stem cell transplantation and hematopoietic stem cell (HSC) transplantation hold promise as a treatment for end-stage liver disease, with the potential to improve clinical indices of liver function.

While preclinical and clinical studies have provided valuable insights into the therapeutic potential of SCs, further research is needed to address remaining challenges and optimize treatment strategies for widespread clinical application (**Figure 7**).

11. Gastrointestinal (GI) tract

The gastrointestinal (GI) tract plays a crucial role in the body's digestive system, and disorders affecting this system can significantly impact an individual's quality of life.

Inflammatory bowel disease (IBD), encompassing conditions such as Crohn's disease and ulcerative colitis, is characterized by chronic inflammation of the GI tract. Conventional treatments often involve immunosuppressive drugs, which may have significant side effects and limited efficacy in some patients. Mesenchymal stem cell (MSCs) therapy has emerged as a promising approach for treating various GI tract disorders due to its regenerative and immunomodulatory properties [9].

Preclinical and clinical studies have demonstrated the potential of MSCs to attenuate inflammation and promote tissue repair in IBD. MSCs exert their therapeutic effects through paracrine signaling, secretion of anti-inflammatory cytokines and modulation of immune cell activity because MSCs possess the ability to migrate to sites of inflammation within the GI tract, enhancing their therapeutic relevance.

A study sponsored by Cellerix, and later adopted by TiGenix, investigated the use of adipose-derived mesenchymal stem cells (MSCs) for the treatment of complex perianal fistulas, focusing on patients without inflammatory bowel disease. In the initial study, autologous adipose MSCs were cultured and injected locally with or without fibrin glue, but they did not demonstrate superiority over fibrin glue alone in achieving sustained closure and healing of fistulas. However, this study informed the design of a subsequent phase III trial sponsored by TiGenix, which employed allogeneic ASCs, increased cell dose to 120 million cells, eliminated fibrin glue matrix and enrolled only Crohn's disease patients. *This clinical trial* (NCT01541579) showed significant superiority of allogeneic ASCs over placebo in treating Crohn's-associated perianal fistulas, with a higher proportion of patients achieving remission and experiencing fewer adverse events. Importantly, the study utilized batch-manufactured allogeneic ASCs that were allowed to fully recover in culture before administration and delivered them locally into the fistulas, which likely enhanced their therapeutic efficacy. This approach holds promise for the clinical utility of MSCs in treating inflammatory disorders.

One notable aspect of MSC therapy in IBD is its safety profile. Clinical trials have reported minimal adverse effects associated with MSC administration, making it a viable option for patients who are intolerant or refractory to conventional therapies.

12. Orthopedics

In the field of orthopedics, the complete restoration of cartilage, bones and tendons is at the stage of research and study. Wound healing is likened to skin tissue healing and muscle recovery can also be likened to cardiac clinical investigations.

As we have already mentioned, mesenchymal cells are characterized by paracrine, anti-inflammatory and immune effects, by the release of several growth factors and cytokines; collecting the sample for intrathecal injections is easy and safe. It also has the potential for the sample to differentiate into cartilage tissue where these characteristics make MSCs of significant importance in the treatment of osteoarthritis [20, 22].

Through clinical studies involving knee osteoarthritis, results such as pain relief, functional improvement and fat regeneration have been successfully demonstrated using intra-articular injection of mesenchymal cells. Successful treatment was shown in a study of patients with knee osteoarthritis (KOA) where by dividing the patients into two subgroups, and they were given two different types of intra-articular injections in each group. The results were encouraging for future treatments and research. One group was given an intra-arthritis injection of autologous bone marrow cells (BMAC), and the other was given an intra-arthritis injection of adipose-derived cells (ADSCs).

Evaluating the Knee Injury and Osteoarthritis Outcome Score (KOOS), the Oxford Knee Score (OKS) and the visual analog scale (VAS), they concluded that both intrathecal injections of BMAC and ADSCs significantly improved both pain and functionality within the 6-month period during which the KOA patients were followed. Patients with mild KOA had a better picture, compared to those with more moderate and severe forms of KOA.

Clinical trials have reported minimal adverse effects associated with MSC administration, making it a viable option for patients (**Figure 8**).

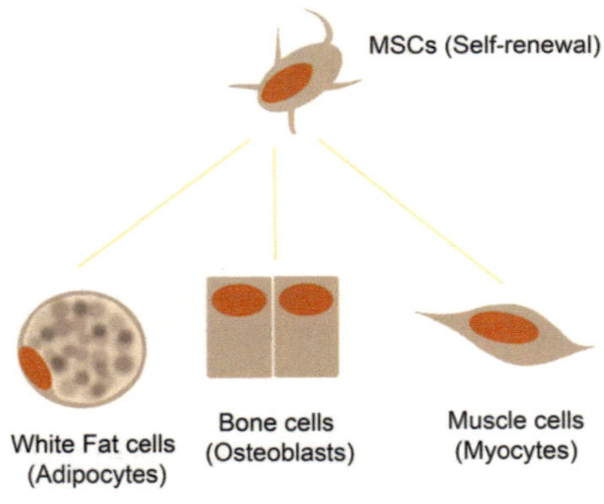

Figure 8.
Schematic representation of MSCs markers.

Author details

Vasiliki E. Kalodimou[1*], Marios Kozakos[2] and Maria Toumazou[1]

1 EUC School of Medicine Frankfurt, Germany

2 Sankt Katharinen Krankenhaus Hospital, Frankfurt, Germany

*Address all correspondence to: v.kalodimou@euc.ac.cy

References

[1] Kusuma GD, Carthew J, Lim R, Frith JE. Effect of the micro-environment on mesenchymal stem cell paracrine signaling: Opportunities to engineer the therapeutic effect. Stem Cells and Development. 2017;**26**:617-631

[2] Caplan AI. Adult mesenchymal stem cells for tissue engineering versus regenerative medicine. Journal of Cellular Physiology. 2007;**213**:341-347

[3] Caplan AI. Why are MSCs therapeutic? New data: New insight. The Journal of Pathology. 2009;**217**:318-324

[4] Charles-de-Sá L, Gontijo-de-Amorim NF, Rigotti G, Sbarbati A, Bernardi P, Benati D, et al. Photoaged skin therapy with adipose-derived stem cells. Plastic and Reconstructive Surgery. 2020;**145**(6):1037e-1049e. DOI: 10.1097/PRS.0000000000006867

[5] Charles-de-Sá L, Gontijo-de-Amorim NF, Takiya CM, et al. Antiaging treatment of the facial skin by fat graft and adipose-derived stem cells. Plastic and Reconstructive Surgery. 2015;**135**:999-1009

[6] Chavez-Munoz C, Nguyen KT, Xu W, Hong SJ, Mustoe TA, Galiano RD. Transdifferentiation of adipose-derived stem cells into keratinocyte-like cells: Engineering a stratified epidermis. PLoS ONE. 2013;**8**(12):e80587. DOI: 10.1371/journal.pone.0080587

[7] Beeravolu N, McKee C, Alamri A, Mikhael S, Brown C, Perez-Cruet M, et al. Isolation and characterization of mesenchymal stromal cells from human umbilical cord and fetal placenta. Journal of Visualized Experiments. 2017;**122**:55224

[8] Benavente CA, Sierralta WD, Conget PA, Minguell JJ. Subcellular distribution and mitogenic effect of basic fibroblast growth factor in mesenchymal uncommitted stem cells. Growth Factors. 2003;**21**:87-94

[9] Averyanov A, Koroleva I, Konoplyannikov M, Revkova V, Lesnyak V, Kalsin V, et al. First-in-human high-cumulative-dose stem cell therapy in idiopathic pulmonary fibrosis with rapid lung function decline. Stem Cells Translational Medicine. 2020;**9**(1):6-16. DOI: 10.1002/sctm.19-0037. Epub 2019 Oct 15

[10] Assmus B, Schachinger V, Teupe C, Britten M, Lehmann R, Dobert N, et al. Transplantation of progenitor cells and regeneration enhancement in acute myocardial infarction (TOPCAPE-AMI). Circulation. 2002;**106**:3009-3017

[11] Baiguera S, Birchall MA, Macchiarini P. Tissue-engineered tracheal transplantation. Transplantation. 2010;**89**:485-491

[12] Can A, Coskun H. The rationale of using mesenchymal stem cells in patients with COVID-19-related acute respiratory distress syndrome: What to expect. Stem Cells Translational Medicine. 2020;**9**:1287-1302

[13] Covas DT, Panepucci RA, Fontes AM, Silva WA Jr, Orellana MD, Freitas MC, et al. Multipotent mesenchymal stromal cells obtained from diverse human tissues share functional properties and gene-expression profile with CD146+ perivascular cells and fibroblasts. Experimental Hematology. 2008;**36**:642-654

[14] Djouad F, Bouffi C, Ghannam S, et al. Mesenchymal stem cells: Innovative

therapeutic tools for rheumatic diseases. Nature Reviews Rheumatology. 2009;**5**:392-399

[15] Hmadcha A, Martin-Montalvo A, Gauthier BR, Soria B, Capilla-Gonzalez V. Therapeutic potential of mesenchymal stemcells for cancer therapy. Frontiers in Bioengineering and Biotechnology. 2020;**8**:43

[16] Horie M, Choi H, Lee RH, et al. Intra-articular injection of human mesenchymal stem cells (MSCs) promote rat meniscal regeneration by being activated to express Indian hedgehog that enhances expression of type II collagen. Osteoarthritis and Cartilage. 2012;**20**:1197-1207

[17] Hu MS, Borrelli MR, Lorenz HP, Longaker MT, Wan DC. Mesenchymal stromal cells and cutaneous wound healing: A comprehensive review of the background, role, and therapeutic potential. Stem Cells International. 2018;**2018**:6901983

[18] Jakobsson L, Franco CA, Bentley K, Collins RT, Ponsioen B, Aspalter IM, et al. Endothelial cells dynamically compete for the tip cell position during angiogenic sprouting. Nature Cell Biology. 2010;**12**:943-953

[19] Kalodimou VE. A Handbook to Mesenchymal Stem Cells in Regenerative Medicine. 2015. ISBN: 978-618-81976-0-2

[20] Acharya C, Adesida A, Zajac P, et al. Enhanced chondrocyte proliferation and mesenchymal stromal cells chondrogenesis in coculture pellets mediate improved cartilage formation. Journal of Cellular Physiology. 2012;**227**:88-97

[21] Kassmeyer S, Plendl J, Custodis P, Bahramsoltani M. New insights in vascular development: Vasculogenesis and endothelial progenitor cells. Anatomia, Histologia, Embryologia. 2009;**38**:1-11

[22] Barry F, Murphy M. Mesenchymal stem cells in joint disease and repair. Nature Reviews Rheumatology. 2013;**9**:584-594

Chapter 15

The Integral Role of Mesenchymal Stem Cells in Stem Cell Transplantation from Promotion of Stem Cell Engraftment to Immunomodulation

Mohini Mendiratta, Meenakshi Mendiratta, Deepak Garg, Sujata Mohanty and Ranjit Kumar Sahoo

Abstract

Graft-versus-host disease (GVHD), a severe complication of allogeneic stem cell transplantation, arises when donor immune cells attack host tissues, triggering a cytokine storm. Mesenchymal stem cells (MSCs) have emerged as a promising therapeutic option due to their immunomodulatory, tissue repair, and regenerative properties. Recent studies highlight the fate of MSCs post-administration, revealing their clearance by host macrophages through efferocytosis following apoptosis. This interaction not only modulates the recipient's immune system but also underscores the role of MSC-derived factors in immune regulation. In addition to mitigating GVHD, MSCs promote engraftment by secreting hematopoietic cytokines and suppressing host immune responses. Despite these advantages, the clinical efficacy of MSCs remains inconsistent, posing a challenge to their widespread application. Notably, MSCs derived from sources such as Wharton's jelly and adipose tissue exhibit superior immunomodulatory capabilities due to their unique gene signatures that enhance immune regulation. This chapter explores the evolving role of MSCs in GVHD prevention and treatment, emphasizing their mechanistic insights, therapeutic limitations, and the potential of tissue-specific MSCs. It also discusses future strategies for optimizing MSC-based therapies to improve clinical outcomes in stem cell transplantation.

Keywords: allogeneic hematopoietic stem cell transplantation, mesenchymal stem cells, engraftment, immunomodulation, graft-versus-host-disease

1. Introduction

Allogeneic hematopoietic stem cell transplantation (Allo-HSCT) is a therapeutic intervention for treating hematologic disorders [1]. Despite its therapeutic potential, Allo-HSCT is associated with significant challenges, primarily graft-versus-host disease

(GVHD) [2] and engraftment failure [3]. GVHD is the leading cause of non-relapse mortality and affects approximately 40–60% of recipients [4]. It occurs when the donor's immune cells recognize the recipient's tissues as foreign and mount an immune response against them. This condition can manifest acutely or chronically, affecting various organs, including the skin, liver, and gastrointestinal tract [5]. Several factors, including the source of stem cells, recipient age, sex mismatch, ABO incompatibility, conditioning regimens, and prophylactic measures, influence the development of GVHD [5].

Engraftment failure, on the other hand, refers to the inability of the transplanted stem cells to establish and proliferate in the recipient's bone marrow [3]. This failure can result from various factors, including insufficient $CD34^+$ stem cell numbers, immunological rejection, or poor marrow niche conditions [6]. Both GVHD and engraftment failure significantly impact patient morbidity and mortality, posing substantial hurdles to the success of Allo-HSCT.

While the advancements in Allo-HSCT have significantly improved patient outcomes, the persistent challenges of GVHD and engraftment failure continue to limit its success. In recent years, research has turned toward innovative solutions to mitigate these complications. One promising approach involves the use of mesenchymal stem cells (MSCs). MSCs, known for their immunomodulatory and regenerative properties, have the potential to enhance engraftment and reduce the incidence and severity of GVHD [7].

Mesenchymal Stem Cells (MSCs), initially isolated from BM by Friedenstein and his colleagues in 1968 [8], are multipotent, non-hematopoietic stem cells with a spindle-shaped morphology [9, 10]. MSCs can also be derived from various sources, including adipose tissue (AT), umbilical cord blood (UCB), amniotic fluid, amniotic membrane, Wharton's Jelly (WJ), dental pulp (DP), menstrual blood (MB), and peripheral blood (PB) [11, 12]. The International Society of Cellular Therapy (ISCT) defines MSCs by their ability to adhere to plastic, marked by expression of mesenchymal markers (CD105, CD73), cell adhesion molecules (CD90), and absence of hematopoietic markers (CD34, CD45, CD14, CD11b, CD19, and CD79α) and differentiate into mesodermal lineages such as osteoblast, adipocytes, and chondroblasts under *in vitro* conditions [13].

Moreover, MSCs exhibit low immunogenicity due to low expression of HLA-I molecules and the absence of HLA-II and costimulatory molecules (CD80, CD86, CD40, and CD40L). Moreover, this characteristic allows for their use in allogeneic settings without eliciting significant immune rejection [10, 14].

MSCs are widely studied for their ability to modulate the immune system, adjusting their activity based on the surrounding immune environment [15]. This adaptability makes them promising for treating immune-related diseases and improving Allo-HSCT outcomes. This chapter focuses on how MSCs help prevent GVHD and support stem cell engraftment in Allo-HSCT, while also discussing their therapeutic potential, current challenges, and future directions for incorporating MSC therapy into routine practice.

2. MSCs in immunomodulation and GVHD prevention

MSCs play a crucial role in immunomodulation and GVHD prevention by secreting anti-inflammatory cytokines, inhibiting activated T-cell proliferation, and promoting regulatory T-cell and macrophage polarization toward M2 phenotype [16]. Both pre-clinical and clinical studies have demonstrated their potential to decrease the incidence and severity of GVHD, offering a promising adjunctive therapy in allogeneic hematopoietic stem cell transplantation.

2.1 Mechanisms of MSC-mediated immunomodulation

MSCs mediate immunomodulation through a multifaceted approach involving direct cell-cell contact and the secretion of a diverse array of soluble immune modulators (metabolites, cytokines, growth factors, and extracellular vesicles). Moreover, these mechanisms work in concert to stimulate metabolism, modulate the immune response, promote tissue repair, and enhance tolerance [17].

2.1.1 Direct cell-cell contact

Research has consistently demonstrated that direct cell-cell contact is pivotal for the immunomodulatory effects of MSCs. Several factors contribute to the immunomodulatory effects of MSCs through direct cell-cell contact, as depicted in **Figure 1**.

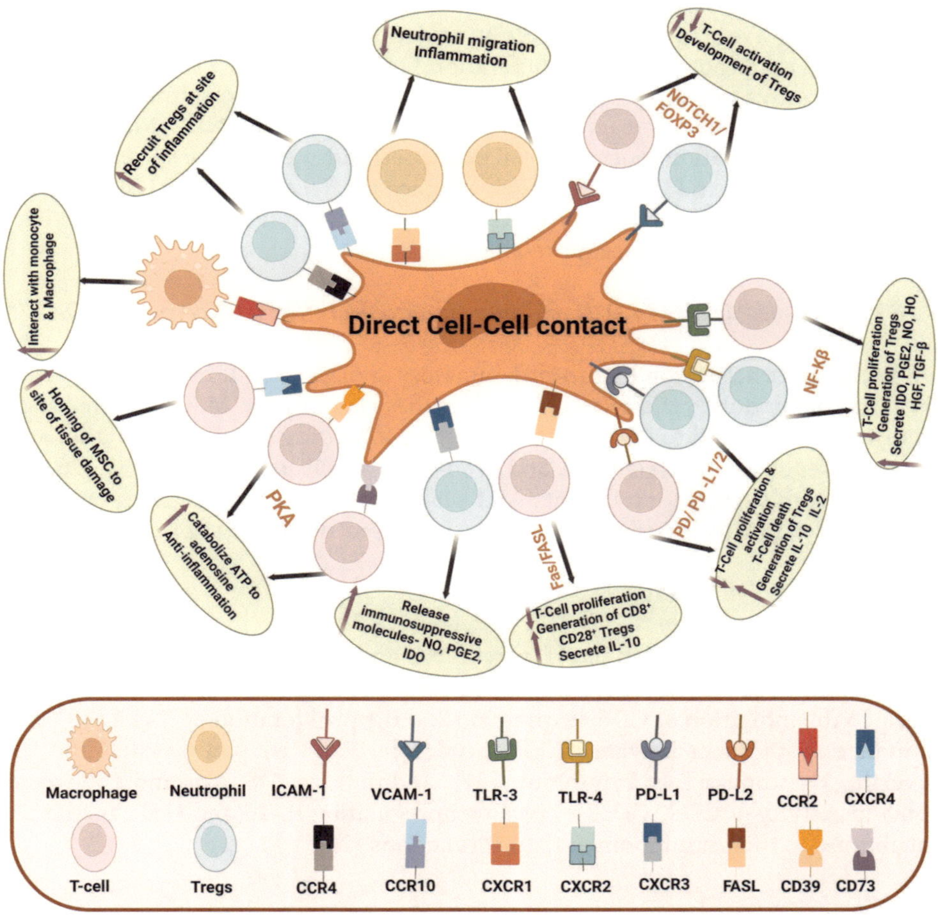

Figure 1.
MSCs-mediated immunomodulation by direct cell-cell contact. This figure demonstrates the critical role of direct cell-cell contact in MSCs-mediated immunomodulation, highlighting the interactions with various immune cells that contribute to their therapeutic potential in GVHD.

2.1.1.1 Cell adhesion molecules

Cell adhesion molecules are critical for direct cell-cell interactions, which are essential for MSCs to modulate immune cells. Cell adhesion molecules like Intercellular Adhesion Molecule 1 (ICAM-1) and Vascular Cell Adhesion Molecule 1 (VCAM-1) on MSCs, bind to LFA-1 (lymphocyte function-associated antigen 1) and VLA-4 (very late antigen-4) on T-cell and other immune cells [18]. This facilitates direct interactions with T-cell, [19], disrupting T-cell activation [20] and promoting the development of Tregs through the Neurogenic locus notch homolog protein 1/ forkhead box protein 3 gene (Notch1/FOXP3) pathway [19]. These molecules are upregulated by inflammatory cytokines (IFN-γ) [21] and GVHD patients-derived plasma [22], resulting in the enhancement of MSCs' immunomodulatory effects.

2.1.1.2 Toll-like receptors (TLR)

MSCs express TLR3 and TLR4, which are key in their immunomodulatory functions. TLR3 activation polarizes MSCs into an anti-inflammatory (MSC2) phenotype [23], promoting Tregs induction, secretion of anti-inflammatory factors like TGF-β and IDO, and inhibition of T-cell proliferation [24]. In contrast, TLR4 activation drives MSCs into a pro-inflammatory (MSC1) phenotype [25], resulting in the secretion of chemokines such as macrophage inflammatory protein (MIP)-1α, MIP-1β, regulated upon activation, normal T-cell expressed and presumably secreted (RANTES), CXCL9, and CXCL10, which enhance the T-cell response [24]. Additionally, TLR-preconditioned MSCs are more resistant to NK cell-mediated cytotoxicity, which enhances their immunosuppressive effects [26].

2.1.1.3 Programmed death 1 (PD-1) and programmed death-ligand 1/ligand 2 (PD-L1/L2)

MSCs require activation by pro-inflammatory signals (e.g., IFN-γ and TNF-α) to exert immunomodulatory effects [27]. Under these conditions, MSCs secrete PD-L1 and PD-L2, which bind to PD-1 on lymphocytes, inhibiting T-cell proliferation and activation, reducing IL-2 secretion, and promoting T-cell anergy and death [26, 28]. Additionally, this interaction inhibits AKT phosphorylation and enhances Foxp3 expression, leading to Tregs generation and increased IL-10 secretion [26, 28].

2.1.1.4 Fas/Fas ligand (FASL)

MSCs express FasL, which interacts with Fas on activated T-cell to exert immunosuppressive effects [29]. Specifically, activated MSCs express both PD-L1 and FasL, leading to the inhibition of CD69 expression and the proliferation of T-cell [30]. MSCs can induce apoptosis in activated T-cell through the Fas/FasL pathway and by converting tryptophan into kynurenine [31]. Additionally, MSCs promote the generation and activity of $CD8^{+}$ $CD28^{-}$ Tregs by upregulating IL-10 and FasL, further contributing to their immunomodulatory functions [26].

2.1.1.5 CD39 and CD73

CD39 catabolizes adenosine triphosphate (ATP) to adenosine monophosphate (AMP), and CD73 further converts AMP to adenosine. While extracellular ATP exerts pro-inflammatory effects, adenosine promotes anti-inflammatory responses *via* the cyclic adenosine monophosphate (cAMP) and protein kinase A (PKA) pathways. Therefore, the activity of CD39 and CD73 on MSCs helps cleave extracellular ATP into adenosine, effectively shifting the immune response from a pro-inflammatory to an anti-inflammatory state [26, 32].

2.1.1.6 Chemokine receptors

Chemokine receptors on MSCs play a pivotal role in their immunosuppressive function by guiding their migration to inflamed tissues [33]. CXCR4-CXCL12 interaction is crucial for MSCs homing to injury sites, where they modulate immune responses [33, 34]. Interestingly, MSCs also express CCR2, facilitating their recruitment to inflamed areas and interaction with monocytes and macrophages [26]. Additionally, CCR4 and CCR10 help recruit Tregs [35, 36], while CXCR3 suppresses effector T-cell [16, 37]. MSCs inhibit the maturation and antigen-presenting capabilities of dendritic cells (DCs) by CCR7-CCL19/CCL21 interaction [33], and CXCR1/CXCR2 responses to Interleukin-8 (IL-8) reduce neutrophil activity, mitigating inflammation and tissue damage [36].

2.1.2 Paracrine effect

The paracrine effects of MSCs are pivotal in their ability to modulate the immune system and to promote tissue repair and homeostasis, making MSCs a promising therapeutic avenue for treating autoimmune diseases, graft-versus-host disease, and other inflammatory conditions. Unlike direct cell-to-cell interactions, the paracrine effect involves the secretion of diverse bioactive molecules that can influence nearby cells and tissues, as shown in **Figure 2**.

2.1.2.1 Indoleamine 2,3 dioxygenase (IDO)

Indoleamine 2,3-dioxygenase (IDO) is a heme enzyme that metabolizes tryptophan into kynurenine and other metabolites *via* the kynurenine pathway [38]. It is primarily expressed by antigen-presenting cells and MSCs under inflammatory conditions [39], which promotes immune tolerance by depleting tryptophan [38], inducing Tregs [38]. IDO expression is upregulated by pro-inflammatory cytokines, acting as a negative feedback signal during inflammation [38, 40], supporting the differentiation of tolerogenic macrophages, and suppressing NK-cell function [38]. In GVHD, IDO is upregulated, and its deficiency leads to increased lethality [38].

2.1.2.2 Prostaglandin E2 (PGE2)

Prostaglandin E2 (PGE2) is synthesized through the cyclooxygenase (COX) pathway [26]. MSCs express COX-2 and produce PGE2, significantly increasing its production in response to inflammatory stimuli [41]. It enhances the differentiation of M2 macrophages [26, 42] and Th2 cells while inhibiting Th1 production [26]. Although it generally has anti-inflammatory effects, its impact can vary with concentration as their high levels inhibit the inflammatory response and promote Foxp3^{+} Treg cell production [43], while low levels can enhance inflammation [44].

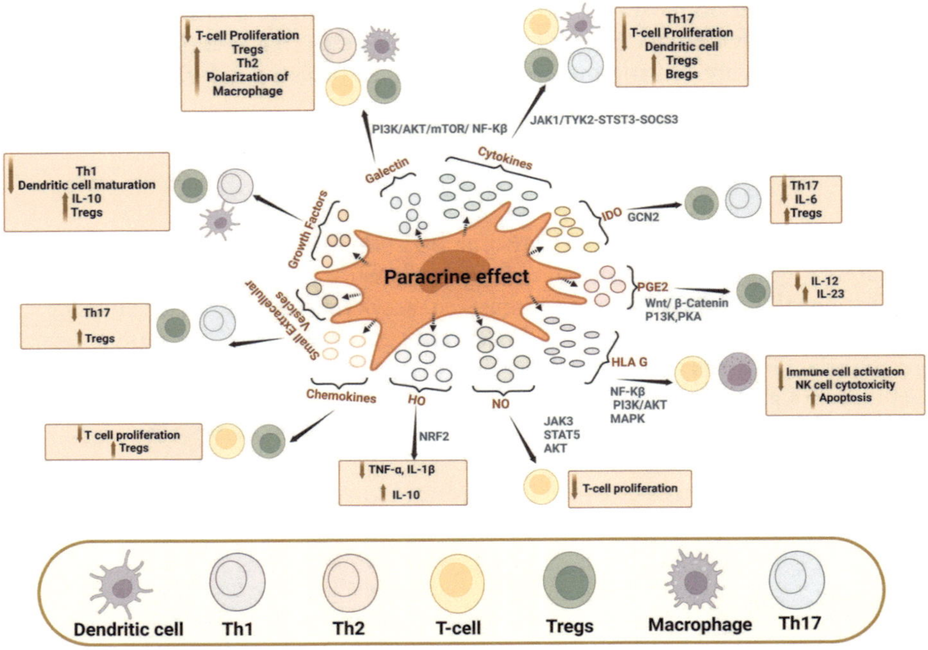

Figure 2.
Immunomodulatory effects of MSCs through paracrine signaling. MSCs secrete various bioactive molecules, including cytokines, chemokines, and growth factors, that modulate immune cells' activity, playing a crucial role in reducing inflammation and promoting immune tolerance in GVHD.

2.1.2.3 Human leukocyte antigen-G5 (HLA-G5)

Human leukocyte antigen-G5 (HLA-G5), a non-classical major histocompatibility complex (MHC) class I molecule [45], secreted by MSCs under the influence of IL-10, IFN-γ, and TNF-α [26] is critical for mediating immune suppression in GVHD. HLA-G5 exerts its immunomodulatory and immunosuppressive properties primarily by engaging with inhibitory receptors on various immune cells, including NK cells, T-cell, DC, macrophages, and monocytes. HLA-G5 engagement with its receptors activates the phosphoinositide 3-kinase (PI3K)/AKT pathway, which promotes cell survival and reduces apoptosis. It also increases levels of cAMP, which activates protein kinase A (PKA). This supports the expansion and suppressive function of Tregs, which are crucial for maintaining immune tolerance and preventing excessive immune responses in GVHD [46].

2.1.2.4 Nitric oxide (NO)

MSCs express inducible nitric oxide synthase (iNOS), which metabolizes L-arginine to produce NO [47]. NO interferes with the IL-2 signaling pathway, affecting key components such as Janus kinase 3 (JAK3), signal transducer and activator of transcription 5 (STAT5), extracellular signal-regulated kinase (ERK), and protein kinase B (AKT). This results in the inhibition of T-cell proliferation and function [26] and also suppresses the expression of MHC II, which is crucial for antigen presentation and subsequent T-cell activation [26]. In the presence of inflammatory stimuli (IL-17), the iNOS gene in MSCs is upregulated, leading to increased production of

NO, resulting in immune suppression [48]. However, in mice, rabbits, rats, and hamsters, MSCs primarily use iNOS to exert their suppressive effects while in human, pigs, and monkeys predominantly rely on IDO for immunosuppression [26, 39, 49].

2.1.2.5 Heme oxygenase-1 (HO)

Heme oxygenase-1 (HO-1) is an intracellular enzyme that plays a crucial role in the metabolism of heme to biliverdin, carbon monoxide (CO), and free divalent iron [47], with antioxidant and anti-inflammation [50]. CO can inhibit the production of pro-inflammatory cytokines (TNF-α, IL-1β, and IL-6) and promote the production of anti-inflammatory cytokines (IL-10) along with reducing leukocyte migration [51]. This shift toward an anti-inflammatory cytokine profile helps to reduce the inflammatory responses characteristic of GVHD. Free iron and biliverdin have strong antioxidant properties [47], and HO can induce a tolerogenic phenotype in DCs, reducing their ability to activate T-cell and promoting the development of Tregs [52].

2.1.2.6 Chemokines

Chemokines, like CXCL9, CXCL10, and CXCL11, play a key role in MSCs-mediated immunomodulation by attracting immune cells, particularly T-cell, *via* the CXCR3 receptor of activated T-cell [26]. This chemokine-mediated recruitment of T-cell to MSCs facilitates immune regulation through various mechanisms, including NO production, which inhibits T-cell proliferation and activation and also promotes Tregs generation, contributing to immune homeostasis [53]. Interestingly, chemokines enhance the efficiency of immune suppression by concentrating T-cell near MSCs, where they encounter other immunosuppressive factors like PGE2, TGF-β, and IDO [17].

2.1.2.7 Cytokines

MSCs secrete key immunoregulatory cytokines like TGF-β and IL-10, which help balance pro- and anti-inflammatory responses [26]. TGF-β is constitutively secreted by MSCs and its production can be further upregulated by inflammatory factors such as IFN-γ and TNF-α [54], inhibiting IL-2, MHC II, and costimulatory factors in DCs and T-cell, suppressing their activation. It also promotes Tregs and regulatory B cell (Bregs), maintaining immune tolerance [55, 56]. Remarkably, autocrine TGF-β can inhibit MSCs immunosuppressive effects by suppressing iNOS and IDO through SMAD3 [33]. IL-10, enhanced by TLR ligands and PGE2, limits APC maturation, suppresses Th17 cells, and promotes Tregs *via* the JAK1-TYK2-STAT3-SOCS3 pathway [26, 57]. IL-1 Receptor Antagonist (IL-1Ra) from MSCs reduces inflammation, promotes M2 macrophages, and accelerates wound healing [26, 58].

2.1.2.8 Growth factors

MSCs secrete several growth factors, such as vascular endothelial growth factor (VEGF), basic fibroblast growth factor (bFGF), keratinocyte growth factor (KGF), insulin-like growth factors (IGF-1 and IGF-2), and HGF, which contribute to their immunomodulatory properties [33].

2.1.2.9 Galectins

MSCs secrete various galectins, such as Galectin-1 (Gal-1) [59], Galectin-3 (Gal-3) [60], and Galectin-9 (Gal-9) [61], which play significant roles in immunosuppression through various mechanisms. Gal-1 binds to specific glycoreceptors on T-cell and activates the ERK/MAPK signaling pathway, which contributes to the inhibition of T-cell activation and proliferation, shifts immune responses to an anti-inflammatory Th2 and Treg response [26, 62], tolerogenic DCs [63], and inhibits NK cell cytotoxicity [64]. Gal-3 inhibits T-cell activation [65], promotes M2 macrophage polarization *via* the STAT3 signaling pathway [66]. It reduces neutrophil recruitment and promotes apoptosis, aiding in the resolution of inflammation by activating the MAPK signaling pathway [67]. Gal-9 interacts with the Tim-3 receptor on T-cell while enhancing Tregs expansion through the PI3K/AKT/mTOR pathway [68].

2.1.2.10 Small extracellular vesicles (exosomes)

MSCs-derived exosomes have therapeutic potential due to their ability to carry microRNAs, lipids, and proteins across tissues and barriers with low immunogenicity, making them ideal for repeated administration [69]. They reduce alloreactive T-cell (Th17) proliferation, promote Treg generation, and help maintain immune tolerance, thus reducing cGVHD severity [70]. Additionally, they aid tissue repair by promoting angiogenesis, reducing apoptosis, and stimulating resident stem cell proliferation and differentiation [71].

2.2 Apoptosis and efferocytosis of MSCs

Recent insights suggest that the immunosuppressive effects of MSCs are driven not by their survival but by their ability to undergo apoptosis post-administration. This regulated cell death, followed by efferocytosis by local myeloid phagocytes, especially macrophages, plays a crucial role in mediating their therapeutic effects [72]. This challenges the long-held belief that MSC viability is essential for their efficacy, shifting the focus to apoptosis as a key mechanism in immunosuppression and tissue repair.

2.2.1 Dynamics of MSCs' fate post-administration

Upon administration, MSCs are exposed to inflammatory environments, which are crucial for their immunosuppressive function [73]. Despite *in vitro* evidence, MSCs rapidly undergo apoptosis *in vivo*, triggering efferocytosis by macrophages. This process reprograms macrophages to an anti-inflammatory M2 phenotype, characterized by reduced TNF-α production and increased IL-10 secretion [72, 74, 75]. Thus, the apoptosis and clearance of MSCs are now seen as central to their therapeutic action, shifting focus from survival and engraftment to immunosuppression driven by cell death.

2.2.1.1 Apoptotic MSCs and the inflammatory environment

MSCs require an inflammatory environment to enhance their immunosuppressive function. Inflammatory signals prime MSCs and optimize the apoptosis-efferocytosis pathway, which leads to an anti-inflammatory response in macrophages.

2.2.1.2 Stromal cell apoptosis and tissue homeostasis

MSCs share similarities with tissue fibroblasts, raising questions about whether stromal cell apoptosis plays a role in tissue repair and homeostasis. This apoptosis may act as an innate mechanism to regulate the inflammatory environment, contributing to tissue healing [72].

2.2.2 Role of host macrophages in clearing apoptotic MSCs

Macrophages are critical in clearing apoptotic MSCs, undergoing efferocytosis post-infusion. This process shifts macrophages from a pro-inflammatory M1 to an anti-inflammatory M2 phenotype. The reprogrammed macrophages secrete anti-inflammatory cytokines like IL-10 [76], while upregulating immunomodulatory molecules such as PGE2 and PD-L1 [77], essential for the therapeutic effects of MSCs.

2.2.3 Contribution to immune regulation

The interaction between apoptotic MSCs and host macrophages creates an immunosuppressive environment. Reprogrammed macrophages release anti-inflammatory molecules like IL-10, TGF-β, and IDO, which suppress pro-inflammatory responses and promote tissue repair. This macrophage reprogramming and cytokine secretion are central to the immune-regulatory benefits of MSC therapy, especially in conditions with excessive inflammation [78]. Thus, the contribution of apoptotic MSCs to immune regulation is a pivotal aspect of their therapeutic mechanism, underscoring the importance of understanding and harnessing this process for effective clinical outcomes.

3. MSCs in promoting stem cell engraftment

MSCs facilitate hematopoietic stem cell (HSCs) engraftment through a multifaceted approach, shown in **Figure 3**, that includes enhancing HSCs homing and survival, promoting hematopoietic reconstitution, and modulating immune responses to prevent graft rejection [9]. While the precise interactions between MSCs and HSCs continue to be studied, the existing evidence underscores the significant potential of MSCs in improving the outcomes of HSCs transplantation.

3.1 Enhancing homing and survival of HSCs

The optimization of self-renewal and proliferation of HSCs depends on their survival and homing into the BM. MSCs significantly improve these processes, thereby enhancing hematopoietic reconstitution and achieving better chimerism compared to conventional transplantation methods. However, the precise interactions between MSCs and HSCs remain partially understood.

Several studies have been conducted to investigate the safety and efficacy of MSC and HSCs co-transplantation. Despite some controversy due to heterogeneous doses, routes, and sources of MSCs and HSCs, evidence suggests that MSCs can improve the homing of HSCs to the BM [79]. For instance, the co-transplantation of MSCs expanded from fetal lung tissue with UC-derived CD34+ cells improved bone marrow engraftment of human HSCs in NOD/SCID mice [80].

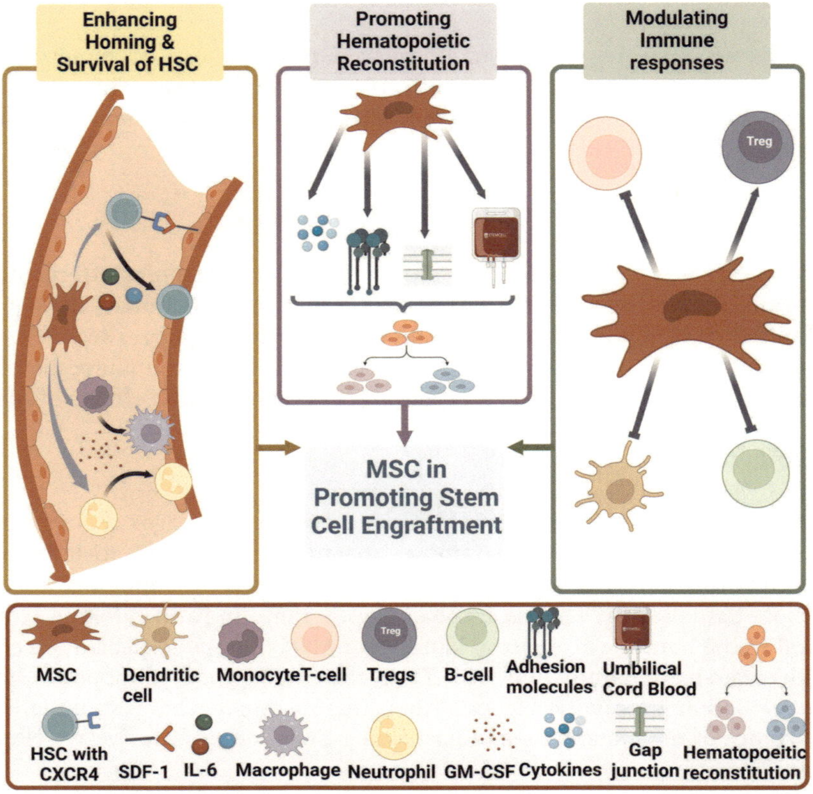

Figure 3.
Mechanism by which MSCs promote engraftment in Allo-HSCT. MSCs facilitate engraftment by creating a supportive microenvironment by secretion of growth factors, cytokines, and extracellular matrix components. They enhance the homing and proliferation of hematopoietic stem cells (HSCs) in the bone marrow niche. Additionally, MSCs modulate the immune response, reducing the risk of graft rejection and GVHD, thereby improving the overall success of Allo-HSCT.

MSCs contribute to the homing of HSCs by secreting various cytokines and growth factors that create a supportive microenvironment. These include:

a. *Granulocyte-macrophage Colony-Stimulating Factor (GM-CSF):* It promotes the differentiation and proliferation of granulocyte and macrophage progenitors [9].

b. *Interleukin-6 (IL-6):* It supports HSC differentiation and maturation [9].

c. *Stromal Cell-Derived Factor-1 (SDF-1):* It facilitates the homing of HSCs to the BM by binding to the CXCR4 receptor on HSCs, thereby guiding them to the niche [81].

3.2 Promoting hematopoietic reconstitution

MSCs play an essential role in promoting hematopoietic reconstitution after HSCs transplantation. This is particularly important given that preconditioning regimens, such as chemotherapy and irradiation, often damage the BM microenvironment, making it challenging for transplanted HSCs to re-establish hematopoiesis [82].

Studies have shown that MSCs can enhance hematopoietic recovery through several mechanisms:

a. *Cytokine secretion:* MSCs secrete a range of hematopoietic cytokines, such as G-CSF, GM-CSF, and IL-6, which support the proliferation and differentiation of HSCs [83]. UC-MSCs expressed mRNA for granulocyte-macrophages and granulocyte colony-stimulating factors, as well as various cell growth factors, indicating strong hematopoietic support [84].

b. *Homing and adhesion molecules:* MSCs produce extracellular matrix components and adhesion molecules that provide a scaffold for HSCs, facilitating their retention in the bone marrow niche. The secretion of SDF-1 by MSCs is particularly important for guiding HSCs to the BM [85].

c. *Direct cell-cell interactions:* MSCs interact directly with HSCs through cell surface molecules and gap junctions, providing physical support and signaling cues essential for HSC maintenance and differentiation [86].

d. Engraftment efficiency in clinical studies: Clinical studies have demonstrated the potential of MSCs to enhance HSC engraftment efficiency. For instance, in a phase 1/2 clinical trial involving breast cancer patients, MSC infusions facilitated rapid hematopoietic recovery post-chemotherapy without observable toxicity [87].

e. *Use in umbilical cord blood transplantation (UCBT):* MSCs have also been studied in the context of UCBT and reported that co-infusion of UCB with MSCs significantly shortened the time for neutrophil and platelet recovery, indicating improved hematopoietic function [88].

3.3 Modulating immune responses

One of the critical roles of MSCs in facilitating engraftment is their ability to modulate immune responses, thereby preventing graft rejection and GVHD. MSCs achieve this through several pathways:

a. *Suppression of T-cell proliferation:* MSCs inhibit the activation and proliferation of T-cell, which are major mediators of graft rejection. They secrete immunosuppressive factors such as TGF-β, HGF, and IDO [89].

b. *Induction of Tregs:* MSCs promote the expansion and function of Tregs, which help maintain immune tolerance by suppressing effector T-cell activity [90].

c. *Inhibition of DC maturation:* MSCs impair the maturation and function of dendritic cells, reducing their ability to activate alloreactive T-cell and initiate immune responses [89].

d. *Modulation of B cell responses:* MSCs inhibit B cell proliferation and differentiation into plasma cells, reducing antibody production that can target transplanted cells, and enhancing the proliferation of IL-10-producing CD5+ regulatory B cells [91].

Clinical studies have confirmed the immunomodulatory effects of MSCs. For example, in a phase 1/2 trial involving pediatric leukemia patients, MSCs infusions were associated with successful neutrophil engraftment and increased survival probability compared to controls [92]. Further research and clinical trials are necessary to optimize MSCs and HSCs co-transplantation protocols and fully harness the therapeutic benefits of MSCs in hematopoietic stem cell engraftment.

4. Clinical insights of MSCs in Allo-HSCT

Several clinical trials have evaluated the safety and efficacy of MSCs in allo-HSCT. Overall, MSCs therapy has been well-tolerated with minimal adverse effects. Clinical studies have shown that co-transplantation of MSCs with HSCs leads to faster hematopoietic recovery and reduced transplant-related mortality. This is particularly beneficial in patients with high-risk hematologic malignancies, where rapid engraftment is crucial for improving survival rates. Moreover, the infusion of MSCs also reduces the incidence and severity of acute and chronic GVHD. A brief overview of clinical trials involving MSCs-based therapy in Allo-HSCT, registered in ClinicalTrials.gov is summarized in **Table 1**.

5. Current challenges in translating MSCs from bench to bedside

Over the past decade, MSCs research has progressed significantly, with promising potential for treating various diseases. However, clinical trials have highlighted discrepancies and unreported adverse outcomes, underscoring the need for a thorough analysis of MSCs treatment risks, depicted in **Figure 4**. This section discusses the regulatory framework for MSCs as Advanced Therapy Medicinal Products (ATMPs), challenges in their use for GVHD treatment, and the inherent heterogeneity of MSCs, which complicates therapeutic consistency and effectiveness.

5.1 Regulatory guidelines and good manufacturing practices (GMP)

MSCs are classified as ATMPs and regulated by stringent Food Drug and Administration (FDA) and the European Medicines Agency (EMA) guidelines under GMP standards. These guidelines ensure the reproducibility, efficiency, and safety of cell culture procedures [21, 117]. A study on bioprocess engineering demonstrated the effective isolation, expansion, and validation of BM-MSCs from 48 iliac crest samples for autologous transplantation. The process, which included steps like isolation, trypsinization, and quality control, confirmed the safety and potency of MSCs for clinical use [118]. However, further validation of isolation and culture protocols is needed to meet GMP requirements and ensure consistent MSC products.

5.2 Challenges associated with the exogenous application of MSCs in GVHD

GVHD is a major complication of Allo-HSCT, with corticosteroids as the standard treatment, though not always effective [119]. MSCs, due to their immunomodulatory properties, have been explored for treating steroid-refractory GVHD [120]. While MSCs have shown potential in reducing inflammation in clinical trials, concerns

Study details	Diagnosis	Source of MSCs	Dose and frequency of MSCs	Outcome
Open-label, multicenter, parallel randomized study; (ChiCTR-IIR-16007806) [93]	Acute leukemia	UC	1*10^6 cells/kg; 4 doses every 2 weeks starting from D + 45 post-transplantation	• Cumulative incidence of chronic GVHD was 5.4% in the MSC group and 17.4% in the control group • No difference in cumulative incidence of leukemia relapse in both MSC and control group • The incidence of grade II-IV aGVHD in the MSCs group was lower than the control
ChiCTR2000035740 [94]	Steroid refractory acute GVHD (Grade II-IV)	UC	1*10^6 cells/kg; twice a week for 4 weeks	The outcome is not publicly available
Phase III, open-label [95]	Steroid-resistant aGVHD	BM	1*10^6 cells/kg; once a week (4 doses)	• 56.6%, 26%, 17.2%, and 82.8% were CR, PR, NR, and OR respectively in the MSC group • 11.3 months was OS in the MSC cohort
Phase III randomized study (NCT00366145 [96]	Steroid refractory aGVHD	BM	2*10^6 cells/kg; twice weekly for 4 consecutive weeks, with infusions at least 3 days apart	• 29% of OR was observed in patients with liver involvement • 58% of OR was observed in grade III-IV aGVHD patients
Phase III, single arm multicentre study (NCT02336230) [97]	Steroid refractory aGVHD	BM	2*10^6 cells/kg; twice weekly for 4 consecutive weeks	• OR at day 28 was 69.1% • 74.5 and 68.5% of patients were alive at days 100 and days 100 respectively
Prospective single-center study [98]	Steroid-resistant GVHD	BM	0.59–1.8*10^6 cells/kg	• 84.6 and 30.8% were OS at 40 days in responders and non-responders respectively • Lymphocytes (NK cells, T-cell) predict a good response to therapy
Multicentre prospective study [99]	Steroid refractory aGVHD	BM	1–2*10^6 cells/kg (n = 20), 3–4*10^6 cells/kg (n = 13)	• 18% was OS at 1-year • 21.9 and 40% were CR and OR respectively
Prospective study [100]	Steroid-refractory aGvHD	BM	1–2*10^6 cells/kg	• CR and PR were 21.09% and 40% respectively • 18% was OS at 1-year
Prospective randomized trial (NCT-02172937) [101]	Severe aGVHD	Placenta derived decidua	1*10^6 cells/kg, dose/week [1 infusion (n = 17), 2 infusions (n = 21)]	All patients achieved CR and PR and the best one-year survival

Study details	Diagnosis	Source of MSCs	Dose and frequency of MSCs	Outcome
Cross-sectional, multicenter, retrospective study [102]	Steroid refractory aGVHD (III/IV)	BM	Cumulative dose: $6.81*10^6$/kg; a median of three doses (range 1–7)	• 6.5, 43.5, and 50% were CR, PR, and OR respectively in the MSC group 34.4, and 7.4% OS at 100-day and 2-year respectively
Retrospective, multicentre study [103]	Steroid-resistant aGVHD	BM	$<1.5*10^6$ cells/kg/infusion: n = 11; $1.5–3.0*10^6$ cells/kg/infusion: n = 32; $>3.0*10^6$ cells/kg/infusion: n = 17	• 53% of patients responded to MSCs • 1·6, 51·6% of patients achieved CR and PR, respectively OS was 104 days
Single center [104]	Steroid refractory GVHD	BM	$2*10^6$ cells/kg; twice a week (aGvHD and cGvHD) or once a week (cGvHD)	• 62% of aGVHD patients achieved a response (27% CR, 15% VGPR, 19% PR) at day 28 post-therapy • 38% of the patients had no response
Phase II/III study [105]	Steroid refractory grade III or IV aGVHD	BM	$2*10^6$ MSCs/kg; twice a week for 4 weeks	• 24 and 36% of patients achieved CR and PR at 4 weeks • Durable CR was achieved in 48% of patients by 24 weeks
Single center [106]	Steroid refractory grade I-IV aGVHD	BM	$0.99*10^6$ MSCs/kg	• 8.62, 8.62, 29.31, 53.44% of patients achieved CR, very good PR, PR, and no response • Overall response rate of 47% was observed
Prospective, open-label, nonrandomized, multicenter study [107]	Refractory aGVHD	BM	1×10^6 cells/kg, weekly until patients achieved CR or received eight doses of MSCs	• OR was 75% after 125 doses (median four doses) of MSCs • CR and 3-year OS was 60% and 46% respectively Tregs and sjTRECs levels were increased in the MSC group at 8 and 12 weeks after treatment
Open-label, non-randomized, phase II study (NCT00827398) [108]	Steroid-resistant aGVHD	BM	$1–2*10^6$ cells/kg	CR was 25 and 44% OS at 1-year
Prospective pilot study [91]	Refractory cGVHD	BM	$1*10^6$ cells/kg/infusion; three doses were given at a 4-week interval	• 86.95% patients achieved CR or PR in a 12 month follow-up • Increased number of IL-10 producing CD5+ regulatory B cells

Study details	Diagnosis	Source of MSCs	Dose and frequency of MSCs	Outcome
Prospective study [109]	Severe aplastic anemia	UC	$5.0*10^5$/kg; BM was infused after infusion of UC-MSCs with a 4-hour interval	• No patients experienced disease relapse or progression • Promotes donor engraftment
Single center study [110]	Hematologic malignancy and Severe aplastic anemia	UC	$5.0*10^5$ cells/kg; BM and PBSCs were infused after infusion of MSCs with a 4-hour interval	• The engraftment of hematopoietic stem cells was enhanced with an incidence of mild GVHD (grade I-II) in the MSC cohort (33.3%) • Hematopoietic engraftment was sustained without an increase in severe acute GVHD (grade II-IV) in the MSC group
Phase I study [111]	Steroid-resistant GVHD (acute/chronic), Grade II to IV	BM	$1 \pm .5*10^6$ cells/kg; 2 infusions with about 5–7 days of interval	• 27.5, 40, and 67.5% were CR, PR, and OR respectively • 50 and 38.6% were OS at 1-year and 2-year respectively
Prospective clinical study [112]	Steroid-resistant aGVHD	BM	1.05×10^6 cells/kg	• Disease-free survival was 56% at 3.6 years. • CR and OR were 34 and 66%, respectively
Multicentre study [113]	Hematologic malignancies	BM	$1.0*10^6$ cells/kg; infusion of MSCs followed by UCB cells with a 4 h interval	• Incidence of grade III and IV aGVHD and TRM was decreased in the MSC group • No significant difference was observed in hematopoietic engraftment in UCBT while accelerating leukocyte recovery without graft rejection in the MSC group
Phase I-II study [92]	Hematologic disorders	BM	$5.0*10^6$ cells/kg; Cohort I received a single dose of MSCs on D0 while cohort II received two doses (D0 and D + 21)	• Hematopoietic engraftment was enhanced with a significant decrease in cumulative incidence of grade II-IV acute GVHD in the MSC group (38%) • Chronic GVHD was not observed
Pilot study [114]	Hematologic disorders	BM	$1.04–2.15*10^6$ immediately after infusion of CB and mobilized hematopoietic stem cells	No significant differences in engraftment of CB and incidence of acute GVHD were observed

Study details	Diagnosis	Source of MSCs	Dose and frequency of MSCs	Outcome
Multicentre study [115]	Steroid refractory aGVHD	BM	$1.4*10^6$ cells/kg; 1 dose (n = 27); 2 doses (n = 22); 3–5 doses (n = 6)	• CR was observed in 54.5% of patients • 49.09% of patients had CR, 3.63% of patients had PR; 47.27% of patients did not respond after one dose • One and six patients developed grade I and II acute GVHD respectively
Phase I multicentre study [116]	Hematologic malignancies	BM	Dose escalation (0, 2.5, or $5.0*10^6$/kg); infusion of MSCs followed by BM/PBSC cells with a 4 h interval	• Hematopoietic recovery was improved with a decrease in grade II-IV acute GVHD in the MSC group (28%) • Chronic GVHD was observed in 61% of patients in the study cohort

Abbreviations: aGVHD: Acute Graft-versus-Host-Disease; UC: Umbilical Cord; UCBT: Umbilical Cord Blood Transplantation; TRM: Transplant-Related Mortality; PBSC: Peripheral Blood Stem Cells; BM: Bone marrow; CB: Cord Blood; CR: Complete Response; PR: Partial Response; VGPR: Very Good Partial Response; cGVHD: Chronic GVHD; OR: Overall Response; NR: No Response.

Table 1.
Clinical trials of MSCs in Allo-HSCT aim to promote engraftment and/or prevent or treat GVHD.

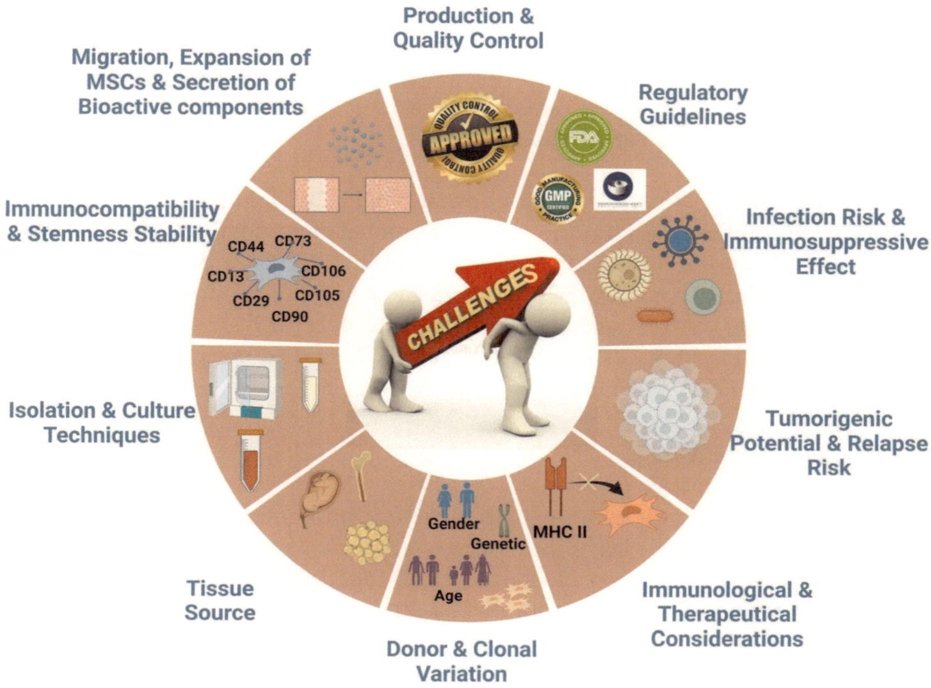

Figure 4.
Challenges Associated with MSCs Therapy in Allo-HSCT: From Bench to Bedside. This figure highlights key challenges in translating MSCs-based therapy from research to clinical application in transplantation. The challenges include MSC heterogeneity, manufacturing scalability, regulatory and quality control issues, immunogenicity risks, limited in vivo *persistence, and the need for standardized clinical trial designs.*

remain regarding their efficacy and safety. One of the key challenges is the increased risk of infections, particularly in patients already on immunosuppressive therapies post-HSCT, which complicates the broader use of MSCs in GVHD treatment.

5.2.1 Infection risk and immunosuppressive effects

MSCs infusion in GVHD patients has been linked to increased infection risks, with studies showing high mortality rates (54%) due to infection-related causes, such as CMV interstitial pneumonia and fungal infections [121]. Patients receiving BM-MSCs post-HSCT had higher incidences of severe infections, including pneumonia and elevated CMV viral loads [122]. While early studies suggested MSCs suppressed cytotoxic T-cell responses to CMV [123], more recent findings indicate that MSCs may act as viral transmitters during infections, raising concerns about their safety in this context [124].

5.2.2 Tumorigenic potential and relapse risk

MSCs can promote tumor growth by modulating the tumor microenvironment. A pilot study found that while MSCs reduced GVHD in hematologic malignancy patients, they increased relapse rates. This protumorigenic effect is likely linked to

MSCs' immunosuppressive properties and potential interaction with tumor stroma [117]. However, no conclusive evidence of malignant transformation of human MSCs has been found yet.

5.2.3 Immunological and therapeutic considerations

Recent studies suggest MSCs are not immunologically silent *in vivo* [117]. This reclassification implies that immune compatibility between donors and recipients must be carefully considered, particularly in long-term treatments involving repeated MSCs infusions. Furthermore, MSCs may lose their immunosuppressive properties in mismatched settings, as demonstrated in murine models [125]. It was observed that grafted BM-MSCs in phase II/III studies for acute GVHD did not prevent the development of chronic GVHD [105].

5.3 Heterogeneity of MSCs

MSCs exhibit significant heterogeneity, which is influenced by numerous factors including donor characteristics, tissue sources, and cell isolation and culture techniques [126]. This heterogeneity can be a double-edged sword; while it offers a range of therapeutic possibilities, it also complicates standardization and consistency in clinical applications.

5.3.1 Donor characteristics

The characteristics of MSCs can vary significantly based on the donor's age, gender, genetic background, and health status [127, 128]. For instance, MSCs from older donors or individuals with conditions like diabetes or rheumatoid arthritis show reduced proliferation and differentiation potential [129]. Additionally, MSCs from obese individuals exhibit significantly impaired capabilities, underscoring the impact of donor health on MSC quality and efficacy [130].

5.3.2 Tissue sources

MSCs can be isolated from various tissues including BM, UC, AT, and synovium. Each tissue source provides MSCs with distinct characteristics. For example, BM-MSCs often have a longer doubling time compared to those derived from AT. Additionally, the immunomodulatory potential and differentiation capacities can vary depending on the tissue source. Moreover, AT-MSCs had more inhibitory effects on B, T-cell, and NK cells, while UC-MSCs had less effect [131].

5.3.3 Isolation and culture techniques

The methods used to isolate and culture MSCs can also influence their properties. Culture conditions such as concentration of fetal bovine serum, seeding density, oxygen concentration, and growth supplements affect the functionality of MSCs. Moreover, cryopreservation and thawing can affect cell purity, sub-populations, and cell viability. For example, MSCs cultured in different conditions such as varying oxygen levels, types of culture surfaces, and passage numbers may exhibit different proliferation rates and differentiation capacities. Optimal culture conditions are crucial to preserve the multipotency of MSCs and to ensure their efficacy in clinical applications [21].

5.3.4 Clonal variations

MSCs exhibit both inter-clonal and intra-clonal variations. This means that even within a single clone, individual cells can have different morphologies and gene expression profiles [126]. For instance, cells located at the outer periphery of a colony may express higher levels of genes related to cell proliferation, while cells in the interior may have higher expression of extracellular matrix genes [132]. These variations can significantly impact the therapeutic potential of MSCs, as different sub-populations may possess distinct biological functions and capabilities.

5.3.5 Immunocompatibility and stemness stability

MSCs were initially considered immune privileged due to low MHC I expression and the absence of MHC II and costimulatory molecules. However, recent studies suggest they can become immunogenic under inflammatory conditions or high cell density, triggering MHC II expression [133]. Therefore, immune compatibility is essential for long-term MSCs infusions.

The maintenance of MSCs stemness and differentiation potential during culture expansion is challenging. Serial passaging and prolonged culture can induce senescence, reducing proliferation and differentiation [134]. Certain stemness genes and markers, such as CD13, CD29, CD44, CD73, CD90, CD105, and CD106, are downregulated during culture, affecting the therapeutic potential of MSCs [135]. Researchers are exploring various strategies to preserve the stemness of MSCs, including optimizing culture conditions and using 3D culture systems.

5.3.6 Migration, expansion of MSCs, and secretion of bioactive components

The therapeutic efficacy of MSCs depends on their migration, homing capacities, and delivery method. Freshly isolated MSCs show better homing, but this decreases with somatic expansion. The expression of chemokine receptors can be enhanced through genetic modification to improve MSCs migration and targeting. Intra-tissue or intra-organ delivery is more effective than systemic delivery [126, 136]. MSCs exert their effects through the secretion of cytokines, growth factors, and extracellular vesicles, with composition varying based on culture conditions and the target tissue's microenvironment [126].

5.3.7 Production and quality control

The manufacturing of GMP-grade MSCs for clinical use is challenging and costly, requiring accredited labs and compliance with strict regulations. Additionally, the collection of BM, a common source of MSCs, involves invasive procedures that can be painful and carry risks such as infection. Quality control is critical, as variability in donor characteristics, tissue sources, and isolation methods can lead to inconsistencies in MSCs properties, such as proliferation, differentiation, and immunomodulation [137].

5.3.7.1 Future strategies and innovations

The future of MSCs therapy is promising, but challenges limit its full potential. Innovative strategies, as depicted in **Figure 5**, are needed to overcome these hurdles to enhance MSCs properties and ensure consistent therapeutic effects. This section explores advanced approaches to maximize MSCs clinical potential, improving patient outcomes and driving progress in regenerative medicine.

5.4 Enhancing MSCs potency and consistency

5.4.1 Genetic modification

Genetic modification of MSCs offers a powerful approach to enhance their therapeutic potency independently of external stimuli. This includes overexpression of genes that promote anti-inflammatory effects, increase homing capabilities to injury sites, or enhance survival and proliferation, thereby improving their efficacy in various clinical applications [138].

5.4.2 Enhancing therapeutic potency

Overexpression of VEGF promotes angiogenesis and improves brain infarction recovery [139], while co-expression with Bcl-2 enhances survival and paracrine effects [140]. To mimic the effects of hypoxic preconditioning, hypoxia-inducible

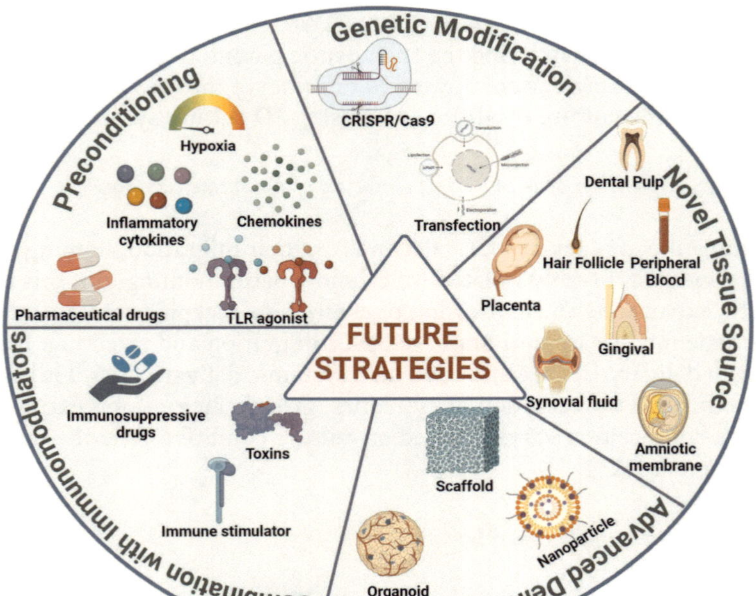

Figure 5.
Future strategies to enhance the efficacy of MSC-based therapy. Various innovative strategies include genetic modification of MSCs, preconditioning with bioactive molecules, combination therapies with other immunomodulators, advanced delivery methods such as targeted delivery and nanotechnology, and personalized treatment plans to boost the efficacy of MSC therapy.

factor 1-alpha (HIF-1α) can be transduced into BM-MSCs, thereby enhancing their therapeutic effects without requiring actual hypoxic exposure [141]. Additionally, overexpression of prostaglandin I synthase (PGIS) in BM-MSCs has been found to protect damaged hearts and restore cardiac function in a myocardial infarction (MI) mouse model [142].

5.4.3 Advanced genetic editing technologies

The advent of advanced genetic editing technologies, such as CRISPR/Cas9 RNA-based nucleases, has revolutionized MSCs modifications by enabling precise genome editing. CRISPR/Cas9 allows for targeted genome editing, enabling MSCs to increase survival rates and alter differentiation preferences [143]. Moreover, CRISPR/Cas9-induced knockout of Keap1 has been shown to enhance the anti-oxidative properties of AT-MSCs [144].

5.4.3.1 Improving migration, adhesion, and survival

Overexpression of chemokine receptors such as CXCR1, CXCR4, and CXCR7 has been shown to improve MSCs migration and proliferation, with significant therapeutic impacts observed in liver, kidney, cerebral ischemia, and myocardial infarction models [138]. The adhesion of MSCs can be enhanced through the expression of integrins and focal adhesion complexes, which have improved their survival and retention in ischemic myocardium models [145].

5.4.3.2 Reducing premature senescence

Genetic modifications involving the overexpression of genes such as Sox2 and Oct4 have been shown to enhance MSCs differentiation and proliferation, as well as their anti-inflammatory effects [146]. Additionally, telomerase reverse transcriptase (TERT) transfection has been reported to prevent senescence and promote higher proliferative capacity and cell cycle-related gene expression [147].

5.4.3.3 Clinical considerations and safety issues

While genetic modification of MSCs holds significant promise, its application in clinical settings raises critical safety concerns. Persistent activation of specific genes may lead to the development of stem cell-derived malignant tumors. Therefore, transient modifications are preferred to enhance therapeutic potential while mitigating safety risks [143].

5.4.4 Preconditioning

Preconditioning of MSCs involves *ex vivo* exposure to chemical and physical factors to enhance the intrinsic therapeutic properties of MSCs against the hostile conditions within their transplanted microenvironment [138].

1. *Improving migration:* MSCs thrive in low oxygen environments (1–7%) [148] *in vivo*, but *in vitro* cultivation typically occurs at higher oxygen levels (20–21%), which can impair their function. Hypoxia preconditioning is a well-established method to enhance MSCs properties, including migration, proliferation, and

survival [138]. It boosts the expression of pro-survival genes and trophic factors, such as SDF-1/CXCR4, promoting migration [149]. Moreover, it also increases lincRNA-p21, CXCR4/7, and HIF-1α, further supporting MSCs migratory capacity [150]. Additionally, preconditioning with deferoxamine [151] and oncostatin M (OSM) enhances migration and proliferation, showing therapeutic potential, such as reducing lung fibrosis in mice [152].

2. *Improving adhesion:* Hypoxic preconditioning enhances MSCs engraftment and survival in target tissues by increasing the expression of SDF-1 receptors (CXCR4 and CXCR7), HIF-1α, and Akt phosphorylation [153]. Preconditioning modifications aimed at inhibiting or knocking down PAI-1, which negatively regulates MSCs survival *in vivo*, produce enhanced MSCs adhesion and autograft survival [154].

3. *Reducing premature senescence:* Hypoxic preconditioned MSCs exert superior protective roles involving HIF1α and Beclin1 signaling pathways [155]. Preconditioning at different hypoxic oxygen concentrations improves MSCs' inherent characteristics and reduces senescence. For instance, 1% O2 enhances proliferation, viability, stemness, and chemokine-related gene expression, including CXCR7 and OCT4, preventing phenotypic changes and senescence-associated-β-gal expression [156]. Furthermore, 2% O2 enhances bioactivity, proliferation, and survival by suppressing cell death signaling pathways and augmenting angiogenic cytokine secretion [148].

Herbal extracts have also been used to precondition MSCs. *Withania somnifera* root and *Tinospora cordifolia* leaf extracts delayed senescence, decreased apoptosis, increased proliferation, and prolonged the G2/M phase of the cell cycle in WJ-MSCs [157].

5.4.5 Combination with immunomodulators

Combining MSCs therapy with immunomodulatory treatments enhances therapeutic outcomes, especially in conditions like GVHD and autoimmune diseases. Co-treatment with immunosuppressants like rapamycin and tacrolimus improves MSCs transplantation survival and reduces medication side effects [143]. Additionally, the use of bioactive reagents, such as the innate immune stimulator MIS416, can enhance MSCs immunomodulation, restoring immune homeostasis and boosting therapeutic function, as shown in experimental colitis [158]. Similarly, a combination of N-acetylcysteine (NAC) and MSCs has shown superior therapeutic effects in resolving inflammation and apoptosis in an interstitial cystitis experimental model compared to both monotherapies [159].

5.5 Exploration of novel tissue sources

Traditionally, MSCs are sourced from BM, AT, and UCB. While these sources have shown significant therapeutic potential, they have limitations such as varying potency and proliferative capacities. As a result, there is a growing interest in exploring novel tissue sources such as postnatal (dental pulp, hair follicle, menstrual blood, synovial fluid, gingival, peripheral blood) and perinatal (placental tissue, amniotic fluid, amniotic membrane, Wharton's Jelly) tissue to enhance the immunomodulatory capabilities of MSCs, which could lead to more effective and reliable treatments [160].

5.6 Advanced delivery methods for enhanced MSC therapies

Advanced delivery methods play a critical role in enhancing the therapeutic efficacy of MSCs. Traditional delivery techniques, such as systemic infusion or local injection, often face challenges like limited cell survival, poor homing efficiency, and uncontrolled differentiation. To address these issues, researchers have been exploring innovative delivery methods, such as targeted delivery, and the application of nanotechnology that can improve the targeting, survival, and functional integration of MSCs [161].

6. Clinical implications and path forward

The transition from preclinical findings to clinical applications of stem cell therapies presents significant challenges, particularly in ensuring safety, efficacy, and standardizing protocols. While promising results have been observed in tissue repair and immunomodulation, translating these successes to clinical practice requires robust clinical trials and an understanding of stem cell mechanisms. Personalized MSCs therapies offer a tailored approach to treatment, optimizing efficacy by considering patient-specific factors such as genetic profiles, disease conditions, and immune responses. Additionally, advances in genomic and proteomic technologies, along with improved methods for cell characterization and manipulation, are pivotal in the development of personalized MSCs treatments. However, ethical concerns such as informed consent and equitable access must be addressed alongside regulatory standards established by agencies like the FDA. These guidelines ensure the safety, sourcing, and processing of stem cells while fostering innovation. Future research should focus on large-scale trials, understanding the immune interaction of MSCs, and developing advanced delivery systems. Collaboration between academia, industry, and regulatory bodies is key to accelerating the clinical translation of stem cell therapies, ensuring safe, effective treatments for patients.

7. Conclusion

The exploration of MSCs in allo-HSCT has revealed their potential to reduce GVHD by secreting anti-inflammatory cytokines, inducing regulatory T-cell, and inhibiting alloreactive T-cell proliferation. MSCs also enhance hematopoietic stem cell engraftment, aiding post-transplant recovery. Advances like genetic modification and targeted homing strategies have improved MSCs efficacy, but challenges remain in standardizing production and ensuring consistent outcomes. Collaborative efforts among researchers, clinicians, regulatory bodies, and industry will be essential to translate scientific breakthroughs into clinical practice, ultimately improving patient outcomes and advancing the field of regenerative medicine. As we continue to unravel the complexities of MSCs biology and refine their clinical applications, MSCs hold the potential to revolutionize the landscape of stem cell transplantation and offer new hope for patients undergoing allo-HSCT.

Acknowledgements

All figures are created using Biorender.com and the details of clinical trials are retrieved from clinicaltrials.gov.

Conflict of interest

The authors declare that there is no conflict of interest.

Abbreviations

Allo-HSCT	allogeneic hematopoietic stem cell transplantation
GVHD	graft-versus-host-disease
MSCs	mesenchymal stem cells
BM	bone marrow
PB	peripheral blood
UCB	umbilical cord blood
ICAM-1	intercellular adhesion molecule 1
VCAM-1	vascular cell adhesion molecule 1
LFA-1	lymphocyte function-associated antigen 1
VLA-4	very late antigen-4
NOTCH1	neurogenic locus notch homolog protein 1
FOXP3	Forkhead box protein 3 gene
TLR	toll-like receptors
TGF-β	transforming growth factor-beta
IDO	indoleamine 2,3 dioxygenase
PGE2	prostaglandin E2
NO	nitric oxide
HO	heme oxygenase
HGF	hepatocyte growth factor
MIP	macrophage inflammatory protein
RANTES	regulated upon activation, normal T Cell expressed and presumably secreted
NK	natural killer
PD-1	programmed death 1
PD-L1/L2	programmed death-ligand 1/ligand 2
TNF-α	tumor necrosis factor-alpha
Fas L	Fas ligand
ATP	adenosine triphosphate
AMP	adenosine monophosphate
cAMP	cyclic adenosine monophosphate
SDF-1	stromal cell-derived factor 1
MCP-1	monocyte chemoattractant protein-1
DCs	dendritic cells
IL	interleukin
COX	cyclooxygenase
HLA-G5	human leukocyte antigen-G5

PKA	protein kinase A
JAK3	Janus kinase 3
STAT5	signal transducer and activator of transcription 5
ERK	extracellular signal-regulated kinase
MAPK	mitogen-activated protein kinase
iNOS	inducible nitric oxide synthase
CO	carbon monoxide
Bregs	regulatory b cells
MHC II	major histocompatibility complex II
VEGF	vascular endothelial growth factor
bFGF	basic fibroblast growth factor
KGF	keratinocyte growth factor
IGF	insulin-like growth factors
MDSCs	myeloid-derived suppressor cells
HSCs	hematopoietic stem cells
GM-CSF	granulocyte-macrophage colony-stimulating factor
SDF-1	stromal cell-derived factor-1
UCBT	umbilical cord blood transplantation
5-aza	5-aza-2′-deoxycytidine
ATMPs	advanced therapy medicinal products
GMP	good manufacturing practices
FDA	food drug and administration
EMA	European Medicines Agency
PGIS	prostaglandin I synthase
MI	myocardial infarction
HIF-1α	hypoxia-inducible factor 1-alpha
OSM	oncostatin M

Author details

Mohini Mendiratta[1], Meenakshi Mendiratta[2], Deepak Garg[1], Sujata Mohanty[2] and Ranjit Kumar Sahoo[1]*

1 Department of Medical Oncology, Dr. B. R. Ambedkar Institute Rotary Cancer Hospital, All India Institute of Medical Sciences, New Delhi, India

2 Stem Cell Facility (DBT-Centre of Excellence for Stem Cell Research), All India Institute of Medical Sciences, New Delhi, India

*Address all correspondence to: drranjitmd@gmail.com

References

[1] Khaddour K, Hana CK, Mewawalla P. Hematopoietic stem cell transplantation. In: StatPearls [Internet]. Treasure Island (FL): StatPearls Publishing; 2024. Available from: http://www.ncbi.nlm.nih.gov/books/NBK536951/ [Accessed: July 23, 2024]

[2] Blazar BR, Murphy WJ, Abedi M. Advances in graft-versus-host disease biology and therapy. Nature Reviews. Immunology. 2012;**12**(6):443-458

[3] Ozdemir ZN, Civriz BS. Graft failure after allogeneic hematopoietic stem cell transplantation. Transfusion and Apheresis Science. 2018;**57**(2):163-167

[4] Murray J, Stringer J, Hutt D. Graft-versus-host disease (GvHD). In: Kenyon M, Babic A, editors. The European Blood and Marrow Transplantation Textbook for Nurses: Under the Auspices of EBMT [Internet]. Cham (CH): Springer; 2018. Available from: http://www.ncbi.nlm.nih.gov/books/NBK543657/ [Accessed: July 23, 2024]

[5] Justiz Vaillant AA, Modi P, Mohammadi O. Graft-Versus-Host Disease. In: StatPearls [Internet]. Treasure Island (FL): StatPearls Publishing; 2024. Available from: http://www.ncbi.nlm.nih.gov/books/NBK538235/ [Accessed: July 23, 2024]

[6] Imamura M. Impaired hematopoiesis after allogeneic hematopoietic stem cell transplantation: Its pathogenesis and potential treatments. Hemato. 2021;**2**(1):43-63

[7] Zhao Q, Ren H, Han Z. Mesenchymal stem cells: Immunomodulatory capability and clinical potential in immune diseases. Journal of Cellular Immunotherapy. 2016;**2**(1):3-20

[8] Friedenstein AJ, Chailakhjan RK, Lalykina KS. The development of fibroblast colonies in monolayer cultures of Guinea-pig bone marrow and spleen cells. Cell Proliferation. 1970;**3**(4):393-403

[9] Burnham AJ, Daley-Bauer LP, Horwitz EM. Mesenchymal stromal cells in hematopoietic cell transplantation. Blood Advances. 2020;**4**(22):5877-5887

[10] Kim EJ, Kim N, Cho SG. The potential use of mesenchymal stem cells in hematopoietic stem cell transplantation. Experimental & Molecular Medicine. 2013;**45**(1):e2

[11] Mendiratta M, Mendiratta M, Mohanty S, Sahoo RK, Prakash H. Breaking the graft-versus-host-disease barrier: Mesenchymal stromal/stem cells as precision healers. International Reviews of Immunology. 2024;**43**(2):95-112

[12] Nagamura-Inoue T, He H. Umbilical cord-derived mesenchymal stem cells: Their advantages and potential clinical utility. World Journal of Stem Cells. 2014;**6**(2):195-202

[13] Dominici M, Le Blanc K, Mueller I, Slaper-Cortenbach I, Marini F, Krause D, et al. Minimal criteria for defining multipotent mesenchymal stromal cells. The International Society for Cellular Therapy position statement. Cytotherapy. 2006;**8**(4):315-317

[14] Li T, Luo C, Zhang J, Wei L, Sun W, Xie Q, et al. Efficacy and safety of mesenchymal stem cells co-infusion in allogeneic hematopoietic stem cell transplantation: A systematic review and meta-analysis. Stem Cell Research & Therapy. 2021;**12**(1):246

[15] Weiss ARR, Dahlke MH. Immunomodulation by mesenchymal stem cells (MSCs): Mechanisms of action of living, apoptotic, and dead MSCs. Frontiers in Immunology [Internet]. 2019;**10**:1191

[16] Kadri N, Amu S, Iacobaeus E, Boberg E, Le Blanc K. Current perspectives on mesenchymal stromal cell therapy for graft versus host disease. Cellular & Molecular Immunology. 2023;**20**(6): 613-625

[17] Wang M, Yuan Q, Xie L. Mesenchymal stem cell-based immunomodulation: Properties and clinical application. Stem Cells International. 2018;**2018**:3057624

[18] Ren G, Roberts AI, Shi Y. Adhesion molecules. Cell Adhesion & Migration. 2011;**5**(1):20-22

[19] Song N, Scholtemeijer M, Shah K. Mesenchymal stem cell immunomodulation: Mechanisms and therapeutic potential. Trends in Pharmacological Sciences. 2020;**41**(9): 653-664

[20] Zheng S, Huang K, Xia W, Shi J, Liu Q, Zhang X, et al. Mesenchymal stromal cells rapidly suppress TCR signaling-mediated cytokine transcription in activated T cells through the ICAM-1/ CD43 interaction. Frontiers in Immunology [Internet]. 2021;**12**:609544

[21] Gao F, Chiu SM, Motan DAL, Zhang Z, Chen L, Ji HL, et al. Mesenchymal stem cells and immunomodulation: Current status and future prospects. Cell Death & Disease. 2016;**7**(1):e2062-e2062

[22] Silva-Carvalho AÉ, Rodrigues LP, Schiavinato JL, Alborghetti MR, Bettarello G, Simões BP, et al. GVHD-derived plasma as a priming strategy of mesenchymal stem cells. Stem Cell Research & Therapy. 2020;**11**(1):156

[23] Zhao X, Liu D, Gong W, Zhao G, Liu L, Yang L, et al. The toll-like receptor 3 ligand, poly(I:C), improves immunosuppressive function and therapeutic effect of mesenchymal stem cells on sepsis via inhibiting MiR-143. Stem Cells. 2014;**32**(2):521-533

[24] Bernardo ME, Fibbe WE. Mesenchymal stromal cells: Sensors and switchers of inflammation. Cell Stem Cell. 2013;**13**(4):392-402

[25] Waterman RS, Tomchuck SL, Henkle SL, Betancourt AM. A new mesenchymal stem cell (MSC) paradigm: Polarization into a pro-inflammatory MSC1 or an immunosuppressive MSC2 phenotype. PLoS One. 2010;**5**(4):e10088

[26] Jiang W, Xu J. Immune modulation by mesenchymal stem cells. Cell Proliferation. 2019;**53**(1):e12712

[27] PD1-mediated mesenchymal stem cells immunomodulation: The two sides of the coin. Pennesi G. International Clinical Pathology Journal. 2018;**6**(4):64-165

[28] Davies LC, Heldring N, Kadri N, Le Blanc K. Mesenchymal stromal cell secretion of programmed death-1 ligands regulates T cell mediated immunosuppression. Stem Cells. 2017;**35**(3):766-776

[29] Vacaru AM, Dumitrescu M, Vacaru AM, Fenyo IM, Ionita R, Gafencu AV, et al. Enhanced suppression of immune cells in vitro by MSC overexpressing FasL. International Journal of Molecular Sciences. 2020;**22**(1):348

[30] Zheng GY, Xue Q, Jing CY, Yu GH, de Qing M, Shen Y, et al. Different roles of PD-L1 and FasL in

immunomodulation mediated by human placenta-derived mesenchymal stem cells. Human Immunology. 2013;**74**(3): 267-276

[31] Akiyama K, Chen C, Wang D, Xu X, Qu C, Yamaza T, et al. Mesenchymal-stem-cell-induced immunoregulation involves FAS-ligand-/FAS-mediated T cell apoptosis. Cell Stem Cell. 2012;**10**(5): 544-555

[32] Kerkelä E, Laitinen A, Räbinä J, Valkonen S, Takatalo M, Larjo A, et al. Adenosinergic immunosuppression by human mesenchymal stromal cells requires co-operation with T cells. Stem Cells. 2016;**34**(3):781-790

[33] Han Y, Yang J, Fang J, Zhou Y, Candi E, Wang J, et al. The secretion profile of mesenchymal stem cells and potential applications in treating human diseases. Signal Transduction and Targeted Therapy. 2022;**7**(1):1-19

[34] Hervás-Salcedo R, Fernández-García M, Hernando-Rodríguez M, Suárez-Cabrera C, Bueren JA, Yáñez RM. Improved efficacy of mesenchymal stromal cells stably expressing CXCR4 and IL-10 in a xenogeneic graft versus host disease mouse model. Frontiers in Immunology. 2023;**14**:1062086

[35] Bogacka J, Pawlik K, Ciapała K, Ciechanowska A, Mika J. CC chemokine receptor 4 (CCR4) as a possible new target for therapy. International Journal of Molecular Sciences. 2022;**23**(24): 15638

[36] Cuesta-Gomez N, Graham GJ, Campbell JDM. Chemokines and their receptors: Predictors of the therapeutic potential of mesenchymal stromal cells. Journal of Translational Medicine. 2021; **19**(1):156

[37] Tokunaga R, Zhang W, Naseem M, Puccini A, Berger MD, Soni S, et al. CXCL9, CXCL10, CXCL11/CXCR3 axis for immune activation – A target for novel cancer therapy. Cancer Treatment Reviews. 2018;**63**:40-47

[38] Dunavin N, Dias A, Li M, McGuirk J. Mesenchymal stromal cells: What is the mechanism in acute graft-versus-host disease? Biomedicine [Internet]. 2017;**5** (3). Available from: https://www.ncbi.nlm.nih.gov/pmc/articles/PMC5618297/

[39] Ren G, Su J, Zhang L, Zhao X, Ling W, L'huillie A, et al. Species variation in the mechanisms of mesenchymal stem cell-mediated immunosuppression. Stem Cells. 2009;**27**(8):1954-1962

[40] Loisel S, Dulong J, Ménard C, Renoud ML, Meziere N, Isabelle B, et al. Brief report: Proteasomal indoleamine 2,3-dioxygenase degradation reduces the immunosuppressive potential of clinical grade-mesenchymal stromal cells undergoing replicative senescence. Stem Cells. 2017;**35**(5):1431-1436

[41] Yudhawati R, Shimizu K. PGE2 produced by exogenous MSCs promotes immunoregulation in ARDS induced by highly pathogenic influenza a through activation of the Wnt-β-catenin signaling pathway. International Journal of Molecular Sciences. 2023;**24**(8):7299

[42] Ylöstalo JH, Bartosh TJ, Coble K, Prockop DJ. Human mesenchymal stem/stromal cells cultured as spheroids are self-activated to produce prostaglandin E2 that directs stimulated macrophages into an anti-inflammatory phenotype. Stem Cells. 2012;**30**(10):2283-2296

[43] Sharma S, Yang SC, Zhu L, Reckamp K, Gardner B, Baratelli F, et al. Tumor cyclooxygenase-2/prostaglandin E2-dependent promotion of FOXP3 expression and CD4+ CD25+ T regulatory cell activities in lung cancer. Cancer Research. 2005;**65**(12):5211-5220

[44] Sreeramkumar V, Fresno M, Cuesta N. Prostaglandin E2 and T cells: Friends or foes? Immunology & Cell Biology. 2012;**90**(6):579-586

[45] Bu X, Zhong J, Li W, Cai S, Gao Y, Ping B. Immunomodulating functions of human leukocyte antigen-G and its role in graft-versus-host disease after allogeneic hematopoietic stem cell transplantation. Annals of Hematology. 2021;**100**(6):1391-1400

[46] Lin A, Yan WH. Heterogeneity of HLA-G expression in cancers: Facing the challenges. Frontiers in Immunology [Internet]. 2018;**9**. DOI: 10.3389/fimmu.2018.02164/full

[47] Liu R, Yang J, Li Y, Xie J, Wang J. Heme oxygenase-1: The roles of both good and evil in neurodegenerative diseases. Journal of Neurochemistry. 2023;**167**(3):347-361

[48] Han X, Yang Q, Lin L, Xu C, Zheng C, Chen X, et al. Interleukin-17 enhances immunosuppression by mesenchymal stem cells. Cell Death and Differentiation. 2014;**21**(11):1758-1768

[49] Su J, Chen X, Huang Y, Li W, Li J, Cao K, et al. Phylogenetic distinction of iNOS and IDO function in mesenchymal stem cell-mediated immunosuppression in mammalian species. Cell Death and Differentiation. 2014;**21**(3):388-396

[50] Iwata M, Inoue T, Asai Y, Hori K, Fujiwara M, Matsuo S, et al. The protective role of localized nitric oxide production during inflammation may be mediated by the heme oxygenase-1/carbon monoxide pathway. Biochemistry and Biophysics Reports. 2020;**23**:100790

[51] Pae HO, Chung HT. Heme oxygenase-1: Its therapeutic roles in inflammatory diseases. Immune Network. 2009;**9**(1):12-19

[52] Wong TH, Chen HA, Gau RJ, Yen JH, Suen JL. Heme oxygenase-1-expressing dendritic cells promote Foxp3 + regulatory T cell differentiation and induce less severe airway inflammation in murine models. PLoS One. 2016;**11**(12):e0168919

[53] Sokol CL, Luster AD. The chemokine system in innate immunity. Cold Spring Harbor Perspectives in Biology. 2015;**7**(5):a016303

[54] English K, Barry FP, Field-Corbett CP, Mahon BP. IFN-gamma and TNF-alpha differentially regulate immunomodulation by murine mesenchymal stem cells. Immunology Letters. 2007;**110**(2):91-100

[55] Najar M, Raicevic G, Fayyad-Kazan H, Bron D, Toungouz M, Lagneaux L. Mesenchymal stromal cells and immunomodulation: A gathering of regulatory immune cells. Cytotherapy. 2016;**18**(2):160-171

[56] Mattar P, Bieback K. Comparing the immunomodulatory properties of bone marrow, adipose tissue, and birth-associated tissue mesenchymal stromal cells. Frontiers in Immunology [Internet]. 2015;**6**. DOI: 10.3389/fimmu.2015.00560/full

[57] Saraiva M, O'Garra A. The regulation of IL-10 production by immune cells. Nature Reviews. Immunology. 2010;**10**(3):170-181

[58] Martino MM, Maruyama K, Kuhn GA, Satoh T, Takeuchi O, Müller R, et al. Inhibition of IL-1R1/MyD88 signalling promotes mesenchymal stem cell-driven tissue regeneration. Nature Communications. 2016;**7**(1):11051

[59] Gieseke F, Böhringer J, Bussolari R, Dominici M, Handgretinger R, Müller I. Human multipotent mesenchymal

stromal cells use galectin-1 to inhibit immune effector cells. Blood. 2010;**116**(19):3770-3779

[60] Liu GY, Xu Y, Li Y, Wang LH, Liu YJ, Zhu D. Secreted galectin-3 as a possible biomarker for the immunomodulatory potential of human umbilical cord mesenchymal stromal cells. Cytotherapy. 2013;**15**(10):1208-1217

[61] Gieseke F, Kruchen A, Tzaribachev N, Bentzien F, Dominici M, Müller I. Proinflammatory stimuli induce galectin-9 in human mesenchymal stromal cells to suppress T cells proliferation. European Journal of Immunology. 2013;**43**(10):2741-2749

[62] Cedeno-Laurent F, Dimitroff CJ. Galectin-1 research in T cell immunity: Past, present and future. Clinical Immunology. 2012;**142**(2):107-116

[63] Thiemann S, Man JH, Chang MH, Lee B, Baum LG. Galectin-1 regulates tissue exit of specific dendritic cell populations. Journal of Biological Chemistry. 2015;**290**(37):22662-22677

[64] Abbasi B, Shamsasenjan K, Ahmadi M, Beheshti SA, Saleh M. Mesenchymal stem cells and natural killer cells interaction mechanisms and potential clinical applications. Stem Cell Research & Therapy. 2022;**13**(1):97

[65] Liu FT, Stowell SR. The role of galectins in immunity and infection. Nature Reviews. Immunology. 2023;**23**(8):479-494

[66] Shirakawa K, Endo J, Kataoka M, Katsumata Y, Yoshida N, Yamamoto T, et al. IL (interleukin)-10–STAT3–Galectin-3 Axis is essential for osteopontin-producing reparative macrophage polarization after myocardial infarction. Circulation. 2018;**138**(18):2021-2035

[67] Fernández GC, Ilarregui JM, Rubel CJ, Toscano MA, Gómez SA, Beigier Bompadre M, et al. Galectin-3 and soluble fibrinogen act in concert to modulate neutrophil activation and survival: Involvement of alternative MAPK pathways. Glycobiology. 2005;**15**(5):519-527

[68] Zhao Y, Yu D, Wang H, Jin W, Li X, Hu Y, et al. Galectin-9 mediates the therapeutic effect of mesenchymal stem cells on experimental endotoxemia. Frontiers in Cell and Development Biology. 2022;**10**:700702

[69] Ma ZJ, Yang JJ, Lu YB, Liu ZY, Wang XX. Mesenchymal stem cell-derived exosomes: Toward cell-free therapeutic strategies in regenerative medicine. World Journal of Stem Cells. 2020;**12**(8):814-840

[70] Lai P, Chen X, Guo L, Wang Y, Liu X, Liu Y, et al. A potent immunomodulatory role of exosomes derived from mesenchymal stromal cells in preventing cGVHD. Journal of Hematology & Oncology. 2018;**11**(1):135

[71] Aguiar Koga BA, Fernandes LA, Fratini P, Sogayar MC, Carreira ACO. Role of MSC-derived small extracellular vesicles in tissue repair and regeneration. Frontiers in Cell and Development Biology. 2023;**10**:1047094

[72] Giacomini C, Granéli C, Hicks R, Dazzi F. The critical role of apoptosis in mesenchymal stromal cell therapeutics and implications in homeostasis and normal tissue repair. Cellular & Molecular Immunology. 2023;**20**(6):570-582

[73] Krampera M. Mesenchymal stromal cell 'licensing': A multistep process. Leukemia. 2011;**25**(9):1408-1414

[74] Elliott MR, Koster KM, Murphy PS. Efferocytosis signaling in the regulation

of macrophage inflammatory responses. The Journal of Immunology. 2017;**198**(4):1387-1394

[75] Saas P, Daguindau E, Perruche S. Concise review: Apoptotic cell-based therapies–rationale, preclinical results and future clinical developments. Stem Cells. 2016;**34**(6):1464-1473

[76] Ghahremani Piraghaj M, Soudi S, Ghanbarian H, Bolandi Z, Namaki S, Hashemi SM. Effect of efferocytosis of apoptotic mesenchymal stem cells (MSCs) on C57BL/6 peritoneal macrophages function. Life Sciences. 2018;**212**:203-212

[77] Cheung TS, Galleu A, von Bonin M, Bornhäuser M, Dazzi F. Apoptotic mesenchymal stromal cells induce prostaglandin E2 in monocytes: Implications for the monitoring of mesenchymal stromal cell activity. Haematologica. 2019;**104**(10):e438-e441

[78] Schrodt MV, Behan-Bush RM, Liszewski JN, Humpal-Pash ME, Boland LK, Scroggins SM, et al. Efferocytosis of viable versus heat-inactivated MSC induces human monocytes to distinct immunosuppressive phenotypes. Stem Cell Research & Therapy. 2023;**14**:206

[79] Lin T, Yang Y, Chen X. A review of the application of mesenchymal stem cells in the field of hematopoietic stem cell transplantation. European Journal of Medical Research. 2023;**28**(1):268

[80] Noort WA, Kruisselbrink AB, In't Anker PS, Kruger M, van Bezooijen RL, de Paus RA, et al. Mesenchymal stem cells promote engraftment of human umbilical cord blood–derived CD34+ cells in NOD/SCID mice. Experimental Hematology. 2002;**30**(8):870-878

[81] Wang Y, Chen X, Cao W, Shi Y. Plasticity of mesenchymal stem cells in immunomodulation: Pathological and therapeutic implications. Nature Immunology. 2014;**15**(11):1009-1016

[82] Sokolov M, Neumann R. Lessons learned about human stem cell responses to ionizing radiation exposures: A long road still ahead of us. International Journal of Molecular Sciences. 2013;**14**(8):15695-15723

[83] Frenette PS, Pinho S, Lucas D, Scheiermann C. Mesenchymal stem cell: Keystone of the hematopoietic stem cell niche and a stepping-stone for regenerative medicine. Annual Review of Immunology. 2013;**31**:285-316

[84] Lu LL, Liu YJ, Yang SG, Zhao QJ, Wang X, Gong W, et al. Isolation and characterization of human umbilical cord mesenchymal stem cells with hematopoiesis-supportive function and other potentials. Haematologica. 2006;**91**(8):1017-1026

[85] Fouillard L, Francois S, Bouchet S, Bensidhoum M, Elm'selmi A, Chapel A. Innovative cell therapy in the treatment of serious adverse events related to both chemo-radiotherapy protocol and acute myeloid Leukemia syndrome: The infusion of mesenchymal stem cells post-treatment reduces hematopoietic toxicity and promotes hematopoietic reconstitution. Current Pharmaceutical Biotechnology. 2013;**14**(9):842-848

[86] Ogawa Y, Akamatsu R, Fuchizaki A, Yasui K, Saino O, Tanaka M, et al. Gap junction–mediated transport of metabolites between stem cells and vascular endothelial cells. Cell Transplantation. 2022;**31**:09636897221136151

[87] Koç ON, Gerson SL, Cooper BW, Dyhouse SM, Haynesworth SE, Caplan AI, et al. Rapid hematopoietic recovery after coinfusion of autologous-blood

stem cells and culture-expanded marrow mesenchymal stem cells in advanced breast cancer patients receiving high-dose chemotherapy. JCO. 2000;**18**(2): 307-307

[88] Wu KH, Sheu JN, Wu HP, Tsai C, Sieber M, Peng CT, et al. Cotransplantation of umbilical cord–derived mesenchymal stem cells promote hematopoietic engraftment in cord blood transplantation: A pilot study. Transplantation. 2013;**95**(5):773

[89] Garrigós MM, de Oliveira FA, Nucci MP, Nucci LP, da Alves AH, OFM D, et al. How mesenchymal stem cell cotransplantation with hematopoietic stem cells can improve engraftment in animal models. World Journal of Stem Cells. 2022;**14**(8):658-679

[90] Zhao J, Chen J, Huang F, Wang J, Su W, Zhou J, et al. Human gingiva tissue-derived MSC ameliorates immune-mediated bone marrow failure of aplastic anemia via suppression of Th1 and Th17 cells and enhancement of CD4+Foxp3+ regulatory T cells differentiation. American Journal of Translational Research. 2019;**11**(12):7627-7643

[91] Peng Y, Chen X, Liu Q, Zhang X, Huang K, Liu L, et al. Mesenchymal stromal cells infusions improve refractory chronic graft versus host disease through an increase of CD5+ regulatory B cells producing interleukin 10. Leukemia. 2015;**29**(3): 636-646

[92] Macmillan ML, Blazar BR, DeFor TE, Wagner JE. Transplantation of ex-vivo culture-expanded parental haploidentical mesenchymal stem cells to promote engraftment in pediatric recipients of unrelated donor umbilical cord blood: Results of a phase I-II clinical trial. Bone Marrow Transplantation. 2009;**43**(6):447-454

[93] Huang R, Chen T, Wang S, Wang J, Su Y, Liu J, et al. Mesenchymal stem cells for prophylaxis of chronic graft-vs-host disease after Haploidentical hematopoietic stem cell transplant: An open-label randomized clinical trial. JAMA Oncol. 2024;**10**(2):220-226

[94] Yang D, Hou X, Qian K, Li Y, Hu L, Li L, et al. Efficacy and safety of human umbilical cord-derived mesenchymal stem cells (hUC-MSC PLEB001) for the treatment of grade II–IV steroid-refractory acute graft-versus-host disease: A study protocol for a multicenter, randomized, double-blind, placebo-controlled, phase II trial. Trials. 2023;**24**(1):306

[95] Zhao K, Lin R, Fan Z, Chen X, Wang Y, Huang F, et al. Mesenchymal stromal cells plus basiliximab, calcineurin inhibitor as treatment of steroid-resistant acute graft-versus-host disease: A multicenter, randomized, phase 3, open-label trial. Journal of Hematology & Oncology. 2022;**15**(1):22

[96] Kebriaei P, Hayes J, Daly A, Uberti J, Marks DI, Soiffer R, et al. A phase 3 randomized study of remestemcel-L versus placebo added to second-line therapy in patients with steroid-refractory acute graft-versus-host disease. Biology of Blood and Marrow Transplantation. 2020;**26**(5):835-844

[97] Kurtzberg J, Abdel-Azim H, Carpenter P, Chaudhury S, Horn B, Mahadeo K, et al. A phase 3, single-arm, prospective study of Remestemcel-L, ex vivo culture-expanded adult human mesenchymal stromal cells for the treatment of Pediatric patients who failed to respond to steroid treatment for acute graft-versus-host disease. Biology of Blood and Marrow Transplantation. 2020;**26**(5):845-854

[98] Hinden L, Avner M, Stepensky P, Or R, Almogi-Hazan O. Lymphocyte counts

may predict a good response to mesenchymal stromal cells therapy in graft versus host disease patients. PLoS One. 2019;**14**(6):e0217572

[99] Servais S, Baron F, Lechanteur C, Seidel L, Selleslag D, Maertens J, et al. Infusion of bone marrow derived multipotent mesenchymal stromal cells for the treatment of steroid-refractory acute graft-versus-host disease: A multicenter prospective study. Oncotarget. 2018;**9**(29):20590-20604

[100] Bader P, Kuçi Z, Bakhtiar S, Basu O, Bug G, Dennis M, et al. Effective treatment of steroid and therapy-refractory acute graft-versus-host disease with a novel mesenchymal stromal cell product (MSC-FFM). Bone Marrow Transplantation. 2018;**53**(7): 852-862

[101] Ringden O, Baygan A, Remberger M, Gustafsson B, Winiarski J, Khoein B, et al. Placenta-derived decidua stromal cells for treatment of severe acute graft-versus-host disease. Stem Cells Translational Medicine. 2018;**7**(4): 325-331

[102] Dotoli GM, De Santis GC, Orellana MD, de Lima PK, Caruso SR, Fernandes TR, et al. Mesenchymal stromal cell infusion to treat steroid-refractory acute GvHD III/IV after hematopoietic stem cell transplantation. Bone Marrow Transplantation. 2017;**52**(6):859-862

[103] Galleu A, Milojkovic D, Deplano S, Szydlo R, Loaiza S, Wynn R, et al. Mesenchymal stromal cells for acute graft-versus-host disease: Response at 1 week predicts probability of survival. British Journal of Haematology. 2019;**185** (1):89-92

[104] Salmenniemi U, Itälä-Remes M, Nystedt J, Putkonen M, Niittyvuopio R, Vettenranta K, et al. Good responses but high TRM in adult patients after MSC therapy for GvHD. Bone Marrow Transplantation. 2017;**52**(4):606-608

[105] Muroi K, Miyamura K, Okada M, Yamashita T, Murata M, Ishikawa T, et al. Bone marrow-derived mesenchymal stem cells (JR-031) for steroid-refractory grade III or IV acute graft-versus-host disease: A phase II/III study. International Journal of Hematology. 2016;**103**(2):243-250

[106] von Dalowski F, Kramer M, Wermke M, Wehner R, Röllig C, Alakel N, et al. Mesenchymal stromal cells for treatment of acute steroid-refractory graft versus host disease: Clinical responses and long-term outcome. Stem Cells. 2016;**34**(2):357-366

[107] Zhao K, Lou R, Huang F, Peng Y, Jiang Z, Huang K, et al. Immunomodulation effects of mesenchymal stromal cells on acute graft-versus-host disease after hematopoietic stem cell transplantation. Biology of Blood and Marrow Transplantation. 2015;**21**(1):97-104

[108] te Boome LCJ, Mansilla C, van der Wagen LE, Lindemans CA, Petersen EJ, Spierings E, et al. Biomarker profiling of steroid-resistant acute GVHD in patients after infusion of mesenchymal stromal cells. Leukemia. 2015;**29**(9):1839-1846

[109] Wu Y, Cao Y, Li X, Xu L, Wang Z, Liu P, et al. Cotransplantation of haploidentical hematopoietic and umbilical cord mesenchymal stem cells for severe aplastic anemia: Successful engraftment and mild GVHD. Stem Cell Research. 2014;**12**(1):132-138

[110] Wu Y, Wang Z, Cao Y, Xu L, Li X, Liu P, et al. Cotransplantation of haploidentical hematopoietic and umbilical cord mesenchymal stem cells with a myeloablative regimen for

refractory/relapsed hematologic malignancy. Annals of Hematology. 2013;**92**(12):1675-1684

[111] Introna M, Lucchini G, Dander E, Galimberti S, Rovelli A, Balduzzi A, et al. Treatment of graft versus host disease with mesenchymal stromal cells: A phase I study on 40 adult and Pediatric patients. Biology of Blood and Marrow Transplantation. 2014;**20**(3):375-381

[112] Resnick IB, Barkats C, Shapira MY, Stepensky P, Bloom AI, Shimoni A, et al. Treatment of severe steroid resistant acute GVHD with mesenchymal stromal cells (MSC). American Journal of Blood Research. 2013;**3**(3):225-238

[113] Bernardo ME, Ball LM, Cometa AM, Roelofs H, Zecca M, Avanzini MA, et al. Co-infusion of ex vivo-expanded, parental MSCs prevents life-threatening acute GVHD, but does not reduce the risk of graft failure in pediatric patients undergoing allogeneic umbilical cord blood transplantation. Bone Marrow Transplantation. 2011;**46**(2):200-207

[114] Gonzalo-Daganzo R, Regidor C, Martín-Donaire T, Rico MA, Bautista G, Krsnik I, et al. Results of a pilot study on the use of third-party donor mesenchymal stromal cells in cord blood transplantation in adults. Cytotherapy. 2009;**11**(3):278-288

[115] Blanc KL, Frassoni F, Ball L, Locatelli F, Roelofs H, Lewis I, et al. Mesenchymal stem cells for treatment of steroid-resistant, severe, acute graft-versus-host disease: A phase II study. The Lancet. 2008;**371**(9624):1579-1586

[116] Lazarus HM, Koc ON, Devine SM, Curtin P, Maziarz RT, Holland HK, et al. Cotransplantation of HLA-identical sibling culture-expanded mesenchymal stem cells and hematopoietic stem cells in hematologic malignancy patients. Biology of Blood and Marrow Transplantation. 2005;**11**(5):389-398

[117] Lukomska B, Stanaszek L, Zuba-Surma E, Legosz P, Sarzynska S, Drela K. Challenges and controversies in human mesenchymal stem cell therapy. Stem Cells International. 2019;**2019**:e9628536. Available from: https://www.hindawi.com/journals/sci/2019/9628536/

[118] Dlouhy BJ, Awe O, Rao RC, Kirby PA, Hitchon PW. Autograft-derived spinal cord mass following olfactory mucosal cell transplantation in a spinal cord injury patient. Journal of Neurosurgery Spine. 2014;**21**(4):618-622

[119] Martin PJ, Rizzo JD, Wingard JR, Ballen K, Curtin PT, Cutler C, et al. First- and second-line systemic treatment of acute graft-versus-host disease: Recommendations of the American Society of Blood and Marrow Transplantation. Biology of Blood and Marrow Transplantation. 2012;**18**(8): 1150-1163

[120] Le Blanc K, Mougiakakos D. Multipotent mesenchymal stromal cells and the innate immune system. Nature Reviews. Immunology. 2012;**12**(5): 383-396

[121] von Bahr L, Sundberg B, Lönnies L, Sander B, Karbach H, Hägglund H, et al. Long-term complications, immunologic effects, and role of passage for outcome in mesenchymal stromal cell therapy. Biology of Blood and Marrow Transplantation. 2012;**18**(4):557-564

[122] Forslöw U, Blennow O, LeBlanc K, Ringdén O, Gustafsson B, Mattsson J, et al. Treatment with mesenchymal stromal cells is a risk factor for pneumonia-related death after allogeneic hematopoietic stem cell transplantation. European Journal of Haematology. 2012; **89**(3):220-227

[123] Karlsson H, Samarasinghe S, Ball LM, Sundberg B, Lankester AC, Dazzi F, et al. Mesenchymal stem cells exert differential effects on alloantigen and virus-specific T cells responses. Blood. 2008;**112**(3):532-541

[124] Thanunchai M, Hongeng S, Thitithanyanont A. Mesenchymal stromal cells and viral infection. Stem Cells International. 2015;**2015**:860950

[125] Badillo AT, Peranteau WH, Heaton TE, Quinn C, Flake AW. Murine bone marrow derived stromal progenitor cells fail to prevent or treat acute graft-versus-host disease. British Journal of Haematology. 2008;**141**(2):224-234

[126] Zhou T, Yuan Z, Weng J, Pei D, Du X, He C, et al. Challenges and advances in clinical applications of mesenchymal stromal cells. Journal of Hematology & Oncology. 2021;**14**(1):24

[127] Dufrane D. Impact of age on human adipose stem cells for bone tissue engineering. Cell Transplantation. 2017; **26**(9):1496-1504

[128] Kokai LE, Traktuev DO, Zhang L, Merfeld-Clauss S, DiBernardo G, Lu H, et al. Adipose stem cell function maintained with age: An intra-subject study of long-term cryopreserved cells. Aesthetic Surgery Journal. 2017;**37**(4):454-463

[129] Pachón-Peña G, Serena C, Ejarque M, Petriz J, Duran X, Oliva-Olivera W, et al. Obesity determines the Immunophenotypic profile and functional characteristics of human mesenchymal stem cells from adipose tissue. Stem Cells Translational Medicine. 2016;**5**(4):464-475

[130] Baptista LS, Silva KR, Borojevic R. Obesity and weight loss could alter the properties of adipose stem cells? World Journal of Stem Cells. 2015;**7**(1):165-173

[131] Ribeiro A, Laranjeira P, Mendes S, Velada I, Leite C, Andrade P, et al. Mesenchymal stem cells from umbilical cord matrix, adipose tissue and bone marrow exhibit different capability to suppress peripheral blood B, natural killer and T cells. Stem Cell Research & Therapy. 2013;**4**(5):125

[132] Andrzejewska A, Dabrowska S, Nowak B, Walczak P, Lukomska B, Janowski M. Mesenchymal stem cells injected into carotid artery to target focal brain injury home to perivascular space. Theranostics. 2020;**10**(15):6615-6628

[133] Chan JL, Tang KC, Patel AP, Bonilla LM, Pierobon N, Ponzio NM, et al. Antigen-presenting property of mesenchymal stem cells occurs during a narrow window at low levels of interferon-γ. Blood. 2006;**107**(12):4817-4824

[134] Ng CP, Mohamed Sharif AR, Heath DE, Chow JW, Zhang CBY, Chan-Park MB, et al. Enhanced *ex vivo* expansion of adult mesenchymal stem cells by fetal mesenchymal stem cell ECM. Biomaterials. 2014;**35**(13):4046-4057

[135] Cao H, Xiao J, Reeves ME, Payne K, Chen CS, Baylink DJ, et al. Discovery of proangiogenic CD44+mesenchymal cancer stem cells in an acute myeloid leukemia patient's bone marrow. Journal of Hematology & Oncology. 2020;**13**(1):63

[136] Dick AJ, Guttman MA, Raman VK, Peters DC, Pessanha BSS, Hill JM, et al. Magnetic resonance fluoroscopy allows targeted delivery of mesenchymal stem cells to infarct Borders in swine. Circulation. 2003;**108**(23):2899-2904

[137] Flamant S, Loinard C, Tamarat R. MSC beneficial effects and limitations, and MSC-derived extracellular vesicles as a new cell-free therapy for tissue

regeneration in irradiated condition. Environmental Advances. 2023;**13**: 100408

[138] Ocansey DKW, Pei B, Yan Y, Qian H, Zhang X, Xu W, et al. Improved therapeutics of modified mesenchymal stem cells: An update. Journal of Translational Medicine. 2020;**18**(1):42

[139] Garcia KD, Ornellas FL, Matsumoto P, Patti CD, Mello LE, Frussa-Filho R, et al. Therapeutic effects of the transplantation of VEGF overexpressing bone marrow mesenchymal stem cells in the hippocampus of murine model of Alzheimer's disease. Frontiers in Aging Neuroscience [Internet]. 2014;**6**. Available from: https://www.frontiersin.org/journals/aging-neuroscience/articles/10.3389/fnagi.2014.00030/full

[140] Ni X, Ou C, Guo J, Liu B, Zhang J, Wu Z, et al. Lentiviral vector-mediated co-overexpression of VEGF and Bcl-2 improves mesenchymal stem cell survival and enhances paracrine effects in vitro. International Journal of Molecular Medicine. 2017;**40**(2):418-426

[141] Hnatiuk AP, Ong S, Olea FD, Locatelli P, Riegler J, Lee WH, et al. Allogeneic mesenchymal stromal cells overexpressing mutant human hypoxia-inducible factor 1-α (HIF1-α) in an ovine model of acute myocardial infarction. Journal of the American Heart Association; 2016;**5**(7):e003714

[142] Lian WS, Cheng WTK, Cheng CC, Hsiao FSH, Chen JJ, Cheng CF, et al. *In vivo* therapy of myocardial infarction with mesenchymal stem cells modified with prostaglandin I synthase gene improves cardiac performance in mice. Life Sciences. 2011;**88**(9):455-464

[143] Lee BC, Kang KS. Functional enhancement strategies for immunomodulation of mesenchymal stem cells and their therapeutic application. Stem Cell Research & Therapy. 2020;**11**(1):397

[144] Hu Y, Liu S, Zhu BM. CRISPR/Cas9-induced loss of Keap1 enhances anti-oxidation in rat adipose-derived mesenchymal stem cells. Frontiers in Neurology [Internet]. 2020;**10**. DOI: 10.3389/fneur.2019.01311/full

[145] Song SW, Chang W, Song BW, Song H, Lim S, Kim HJ, et al. Integrin-linked kinase is required in hypoxic mesenchymal stem cells for strengthening cell adhesion to ischemic myocardium. Stem Cells. 2009;**27**(6):1358-1365

[146] Han SM, Han SH, Coh YR, Jang G, Chan Ra J, Kang SK, et al. Enhanced proliferation and differentiation of Oct4- and Sox2-overexpressing human adipose tissue mesenchymal stem cells. Experimental & Molecular Medicine. 2014;**46**(6):e101-e101

[147] Lee HJ, Choi JH, Jung J, Kim JK, Lee SS, Kim GJ. Changes in PTTG1 by human TERT gene expression modulate the self-renewal of placenta-derived mesenchymal stem cells. Cell and Tissue Research. 2014;**357**(1):145-157

[148] Lee JH, Yoon YM, Lee SH. Hypoxic preconditioning promotes the bioactivities of mesenchymal stem cells via the HIF-1α-GRP78-Akt Axis. International Journal of Molecular Sciences. 2017;**18**(6):1320

[149] Li L, Wu S, Li P, Zhuo L, Gao Y, Xu Y. Hypoxic preconditioning combined with microbubble-mediated ultrasound effect on MSCs promote SDF-1/CXCR4 expression and its migration ability: An In vitro study. Cell Biochemistry and Biophysics. 2015;**73**(3):749-757

[150] Meng SS, Xu XP, Chang W, Lu ZH, Huang LL, Xu JY, et al. LincRNA-p21

promotes mesenchymal stem cell migration capacity and survival through hypoxic preconditioning. Stem Cell Research & Therapy. 2018;**9**(1):280

[151] Najafi R, Sharifi AM. Deferoxamine preconditioning potentiates mesenchymal stem cell homing in vitro and in streptozotocin-diabetic rats. Expert Opinion on Biological Therapy. 2013;**13**(7):959-972

[152] Lan YW, Theng SM, Huang TT, Choo KB, Chen CM, Kuo HP, et al. Oncostatin M-preconditioned mesenchymal stem cells alleviate bleomycin-induced pulmonary fibrosis through paracrine effects of the hepatocyte growth factor. Stem Cells Translational Medicine. 2017;**6**(3):1006-1017

[153] Liu H, Xue W, Ge G, Luo X, Li Y, Xiang H, et al. Hypoxic preconditioning advances CXCR4 and CXCR7 expression by activating HIF-1α in MSCs. Biochemical and Biophysical Research Communications. 2010;**401**(4):509-515

[154] Copland IB, Lord-Dufour S, Cuerquis J, Coutu DL, Annabi B, Wang E, et al. Improved autograft survival of mesenchymal stromal cells by plasminogen activator inhibitor 1 inhibition. Stem Cells. 2009;**27**(2):467-477

[155] Lu N, Li X, Tan R, An J, Cai Z, Hu X, et al. HIF-1α/Beclin1-mediated autophagy is involved in neuroprotection induced by hypoxic preconditioning. Journal of Molecular Neuroscience. 2018;**66**(2):238-250

[156] Kim DS, Ko YJ, Lee MW, Park HJ, Park YJ, Kim DI, et al. Effect of low oxygen tension on the biological characteristics of human bone marrow mesenchymal stem cells. Cell Stress and Chaperones. 2016;**21**(6):1089-1099

[157] Sanap A, Chandravanshi B, Shah T, Tillu G, Dhanushkodi A, Bhonde R, et al. Herbal pre-conditioning induces proliferation and delays senescence in Wharton's jelly mesenchymal stem cells. Biomedicine & Pharmacotherapy. 2017; **93**:772-778

[158] Lee BC, Shin N, Lee JY, Kang I, Kim JJ, Lee SE, et al. MIS416 enhances therapeutic functions of human umbilical cord blood-derived mesenchymal stem cells against experimental colitis by modulating systemic immune milieu. Frontiers in Immunology [Internet]. 2018;**9**:1078

[159] Shin JH, Ryu CM, Ju H, Yu HY, Song S, Shin DM, et al. Synergistic effects of N-Acetylcysteine and mesenchymal stem cell in a lipopolysaccharide-induced interstitial cystitis rat model. Cells. 2020;**9**(1):86

[160] Bhonde RR, Sheshadri P, Sharma S, Kumar A. Making surrogate β-cells from mesenchymal stromal cells: Perspectives and future endeavors. The International Journal of Biochemistry & Cell Biology. 2014;**46**:90-102

[161] Samuel P, Sundarraj S, Sudarmani DNP, Samuel P, Sundarraj S, Sudarmani DNP. Nanotechnology-based stem cell therapy: Current status and perspectives. In: Possibilities and Limitations in Current Translational Stem Cell Research [Internet]. London: IntechOpen. DOI: 10.5772/intechopen.109275

Chapter 16

Schwann-Like Cells Derived from Mesenchymal Stem Cells: Their Potential for Peripheral Nerve Regeneration

Rogério Martins Amorim
and Lucas Vinícius de Oliveira Ferreira

Abstract

Peripheral nervous system (PNS) injuries pose a significant clinical challenge, often resulting in motor, sensory, or autonomic dysfunction that impacts patients' quality of life. Despite the PNS's capacity for regeneration, outcomes are not always satisfactory. In response to these challenges, new research is encouraged to provide more effective therapeutic approaches. In this context, cellular therapy emerges as a promising alternative. Evidence of the therapeutic potential of Schwann cells (SCs) in PNS injuries has been observed, yet their clinical application faces significant limitations. To address these difficulties, several studies have highlighted the ability of mesenchymal stem cells (MSCs) to transdifferentiate into Schwann-like cells (SLCs), holding the potential for treating peripheral nerve injuries. Therefore, this chapter not only reviews the involvement of SCs in peripheral nerve regeneration but also provides an overview of recent advancements in developing SLCs derived from MSCs and their therapeutic potential in peripheral nerve injuries. Additionally, it explores the future perspective of manufacturing nerve guidance conduits (NGCs).

Keywords: nerve injury, neuroregeneration, cell-based therapy, nerve guidance conduits, regenerative medicine

1. Introduction

Injuries to the peripheral nervous system (PNS) are prevalent among both humans and animals, presenting a significant clinical challenge [1, 2]. These injuries stem from various causes, frequently associated with traumatic incidents [3, 4]. The resulting motor and sensory impairments are the foremost consequences, often leading to functional deficits and detrimentally impacting the patient's quality of life [5, 6].

While the PNS demonstrates an innate ability for regeneration, primarily attributed to Schwann cells (SCs), the results frequently fall short of expectations [7, 8]. Given the complexities associated with peripheral nerve injuries, new research is encouraged to develop more efficacious therapeutic strategies.

In this context, cellular therapy emerges as a promising alternative. Evidence of Schwann cells' therapeutic potential in treating PNS injuries has been observed [9, 10]. Nevertheless, its clinical application faces substantial limitations, including the necessity to sacrifice a functional nerve to obtain the cells and the challenge of cell expansion [11, 12].

To address these challenges, several studies have highlighted the potential of mesenchymal stem cells (MSCs) sourced from diverse sources and species to transdifferentiate into Schwann-like cells (SLCs), offering promising avenues for treating peripheral nerve injuries [13–15].

Therefore, this chapter reviews the role of SCs in peripheral nerve regeneration and offers insights into recent progress in creating SLCs from MSCs and their therapeutic potential in peripheral nerve injuries. Furthermore, it explores the future perspectives of fabricating nerve guidance conduits (NGCs).

2. Anatomy of the peripheral nerve

The PNS is composed of neurons, glial cells (including SCs), and stromal cells [16, 17]. This system consists of a combination of motor, sensory, and autonomic nerves responsible for connecting tissues and organs to the central nervous system (CNS) [18, 19]. Each peripheral nerve is formed by various bundles of fibers known as nerve fascicles or nerve bundles [20, 21]. The nerve fibers consist of axons surrounded by SCs, with differences in their diameter [22]. Axons with larger diameters are considered myelinated, as they are covered by the myelin sheath [23]. This is discontinuous, and the small areas between two segments of myelin, the unmyelinated space, are called Ranvier's nodes, whose function is to allow the saltatory conduction of nerve impulses from one node to another, speeding up their transmission [22, 24].

Histologically, three layers of connective tissue support the peripheral nerves (**Figure 1**). The epineurium is the outermost layer comprising numerous nerve fascicles, blood vessels, lymphocytes, and fibroblasts, and is formed by loose connective tissue, whose function is to provide support to nerve bundles, mechanical protection, and isolation from the external environment [20, 25, 26]. The perineurium is the intermediate region that surrounds each nerve fascicle, providing tensile strength to the nerves and acting as a barrier regulating the entry and exit of substances from the nerve [25, 27]. The endoneurium is the innermost layer composed of loose collagen

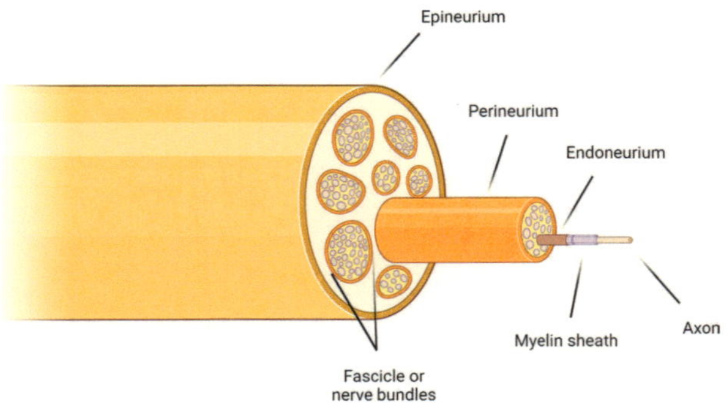

Figure 1.
Schematic representation of peripheral nerve anatomy. Created with Biorender.com

that surrounds both myelinated and unmyelinated nerve fibers, responsible for providing protection and nourishment to the individual axons [26, 28].

3. Peripheral nervous system injuries

Injuries to the PNS hold considerable clinical significance for medicine. These injuries are prevalent and can result in chronic pain, as well as loss of motor, sensory, or autonomic function in affected body segments, thereby diminishing the quality of life for both animals and humans [24, 29–31].

This condition affects a global population of at least 2 million individuals [2]. Epidemiological studies in developed countries indicate that peripheral nerve injuries occur at an estimated rate of 13–23 cases per 100,000 individuals annually [32]. The outcome of peripheral nerve repair after an injury is often unsatisfactory. It is estimated that only 3% of patients regain sensitivity, while motor function is regained by less than 25% of patients [33].

In veterinary medicine, studies demonstrate that occurrences of brachial plexus or sciatic nerve injuries are frequent in dogs and cats [34, 35]. Moreover, a study revealed that direct trauma is the most common cause of peripheral neuropathies in the thoracic limbs of horses [36]. Additionally, facial nerve paralysis is one of the most common neuropathies in horses, often attributed to traumatic events [37].

4. Classification of peripheral nerve injuries

The researcher Herbert Seddon first introduced the classification of peripheral nerve injuries in 1943, suggesting three distinct grades of injury [38]. The mildest grade, neuropraxia, involves focal demyelination without damage to the axons or surrounding connective tissues. In second-degree injuries, termed axonotmesis, there is a loss of axon continuity beyond focal demyelination. Finally, neurotmesis, the most severe grade, involves complete nerve sectioning.

In 1951, Sydney Sunderland expanded the classification to five grades (I–V), aiming to differentiate the extent of injuries to the connective tissues [39]. Grades I and V correspond to neuropraxia and neurotmesis, respectively, as classified by Seddon. Grades II and IV represent subdivisions of axonotmesis, differing in the involvement of connective tissue. In grade II, axon injury occurs without connective tissue involvement; in grade III, both axon and endoneurium are injured, while in grade IV, damage extends to axons, endoneurium, and perineurium. A new degree of injury was proposed by Mackinnon and Dellon in 1988, although its usage is not widely adopted, it is characterized by the occurrence of mixed lesions (**Table 1**) [40].

Seddon	Sunderland	Injury
Neuropraxia	I	Focal demyelination
Axonotmesis	II	Axon damaged without the involvement of the connective tissue
	III	Axon and endoneurium damaged
	IV	Axon, endoneurium, and perineurium damaged
Neurotmesis	V	Complete section of the nerve

Table 1.
Summary of the classification of peripheral nerve injuries. Adapted from Menorca et al. [18].

5. Peripheral nerve degeneration

After PNS injury, Wallerian degeneration ensues, marked by distal degeneration of the injured nerve, clearance of inhibitory cellular debris, and proliferation of SCs, establishing a pro-regenerative microenvironment [41]. The axonal microtubules start to disorganize, initiating the dissolution of the axonal cytoskeleton [22]. During the initial days following injury, an influx of calcium ions triggers the activation of calpain proteases, which proceed to degrade axonal neurofilaments, resulting in the release of granular debris [42–44].

Additionally, SCs convert into a repair phenotype, characterized by a specific profile that facilitates the regeneration process [45]. These cells assist in the degradation of cellular debris and myelin residues, an essential process, considering that myelin contains molecules that inhibit axonal growth [24, 46].

The expression of cytokines such as tumor necrosis factor (TNF)-α, interleukin (IL)-1α, -1β, leukemia inhibitory factor (LIF), and monocyte chemoattractant protein (MCP)-1 is positively regulated [47]. This process leads to SC proliferation and the recruitment of macrophages and other immune cells to the injury site [48]. The macrophages, attracted to the site of injury along with the SCs, are essential for neuroregeneration, as they control the inflammatory process and degrade myelin debris, thus promoting a favorable microenvironment for regeneration [49]. After phagocytizing myelin debris and other waste, macrophages are either eliminated through local apoptosis or re-enter circulation [22].

In the proximal segment of the injured nerve, to a lesser extent, the process of retrograde degeneration occurs until the next preserved Ranvier node [18, 22, 50]. The alterations in this segment differ depending on the severity and location of the injury. Apoptosis typically occurs in lesions near the neuronal cell body, while reversible chromatolysis may occur in certain cases [18, 51].

6. Involvement of Schwann cells in peripheral nerve regeneration

After injury, the PNS exhibits a remarkable capacity for regeneration [52]. Concurrently with the degeneration process, a series of cellular and molecular events are initiated to facilitate nerve repair. Essential to this process is the integrity of both the neuronal body and SCs [53]. The SCs promote regeneration through proliferation, secretion of neurotrophic factors, cytokines, and phagocytosis of myelin debris [46, 54]. These cells proliferate, undergo elongation, and align within their basal lamina to form Büngner bands, which serve as guidance and support for axonal growth [23, 44].

Furthermore, SCs play a pivotal role in synthesizing neurotrophic factors crucial for axonal guidance and growth. These factors include brain-derived neurotrophic factor (BDNF), ciliary neurotrophic factor (CNTF), fibroblast growth factor 2 (FGF-2), glial-derived neurotrophic factor (GDNF), nerve growth factor (NGF), and neurotrophins -3, -4, and -5 (NT-3/4/5), which selectively interact with tropomyosin receptor kinases (TrkA, -B, and -C) and p75 neurotrophin receptors [50, 55].

SCs play a pivotal role in creating a conducive environment post-injury, employing autophagy/myelinophagy mechanisms and orchestrating macrophage recruitment [23]. Macrophages, in turn, release molecules associated with neovascularization, such as vascular endothelial growth factor A (VEGF-A), which aids in oxygen supply to the injury site and guides repair SCs during the regenerative phase [49]. Additionally,

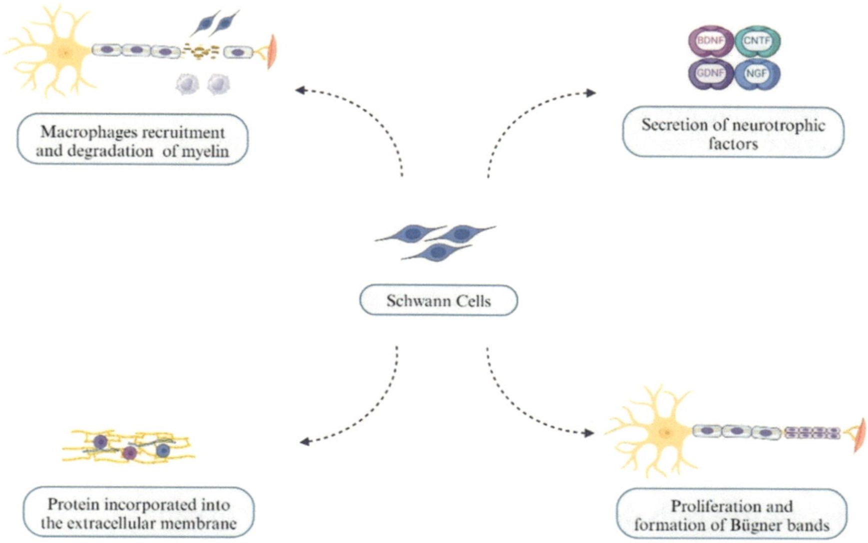

Figure 2.
Schematic representation of participation of SCs in peripheral nerve regeneration after injury. The SCs promote regeneration through proliferation, recruitment of macrophages, phagocytosis of myelin debris, secretion of neurotrophic factors, incorporation of proteins into the extracellular membrane, and the formation of Büngner bands. Created with Biorender.com

SCs produce proteins such as laminin and fibronectin that are incorporated into the extracellular matrix and promote the extension of regenerating axons [56].

Although the PNS exhibits intrinsic repair capabilities, several factors may hinder this process, such as the disorganized growth of regenerating axons, which may fail to reach their target organs, difficulty in obtaining a large number of axons from the proximal to the distal stump, loss of nerve-muscle connections leading to muscle fiber atrophy. Furthermore, when there is chronic loss of contact between Schwann cells and axons, the basal laminae and Büngner bands are not maintained, creating an environment that does not support regeneration (**Figure 2**) [23, 48, 53].

7. Mesenchymal stem cells

MSCs are multipotent progenitor cells of non-hematopoietic origin isolated from various adult tissues. They are characterized by their ability to proliferate and differentiate *in vitro* into various mesenchymal lineages in response to appropriate stimuli, including adipocytes, chondrocytes, and osteocytes [57, 58].

Although it is a widely debated topic, studies suggest that MSCs, under specific conditions, possess the ability to differentiate into cells derived from all three germ layers, including neurons, glial cells, and hepatocytes [15, 59–61].

The International Society for Cellular Therapy has defined minimum criteria for characterizing MSCs in humans. These criteria include adherence to plastic, expression or absence of specific surface markers (Cluster of Differentiation), and the capacity to differentiate into the three mesodermal lineages, osteogenic, adipogenic, and chondrogenic, when subjected to appropriate stimuli [62].

Cellular therapy based on MSCs holds promise in regenerative medicine, especially for treating nerve injuries in both humans and veterinary medicine. MSCs have the potential to differentiate into various cell types and demonstrate immunomodulatory, neuroprotective, anti-inflammatory, antiapoptotic, and angiogenic properties [61, 63].

The initial source of MSCs was discovered in bone marrow, and despite requiring invasive harvesting procedures, it continues to be extensively investigated [64]. The most commonly utilized MSCs in clinical trials are derived from adipose tissue, bone marrow, and umbilical cord [65]. Adipose tissue-derived MSCs offer several advantages, including easy isolation, high availability, and a high rate of *in vitro* proliferation [64, 66]. Additionally, recent studies are exploring MSCs from other sources, such as those derived from umbilical cord, due to their non-invasive obtainment, abundance, and easy accessibility [67]. Furthermore, they can be easily obtained after birth and raise fewer ethical concerns [68].

MSCs can also be isolated from a variety of tissues, including Wharton's jelly, amniotic fluid, endometrium, placenta, hair follicles, skin, peripheral blood, synovium, and synovial fluid [63].

8. Schwann-like cells derived from mesenchymal stem cells

SCs play a crucial role in peripheral nerve regeneration, as previously discussed. However, their clinical application involves sacrificing a healthy nerve from the donor, and their proliferation in culture is time-consuming [69]. Due to these limitations, transdifferentiating MSCs into SLC cells has been proposed as an alternative approach for cell-based therapy in peripheral nerve injuries.

The first protocol developed to induce the transdifferentiation of MSCs into SLCs *in vitro* was outlined by Dezawa et al. [70]. In this study, rat bone marrow-derived MSCs were exposed to a combination of factors, including beta-mercaptoethanol, retinoic acid, forskolin, basic fibroblast growth factor (bFGF), platelet-derived growth factor (PDGF), and Heregulin (HRG) [70]. Based on this method, several other approaches have been developed over time, utilizing MSCs from different sources and species.

The first documented success in transdifferentiating adipose tissue-derived MSCs into SLCs *in vitro* was reported by Kingham and colleagues using rat MSCs [71]. The researchers achieved this by extending the induction period, increasing the concentration of FSK, and incorporating glial growth factor-2 (GGF2) into the protocol initially established by Dezawa et al. [70, 71].

Other alternative strategies for transdifferentiation include the addition of neurotrophic factors such as BDNF [72] and NGF [73], as well as glucocorticoids, insulin and progesterone to the differentiation medium [74]. Additionally, methods such as electrostimulation [75], intermittent induction [76], cell imprinting [77], and the use of conditioned culture medium containing factors secreted from peripheral nerves of rats [78–80] and equine [15] have been reported.

Most studies assess the success of *in vitro* transdifferentiation based on morphological changes and the expression of specific markers of SCs [15, 78, 79, 81]. These morphological changes typically involve cells adopting a bipolar and/or tripolar shape, featuring large nuclei and fine cytoplasmic processes [15, 78, 80]. Regarding markers, S100 calcium-binding protein (S100), glial fibrillary acidic protein (GFAP), and p75 are commonly used for the evaluation of transdifferentiation. However, there is no consensus on which ones to use (**Table 2**). However, the expression of glial markers is considered the most important criterion for identifying transdifferentiation [126].

Schwann cell markers	References
GFAP	[83]
S100	[84–86]
S100, GFAP	[14, 30, 78–80, 87–92]
S100, p75 NGFR	[75, 93–96]
S100, GFAP, p75 NGFR	[13, 71, 73, 81, 97–109]
S100, GFAP, Nestin	[110, 111]
S100, GFAP, MBP	[112–114]
S100, p75 NGFR, Oct-6	[115]
S100, p75 NGFR, Integrin β4	[116]
S100, GFAP, p75 NGFR, O4	[70]
S100, GFAP, p75 NGFR, erbB3	[117]
S100, GFAP, p75 NGFR, Nestin	[118]
S100, p75 NGFR, PMP-22	[119]
S100, GFAP, MBP, Nestin	[120]
S100, p75 NGFR, P0, MAG	[121]
S100, GFAP, P0, PMP-22	[122]
S100, GFAP, p75 NGFR, MBP	[15]
S100, GFAP, p75 NGFR, P0, L1	[123]
S100, GFAP, p75 NGFR, MBP, P0	[124]
S100, GFAP, p75 NGFR, P0, O4, Sox10, Krox20	[125]

Abbreviation: glial fibrillary acidic protein (GFAP); S100 calcium-binding protein (S100); nerve growth factor receptor (NGFR); myelin basic protein (MBP); oligodendrocyte 4 (O4); epidermal growth factor receptor 3 (erbB3); peripheral myelin protein 22 (PMP-22); protein zero (P0); myelin-associated glycoprotein (MAG).

Table 2.
Different markers used to characterize transdifferentiation in SLCs. Adapted from Ramli et al. [82].

Western blot, immunofluorescence, and PCR are the primary methods utilized to verify their expression [127]. Additionally, other assessments such as evaluating the secretion of neurotrophic factors, co-culturing with dorsal root ganglion neurons and assessing myelination can be conducted [126, 128].

Studies involving MSCs from rats and humans have demonstrated an increase in the secretion of neurotrophic factors, including BDNF, GDNF, NGF, and VEGF-A, following transdifferentiation [129, 130]. Furthermore, a significant upregulation in the expression of BDNF and GDNF was observed after *in vitro* transdifferentiation using human [131] and equine [15] MSCs.

9. Therapeutic potential of Schwann-like cells derived from mesenchymal stem cells for peripheral nerve injuries

The evaluation of SLCs transplantation in peripheral nerve injuries is constantly evolving, and the promising results have been reported. In this topic, assessments have been divided, considering the different sources of MSCs to obtain SLCs.

9.1 Bone marrow MSCs-derived SLCs

Bone marrow-derived MSCs (BM-MSCs) have been extensively researched for peripheral nerve regeneration. Studies highlight the promising results achieved with BM-MSCs-derived SLCs. In models of sciatic nerve transection in rats with a 12 mm gap, BM-MSCs derived SLCs from rats were transferred to hollow fibers and cellular transplantation was performed [97]. This intervention resulted in significant improvements in motor nerve conduction velocity, sciatic nerve function, reconstruction of Ranvier's nodes, and myelination [97]. Additionally, no tumor formations were detected within six months after transplantation [97]. Another study, also employing a rat model with a 12-mm gap lesion, utilized chitosan nerve conduits seeded with SLCs, resulting in improvements in remyelination and axonal regrowth [94]. However, no significant results were observed compared to treatment with SCs derived from the sciatic nerve [94]. Using human BM-MSCs, a study conducted the transplantation of hollow conduits seeded with SCLs and treated with tacrolimus in rats with a 10 mm gap lesion, resulting in functional recovery of the sciatic nerve [123].

Moreover, in a model of transection of the buccal branch of the rabbit facial nerve with a 1-cm gap, it was evidenced that rabbit SLCs within the autologous vein graft accelerate axonal regeneration and promote better remyelination [103]. Another study, utilizing rat SLCs incorporated into a polyglycolic acid tube, revealed promising results in nerve regeneration following neurotmesis of the mandibular branch of the rat facial nerve [115].

9.2 Adipose tissue MSCs-derived SLCs

Adipose tissue-derived MSCs (AT-MSCs) have garnered interest in regenerative medicine due to their widespread availability and ease of obtainment. In experimental models of rat sciatic nerve transection with a 10-mm gap [116, 130, 132, 133] and 15-mm gap [134], the transplantation of AT-MSCs transdifferentiated into SLCs, from both rats [116, 132–134] and humans [130], into nerve fibrin conduits [130, 132, 133], aligned collagen matrix [134] and silicone [116] has shown improvements in axonal regeneration [116, 130, 132–134], increased neurotrophic factors, vascularization [130], myelination [116] and reduction in muscle atrophy [133].

A study compared the transplantation of rat AT-MSCs, SLCs, and SCs incorporated into silicone nerve conduits with collagen gel in a 7-mm gap model of rat facial nerve and concluded that the groups exhibited a similar potential for nerve regeneration [108].

Additionally, the use of sheep as an animal model showed promising results by employing acellular nerve allografts recellularized with ovine SLCs after experimentally inducing a 30-mm gap in the animals' peroneal nerve [86].

In crush models, positive results were also observed. The capacity for myelination formation was demonstrated after the transplantation of human SLCs in an experiment involving tibial nerve crush injury in athymic nude rats [135]. Another study demonstrated that the transplantation of SLCs from rats obtained through the approach utilizing a conditioned medium with factors secreted by the rat sciatic nerve assisted in peripheral nerve regeneration in a rat sciatic nerve crush injury model [79].

9.3 Wharton's jelly MSCs-derived SLCs

Wharton's jelly-derived MSCs can be easily obtained in abundance and have been widely investigated [136]. The use of human MSCs transdifferentiated into SLCs, combined with Matrigel grafting in a rat sciatic nerve transection model with an 8-mm gap, promoted myelination and enhanced nerve regeneration [125]. Moreover, acellular nerve grafts combined with SLCs have been shown to promote peripheral nerve regeneration in a rat sciatic nerve injury model with 6-mm nerve defects [12].

9.4 Umbilical cord blood MSCs-derived SLCs

Umbilical cord blood-derived MSCs (UCB-MSCs) are widely investigated due to their extensive availability and non-invasive collection process [137]. Transplantation of human UCB-MSCs-derived SLCs, organized into three-dimensional (3D) cellular spheroids using a methylcellulose hydrogel system, in a rat sciatic nerve crush injury model, stimulated nerve structure regeneration and promoted motor function recovery [138].

10. Future perspectives of fabricating nerve guidance conduits (NGCs) for peripheral nerve injuries

To promote the regeneration of peripheral nerve injuries, various strategies are being investigated, including the use of NGCs in defects with long gaps [139]. These tubular structures provide a suitable microenvironment between the ends of the injured nerve, favoring neuroregeneration [43]. NGCs can be constructed from non-biological materials, as well as autogenous and allogeneic biological materials [140].

NGCs may originate from synthetic or natural origin, a blend of both, and exhibit structural variations related to porosity and filament presence [25]. The material is chosen based on its biocompatibility, permeability, resorption time, strength, and surface characteristics [141, 142]. NGCs are designed and manufactured to provide different mechanical, biological, and biochemical stimuli (**Figure 3**) [25, 143].

Several techniques, such as micro-patterning, injection molding, unidirectional freezing, electrospinning, and 3D printing, have been employed in the fabrication of NGCs [25]. The emergence of 3D bioprinting allows for the construction of NGCs aimed at regenerating tissues with complex architectures and specific production for each case [139]. In bioprinting, bioink is prepared with living cells that are employed as fundamental elements, while support materials serve as a base structure, enabling the creation of 3D tissue structures [144].

The use of 3D bioprinting allows for greater mechanical stability, high porosity, reproducibility, and adaptability to various polymers, mimicking the structural details of the nerve region to be replaced [144–146]. Bioprinting shows great potential; however, it faces some challenges that need to be overcome to acquire the precise architecture of nervous tissue. This involves reproducing nerve microanatomy, precisely depositing different types of cells in a controlled environment within the hydrogel, in addition to the composition and mechanical properties of the bioink [144]. A futuristic area is the use of the four-dimensional (4D) technology approach, which may allow for material adaptation in response to stimuli, mimicking physiological variations of the body [25, 147]. The primary goal of future research on NGCs is to make them functional, and the combination with SLCs represents potential future research that could further enhance the neureregenerative capacity after injury.

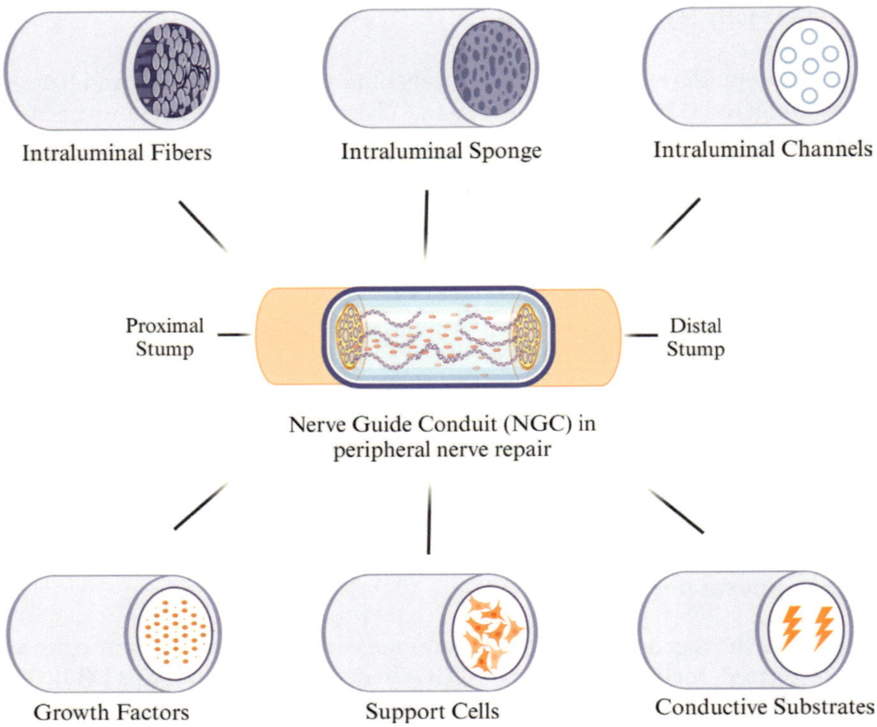

Figure 3.
Schematic illustration of NGC fabrication featuring various configurations and integrated with biochemical, biological, and mechanical cues for peripheral nerve regeneration. Created with Biorender.com

11. Conclusions

The regeneration of the PNS remains a major challenge. SCs are the main glial cells present in the PNS and are of the utmost importance in its regeneration. SC transplantation has emerged as a promising cellular therapy for the treatment of peripheral nerve injuries. However, due to the limitations associated with its use, research has advanced toward alternative approaches to overcome these constraints, such as the transdifferentiation of MSCs into SLCs. Several sources of MSCs, from different species, have been transdifferentiated into SLCs and have become a great promise as substitutes for SCs in nerve regeneration.

Several studies have provided evidence of the beneficial effects of SLC transplantation for treating peripheral nerve injuries. However, there are still significant issues to address. Further research is needed to investigate different doses, application frequencies, long-term assessments post-transplantation, potential immune responses, transplantation in chronic injuries, and the use of several biomaterials as cell carriers. Therefore, conducting more preclinical studies is crucial before advancing to clinical application. SC-based therapy has the potential to become an even more attractive and effective strategy for treating PNS injuries.

Conflict of interest

The authors have no conflict of interest.

Acronyms and abbreviations

3D	three-dimensional
4D	four-dimensional
AT-MSCs	adipose tissue-derived mesenchymal stem cells
BDNF	brain-derived neurotrophic factor
bFGF	basic fibroblast growth factor
BME	beta-mercaptoethanol
BM-MSCs	bone marrow-derived mesenchymal stem cells
CNS	central nervous system
CNTF	ciliary neurotrophic factor
ErbB3	epidermal growth factor receptor 3
FGF-2	fibroblast growth factor 2
FSK	forskolin
GDNF	glial-derived neurotrophic factor
GFAP	glial fibrillary acidic protein
GGF2	glial growth factor-2
HGF	hepatocyte growth factor
HRG	heregulin
IDO	indoleamine 2,3-dioxygenase
IGF	insulin-like growth factor
IL	interleukin
LIF	leukemia inhibitory factor
MAG	myelin-associated glycoprotein
MBP	myelin basic protein
MCP-1	monocyte chemoattractant protein-1
MSCs	mesenchymal stem cells
NGC	nerve guidance conduits
NGF	nerve growth factor
NGFR	nerve growth factor receptor
NT	neurotrophins
O4	oligodendrocyte 4
P0	protein zero
PDGF	platelet-derived growth factor
PGE2	prostaglandin E2
PMP-22	peripheral myelin protein 22
PNS	peripheral nervous system
RA	retinoic acid
S100	S100 calcium-binding protein
SCs	Schwann cells
SLCs	Schwann-like cells
TGF-β	transforming growth factor beta
TNF-α	tumor necrosis factor
UCB-MSCs	umbilical cord blood-derived mesenchymal stem cells
VEGF-A	vascular endothelial growth factor A

Author details

Rogério Martins Amorim and Lucas Vinícius de Oliveira Ferreira*
Department of Veterinary Clinic, School of Veterinary Medicine and Animal Science, São Paulo State University (UNESP), Botucatu, SP, Brazil

*Address all correspondence to: lv.ferreira@unesp.br

References

[1] Sivanarayanan TB, Bhat IA, Sharun K, Palakkara S, Singh R, Remya V, et al. Allogenic bone marrow-derived mesenchymal stem cells and its conditioned media for repairing acute and sub-acute peripheral nerve injuries in a rabbit model. Tissue and Cells. 2023;**82**:102053. DOI: 10.1016/j.tice.2023.102053

[2] Shan Y, Xu L, Cui X, Wang E, Jiang F, Li J, et al. A responsive cascade drug delivery scaffold adapted to the therapeutic time window for peripheral nerve injury repair. Materials Horizons. 2024;**11**:1032-1045. DOI: 10.1039/d3mh01511d

[3] Piovesana R, Faroni A, Tata AM, Reid AJ. Schwann-like adipose-derived stem cells as a promising therapeutic tool for peripheral nerve regeneration: Effects of cholinergic stimulation. Neural Regeneration Research. 2021;**16**:1218-1220. DOI: 10.4103/1673-5374.300433

[4] Dong Y, Alhaskawi A, Zhou H, Zou X, Liu Z, Ezzi SHA, et al. Imaging diagnosis in peripheral nerve injury. Frontiers in Neurology. 2023;**14**:1250808. DOI: 10.3389/fneur.2023.1250808

[5] Luo L, He Y, Jin L, Zhang Y, Guastaldi FP, Albashari AA, et al. Application of bioactive hydrogels combined with dental pulp stem cells for the repair of large gap peripheral nerve injuries. Bioactive Materials. 2021;**6**:638-654. DOI: 10.1016/j.bioactmat.2020.08.028

[6] Zhang RC, Du WQ, Zhang JY, Yu SX, Lu FZ, Ding HM, et al. Mesenchymal stem cell treatment for peripheral nerve injury: A narrative review. Neural Regeneration Research. 2021;**16**:2170-2176. DOI: 10.4103/1673-5374.310941

[7] Alvites R, Lopes B, Coelho A, Maurício AC. Peripheral nerve regeneration: A challenge far from being overcome. Regenerative Medicine. 2024;**19**:155-159. DOI: 10.2217/rme-2023-0072

[8] Huang Y, Wu L, Zhao Y, Guo J, Li R, Ma S, et al. Schwann cell promotes macrophage recruitment through IL-17B/IL-17RB pathway in injured peripheral nerves. Cell Reports. 2024;**43**:113753. DOI: 10.1016/j.celrep.2024.113753

[9] Zhang YL, Chen DJ, Yang BL, Liu TT, Li JJ, Wang XQ, et al. Microencapsulated Schwann cell transplantation inhibits P2X3 receptor expression in dorsal root ganglia and neuropathic pain. Neural Regeneration Research. 2018;**13**:1961-1967. DOI: 10.4103/1673-5374.238715

[10] Han GH, Peng J, Liu P, Ding X, Wei S, Lu S, et al. Therapeutic strategies for peripheral nerve injury: Decellularized nerve conduits and Schwann cell transplantation. Neural Regeneration Research. 2019;**14**:1343-1351. DOI: 10.4103/1673-5374.253511

[11] Liu Y, Dong R, Zhang C, Yang Y, Xu Y, Wang H, et al. Therapeutic effects of nerve leachate-treated adipose-derived mesenchymal stem cells on rat sciatic nerve injury. Experimental and Therapeutic Medicine. 2020;**19**:223-231. DOI: 10.3892/etm.2019.8203

[12] Choi SJ, Park SY, Shin YH, Heo SH, Kim KH, Lee H, et al. Mesenchymal stem cells derived from Wharton's Jelly can differentiate into Schwann cell-like cells and promote peripheral nerve regeneration in acellular nerve grafts. Tissue Engineering and Regenerative Medicine. 2021;**18**:467-478. DOI: 10.1007/s13770-020-00329-6

[13] Park S, Jung N, Myung S, Choi Y, Chung KW, Choi BO, et al. Differentiation of human tonsil-derived mesenchymal stem cells into Schwann-like cells improves neuromuscular function in a mouse model of Charcot-Marie-tooth disease type 1A. International Journal of Molecular Sciences. 2018;**19**:2393. DOI: 10.3390/ijms19082393

[14] Hu X, Wang X, Xu Y, Li L, Liu J, He Y, et al. Electric conductivity on aligned nanofibers facilitates the transdifferentiation of mesenchymal stem cells intro Schwann cells and regeneration of injured peripheral nerve. Advanced Healthcare Materials. 2020;**9**:e1901570. DOI: 10.1002/adhm.201901570

[15] Ferreira LVO, Kamura BC, Oliveira JPM, Chimenes ND, Carvalho M, Santos LA, et al. In vitro transdifferentiation potential of equine mesenchymal stem cells into Schwann-like cells. Stem Cells and Development. 2023;**32**:422-432. DOI: 10.1089/scd.2022.0274

[16] Leberfinger AN, Ravnic DJ, Payne R, Rizk E, Koduru SV, Hazard SW. Adipose-derived stem cells in peripheral nerve regeneration. Current Surgery Reports. 2017;**5**:1-9. DOI: 10.1007/s40137-017-0169-2

[17] Sumarwoto T, Suroto H, Mahyudin F, Utomo DN, Romaniyanto TD, et al. Role of adipose mesenchymal stem cells and secretome in peripheral nerve regeneration. Annals of Medicine and Surgery. 2021;**67**:102482. DOI: 10.1016/j.amsu.2021.102482

[18] Menorca RMG, Fussell TS, Elfar JC. Nerve physiology: Mechanisms of injury and recovery. Hand Clinics. 2013;**29**:317-330. DOI: 10.1016/j.hcl.2013.04.002

[19] Lanigan LG, Russell DS, Woolard KD, Pardo ID, Godfrey V, Jortner BS, et al. Comparative pathology of the peripheral nervous system. Veterinary Pathology. 2021;**58**:10-33. DOI: 10.1177/0300985820959231

[20] Liu Q, Wang X, Yi S. Pathophysiological changes of physical barriers of peripheral nerves after injury. Frontiers in Neuroscience. 2018;**12**:597. DOI: 10.3389/fnins.2018.00597

[21] Wang ML, Rivlin M, Graham JG, Beredjiklian PK. Peripheral nerve injury, scarring, and recovery. Connective Tissue Research. 2019;**60**:3-9. DOI: 10.1080/03008207.2018.1489381

[22] Alvites RD, Branquinho MV, Caseiro AR, Pedrosa SS, Luís AL, Geuna S, et al. . In: Turker H, Benavides LG, Gallardo GR, Villar MM, editors. Peripheral Nerve Disorders and Treatment [Internet]. London, UK: IntechOpen; 2020. pp. 67-115. DOI: 10.5772/intechopen.78159

[23] Jessen KR, Mirsky R. The success and failure of the Schwann cell response to nerve injury. Frontiers in Cellular Neuroscience. 2019;**13**:33. DOI: 10.3389/fncel.2019.00033

[24] Riccio M, Marchesini A, Pugliese P, Francesco F. Nerve repair and regeneration: Biological tubulization limits and future perspectives. Journal of Cellular Physiology. 2019;**234**:3362-3375. DOI: 10.1002/jcp.27299

[25] Vijayavenkataraman S. Nerve guide conduits for peripheral nerve injury repair: A review on design, materials and fabrication methods. Acta Biomaterialia. 2020;**106**:54-69. DOI: 10.1016/j.actbio.2020.02.003

[26] Wang Y, Zhang Y, Li X, Zhang Q. The progress of biomaterials in peripheral nerve repair and regeneration. Journal of Neurorestoratology. 2020;**8**:252-269. DOI: 10.26599/JNR.2020.9040022

[27] Kim J. Neural reanimation advances and new technologies. Facial Plastic Surgery Clinics of North America. 2016;**24**:71-84. DOI: 10.1016/j.fsc.2015.09.006

[28] Houdek MT, Shin AY. Management and complications of traumatic peripheral nerve injuries. Hand Clinics. 2015;**31**:151-163. DOI: 10.1016/j.hcl.2015.01.007

[29] Campbell WW. Evaluation and management of peripheral nerve injury. Clinical Neurophysiology. 2008;**119**:1951-1965. DOI: 10.1016/j.clinph.2008.03.018

[30] Villagrán CC, Schumacher J, Donnell R, Dhar MS. A novel model for acute peripheral nerve injury in the horse and evaluation of the effect of mesenchymal stromal cells applied in situ on nerve regeneration: A preliminary study. Frontiers in Veterinary Science. 2016;**3**:80. DOI: 10.3389/fvets.2016.00080

[31] Bolívar S, Navarro X, Udina E. Schwann cell role in selectivity of nerve regeneration. Cells. 2020;**9**:2131. DOI: 10.3390/cells9092131

[32] Li R, Liu Z, Pan Y, Chen L, Zhang Z, Lu L. Peripheral nerve injuries treatment: A systematic review. Cell Biochemistry and Biophysics. 2014;**68**:449-454. DOI: 10.1007/s12013-013-9742-1

[33] Houshyar S, Bhattacharyya A, Shanks R. Peripheral nerve conduit: Materials and structures. ACS Chemical Neuroscience. 2019;**10**:3349-3365. DOI: 10.1021/acschemneuro.9b00203

[34] Forterre F, Tomek A, Rytz U, Brunnberg L, Jaggy A, Spreng D. Iatrogenic sciatic nerve injury in eighteen dogs and nine cats (1997-2006). Veterinary Surgery. 2007;**36**:464-471. DOI: 10.1111/j.1532-950X.2007.00293.x

[35] Soens IV, Struys MM, Polis IE, Bhatti SF, Meervenne SAV, Martlé VA, et al. Magnetic stimulation of the radial nerve in dogs and cats with brachial plexus trauma: A report of 53 cases. The Veterinary Journal. 2009;**182**:108-113. DOI: 10.1016/j.tvjl.2008.05.007

[36] Emond AL, Bertoni L, Seignour M, Coudry V, Denoix JM. Peripheral neuropathy of a forelimb in horses: 27 cases (2000-2013). Journal of the American Veterinary Medical Association. 2016;**249**:1187-1195. DOI: 10.2460/javma.249.10.1187

[37] Boorman S, Scherrer NM, Stefanovski D, Johnson AL. Facial nerve paralysis in 64 equids: Clinical variables, diagnosis, and outcome. Journal of Veterinary Internal Medicine. 2020;**34**:1308-1320. DOI: 10.1111/jvim.15767

[38] Seddon HJ. Three types of nerve injury. Brain. 1943;**66**:237-288. DOI: 10.1093/brain/66.4.237

[39] Sunderland S. A classification of peripheral nerve injuries producing loss of function. Brain. 1951;**74**:491-516. DOI: 10.1093/brain/74.4.491

[40] Mackinnon S, Dellon A. Diagnosisof nerve injury. In: Surgery of the Peripheral Nerve. New York: Thieme; 1988. pp. 74-79

[41] Lin YF, Xie Z, Zhou J, Yin G, Lin HD. Differential gene and protein expression between rat tibial nerve and common peroneal nerve during Wallerian degeneration. Neural Regeneration Research. 2019;**14**:2183-2191. DOI: 10.4103/1673-5374.262602

[42] Salvadores N, Gerónimo-Olvera C, Court FA. Axonal degeneration in AD: The contribution of Aβ and Tau. Frontiers in Aging Neuroscience.

2020;**12**:581767. DOI: 10.3389/fnagi.2020.581767

[43] Lopes B, Sousa P, Alvites R, Branquinho M, Sousa AC, Mendonça C, et al. Peripheral nerve injury treatments and advances: One health perspective. International Journal of Molecular Science. 2022;**23**:918. DOI: 10.3390/ijms23020918

[44] Wan T, Zhang FS, Qin MY, Jiang HR, Zhang M, Qu Y. Growth factors: Bioactive macromolecular drugs for peripheral nerve injury treatment – Molecular mechanisms and delivery platforms. Biomedicine & Pharmacotherapy. 2024;**170**:116024. DOI: 10.1016/j.biopha.2023.116024

[45] Wilcox M, Gregory H, Powell R, Quick TJ, Phillips JB. Strategies for peripheral nerve repair. Current Tissue Microenvironment Reports. 2020;**1**:49-59. DOI: 10.1007/s43152-020-00002-z

[46] Rao Z, Lin Z, Song P, Quan D, Bai Y. Biomaterial-based Schwann cell transplantation and Schwann cell-derived biomaterials for nerve regeneration. Frontiers in Cellular Neuroscience. 2022;**16**:926222. DOI: 10.3389/fncel.2022.926222

[47] Pandey S, Mudgal J. A review on the role of endogenous neurotrophins and Schwann cells in axonal regeneration. Journal of Neuroimmune Pharmacology. 2022;**17**:398-408. DOI: 10.1007/s11481-021-10034-3

[48] Rhode SC, Beier JP, Ruhl T. Adipose tissue stem cells in peripheral nerve regeneration—In vitro and in vivo. Journal of Neuroscience Research. 2021;**99**:545-560. DOI: 10.1002/jnr.24738

[49] Wu G, Wen X, Kuang R, Lui KHW, He B, Li G, et al. Roles of macrophages and their interactions with Schwann cells after peripheral nerve injury. Cellular and Molecular Neurobiology. 2024;**44**:11. DOI: 10.1007/s10571-023-01442-5

[50] Bhandari PS. Management of peripheral nerve injury. Journal of Clinical Orthopaedics and Trauma. 2019;**10**:862-866. DOI: 10.1016/j.jcot.2019.08.003

[51] Rishal I, Fainzilber M. Axon-soma communication in neuronal injury. Nature Reviews Neuroscience. 2014;**15**:32-42. DOI: 10.1038/nrn3609

[52] Lee D, Tran HQ, Dudley AT, Yang K, Yan Z, Xie J. Advancing nerve regeneration: Peripheral nerve injury (PNI) chip empowering high-speed biomaterial and drug screening. Chemical Engineering Journal. 2024;**486**:150210. DOI: 10.1016/j.cej.2024.150210

[53] Scheib J, Höke A. Advances in peripheral nerve regeneration. Nature Reviews Neuroscience. 2013;**9**:668-676. DOI: 10.1038/nrneurol.2013.227

[54] Sullivan R, Dailey T, Duncan K, Abel N, Borlongan CV. Peripheral nerve injury: Stem cell therapy and peripheral nerve transfer. International Journal of Molecular Sciences. 2016;**17**:2101. DOI: 10.3390/ijms17122101

[55] Gonçalves NP, Mohseni S, El Soury M, Ulrichsen M, Richner M, Xiao J, et al. Peripheral nerve regeneration is independent from Schwann cell p75NTR expression. Frontiers in Cellular Neuroscience. 2019;**13**:235. DOI: 10.3389/fncel.2019.00235

[56] Caillaud M, Richard L, Vallat JM, Desmoulière A, Billet F. Peripheral nerve regeneration and intraneural revascularization. Neural Regeneration Research. 2019;**14**:24-33. DOI: 10.4103/1673-5374.243699

[57] Barberini DJ, Freitas NPP, Magnoni MS, Maia L, Listoni AJ, Heckler MC, et al. Equine mesenchymal stem cells from bone marrow, adipose tissue and umbilical cord: Immunophenotypic characterization and differentiation potential. Stem Cell Research & Therapy. 2014;**5**:25. DOI: 10.1186/scrt414

[58] MacDonald E, Barrett JG. The potential of mesenchymal stem cells to treat systemic inflammation in horses. Frontiers in Veterinary Science. 2020;**6**:507. DOI: 10.3389/fvets.2019.00507

[59] Villagrán CC, Amelse L, Neilsen N, Dunlap J, Dhar M. Differentiation of equine mesenchymal stromal cells into cells of neural lineage: Potential for clinical applications. Stem Cells International. 2014;**2014**:891518. DOI: 10.1155/2014/891518

[60] Pennington MR, Curtis TM, Divers TJ, Wagner B, Ness SL, Tennant BC, et al. Equine mesenchymal stromal cells from different sources efficiently differentiate into hepatocyte-like cells. Tissue Engineering Part C: Methods. 2016;**22**:596-607. DOI: 10.1089/ten.TEC.2015.0403

[61] Andrzejewska A, Lukomska B, Janowski M. Concise review: Mesenchymal stem cells: From roots to boost. Stem Cells. 2019;**37**:855-864. DOI: 10.1002/stem.3016

[62] Dominici M, Blanc KL, Mueller I, Cortenbach IS, Marini FC, Krause D, et al. Minimal criteria for defining multipotent mesenchymal stem stromal cells. The international society for cellular therapy position statement. Cytotherapy. 2006;**8**:315-317. DOI: 10.1080/14653240600855905

[63] Maličev E, Jazbec K. An overview of mesenchymal stem cell heterogeneity and concentration. Pharmaceuticals. 2024;**17**:350. DOI: 10.3390/ph17030350

[64] Hopf A, Schaefer DJ, Kalbermatten DF, Guzman R, Madduri S. Schwann cell-like cells: Origin and usability for repair and regeneration of the peripheral and central nervous system. Cells. 2020;**9**:1990. DOI: 10.3390/cells9091990

[65] Biglari N, Mehdizadeh A, Mastanabad MV, Gharaeikhezri MH, Afrakoti LGMP, Pourbala H, et al. Application of mesenchymal stem cells (MSCs) in neurodegenerative disorders: History, findings, and prospective challenges. Pathology, Research and Practice. 2023;**247**:154541. DOI: 10.1016/j.prp.2023.154541

[66] Qomi RT, Sheykhhasan M. Adipose-derived stromal cell in regenerative medicine: A review. World Journal of Stem Cells. 2017;**9**:107-117. DOI: 10.4252/wjsc.v9.i8.107

[67] Merlo B, Teti G, Mazzotti E, Ingrà L, Salvatore V, Buzzi M, et al. Wharton's Jelly derived mesenchymal stem cells: Comparing human and horse. Stem Cell Reviews and Reports. 2018;**14**:574-584. DOI: 10.1007/s12015-018-9803-3

[68] Jiang L, Jones S, Jia X. Stem cell transplantation for peripheral nerve regeneration: Current options and opportunities. International Journal of Molecular Sciences. 2017;**18**:94. DOI: 10.3390/ijms18010094

[69] Saffari S, Saffari TM, Ulrich DJO, Hovius SER, Shin AY. The interaction of stem cells and vascularity in peripheral nerve regeneration. Neural Regeneration Research. 2021;**16**:1510-1517. DOI: 10.4103/1673-5374.303009

[70] Dezawa M, Takahashi I, Esaki M, Takano M, Sawada H.

Sciatic nerve regeneration in rats induced by transplantation of in vitro differentiated bone-marrow stromal cells. The European Journal of Neuroscience. 2001;**14**:1771-1776. DOI: 10.1046/j.0953-816x.2001.01814.x

[71] Kingham PJ, Kalbermatten DF, Mahay D, Armstrong SJ, Wiberg M, Terenghi G. Adipose-derived stem cells differentiate into a Schwann cell phenotype and promote neurite outgrowth in vitro. Experimental Neurology. 2007;**207**:267-274. DOI: 10.1016/j.expneurol.2007.06.029

[72] Gao S, Zheng Y, Cai Q, Wu X, Yao W, Wang J. Different methods for inducing adipose-derived stem cells to differentiate into Schwann-like cells. Archives of Medical Science: AMS. 2015;**11**:886-892. DOI: 10.5114/aoms.2015.53310

[73] Xiao YZ, Wang S. Differentiation of Schwann-like cells from human umbilical cord blood mesenchymal stem cells in vitro. Molecular Medicine Reports. 2015;**11**:1146-1152. DOI: 10.3892/mmr.2014.2840

[74] Kang Y, Liu Y, Liu Z, Ren S, Xiong H, Chen J, et al. Differentiated human adipose-derived stromal cells exhibits the phenotypic and functional characteristics of mature Schwann cells through a modified approach. Cytotherapy. 2019;**21**:987-1003. DOI: 10.1016/j.jcyt.2019.04.061

[75] Das SR, Uz M, Ding S, Lentner MT, Hondred JA, Cargill AA, et al. Electrical differentiation of mesenchymal stem cells into Schwann-cell like phenotypes using inkjet printed graphene circuits. Advanced Healthcare Materials. 2017;**6**:1601087. DOI: 10.1002/adhm.201601087

[76] Sun X, Zhu Y, Yin HY, Guo ZY, Xu F, Xiao B, et al. Differentiation of adipose-derived stem cells into Schwann cell-like cells through intermittent induction: Potential advantage of cellular transient memory function. Stem Cell Research & Therapy. 2018;**9**:133. DOI: 10.1186/s13287-018-0884-3

[77] Moghaddam MM, Bonakdar S, Shokrgozar MA, Zaminy A, Vali H, Faghihi S. Engineered substrates with imprinted cell-like topographies induce direct differentiation of adipose-derived mesenchymal stem cells into Schwann cells. Artificial Cells, Nanomedicine, and Biotechnology. 2019;**47**:1022-1035. DOI: 10.1080/21691401.2019.1586718

[78] Liu Y, Zhang Z, Qin Y, Wu H, Lv Q, Chen X, et al. A new method for Schwann-like cells differentiation of adipose derived stem cells. Neuroscience Letters. 2013;**551**:79-83. DOI: 10.1016/j.neulet.2013.07.012

[79] Tawab SA, Omar SMM, Zeid AAA, Saba C. Role of adipose tissue-derived stem cells versus differentiated Schwann-like cells transplantation on the regeneration of crushed sciatic nerve in rats. A histological study. International Journal of Stem Cell Research and Therapeutics. 2018;**1**:1-10

[80] Galhom RA, Raouf HHHA, Ali MHM. Role of bone marrow derived mesenchymal stromal cells and Schwann-like cells transplantation on spinal cord injury in adult male albino rats. Biomedicine & Pharmacotherapy. 2018;**108**:1365-1375. DOI: 10.1016/j.biopha.2018.09.131

[81] Mathot F, Shin AY, Wijnen AJV. Targeted stimulation of MSCs in peripheral nerve repair. Gene. 2019;**710**:17-23. DOI: 10.1016/j.gene.2019.02.078

[82] Ramli K, Gasim IA, Ahmad AA, Hassan S, Law ZK, Tan GC, et al. Human bone marrow-derived MSCs

spontaneously express specific Schwann cell markers. Cell Biology International. 2019;**43**:233-252. DOI: 10.1002/cbin.11067

[83] Faroni A, Rothwell SW, Grolla AA, Terenghi G, Magnaghi V, Verkhratsky A. Differentiation of adipose-derived stem cells into Schwann cell phenotype induces expression of P2X receptors that control cell death. Cell Death & Disease. 2013;**4**:e743. DOI: 10.1038/cddis.2013.268

[84] Han IH, Sun F, Choi YJ, Zou F, Nam KH, Cho WH, et al. Cultures of Schwann-like cells differentiated from adipose-derived stem cells on PDMS/MWNT sheets as a scaffold for peripheral nerve regeneration. Journal of Biomedical Materials Research Part A. 2015;**103**:3642-3648. DOI: 10.1002/jbm.a.35488

[85] Zhu H, Yang A, Du J, Li D, Liu M, Ding F, et al. Basic fibroblast growth factor is a key factor that induces bone marrow mesenchymal stem cells towards cells with Schwann cell phenotype. Neuroscience Letters. 2014;**559**:82-87. DOI: 10.1016/j.neulet.2013.11.044

[86] Tamez-Mata Y, Montoya FEP, Rodríguez HGM, Pérez MMG, Cantú AAR, Flores JRG, et al. Nerve gaps repaired with acellular nerve allografts recellularized with Schwann-like cells: Preclinical trial. Journal of Plastic, Reconstructive & Aesthetic Surgery. 2022;**75**:296-306. DOI: 10.1016/j.bjps.2021.05.066

[87] Jiang TM, Yang ZJ, Kong CZ, Zhang HT. Schwann-like cells can be induction from human nestin-positive amniotic fluid mesenchymal stem cells. In Vitro Cellular & Developmental Biology Animal. 2010;**46**:793-800. DOI: 10.1007/s11626-010-9335-x

[88] Razavi S, Ahmadi N, Kazemi M, Mardani M, Esfandiari E. Efficient transdifferentiation of human adipose-derived stem cells into Schwann-like cells: A promise for treatment of demyelinating diseases. Advanced Biomedical Research. 2012;**1**:12. DOI: 10.4103/2277-9175.96067

[89] Younesi E, Bayati V, Hashemitabar M, Azandeh SS, Bijannejad D, Bahreini A. Differentiation of adipose-derived stem cells into Schwann-like: Fetal bovine serum or human serum? Anatomy & Cell Biology. 2015;**48**:170-176. DOI: 10.5115/acb.2015.48.3.170

[90] Halselbach D, Raffoul W, Larcher L, Tremp M, Kalbermatten DF, Summa PG. Regeneration patterns influence hindlimb automutilation after sciatic nerve repair using stem cells in rats. Neuroscience Letters. 2016;**634**:153-159. DOI: 10.1016/j.neulet.2016.10.024

[91] Xie S, Lu F, Han J, Tao K, Wang H, Simental A, et al. Efficient generation of functional Schwann cells from adipose-derived stem cells in defined conditions. Cell Cycle. 2017;**16**:841-851. DOI: 10.1080/15384101.2017.1304328

[92] Fu XM, Wang Y, Fu WL, Liu DH, Zhang CY, Wang QL, et al. The combination of adipose-derived Schwann-like cells and acellular nerve allografts promotes sciatic nerve regeneration and repair through the JAK2/STAT3 signaling pathway in rats. Neuroscience. 2019;**422**:134-145. DOI: 10.1016/j.neuroscience.2019.10.018

[93] Shea GKH, Tsui AYP, Chan YS, Shum DKY. Bone marrow-derived Schwann cells achieve fate commitment—A prerequisite for remyelination therapy. Experimental Neurology. 2010;**224**:448-458. DOI: 10.1016/j.expneurol.2010.05.005

[94] Ao Q, Fung CK, Tsui AYP, Cai S, Zuo HC, Chan YS, et al. The regeneration

of transected sciatic nerves of adult rats using chitosan nerve conduits seeded with bone marrow stromal cell-derived Schwann-cells. Biomaterials. 2011;**32**:787-796. DOI: 10.1016/j.biomaterials.2010.09.046

[95] Rosa MB, Sharma AD, Mallapragada SK, Sakaguchi DS. Transdifferentiation of brain-derived neurotrophic factor (BDNF)—Secreting mesenchymal stem cells significantly enhance BDNF secretion and Schwann cell marker proteins. Journal of Bioscience and Bioengineering. 2017;**124**:572-582. DOI: 10.1016/j.jbiosc.2017.05.014

[96] Uz M, Büyüköz M, Sharma AD, Sakaguchi DS, Altinkaya SA, Mallapragada SK. Gelatin-based 3D conduits for transdifferentiation of mesenchymal stem cells into Schwann cell-like pehnotypes. Acta Biomaterialia. 2017;**53**:293-306. DOI: 10.1016/j.actbio.2017.02.018

[97] Mimura T, Dezawa M, Kanno H, Sawada H, Yamamoto I. Peripheral nerve regeneration by transplantation of bone marrow stromal cell-derived Schwann cells in adults rats. Journal of Neurosurgery. 2004;**101**:806-812. DOI: 10.3171/jns.2004.101.5.0806

[98] Caddik J, Kingham PJ, Gardiner NJ, Wiberg M, Terenghi G. Phenotypic and functional characteristics of mesenchymal stem cells differentiated along a Schwann cell lineage. Glia. 2006;**54**:840-849. DOI: 10.1002/glia.20421

[99] Lin WW, Chen X, Wang XD, Liu J, Gu XS. Adult rat bone marrow stromal cells differentiate into Schwann cell-like cells in vitro. In Vitro Cellular & Developmental Biology Animal. 2008;**44**:31-40. DOI: 10.1007/s11626-007-9064-y

[100] Radtke C, Schmitz B, Spies M, Kocsis JD, Vogt PM. Peripheral glial cell differentiation from neurospheres derived from adipose mesenchymal stem cells. International Journal of Development Neuroscience. 2009;**27**:817-823. DOI: 10.1016/j.ijdevneu.2009.08.006

[101] Wei Y, Gong K, Zheng Z, Liu L, Wang A, Zhang L, et al. Schwann-like cell differentiation of rat adipose-derived stem cells by indirect co-culture with Schwann cells in vitro. Cell Proliferation. 2010;**43**:606-616. DOI: 10.1111/j.1365-2184.2010.00710.x

[102] Ladak A, Olson J, Tredget EE, Gordon T. Differentiation of mesenchymal stem cells to support peripheral nerve regeneration in a rat model. Experimental Neurology. 2011;**228**:242-252. DOI: 10.1016/j.expneurol.2011.01.013

[103] Wang X, Luo E, Li Y, Hu J. Schwann-like mesenchymal stem cells within vein graft facilitate facial nerve regeneration and remyelination. Brain Research. 2011;**1383**:71-80. DOI: 10.1016/j.brainres.2011.01.098

[104] Kaewkhaw R, Scutt AM, Haycock JW. Anatomical site influences the differentiation of adipose-derived stem cells for Schwann-cell phenotype and function. Glia. 2011;**59**:734-749. DOI: 10.1002/glia.21145

[105] Sharma AD, Zbarska S, Petersen EM, Marti ME, Mallapragada SK, Sakaguchi DS. Oriented growth and transdifferentiation of mesenchymal stem cells towards a Schwann cell fate on micropatterned substrates. Journal of Bioscience and Bioengineering. 2016;**121**:325-335. DOI: 10.1016/j.jbiosc.2015.07.006

[106] Najafabadi MM, Bayati V, Orazizadeh M, Hashmitabar M,

Absalan F. Impact of cell density on differentiation efficiency of rat adipose-derived stem cells into Schwann-like cells. International Journal of Stem Cell. 2016;**9**:213-220. DOI: 10.15283/ijsc16031

[107] Fu X, Tong Z, Li Q, Niu Q, Zhang Z, Tong X, et al. Induction of adipose-derived stem cells into Schwann-like cells and observation of Schwann-like cell proliferation. Molecular Medicine Reports. 2016;**14**:1187-1193. DOI: 10.3892/mmr.2016.5367

[108] Watanabe Y, Sasaki R, Matsumine H, Yamato M, Okano T. Undifferentiated and differentiated adipose-derived stem cells improve nerve regeneration in a rat model of facial nerve defect. Journal of Tissue Engineering and Regenerative Medicine. 2017;**11**:362-374. DOI: 10.1002/term.1919

[109] Ouasti S, Faroni A, Kingham PJ, Ghibaudi M, Reid AJ, Tirelli N. Hyaluronic acid (HA) receptors and the motility of Schwann cell(-like) phenotypes. Cells. 2020;**9**:1477. DOI: 10.3390/cells9061477

[110] Li X, Dapeng L, Ping G, Quan Y, Zhen T. Neural differentiation of adipose-derived stem cells by indirect co-culture with Schwann cells. Archives of Biological Sciences. 2009;**61**:703-711. DOI: 10.2298/ABS0904703L

[111] Liao D, Gong P, Li X, Tan Z, Yuan Q. Co-culture with Schwann cells is an effective way for adipose-derived stem cells neural transdifferentiation. Archives of Medical Sciences. 2010;**6**:145-151. DOI: 10.5114/aoms.2010.13885

[112] Montzka K, Lassonczyk N, Tschöke B, Neuss S, Führmann T, Franzen R, et al. Neural differentiation potential of human bone marrow-derived mesenchymal stromal cells: Misleading marker gene expression. BMC Neuroscience. 2009;**10**:16. DOI: 10.1186/1471-2202-10-16

[113] Razavi S, Esfahani HZ, Morshed M, Vaezifar S, Karbasi S, Golozar MA. Nanobiocomposite of poly(lactide-co-glycolide)/chitosan electrospun scaffold can promote proliferation and transdifferentiation of Schwann-like cells from human adipose-derived stem cells. Journal of Biomedical Materials Research Part A. 2015;**103**:2628-2634. DOI: 10.1002/jbm.a.35398

[114] Zarinfard G, Tadjalli M, Razavi S, Kazemi M. Effect of laminin on neurotrophic factors expression in Schwann-like cells induced from human adipose-derived stem cells in vivo. Journal of Molecular Neuroscience. 2016;**60**:465-473. DOI: 10.1007/s12031-016-0808-6

[115] Costa HJZR, Bento RF, Salomone R, Nogueira DA, Zanatta DB, Costa MP, et al. Mesenchymal bone marrow stem cells within polyglycolic acid tube observed in vivo after six weeks enhance facial nerve regeneration. Brain Research. 2013;**1510**:10-21. DOI: 10.1016/j.brainres.2013.03.025

[116] Orbay H, Uysal AC, Hyakusoku H, Mizuno H. Differentiated and undifferentiated adipose-derived stem cells improve function in rats with peripheral nerve gaps. Journal of Plastic, Reconstructive & Aesthetic Surgery. 2012;**65**:657-664. DOI: 10.1016/j.bjps.2011.11.035

[117] Brohlin M, Mahay D, Novikov LN, Terenghi G, Wiberg M, Shawcross SG, et al. Characterisation of human mesenchymal stem cells following differentiation into Schwann cell-like cells. Neuroscience Research. 2009;**64**:41-49. DOI: 10.1016/j.neures.2009.01.010

[118] Hassan NH, Sulong AF, Ng MH, Htwe O, Idrus RBH, Roohi S, et al. Neural-differentiated mesenchymal stem cells incorporated into muscle stuffed vein scaffold forms a stable living nerve conduit. Journal of Orthopaedic Research. 2012;**30**:1674-1681. DOI: 10.1002/jor.22102

[119] Zaminy A, Shokrgozar MA, Sadeghi Y, Noroozian M, Heidari MH, Piryaei A. Mesenchymal stem cells as an alternative for Schwann cells in rat spinal cord injury. Iran Biomedical Journal. 2013;**17**:113-122. DOI: 10.6091/ibj.1121.2013

[120] Razavi S, Mardani M, Kazemi M, Esfandiari E, Narimani M, Esmaeili A, et al. Effect of leucemia inhibitory fator on the myelinogenic ability of Schwann-like cells induced from human adipose-derived stem cells. Cell and Molecular Neurobiology. 2013;**33**:283-289. DOI: 10.1007/s10571-012-9895-2

[121] Schuh CMAP, Morton TJ, Banerjee A, Grasl C, Schima H, Schmidhammer R, et al. Activated Schwann cell-like on aligned fibrin-poly(lactic-co-glycolic acid) structures: A novel construct for application in peripheral nerve regeneration. Cells, Tissue, Organs. 2015;**200**:287-299. DOI: 10.1159/000437091

[122] Liu Y, Chen J, Liu W, Lu X, Liu Z, Zhao X, et al. A modified approach to inducing bone marrow stromal cells to differentiate into cells with mature Schwann cell phenotypes. Stem Cells and Development. 2016;**25**:347-359. DOI: 10.1089/scd.2015.0295

[123] Shimizu S, Kitada M, Ishikawa H, Itokazu Y, Wakao S, Dezawa M. Peripheral nerve regeneration by the in vitro differentiated-human bone marrow stromal cells with Schwann cell property. Biochemical and Biophysical Research Communications. 2007;**359**:915-920. DOI: 10.1016/j.bbrc.2007.05.212

[124] Peng J, Wang Y, Zhang L, Zhao B, Zhao Z, Chen J, et al. Human umbilical cord Wharton's jelly-derived mesenchymal stem cells differentiate into a Schwann-cell phenotype and promote neurite outgrowth in vitro. Brain Research Bulletin. 2011;**84**:235-243. DOI: 10.1016/j.brainresbull.2010.12.013

[125] Matsuse D, Kitada M, Kohama M, Nishikawa K, Makinoshima H, Wakao S, et al. Human umbilical cord-derived mesenchymal stromal cells differentiate into functional Schwann cells that sustain peripheral nerve regeneration. Journal of Neurophathology & Experimental Neurology. 2010;**69**:973-985. DOI: 10.1097/NEN.0b013e3181eff6dc

[126] Liu Z, Jin YQ, Chen L, Wang Y, Yang X, Cheng J, et al. Specific marker expression and cell state of Schwann cells during culture in vitro. PLoS One. 2015;**10**:e0123278. DOI: 10.1371/journal.pone.0123278

[127] Hörner SJ, Couturier N, Gueiber DC, Hafner M, Rudolf R. Development and in vitro differentiation of Schwann cells. Cells. 2022;**11**:3753. DOI: 10.3390/cells11233753

[128] Pan Y, Cai S. Current state of the development of mesenchymal stem cells into clinically applicable Schwann cell transplants. Molecular and Cellular Biochemistry. 2012;**368**:127-135. DOI: 10.1007/s11010-012-1351-6

[129] Tomita K, Madura T, Mantovani C, Terenghi G. Differentiated adipose-derived stem cells promote myelination and enhance functional recovery in a rat model of chronic denervation. Journal of Neuroscience Research. 2012;**90**:1392-1402. DOI: 10.1002/jnr.23002

[130] Kingham PJ, Kolar MK, Novikova LN, Novikov LN, Wiberg M. Stimulating the neurotrophic and angiogenic properties of human adipose-derived stem cells enhances nerve repair. Stem Cells and Development. 2014;**23**:741-754. DOI: 10.1089/scd.2013.0396

[131] Faroni A, Smith RJP, Lu L, Reid AJ. Human Schwann-like cells derived from adipose-derived mesenchymal stem cells rapidly de-differentiate in the absence of stimulating medium. The European Journal of Neuroscience. 2016;**43**:417-430. DOI: 10.1111/ejn.13055

[132] di Summa PG, Kingham PJ, Raffoul W, Wiberg M, Terenghi G, Kalbermatten DF. Adipose-derived stem cells enhance peripheral nerve regeneration. Journal of Plastic, Reconstructive & Aesthetic Surgery. 2010;**63**:1544-1552. DOI: 10.1016/j.bjps.2009.09.012

[133] di Summa PG, Kalbermatten DF, Pralong E, Raffoul W, Kingham PJ, Terenghi G. Long-term in vivo regeneration of peripheral nerves through bioengineered nerve grafts. Neuroscience. 2011;**181**:278-291. DOI: 10.1016/j.neuroscience.2011.02.052

[134] Georgiou M, Golding JP, Loughlin AJ, Kingham PJ, Phillips JB. Engineered neural tissue with aligned, differentiated adipose-derived stem cells promotes peripheral nerve regeneration across a critical sized defect in rat sciatic nerve. Biomaterials. 2015;**37**:242-251. DOI: 10.1016/j.biomaterials.2014.10.009

[135] Tomita K, Madura T, Sakai Y, Yano K, Terenghi G, Hosokawa K. Glial differentiation of human adipose-derived stem cells: Implications for cell-based transplantation therapy. Neuroscience. 2013;**236**:55-65. DOI: 10.1016/j.neuroscience.2012.12.066

[136] Liau LL, Ruszymah BHI, Ng MH, Law JX. Characteristics and clinical applications of Wharton's jelly-derived mesenchymal stromal cells. Current Research in Translational Medicine. 2020;**68**:5-16. DOI: 10.1016/j.retram.2019.09.001

[137] Sibov TT, Severino P, Marti LC, Pavon LF, Oliveira DM, Tobo PR, et al. Mesenchymal stem cells from umbilical cord blood: Parameters for isolation, characterization and adipogenic differentiation. Cytotechnology. 2012;**64**:511-521. DOI: 10.1007/s10616-012-9428-3

[138] Lin YJ, Lee YW, Chang CW, Huang CC. 3D spheroids of umbilical cord blood MSC-derived Schwann cells promote peripheral nerve regeneration. Frontiers in Cell and Developmental Biology. 2020;**8**:604946. DOI: 10.3389/fcell.2020.604946

[139] Zhang C, Gong J, Zhang J, Zhu Z, Qian Y, Lu K, et al. Three potential elements of developing nerve guidance conduit for peripheral nerve regeneration. Adavanced Functional Materials. 2023;**33**:2302251. DOI: 10.1002/adfm.202302251

[140] Colen KL, Choi M, Chiu DTW. Nerve grafts and conduits. Plastic and Reconstructive Surgery. 2009;**124**:e386-e394. DOI: 10.1097/PRS.0b013e3181bf8430

[141] Moskow J, Ferrigno B, Mistry N, Jaiswal D, Bulsara K, Rudraiah S. Review: Bioengineering approach for the repair and regeneration of peripheral nerve. Bioactive Materials. 2019;**4**:107-113. DOI: 10.1016/j.bioactmat.2018.09.001

[142] Liu Y, Hsu SH. Biomaterials and neural regeneration. Neural Regeneration Research. 2020;**15**:1243-1244. DOI: 10.4103/1673-5374.272573

[143] Gu X, Ding F, Williams DF. Neural tissue engineering options for peripheral nerve regeneration. Biomaterials. 2014;**35**:6143-6156. DOI: 10.1016/j.biomaterials.2014.04.064

[144] Soman SS, Vijayavenkataraman S. Perspectives on 3D bioprinting of peripheral nerve conduits. International Journal of Molecular Sciences. 2020;**21**:5792. DOI: 10.3390/ijms21165792

[145] Pedde RD, Mirani B, Navaei A, Styan T, Wong S, Mehrali M, et al. Emerging biofabrication strategies for engineering complex tissue constructs. Advanced Materials. 2017;**29**:1606061. DOI: 10.1002/adma.201606061

[146] Zhu W, Tringale KR, Woller SA, You S, Johnson S, Shen H, et al. Rapid continuous 3D printing of customizable peripheral nerve guidance conduits. Materials Today. 2018;**21**:951-959. DOI: 1016/j.mattod.2018.04.001

[147] Yang GH, Yeo M, Koo YW, Kim GH. 4D bioprinting: Technological advances in biofabrication. Macromolecular Bioscience. 2019;**19**:e1800441. DOI: 10.1002/mabi.201800441

Chapter 17

The Function and Effectiveness of MSCs in Preclinical Models of Diseases

Zhilang Xie, Huayin Zhou, TingTing Ou, Wendan Pei and Xingliang Fan

Abstract

Mesenchymal stem cells (MSCs) have shown immense promise across diverse fields, particularly in treating various diseases. A multitude of preclinical investigations, encompassing both laboratory cellular-level studies and animal trials, have demonstrated the significant effectiveness of MSCs in addressing a broad spectrum of diseases. These studies collectively highlight the enormous potential of therapies utilizing MSCs across a range of medical scenarios. From neurological and cardiovascular disorders to autoimmune and musculoskeletal disorders, including recent challenges like COVID-19, MSCs present therapeutic possibilities due to their versatile capabilities. These discoveries lay a robust foundation for further clinical research and the eventual translation of MSC-based treatments into mainstream clinical practice, offering renewed optimism for patients contending with complex medical conditions.

Keywords: mesenchymal stem cells, stem cell transplantation, preclinical model, autoimmune diseases, neurological disorders, cardiovascular disorders, COVID-19, musculoskeletal disorders

1. Introduction

Mesenchymal stem cells (MSCs) are multipotent stem cells derived from mesoderm, which have the basic characteristics of self-renewal and multidirectional differentiation. Furthermore, MSCs demonstrate the capacity for immune regulation and tissue repair and regeneration, thus presenting extensive prospects in regenerative medicine, disease treatment, tissue engineering, and related domains.

Preclinical research has robustly demonstrated the efficacy and significance of MSCs in disease treatment, providing a sound theoretical and practical basis for their clinical application [1, 2]. Regarding immune regulation, these studies have confirmed that MSCs can stimulate the secretion of anti-inflammatory factors, consequently mitigating inflammatory responses and tissue damage. Additionally, concerning tissue repair and regeneration, preclinical investigations have highlighted MSCs' potential for the repair of tissue defects.

This chapter presents an overview of the characteristics and therapeutic mechanisms of MSCs, with a specific emphasis on their applications and challenges in treating autoimmune diseases, neurological diseases, cardiovascular diseases, and COVID-19, alongside their therapeutic efficacy in preclinical experimental settings. It is anticipated that through continued research into the characteristics and therapeutic mechanisms of MSCs, their utilization in clinics will deepen, offering promising prospects for an expanding number of patients.

2. Therapeutic potential of MSCs in different disease categories

2.1 Neurological disorders

Nervous system diseases comprise conditions that impair the function of the nervous system. They encompass various types such as neurodegenerative diseases, cerebrovascular diseases, and neurodevelopmental disorders. Among them, Parkinson's disease (PD) and Alzheimer's disease (AD) stand out as the most prevalent senile neurodegenerative diseases (ND), both characterized by neuroinflammation [3]. PD is primarily marked by the progressive loss of dopaminergic (DA) neurons in the midbrain substantia nigra (SN) [4]. Currently, the primary treatment for PD involves L-DOPA but prolonged usage often leads to L-DOPA-induced dyskinesia (LID), significantly impacting patients' quality of life [5]. AD, on the other hand, is characterized by the accumulation of amyloid beta (Aβ) plaques in the extracellular space and the formation of neurofibrillary tangles within cells. Presently, Aβ targeted therapy for AD is costly and associated with side effects [6].

Cerebrovascular disease (CVD) is characterized by clinical manifestations such as hemiplegia, speech disorders, and coma, resulting from cerebral ischemia, hypoxia, edema, necrosis, and functional impairment due to lesions affecting the arteries supplying or innervating the brain. Clinically, CVD can be categorized into two main types based on lesion nature: ischemic and hemorrhagic strokes. Ischemic stroke ranks among the top three leading causes of death and disability, yet effective treatment strategies remain elusive in clinical practice [7]. Hypoxic-ischemic encephalopathy (HIE), a possible consequence of perinatal asphyxia, stands as the primary cause of neonatal cerebral palsy. It significantly impacts both the physical and mental well-being of children and imposes a substantial burden on families [8].

2.1.1 Mechanisms of MSC therapy

MSCs exhibit significant potential and therapeutic efficacy in addressing nervous system diseases, leveraging their distinctive biological traits including paracrine function, multidirectional differentiation capacity, immunomodulatory ability, and promotion of angiogenesis. This offers a novel approach to treating nervous system diseases (**Figure 1**).

2.1.1.1 Paracrine function of MSCs

MSCs possess paracrine effects, facilitating nerve tissue repair and regeneration, apoptosis inhibition, and enhancement of the local brain microenvironment through the secretion of various neurotrophic factors and bioactive agents [9]. These paracrine effects are mediated, in part, by the release of extracellular vesicles (EVs) from MSCs. Notably, miRNAs within MSC-derived EVs, such as miR-125a,

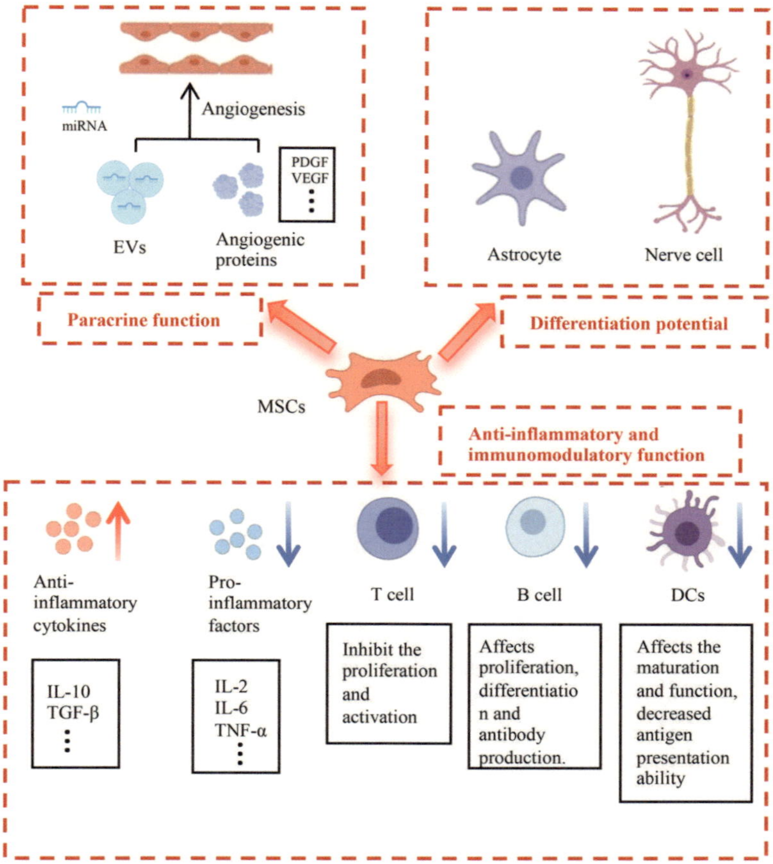

Figure 1.
Mechanism of MSCS in the treatment of nervous system diseases. (1) MSC promotes the repair and regeneration of nerve tissue by secreting a variety of neurotrophic factors and bioactive factors, and directly acts on endothelial cells by secreting angiogenic proteins to promote angiogenesis. (2) MSCs can differentiate into neurons or glial cells under specific conditions to replace damaged nerve tissue. (3) By secreting anti-inflammatory cytokines, MSC inhibits the expression of pro-inflammatory factors and alleviates neuroinflammation; In addition, they regulate the activity of immune cells, including inhibiting the function of T and B cells, as well as regulating the maturation and antigen presentation capacity of dendritic cells, reducing oxidative stress, and protecting nerve cells from further damage.

miR-423-5p, and miR-181b, target relevant genes like DLL4, Sufu, TRPM7, HIF-1α, and VEGF to promote endothelial cell angiogenesis [10, 11]. Moreover, MSC-derived EVs directly stimulate endothelial cells through the secretion of angiogenic proteins like platelet-derived growth factor (PDGF) and vascular endothelial growth factor (VEGF), enhancing their proliferation and migration to facilitate angiogenesis [12–14]. Braun et al. demonstrated that the angiogenic mechanism of MSC-derived EVs partly involves VEGF, as evidenced by experiments utilizing MSC-derived EVs and anti-VEGF antibodies on human umbilical vein endothelial cells (HUVECs) [13]. Gregorius et al. revealed that MSC-derived EVs regulate the expression of miRNAs associated with angiogenesis in hCMEC/D3 cells under hypoxic conditions, suggesting their potential to promote angiogenesis and improve neural function recovery [15]. Additionally, specific components within EVs derived from human umbilical cord MSCs (hUC-MSCs), such as adaptor protein complex 2 subunit α-1 (AP2A1) and adaptor protein complex 2 subunit β-1 (AP2B1), may play a crucial role in modulating the synaptic vesicular circulation signaling pathway for AD treatment [16]. Current research suggests that stroke treatment can involve modulating specific EV-derived miRNAs in acute ischemic stroke (AIS), with the enhanced functionality of MSCs post-stroke potentially linked to blood-brain barrier (BBB) permeability.

2.1.1.2 Differentiation potential of MSC

MSCs possess the capacity for multidirectional differentiation, including the ability to differentiate into nerve cells. Induction factors play a crucial role in directing the differentiation of MSCs by activating or inhibiting specific signaling pathways and modulating gene expression. When exposed to inducers such as NGF, GDNF, WNT1, and LMX1a, MSCs can be guided to differentiate into dopaminergic neurons through the activation of specific signaling pathways such as TRK receptors, PI3K/Akt, and MAPK, which holds significance in PD treatment. Khademizadehd et al. isolated MSCs from adipose tissue obtained from healthy volunteers and cultured them in a neural base medium supplemented with differentiation factors, leading to the expression of neuronal marker NeuN and dopaminergic marker tyrosine hydroxylase (TH), indicating the successful differentiation of dopaminergic neurons [17]. Studies have indicated that MSC transplantation therapy can mitigate brain tissue damage resulting from ischemic stroke through the directed proliferation and differentiation of nerve cells and alternative pathways [18]. This therapy bypasses limitations such as ethical concerns, isolation and culture complexities, immune rejection, and post-transplantation complications. Directed induction of MSC differentiation into neurons holds promise in effectively ameliorating nerve function impairments caused by cerebrovascular diseases. Furthermore, upon transplantation into damaged nervous systems, MSCs can differentiate into glial cells, which contribute to nervous system repair and regeneration by providing support and protection, participating in tissue regeneration, facilitating substance metabolism, supplying nutrients, and modulating immune responses.

2.1.1.3 Immunomodulatory function of MSCs

The utilization of MSCs in treating neurological diseases primarily hinges on their anti-inflammatory and immunomodulatory capabilities. Research indicates that MSCs can suppress the expression of pro-inflammatory factors like IL-2, IL-6, and TNF-α by secreting anti-inflammatory cytokines such as IL-10 and TGF-β, thereby mitigating neuroinflammation. Che et al. demonstrated that hUC-MSC-derived EVs effectively reduced lipopolysaccharide (LPS)- or H_2O_2-induced oxidative stress and pro-inflammatory cytokine expression. Furthermore, they noted that inhibiting the microglial NRF2/NF-κB/NLRP3 signaling pathway improved LPS/H_2O_2-induced neuroinflammation and oxidative stress [19]. By downregulating IL-1β and TNF-α, MSC-derived EVs promote the polarized development of M2 microglia and enhance A2 astrocyte activity, thus dampening their pro-inflammatory responses. This, in turn, alleviates ischemic stroke cell damage, enhances their protective effects on nerve cells, and accelerates nerve tissue functional recovery. Additionally, MSCs regulate immune cell activity by inhibiting T cell proliferation and function, decreasing B cell antibody production, and modulating dendritic cell (DC) maturation and antigen presentation capacity, thereby safeguarding against neuronal damage [20]. MSCs also can scavenge harmful free radicals, diminish oxidative stress, and shield nerve cells from further harm.

2.1.2 Preclinical application of MSC therapy

In order to evaluate the potential and efficacy of MSCs in the treatment of neurological diseases in real organisms, MSCs were utilized in animal model studies in the treatment of PD, AD, cerebrovascular diseases, and pediatric cerebral palsy. These studies have provided the scientific basis and therapeutic strategies for future clinical applications of MSC application in the human body (**Figure 2**).

2.1.2.1 MSC treatment of Parkinson's disease

Transplantation of MSCs effectively ameliorated motor and memory deficits as well as oxidative stress in rat models of PD via both intravenous and intracranial administration routes. In a PD rat model induced by rotenone (ROT), Essawy et al. conducted MSC transplantation utilizing two delivery methods: intracranial and intravenous administration [21, 20]. They observed restoration of superoxide dismutase (SOD) and catalase (CAT) activities post-transplantation. Moreover, MSC

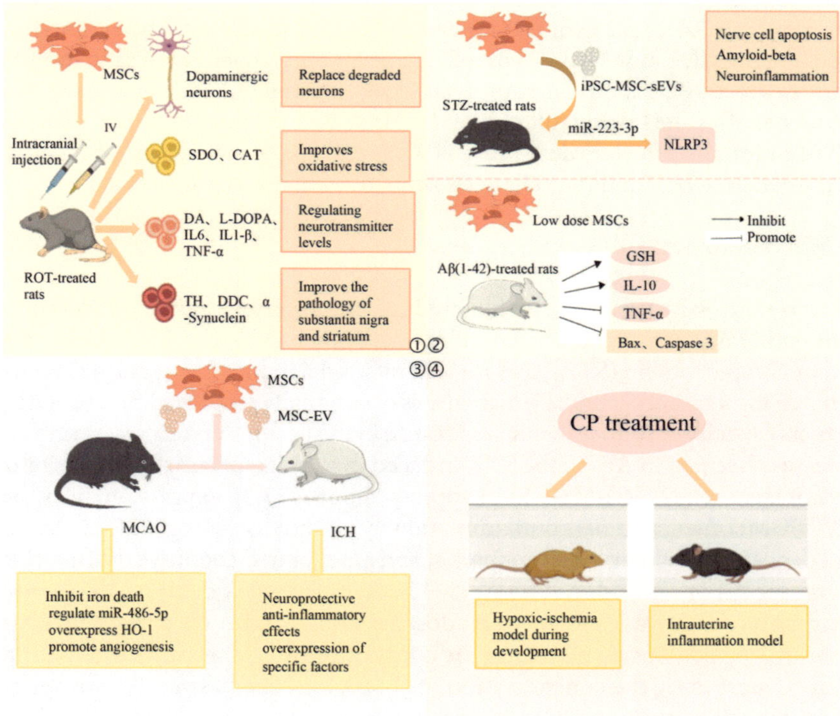

Figure 2.
Preclinical model of nervous system diseases and MSC treatment. MSC therapy can restore the activity of superoxide dismutase (SOD) and catalase (CAT), regulate the levels of dopamine (DA) L-Dopa, IL-6, IL-1β and TNF-α, and reverse the pathological changes induced by rotenone (ROT). The expression levels of tyrosine hydroxylase, dopa decarboxylase and α-synuclein were normalized in the substantia nigra and striatum of male rats. MSC-derived EVs can reduce NLRP3/GSDMD-mediated neuroinflammation and reduce amyloid deposition. Transplantation of low-dose MSCs increased the concentration of the antioxidant glutathione (GSH), increased the concentration of IL-10, and decreased the concentration of TNF-α and the expression of apoptosis factors (BAX and CASPASE 3). MSC-EVs can inhibit iron death, increase miR-486-5p content, overexpress HO-1 and specific factors, and play neuroprotective and anti-inflammatory effects. The two most common animal models for cerebral palsy treatment are hypoxic-ischemia model during development and intrauterine inflammation model.

transplantation regulated dopamine (DA), L-Dopa (L-DOPA), IL-6, IL1-β, and TNF-α levels, reversed ROT-induced pathological alterations, and normalized expression levels of tyrosine hydroxylase, dopa decarboxylase, and α-synuclein in the substantia nigra and striatum tissues of male rats. Research indicates that both delivery routes of MSCs improve motor function, protect the nigrostriatal system, and enhance striatal dopamine release. While neither route demonstrates significant superiority over the other in treatment outcomes, intravenous administration appears to be safer and more feasible. Besides, EVs derived from MSCs have exhibited superior neuroprotective efficacy compared to L-Dopa in Parkinson's disease (PD) patients. Mohamed et al. conducted a study using a rat model of PD to assess the inhibitory effects of MSC-derived EVs *versus* L-Dopa on neurodegeneration [22]. Their findings revealed that the MSC-derived EV treatment group displayed enhancements in motor function, histopathology, α-synuclein, PARKIN, and DJ-1 levels compared to the L-Dopa treatment group. Notably, MSC-derived EVs demonstrated greater improvements in miRNA-34b and circRNA.2837 levels compared to the L-Dopa treatment group.

Furthermore, it was found that human embryonic stem cell (HESC)-derived dopaminergic progenitor cell transplantation is a promising strategy for the treatment of PD, which has been approved by the Korean Ministry of Food and Drug Safety and has begun clinical trials [23]. Park et al. evaluated the toxicity, biodistribution, and tumor-forming potential of clinically grade HESC-derived dopaminergic progenitor cells in immunodeficient rats. In a semi-PD rat model, significant dose-dependent behavioral improvements were observed after the transplantation of dopaminergic progenitor cells at different doses. The minimum effective dose range is 5000–10,000 dopaminergic progenitor cells, with a low cell dose (3.15 million cells) determined for human clinical trials.

2.1.2.2 MSC treatment of Alzheimer's disease

Numerous investigations have highlighted the use of various animal models in experiments targeting AD treatment. These models primarily include the AD mouse model induced by Streptozotocin (STZ), and the AD rat model induced by Aβ(1-42) peptide. STZ, known for its ability to induce diabetes, is occasionally employed to create AD models due to its capacity to induce insulin resistance in the brain, mirroring certain pathological processes observed in AD. In the STZ-induced model, intracranial injection of induced pluripotent stem cell (iPSC)-MSC-EVs suppressed miR-223-3p expression, mitigated NLRP3/GSDMD-mediated neuroinflammation by directly targeting NLRP3, decreased amyloid deposition and nerve cell apoptosis, and ameliorated cognitive dysfunction [24].

In a rat model of AD induced by Aβ(1-42) peptide, MSCs exhibited a neuroprotective effect that was dependent on dosage. Transplantation of low-dose MSCs improved AD symptoms by elevating the concentration of the antioxidant glutathione (GSH), increasing the concentration of IL-10, and decreasing the concentration of TNF-α, as well as reducing the expression levels of apoptosis factors (BAX and CASPASE 3). These effects were attributed to the neuroprotective, anti-inflammatory, and antioxidant properties of MSCs. However, increasing the dosage of MSCs resulted in a diminished therapeutic effect in treating AD. In a comparative experiment involving untreated rats, rats treated with Aβ to simulate the pathological state of Alzheimer's disease (AD group), and rats treated with MSCs (AD+MSC group) to simulate MSC treatment following Aβ treatment, the expression levels of peroxisome proliferator-activated receptor γ (PPARγ), lncRNATUSC7, and CD36 genes were found to be enhanced in the AD+MSC group compared to the AD group. These genes were observed to act through non-coding RNAs (ncRNAs), potentially leading to a reduction in plaque formation in cell receptors. The study's findings suggest that MSC treatment may enhance the cognitive function of AD rat models by modulating specific molecular pathways. This sheds light on potential molecular mechanisms and therapeutic strategies for the application of MSCs in AD treatment [25].

2.1.2.3 MSC treatment of cerebrovascular disease

When exploring the clinical application of MSC in cerebrovascular diseases, researchers commonly use middle-cerebral artery occlusion (MCAO) model and intracerebral hemorrhage (ICH) model.

In the mouse MCAO model, Deng et al. discovered that MSC-EV overexpression of steroid receptor coactivator 3 (SRC-3) can significantly inhibit lipid peroxidation and ferroptosis of neurons, reduce the production of inflammatory factors, brain tissue water content, and infarct volume [26]. This also inhibited the activation of microglia and astrocytes, leading to improved neural function in MCAO mice. Zhu et al. found that miR-486-5p within MSC-EVs could alleviate cerebral ischemia and reperfusion (I/R) injury and improve neurological functional deficits [27]. Yang et al. showed that MSCs overexpressing HO-1 had a significant therapeutic effect on excessive inflammatory damage post-stroke in mice, ultimately promoting recovery after ischemic stroke [28]. In the MCAO-treated C57BL/6 mouse model, Lu et al. [28] demonstrated that human-induced multipotent stem cell-derived MSC exosomes (hiPS-MSC-EV) promoted the proliferation of HT22 cells, reduced infarct volume, and improved the spontaneous motor ability of mice [29]. Additionally, the expression of VEGF and CXCR4 protein enhanced angiogenesis in the infarct hemisphere of MCAO-treated mouse models. These studies collectively indicate that MSCs and MSC-EVs hold significant therapeutic potential for treating cerebrovascular diseases, especially in the MCAO model. Through mechanisms such as inhibiting ferroptosis, regulating the miR-486-5p and PTEN axes, overexpressing HO-1, or promoting angiogenesis, MSCs and their exosomes can alleviate ischemic damage and enhance neural function.

In mouse ICH models, MSCs and MSC-EVs function through various mechanisms, including neuroprotection, anti-inflammatory effects, regulation of gut microbiota and metabolism, and modulation of specific molecular pathways like the TRAF6/NF-κB axis, demonstrating potential for treating cerebrovascular diseases. Furthermore, the therapeutic effect of MSCs can be enhanced by the overexpression of specific factors (such as GDNF) or specific miRNAs (such as miR-150-3p) [30–32].

2.1.2.4 MSC treatment of cerebral palsy in children

In the study of cerebral palsy treatment, the two most common animal models are the developmental hypoxic-ischemia model and the intrauterine inflammation model. Developmental hypoxic-ischemia models are created by inducing hypoxic-ischemic conditions in newborn or developing animals (usually rats or mice) to simulate perinatal brain damage, which is a common cause of cerebral palsy. This model effectively replicates the neurological damage and dysfunction seen in cerebral palsy and is widely used to study the pathological mechanisms of the disease and to evaluate treatments, including MSC therapy. Studies have shown that human umbilical cord-derived MSC (hUC-MSC) transplantation is a promising strategy for treating cerebral palsy, and that the route of administration is crucial for hUC-MSC to provide adequate neuroprotection [33].

The intrauterine inflammation model simulates intrauterine infection by injecting LPS or other inflammatory inducers into pregnant animals, replicating a known risk factor for cerebral palsy. Intrauterine inflammation can lead to fetal brain damage, resulting in cerebral palsy-like symptoms in animals after birth. This model is valuable for studying the effects of inflammation on brain development and the role of MSCs in providing anti-inflammatory and neuroprotective benefits. Currently, the literature on the effects of MSCs in the intrauterine inflammation model is limited. Many research teams are actively conducting preclinical studies to evaluate the safety and efficacy of MSCs for treating pediatric cerebral palsy using animal models. These studies explore various potential mechanisms of MSCs, typically involving MSC transplantation, assessment of different administration routes, and evaluation of motor function, nerve regeneration, immunomodulatory, and neuroprotective effects in animal models to determine the most effective treatment approaches.

2.1.3 Summary

To sum up, in the treatment of PD, MSCs can alleviate symptoms by differentiating into dopamine neurons or promoting the regeneration of dopamine neurons. In AD treatment, MSCs can repair brain damage by enhancing autophagy, and their EVs have been shown to reduce disease progression. For stroke treatment, MSCs enhance therapeutic effects by reducing inflammation through immune regulation, releasing nutritional factors, and inducing angiogenesis and neurogenesis. In treating children with cerebral palsy, MSCs are considered a potentially effective method due to their multidirectional differentiation potential and immunoregulatory properties.

It is important to note that although MSCs show promise in treating neurological diseases, current research is still in its early stages. Further research and clinical trials are necessary to determine the optimal application, dosage, and long-term effects. Therefore, in practical applications, it is essential to carefully evaluate the patient's condition and follow medical advice and guidance.

2.2 Cardiovascular diseases

Currently, common cardiovascular diseases include coronary heart disease, myocardial infarction, cardiomyopathy, heart failure (HF), and arrhythmias. MSCs have shown potential in treating cardiovascular diseases such as myocardial infarction and heart failure. The occurrence of cardiovascular diseases is often associated with aging and frailty. As a candidate for tissue regeneration, MSCs hold significant promise in treating these conditions. However, MSCs face challenges such as replicative senescence, limiting their large-scale expansion. Additionally, MSCs are difficult to extract from patients with systemic diseases, and the extraction process is time-consuming [34]. Moreover, long-term *in vitro* expansion and implantation of MSCs may pose a risk of tumor formation [35]. Despite these challenges, advancements in MSC treatment and medical technology have increased the overall survival and longevity of heart failure patients, underscoring the considerable potential of MSCs in treating cardiovascular diseases.

2.2.1 Mechanisms of MSC therapy

2.2.1.1 MSC effects in myocardial infarction

Myocardial infarction (MI), characterized by myocardial necrosis due to acute coronary artery occlusion and insufficient blood supply to the affected myocardium, is the leading cause of heart failure (HF) and death globally [36]. In all living tissues, the blood supply must meet the oxygen demand of the muscles [37]. When the heart rate is too fast or blood pressure drops, the imbalance between blood supply and oxygen demand can result in myocardial damage.

MSCs can mitigate myocardial damage by reducing inflammatory cell infiltration and the release of inflammatory cytokines (**Figure 3**). Recent research showed that MSC-EVs or allogeneic peripheral blood mononuclear cells (PBMCs) inhibit the proliferation of consubstantial PBMCs and decrease the secretion of proinflammatory cytokines such as IL-12p40 and TNF-α. Additionally, MSCs reduce T cell infiltration and promote the transcription of the anti-inflammatory factorIL-10 in the infarction core area [38]. T cell infiltration indicates an ongoing immune response, so a decrease in T cell infiltration highlights the anti-inflammatory effect of MSCs.

MSCs can also secrete various growth factors and cytokines to stimulate angiogenesis in damaged myocardium and treat damage caused by MI. For instance, MSCs secrete insulin-like growth factor 1 (IGF-1) and VEGF, which can induce blood vessel and cell formation [39–41]. Notably, IGF-1 plays a crucial role in cardiac muscle growth. One month after delivering Ad-hVEGF165 (a type of VEGF) to damaged heart tissue, the microvessel density in the tissue significantly increased, confirming that MSCs can promote angiogenesis to treat damaged myocardium [42]. Other studies suggest that endothelial cells (OECs) incorporated with VEGF or MSC-EV migrate actively to agarose sites containing these factors, with a thickened vascular network forming over time [38]. These experiments demonstrate that MSCs can enhance angiogenesis and improve cardiac function by recruiting VEGF and other regenerative cells.

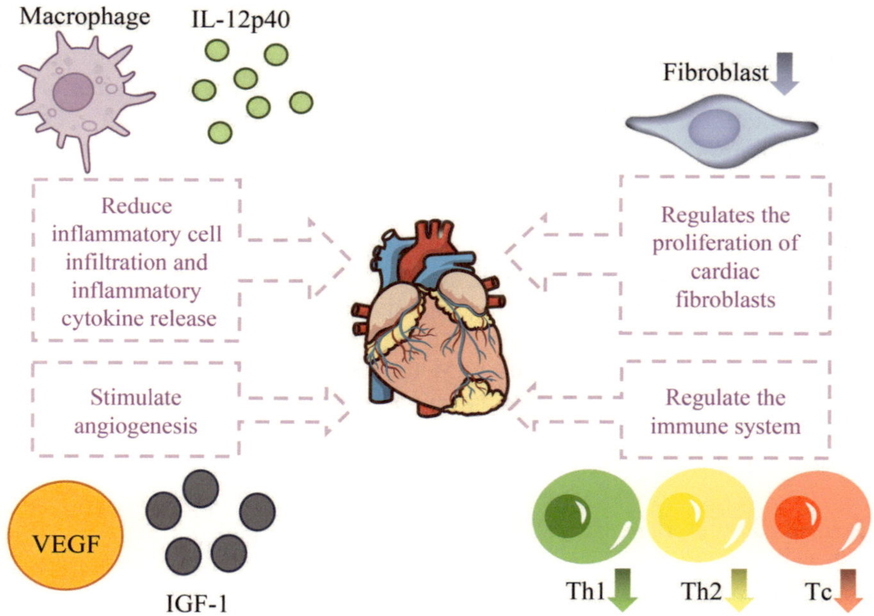

Figure 3.
Mechanism of MSCS in the treatment of cardiovascular diseases. First, MSCs can achieve therapeutic purposes by reducing inflammatory cell infiltration and the release of inflammatory factors, such as macrophages and inflammatory factors IL-12p40 and TNF-α. Secondly, MSCs can stimulate angiogenesis through growth factors such as IGF-1 and VEGF. In addition, the inhibition of proliferation of T cells such as type 1helper T (Th1) cells, type 2 helper T (Th2) cells, and cytotoxic T cells (Tc) after co-culture with MSCs also indicated the therapeutic effect of MSCs. Finally, MSCs may reduce fibrosis by reducing the proliferation of fibroblasts.

2.2.1.2 MSCs effects in heart failure

HF is a clinical condition characterized by tissue hypoperfusion and congestion in the systemic or pulmonary circulation, resulting from impaired heart function and reduced ejection and filling capacity. It represents the final stage of various cardiovascular diseases, including congenital heart disease and hypertension [43]. Studies have shown that hypoxia and mitochondria-induced apoptosis are crucial mechanisms in HF. Under hypoxic conditions, the heart activates hypoxia-inducible factor (HIF-1α) to induce angiogenesis during the compensatory period, and it undergoes adaptive hypertrophy to maintain blood supply [44]. During decompensation, angiogenesis is suppressed, and the blood supply fails to meet the needs of the heart muscle, leading to a deterioration in heart function, which can eventually result in HF. Mitochondrial metabolism is also closely linked to HF. When pyruvate oxidation in mitochondria decreases and lactic acid production increases, the protective effect of plasma amino acid metabolism on the failing heart is enhanced. MSCs are also highly efficient mitochondrial donors, making them promising for HF treatment [45].

First of all, MSC protects cardiomyocytes from further damage through its anti-fibrotic effect. MSCs were found to inhibit fibroblast proliferation by enhancing fibroblast to express a large number of factors such as myocardium and elastin [39, 46]. Moreover, MSC conditioned medium inhibited the translation and transcription of type I and type III collagen genes (COl1a1 and COl3a1) in fibroblasts, suggesting that MSCs can regulate the proliferation of cardiac fibroblasts and down-regulate the anti-fibrotic effects of type I and type III collagen synthesis, thereby safeguarding cardiomyocytes.

Additionally, MSC possesses the ability to secrete vascular growth factors, promoting angiogenesis, enhancing myocardial blood supply, and aiding in myocardial repair and functional restoration. Through the secretion of VEGF, indoleamine 2,3-dioxygenase (IDO), and Hepatocyte Growth Factor (HGF), MSC can enhance myocardial function and stimulate angiogenesis. Notably, the expression of HGF in hUC-MSCs was approximately 55 times greater than that observed in bone marrow-derived MSCs, indicating a significant potential for therapeutic efficacy [47].

Furthermore, MSCs exhibit an immunomodulatory role, regulating the immune system's response to prevent excessive damage to the cardiac muscle caused by an exaggerated immune reaction. Regarding anti-inflammatory mechanisms, MSC can modulate interleukins such as IL-6, along with anti-inflammatory factors like TGF-β1 and TGF-β3, as well as proinflammatory factors like Prostaglandin E2 (PGE2) and other secretions, to achieve immune regulation. Moreover, co-culturing with MSC resulted in the inhibition of the proliferation of various T cells, including helper T cells 1, helper T cells 2, and cytotoxic T cells, suggesting MSC's role in promoting immune system reconstruction and balance, ultimately inhibiting excessive immune responses [47].

2.2.2 Preclinical application of MSC therapy

2.2.2.1 MSCs treatment of myocardial infarction

Previous experiments have demonstrated that MSCs can enhance heart function. Shake et al. performed MSC injections on 12 pigs with induced myocardial infarction (MI) and found that, compared to the control group, MSC implantation significantly improved systolic wall thickening by preventing abnormal systolic wall thinning [48]. This provides compelling evidence for the potential application of MSCs in treating MI.

Another group of researchers further confirmed the efficacy of MSCs in treating MI. Schuleri et al. treated pigs with heart infarction and then randomly assigned them to a placebo group and an MSC treatment group. Over time, it was observed that while the scar size remained unchanged in the placebo group, the MSC treatment group exhibited a tendency to reduce scar volume [49]. More importantly, the treatment group showed a reduction in infarct size, an increase in systolic force and perfusion, and an improvement in overall left ventricular function compared to the placebo group.

In 2018, a study involving miniature pigs with MI treated with intravenous MSCs found that, compared to the control group injected with phosphate-buffered solution (PBS), the MSC treatment group showed an increase in left ventricular ejection fraction (LVEF) and a reduction in infarction area [50]. Additionally, the study revealed that the expression of TNF-α and IL-6 genes in the infarct and border areas was significantly lower in the MSC group than in the PBS group. The mRNA expression level of VEGF in the infarct border area was also increased. This indicates that MSC injection promotes angiogenesis and inhibits inflammation in the infarct area. These results suggest that MSCs can enhance the production of angiogenic factors and reduce inflammation.

In another study using rat models to simulate MI, researchers ligated the left coronary artery and randomly divided the rats into two groups. One group received an intravenous injection of MSCs 3 hours later, while the control group received an intravenous injection of PBS. The results showed a significant reduction in MI area and an increase in capillary density in the MSC treatment group compared to the control group. Additionally, the treatment group exhibited decreased left ventricular end-diastolic pressure and an increased isovolumic systolic period [51]. These findings indicate that MSCs enhance cardiac function after MI by promoting angiogenesis in the ischemic myocardium, suggesting a potential therapeutic strategy for MI treatment.

2.2.2.2 MSC treatment of heart failure

Direct epicardial injection of allogeneic MSCs has been shown to increase blood vessel density and enhance myocardial function in canine models of ischemic HF. Additionally, MSCs have been confirmed to support the formation of new blood vessels from endothelial colony-derived cells in other experiments [52]. However, despite the promising results from numerous trials, the efficacy of MSCs in treating HF patients remains controversial. Future experiments are needed to further validate the role of MSCs in improving HF through angiogenesis stimulation and immune regulation.

2.2.3 MSC-EVs in cardiovascular therapy

MSC can treat MI by secreting EVs. EVs secreted by MSCs from adipose tissue represent a promising therapeutic approach for cardiac healing post-MI. Hypoxia-stimulated MSC-derived EVs have been shown to reduce cardiac apoptosis after infarction and enhance cardiomyocyte survival under hypoxic conditions.

In an experiment involving male Sprague-Dawley rats, the experimental group underwent coronary artery ligation after anesthesia, while the control group received a sham ligation. Subsequently, PBS was injected into the myocardium near the ligation area in all rats, and EVs from various MSC sources were injected into the experimental group. After a designated period, the rats were euthanized, and cell viability was assessed using transmission electron microscopy, real-time quantitative PCR, and other techniques. The results demonstrated that the number of apoptotic cardiomyocytes significantly decreased in the animal tissue samples treated with EVs compared to those treated with PBS [53]. This indicates that MSC-derived EVs have substantial potential in treating MI and reducing cardiac scarring.

In another study, Monguio et al. further explored the role of MSC-EVs in MI repair. They implanted a heart stent rehydrated with a hydrogel containing MSC-EVs or buffer into a pig model of MI, securing it to the ischemic heart muscle with surgical glue. Compared to the untreated group, the group treated with the MSC-EV-implanted stent showed a significant increase in vascularization in the infarction core area and a reduction in distal myocardial fibrosis [38]. This suggests that the MSC-EV-implanted stent has the potential to promote angiogenesis. Additionally, this animal study confirmed that MSCs can reduce cardiac fibrosis by increasing the transcription of fibrosis regulatory factors, thereby aiding in MI treatment. Monguio et al. conducted another experiment involving a pig model with permanent coronary artery ligation to simulate MI. The pigs were divided into two groups: untreated and treated. In the treatment group, acellular pericardial scaffolds filled with peptide hydrogels and MSC-EVs were placed on the myocardium of post-infarction pigs, while the control group received only buffers. After 30 days, the treated animals showed improved cardiac function, demonstrated by a significantly increased ventricular ejection fraction, smaller scar size, and reduced myocardial fibrosis [54]. These results indicate that MSC-EVs have the potential to regulate ischemic injury and aid in post-MI repair.

2.2.4 Summary

In summary, MSCs and their bioactive secretions play a significant role in the treatment of cardiovascular diseases. MSCs hold significant promise for the treatment of cardiovascular diseases. Numerous experiments have demonstrated their capabilities in anti-fibrosis, promoting angiogenesis, and immune regulation. Furthermore, the use of MSC secretions is emerging as a safer therapeutic approach. As research continues to delve into the mechanisms by which MSCs and their secretions treat cardiovascular diseases, the goal is to develop safer and more effective treatment protocols. These advancements aim to enhance clinical outcomes, reduce treatment costs, and bring hope to more patients suffering from cardiovascular conditions.

2.3 Autoimmune diseases

Autoimmune diseases are a complex group of conditions characterized by the immune system mistakenly attacking the body's own tissues and organs. These diseases involve various immune cell populations, including T lymphocytes, B lymphocytes, and monocytes/macrophages, which are crucial in the onset and persistence of the disease [55]. Autoimmune diseases are marked by chronic inflammation and tissue damage, often resulting from the abnormal recognition and response of immune cells to self-antigens.

Rheumatoid arthritis (RA) and systemic lupus erythematosus (SLE) are notable examples of autoimmune diseases where the abnormal activation of T and B lymphocytes and the production of autoantibodies play a pivotal role in disease progression [55]. However, RA and SLE differ in their clinical manifestations and the organs they affect. RA primarily targets the joints, causing inflammation and erosion that impair joint function. In contrast, SLE affects multiple organs and systems, producing a broad range of autoantibodies, particularly against nuclear components [56]. These antibodies form immune complexes with autoantigens, which are deposited in the skin, joints, kidneys, and other tissues, triggering inflammatory responses and resulting in multisystem damage [57]. Type 1 diabetes (T1D) is another prevalent autoimmune disease, typically treated with exogenous insulin injections to maintain physiological blood glucose levels. However, this approach can lead to long-term complications [58]. In addition, autoimmune hepatitis (AIH) and Crohn's disease (CD) are significant autoimmune conditions. AIH is characterized by the presence of autoantibodies and elevated serum levels of immunoglobulin G (IgG) [59]. CD is an intestinal disease marked by immune-mediated chronic inflammation, with pathogenesis involving multiple factors [60, 61].

Autoimmune diseases exhibit complex and diverse clinical features, involving abnormal immune system responses and recognition of autoantigens. Despite the shared characteristics, each disease affects different organs, presents distinct clinical manifestations, and requires unique treatment strategies. Understanding these commonalities and differences is crucial for comprehending their pathogenesis and developing new therapeutic approaches. MSCs offer promising potential for targeting these mechanisms, providing more effective treatment options and opportunities for clinical management of these diseases.

2.3.1 Mechanisms of MSC therapy

The mechanisms and effects of MSCs in treating autoimmune diseases have been extensively studied. These studies emphasize the crucial role of MSCs in immune regulation and highlight the therapeutic potential of MSCs and their derivatives.

MSCs are widely recognized for their immunomodulatory properties and their ability to modulate immune responses through various mechanisms. First, MSCs can inhibit the proliferation and activation of T cells by secreting several soluble cytokines, including nitric oxide, TGF-β, and PGE2 [62–64]. They also influence the expression of cell surface markers, thereby affecting the proliferation and function of specific cell types. Additionally, MSCs negatively regulate B cells either through direct contact or by secreting cytokines, which inhibits B cell proliferation and activation, thus reducing autoantibody production [65]. When co-cultured with DCs, MSCs can inhibit the differentiation and maturation of DCs, thereby modulating their immune activity [66].

Furthermore, MSC derivatives, such as EVs, have shown therapeutic potential. These EVs carry a variety of active substances, including cytokines, growth factors, and miRNAs, which influence the function of immune cells through intercellular communication [67]. MSC-EVs exhibit low immunogenicity and can cross the blood-brain barrier, offering potential advantages in treating autoimmune diseases that involve the central nervous system. Specific miRNAs and other non-coding RNAs enriched in MSC-EVs can regulate immune cell responses and promote the repair of damaged tissues, making them promising candidates for treating autoimmune diseases [68].

2.3.1.1 MSC effects on DCs

The effects of MSCs on DCs in autoimmune diseases share several similarities. Generally, MSCs inhibit the maturation, migration, and antigen presentation of DCs while promoting the production of regulatory T cells (Treg cells) [69]. Additionally, MSCs can prevent the differentiation of monocytes into DCs by secreting various factors such as IL-6, or they may promote differentiation into macrophages rather than antigen-presenting DCs, and inhibit the production of pro-inflammatory cytokines like IL-1Ra (**Figure 4**) [70, 71]. The role of MSCs can vary in specific disease contexts.

In RA, synovial cells or MSCs act as antigen-presenting cells, activating T cells through interactions with T lymphocytes and promoting their proliferation. MSCs are also involved in the pathogenesis of RA by activating pattern recognition receptors such as TLR2 and TLR4 [72, 73]. In SLE, plasmacytoid dendritic cells (pDCs) are the main producers of IFN-α, inducing immature B lymphocytes to differentiate into plasma cells without promoting regulatory B cells (Breg cells) to secrete IFN-α [74]. While IFN-α has little direct effect on MSCs, it can inhibit cell proliferation and osteogenic differentiation potential when MSCs are treated *in vitro* [75].

In T1D, MSCs exert immunosuppressive effects by expressing inhibitory molecules such as IDO, PGE2, and TGF-β, which affect the maturation and function of DCs and inhibit the activation and proliferation of T cells [76]. In CD, MSCs inhibit the maturation of DCs *in vitro*, impair their antigen-presenting and migration functions, and induce their differentiation into regulatory dendritic cells through the Notch signaling pathway [77].

2.3.1.2 MSC effects on macrophages

In autoimmune diseases, the interaction between macrophages and MSCs is crucial for disease progression and tissue repair. This interaction is essential for maintaining the inflammatory microenvironment and tissue balance (**Figure 4**). Generally, MSCs can inhibit the pro-inflammatory state of macrophages by secreting factors such as IDO and COX-2/PGE2 [78]. They promote the transformation of macrophages to an anti-inflammatory state, regulate the levels of overproduced inflammatory mediators (e.g., reducing IL-12 and TNF-α), and enhance their anti-inflammatory functions [79]. This modulation has positive implications for alleviating the inflammatory response in RA. However, the role of MSCs can vary across different autoimmune diseases. In CD, MSCs mainly exert their therapeutic effects by inhibiting the maturation and function of macrophages and influencing their production of regulatory cytokines [61]. In AIH, MSCs alleviate the disease's pathological process by regulating macrophage polarization and inhibiting their excessive production of inflammatory mediators [59]. These differences highlight that, while the regulatory interactions between MSCs and macrophages are commonly beneficial across various autoimmune diseases, the specific mechanisms may differ depending on the disease.

2.3.1.3 MSC effects on T-lymphocytes

In RA, the interaction between T cells and MSCs is pivotal in both disease development and treatment. T cells influence the function and differentiation potential of MSCs through the secretion of pro-inflammatory factors like IFN-γ and TNF-α. Conversely, MSCs regulate the inflammatory response by inhibiting T cell proliferation and secreting regulatory cytokines (**Figure 4**). The secretions from Th17 cells impact the chondrogenesis and differentiation of MSCs, while MSCs dampen T cell reactivity and promote differentiation into Treg cells by expressing molecules such as HLA-G5 [55]. In AIH, MSCs impede the proliferation and activation of T cells and alter the differentiation direction of T cells by secreting a range of anti-inflammatory molecules, affecting T cell subsets including CD4+ T cells and CD8+ T cells [59]. In CD, MSCs exert an immunomodulatory role by suppressing the proliferation of Th1/Th17 cells, facilitating the recruitment of Treg cells, and inducing antigen-presenting cells to acquire regulatory properties, thereby mitigating autoimmune reactions [61].

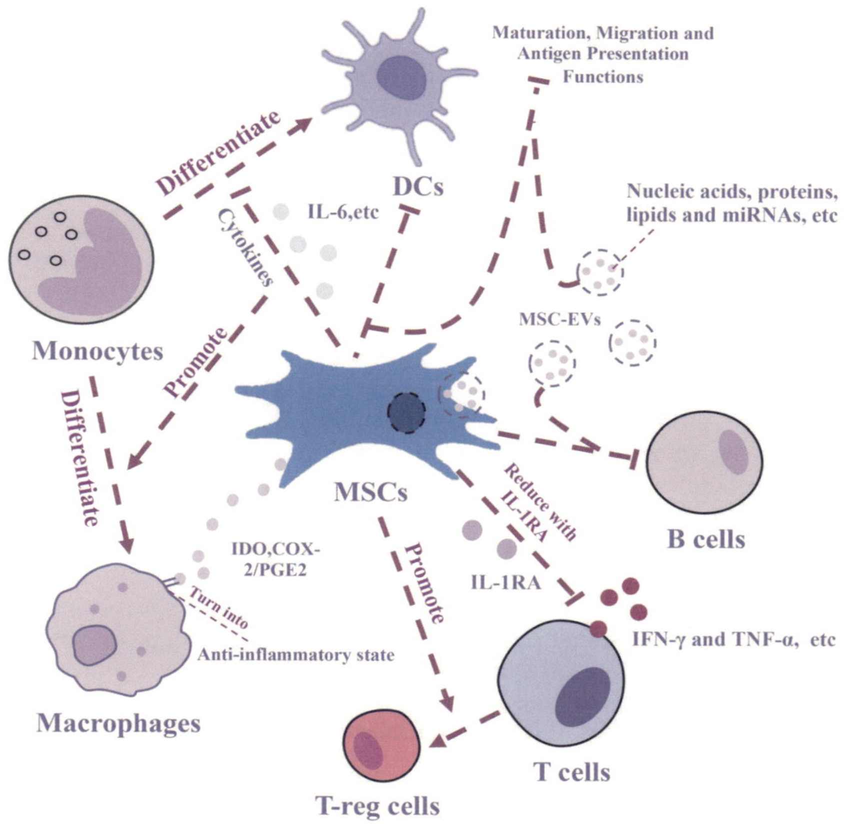

Figure 4.
Mechanism of MSCs therapy on the immune cells in autoimmune diseases. MSCs can inhibit the immune function of DCs and B cells through direct secretion of factors or other pathways, and they can also inhibit the differentiation of monocytes into DCs into situational cells through the secretion of cytokines, such as IL-6, IDO, and COX-2/PGE2. The secretion of IDO, COX-2/PGE2 can also promote the transformation of macrophages to anti-inflammatory state. In addition, MSCs will also secrete anti-inflammatory factor to inhibit the level of T cell secretion of IFN-γ and TNF-α, and at the same time, promote the transformation of T cells to Treg cells. EVs secreted by MSCs also have a good function in regulating a variety of immune cells, which broaden the immunoregulatory ability of MSCs.

2.3.1.4 MSC effects on B-lymphocytes

In autoimmune diseases, the interactions between MSCs and B lymphocytes share certain common characteristics. Generally, MSCs possess the ability to suppress the proliferation and activity of B lymphocytes (**Figure 4**). However, in diseases like RA and SLE, this inhibitory effect may be influenced, leading to a reprogramming of MSCs to support the pro-inflammatory function of B lymphocytes rather than regulating the immune response [55]. Although our understanding of the mechanisms underlying these interactions is currently limited, further research in this area could contribute to a deeper comprehension of autoimmune pathological processes and offer optimized directions for MSC-based treatment strategies. In T1D, MSCs impact the proliferation and antibody production of B cells, hinder the differentiation of B cells into plasma cells, enhance the proliferation of natural killer (NK) cells by up-regulating CD73 expression, and diminish the production of IFN-γ and the cytotoxicity of NK cells through mechanisms involving IDO and PGE2 [80]. In AIH, MSCs impede the proliferation of B cells and dampen B cell activity by activating the PD-1/PD-L1 pathway. They also inhibit signal transduction and activation of transcription factor 3 (STAT-3) in B cells by secreting CCL2, promoting the expression of pD-5 (PAX5), and suppressing the production of immunoglobulins in plasma cells [59].

2.3.1.5 MSC-EV effects in autoimmune diseases

MSC-EVs exhibit similar therapeutic characteristics across autoimmune diseases (**Figure 4**). First, they actively participate in and regulate chronic inflammatory and immune processes by delivering a diverse array of biomolecules, including nucleic acids, proteins, and lipids. Through this mode of intercellular communication, MSC-EVs can remotely modulate the function of immune cells, thereby influencing disease development and pathological processes. Second, MSC-EVs have the ability to inhibit T cell proliferation, promote apoptosis, and impact the activity of DCs and macrophages, thereby achieving immune balance and regulation. Moreover, MSC-EVs carry molecules such as microRNAs, which can target and down-regulate key inflammatory and angiogenesis-related gene pathways, thereby impeding the progression of pathological processes [67]. In conditions like RA and AIH, MSC-EVs play a crucial regulatory role by transporting specific miRNAs, such as miR-150-5p, miR-192-5p, miR-124a, etc. These miRNAs can negatively regulate multiple target genes at the post-transcriptional level, thereby influencing immune cell function and inflammatory processes [81, 82].

MSC-EVs not only mitigate the inflammatory response but also facilitate the repair and regeneration of damaged tissues, primarily by modulating the activity of immune cells. Specifically, MSC-EVs suppress the overactivation of T cells and enhance the population of Treg cells, thereby dampening the inflammatory response and promoting joint tissue repair. Furthermore, they modulate the activity of DCs and macrophages, thereby regulating the differentiation and function of immune cells [68]. Overall, MSC-EVs demonstrate potential advantages and promise as a cell-free therapy for various autoimmune diseases, though further research is required to address the challenges and limitations associated with their clinical application.

2.3.2 Preclinical application of MSC therapy

2.3.2.1 MSCs treatment of rheumatoid arthritis

In the realm of RA research, the collagen-induced arthritis (CIA) model stands out as the primary experimental model for investigating MSC therapy. Among the selection of experimental animals, mice are the most commonly utilized [55]. The CIA model is established by injecting type II collagen into mice at various sites, disrupting immune tolerance and instigating systemic autoantibody-driven arthritis. Additionally, researchers may opt for alternative models such as the rat adjuvant-induced arthritis model (rAIA) or the mouse antigen-induced arthritis model (mAIA) depending on specific research requirements (**Table 1**) [55].

In preclinical RA model treatments involving MSCs, intravenous or intraperitoneal injections are the predominant modes of administration [83]. However, numerous preclinical studies indicate that only a small fraction of administered MSCs migrate to inflamed joints, with no consistent correlation between their biological distribution and therapeutic efficacy. Alternative administration routes such as periarticular and intramuscular injections have been explored, albeit with no significant enhancement in effectiveness [84].

Model	Specie	Immunogen and protocol
CIA	Mouse	Induction of arthritis by one/multiple subcutaneous (SC), ID, IP, IA, injections of CII, which can be derived from chicken or bovine sources emulsified in complete Freund's adjuvant (CFA) containing *Mycobacterium tuberculosis*, with injections given at intervals of 1, 2, or 3 weeks.
rAIA	Rat	SC-injected *Mycobacterium butyricum* in IFA -SC-injected mBSA in CFA & IAinjected mBSA -One or two IA injections of CFA 5 or 14 days apart.
mAIA	Mouse	SC/IP-injected mBSA in/or CFA with Bordetella pertussis toxin and IA-injected mBSA three weeks later.
SLE model	Mouse(MRL/lpr)	The lpr gene in mice undergoes a spontaneous mutation, leading to a defect in the Fas protein. A large number of lymphocytes are unable to undergo normal apoptosis, thereby triggering autoimmune diseases.
SLE model	Mouse(NZB/WF1)	The mice produced by crossing NZB with SWR/J mice develop a strain that exhibits diseases similar to SLE.
T1D model	Mouse	S100 is administered intravenously via the tail vein into male C57BL/6 J mice to establish an autoimmune hepatitis model.
AIH model	Mouse	Streptozotocin is typically dissolved in citrate buffer or other suitable solvents and is administered intraperitoneally. After the injection of STZ, the mice are provided with a high-fat, high-sugar diet to promote the development of diabetes
CD model	Mouse	A mixed solution of DSS (Dextran Sulfate Sodium) and PBS (Phosphate-Buffered Saline), along with MSCs (Mesenchymal Stem Cells) used in the experiment, is injected around the intestine via intraperitoneal injection.

AIH: autoimmune hepatitis; CD: Crohn's disease; CFA: complete Freund's adjuvant; CIA: Collagen-induced arthritis model; IA: intra-articular; ID: intradermal; IFA: incomplete Freund's adjuvant; IP: intraperitoneal; lpr: lymphoproliferation; mAIA: Mouse antigen-induced arthritis model; RA: rheumatoid arthritis; rAIA: Rat adjuvant-induced arthritis model; SC: subcutaneous; SLE: systemic lupus erythematosus; T1D: Type 1 diabetes.

Table 1.
Common preclinical models for autoimmune diseases.

Evaluation of MSC therapeutic efficacy in animal models of RA entails various indicators, including paw swelling, histological analysis, serum levels of anti-type II collagen antibodies and rheumatoid factors (such as TNF-α and IL-1β) [84, 85]. Animal experiments not only corroborate findings from *in vitro* studies but also underscore the significant potential of MSCs in mitigating RA symptoms across diverse indicators. Notably, joint inflammation, bone erosion, and inflammatory marker levels in experimental mice treated with MSCs exhibit notable reductions. Furthermore, MSCs derived from different sources exhibit varying therapeutic effects, with UC-MSCs demonstrating superior efficacy [85].

Despite promising results from preclinical studies suggesting the therapeutic potential of MSCs in RA treatment, several challenges persist. These include potential functional alterations of MSCs in chronic inflammatory environments and the need to enhance MSC persistence and migration within the body. Moreover, the underlying mechanisms remain incompletely understood, and optimal treatment regimens, dosages, and administration routes for clinical applications are still subjects of ongoing investigation.

2.3.2.2 MSCs treatment of systemic lupus erythematosus

In preclinical investigations into the therapeutic efficacy of MSCs for SLE, researchers predominantly rely on two mouse models: MRL/lpr and NZB/WF1, both acknowledged for their ability to replicate key pathological features of human SLE. In these studies, MSCs are typically administered to mice via tail intravenous injection to enable their immunomodulatory and anti-inflammatory effects (**Table 1**). Beyond standard histological analysis, assessments to gauge MSC efficacy encompass NF-κB and TNF-α levels, pivotal in SLE pathogenesis. Additionally, metrics such as anti-double-stranded DNA (dsDNA) antibody levels, proteinuria, antinuclear antibody (ANA) levels, and kidney scores offer comprehensive insights into SLE disease activity and severity, thus serving as crucial indicators of MSC therapeutic efficacy [83, 86, 87].

Cumulative literature findings underscore MSCs' favorable impact on regulating multiple SLE indicators. MSCs effectively modulate key pathological processes of SLE, as evidenced by tissue score and cytokine level regulation. Notably, MSCs demonstrate significant therapeutic benefits in reducing anti-dsDNA antibody levels, diminishing proteinuria, enhancing ANA levels, and improving kidney scores. These observations affirm MSCs' capacity to regulate the immune system and mitigate inflammation, signaling promising potential and clinical applicability in SLE treatment.

In summary, MSCs, as an emerging cell therapy, have shown promising strides in SLE treatment. Future investigations should aim to optimize MSC sourcing, dosage, administration routes, and combinations with other therapies to maximize therapeutic outcomes and facilitate their translation into effective clinical practices.

2.3.2.3 MSCs treatment of type 1 diabetes

When investigating the *in vivo* experimental aspect of MSC treatment for T1D, the prevailing approach typically involves utilizing streptozotocin-induced diabetic C57BL/6 mice. Additionally, primate models, including lean monkeys with STZ-induced diabetes, are frequently employed in such *in vivo* studies (**Table 1**). Standard treatment protocols entail co-culturing model islets with human MSCs from various sources for a specified duration, followed by diverse transplantation methods. The study objectives often revolve around comparing tissue scores and cytokine levels. Besides the conventional cytokines associated with autoimmune diseases, assessing blood glucose levels and insulin secretion function in experimental animals is imperative for evaluating T1D treatment with MSCs.

Following the transplantation of islets co-cultured with MSCs into mice, notable improvements have been observed. These include significantly enhanced transplantation efficiency and islet survival rates, accelerated blood glucose level recovery in diabetic mice, and enhanced tissue structure and vascularization of islets [57, 86]. Furthermore, studies have revealed that MSCs enhance the survival and function of islets through intercellular contact mediated by N-cadherin, with co-encapsulated islets maintaining normal blood glucose levels in diabetic mice for an extended duration [87]. Particularly noteworthy is the finding from primate experiments that MSC therapy not only augments islet implantation and function but is also associated with an increase in the number of Treg cells [83]. Collectively, these findings underscore the potential utility of MSCs in improving therapeutic outcomes for T1D.

2.3.2.4 MSCs treatment of autoimmune hepatitis

In preclinical investigations of the therapeutic efficacy of MSCs on AIH, several animal models are available for selection, such as the establishment of an AIH model using S100-induced C57BL/6 J male mice (**Table 1**) [88]. Similar to their administration in other autoimmune diseases, MSCs are typically delivered intravenously in mice. Various indicators are utilized to assess the therapeutic efficacy of MSCs in this disorder, primarily including liver histopathological injury, serum ALT and AST levels, and the levels of various T cell-related markers and inflammatory factors [82, 88].

Preclinical studies have revealed that MSCs derived from different sources, such as hUC-MSCs, can suppress immune responses in immune-mediated liver diseases [88]. Additionally, their EVs have been demonstrated to protect against oxidative stress-induced bile duct cell senescence and reduce aging-related secretory phenotype (SASP). Moreover, MSC-derived EVs, including those derived from bone marrow, exhibit the ability to mitigate liver ischemia/reperfusion injury by modulating the inflammatory response, thereby offering a novel therapeutic approach for liver injury treatment. In summary, findings from preclinical experiments underscore the significant therapeutic potential of MSCs in the management of AIH.

2.3.2.5 MSCs treatment of Crohn's disease

In the selection of animal models for studying CD, the mouse model of dextran sulfate sodium-induced ulcerative colitis is frequently employed (**Table 1**) [89]. Additionally, reagents like 2,4,6-Trinitrobenzene sulfonic acid are commonly utilized for induction. Similar to several other diseases mentioned earlier, MSCs are typically administered to laboratory mice via the tail vein. The therapeutic efficacy of MSCs on CD is assessed by comparing histopathological scores, expressions of pro-inflammatory and anti-inflammatory factors, and CD-specific indicators such as colon length and intestinal cell markers.

Preclinical investigations have revealed that intestinal epithelial cell proliferation, the Wnt signaling pathway, and T cell immunity play pivotal roles in CD pathogenesis, with MSCs demonstrating effective regulatory capabilities in these aspects [90]. Furthermore, MSCs, when stimulated by different agents like IFN-γ, exhibit enhanced therapeutic effects [89]. Additionally, MSC-induced regulatory dendritic cells (rDCs) have shown therapeutic potential in modulating inflammatory responses and promoting the activation of Treg cells, thereby offering novel therapeutic avenues for CD treatment.

2.3.3 Summary

MSCs have exhibited considerable promise in addressing autoimmune diseases. Both *in vitro* cellular experiments and animal model investigations have corroborated the capacity of MSCs to modulate immune responses through diverse mechanisms. These include curtailing the proliferation and activation of T and B cells, as well as regulating the maturation and functionality of DCs and macrophages. Furthermore, MSC-derived EVs have shown potential in immune modulation and facilitating tissue repair in damaged areas.

Despite the encouraging outcomes of MSC therapy observed in animal models, several challenges need to be addressed before these treatments can be effectively translated into clinical practice. These challenges encompass potential alterations in MSC functionality within a chronic inflammatory milieu and the enhancement of their persistence and migratory capability within the body. To overcome these hurdles, future research endeavors should focus on refining MSC treatment strategies, including optimizing the source, dosage, and route of administration.

2.4 COVID-19

Severe acute respiratory syndrome coronavirus type 2 (SARS-CoV-2) is a pathogenic β coronavirus that causes coronaviruses 2019 (COVID-19), profoundly disrupting the normal course of human life [91]. SARS-CoV-2 is characterized by a single plus-stranded RNA genome [92]. The virus enters host cells through the interaction of its spike protein with the angiotensin-converting enzyme 2 (ACE2) receptor situated on the surface of target cells [93]. Notably, ACE2 receptors are predominantly found in lung AT2 cells, alveolar macrophages, and capillary endothelial cells, making the respiratory system particularly vulnerable to this viral infection [94].

SARS-CoV-2 infects ACE2-expressing AT2 cells and resident alveolar macrophages, prompting the secretion of chemokines such as CXCL8, CXCL9C, XCL10, CCL2, and CCL3. These chemokines attract inflammatory and immune cells into the infected alveoli [95]. The influx of inflammatory cells leads to the release of numerous pro-inflammatory cytokines including IL-1β, IL-6, IL-7, IL-12, granulocyte colony-stimulating factor (G-CSF), IFN-γ, and TNF-α, precipitating a "cytokine storm" [96]. This hyperactive inflammatory response elevates reactive oxygen species ROS production, disrupting redox homeostasis, inducing oxidative stress and inflammation, inflicting damage on lung tissue, and culminating in acute respiratory distress syndrome (ARDS) [97]. Endothelial cells of capillaries, typically linked by intercellular junctions regulating the flow of inflammatory cells and fluid into the interstitial space, are affected by SARS-CoV-2 infection. Migrating neutrophils and monocyte-derived macrophages release toxic mediators, causing damage to endothelial and epithelial cells, increasing permeability, and facilitating the migration of inflammatory cells. This is characterized by intercellular junction damage, loss of adhesion to the basement membrane, and cellular swelling [98]. Consequently, the air cavities fill with fluid (alveolar edema), causing breathing difficulties.

Severe respiratory conditions like acute lung injury (ALI), pulmonary edema, and ARDS are the primary manifestations of SARS-CoV-2 infection [99]. MSCs possess immune regulatory capabilities, multi-directional differentiation potential, and regenerative effects. Multiple preclinical studies have demonstrated the promising potential of MSCs as an emerging treatment for ARDS and ALI, severe clinical consequences of COVID-19 [100].

2.4.1 Mechanisms of MSC therapy

2.4.1.1 Immunomodulation

MSCs possess the ability to modulate the immune response, which is crucial in the treatment of COVID-19 due to the tendency of viral infections to trigger excessive immune activation and inflammatory responses. MSCs exert their regulatory influence on immune cells through both direct contact and the secretion of anti-inflammatory cytokines like HGF, VEGF, and keratinocyte growth factor (KGF). These factors are instrumental in maintaining immune balance and mitigating inflammatory damage [101]. Upon receiving activation signals from the inflamed environment, MSCs migrate to the site of inflammation facilitated by inflammatory mediators (cytokines and chemokines) produced during infection or tissue damage.

MSCs modulate innate immune responses by affecting DCs, neutrophils, macrophages, and NK cells. Particularly, macrophages play a pivotal role in the wound healing process, making them prime targets for therapeutic tissue regeneration strategies [102]. Upon reaching the site of infection or injury, MSCs can induce a shift in M1 macrophages (pro-inflammatory) to M2 macrophages (anti-inflammatory) by secreting IDO and PGE2, thereby reducing inflammation and enhancing phagocytic activity [103]. MSCs also counteract neutrophil respiratory bursts and prevent neutrophil death through an IL-6-dependent mechanism [104]. By secreting PGE2 and IL-10, MSCs decrease the release of proinflammatory cytokines, diminish antigen presentation capability, and impede the differentiation and maturation of DCs. Furthermore, MSCs dampen the cytotoxic activity of resting NK cells by reducing the production of natural cytotoxic receptor 3 (NKp30) and natural killer group 2 member D (NKG2D), both involved in NK cell activation and target cell killing [105].

In addition to modulating innate immunity, MSCs also regulate adaptive immunity, including T cells and B cells. They inhibit the proliferation of antigen-activated T cells, induce the polarization of pro-inflammatory Th1 type $CD4^+$ T cells to anti-inflammatory Th2 type, and down-regulate $CD8^+$ T cell-mediated cytotoxicity. MSCs further inhibit B cell proliferation *in vitro* and enhance the frequency and activity of Breg cells by secreting IL-10, TGF-β, and IDO (**Figure 5**) [103].

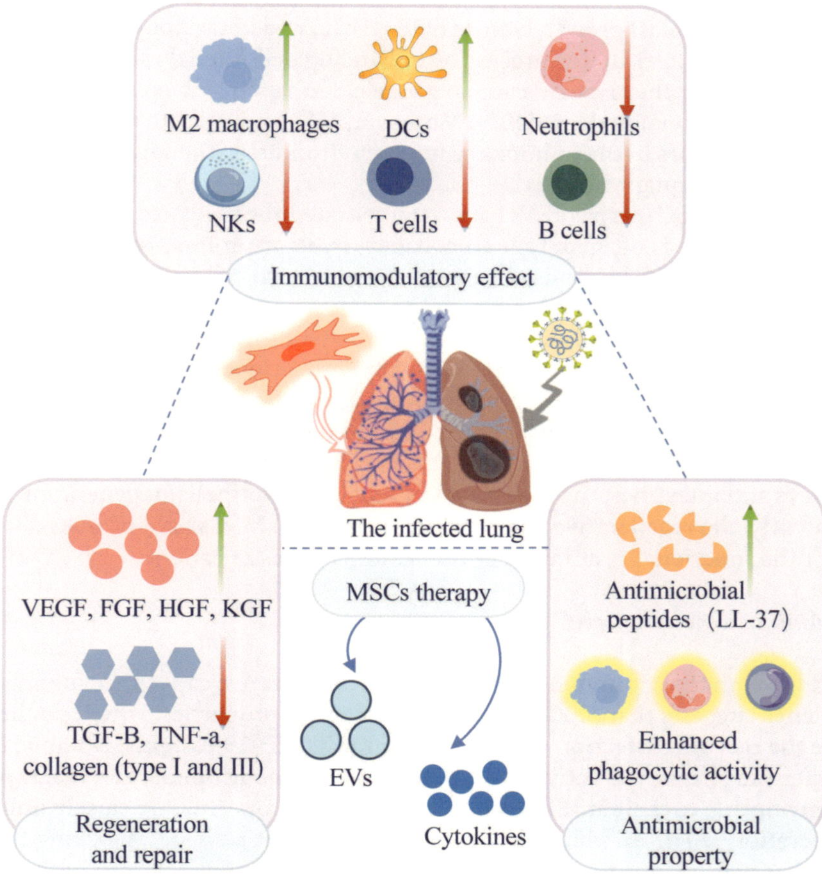

Figure 5.
Mechanisms of MSC therapy on COVID-19. MSCs mainly play roles in immune regulation, regeneration and repair, as well as antibacterial properties in COVID-19 treatment by secreting cytokines and EVs. MSCs maintain immune balance and reduce cytokine storms caused by viral infections by secreting anti-inflammatory cytokines and influencing immune cells such as dendritic cells, neutrophils, macrophages, and natural killer cells. MSCs release excessive regenerative factors in response to damage signals that promote tissue repair, such as VEGF, FGF, HGF, and KGF. In addition, MSCs also exhibit antibacterial properties, enhance the phagocytic activity of host cells, and resist viruses by secreting antimicrobial peptides and other substances, reducing the risk of secondary infection.

Overall, MSC-mediated immune regulation can mitigate the pro-inflammatory factors contributing to cytokine storms while boosting anti-inflammatory factors, thus playing a crucial role in promoting lung damage repair and lung function recovery in COVID-19 patients.

2.4.1.2 Regeneration and repair

As discussed previously, MSCs exhibit an immunomodulatory role in rebalancing the immune system during inflammation in COVID-19 patients. Beyond this function, MSCs possess regenerative capabilities that facilitate the repair of damaged cells and tissues. Pathological inflammatory processes induce extensive tissue damage characterized by compromised integrity of alveolar capillary membranes, resulting in pulmonary edema, lung tissue degeneration, and fibrosis, all of which can be alleviated in the presence of MSCs [106]. In response to injury signals, MSCs release an array of regenerative factors that promote tissue repair. Notably, growth factors such as VEGF, fibroblast growth factor (FGF), HGF, and keratinocyte growth factor (KGF) play pivotal roles in this process by enhancing endothelial cell stability, thus facilitating repair of the alveolar-capillary barrier, enhancing alveolar fluid clearance, and reducing pulmonary edema [107]. Moreover, MSCs contribute to the reduction of pro-fibrotic factors in the pulmonary microenvironment, thereby aiding in the prevention of pulmonary fibrosis.

Interestingly, MSC-derived EVs have demonstrated the ability to diminish levels of TGF-β, TNF-α, type I collagen, type III collagen, hydroxyproline, and serum ceruloplasmin in lung tissue. This leads to decreased endothelial cell apoptosis, suppressed myofibroblast proliferation, and enhanced regeneration of alveolar epithelial cells [108]. Furthermore, MSCs facilitate alveolar cell repair through mitochondrial transfer via direct cell-to-cell contact, thereby boosting their bioenergy levels. Additionally, MSCs safeguard lung cells by promoting autophagy through the phosphoinositide 3-kinase/protein kinase-B pathway [109]. Beyond their regenerative effect on lung tissue, MSCs also exhibit this reparative role in other affected organs, including the kidneys and intestines, through mucosal repair and epithelial regeneration. Consequently, the regenerative and reparative effects of MSCs hold profound significance for the functional restoration of tissues and organs in patients with COVID-19.

2.4.1.3 Antimicrobial properties

In addition to their immunomodulatory and regenerative capacities, mounting evidence suggests that MSCs possess antimicrobial properties, which can help mitigate the risk of reinfection often observed in COVID-19 patients. Notably, MSCs can bolster the phagocytic activity of host monocytes, neutrophils, and macrophages, thus contributing to their antibacterial effects. These properties are closely linked to the secretion of HL-37, Lipocalin-2, and FGF-7 by MSCs. In fact, blocking these peptides has been shown to deactivate the protective effects of MSCs in both *in vivo* and *in vitro* models of lung injury [110]. Among these, antimicrobial peptides (AMPs) represent a class of widely distributed low molecular weight proteins that serve as the first line of defense against various pathogens, including bacteria, fungi, and viruses. Notably, studies indicate that the antimicrobial peptide LL-37 primarily mediates the antibacterial effect of MSCs. Well-designed experiments have demonstrated that both MSCs and MSC-conditioned media inhibit the growth of both Gram-negative and Gram-positive bacteria *in vitro* [91].

Moreover, the transfer of mitochondria from MSCs to macrophages, facilitated either through tunneling nanotube structures or EVs, also contributes significantly to the enhanced phagocytic activity of macrophages in ARDS [111]. While there is relatively limited research on the specific antiviral activity of MSCs compared to antibacterial studies, evidence suggests that MSC-derived exosomal microRNAs exhibit inhibitory effects on hepatitis C virus infection *in vitro* (**Figure 5**) [112]. These findings are particularly crucial in the context of injuries caused by the COVID-19 virus, as they offer insights into strategies for preventing or treating opportunistic bacterial infections.

2.4.2 Preclinical application of MSC therapy

2.4.2.1 Immunomodulation of MSCs

It is demonstrated that the administration of MSCs leads to immunomodulation, resulting in improved inflammatory responses observed in animal models associated with COVID-19. For instance, in mice with LPS-induced ARDS, MSC treatment has been shown to decrease neutrophil infiltration, collagen deposition, and fibrosis [113]. In a similar model, the administration of MSCs shielded mice from LPS-induced ALI by suppressing macrophages, a process reliant on the secretion of PGE2, granulocyte-macrophage colony-stimulating factor (GM-CSF), and IL-13 [112]. Apart from their own secretion of anti-inflammatory molecules, MSCs can also counteract the release of certain pro-inflammatory cytokines. Research indicates that in a mouse model of ALI induced by oleic acid, intravenous transplantation of MSCs lowers the concentration of pro-inflammatory TNF-α in plasma and lung tissue extracts, while enhancing the levels of anti-inflammatory IL-10 following MSC treatment [114].

Moreover, MSCs regulate the recruitment of specific immune cell types to the site of injury, favoring the infiltration of tolerant cell types such as Treg cells. Studies have demonstrated that MSCs ameliorate LPS-induced ALI by modulating the balance between Treg cells and Th17 cells, with an observed increase in Treg cells in a mouse lung contusion model treated with MSCs [115]. The efficacy of MSCs in mediating intrapulmonary anti-inflammatory responses has also been replicated in large animal models. Investigations in a sheep model of bacterial pneumonia revealed that the severity of ALI was reduced after intravenous injection of MSCs, leading to improved oxygenation [116]. These findings collectively highlight the therapeutic potential of MSCs for mitigating COVID-19-related inflammation in animal models through diverse mechanisms, providing promising early evidence for the use of MSCs in human COVID-19 treatment.

2.4.2.2 Regeneration of MSCs

While MSCs' immunomodulatory function is crucial for addressing an exacerbated immune response, in the context of ARDS, there is also a necessity for cell and tissue repair. As previously mentioned, MSCs secrete various regenerative factors pivotal in tissue repair and regulation. Studies have shown that inhibiting key regenerative mediators in MSCs can nullify their ability to mitigate lung injury *in vivo*. For instance, MSC transplantation in mice reduced lung permeability, shielded lung epithelium from apoptosis, facilitated the restoration of vascular endothelial-cadherin, and alleviated lung injury. However, these beneficial effects significantly diminished when VEGF expression was inhibited in transplanted MSCs [117]. In a bleomycin-induced mouse model of pulmonary fibrosis, decreased HGF expression in MSCs compromised their protective effect against pulmonary fibrosis [112].

Apart from their inherent secretion of regenerative molecules, MSCs can also induce the expression of regenerative factors in specific cells and tissues. In the LPS-induced ALI mouse model, intravenous injection of MSCs led to the upregulation of KGF, a crucial mediator of alveolar repair [118]. In rodent models of radiation-induced pulmonary fibrosis, MSC treatment preserved the structure of irradiated lungs, partly mediated by endogenous HGF expression [119]. In rodent models of phosgene-induced lung injury, MSCs facilitated the repair of the lung blood-air barrier [120]. Furthermore, in a recent study, hUC-MSCs were successfully induced to differentiate into alveolar epithelial cells, consequently reducing mortality due to pulmonary fibrosis in mice [121].

These findings underscore the significant potential of MSCs in treating COVID-19-mediated ARDS and pulmonary fibrosis, laying a solid scientific foundation for clinical interventions.

2.4.2.3 Antimicrobial properties of MSCs

The antimicrobial effect of MSCs primarily hinges on the action of the antimicrobial peptide LL-37, a mechanism substantiated across various animal models. In a mouse model of *Escherichia coli* (*E. coli*)-induced pneumonia, the administration of MSCs led to a reduction in bacterial growth in bronchoalveolar lavage fluid (BALF) and lung homogenate. This effect was notably diminished when neutralizing antibodies against LL-37 were introduced [110]. Similarly, in another study investigating *E. coli*-induced pneumonia, MSC treatment decreased the bacterial load in the lungs, concomitant with an elevation in LL-37 concentration in the alveoli [122]. The LL-37-mediated bacterial clearance induced by MSCs was also evident in a mouse model of cystic fibrosis [123]. Moreover, in a rodent model of *E. coli*-induced ARDS, MSC administration correlated with a decreased formation of *E. coli* colony-forming units in BALF [124]. Additionally, β-defensin 2 (BD2), another antimicrobial peptide secreted by MSCs, was found to contribute to bacterial clearance. In a mouse model of *E. coli*-induced ALI, MSC administration resulted in enhanced bacterial clearance, consistent with increased secretion of BD2 [112]. These findings provide further evidence of the therapeutic potential of MSCs in bolstering pathogen control.

2.4.3 Summary

In individuals experiencing severe manifestations of COVID-19, life-threatening complications like ARDS are mediated by widespread inflammation, cytokine storm syndrome, and dysregulated immune responses. Leveraging their immunomodulatory, regenerative repair, and antimicrobial properties, MSCs have demonstrated therapeutic promise in preclinical studies targeting ARDS and ALI in severe COVID-19 cases. Most preclinical investigations focusing on MSC therapy have indicated that MSCs exhibit minimal immunogenicity upon administration to patients, without exacerbating inflammation or worsening cytokine storms. Although numerous clinical studies have affirmed the safety profile of mesenchymal stem cells, evidence regarding their efficacy remains less conclusive. To enhance COVID-19 treatment strategies sensibly, further research is imperative to elucidate the mechanisms underlying the immunomodulatory, regenerative, and antimicrobial capacities of MSCs, while identifying the microenvironment nuances inherent in different SARS-CoV-2 scenarios.

3. Clinical exploration and translation of MSC-based therapies

In the realm of medical challenges encompassing neurological diseases, cardiovascular diseases, autoimmune diseases, and COVID-19, MSCs have emerged as a promising treatment avenue, showcasing their unique potential and promise. However, this field also presents a spectrum of challenges and opportunities throughout the clinical exploration and translational process.

The multidirectional differentiation capabilities and immunomodulatory properties of MSCs offer renewed optimism for patients grappling with nervous system diseases. Nevertheless, navigating the clinical translation journey entails grappling with hurdles such as the intricate mechanisms underlying MSCs' differentiation, the risk of immune rejection, stringent adherence to safety and ethical standards, the rigorous approval process for clinical trials, and the substantial cost of treatment. Despite these challenges, advancements in technology and bolstered policy support augur well for the future application of MSC therapy in neurological diseases. Moreover, the intersection of interdisciplinary collaboration and the burgeoning demand for personalized treatment avenues innovative approaches for stem cell therapy.

In the realm of cardiovascular and cerebrovascular diseases, MSCs have demonstrated remarkable efficacy in both experimental settings and laboratory studies. Nonetheless, a significant technical challenge lies in ensuring their precise migration to damaged cardiovascular tissues to exert therapeutic effects. Moreover, the use of MSC therapy poses potential risks such as immune reactions and tumor formation, necessitating thorough safety assessments. Ethical debates, as well as the costs associated with preparation and transportation, further impede widespread adoption. However, with a rising prevalence of cardiovascular diseases, the market demand for MSC-based treatments is substantial, and advancements in technology along with supportive policy frameworks offer avenues for progress in MSC therapy.

In the domain of autoimmune disease treatment, numerous challenges must be addressed, including the heterogeneity of MSCs, their mechanisms of action, discrepancies between preclinical models and human diseases, persistence and migration within the body, optimal dosage and administration routes, long-term safety considerations, cost-effectiveness, and ethical and legal considerations. Each autoimmune disease's unique characteristics necessitate tailored treatment strategies. Nonetheless, the pivotal role of MSCs in immune modulation, coupled with the potential of MSC-EVs as a cell-free therapeutic approach, presents novel prospects for addressing autoimmune diseases.

Preclinical studies have underscored the potential of MSCs in treating COVID-19, yet the precise mechanisms and optimal delivery routes remain unclear. Key challenges include MSC product variability, intravascular entrapment, and low cell survival rates. Additionally, intravenous administration of MSCs may lead to poor homing efficiency and clotting risks. To address these issues, further investigations into MSC mechanisms of action, strategies to enhance *in vivo* survival and persistence, exploration of alternative delivery routes, and mitigation of clotting risks are essential. By advancing our understanding and refining delivery methods, more effective and safer treatment modalities can be developed for COVID-19 patients.

In summary, MSCs represent a promising therapeutic avenue for various medical conditions. Despite existing challenges, ongoing research efforts and technological advancements offer hope for the clinical application of MSC therapy. With continued innovation and supportive policies, it is anticipated that these challenges will be surmounted, leading to the widespread adoption of MSC therapy and providing patients with more efficacious and safer treatment options.

4. Conclusion

In conclusion, the multifaceted capabilities of MSCs, including their ability to differentiate into various cell types, modulate immune responses, and facilitate tissue repair, hold significant promise for regenerative medicine and disease treatment.

Preclinical studies have demonstrated the immunomodulatory effects of MSCs, highlighting their ability to regulate immune cell function and promote tissue healing through the secretion of EVs. Despite promising results in animal models, challenges remain in translating MSC therapy into clinical practice. Addressing issues such as the behavior of MSCs in inflammatory environments and optimizing administration protocols are crucial for successful implementation.

Furthermore, the ideal source, dosage, and administration route of MSCs need further refinement, especially in the context of treating COVID-19-related conditions such as ARDS and ALI. Understanding the specific mechanisms underlying MSC therapy and addressing concerns such as product variability are essential for maximizing its therapeutic potential.

Overall, the compelling evidence from preclinical studies underscores the potential of MSC therapy in addressing a wide range of diseases. Continued research efforts and technological advancements are expected to enhance the clinical application of MSCs, offering new treatment options and renewed hope for patients.

Acknowledgements

This work was supported by grants from NSFC (81900919), the Natural Science Foundation of Guangdong Province (2022A1515011817), the Guangdong Provincial Department of Science and Technology Characteristic Innovation Project for Universities (2023KTSCX097), the Special Fund for Science and Technology Innovation Strategy of Guangdong Province (pdjh2023b0388) and the National College Student Innovation and Entrepreneurship Training Program (202414278010).

Conflict of interest

The authors declare no conflict of interest.

Author details

Zhilang Xie, Huayin Zhou, TingTing Ou, Wendan Pei and Xingliang Fan*
Institute of Biological and Food Engineering, Guangdong University of Education, Guangzhou, P.R. China

*Address all correspondence to: fanxl84@gmail.com

References

[1] Han Y, Li X, Zhang Y, Han Y, Chang F, Ding J. Mesenchymal stem cells for regenerative medicine. Cells. 2019;**8**(8):886

[2] Jovic D, Yu Y, Wang D, Wang K, Li H, Xu F, et al. A brief overview of global trends in MSC-based cell therapy. Stem Cell Reviews and Reports. 2022;**18**(5):1525-1545

[3] Garfias S, Tamaya Domínguez B, Toledo Rojas A, Arroyo M, Rodríguez U, Boll C, et al. Peripheral blood lymphocyte phenotypes in Alzheimer and Parkinson's diseases. Neurología (English Edition). 2022;**37**(2):110-121

[4] Ng XY, Cao M. Dysfunction of synaptic endocytic trafficking in Parkinson's disease. Neural Regeneration Research. 2024;**19**(12):2649-2660

[5] Johnston TH, Lacoste AMB, Ravenscroft P, Su J, Tamadon S, Seifi M, et al. Using artificial intelligence to identify drugs for repurposing to treat l-DOPA-induced dyskinesia. Neuropharmacology. 2024;**248**:109880

[6] Ye J, Wan H, Chen S, Liu GP. Targeting tau in Alzheimer's disease: From mechanisms to clinical therapy. Neural Regeneration Research. 2024;**19**(7):1489-1498

[7] Zhang GL, Zhu ZH, Wang YZ. Neural stem cell transplantation therapy for brain ischemic stroke: Review and perspectives. World Journal of Stem Cells. 2019;**11**(10):817-830

[8] Herron MS, Wang L, von Bartheld CS. Prevalence and types of strabismus in cerebral palsy: A global and historical perspective based on a systematic review and meta-analysis. Ophthalmic Epidemiology. 2024:1-18

[9] Heris RM, Shirvaliloo M, Abbaspour-Aghdam S, Hazrati A, Shariati A, Youshanlouei HR, et al. The potential use of mesenchymal stem cells and their exosomes in Parkinson's disease treatment. Stem Cell Research & Therapy. 2022;**13**(1):371

[10] Xu F, Xiang Q, Huang J, Chen Q, Yu N, Long X, et al. Exosomal miR-423-5p mediates the proangiogenic activity of human adipose-derived stem cells by targeting Sufu. Stem Cell Research & Therapy. 2019;**10**(1):106

[11] Yang Y, Cai Y, Zhang Y, Liu J, Xu Z. Exosomes secreted by adipose-derived stem cells contribute to angiogenesis of brain microvascular endothelial cells following oxygen-glucose deprivation in vitro through MicroRNA-181b/TRPM7 Axis. Journal of Molecular Neuroscience. 2018;**65**(1):74-83

[12] Babaei M, Rezaie J. Application of stem cell-derived exosomes in ischemic diseases: Opportunity and limitations. Journal of Translational Medicine. 2021;**19**(1):196

[13] BraunRK,ChettyC,BalasubramaniamV, Centanni R, Haraldsdottir K, Hematti P, et al. Intraperitoneal injection of MSC-derived exosomes prevent experimental bronchopulmonary dysplasia. Biochemical and Biophysical Research Communications. 2018;**503**(4):2653-2658

[14] Zhang J, Zhang H, Chen Y, Fu J, Lei Y, Sun J, et al. Platelet-derived growth factor D promotes the angiogenic capacity of endothelial progenitor cells. Molecular Medicine Reports. 2019;**19**(1):125-132

[15] Gregorius J, Wang C, Stambouli O, Hussner T, Qi Y, Tertel T, et al. Small

extracellular vesicles obtained from hypoxic mesenchymal stromal cells have unique characteristics that promote cerebral angiogenesis, brain remodeling and neurological recovery after focal cerebral ischemia in mice. Basic Research in Cardiology. 2021;**116**(1):40

[16] Li S, Zhang J, Liu X, Wang N, Sun L, Liu J, et al. Proteomic characterization of hUC-MSC extracellular vesicles and evaluation of its therapeutic potential to treat Alzheimer's disease. Scientific Reports. 2024;**14**(1):5959

[17] Khademizadeh M, Messripour M, Ghasemi N, Momen Beik F, Movahedian AA. Differentiation of adult human mesenchymal stem cells into dopaminergic neurons. Research in Pharmaceutical Sciences. 2019;**14**(3):209-215

[18] Zhou L, Wang J, Huang J, Song X, Wu Y, Chen X, et al. The role of mesenchymal stem cell transplantation for ischemic stroke and recent research developments. Frontiers in Neurology. 2022;**13**:1000777

[19] Che J, Wang H, Dong J, Wu Y, Zhang H, Fu L, et al. Human umbilical cord mesenchymal stem cell-derived exosomes attenuate neuroinflammation and oxidative stress through the NRF2/NF-κB/NLRP3 pathway. CNS Neuroscience & Therapeutics. 2024;**30**(3):e14454

[20] Zhou L, Liang J, Xiong T. Research progress of mesenchymal stem cell-derived exosomes on inflammatory response after ischemic stroke. Zhejiang Da Xue Xue Bao. Yi Xue Ban. 2022;**51**(4):500-506

[21] Essawy Essawy A, Abou-ElNaga OA, Mehanna RA, Badae NM, Elsawy ES, Soffar AA. Comparing the effect of intravenous versus intracranial grafting of mesenchymal stem cells against parkinsonism in a rat model: Behavioral, biochemical, pathological and immunohistochemical studies. PLoS One. 2024;**19**(2):e0296297

[22] Mohamed AS, Abdel-Fattah DS, Abdel-Aleem GA, El-Sheikh TF, Elbatch MM. Biochemical study of the effect of mesenchymal stem cells-derived exosome versus L-Dopa in experimentally induced Parkinson's disease in rats. Molecular and Cellular Biochemistry. 2023;**478**(12):2795-2811

[23] Park S, Park CW, Eom JH, Jo MY, Hur HJ, Choi SK, et al. Preclinical and dose-ranging assessment of hESC-derived dopaminergic progenitors for a clinical trial on Parkinson's disease. Cell Stem Cell. 2024;**31**(1):25-38.e8

[24] Lin L, Huang L, Huang S, Chen W, Huang H, Chi L, et al. MSC-derived extracellular vesicles alleviate NLRP3/GSDMD-mediated Neuroinflammation in mouse model of sporadic Alzheimer's disease. Molecular Neurobiology. 2024. DOI: 10.1007/s12035-024-03914-1 [Epub ahead of print]

[25] Pezeshki SP, Karimi Darabi M, Nazeri Z, Sarkaki A, Rashidi M, Babaahmadi-Rezaei H, et al. Mesenchymal stem cells therapy led to the improvement of spatial memory in rats with Alzheimer's disease through changing the expression of LncRNA TUSC7/miR-449a/PPARγ and CD36 genes in the brain tissue. International Journal of Molecular and Cellular Medicine. 2023;**12**(2):108-119

[26] Deng M, Hou Y, Liu J, He J, Lan Z, Xiao H. Mesenchymal stem cell-derived exosomes overexpressing SRC-3 protect mice from cerebral ischemia by inhibiting ferroptosis. Brain Research Bulletin. 2024;**211**:110948

[27] Zhu G, Jiang L, Tan K, Li Y, Hu M, Zhang S, et al. MSCs-derived exosomes

containing miR-486-5p attenuate cerebral ischemia and reperfusion (I/R) injury. Gene. 2024;**906**:148262

[28] Yang Y, Liu Q, Deng S, Shao Q, Peng L, Ling Y, et al. Human umbilical cord derived mesenchymal stem cells overexpressing HO-1 attenuate neural injury and enhance functional recovery by inhibiting inflammation in stroke mice. CNS Neuroscience & Therapeutics. 2024;**30**(2):e14412

[29] Lu G, Su X, Wang L, Leung CK, Zhou J, Xiong Z, et al. Neuroprotective effects of human-induced pluripotent stem cell-derived mesenchymal stem cell extracellular vesicles in ischemic stroke models. Biomedicine. 2023;**11**(9):2550

[30] Jiang X, Zhou L, Sun Z, Xie B, Lin H, Gao X, et al. MSCs overexpressing GDNF restores brain structure and neurological function in rats with intracerebral hemorrhage. Molecular Biomedicine. 2023;**4**(1):43

[31] Piede N, Bremm M, Farken A, Pfeffermann LM, Cappel C, Bonig H, et al. Validation of an ICH Q2 compliant flow cytometry-based assay for the assessment of the inhibitory potential of mesenchymal stromal cells on T cell proliferation. Cells. 2023;**12**(6):850

[32] Sun J, Xu G. Mesenchymal stem cell-derived Exosomal miR-150-3p affects intracerebral Hemorrhage by regulating TRAF6/NF-κB Axis, gut microbiota and metabolism. Stem Cell Reviews and Reports. 2023;**19**(6):1907-1921

[33] Huang S, Liu L, Huang Y, Fu C, Peng T, Yang X, et al. Potential optimized route for mesenchymal stem cell transplantation in a rat model of cerebral palsy. Experimental Cell Research. 2023;**430**(2):113734

[34] Zhu Y, Ge J, Huang C, Liu H, Jiang H. Application of mesenchymal stem cell therapy for aging frailty: From mechanisms to therapeutics. Theranostics. 2021;**11**(12):5675-5685

[35] Bytyci I, Bajraktari G. Mortality in heart failure patients. Anatolian Journal of Cardiology. 2015;**15**(1):63-68

[36] Bhaskara M, Anjorin O, Wang M. Mesenchymal stem cell-derived Exosomal microRNAs in cardiac regeneration. Cells. 2023;**12**(24):2815

[37] Saleh M, Ambrose JA. Understanding myocardial infarction. F1000Res. 2018;7:F1000. Faculty Rev-1378

[38] Monguio-Tortajada M, Prat-Vidal C, Moron-Font M, Clos-Sansalvador M, Calle A, Gastelurrutia P, et al. Local administration of porcine immunomodulatory, chemotactic and angiogenic extracellular vesicles using engineered cardiac scaffolds for myocardial infarction. Bioactive Materials. 2021;**6**(10):3314-3327

[39] Nagaya N, Kangawa K, Itoh T, Iwase T, Murakami S, Miyahara Y, et al. Transplantation of mesenchymal stem cells improves cardiac function in a rat model of dilated cardiomyopathy. Circulation. 2005;**112**(8):1128-1135

[40] Silva GV, Litovsky S, Assad JA, Sousa AL, Martin BJ, Vela D, et al. Mesenchymal stem cells differentiate into an endothelial phenotype, enhance vascular density, and improve heart function in a canine chronic ischemia model. Circulation. 2005;**111**(2):150-156

[41] Sanganalmath SK, Bolli R. Cell therapy for heart failure: A comprehensive overview of experimental and clinical studies, current challenges, and future directions. Circulation Research. 2013;**113**(6):810-834

[42] Gao F, He T, Wang H, Yu S, Yi D, Liu W, et al. A promising strategy for

the treatment of ischemic heart disease: Mesenchymal stem cell-mediated vascular endothelial growth factor gene transfer in rats. Canadian Journal of Cardiology. 2007;**23**(11):891-898

[43] Katagiri M, Yamada S, Katoh M, Ko T, Ito M, Komuro I. Heart failure pathogenesis elucidation and new treatment method development. JMA Journal. 2022;**5**(4):399-406

[44] Yu H, Yu M, Li Z, Zhang E, Ma H. Identification and analysis of mitochondria-related key genes of heart failure. Journal of Translational Medicine. 2022;**20**(1):410

[45] Liang X, Zhang Y, Lin F, Li M, Li X, Chen Y, et al. Direct administration of mesenchymal stem cell-derived mitochondria improves cardiac function after infarction via ameliorating endothelial senescence. Bioengineering & Translational Medicine. 2023;**8**(1):e10365

[46] Ohnishi S, Sumiyoshi H, Kitamura S, Nagaya N. Mesenchymal stem cells attenuate cardiac fibroblast proliferation and collagen synthesis through paracrine actions. FEBS Letters. 2007;**581**(21):3961-3966

[47] Bartolucci J, Verdugo FJ, Gonzalez PL, Larrea RE, Abarzua E, Goset C, et al. Safety and efficacy of the intravenous infusion of umbilical cord mesenchymal stem cells in patients with heart failure: A phase 1/2 randomized controlled trial (RIMECARD trial [randomized clinical trial of intravenous infusion umbilical cord mesenchymal stem cells on cardiopathy]). Circulation Research. 2017;**121**(10):1192-1204

[48] Shimizu Y, Ntege EH, Azuma C, Uehara F, Toma T, Higa K, et al. Management of Rheumatoid Arthritis: Possibilities and challenges of mesenchymal stromal/stem cell-based therapies. Cells. 2023;**12**(14):1905

[49] Schuleri KH, Feigenbaum GS, Centola M, Weiss ES, Zimmet JM, Turney J, et al. Autologous mesenchymal stem cells produce reverse remodelling in chronic ischaemic cardiomyopathy. European Heart Journal. 2009;**30**(22):2722-2732

[50] Lim M, Wang W, Liang L, Han ZB, Li Z, Geng J, et al. Intravenous injection of allogeneic umbilical cord-derived multipotent mesenchymal stromal cells reduces the infarct area and ameliorates cardiac function in a porcine model of acute myocardial infarction. Stem Cell Research & Therapy. 2018;**9**(1):129

[51] Nagaya N, Fujii T, Iwase T, Ohgushi H, Itoh T, Uematsu M, et al. Intravenous administration of mesenchymal stem cells improves cardiac function in rats with acute myocardial infarction through angiogenesis and myogenesis. American Journal of Physiology. Heart and Circulatory Physiology. 2004;**287**(6):H2670-H2676

[52] Roubelakis MG, Tsaknakis G, Pappa KI, Anagnou NP, Watt SM. Spindle shaped human mesenchymal stem/stromal cells from amniotic fluid promote neovascularization. PLoS One. 2013;**8**(1):e54747

[53] Cheng H, Chang S, Xu R, Chen L, Song X, Wu J, et al. Hypoxia-challenged MSC-derived exosomes deliver miR-210 to attenuate post-infarction cardiac apoptosis. Stem Cell Research & Therapy. 2020;**11**(1):224

[54] Monguio-Tortajada M, Prat-Vidal C, Martinez-Falguera D, Teis A, Soler-Botija C, Courageux Y, et al. Acellular cardiac scaffolds enriched with MSC-derived extracellular vesicles limit ventricular remodelling and exert local

and systemic immunomodulation in a myocardial infarction porcine model. Theranostics. 2022;**12**(10):4656-4670

[55] El-Jawhari JJ, El-Sherbiny Y, McGonagle D, Jones E. Multipotent mesenchymal stromal cells in rheumatoid arthritis and systemic lupus erythematosus; from a leading role in pathogenesis to potential therapeutic saviors? Frontiers in Immunology. 2021;**12**:643170

[56] Marion TN, Postlethwaite AE. Chance, genetics, and the heterogeneity of disease and pathogenesis in systemic lupus erythematosus. Seminars in Immunopathology. 2014;**36**(5):495-517

[57] Li A, Guo F, Pan Q, Chen S, Chen J, Liu HF, et al. Mesenchymal stem cell therapy: Hope for patients with systemic lupus erythematosus. Frontiers in Immunology. 2021;**12**:728190

[58] Gamble A, Pawlick R, Pepper AR, Bruni A, Adesida A, Senior PA, et al. Improved islet recovery and efficacy through co-culture and co-transplantation of islets with human adipose-derived mesenchymal stem cells. PLoS One. 2018;**13**(11):e0206449

[59] He C, Yang Y, Zheng K, Chen Y, Liu S, Li Y, et al. Mesenchymal stem cell-based treatment in autoimmune liver diseases: Underlying roles, advantages and challenges. Therapeutic Advances in Chronic Disease. 2021;**12**:2040622321993442

[60] Wang R, Yao Q, Chen W, Gao F, Li P, Wu J, et al. Stem cell therapy for Crohn's disease: Systematic review and meta-analysis of preclinical and clinical studies. Stem Cell Research & Therapy. 2021;**12**(1):463

[61] Grégoire C, Lechanteur C, Briquet A, Baudoux É, Baron F, Louis E, et al. Review article: Mesenchymal stromal cell therapy for inflammatory bowel diseases. Alimentary Pharmacology & Therapeutics. 2017;**45**(2):205-221

[62] Eom YW, Kim G, Baik SK. Mesenchymal stem cell therapy for cirrhosis: Present and future perspectives. World Journal of Gastroenterology. 2015;**21**(36): 10253-10261

[63] Sharma RR, Pollock K, Hubel A, McKenna D. Mesenchymal stem or stromal cells: A review of clinical applications and manufacturing practices. Transfusion. 2014;**54**(5):1418-1437

[64] Cipriani P, Ruscitti P, Di Benedetto P, Carubbi F, Liakouli V, Berardicurti O, et al. Mesenchymal stromal cells and rheumatic diseases: New tools from pathogenesis to regenerative therapies. Cytotherapy. 2015;**17**(7):832-849

[65] Zeng L, Yu G, Yang K, Xiang W, Li J, Chen H. Efficacy and safety of mesenchymal stem cell transplantation in the treatment of autoimmune diseases (rheumatoid arthritis, systemic lupus erythematosus, inflammatory bowel disease, multiple sclerosis, and ankylosing spondylitis): A systematic review and meta-analysis of randomized controlled trial. Stem Cells International. 2022;**2022**:9463314

[66] Zaripova LN, Midgley A, Christmas SE, Beresford MW, Pain C, Baildam EM, et al. Mesenchymal stem cells in the pathogenesis and therapy of autoimmune and autoinflammatory diseases. International Journal of Molecular Sciences. 2023;**24**(22):16040

[67] Liu H, Chen Y, Yin G, Xie Q. Therapeutic prospects of MicroRNAs carried by mesenchymal stem cells-derived extracellular vesicles in autoimmune diseases. Life Sciences. 2021;**277**:119458

[68] Martinez-Arroyo O, Ortega A, Forner MJ, Cortes R. Mesenchymal stem cell-derived extracellular vesicles as non-coding RNA therapeutic vehicles in autoimmune diseases. Pharmaceutics. 2022;**14**(4):733

[69] Cahill EF, Tobin LM, Carty F, Mahon BP, English K. Jagged-1 is required for the expansion of CD4+ CD25+ FoxP3+ regulatory T cells and tolerogenic dendritic cells by murine mesenchymal stromal cells. Stem Cell Research & Therapy. 2015;**6**(1):19

[70] Djouad F, Charbonnier LM, Bouffi C, Louis-Plence P, Bony C, Apparailly F, et al. Mesenchymal stem cells inhibit the differentiation of dendritic cells through an interleukin-6-dependent mechanism. Stem Cells (Dayton, Ohio). 2007;**25**(8):2025-2032

[71] Kikodze N, Pantsulaia I, Chikovani T. The role of T regulatory and TH17 cells in the pathogenesis of rheumatoid arthritis (review). Georgian Medical News. 2016;**261**:62-68

[72] Roelofs MF, Boelens WC, Joosten LA, Abdollahi-Roodsaz S, Geurts J, Wunderink LU, et al. Identification of small heat shock protein B8 (HSP22) as a novel TLR4 ligand and potential involvement in the pathogenesis of rheumatoid arthritis. Journal of Immunology (Baltimore, Md: 1950). 2006;**176**(11):7021-7027

[73] Schaefer L, Babelova A, Kiss E, Hausser HJ, Baliova M, Krzyzankova M, et al. The matrix component biglycan is proinflammatory and signals through toll-like receptors 4 and 2 in macrophages. The Journal of Clinical Investigation. 2005;**115**(8):2223-2233

[74] Menon M, Blair PA, Isenberg DA, Mauri C. A regulatory feedback between plasmacytoid dendritic cells and regulatory B cells is aberrant in systemic lupus erythematosus. Immunity. 2016;**44**(3):683-697

[75] Ganguly P, Burska AN, Davis CLM, El-Jawhari JJ, Giannoudis PV, Jones EA. Intrinsic type 1 interferon (IFN1) profile of uncultured human bone marrow CD45(low)CD271(+) multipotential stromal cells (BM-MSCs): The impact of donor age, culture expansion and IFNα and IFNβ stimulation. Biomedicine. 2020;**8**(7):214

[76] Koehler N, Buhler L, Egger B, Gonelle-Gispert C. Multipotent mesenchymal stromal cells interact and support islet of Langerhans viability and function. Frontiers in Endocrinology. 2022;**13**:822191

[77] Li YP, Paczesny S, Lauret E, Poirault S, Bordigoni P, Mekhloufi F, et al. Human mesenchymal stem cells license adult CD34+ hemopoietic progenitor cells to differentiate into regulatory dendritic cells through activation of the notch pathway. Journal of Immunology (Baltimore, Md: 1950). 2008;**180**(3):1598-1608

[78] Pajarinen J, Lin T, Gibon E, Kohno Y, Maruyama M, Nathan K, et al. Mesenchymal stem cell-macrophage crosstalk and bone healing. Biomaterials. 2019;**196**:80-89

[79] Kim J, Hematti P. Mesenchymal stem cell-educated macrophages: A novel type of alternatively activated macrophages. Experimental Hematology. 2009;**37**(12):1445-1453

[80] Refaie AF, Elbassiouny BL, Kloc M, Sabek OM, Khater SM, Ismail AM, et al. From mesenchymal stromal/stem cells to insulin-producing cells: Immunological considerations. Frontiers in Immunology. 2021;**12**:690623

[81] Qiu B, Xu X, Yi P, Hao Y. Curcumin reinforces MSC-derived exosomes in attenuating osteoarthritis via modulating the miR-124/NF-kB and miR-143/ROCK1/TLR9 signalling pathways. Journal of Cellular and Molecular Medicine. 2020;**24**(18):10855-10865

[82] Fang X, Gao F, Yao Q, Xu H, Yu J, Cao H, et al. Pooled analysis of mesenchymal stromal cell-derived extracellular vesicle therapy for liver disease in preclinical models. Journal of Personalized Medicine. 2023;**13**(3):441

[83] Shandil RK, Dhup S, Narayanan S. Evaluation of the therapeutic potential of mesenchymal stem cells (MSCs) in preclinical models of autoimmune diseases. Stem Cells International. 2022;**2022**:6379161

[84] Lopez-Santalla M, Bueren JA, Garin MI. Mesenchymal stem/stromal cell-based therapy for the treatment of rheumatoid arthritis: An update on preclinical studies. eBioMedicine. 2021;**69**:103427

[85] Zhang Q, Li Q, Zhu J, Guo H, Zhai Q, Li B, et al. Comparison of therapeutic effects of different mesenchymal stem cells on rheumatoid arthritis in mice. PeerJ. 2019;**7**:e7023

[86] Zhou T, Liao C, Li HY, Lin W, Lin S, Zhong H. Efficacy of mesenchymal stem cells in animal models of lupus nephritis: A meta-analysis. Stem Cell Research & Therapy. 2020;**11**(1):48

[87] Liu J, Lu X, Lou Y, Cai Y, Cui W, Wang J, et al. Xenogeneic transplantation of human placenta-derived mesenchymal stem cells alleviates renal injury and reduces inflammation in a mouse model of lupus nephritis. BioMed Research International. 2019;**2019**:9370919

[88] Wei X, Cheng X, Luo Y, Li X. Umbilical cord-derived mesenchymal stem cells attenuate S100-induced autoimmune hepatitis via modulating Th1 and Th17 cell responses in mice. Stem Cells International. 2023;**2023**:9992207

[89] Lim JY, Kim BS, Ryu DB, Kim TW, Park G, Min CK. The therapeutic efficacy of mesenchymal stromal cells on experimental colitis was improved by the IFN-γ and poly(I:C) priming through promoting the expression of indoleamine 2,3-dioxygenase. Stem Cell Research & Therapy. 2021;**12**(1):37

[90] Gao JG, Yu MS, Zhang MM, Gu XW, Ren Y, Zhou XX, et al. Adipose-derived mesenchymal stem cells alleviate TNBS-induced colitis in rats by influencing intestinal epithelial cell regeneration, Wnt signaling, and T cell immunity. World Journal of Gastroenterology. 2020;**26**(26):3750-3766

[91] KirkhamAM,MonaghanM,BaileyAJM, Shorr R, Lalu MM, Fergusson DA, et al. Mesenchymal stem/stromal cell–based therapies for COVID-19: First iteration of a living systematic review and meta-analysis. Cytotherapy. 2022;**24**(6):639-649

[92] Yesudhas D, Srivastava A, Gromiha MM. COVID-19 outbreak: History, mechanism, transmission, structural studies and therapeutics. Infection. 2020;**49**(2):199-213

[93] Hirano T, Murakami M. COVID-19: A new virus, but a familiar receptor and cytokine release syndrome. Immunity. 2020;**52**(5):731-733

[94] Zhou P, Yang X-L, Wang X-G, Hu B, Zhang L, Zhang W, et al. A pneumonia outbreak associated with a new coronavirus of probable bat origin. Nature. 2020;**579**(7798):270-273

[95] Al-Khawaga S, Abdelalim EM. Potential application of mesenchymal stem cells and their exosomes in lung

injury: An emerging therapeutic option for COVID-19 patients. Stem Cell Research & Therapy. 2020;**11**(1):437

[96] Samudrala PK, Kumar P, Choudhary K, Thakur N, Wadekar GS, Dayaramani R, et al. Virology, pathogenesis, diagnosis and in-line treatment of COVID-19. European Journal of Pharmacology. 2020;**883**:173375

[97] Unutmaz D, Waterman RS, Tomchuck SL, Henkle SL, Betancourt AM. A new mesenchymal stem cell (MSC) paradigm: Polarization into a pro-inflammatory MSC1 or an immunosuppressive MSC2 phenotype. PLoS One. 2010;**5**(4):e10088

[98] Ackermann M, Verleden SE, Kuehnel M, Haverich A, Welte T, Laenger F, et al. Pulmonary vascular Endothelialitis, thrombosis, and angiogenesis in Covid-19. New England Journal of Medicine. 2020;**383**(2):120-128

[99] Liu L, Wong CW, Han M, Farhoodi HP, Liu G, Liu Y, et al. Meta-analysis of preclinical studies of mesenchymal stromal cells to treat rheumatoid arthritis. eBioMedicine. 2019;**47**:563-577

[100] Tsujimaru K, Takanashi M, Sudo K, Ishikawa A, Mineo S, Ueda S, et al. Extracellular microvesicles that originated adipose tissue derived mesenchymal stem cells have the potential ability to improve rheumatoid arthritis on mice. Regenerative Therapy. 2020;**15**:305-311

[101] Shi L, Wang L, Xu R, Zhang C, Xie Y, Liu K, et al. Mesenchymal stem cell therapy for severe COVID-19. Signal Transduction and Targeted. Therapy. 2021;**6**(1):339

[102] Saldaña L, Bensiamar F, Vallés G, Mancebo FJ, García-Rey E, Vilaboa N. Immunoregulatory potential of mesenchymal stem cells following activation by macrophage-derived soluble factors. Stem Cell Research & Therapy. 2019;**10**(1):58

[103] Lee DK, Song SU. Immunomodulatory mechanisms of mesenchymal stem cells and their therapeutic applications. Cellular Immunology. 2018;**326**:68-76

[104] Raffaghello L, Bianchi G, Bertolotto M, Montecucco F, Busca A, Dallegri F, et al. Human mesenchymal stem cells inhibit neutrophil apoptosis: A model for neutrophil preservation in the bone marrow niche. Stem Cells (Dayton, Ohio). 2008;**26**(1):151-162

[105] Moretta A, Bottino C, Vitale M, Pende D, Cantoni C, Mingari MC, et al. Activating receptors and coreceptors involved in human natural killer cell-mediated cytolysis. Annual Review of Immunology. 2001;**19**:197-223

[106] Loke XY, Imran SAM, Tye GJ, Wan Kamarul Zaman WS, Nordin F. Immunomodulation and regenerative capacity of MSCs for Long-COVID. International Journal of Molecular Sciences. 2021;**22**(22):12421

[107] Li Z, Niu S, Guo B, Gao T, Wang L, Wang Y, et al. Stem cell therapy for COVID-19, ARDS and pulmonary fibrosis. Cell Proliferation. 2020;**53**(12):e12939

[108] Rossello-Gelabert M, Gonzalez-Pujana A, Igartua M, Santos-Vizcaino E, Hernandez RM. Clinical progress in MSC-based therapies for the management of severe COVID-19. Cytokine & Growth Factor Reviews. 2022;**68**:25-36

[109] Barros I, Silva A, de Almeida LP, Miranda CO. Mesenchymal stromal cells to fight SARS-CoV-2: Taking advantage

of a pleiotropic therapy. Cytokine & Growth Factor Reviews. 2021;**58**:114-133

[110] Krasnodembskaya A, Song Y, Fang X, Gupta N, Serikov V, Lee JW, et al. Antibacterial effect of human mesenchymal stem cells is mediated in part from secretion of the antimicrobial peptide LL-37. Stem Cells (Dayton, Ohio). 2010;**28**(12):2229-2238

[111] Jackson MV, Krasnodembskaya AD. Analysis of mitochondrial transfer in direct Co-cultures of human monocyte-derived macrophages (MDM) and mesenchymal stem cells (MSC). Bio-Protocol. 2017;**7**(9):e2255

[112] Durand N, Mallea J, Zubair AC. Insights into the use of mesenchymal stem cells in COVID-19 mediated acute respiratory failure. NPJ Regenerative Medicine. 2020;**5**(1):17

[113] Jung YJ, Park YY, Huh JW, Hong SB. The effect of human adipose-derived stem cells on lipopolysaccharide-induced acute respiratory distress syndrome in mice. Annals of Translational Medicine. 2019;**7**(22):674

[114] Xu YL, Liu YL, Wang Q, Li G, Lü XD, Kong B. Intravenous transplantation of mesenchymal stem cells attenuates oleic acid induced acute lung injury in rats. Chinese Medical Journal. 2012;**125**(11):2012-2018

[115] Gore AV, Bible LE, Song K, Livingston DH, Mohr AM, Sifri ZC. Mesenchymal stem cells increase T-regulatory cells and improve healing following trauma and hemorrhagic shock. The Journal of Trauma and Acute Care Surgery. 2015;**79**(1):48-52. discussion

[116] Asmussen S, Ito H, Traber DL, Lee JW, Cox RA, Hawkins HK, et al. Human mesenchymal stem cells reduce the severity of acute lung injury in a sheep model of bacterial pneumonia. Thorax. 2014;**69**(9):819-825

[117] Yang Y, Hu S, Xu X, Li J, Liu A, Han J, et al. The vascular endothelial growth factors-expressing character of mesenchymal stem cells plays a positive role in treatment of acute lung injury In vivo. Mediators of Inflammation. 2016;**2016**:2347938

[118] Xiang B, Chen L, Wang X, Zhao Y, Wang Y, Xiang C. Transplantation of menstrual blood-derived mesenchymal stem cells promotes the repair of LPS-induced acute lung injury. International Journal of Molecular Sciences. 2017;**18**(4):689

[119] Dong LH, Jiang YY, Liu YJ, Cui S, Xia CC, Qu C, et al. The anti-fibrotic effects of mesenchymal stem cells on irradiated lungs via stimulating endogenous secretion of HGF and PGE2. Scientific Reports. 2015;**5**:8713

[120] Chen J, Shao Y, Xu G, Lim C, Li J, Xu D, et al. Bone marrow-derived mesenchymal stem cells attenuate phosgene-induced acute lung injury in rats. Inhalation Toxicology. 2015;**27**(5):254-261

[121] Samanipour R, Tabatabaee S, Delyanee M, Tavakoli A. The promising approach of MSCs therapy for COVID-19 treatment. Cell and Tissue Banking. 2023;**24**(3):597-612

[122] Devaney J, Horie S, Masterson C, Elliman S, Barry F, O'Brien T, et al. Human mesenchymal stromal cells decrease the severity of acute lung injury induced by *E. Coli* in the rat. Thorax. 2015;**70**(7):625-635

[123] Sutton MT, Fletcher D, Ghosh SK, Weinberg A, van Heeckeren R, Kaur S, et al. Antimicrobial properties of mesenchymal stem cells: Therapeutic

potential for cystic fibrosis infection, and treatment. Stem Cells International. 2016;**2016**:5303048

[124] Curley GF, Jerkic M, Dixon S, Hogan G, Masterson C, O'Toole D, et al. Cryopreserved, Xeno-free human umbilical cord mesenchymal stromal cells reduce lung injury severity and bacterial burden in rodent Escherichia coli-induced acute respiratory distress syndrome. Critical Care Medicine. 2017;**45**(2):e202-ee12

Chapter 18

An *In Vitro* Model for the Study of Itch Using Human iPS Cell-Derived Sensory Neurons

Sumika Toyama, Mitsutoshi Tominaga and Kenji Takamori

Abstract

Neuronal morphological changes in the epidermis contribute to abnormal skin sensations in dry skin-based skin diseases. Intractable itch significantly reduces the quality of life, making it essential to understand the fundamental mechanisms of itch for developing antipruritic treatments. Itch is mediated by peripheral sensory neurons (SN) that originate from the neural crest (NC) during development. Itch-associated signaling molecules have been identified in genetically engineered animals and in cultures of peripheral neurons from dorsal root ganglia (DRG). However, ethical difficulties in collecting peripheral neurons from human DRG have limited the analysis of itch in humans. This chapter describes a method for differentiating peripheral neurons from human induced pluripotent stem cells (hiPSC) and a method for physiological studies of itch using hiPSC-derived peripheral neurons. It also includes an analysis of their interactions with non-neuronal cells.

Keywords: iPS cell, itch, nerve, peripheral neuron, skin

1. Introduction

Sensory neurons (SN) detect and transmit signals related to nociception, like itch, pain, and temperature; mechanoreception, including touch and vibration; and proprioception, such as pressure and tension. Somatic sensations, like itch and pain, are mediated by primary SN with cell bodies located in the dorsal root ganglion (DRG) or trigeminal ganglia [1, 2]. Primary SNs transmit somatosensory information to the central nervous system. The cell bodies of these neurons vary in size, expression of ion channels and receptors, innervating regions, and electrophysiological properties [1]. Nociceptors are activated by potentially noxious thermal, mechanical, or chemical stimuli and may convey the sensation of pain. These relatively small- to medium-diameter cells are classified as thin myelinated (Aδ) or unmyelinated axons (C fibers) [2, 3]. C fibers, which are the most common type of nociceptor, respond to both peptidergic and non-peptidergic neurotransmitters, and they are particularly sensitive to peptidergic neurotransmitters [3]. DRG neurons, which include these C fibers, are primary afferent neurons induced by both non-nociceptive and nociceptive stimuli. DRG neurons are the primary mediators of itch and pain [3–5].

Studies in genetically modified animals and rodent DRG neuron cultures have identified signaling molecules and receptors involved in the transmission of sensory stimuli. However, these mechanisms markedly differ between humans and rodents. Detailed functional analysis of SN responses to biological, biochemical, and physiological stimuli is essential to advance medical technology and understanding of biology and pathology. The collection and analysis of peripheral neurons from human DRG is limited by the available number of cells and ethics regarding invasive collection techniques. Thus, peripheral SNs derived from human cell resources are used for functional analysis.

In recent years, several methods have been developed to induce human peripheral neurons from human pluripotent stem cells (hPSC) [6–8]. These approaches are valuable for modeling human pathologies and developing therapeutic strategies. Moreover, they facilitate the analysis of interactions between human SN and non-neuronal cells, such as immune cells and epithelial cells, in co-culture systems.

This chapter presents recent studies on hPSC-derived SN (hPSC-SN) which facilitate our understanding of human sensory biology and diseases.

2. Methods and applications for inducing human peripheral sensory neurons from pluripotent stem cells

2.1 Induction of human peripheral sensory neurons from pluripotent stem cells

Neural crest cells (NC) are a pluripotent and migratory cell population in the developing embryo that contribute to forming various tissues. During vertebrate development, peripheral SNs are also derived from NC. NCs emerge at the interface between the neural and non-neural ectoderm and differentiate into various cell types, including neurons, glial cells, cranial bone, cartilage, smooth muscle, and melanocytes [8–10].

In recent years, several differentiation protocols for generating NC and SN from hPSC have been established (**Table 1**) [6, 7, 11, 14, 18]. About 30 years ago, it was reported that retinoic acid (RA) induced neuronal differentiation from mouse embryonic stem (ES) cells via embryoid bodies [24]. In mice, Kawasaki et al. established the method of stromal cell-derived inducing activity to induce dopamine neurons using PA6 cells, which support differentiation of hematopoietic stem cells, osteoclasts, and preadipocytes, as feeders [25]. Tropepe et al. established a method to form neurospheres by culturing ES cells at low density in a culture medium containing leukemia inhibitory factor [19]. Six years later, in 2007, human iPSC were reported, and it was found that neural differentiation could be achieved using a method similar to that for human embryonic stem cells [26].

Surmacz et al. established the LSB (*L*DN193189, *SB*43152) protocol using dual SMAD inhibition. According to this protocol, NCs are induced by inhibiting glycogen synthase kinase (GSK)-3 (CHIR99021), notch inhibition; DAPT (γ-gamma secretase), and vascular endothelial growth factor receptor/fibroblast growth factor receptor (SU5402), achieving an induction efficiency of 80% [6, 11]. At this stage, sensory neuron fate specification is possible, followed by neuronal maturation with growth factors such as brain-derived neurotrophic factor (BDNF), glial-derived neurotrophic factor (GDNF), nerve growth factor(NGF), and neurotrophin-3 (NT-3) supplemented with ascorbic acid. Neurons induced by this protocol express standard markers of SN, including TuJ-1 (β-III tubulin), TAC1 (substance P/neurokinin A

LSB					
	Chambers et al. [11]	Young et al. [12]		Dionisi et al. [13]	
NC induction efficiency	>95%	75%		?	
SN induction efficiency	75%	–		50%	
NSB					
	Chambers et al. [14]	Lee et al. [8]	Surmacz et al. [6]	Umehara et al. [15]	
NC induction efficiency	>80%	15–40%	79 ± 3.1%	>90%	
SN induction efficiency	?	?	–	?	
CSB					
	Fukuta et al. [7]	Kirino et al. [16]	Alshawaf et al. [17]	Saito-Diaz et al. [18]	
NC induction efficiency	70–80%	Approx. 45%	Approx. 11%	<80%	
SN induction efficiency	–	>85%	?	<99%	
Neurosphere					
	Tropepe et al. [19]	Ebert et al. [20]	Nori et al. [21]	Imaizumi et al. [22]	Matsumoto et al. [23]
Neurosphere induction efficiency	40–50%	?	?	<65%	100%
SN induction efficiency	26%	Approx. 15%	<50%	Approx. 20%	>45%

Table 1.
NC and SN induction efficiency by each NC induction method.

precursor), peripherin, Islet-1, brain-specific homeobox/POU domain protein 3A (Brn3A (POU4F1)), SLC17A6 (VGLUT2), SCN9A (Nav1.7), SCN10A (Nav1.8), and the transcription factor RUNX1. Electrophysiological recordings revealed that ion channels Nav1.8 and P2X3, associated with nociception, are also expressed [6, 11, 12]. Using a multi-electrode array (MEA), Hiranuma et al. demonstrated that SNs induced by the LSB protocol express genes and proteins related to nociceptors, mechanoreceptors, and self-receptors, although these sensory neurons varied in maturity [27].

Fukuta et al. established the CSB system using *C*HIR99021 and *SB*431542 [7]. This protocol induced NCs (70–80%) from hPSC regardless of the type (human embryonic stem cells (hESC) vs. hiPSC) or generating method (viral-integrated vs. plasmid-episomal). Additionally, they successfully induced not only NCs but also mesenchymal stem cells and osteogenic cells using iPS cells produced and maintained in feeder-free and xeno-free culture systems and chemically defined media. In this protocol, after NCs are sorted, BDNF, GDNF, NT-3, and NGF can be added to induce TuJ-1 and peripherin-positive neurons [7]. To improve induction efficiency, they used the PHOX2B::eGFP reporter hPSC line to optimize each differentiation stage and established a robust protocol to induce functional SNs via neural mesoderm progenitor-like cells and somatic NC cells [16]. Saito-Diaz et al. examined the timing of differentiation and concentrations of CHIR99021, establishing a protocol with 80% efficiency in inducing NCs [18]. To increase the efficiency and speed of differentiation into SNs, SU-5402 and DAPT were used, following the method established by Lee et al. [8–10]. The results showed that 99% of all cells were TuJ-1 positive and 60–70% expressed the pan-sensory nerve marker BRN3A at 20 days in culture. Furthermore, 30% of the induced sensory nerves were $TRKB^+/TRKC^+$ mechanoreceptors and $TRKC^+$ proprioceptors. Common markers of sensory nerve maturation (VGLUT1/2/3), nociceptors (somatostatin, PLXNC1, TAC1), and temperature and pain sensors transient receptor potential (TRP) V1 and TRPV2, TRPM8, and TRPA1 were present on day 30. Additionally, the genes encoding the Na^+ channels Nav1.6 (SCN8A), Nav1.7 (SCN9A), SCN10A, and Nav1.9 (SCN11A), as well as the ATP-activated receptor P2X3, were expressed along with mechanoreceptors and intrinsic receptors [18].

Lee et al. and Chamber et al. established the NSB system using *N*oggin and *SB*431542 for NC induction [8, 14]. This protocol can further induce Brn3a + and peripherin + peripheral nerves with the addition of brain-derived neurotrophic factor, AA, GDNF, NGF, NT-3, and cyclic AMP [8, 14]. However, the NC induction efficiency of this protocol is less than 40% and requires sorting [8]. Umehara et al. modified the NSB system to establish a highly efficient robust protocol for NC induction from hiPSC, achieving over 90% double positive cells for p75 and carbohydrate chain HNK1, making it more efficient compared to the LSB and CSB protocols. Additionally, these neurons responded well to itch and pain stimuli [15].

Reynolds et al. developed a neurosphere (NS) method to grow neural stem cells in a floating culture using self-renewal of cells extracted from the mouse striatum in a serum-free medium [28]. Furthermore, Nori et al. established a method to induce differentiation from hiPSC-NS into neurons [21]. Based on this method, Matsumoto et al. developed a technique to culture NS under low cell density and hypoxic conditions, efficiently and stably differentiating them into mature, functional neurons [23]. The efficiency of this method is comparable to when NS are induced by the NSB protocol.

2.2 Functional analyses using human peripheral neurons derived from human pluripotent stem cells

Functional analyses using sensory neurons derived from human pluripotent stem cells are in progress (**Table 2**). Patch-clamp experiments and calcium imaging studies have shown that SNs induced by LSB protocol exhibit a nociceptive sensory phenotype [11, 12]. Microarray and single-cell qPCR data analyses revealed that hPSC-SN expresses more than 80% of the ion channel genes found in hDRG samples, indicating a high degree of similarity between hPSC-SN and hDRG [12]. Dionisi et al. improved the LSB protocol, establishing a more efficient and robust induction protocol [13]. They successfully induced SN from hematopoietic stem cells derived from patients with Friedreich's ataxia (FRDA), and these SNs mimicked this pathology.

Electrical recording of neuronal activity has traditionally been the preferred choice for analyzing single neurons and neuronal networks. Recently, MEA has been developed to simultaneously record local field potentials and extracellular action potentials over long periods. Analysis of SN induced by the LSB protocol with MEA by Hiranuma et al. suggests that the generated hiPSC-SN contains nociceptors that respond to nociceptive stimuli such as capsaicin, menthol, bradykinin, nociceptive heat (>43°C), and mechanoreceptors that respond to non-nociceptive stimuli (41°C) [27]. This SN has been applied to studies of the effects of chemicals [29].

LSB				
	Chambers et al. [11]	Young et al. [12]	Dionisi et al. [13]	Hiranuma et al. [27]
Method	Patch-clamp, Ca imaging	Patch-clamp	Patch-clamp, patient-derived iPS-SN	MEA
Function	Nociceptors	Nociceptors express GABAAR, HCN ion channel, KCNQ2/3 channels, and acid-sensing ion channels	Neural firing. FRDA patient-derived iPS-SNs reflect pathological conditions	Includes nociceptors and mechanoreceptors
NSB				
	Umehara et al. [15]			
Method	Calcium imaging			
Function	SN involved in the transmission of itch and pain.			
CSB				
	Alshawaf et al. [17]		Saito-Diaz et al. [18]	
Method	MEA		MEA	
Function	Response to heat, capsaicin, and hypoosmotic stimuli		Response to capsaicin and hypoosmotic stimuli	
Neurosphere				
	Ebert et al. [20]		Imaizumi et al. [22]	
Method	Patient-derived iPS-SN		Patient-derived iPS-SN	
Function	Spinal muscular atrophy patient-derived iPS-SNs reflect pathological conditions		ALS-derived iPS-SNs reflect pathological conditions	

Table 2.
Functional analysis of induced SN.

SN induced by the CSB protocol has also been subjected to functional analysis using MEA. The analyses found spontaneous neural activity in hiPS-SN [18, 17], and hiPS-SN induced by the CSB protocol is neurofired by nociceptors (capsaicin, heat (45°C)) and mechanoreceptors (hypoosmotic pressure) [18, 17].

Umehara et al. performed a functional analysis of SN induced by the NSB protocol by means of calcium imaging. They reported that some neurons responded to capsaicin, bovine adrenal medulla peptide (BAM)8-22, and allyl isothiocyanate (AITC), while others did not. The results suggest that calcium influx is induced independently of TRPV1 and TRPA1. Some neurons responded to capsaicin but not histamine, and others responded to AITC but not BAM8-22, suggesting that these neurons may be involved in pain transmission. Although IL-31 directly stimulates IL-31R-expressing sensory neurons via TRPV1 and TRPA1 [30], some neurons induced by this protocol responded to IL-31 and not to AITC. It is thought that calcium influx is induced in these neurons by TRPV1. In contrast, other neurons respond to IL-31 and AITC, which may have calcium influx induced by TRPA1 [15]. These results suggest that peripheral neurons from hiPSC-NC cells react with various stimuli associated with itch, and may be useful in studies investigating the fundamental mechanisms of itch elicitation in human DRG neurons.

Induction of neurons in iPS cells derived from patients with spinal muscular atrophy using RA confirmed the phenotype that patient-derived cells are fragile [20]. When neurology was induced by the neurosphere method using iPS cells derived from Alzheimer's disease and amyotrophic lateral sclerosis (ALS)patients, the pathological phenotype could be observed [22].

Based on the above, the expression levels of sensory neuron-related genes and proteins in hiPSC-derived sensory neurons are close to those of DRG, which may be valuable in studying the basic mechanisms of human DRG neurons and as a research tool for drug discovery.

3. Conclusion

In conclusion, various approaches to producing human peripheral SN are still being refined, providing valuable opportunities to study human somatosensory biology. These advancements will advance research in somatosensory physiology and may prove useful for drug screening, disease modeling, and tissue regeneration.

Acknowledgements

This research was partly supported by a Grant-in-Aid for Research Activity Start-up (Grant number 17H07096), a Grant-in-Aid for Early-Career Scientists (Grant number 19 K17817 and 21 K16307), a Grant-in-Aid for Scientific Research (B) (Grant number 20H03568, 22H02956, 23 K20320, and 23 K24217). We thank David Price of English Services for Scientists for English proofreading.

Conflict of interest

The authors declare no conflict of interest.

Author details

Sumika Toyama[1,2], Mitsutoshi Tominaga[1,2] and Kenji Takamori[1,2,3*]

1 Juntendo Itch Research Center (JIRC), Institute for Environmental and Gender-Specific Medicine, Juntendo University Graduate School of Medicine, Urayasu, Japan

2 Department of Dermatological Anti-Aging and Regenerative Medicine (DARM), Juntendo University Graduate School of Medicine, Urayasu, Japan

3 Department of Dermatology, Juntendo University Urayasu Hospital, Urayasu, Japan

*Address all correspondence to: ktakamor@juntendo.ac.jp

References

[1] Ikoma A, Steinhoff M, Stander S, Yosipovitch G, Schmelz M. The neurobiology of itch. Nature Reviews. Neuroscience. 2006;**7**(7):535-547. DOI: 10.1038/nrn1950

[2] Belmonte C, Viana F. Molecular and cellular limits to somatosensory specificity. Molecular Pain. 2008;**4**:14. DOI: 10.1186/1744-8069-4-14

[3] Le Pichon CE, Chesler AT. The functional and anatomical dissection of somatosensory subpopulations using mouse genetics. Frontiers in Neuroanatomy. 2014;**8**:21. DOI: 10.3389/fnana.2014.00021

[4] Lallemend F, Ernfors P. Molecular interactions underlying the specification of sensory neurons. Trends in Neurosciences. 2012;**35**(6):373-381. DOI: 10.1016/j.tins.2012.03.006

[5] Poole K, Herget R, Lapatsina L, Ngo HD, Lewin GR. Tuning piezo ion channels to detect molecular-scale movements relevant for fine touch. Nature Communications. 2014;**5**:3520. DOI: 10.1038/ncomms4520

[6] Surmacz B, Fox H, Gutteridge A, Fish P, Lubitz S, Whiting P. Directing differentiation of human embryonic stem cells toward anterior neural ectoderm using small molecules. Stem Cells. 2012;**30**(9):1875-1884. DOI: 10.1002/stem.1166

[7] Fukuta M, Nakai Y, Kirino K, Nakagawa M, Sekiguchi K, Nagata S, et al. Derivation of mesenchymal stromal cells from pluripotent stem cells through a neural crest lineage using small molecule compounds with defined media. PLoS One. 2014;**9**(12):e112291. DOI: 10.1371/journal.pone.0112291

[8] Lee G, Chambers SM, Tomishima MJ, Studer L. Derivation of neural crest cells from human pluripotent stem cells. Nature Protocols. 2010;**5**(4):688-701. DOI: 10.1038/nprot.2010.35

[9] Srinivasan A, Toh YC. Human pluripotent stem cell-derived neural crest cells for tissue regeneration and disease modeling. Frontiers in Molecular Neuroscience. 2019;**12**:39. DOI: 10.3389/fnmol.2019.00039

[10] Dupin E, Calloni GW, Coelho-Aguiar JM, Le Douarin NM. The issue of the multipotency of the neural crest cells. Developmental Biology. 2018;**444**(Suppl. 1):S47-S59. DOI: 10.1016/j.ydbio.2018.03.024

[11] Chambers SM, Qi Y, Mica Y, Lee G, Zhang XJ, Niu L, et al. Combined small-molecule inhibition accelerates developmental timing and converts human pluripotent stem cells into nociceptors. Nature Biotechnology. 2012;**30**(7):715-720. DOI: 10.1038/nbt.2249

[12] Young GT, Gutteridge A, Fox H, Wilbrey AL, Cao L, Cho LT, et al. Characterizing human stem cell-derived sensory neurons at the single-cell level reveals their ion channel expression and utility in pain research. Molecular Therapy. 2014;**22**(8):1530-1543. DOI: 10.1038/mt.2014.86

[13] Dionisi C, Rai M, Chazalon M, Schiffmann SN, Pandolfo M. Primary proprioceptive neurons from human induced pluripotent stem cells: A cell model for afferent ataxias. Scientific Reports. 2020;**10**(1):7752. DOI: 10.1038/s41598-020-64831-6

[14] Chambers SM, Fasano CA, Papapetrou EP, Tomishima M,

Sadelain M, Studer L. Highly efficient neural conversion of human ES and iPS cells by dual inhibition of SMAD signaling. Nature Biotechnology. 2009;**27**(3):275-280. DOI: 10.1038/nbt.1529

[15] Umehara Y, Toyama S, Tominaga M, Matsuda H, Takahashi N, Kamata Y, et al. Robust induction of neural crest cells to derive peripheral sensory neurons from human induced pluripotent stem cells. Scientific Reports. 2020;**10**(1):4360. DOI: 10.1038/s41598-020-60036-z

[16] Kirino K, Nakahata T, Taguchi T, Saito MK. Efficient derivation of sympathetic neurons from human pluripotent stem cells with a defined condition. Scientific Reports. 2018;**8**(1):12865. DOI: 10.1038/s41598-018-31256-1

[17] Alshawaf AJ, Viventi S, Qiu W, D'Abaco G, Nayagam B, Erlichster M, et al. Phenotypic and functional characterization of peripheral sensory neurons derived from human embryonic stem cells. Scientific Reports. 2018;**8**(1):603. DOI: 10.1038/s41598-017-19093-0

[18] Saito-Diaz K, Street JR, Ulrichs H, Zeltner N. Derivation of peripheral nociceptive, mechanoreceptive, and proprioceptive sensory neurons from the same culture of human pluripotent stem cells. Stem Cell Reports. 2021;**16**(3):446-457. DOI: 10.1016/j.stemcr.2021.01.001

[19] Tropepe V, Hitoshi S, Sirard C, Mak TW, Rossant J, van der Kooy D. Direct neural fate specification from embryonic stem cells: A primitive mammalian neural stem cell stage acquired through a default mechanism. Neuron. 2001;**30**(1):65-78. DOI: Doi 10.1016/S0896-6273(01)00263-X

[20] Ebert AD, Yu J, Rose FF Jr, Mattis VB, Lorson CL, Thomson JA, et al. Induced pluripotent stem cells from a spinal muscular atrophy patient. Nature. 2009;**457**(7227):277-280. DOI: 10.1038/nature07677

[21] Nori S, Okada Y, Yasuda A, Tsuji O, Takahashi Y, Kobayashi Y, et al. Grafted human-induced pluripotent stem-cell-derived neurospheres promote motor functional recovery after spinal cord injury in mice. Proceedings of the National Academy of Sciences of the United States of America. 2011;**108**(40):16825-16830. DOI: 10.1073/pnas.1108077108

[22] Imaizumi K, Sone T, Ibata K, Fujimori K, Yuzaki M, Akamatsu W, et al. Controlling the regional identity of hPSC-derived neurons to uncover neuronal subtype specificity of neurological disease phenotypes. Stem Cell Reports. 2015;**5**(6):1010-1022. DOI: 10.1016/j.stemcr.2015.10.005

[23] Matsumoto T, Fujimori K, Andoh-Noda T, Ando T, Kuzumaki N, Toyoshima M, et al. Functional neurons generated from T cell-derived induced pluripotent stem cells for neurological disease modeling. Stem Cell Reports. 2016;**6**(3):422-435. DOI: 10.1016/j.stemcr.2016.01.010

[24] Bain G, Kitchens D, Yao M, Huettner JE, Gottlieb DI. Embryonic stem cells express neuronal properties in vitro. Developmental Biology. 1995;**168**(2):342-357. DOI: 10.1006/dbio.1995.1085

[25] Kawasaki H, Mizuseki K, Nishikawa S, Kaneko S, Kuwana Y, Nakanishi S, et al. Induction of midbrain dopaminergic neurons from ES cells by stromal cell-derived inducing activity. Neuron. 2000;**28**(1):31-40. DOI: 10.1016/s0896-6273(00)00083-0

[26] Okita K, Ichisaka T, Yamanaka S. Generation of germline-competent induced pluripotent stem cells. Nature. 2007;**448**(7151):313-317. DOI: 10.1038/nature05934

[27] Hiranuma M, Okuda Y, Fujii Y, Richard JP, Watanabe T. Characterization of human iPSC-derived sensory neurons and their functional assessment using multi electrode array. Scientific Reports. 2024;**14**(1):6011. DOI: 10.1038/s41598-024-55602-8

[28] Reynolds BA, Weiss S. Generation of neurons and astrocytes from isolated cells of the adult mammalian central nervous system. Science. 1992;**255**(5052):1707-1710. DOI: 10.1126/science.1553558

[29] Satoh M, Suzuki T, Sakurai T, Toyama S, Kamata Y, Kondo S, et al. A novel in vitro assay using human iPSC-derived sensory neurons to evaluate the effects of external chemicals on neuronal morphology: Possible implications in the prediction of abnormal skin sensation. International Journal of Molecular Sciences. 2021;**22**(19):10525. DOI: 10.3390/ijms221910525

[30] Cevikbas F, Wang X, Akiyama T, Kempkes C, Savinko T, Antal A, et al. A sensory neuron-expressed IL-31 receptor mediates T helper cell-dependent itch: Involvement of TRPV1 and TRPA1. The Journal of Allergy and Clinical Immunology. 2014;**133**(2):448-460. DOI: 10.1016/j.jaci.2013.10.048

Section 3

Stem Cell Transplant for Non-Neoplastic Diseases

Chapter 19

Advancements of Stem Cell Transplantation in Myocardial Infarction and Ischemic Heart Failure

Rizaev Jasur Alimdjanovich and
Ziyadullaev Shukhrat Khudayberdiyevich

Abstract

Myocardial infarction (MI), or heart attack, is a major cause of death globally, characterized by the death of heart muscle cells due to insufficient blood supply. Traditional treatments often fail to regenerate damaged heart tissue effectively. Therefore, stem cell transplantation has emerged as a promising therapeutic approach to restore cardiac function and prevent adverse remodeling. This article reviews the role of stem cell therapy in managing MI, highlighting various stem cell types, mechanisms of action, and clinical outcomes. Stem cells, including skeletal myoblasts, bone marrow-derived mononuclear cells, hematopoietic stem cells, endothelial progenitor cells, mesenchymal stem cells, cardiac stem cells, and embryonic stem cells, offer potential for cardiac regeneration. They can differentiate into specialized cells, promote angiogenesis, modulate inflammation, and recruit repair mechanisms. Clinical trials have shown the safety and feasibility of stem cell therapy, with varying efficacy in improving heart function and patient outcomes. The application of stem cell therapy in MI treatment requires careful consideration of patient selection, timing, administration route, and stem cell type. Ongoing research focuses on optimizing treatment protocols and exploring innovative approaches to enhance therapeutic efficacy. In conclusion, stem cell transplantation holds promise for improving myocardial repair, functional recovery, and prognosis in MI patients. Collaborative efforts among researchers, clinicians, and regulatory bodies are crucial to fully realize the potential of stem cell therapy in cardiovascular medicine.

Keywords: myocardial infarction, stem cell therapy, cardiac regeneration, clinical outcomes, therapeutic approaches

1. Introduction

In today's contemporary world, myocardial infarction (MI) is one of the leading causes of death worldwide [1]. MI also known as "heart attack" can be caused by decreased or complete cessation of blood flow from coronary arteries in the myocardium. After myocardial infarction (MI), there is an irreversible loss of cardiomyocytes and the creation of scar tissue, leading to tissue damage or necrosis. Despite advancements in cardiovascular medicine, MI is still one of the medical phenomena that was not solved. Conventional treatments such as reperfusion therapies (e.g., angioplasty and thrombolytic therapy) and pharmacotherapy aim to restore blood flow and alleviate symptoms. However, these approaches often fall short in promoting substantial regeneration of damaged heart tissue, leaving patients vulnerable to adverse cardiac events and progressive heart failure. For this reason, cell transplantation has emerged as a potential therapy, and stem cells may be an important and powerful cellular source. Stem cells possess unique properties, including self-renewal and the capacity to differentiate into various cell types, making them invaluable in repairing damaged tissues. By harnessing the regenerative potential of stem cells, researchers and clinicians aim to restore cardiac function, attenuate adverse remodeling, and enhance the overall prognosis for individuals affected by myocardial infarction. The purpose of this part of the book is to overview the role of stem cell transplantation in the management of myocardial infarction, by exploring various mechanisms, therapeutic approaches, clinical outcomes, challenges, and future prospects associated with stem cell therapy. By understanding this field, clinicians and researchers can find a weapon against one of the lethal diseases named myocardial infarction.

1.1 Background on myocardial infarction

Myocardial infarction (MI) occurs when there is necrosis or death of heart muscle cells due to prolonged and severe lack of blood supply. Typically associated with atherosclerosis-related coronary heart disease, MI can also be caused by other factors that obstruct blood flow, such as arterial spasms. Atherosclerosis, characterized by plaque buildup in arteries, is always involved in the formation of these obstructions. The cause of this plaque ranges from genetics to environmental factors, which include the following:

1. Smoking
2. Abnormal lipid profile/blood apolipoprotein (raised ApoB/ApoA1)
3. Hypertension
4. Diabetes mellitus
5. Abdominal obesity (waist/hip ratio) (greater than 0.90 for males and greater than 0.85 for females)
6. Psychosocial factors such as depression, loss of the locus of control, global stress, financial stress, and life events including marital separation, job loss, and family conflicts
7. Lack of daily consumption of fruits or vegetables
8. Lack of physical activity
9. Alcohol consumption [2].

Therefore, this can be main etiological factors that can lead to obstruction of coronary arteries that potentially cause MI. The clinical presentation of myocardial infarction varies depending on the extent, location, and duration of ischemia. Common symptoms include pectoralis angina or discomfort, which may radiate to the arms, shoulders, neck, jaw, or back. In order to diagnose MI, ECG, cardiac biomarker testing (e.g., troponin), and imaging modalities (e.g., echocardiography and coronary angiography) can be used. After MI, clinicians often worried about heart failure. Since the cardiomyocytes are damaged, this area will face necrosis. In order to decrease the ischemic zone, clinicians prescribe adjunctive pharmacotherapy, such as antiplatelet agents, anticoagulants, beta-blockers, angiotensin-converting enzyme (ACE) inhibitors, and statins. Traditionally, the heart has been viewed as a non-dividing organ due to the limited capacity of its cardiomyocytes to proliferate after birth. When faced with injury or increased workload in disease states, cardiomyocytes in the border region of the damaged heart muscle typically undergo hypertrophy instead of re-entering the cell cycle to replace lost cells. As a result, the loss of functional cardiomyocytes in the damaged heart tissue is often seen as irreversible mainly because cardiomyocytes are considered terminally differentiated, and post-infarction repair mechanisms are inadequate following myocyte apoptosis [3]. So, this was brief background about MI, and the next part of the chapter will consider stem cell transplantation in the infarction zone.

2. Stem cell transplantation in MI

Stem cells are undifferentiated cells characterized by their ability to self-renew and differentiate into specialized cell types. They serve as the building blocks for tissue regeneration and repair throughout life [4]. Stem cells possess two distinct characteristics: (i) the ability to replicate themselves, producing daughter cells that are identical to the parent cell during division, and (ii) the capacity to develop into specialized cells through differentiation [5]. In broad terms, stem cells are categorized into 'somatic' or 'embryonic' stem cells. However, if delve deeper, they are divided based on their potency and differentiation capacity, ranging from totipotent (able to differentiate into any cell type) to pluripotent (able to differentiate into cells of all three germ layers) to multipotent (able to differentiate into a limited range of cell types) (**Figure 1**).

Numerous research endeavors have demonstrated the positive impact of human stem cells in promoting cardiac regeneration. Our attention is directed toward those cell types exhibiting the most encouraging prospects, including skeletal myoblasts (SMs), mononuclear cells derived from bone marrow (BMMNCs), hematopoietic stem cells (HSCs), endothelial progenitor cells (EPCs), mesenchymal stem cells (MSCs), cardiac stem cells (CSCs), and embryonic stem cells (ESCs).

2.1 Skeletal myoblast

Skeletal myoblasts, found between the basal lamina and sarcolemma in mature skeletal muscle, play a crucial role in muscle repair and regeneration. They originate from the mesodermal layer during embryonic development and are responsible for the formation of skeletal muscle fibers through the process of myogenesis. Skeletal myoblasts express various markers and undergo activation in response to muscle damage or disease. These cells possess a high proliferative capacity and maintain their undifferentiated state in appropriate culture conditions. Skeletal myoblasts are

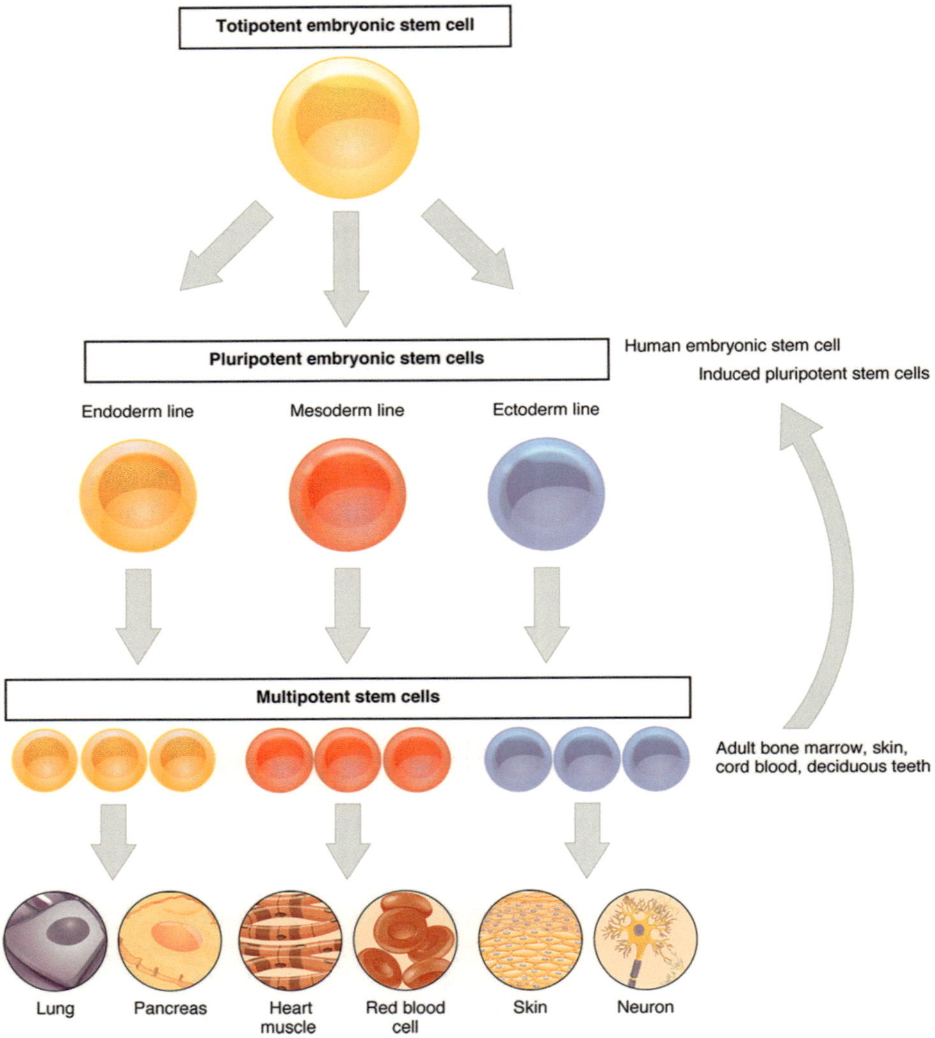

Figure 1.
Progression and potency of stem cells: from totipotent embryonic cells to differentiated adult cells. Source: https://commons.wikimedia.org/wiki/File:422_Feature_Stem_Cell_new.png.

resistant to ischemic stress and commit to a myogenic lineage upon differentiation. Importantly, transplantation of autologous skeletal myoblasts carries low risk of tumorigenicity and eliminates the need for immunosuppression, making them ethically favorable for therapeutic use in muscle regeneration. When it comes to clinical trials, there are some studies that tested skeletal myoblasts cell transplantation to animals such as mouse [6], sheep [7], and pig [8], and all these investigations have shown that skeletal myoblasts successfully halted left ventricular remodeling and maintained overall metrics (such as ejection fraction and left ventricular pressure), along with specific metrics (such as regional stroke work at the site of infarction, thickness of the wall, and velocities of myocardial tissue), indicative of myocardial contractility. However, when it comes to clinical trials in real patient result was different than expected because of variety of patient that was used in trials. For example, in

2003, the first Phase I study on skeletal myoblast transplantation for severe ischemic cardiomyopathy patients (ejection fraction < or = 35%) showed feasibility and safety. It involved injecting autologous skeletal myoblasts in the dose of 871 x 10(6) cells (86% of myoblasts) into scar areas, resulting in improved clinical symptoms and left ventricular function and ejection fraction over time [9]. Moreover, Pagani et al. and Dib et al. demonstrated the viability, feasibility, and safety of skeletal myoblast transplantation in patients with ischemic heart disease, suggesting it as a potential treatment for end-stage heart failure [10]. On the other hand, the first Phase II trial, the MAGIC (myoblast autologous grafting in ischemic cardiomyopathy) trial, showed that skeletal myoblast transplantation alongside CABG (coronary artery bypass graft surgery) did not significantly improve heart function compared to controls. Although high-dose injections reduced left ventricular volumes, they also increased the incidence of arrhythmic events, warranting further investigation. Moreover the electrical conduction of skeletal myoblast is also remains issue, as arrhythmic events happening after transplantation [11]. Other trials such as, SEISMIC, and CAuSMIC, explored catheter-based delivery of autologous skeletal myoblasts in heart failure patients but failed because of relatively small participant numbers. Overall, it could be said that skeletal myoblast is one of the good options in stem cell transplantation in ischemic heart failure, but this method needs more trials in order to fully understand mechanics and synchronization between skeletal myoblasts and cardiomyocytes in terms of electrical conduction. Furthermore, age-related diminution is another area of problem. Aging diminishes the growth and differentiation potential of skeletal myoblasts, reducing their therapeutic effectiveness. Young skeletal myoblasts show better proliferation and differentiation compared to aging ones. This poses challenges for using autologous skeletal myoblasts, especially in elderly patients.

2.2 Bone marrow-derived mononuclear cells

Bone marrow-derived mononuclear cells (BMMNCs) have emerged as promising candidates for cell replacement therapy due to their multifaceted mechanisms of action and the potential to promote cardiac repair and regeneration. To be clear about this point, first, we should analyze the mechanism of action of BMMNCs in cardiac repair. Firstly, BMMNCs contribute to angiogenesis through the secretion of pro-angiogenic factors such as vascular endothelial growth factor (VEGF), fibroblast growth factor (FGF), and hepatocyte growth factor (HGF). These factors stimulate endothelial cell proliferation, migration, and tube formation, leading to the formation of new blood vessels in ischemic myocardium (**Figure 2**).

Additionally, BMMNCs secrete a plethora of bioactive molecules including cytokines, chemokines, growth factors, and extracellular vesicles, which collectively exert paracrine effects on neighboring cardiac cells. These paracrine factors modulate various cellular processes such as cell survival, proliferation, apoptosis, and extracellular matrix remodeling. Furthermore, they mitigate oxidative stress, suppress inflammation, and promote tissue repair and regeneration, thereby enhancing cardiac function and attenuating adverse remodeling in ischemic myocardium. In theory, it sounds well, right? So, in order to clarify this debate, let us delve into some data that were tested by researchers. Numerous preclinical animal studies have provided compelling evidence supporting the efficacy of bone marrow-derived mononuclear cell (BMMNC) therapy in ischemic heart failure. These studies, conducted in various animal models of myocardial infarction or ischemic cardiomyopathy, consistently demonstrate significant improvements in cardiac function following BMMNC

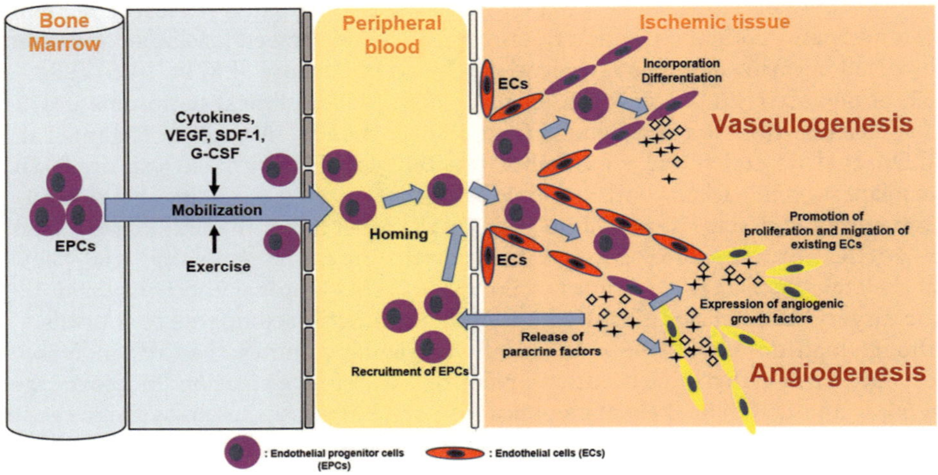

Figure 2.
Mechanisms of endothelial progenitor cell (EPC) mobilization and role in vascular repair and angiogenesis. Source: https://doi.org/10.3400/avd.ra.20-00002.

transplantation. Functional parameters such as left ventricular ejection fraction (LVEF), fractional shortening (FS), and myocardial perfusion are commonly assessed to evaluate treatment outcomes. On the other hand, in real patients, these data vary. Perin et al. demonstrated improvements in left ventricular ejection fraction (LVEF), reduced scar size, and enhanced exercise capacity following BMMNC therapy [12]. Additionally, several phase I/II clinical trials have been conducted to evaluate the safety, feasibility, and efficacy of BMMNC therapy in patients with ischemic heart failure. These trials typically involve the intracoronary or intramyocardial delivery of autologous BMMNCs harvested from the patient's own bone marrow. Safety assessments focus on adverse events, including arrhythmias, myocardial perforation, and immune reactions, while efficacy endpoints encompass improvements in cardiac function, exercise capacity, quality of life, and clinical outcomes such as mortality and hospitalization rates. Although initial trials have demonstrated the safety and feasibility of BMMNC therapy, results regarding efficacy have been heterogeneous, with some studies reporting modest functional improvements while others showing no significant benefit [13]. Overall, the collective evidence from animal studies, clinical trials, and meta-analyses underscores the potential of BMMNC therapy as a promising therapeutic approach for ischemic heart failure. However, in acute myocardial infarction, it shows less result [14]. From this perspective, we can conclude that it is a very effective method in ischemic heart failure but less effective in acute myocardial infarction. Moreover, in the future perspective, combining BMMNC therapy with growth factors or biomaterials represents a promising approach to enhance therapeutic efficacy. All in all, nowadays, we do not have full information about bone marrow-derived mononuclear cells, but in future, it shows promising results.

2.3 Hematopoietic stem cells (HSCs)

Hematopoietic stem cells (HSCs) possess a remarkable ability to migrate toward sites of tissue injury, including the myocardium following myocardial infarction (MI). This process, known as homing, is orchestrated by a complex interplay of chemokines, adhesion molecules, and growth factors. Upon release from the bone marrow niche, HSCs respond to signals such as stromal cell-derived factor-1 (SDF-1) and its receptor CXCR4, which are upregulated in the ischemic myocardium. Through a series of adhesion molecules interactions, including selectins and integrins, HSCs adhere to the endothelium and extravasate into the damaged tissue, where they participate in the reparative processes [15]. In addition to direct cell replacement, HSCs exert profound paracrine effects that contribute to myocardial repair and regeneration. Following engraftment within the infarcted myocardium, HSCs secrete a myriad of bioactive molecules, including cytokines, growth factors, and extracellular vesicles. These factors modulate the inflammatory response, promote angiogenesis, inhibit apoptosis, and stimulate endogenous cardiac progenitor cells, thereby creating a conducive microenvironment for tissue healing. Notably, the paracrine activity of HSCs has been shown to exert beneficial effects on cardiac function even in the absence of significant cell engraftment, underscoring the importance of their secretome in therapeutic efficacy. Although the extent of direct transdifferentiation of HSCs into mature cardiomyocytes remains a subject of debate, emerging evidence suggests that a subset of HSCs possesses multilineage differentiation potential. Under appropriate microenvironmental cues, HSCs can differentiate into cardiomyocyte-like cells, smooth muscle cells, and endothelial cells, thereby contributing to neovascularization and tissue remodeling. While the frequency of spontaneous cardiomyogenesis from HSCs appears to be low, strategies aimed at enhancing their cardiogenic potential, such as genetic modification or co-culture with cardiac cells, hold promise for augmenting their therapeutic efficacy in myocardial repair [16]. So, clinical trails in animals demonstrated the efficacy of hematopoietic stem cell (HSC) transplantation in promoting myocardial repair and functional recovery. These animal models, including rodents and larger mammals such as pigs and sheep, allow researchers to mimic the pathophysiological processes of MI and assess the therapeutic potential of HSCs. Transplantation of HSCs, either directly into the myocardium or via systemic administration, has been shown to attenuate myocardial remodeling, reduce infarct size, and improve cardiac function. Histological analyses often reveal increased angiogenesis, reduced fibrosis, and enhanced cardiomyocyte survival in HSC-treated animals compared to controls, highlighting the regenerative effects of HSCs on the injured myocardium [17]. Several clinical trials have been conducted to assess the safety, feasibility, and efficacy of hematopoietic stem cell (HSC) transplantation in patients with myocardial infarction (MI). These trials vary in design, including randomized controlled trials, open-label studies, and single-arm pilot trials. Trials have investigated different sources of HSCs, such as bone marrow-derived HSCs, mobilized peripheral blood stem cells, and umbilical cord blood-derived stem cells. Additionally, various delivery methods, including intracoronary infusion, intramyocardial injection, and intravenous administration, have been explored. Trials typically evaluate clinical endpoints such as mortality, major adverse cardiac events (MACE), left ventricular function, myocardial perfusion, and quality of life [14]. Immunological rejection represents a challenge in HSCs transplantation, particularly in allogeneic cell-based therapies. Host immune responses, including cytotoxic T-cell-mediated rejection and humoral immune reactions, can compromise the survival and function of transplanted cells. Immunosuppressive regimens may mitigate rejection but pose risks of infection and other adverse effects. Strategies to induce immune tolerance or utilize autologous cells to circumvent immune rejection are areas of ongoing research. The exploration of hematopoietic stem cell (HSC) transplantation in myocardial infarction (MI) therapy has yielded significant insights into its regenerative potential. Preclinical studies have demonstrated the ability of HSCs to home to the injured myocardium, exert paracrine effects, and potentially differentiate into cardiomyocytes and vascular cells, thereby promoting tissue repair and functional recovery. Clinical trials have provided evidence supporting the safety, feasibility, and potential efficacy of HSCs transplantation in improving cardiac function and clinical outcomes in MI patients. Despite challenges such as immune rejection and optimization of delivery methods, HSCs therapy holds promise as a novel therapeutic approach for MI treatment.

2.4 Endothelial progenitor cells (EPCs) and mesenchymal stem cells (MSCs) in myocardial infarction

Now let us delve deeper into one of the best methods in stem cell transplantation in MI. Endothelial progenitor cells (EPCs), mesenchymal stem cells (MSCs) separately and dual transplantation in medicine right now counts as one of the best methods in Myocardial infarction. First of all, endothelial progenitor cells (EPCs) represent a subpopulation of stem cells characterized by their ability to differentiate into mature endothelial cells, which form the inner lining of blood vessels. These cells play a crucial role in vascular homeostasis and repair following injury. EPCs are identified by the expression of specific surface markers, including CD34, CD133, and vascular endothelial growth factor receptor 2 (VEGFR-2). Additionally, they possess the capacity for self-renewal and migration to sites of vascular injury or ischemia. One of the primary functions of EPCs is to promote angiogenesis, the process of new blood vessel formation from pre-existing vasculature (**Figure 3**).

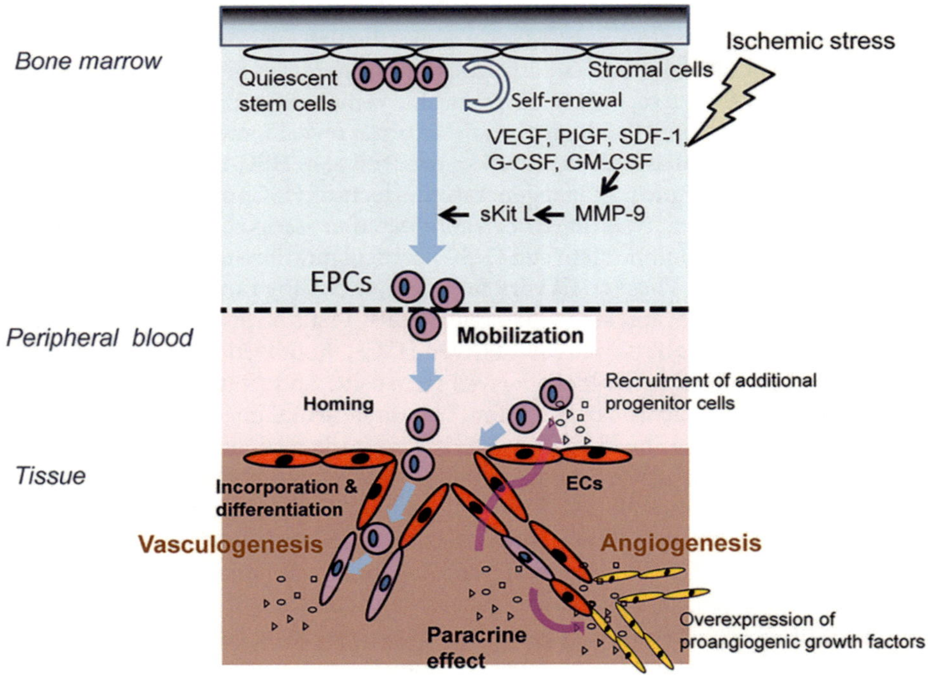

Figure 3.
Pathways of endothelial progenitor cell mediated vascular regeneration. Source: https://doi.org/10.3400/avd.ra.22-00086.

Upon recruitment to sites of tissue ischemia or injury, EPCs contribute to angiogenesis by several mechanisms. Firstly, they directly incorporate into newly forming blood vessels, where they differentiate into endothelial cells and participate in vessel stabilization and maturation. Secondly, EPCs secrete various growth factors, cytokines, and extracellular matrix components that facilitate endothelial cell proliferation, migration, and tube formation, thereby promoting angiogenesis. Additionally, EPCs possess paracrine effects, modulating the function of neighboring cells and enhancing vascular repair processes. EPCs can be derived from various sources, including bone marrow and peripheral blood. Bone marrow-derived EPCs are typically isolated from bone marrow aspirates or by culture of mononuclear cells obtained from bone marrow samples. The findings from clinical trials evaluating EPC transplantation in MI patients have been mixed, with some studies reporting improvements in cardiac function, myocardial perfusion, and clinical symptoms, while others have shown no significant benefits. Meta-analyses of these trials suggest that EPC transplantation may lead to modest improvements in left ventricular ejection fraction and reduction in infarct size. In the studies, only long-term efficiency was unknown, but short-term perspective demonstrates one of the best improvements compared to other stem cell types (**Figure 4**).

Study conducted by Kawamoto et al. showed that when human endothelial progenitor cells (hEPCs) were expanded outside the body and then transplanted, they

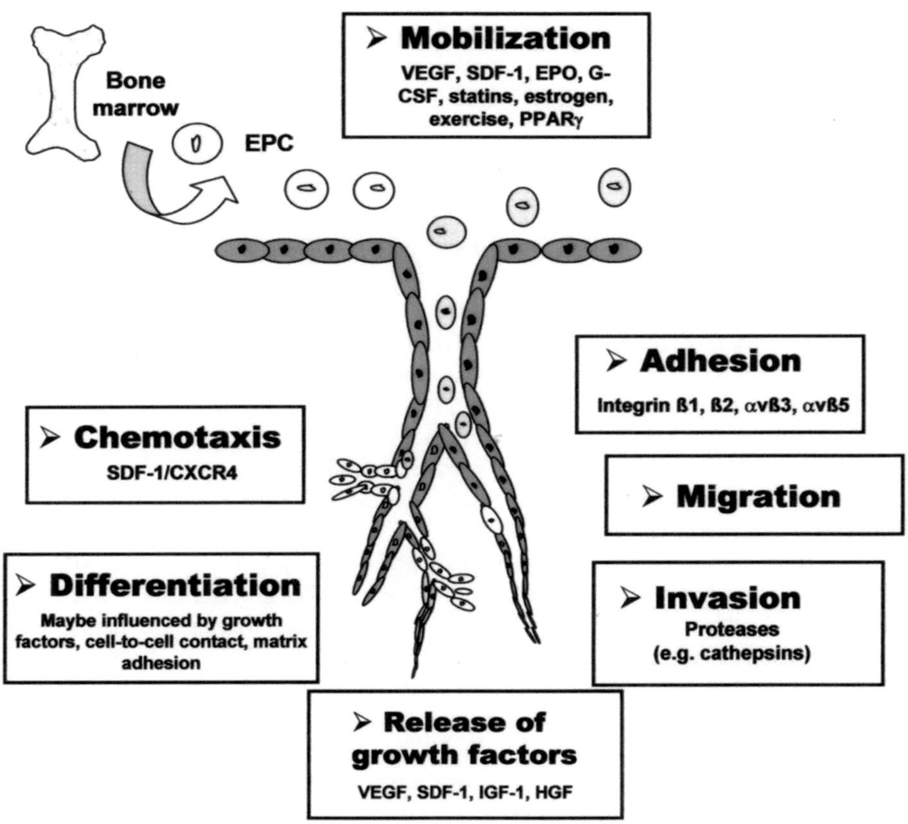

Figure 4.
Endothelial progenitor cell contributions to cardiovascular tissue repair. Source: https://www.ahajournals.org/doi/10.1161/01.RES.0000137877.89448.78.

positively affected the maintenance of left ventricular function [18]. Following the induction of myocardial ischemia, labeled hEPCs were administered intravenously. These cells were observed to gather in the ischemic region and contribute to the formation of new blood vessels in the heart muscle. Echocardiography indicated smaller ventricular dimensions and increased fractional shortening in the group that received hEPC transplantation. Kocher et al. also reported similar results following the transplantation of G-CSF-mobilized CD34+ human cells containing both hematopoietic stem cells (HSCs) and endothelial progenitor cells (EPCs) [19]. This transplantation was found to enhance heart function, prevent cell death in heart muscle cells, and limit changes in heart structure. New blood vessel formation was observed both within the area of heart muscle damage and around its edges. Notably, endothelial cells originating from humans tended to cluster at the center of the damaged area, while those from rats contributed to the formation of new vessels at its edges. This indicates that bone marrow-derived progenitor cells may directly aid in forming new vessels within the core of the damaged tissue and stimulate the growth of blood vessels from existing endothelial cells at the edges of the damaged area. With the promising results from both animal studies and early human clinical trials, the future for therapeutic neovascularization mediated by EPC transplantation appears bright. Despite the promising preclinical data and some encouraging results from clinical trials, several challenges hinder the clinical translation of EPC transplantation for MI patients. One major challenge is the variability in cell preparation techniques, including isolation methods, cell culture protocols, and quality control measures, which can impact the potency and consistency of EPCs used for transplantation. Standardization of cell manufacturing processes and establishment of Good Manufacturing Practice (GMP) guidelines are crucial steps to address this issue.

On the other side, mesenchymal stem cells (MSCs) have also emerged as promising candidates for myocardial infarction (MI) therapy due to their multifaceted mechanisms of action. Understanding these mechanisms is crucial for elucidating their therapeutic potential. Mesenchymal stem cells (MSCs) possess strong immunomodulatory abilities, which are crucial following myocardial infarction (MI) where inflammatory processes contribute to tissue damage. MSCs achieve this by secreting anti-inflammatory molecules like IL-10, TGF-β, and PGE2, creating a less inflammatory environment in the affected heart tissue. Additionally, MSCs interact with immune cells like macrophages, T cells, and NK cells, promoting an anti-inflammatory and tissue-reparative phenotype, thereby reducing damage and fostering healing. In addition, the paracrine effects of mesenchymal stem cells (MSCs) are crucial for tissue repair after myocardial infarction (MI). MSCs secrete bioactive molecules including growth factors like VEGF, HGF, and FGF-2, which aid in angiogenesis, tissue remodeling, and cardiomyocyte survival. Additionally, MSC-derived extracellular vesicles (EVs) contain various bioactive molecules, facilitating intercellular communication and promoting processes like angiogenesis and anti-apoptosis, further supporting tissue repair in the injured myocardium (**Figure 5**).

The clinical use of MSC started between 1995 and 2000, especially for the treatment of cancer and bone/cartilage disease. However, now many clinical trials in cardiovascular diseases also can be found. About 13 trials including 956 patients (468 and 488 in the intervention and control arms) were enrolled. After excluding the biased data, LVEF was significantly increased compared to the baseline among those who received MSC (WMD = 3.78%, 95% CI: 2.14 to 5.42, $p < 0.001$, I2 = 90.2%) with more pronounced effect if the transplantation occurred within the first week after AMI (MD = 5.74%, 95%CI: 4.297 to 7.183; I2 = 79.2% $p < 0.001$). The efficacy of trans-endocardial injection

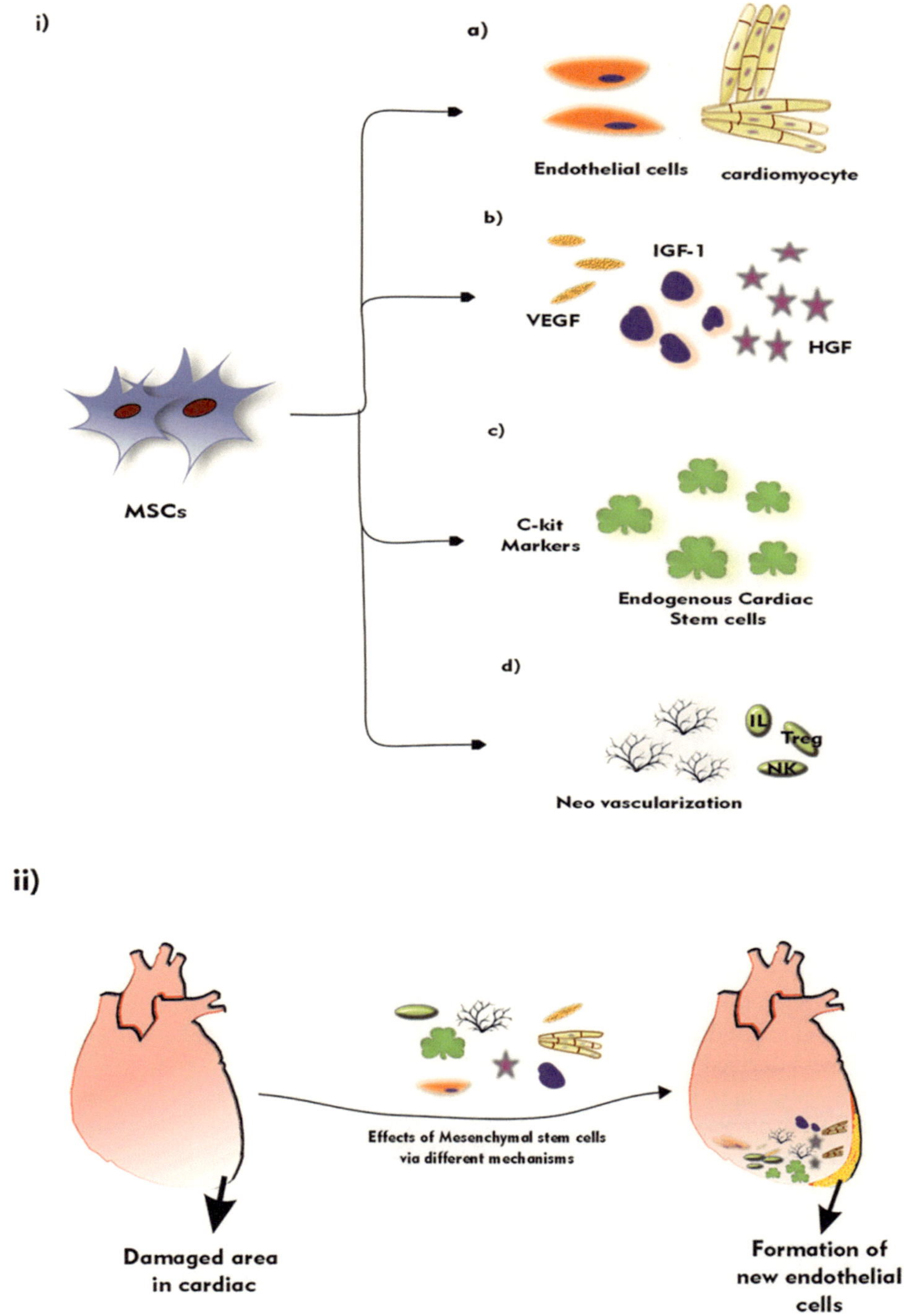

Figure 5.
Mesenchymal stem cell mechanisms in cardiac regeneration and repair. Source: https://doi.org/10.5334/gh.1098.

was similar to that of intracoronary infusion (4% [95%CI: 2.741 to 5.259, $p < 0.001$] vs. 3.565% [95%CI: 1.912 to 5.218, $p < 0.001$], respectively). MSC doses of lower and higher than 107 cells did not improve LVEF differently (5.24% [95%CI: 2.06 to 8.82, $p = 0.001$]

vs. 3.19% [95%CI: 0.17 to 6.12, p = 0.04], respectively) [20]. So, from this, we can conclude that mesenchymal stem cells (MSCs) can be effective in myocardial infarction.

The rationale behind dual transplantation of EPCs and MSCs lies in their complementary mechanisms of action. EPCs promote angiogenesis and neovascularization, enhancing blood supply to the infarcted area. Meanwhile, MSCs exert paracrine effects, modulating the inflammatory response, promoting tissue repair, and reducing fibrosis. By combining these two cell types, researchers aim to achieve synergistic effects that maximize cardiac repair and functional recovery post-MI. Numerous preclinical studies have demonstrated the efficacy of dual transplantation of EPCs and MSCs in animal models of MI. These studies have shown improvements in cardiac function, reduction in infarct size, increased neovascularization, and attenuated adverse remodeling. Additionally, histological analysis has revealed enhanced tissue regeneration and decreased fibrosis in the dual-transplanted hearts compared to single-cell therapy or control groups [15]. While preclinical studies have provided promising results, translating these findings into clinical practice poses several challenges. Clinical trials investigating the safety and efficacy of dual transplantation of EPCs and MSCs in MI are ongoing, with varying outcomes. Challenges such as cell delivery methods, optimal cell dosing, timing of transplantation, immune rejection, and long-term engraftment need to be addressed to maximize therapeutic benefits and ensure patient safety. Ultimately, dual transplantation of EPCs and MSCs holds the potential to revolutionize the treatment paradigm for MI, offering patients a novel therapeutic option that promotes cardiac regeneration, improves outcomes, and enhances quality of life.

2.5 Cardiac stem cells (CSCs)

Cardiac stem cells (CSCs) are a unique population of cells residing within the heart with the capability to self-renew and differentiate into various cardiac cell types. These cells possess stem-like properties, including the ability to proliferate and differentiate into cardiomyocytes, endothelial cells, and smooth muscle cells, thus contributing to myocardial tissue regeneration and repair. They can be found within specialized niches in the heart, such as the atria, ventricles, and cardiac apex, which provide a conducive microenvironment for their maintenance and regulation. Additionally, resident progenitor cells within the heart, including c-kit+, Sca-1+, and Isl1+ cells, have been identified as potential sources of CSCs. Moreover, induced pluripotent stem cells (iPSCs) derived from somatic cells offer a virtually limitless supply of CSCs by differentiating them into cardiac lineage cells. There is no evidence of human cardiac stem cell transplantation, but animal models, including rodents and large animals, have been employed to investigate the safety, efficacy, and mechanisms of action of CSC-based interventions. These studies often involve the transplantation of CSCs into infarcted hearts, either directly or via various delivery methods, such as intramyocardial injection or intracoronary infusion. So, these cells should be analyzed further.

2.6 Embryonic stem cells (ESCs)

Embryonic stem cells (ESCs) are undifferentiated cells derived from the inner cell mass of a blastocyst, an early-stage embryo. These cells possess remarkable properties, notably pluripotency, which means they can differentiate into any cell type in the body. Kehat et al. have succeeded in growing the heart cells from human ESCs [21]. Caspi et al. evaluated the ability of human ESCs and their cardiomyocyte derivatives to engraft and improve myocardial performance in the rat chronic infarction model. The results of this study showed the formation of stable cardiomyocyte grafts, attenuation of the remodeling process, and functional benefit, highlighting their potential for myocardial cell therapy strategies [22]. These methods seem like novel therapy in future as growing whole heart from stem cells have some ethical challenges.

3. Ethical consideration

Stem cell research, particularly involving embryonic stem cells (ESCs), has been a subject of ethical debates due to concerns about the origin, misuse, safety, informed consent, and patient rights. The derivation of ESCs from human embryos raises ethical concerns about the destruction of embryos, leading to objections from various religious and ethical perspectives. Induced pluripotent stem cells (iPSCs) also pose ethical dilemmas as their generation involves genetic manipulation and reprogramming of somatic cells, raising questions about patient privacy, autonomy, and long-term implications of genetic modification. Advancements in stem cell research have led to the development of alternative approaches, such as adult stem cells and iPSCs, which mitigate some of the ethical concerns associated with ESCs. Ethical oversight and adherence to established guidelines are essential to ensure ethical and responsible conduct of stem cell research. Regulatory frameworks and guidelines are established by agencies like the Food and Drug Administration (FDA) and the European Medicines Agency (EMA) to safeguard the welfare of research participants and ensure compliance with ethical principles and legal requirements. These frameworks encompass various aspects of stem cell research, including preclinical studies, clinical trial design, manufacturing practices, product quality control, and post-market surveillance. Informed consent is a fundamental ethical principle in stem cell research and clinical practice, ensuring that individuals are adequately informed about the nature of the research, potential risks and benefits, alternative treatment options, and their rights as research participants or patients. Patient education plays a crucial role in facilitating informed decision-making and promoting transparency in stem cell therapy.

4. Conclusion

In summary, stem cell transplantation holds immense promise as a therapeutic strategy for myocardial infarction, leveraging the regenerative potential of stem cells to repair damaged cardiac tissue and improve clinical outcomes. Key findings from research and clinical trials underscore the multifaceted mechanisms of action of stem cells, including their ability to differentiate into cardiac cell types, modulate the inflammatory response, promote angiogenesis, and recruit endogenous repair mechanisms. While numerous studies have demonstrated the safety and feasibility of stem cell therapy, the efficacy of these interventions in achieving meaningful improvements in myocardial function and patient outcomes remains variable.

The implications of stem cell transplantation for clinical practice are profound, offering new avenues for the management of myocardial infarction and other cardiovascular diseases. Integration of stem cell therapy into the therapeutic armamentarium for myocardial infarction requires careful consideration of patient selection criteria, optimal timing and route of administration, choice of stem cell type, and standardization of treatment protocols. Healthcare professionals must stay abreast of the latest developments in stem cell research and adhere to regulatory guidelines and ethical principles to ensure safe and effective implementation of stem cell-based interventions in clinical practice.

Looking ahead, ongoing research endeavors hold the potential to address existing challenges and further advance the field of stem cell transplantation for myocardial infarction. Future studies should focus on elucidating the optimal dosing, delivery, and timing of stem cell therapy, as well as identifying biomarkers predictive of treatment response and long-term outcomes. Furthermore, comparative effectiveness trials and large-scale multicenter studies are needed to establish the comparative efficacy of different stem cell types and treatment strategies. Additionally, innovative approaches, such as genetic engineering, tissue engineering, and combination therapies, warrant exploration to enhance the regenerative capacity and therapeutic efficacy of stem cell-based interventions. In conclusion, stem cell transplantation represents a promising frontier in the management of myocardial infarction, offering potential benefits for myocardial repair, functional recovery, and long-term prognosis. By addressing ethical, regulatory, and scientific challenges, the integration of stem cell therapy into clinical practice has the potential to revolutionize the treatment paradigm for myocardial infarction and improve the quality of life for affected individuals. Continued collaboration among researchers, clinicians, regulatory agencies, and stakeholders is essential to realize the full potential of stem cell transplantation in cardiovascular medicine.

Acknowledgements

The author acknowledges the use of “Grammarly” for checking grammar mistakes and changing them in the manuscript to make it more grammatically cohesive.

Author details

Rizaev Jasur Alimdjanovich* and Ziyadullaev Shukhrat Khudayberdiyevich
Independent Researcher, Samarkand, Uzbekistan

*Address all correspondence to: dr.jasur@gmail.com

References

[1] Dyrbuś K et al. The prevalence and management of familial hypercholesterolemia in patients with acute coronary syndrome in the polish tertiary Centre: Results from the TERCET registry with 19,781 individuals. Atherosclerosis. 2019;**288**:33-41

[2] Ojha N, Dhamoon AS. Myocardial Infarction. In: StatPearls [Internet]. StatPearls Publishing; 2023

[3] Soonpaa MH, Field LJ. Assessment of cardiomyocyte DNA synthesis in normal and injured adult mouse hearts. American Journal of Physiology-Heart and Circulatory Physiology. 1997;**1**(272):H220-H226

[4] Turner D et al. Clinical-based cell therapies for heart disease-current and future state. Rambam Maimonides Medical Journal. 2020;**2**(11):e0015

[5] Melton D. 'Stemness': Definitions, Criteria, and Standards. Essentials of stem cell biology. Academic Press, Elsevier; 2014. pp. 7-17

[6] Khan M et al. Skeletal myoblasts transplanted in the ischemic myocardium enhance in situ oxygenation and recovery of contractile function. American Journal of Physiology-Heart and Circulatory Physiology. 2007;**4**(293):H2129-H2139

[7] Brasselet C et al. Skeletal myoblast transplantation through a catheter-based coronary sinus approach: An effective means of improving function of infarcted myocardium. European Heart Journal. 2005;**15**(26):1551-1556

[8] Ye L et al. Improved angiogenic response in pig heart following ischaemic injury using human skeletal myoblast simultaneously expressing VEGF 165 and angiopoietin-1. European Journal of Heart Failure. 2007;**1**(9):15-22

[9] Menasché P et al. Autologous skeletal myoblast transplantation for severe postinfarction left ventricular dysfunction. Journal of the American College of Cardiology. 2003;**7**(41):1078-1083

[10] Pagani FD et al. Autologous skeletal myoblasts transplanted to ischemia-damaged myocardium in humans. Histological analysis of cell survival and differentiation. Journal of the American College of Cardiology. 2003;**5**(41):879-888

[11] Menasché P et al. The myoblast autologous grafting in ischemic cardiomyopathy (MAGIC) trial: First randomized placebo-controlled study of myoblast transplantation. Circulation. 2008;**9**(117):1189-1200

[12] Perin EC et al. Transendocardial, autologous bone marrow cell transplantation for severe, chronic ischemic heart failure. Circulation. 2003;**18**(107):2294-2302

[13] Zakrzewski W et al. Stem cells: Past, present, and future. Stem Cell Research & Therapy. 2019;**1**(10):68

[14] Tang Y-H et al. Opportunities and challenges: Stem cell-based therapy for the treatment of ischemic stroke. CNS Neuroscience & Therapeutics. 2015;**4**(21):337-347

[15] Taghavi S, George JC. Homing of stem cells to ischemic myocardium. American Journal of Translational Research. 2013;**4**(5):404-411

[16] Zhang N et al. Blood-borne stem cells differentiate into vascular and cardiac

lineages during normal development. Stem Cells and Development. 2006;**1**(15):17-28

[17] Shah VK, Shalia KK. Stem cell therapy in acute myocardial infarction: A pot of gold or Pandora's box. Stem Cells International. 2011;**2011**:536758

[18] Kawamoto A et al. Therapeutic potential of ex vivo expanded endothelial progenitor cells for myocardial ischemia. Circulation. 2001;**5**(103):634-637

[19] Kocher AA et al. Neovascularization of ischemic myocardium by human bone-marrow–derived angioblasts prevents cardiomyocyte apoptosis, reduces remodeling and improves cardiac function. Nature Medicine. 2001;**4**(7):430-436

[20] Attar A et al. Mesenchymal stem cell transplantation after acute myocardial infarction: A meta-analysis of clinical trials. Stem Cell Research & Therapy. 2021;**1**(12):600

[21] Kehat I et al. Human embryonic stem cells can differentiate into myocytes with structural and functional properties of cardiomyocytes. The Journal of Clinical Investigation. 2001;**3**(108):407-414

[22] Caspi O et al. Transplantation of human embryonic stem cell-derived cardiomyocytes improves myocardial performance in infarcted rat hearts. Journal of the American College of Cardiology. 2007;**19**(50):1884-1893

Chapter 20

Regenerative Medicine for the Hypothalamic-Pituitary Axis: Current Progress and Future Challenges

Miho Kawata and Hiroshi Nagasaki

Abstract

The hypothalamus is composed of heterogenous cell populations and contains highly complex neural circuits that regulate the autonomic nervous system. In particular, the hypothalamic-pituitary axis is the command center for the secretion of hormones that act throughout the body, and its dysfunction leads to severe endocrine disorders. Stem cell therapy is a promising tool for the treatment of hypothalamic-pituitary disorders. To date, several groups have reported the differentiation of hypothalamic neurons from mouse embryonic stem cells (mESCs) or pituitary cells from mESCs or human pluripotent stem cells (hPSCs). In addition, several studies have shown that ectopically transplanted *in vitro*-generated hypothalamic neurons and pituitary cells function properly *in vivo*. This paper reviews the current status and future prospects of hypothalamic and pituitary transplantation, including the results of our hypothalamic organoid transplantation.

Keywords: hypothalamus, pituitary, stem cell, differentiation, transplantation

1. Introduction

The hypothalamus is an evolutionarily ancient region of the brain and that is essential for maintaining homeostasis. For example, energy balance and thermoregulation [1], circadian rhythms [2], water balance [3], sleep-wake cycle [4], feeding behavior [5], sexual behavior [6], social behavior [7, 8], and stress response and memory [9–11] are regulated by the individual nuclei present in the hypothalamus. The hypothalamus is divided into four major divisions from rostral to caudal: preoptic, anterior, tuberal, and mammillary regions. The preoptic area (POA) contains the median (MnPO), medial (MPO), ventrolateral (VLPO), anteroventral periventricular (AVPV), and periventricular (PVpo) preoptic nuclei. The anterior region includes the anterior hypothalamus (AH) and the supraoptic (SON), suprachiasmatic (SCN), and paraventricular (PVN) nuclei. The tuberal area consists of the dorsomedial

medial hypothalamic nucleus (DMH), ventral medial hypothalamic nucleus (VMH), arcuate hypothalamic nucleus (ARC), tuberal hypothalamic nucleus (TuN), and the lateral hypothalamic region (LH), the boundaries of which are not clearly defined. The mammilary region includes the premammillary nucleus (PM), supramammillary nucleus (SuM), tuberomammillary (TM), and posterior hypothalamic (PH) nuclei [12].

Precise functional compartmentalization is difficult because these hypothalamic nuclei often have overlapping functions each other, but some of the nuclei have been identified as being primarily involved in specific physiological functions. For example, the ARC is a major nucleus for the regulation of feeding behavior, which includes the pro-opiomelanocortin (POMC) or agouti-related peptide (AgRP)/neuropeptide Y (NPY) neurons [13, 14]. The PVN, which is composed of corticotropin-releasing hormone (CRH), oxytocin (OXT), and vasopressin (AVP), is involved in the stress and social behavioral responses [15, 16]. The SON is also composed of OXT and AVP neurons, and these neurons release the peptide into vessels of the posterior pituitary [7]. The LH contains the melanin-concentrating hormone (MCH) and orexin neurons and is involved in the sleep-wake cycle [17]. The SCN contains the AVP and vasoactive intestinal polypeptide (VIP) neurons and is the main regulator of the circadian rhythm [18].

In addition to the above, the hypothalamus also has a role in controlling pituitary hormone secretion. The hypothalamus secretes releasing hormones to control pituitary hormone secretion, and the hormones released by the hypothalamus are transported by the blood to the pituitary gland. After receiving the releasing hormones, the pituitary gland secretes the following six hormones (adrenocorticotropic hormone (ACTH), thyroid stimulating hormone (TSH), prolactin (PRL), growth hormone (GH), follicle stimulating hormone (FSH), and luteinizing hormone (LH)). These six hormones are then transported by the blood to the target glands (adrenal, gonadal, thyroid, etc.) for action. The target gland provides constant feedback to the hypothalamus and pituitary gland to maintain appropriate hormone levels. In most cases, a negative feedback mechanism causes the hypothalamus and/or pituitary gland to stop releasing the hormone and turn off the cascade when the hormone reaches a certain blood level. (In some cases, short-loop feedback also occurs, where pituitary hormones act directly on the hypothalamus) [19]. Thus, the hypothalamus and pituitary play an important role in linking the nervous and endocrine systems and maintaining a balance between the external and internal environments. Therefore, disorders of the hypothalamus-pituitary system cause intractable diseases and significantly reduce the quality of life of patients in the long term.

Table 1 shows a list of the etiologies of the hypothalamus-pituitary dysfunction and potential target disorders for the regenerative medicine. The following eight main etiologies are known to cause hypothalamus-pituitary dysfunction: developmental/structural [20–30], traumatic [31–34], neoplastic [35–47], infiltrative/inflammatory [48–56], vascular [57–62], infections [63–66], autoimmune [67], and drug [68]. The diseases included in each group are numerous and very diverse. Surgery or radiotherapy are often performed for all of these diseases, but they can damage the hypothalamus-pituitary system and often do not restore hypothalamic-pituitary dysfunction even after surgery (or even cause new dysfunction after surgery). As a result, patients require lifelong medical therapy, mainly hormone replacement, for the rest of their lives after surgery.

These situations have increased the demand for the regenerative medicine for the hypothalamic disorders using the pluripotent stem cells. Indeed, recent studies have

Etiology	Examples of specific disorder	Treatment options	Ref.
Developmental/ structural	Transcription factor defect	• Surgery • Medical therapy • Radiotherapy	[20–30]
	Pituitary dysplasia/aplasia		
	Congenital CNS mass, encephalocele		
	Primary empty sella		
	Congenital hypothalamic disorders (septo-optic dysplasia, Prader–Willi, Laurence–Moon–Biedl syndrome, Kallmann syndrome)		
	Familial dwarfism		
	Familial neurohypophyseal diabetes insipidus (FNDI)		
	Carney complex		
	Multiple Endocrine Neoplasia type 1:MEN1		
Traumatic	Surgical resection	• Medical therapy	[31–34]
	Radiation damage		
	Head injury		
	Pituitary stalk interruption syndrome (PSIS) (perinatal injury)		
Neoplastic	Pituitary adenoma	• Surgery • Medical therapy • Radiotherapy	[35–47]
	Parasellar mass (germinoma, ependymoma, glioma)		
	Rathke's cyst		
	Craniopharyngioma		
	Hypothalamic hamartoma, gangliocytoma		
	Pituitary metastases (breast, lung, and colon carcinoma)		
	Lymphoma and leukemia		
	Meningioma		
	Teratoma		
Infiltrative/ inflammatory	Lymphocytic hypophysitis	• Surgery • Radiotherapy • Medical therapy	[48–56]
	Hemochromatosis		
	Sarcoidosis		
	Langerhans cell histiocytosis (LCH)		
	Granulomatous hypophysitis		
	meningitis		
	encephalitis		
	IgG4-related hypophysitis		
	Lymphocytic infundibuloneurohypophysitis		
Vascular	Pituitary apoplexy	• Surgery • Radiotherapy • Medical therapy	[57–62]
	Pregnancy-related (infraction with diabetes; postpartum necrosis)		
	Sickle cell disease		
	Arteritis		
	Sheehan's syndrome		
	Aneurysms		

Etiology	Examples of specific disorder	Treatment options	Ref.
Infections	Fungal (histoplasmosis)	• Surgery	[63–66]
	Parasitic (toxoplasmosis)	• Medical therapy	
	Tuberculosis		
	Pneumocystis carinii		
	Pituitary abscess		
Autoimmune	Anti-PIT-1 antibody syndrome	• Medical therapy	[67]
Drug	Pituitary dysfunction induced by immune checkpoint inhibitor (Nivolumab, Ipilimumab)	• Medical therapy	[68]

Table 1.
Etiology causing the hypothalamus-pituitary dysfunction and its treatment.

shown that the various brain region-specific organoids that resemble endogenous tissues, including the hypothalamus, can be induced *in vitro* [69]. For example, disease model brain organoids of neurodevelopmental and neurodegenerative diseases, such as autism spectrum disorder (ASD), Rett syndrome, Zika virus-related microcephaly, Alzheimer's disease (AD), Parkinson's disease (PD), and Huntington's disease (HD) have been established based on these techniques [70]. The application of the electrophysiological studies, genomic analysis and drug screenings, etc., to these brain organoids may bridge the gap between results obtained from animal models and human patients and facilitate the development of personalized regenerative therapies [71, 72]. In particular, the hypothalamus-pituitary region is less well understood in terms of developmental mechanisms and pathophysiology than other brain regions because it is located very deep in the brain, making it difficult to access and sample *in vivo*. Current organoid research has the potential to overcome this problem.

In this chapter, we first review *in vitro* techniques for the differentiation of ES and iPS cells into the hypothalamic and pituitary cells. We then describe the transplantation studies of ES or iPS cell-derived hypothalamic and pituitary cells. In addition, we present our recent work on orthotopic transplantation of mouse ES cell-derived hypothalamic neurons into the adult mouse brain. Finally, we discuss the prospects for regenerative medicine of the hypothalamus-pituitary system.

2. Differentiation of hypothalamus and pituitary from stem cells

Currently, two methods are used to differentiate the central nervous system: 2D or 3D culture. In 2D cultures, two inhibitors of SMAD signaling, Noggin and SB431542, are added to the adherent stem cells, a process called dual SMAD inhibition [73]. This method allowed the efficient neuronal induction by suppressing the TGF-β/NODAL/activin and BMP signaling pathway through inhibition of the intracellular signaling transducer, SMAD. Using these methods, Wang et al. showed that the early activation of sonic hedgehog (SHH) signaling followed by timed NOTCH inhibition in human ESCs/iPSCs induces efficient conversion into hypothalamic NKX2.1^{+} precursors. Furthermore, the application of brain-derived neurotrophic factor (BDNF) in addition to NOTCH inhibitors induced the cells to express the hypothalamic neuron ARC markers, such as POMC, NPY, AGRP, somatostatin, and dopamine [74]. Rajamani et al., then performed combinatorial screening and found that 0 μM RA, 4 μg/

mL insulin, and 10 μM IWR-1-endo were the most efficient combination to induce early hypothalamic progenitor cells from hiPSCs between days 0 and 9 in 2D culture method [75]. Timely withdrawal of the BMP inhibitor Noggin was shown to be necessary for pituitary induction. Noggin withdrawal on day 3 induced 71% $SIX1^+$ putative cranial placode precursor cells [76].

As for the 3D culture method, the serum-free floating culture of embryoid body-like aggregates with quick aggregation (SFEBq), a culture method that can induce telencephalic precursors with high efficiency, is well-known [77, 78]. However, although the SFEBq method efficiently produced cerebral neurons, it was unable to differentiate the hypothalamic neurons. In 2008, Wataya et al., found that insulin, which is normally added to cells to promote cell proliferation and survival, inhibited the differentiation of ES cells into the hypothalamic neurons. Therefore, they cultured the mouse ES cells (mESCs) in growth factor-free chemically defined medium (gfCDM). This SFEBq/gfCDM method can efficiently differentiate the hypothalamic neurons from mESCs. Interestingly, this method also showed that ES cells are automatically induced to differentiate into the hypothalamic neurons without any specific external stimulus [79]. And then, in 2011, pituitary glands were generated from mESCs by modifying the SFEBq method [80]. To achieve the pituitary differentiation, Suga et al. mimicked the pituitary development process of mammalian fetuses *in vitro.* The pituitary gland begins to develop when the oral ectoderm contacts the mesencephalic tissue that will later become the hypothalamus region and receives the signals from the future hypothalamus that promote differentiation. Therefore, while 3000 ES cells per aggregate were used to induce hypothalamus differentiation, in this experiment 10,000 ES cells per aggregate, three times as many, were used in this experiment to differentiate the oral ectoderm and hypothalamus simultaneously. The group subsequently succeeded in inducing pituitary differentiation from human ES cells using the same technique described above. However, in the case of human cells, it was found that early exposure to BMP4 was required for the induction of pituitary differentiation. Furthermore, the subsequent addition of FGF was found to increase the differentiation efficiency of pituitary primordium formation [81].

In *in vitro* culture systems, external factors such as BMP4 and FGF play an important role in determining the direction of differentiation. For hypothalamic neurons, SHH plays an important role in determining the direction of differentiation. Our group has shown that hedgehog (Hh) signaling controls the presence or absence of cocaine-amphetamine-regulated transcripts (CART) in MCH neurons. We also found that Hh signaling supports the differentiation of mESCs into orexin/hypocretin neurons [82]. Another group examined the expression variation of transcription factors (Nkx2.1, Nkx2.2, Pax6, and RAX) by the intensity of Hh signal concentration and showed a transcriptional network that induces differentiation into VMH, dVMH, and ARC cell types [83].

With the development of the differentiation methods described above, it became possible to generate hypothalamus or pituitary organoids. Qian et al. successfully generated the hypothalamus organoids by transferring hiPSCs induced by a 2D dual SMAD inhibition protocol into a small spinning bioreactor (SpinΩ) from day 7 of culture [84, 85]. On the other hand, Kasai et al. modified the 3D culture method, SFEBq, and then successfully generated the hypothalamus-pituitary hybrid organoid. They found that the early pituitary marker LHX3 expression appeared to be higher using 10,000 cells per well, 10% KSR, 10 nM BMP4 during days 6–15, and 2 μM SAG. These organoids increased the ACTH secretion under low glucose conditions, suggesting that the CRH-ACTH pathway is established within the organoids [86]. In addition, congenital pituitary hypoplasia (CPH) model organoids were established by using the patient-derived iPSCs [87]. These organoids demonstrated that the possible mechanism of CPH is that OTX2 gene mutation reduces FGF10 in the hypothalamus, resulting in reduced LHX3 expression in the oral ectoderm and cell death by apoptosis.

As for the hypothalamus, ARC-specific organoids (ARCO) were established in 2021 [88]. These organoids were generated by the modified dual SMAD inhibition protocol, which considers the expression pattern of morphogenetic factors during hypothalamic ARC development and the need for SHH signal activation and Wnt inhibition [89–91]. Single-cell transcriptomic analysis showed that ARCOs have similar molecular signatures to human ARC. In addition, Prader-Willi syndrome (PWS) patient-derived ARCO were also generated, and it was shown that the cellular differentiation deficits and exacerbated inflammatory responses in PWS ARCOs.

Both 2D and 3D methods can be used together to induce hypothalamus or pituitary differentiation (**Table 2**). However, these differentiation methods have several limitations. The most significant issue is the instability of the differentiation efficiency. Merkle et al. differentiated the hypothalamic neurons from human PSCs by using both 2D and 3D methods and counted the number of neuropeptidergic neurons that appeared. The results showed that both methods induced the α-melanin-stimulating hormone (αMSH/POMC), AGRP, hypocretin A/orexin (HCRTA), MCH, CRH or thyrotropin-releasing hormone (TRH) neurons, but the expression of growth hormone releasing hormone (GHRH), kisspeptin (KISS1), or gonadotropin-releasing hormone (GnRH/LHRH) were not observed by either method. Interestingly, OXT or AVP neurons observed in 3D cultures did not appear in 2D cultures [94]. Thus, the efficiency of differentiation of individual neurons varies with current differentiation

Year	Protocol for differentiation	PSC type	Target cell or tissue	Ref.
2008	3D	mESC	Hypothalamic neurons	[79]
2011	3D	mESC	Pituitary	[80]
2013	2D	hiPSC, hESC	Cranial placode	[76]
2015	2D	hiPSC, hESC	ARC hypothalamic neurons (POMC, NPY, AgRP, somatostatin, and dopamine)	[74]
2016	3D	hESC	Pituitary	[81]
2016	2D + spinning bioreactors (SpinΩ)	hiPSC	Hypothalamus organoids	[84]
2018	3D	hESC	AVP neurons	[92]
2018	2D	hiPSC	Hypothalamic neurons	[75]
2019	3D	miPSC	AVP neurons	[93]
2020	3D	hiPSC	Hypothalamus-pituitary hybrid organoid	[86]
2020	3D	hiPSC	Congenital pituitary hypoplasia (CPH) model organoid	[87]
2021	2D + orbital shaker	hiPSC	ARC organoid	[88]
2022	3D	mESC	MCH neuron	[82]

Table 2.
Protocols for differentiation of hypothalamic neurons or pituitary in 2D and 3D.

methods, and it is also not possible to induce only the target cell type with high efficiency.

One approach to address the above issue is to modify the culture conditions. In the case of AVP neurons, differentiation efficiency has been improved by changing the concentration and timing of the addition of BMP4 and SAG [92] or by FGF8 treatment, which is thought to be involved in AVP neuron differentiation *in vivo* but has not been used in conventional differentiation methods [93]. Another approach is to sort the target cells only from the differentiated organs. Our group has established a method to purify only pituitary cells from hiPSC-derived hypothalamic-pituitary organoids. By screening for cell surface markers, we found that EpCAM is specifically expressed in anterior pituitary cells and their ectodermal precursors. Indeed, magnetic separation of EpCAM-positive cells resulted in enrichment of anterior pituitary cells ($ACTH^+$ or $PITX1^+/LHX3^+$) [95]. Using this technique, it will be possible to enrich only the target cell populations from hypothalamus and pituitary organoids by applying this technique. In addition, a method for direct programming of hypothalamic neurons or pituitary cells will be established in the near future. In 2010, Vierbuchen et al. succeeded in directly converting fetal mouse fibroblasts into neurons by co-expressing the neuron-specific transcription factors Brn2, Ascl1, and Myt1l [96]. Using this technique, it will be possible to produce only the target cells in the hypothalamus or pituitary gland. However, to achieve this, it is necessary to understand both the combination and the timing of the expression of transcription factors involved in the target cell development. Single cell RNA-seq data provide useful information in this regard. For example, recent single-cell transcriptomics of the prenatal and adult human hypothalamus revealed a set of transcription factors involved in the maturation of $KISS^+$ neurons within the POMC expression lineage, OXT and AVP-expressing neurons in the PVN [97]. Further development of the above approaches may pave the way for the personalized regenerative medicine for the hypothalamic-pituitary diseases and also provide the new insights into the development of the hypothalamic-pituitary axis.

3. Transplantation of *in vitro*-derived hypothalamus and pituitary

The next consideration for the realization of regenerative medicine is whether cells and tissues differentiated *in vitro* are functional *in vivo* after transplantation. Although there are numerous reports of hypothalamic or pituitary differentiation *in vitro*, there are few reports of *in vivo* transplantation. In addition, most of the reported cases have performed ectopic transplantation (**Table 3**).

As for the pituitary to our knowledge, there are no reports of orthotopic transplantation of differentiated cells. However, several studies have been reported on the ectopic transplantation of differentiated pituitary organs, with good results. The most commonly reported case is the transplantation of organoids under the kidney capsule. In both mESC and hESC-derived pituitary organoids, transplantation under the kidney capsule of hypopituitary mice created by surgical removal of the pituitary resulted in elevated blood ACTH and corticosterone levels compared to the non-transplanted control group. In addition, while hypopituitary mice typically show decreased spontaneous locomotion and die within a few weeks, the organoid-transplanted mice showed a recovery of locomotion and improved survival rate [80, 81]. More recently, hPSC-derived pituitary organoids, enriched for the hormone-producing cells by cell sorting using EpCAM, were transplanted under the kidney capsule of

Year	Graft type	Cell source	Cell type to graft	Host animal	Graft site	Outcome	Ref.
2011	Ectopic	mESC	Pituitary	Hypopituitary mice (ICR background; 10-week-old)	Under kidney capsules	• Increase in blood ACTH and glucocorticoid levels • Increased spontaneous locomotor activity • Improved survival	[80]
2014	Orthotopic	miPSC	Neural stem cell	Adult C57BL/6 mice	Mediobasal hypothalamus	• Induction of neurogenesis • Generation of POMC neurons	[98]
	Ectopic	miPSC	Neural stem cell with dominant-negative IκBα (^{DN}IκBα-NSCs^{iPS})	High-fat diet (HFD) mice (adult C57 BL/6 mice)	Carotid artery injection	• Induction of neurogenesis • Generation of POMC neurons • Reduction in HFD intake • Recovery of leptin action • Improvement of glucose tolerance	
2015	Ectopic	hESC	Hypothalamic neurons	Neonatal C57Bl/6 or Il2rg$^{-/-}$Rag2$^{-/-}$ mice	Lateral ventricle or brain parenchyma	• Long-term survival (18 months) • Synapse formation between host and graft	[94]
2016	Ectopic	hESC	Pituitary	Hypopituitary mice (SCID; 10-week-old)	Under kidney capsules	• Increase in blood ACTH and glucocorticoid levels • Increased spontaneous locomotor activity • Improved survival • Formation of new blood vessels in the graft	[81]
2016	Ectopic	hESC	Pituitary	Hypopituitary rats (Athymic nude rats; > 8-week-old)	Subcutaneous	• Increase in plasma ACTH, GH, and corticosterone levels	[99]

DOI: http://dx.doi.org/10.5772/intechopen.1006504

Year	Graft type	Cell source	Cell type to graft	Host animal	Graft site	Outcome	Ref.
2022	Orthotopic	mESC	Hypothalamic neurons	SCID mice (6–14 weeks old)	SON	• Exhibit axonal projection patterns similar to those of host hypothalamic neurons	[100]
2023	Orthotopic	hESC	Neural stem cell (tanycyte-like cell)	SCID mice (8-week-old)	Ventral hypothalamus	• Differentiation into the hypothalamic neurons	[101]
2023	Ectopic	hESC	Pituitary	Hypopituitary mice (SCID; > 8–9-week-old)	Inguinal subcutaneous white adipose tissue (ISWAT)	• Increase in plasma ACTH levels • Increased spontaneous locomotor activity • Promoting vascularization	[102]
2023	Ectopic	hESC	EpCAM-sorted pituitary organoid	Hypopituitary mice (SCID; 10–17-week-old)	Under kidney capsules	• Long-term increase in blood ACTH levels (24 weeks) • Suppression of ACTH secretion by dexamethasone treatment • LPS-stimulated increase in ACTH secretion	[103]

Table 3.
Protocols for transplantation of stem cell-derived hypothalamic neurons or pituitary.

hypopituitary mice [103]. As a results, long-term improvement in blood ACTH levels of hypopituitary mice was observed up to at least 24 weeks post-transplantation. In addition, it has been suggested that the transplanted organoids may also be responsive to their surrounding environment. For example, transplanted cells were properly suppressed ACTH secretion in the glucocorticoid (dexamethasone)-treated condition. However, when lipopolysaccharide (LPS) was administered as sepsis model, ACTH secretion increased in the transplanted group.

However, transplanting organoids under the kidney capsule is too invasive for clinical use in human. In addition, if a tumor develops from the organoids, it is difficult to remove them from under the kidney capsule. Considering that, hESC-derived pituitary organoids were transplanted into the inguinal subcutaneous white adipose tissue (ISWAT) of mice as a relatively less invasive and easy to remove subcutaneous transplantation method. The results suggest that ACTH is released more efficiently in the group that received organoids transplanted into ISWAT, the richly vascularized area, than in the group that received organoids transplanted into the non-vascularized area [102]. Subcutaneous organoid transplants were also performed in hypophysectomized rats. ACTH levels remained higher in the transplanted group than in the control group throughout the 7-week experimental period. GH and LH secretion were also been examined, and GH levels were significantly increased at 3 and 7 weeks after transplantation, but LH levels were not significantly changed. Thus, although not all hormones were significantly increased in the transplant group compared with the control group, but immunohistochemical staining revealed cells expressing each of the six hormones (ACTH, TSH, PRL, GH, FSH, and LH) in the grafts at 7 weeks after transplantation. These results suggest that the grafts survived and, at least in part, induced further *in vivo* differentiation and maturation in host tissues [99]. Taken together, these results suggest that ectopically implanted pituitary organoids may ameliorate hypopituitarism by sensing *in vivo* conditions and appropriately regulating hormone secretion.

Compared to the pituitary organoids, there have been very few studies on the transplantation of stem cell-derived hypothalamic organoids. In 2015, the transplantation of hPSC-derived hypothalamic neurons into the mouse brain was reported for the first time. hPSC-derived hypothalamic neurons were transplanted into the lateral ventricle or brain parenchyma of newborn mice. The grafts successfully survived for at least 18 months, and they were integrated into the host brain by forming of synapses. In addition, the expression of neuropeptides, such as HCRT and MCH, were found in the graft [94]. Although it was shown that the *in vitro*-differentiated hypothalamic neurons could be transplanted into the brain, there have been no further studies of orthotopic transplantation since this study. However, there have been several studies in which *in vitro*-derived neural stem cells, rather than differentiated cells, have been transplanted into the hypothalamus. In rodents, cells around the third ventricle, called tanycytes, are known to have the properties of hypothalamic neural stem cells, and RAX has been identified as a marker gene for tanycytes [104, 105]. Miwata et al., generated the hypothalamic organoids from RAX::VENUS knock in human ESCs and then sorted RAX^{+} cells from the generated organoids to create tanycyte-like neurospheres. These neurospheres were transplanted into the ventral hypothalamus of immunodeficient mice. As a result, it was confirmed that the neurospheres were survived in the mouse brain and some cells differentiated into the hypothalamic neurons 1 week after transplantation [101]. Li et al. conducted a study in which mouse iPSCs-derived neural stem cells ($NSCs^{iPS}$) were not only transplanted directly into the brain by stereotactic surgery, but also delivered to the hypothalamus by carotid artery injection. First, they transplanted NSCsiPS into the mediobasal hypothalamus (MBH) of wild-type mice. The results showed that NSCsiPS not only induced neurogenesis but also generated POMC cells. Next, they attempted to transplant $NSCs^{iPS}$ into the obesity model. Since obesity-induced hypothalamic IKKβ/NF-κB activation is known to destroy hypothalamic NSCs, they generated a lentivirus-induced NSCsiPS strain (DNIκBα-NSCsiPS) with dominant-negative IκBα (DNIκBα). Carotid injection of DNIκBα-NSCsiPS was then performed with the expectation that the cells would reach the MBH via carotid injection, as the blood–brain barrier (BBB) is more leaky under conditions of obesity and diabetes. As a results, DNIκBα-NSCsiPS were found significantly in MBH of mice fed a high-fat diet (HFD) to induce obesity. Furthermore, it was confirmed that DNIκBα-NSCsiPS induced neurogenesis and led to the generation of POMC cells in the MBH. Restoration of the anorexigenic effect of leptin and improvement of glucose tolerance were also observed [98].

To summarize this section, it has become possible to transplant stem cell-derived hypothalamic neurons into the neonatal brain or to transplant neural stem cells such as tanycytes into the hypothalamic region, and regarding neural stem cell transplantation, it has been shown to improve the obesity. However, the following issues remain to be addressed in these studies. (1) For clinical application, it is necessary to investigate the transplantation not only into the brain of neonates but also in adults, since some patients develop acquired the hypothalamus-pituitary diseases due to the trauma or tumors. (2) The neural stem cell transplantation requires time for the neurons from the grafts to differentiate and mature in the host brain. (3) It has not been determined whether the transplanted neural stem cells will necessarily give rise to the desired cells. To address the above issues, we attempted to transplant mESC-derived mature hypothalamic neurons into the adult mouse brain.

4. Orthotopic transplantation of mESC-derived hypothalamic neurons into the adult mouse brain

We aimed to establish the orthotopic transplantation protocol of *in vitro*-derived hypothalamic neurons into the adult brain. mESCs were differentiated into the hypothalamic neurons using the SFEBq-based methods (**Figure 1**). We confirmed that these differentiated neurons expressed the major hypothalamic peptide hormones such as OXT, AVP, POMC, NPY, AgRP, and MCH after 28 days of culture. In addition, whole-cell patch-clamp studies showed that these neurons were responsive to the external glutamate or GABA. We labeled the mESC-derived hypothalamic neurons by the transfection with AAV-CAG-tdTomato vector and then transplanted the neurons into the SON of the adult immunodeficiency (SCID) mice. As a results, grafts survived until for at least 3 months after transplantation. Although the axons of the grafts projected to the various regions of the brain, it was found that some of the grafts were leaked into the ventricles; therefore, some axons were projected from ventricles rather than from the target SON. Because of these problems, we re-examined the axon projections. It was found that the axons from the grafts projected to the diagonal band nucleus (NDB) and central amygdala nucleus (CEA), both of which are known to be the projection target regions of SON-OXT and AVP neurons [106, 107]. These results suggested that grafts can project the axons to the similar targets as endogenous hypothalamic neurons. Furthermore, in regions rich in host AVP^+ and OXT^+ axon bundles projecting to the posterior pituitary, graft-derived axons were observed to extend parallel to these host axon bundles. These results suggested that graft-derived axons could extend into the posterior pituitary, so we next observed the

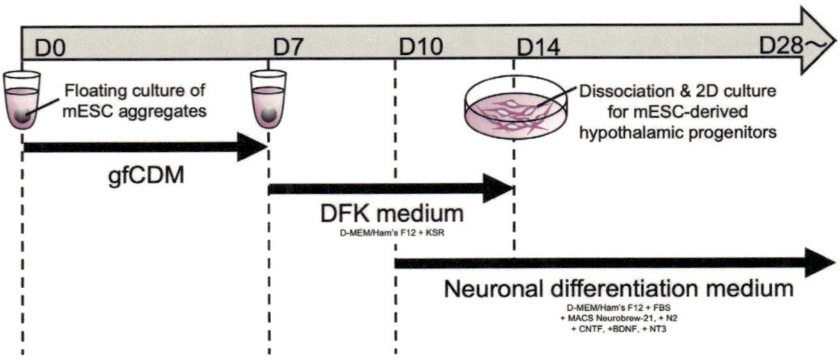

Figure 1.
Schematic of hypothalamic differentiation from mESCs (Modified from [100]).

pituitary. As a result, tdTomato responses of graft-derived axons were observed only in the posterior pituitary, indicating that graft-derived axons project not only into the brain but also into the posterior pituitary (**Figure 2**). This phenomenon was only seen 3 months after transplantation, and we did not observe the axons present in the posterior pituitary 1 or 2 months after transplantation. Considering that only AVP, OXT, or NPY neurons are known to project their axons to the posterior pituitary, grafts have similar characteristics to the endogenous hypothalamic neurons in the host brain. To investigate whether axon projection is altered when the same grafts are transplanted into another areas, we transplanted the mESC-derived hypothalamic neurons into the substantia nigra reticular part (SNr) of adult SCID mouse. As a result, axon projections were changed, and no axons were observed in the posterior pituitary 3 months after transplantation. These results suggest that the match between the cell identity

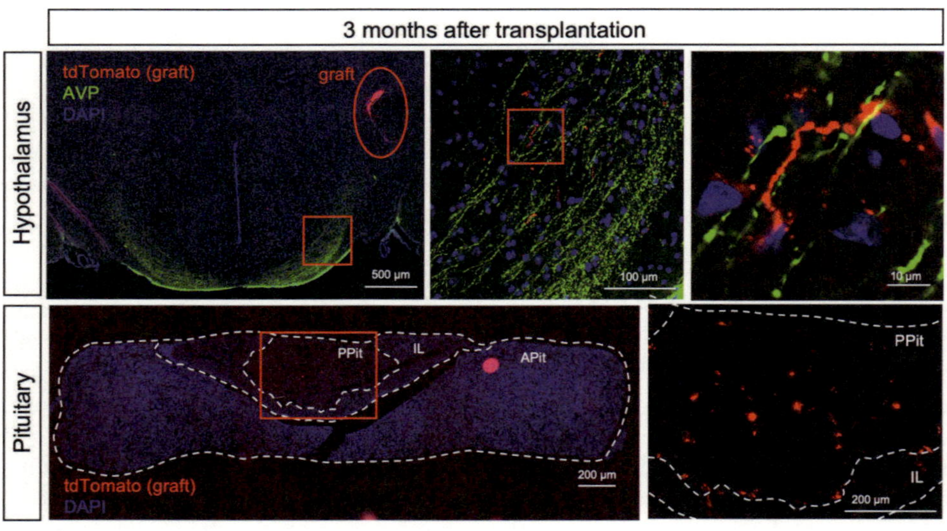

Figure 2.
Projection of the graft-derived axons 3 months after transplantation. The graft-derived axons extending along the host AVP⁺ axon bundles were observed within the host hypothalamus. Signaling of the graft-derived axons was also observed only in the posterior pituitary (PPit). IL, intermediate lobe; APit, anterior pituitary (Modified from [100]).

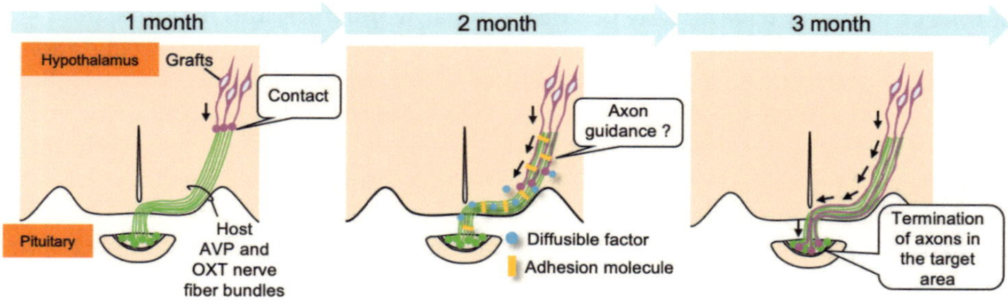

Figure 3.
Hypothetical mechanism of the graft-derived axons elongation. The interaction between the graft-derived axons and the host-derived axons and adhesion factors, as well as factors diffusing into the surrounding area, are assumed to guide the graft-derived axons to their target organs (here, the posterior pituitary) (Modified from [100]).

of the graft and graft area is important for the successful integration of the graft-host connection [100].

As discussed in Section 3, previous studies are establishing the therapeutic approach for the anterior pituitary through the implantation of *in vitro*-derived pituitary organoids. However, a specific treatment for the posterior pituitary disorders has not yet been established. Our research may contribute to establish the treatments for the posterior pituitary disorders such as the central diabetes insipidus (CDI). Although the axon projection of the grafts was similar to that of endogenous hypothalamic neurons, it took more than 3 months for the axons to project into the posterior pituitary. For clinical application in humans, the time required for the axonal projection of the grafts needs to be reduced. Therefore, we are now focusing on axon elongation factors of hypothalamic neurons (**Figure 3**). We observed that the axons derived from the grafts elongated together with the host AVP or OXT axon bundles. These results suggest that the host-graft interaction may be required for proper axon projection. Axon elongation mechanisms similar to our study have been observed in immature cortical neurons, and the cell adhesion molecule TAG-1 has been found to play an important role in axon formation [108]. Perhaps these adhesion factors could promote axon elongation of hypothalamic neurons. Previous studies have provided a mechanism for axon guidance of hypothalamic neurons by diffusible but not adhesive factors. For example, using *ex vivo* analyses in chick, Fang et al., showed that Fgf10 and Fgf3 secreted from the developing neurohypophysis exert direct chemoattractant effects on hypothalamic neurosecretory axons. However, it has also been shown that Fgf10 has a concentration-dependent effect, stimulating axon elongation at low concentrations but possibly inhibiting it at high concentrations [109]. Another genetic study of severely obese patients has identified a rare variant of the diffusible axon guidance factor semaphorin 3 (SEMA3) signaling, and studies using *ex vivo* cultures of mice have shown that SEMA3 promotes axonal projections of POMC neurons from the ARC to the PVN by signaling through neuropilin-2 (NRP2) receptors [110]. We assume that manipulating the expression of the adhesion molecules and diffusible axon guidance factors described above may promote axon elongation in transplanted stem cell-derived hypothalamic neurons.

5. Conclusions

In summary, this chapter has reviewed recent studies on the regeneration of the hypothalamic-pituitary axis. *In vitro* differentiation methods of hypothalamic neurons or pituitary have already become popular worldwide, and hypothalamic and pituitary organoids have been successfully developed as endogenous cells and tissues in recent years. Based on recent results from transcriptome analysis and direct reprogramming studies, it will be possible in the future to induce the differentiation of only specific target cells and tissues by modifying current culture systems in the future. As for *in vivo* studies, many cases of ectopic transplantation of stem cell-derived hypothalamus and pituitary have been reported, but more recently orthotopic transplantation methods have also been reported. By combining the results of *in vitro* and *in vivo* experiments, it is hoped that regenerative medicine will be able to reverse hypothalamic-pituitary disorders. However, many aspects still need to be improved for clinical application, such as the selection of the transplantation site, the choice of the cells to be transplanted (neural stem cell status or mature cells), and the too long time required for the transplanted cells to project their axons in the host brain. Further basic research is needed to accomplish these issues.

Acknowledgements

The work in our lab was supported by Japan Society for the Promotion of Science (JSPS) KAKENHI Grant Number JP20K15917, JP20K09838 and Japan Agency for Medical Research and Development (AMED) under Grant Number JP20bm0404036, JP21ek0109524, JP24ek0109702h001.

Conflict of interest

The authors declare no conflict of interest.

Author details

Miho Kawata* and Hiroshi Nagasaki
Department of Physiology, Fujita Health University School of Medicine, Toyoake, Aichi, Japan

*Address all correspondence to: mkawata@fujita-hu.ac.jp

References

[1] Tran LT, Park S, Kim SK, Lee JS, Kim KW, Kwon O. Hypothalamic control of energy expenditure and thermogenesis. Experimental and Molecular Medicine. 2022;**54**(4):358-369. DOI: 10.1038/s12276-022-00741-z. Epub 2022 Mar 17

[2] Huang W, Ramsey KM, Marcheva B, Bass J. Circadian rhythms, sleep, and metabolism. The Journal of Clinical Investigation. 2011;**121**(6):2133-2141. DOI: 10.1172/JCI46043. Epub 2011 Jun 1

[3] Ball SG. Vasopressin and disorders of water balance: The physiology and pathophysiology of vasopressin. Annals of Clinical Biochemistry. 2007;**44**(Pt 5):417-431. DOI: 10.1258/000456307781646030

[4] Mignot E, Taheri S, Nishino S. Sleeping with the hypothalamus: Emerging therapeutic targets for sleep disorders. Nature Neuroscience. 2002;**5**(Suppl):1071-1075. DOI: 10.1038/nn944

[5] Azevedo EP, Ivan VJ, Friedman JM, Stern SA. Higher-order inputs involved in appetite control. Biological Psychiatry. 2022;**91**(10):869-878. DOI: 10.1016/j.biopsych.2021.07.015. Epub 2021 Jul 24

[6] Yang T, Shah NM. Molecular and neural control of sexually dimorphic social behaviors. Current Opinion in Neurobiology. 2016;**38**:89-95. DOI: 10.1016/j.conb.2016.04.015. Epub 2016 May 7

[7] Rigney N, de Vries GJ, Petrulis A, Young LJ. Oxytocin, vasopressin, and social behavior: From neural circuits to clinical opportunities. Endocrinology. 2022;**163**(9):bqac111. DOI: 10.1210/endocr/bqac111

[8] Mei L, Osakada T, Lin D. Hypothalamic control of innate social behaviors. Science. 2023;**382**(6669):399-404. DOI: 10.1126/science.adh8489. Epub 2023 Oct 26

[9] O'Connor TM, O'Halloran DJ, Shanahan F. The stress response and the hypothalamic-pituitary-adrenal axis: From molecule to melancholia. QJM. 2000;**93**(6):323-333. DOI: 10.1093/qjmed/93.6.323

[10] Kosse C, Burdakov D. Natural hypothalamic circuit dynamics underlying object memorization. Nature Communications. 2019;**10**(1):2505. DOI: 10.1038/s41467-019-10484-7

[11] Hasan MT, Althammer F, Silva da Gouveia M, Goyon S, Eliava M, Lefevre A, et al. A fear memory engram and its plasticity in the hypothalamic oxytocin system. Neuron. 2019;**103**(1):133-146.e8. DOI: 10.1016/j.neuron.2019.04.029. Epub 2019 May 16

[12] Fong H, Zheng J, Kurrasch D. The structural and functional complexity of the integrative hypothalamus. Science. 2023;**382**(6669):388-394. DOI: 10.1126/science.adh8488. Epub 2023 Oct 26

[13] Deem JD, Faber CL, Morton GJ. AgRP neurons: Regulators of feeding, energy expenditure, and behavior. The FEBS Journal. 2022;**289**(8):2362-2381. DOI: 10.1111/febs.16176. Epub 2021 Sep 13

[14] Vohra MS, Benchoula K, Serpell CJ, Hwa WE. AgRP/NPY and POMC neurons in the arcuate nucleus and their potential role in treatment of obesity. European Journal of Pharmacology. 2022;**915**:174611. DOI: 10.1016/j.ejphar.2021.174611. Epub 2021 Nov 17

[15] Füzesi T, Daviu N, Wamsteeker Cusulin JI, Bonin RP, Bains JS. Hypothalamic CRH neurons orchestrate complex behaviours after stress. Nature Communications. 2016;**7**:11937. DOI: 10.1038/ncomms11937

[16] Veenema AH, Neumann ID. Central vasopressin and oxytocin release: Regulation of complex social behaviours. Progress in Brain Research. 2008;**170**:261-276. DOI: 10.1016/S0079-6123(08)00422-6

[17] Bouâouda H, Jha PK. Orexin and MCH neurons: Regulators of sleep and metabolism. Frontiers in Neuroscience. 2023;**17**:1230428. DOI: 10.3389/fnins.2023.1230428

[18] Ono D, Honma KI, Honma S. Roles of neuropeptides, VIP and AVP, in the mammalian central circadian clock. Frontiers in Neuroscience. 2021;**15**:650154. DOI: 10.3389/fnins.2021.650154. Erratum in: Front Neurosci. 2021 Dec 07;15:810796

[19] Hiller-Sturmhöfel S, Bartke A. The endocrine system: An overview. Alcohol Health and Research World. 1998;**22**(3):153-164

[20] Sobrier ML, Maghnie M, Vié-Luton MP, Secco A, di Iorgi N, Lorini R, et al. Novel HESX1 mutations associated with a life-threatening neonatal phenotype, pituitary aplasia, but normally located posterior pituitary and no optic nerve abnormalities. The Journal of Clinical Endocrinology and Metabolism. 2006;**91**(11):4528-4536. DOI: 10.1210/jc.2006-0426. Epub 2006 Aug 29

[21] Spacca B, Amasio ME, Giordano F, Mussa F, Busca G, Donati P, et al. Surgical management of congenital median perisellar transsphenoidal encephaloceles with an extracranial approach: A series of 6 cases. Neurosurgery. 2009;**65**(6):1140-1145; discussion 1145-6. DOI: 10.1227/01.NEU.0000351780.23357.F5

[22] De Marinis L, Bonadonna S, Bianchi A, Maira G, Giustina A. Primary empty sella. The Journal of Clinical Endocrinology and Metabolism. 2005;**90**(9):5471-5477

[23] Gregory LC, Dattani MT. The molecular basis of congenital hypopituitarism and related disorders. The Journal of Clinical Endocrinology and Metabolism. 2020;**105**(6):dgz184. DOI: 10.1210/clinem/dgz184

[24] Cassidy SB, Schwartz S, Miller JL, Driscoll DJ. Prader-Willi syndrome. Genetics in Medicine. 2012;**14**(1):10-26. DOI: 10.1038/gim.0b013e31822bead0. Epub 2011 Sep 26

[25] Kumar A, Husain A Sr, Saleem A, Khawaja UA, Virani S. Laurence-Moon-Bardet-Biedl syndrome: A rare case with a literature review. Cureus. 2020;**12**(11):e11355. DOI: 10.7759/cureus.11355

[26] Kim SH. Congenital hypogonadotropic hypogonadism and Kallmann syndrome: Past, present, and future. Endocrinology and Metabolism (Seoul). 2015;**30**(4):456-466. DOI: 10.3803/EnM.2015.30.4.456

[27] Salvatori R, Hayashida CY, Aguiar-Oliveira MH, Phillips JA 3rd, Souza AH, Gondo RG, et al. Familial dwarfism due to a novel mutation of the growth hormone-releasing hormone receptor gene. The Journal of Clinical Endocrinology and Metabolism. 1999;**84**(3):917-923. DOI: 10.1210/jcem.84.3.5599

[28] Arima H, Azuma Y, Morishita Y, Hagiwara D. Central diabetes insipidus.

Nagoya Journal of Medical Science. 2016;**78**(4):349-358. DOI: 10.18999/nagjms.78.4.349

[29] Lonser RR, Mehta GU, Kindzelski BA, Ray-Chaudhury A, Vortmeyer AO, Dickerman R, et al. Surgical management of carney complex-associated pituitary pathology. Neurosurgery. 2017;**80**(5):780-786. DOI: 10.1227/NEU.0000000000001384

[30] Thakker RV, Newey PJ, Walls GV, Bilezikian J, Dralle H, Ebeling PR, et al. Clinical practice guidelines for multiple endocrine neoplasia type 1 (MEN1). The Journal of Clinical Endocrinology and Metabolism. 2012;**97**(9):2990-3011. DOI: 10.1210/jc.2012-1230. Epub 2012 Jun 20

[31] Prete A, Corsello SM, Salvatori R. Current best practice in the management of patients after pituitary surgery. Therapeutic Advances in Endocrinology and Metabolism. 2017;**8**(3):33-48. DOI: 10.1177/2042018816687240. Epub 2017 Mar 1

[32] Feigl GC, Bonelli CM, Berghold A, Mokry M. Effects of gamma knife radiosurgery of pituitary adenomas on pituitary function. Journal of Neurosurgery. 2002;**97**(Suppl. 5):415-421. DOI: 10.3171/jns.2002.97.supplement

[33] Javed Z, Qamar U, Sathyapalan T. Pituitary and/or hypothalamic dysfunction following moderate to severe traumatic brain injury: Current perspectives. Indian Journal of Endocrinology and Metabolism. 2015;**19**(6):753-763. DOI: 10.4103/2230-8210.167561

[34] Wang CZ, Guo LL, Han BY, Su X, Guo QH, Mu YM. Pituitary stalk interruption syndrome: From clinical findings to pathogenesis. Journal of Neuroendocrinology. 2017;**29**(1). DOI: 10.1111/jne.12451

[35] Lake MG, Krook LS, Cruz SV. Pituitary adenomas: An overview. American Family Physician. 2013;**88**(5):319-327

[36] Molitch ME. Diagnosis and treatment of pituitary adenomas: A review. JAMA. 2017;**317**(5):516-524

[37] van Iersel L, van Santen HM, Potter B, Li Z, Conklin HM, Zhang H, et al. Clinical impact of hypothalamic-pituitary disorders after conformal radiation therapy for pediatric low-grade glioma or ependymoma. Pediatric Blood and Cancer. 2020;**67**(12):e28723

[38] Han SJ, Rolston JD, Jahangiri A, Aghi MK. Rathke's cleft cysts: Review of natural history and surgical outcomes. Journal of Neuro-Oncology. 2014;**117**:197-203

[39] Fan MC, Wang QL, Wang JF, Deng WS, Li LD, Wang ZH, et al. Surgical treatment of symptomatic Rathke's cleft cysts: Clinical features, therapy considerations and outcomes. Chinese Medical Journal. 2012;**125**(16):2919-2924

[40] Garnett MR, Puget S, Grill J, Sainte-Rose C. Craniopharyngioma. Orphanet Journal of Rare Diseases. 2007;**2**:1-7

[41] Müller HL. Craniopharyngioma. Endocrine Reviews. 2014;**35**(3):513-543. DOI: 10.1210/er.2013-1115. Epub 2014 Jan 27

[42] Jensterle M, Jazbinsek S, Bosnjak R, Popovic M, Zaletel LZ, Vesnaver TV, et al. Advances in the management of craniopharyngioma in children and adults. Radiology and Oncology. 2019;**53**(4):388-396. DOI: 10.2478/raon-2019-0036

[43] Arita K, Kurisu K, Kiura Y, Iida K, Otsubo H. Hypothalamic hamartoma. Neurologia Medico-Chirurgica (Tokyo). 2005;**45**(5):221-231. DOI: 10.2176/nmc.45.221

[44] Shimon I. Metastatic spread to the pituitary. Neuroendocrinology. 2020;**110**(9-10):805-808. DOI: 10.1159/000506810. Epub 2020 Feb 27

[45] Wiens AL, Hagen MC, Bonnin JM, Rizzo KA. T-cell lymphoblastic lymphoma/leukemia presenting as a pituitary mass lesion: A case report and review of the literature. Neuropathology. 2012;**32**(6):668-674. DOI: 10.1111/j.1440-1789.2012.01314.x. Epub 2012 Mar 21

[46] Martinez-Perez R, Kortz MW, Florez-Perdomo W, Ung TH, Youssef AS. Endocrinological outcomes after transcranial resection of tuberculum sellae meningiomas: A systematic review and meta-analysis. Neurosurgical Review. 2022;**45**(3):1965-1975. DOI: 10.1007/s10143-022-01744-0. Epub 2022 Feb 12

[47] Nishioka H, Ito H, Haraoka J, Akada K. Immature teratoma originating from the pituitary gland: Case report. Neurosurgery. 1999;**44**(3):644-647; discussion 647-8. DOI: 10.1097/00006123-199903000-00115

[48] Lury KM. Inflammatory and infectious processes involving the pituitary gland. Topics in Magnetic Resonance Imaging. 2005;**16**(4):301-306. DOI: 10.1097/01.rmr.0000224686.21748.ea

[49] Ucler R, Kara E, Atmaca M, Olmez S, Alay M, Dirik Y, et al. A rare presentation of transfusional hemochromatosis: Hypogonadotropic hypogonadism. Case Reports in Endocrinology. 2015;**2015**:493091. DOI: 10.1155/2015/493091. Epub 2015 Jul 21

[50] Bihan H, Christozova V, Dumas JL, Jomaa R, Valeyre D, Tazi A, et al. Sarcoidosis: Clinical, hormonal, and magnetic resonance imaging (MRI) manifestations of hypothalamic-pituitary disease in 9 patients and review of the literature. Medicine (Baltimore). 2007;**86**(5):259-268. DOI: 10.1097/MD.0b013e31815585aa

[51] Tan H, Yu K, Yu Y, An Z, Li J. Isolated hypothalamic-pituitary langerhans' cell histiocytosis in female adult: A case report. Medicine (Baltimore). 2019;**98**(2):e13853. DOI: 10.1097/MD.0000000000013853

[52] Kaltsas GA, Powles TB, Evanson J, Plowman PN, Drinkwater JE, Jenkins PJ, et al. Hypothalamo-pituitary abnormalities in adult patients with langerhans cell histiocytosis: Clinical, endocrinological, and radiological features and response to treatment. The Journal of Clinical Endocrinology and Metabolism. 2000;**85**(4):1370-1376. DOI: 10.1210/jcem.85.4.6501

[53] Tsiakalos A, Xynos ID, Sipsas NV, Kaltsas G. Pituitary insufficiency after infectious meningitis: A prospective study. The Journal of Clinical Endocrinology and Metabolism. 2010;**95**(7):3277-3281. DOI: 10.1210/jc.2010-0144. Epub 2010 Apr 28

[54] Vedeler CA, Storstein A. Autoimmune limbic encephalitis. Acta Neurologica Scandinavica. Supplementum. 2009;**189**:63-67. DOI: 10.1111/j.1600-0404.2009.01204.x

[55] Bhargava R, Hussein Z, Dorward NL, Grieve JP, Jaunmuktane Z, Marcus HJ, et al. IgG4-related hypophysitis: A retrospective cohort study. Acta

Neurochirurgica. 2022;**164**(8):2095-2103. DOI: 10.1007/s00701-022-05231-9. Epub 2022 May 7

[56] Inaba H, Suzuki S, Shigematsu S, Takei M, Miyamoto T, Takeda T, et al. Lymphocytic infundibuloneurohypophysitis: Long-term follow-up of a case cured with glucocorticoid. Medical Principles and Practice. 2010;**19**(1):79-81. DOI: 10.1159/000252841. Epub 2009 Dec 9

[57] Donegan D, Erickson D. Revisiting pituitary apoplexy. Journal of the Endocrine Society. 2022;**6**(9):bvac113. DOI: 10.1210/jendso/bvac113

[58] Karaca Z, Tanriverdi F, Unluhizarci K, Kelestimur F. Pregnancy and pituitary disorders. European Journal of Endocrinology. 2010;**162**(3):453-475

[59] Huang AW, Muneyyirci-Delale O. Reproductive endocrine issues in men with sickle cell anemia. Andrology. 2017;**5**(4):679-690. DOI: 10.1111/andr.12370. Epub 2017 Jun 29

[60] Jamilloux Y, Liozon E, Pugnet G, Nadalon S, Heang Ly K, Dumonteil S, et al. Recovery of adrenal function after long-term glucocorticoid therapy for giant cell arteritis: A cohort study. PLoS One. 2013;**8**(7):e68713. DOI: 10.1371/journal.pone.0068713

[61] Kovacs K. Sheehan syndrome. Lancet. 2003;**361**(9356):520-522. DOI: 10.1016/S0140-6736(03)12490-7

[62] Can A, Gross BA, Smith TR, Dammers R, Dirven CM, Woodmansee WW, et al. Pituitary dysfunction after aneurysmal subarachnoid Hemorrhage: A systematic review and meta-analysis. Neurosurgery. 2016;**79**(2):253-264. DOI: 10.1227/NEU.0000000000001157

[63] Sunil K, Menon R, Goel N, Sanghvi D, Bandgar T, Joshi SR, et al. Pituitary tuberculosis. The Journal of the Association of Physicians of India. 2007;**55**:453-456

[64] Mageshkumar S, Patil DV, Philo AJ, Madhavan K. Hypopituitarism as unusual sequelae to central nervous system tuberculosis. Indian Journal of Clinical Endocrinology and Metabolism. 2011;**15**(Suppl_3):S259-S262. DOI: 10.4103/2230-8210.84881

[65] Kumar S, Kumar KH. Pituitary diseases in the tropics. In: Feingold KR, Anawalt B, Boyce A, Chrousos G, de Herder WW, Dhatariya K, et al., editors. Endotext. South Dartmouth MA: MDText.com, Inc. 2000. Available from: http://www.ncbi.nlm.nih.gov/books/NBK568567/

[66] Agyei JO, Lipinski LJ, Leonardo J. Case report of a primary pituitary abscess and systematic literature review of pituitary abscess with a focus on patient outcomes. World Neurosurgery. 2017;**101**:76-92. DOI: 10.1016/j.wneu.2017.01.077. Epub 2017 Jan 31

[67] Yamamoto M, Iguchi G, Bando H, Kanie K, Hidaka-Takeno R, Fukuoka H, et al. Autoimmune pituitary disease: New concepts with clinical implications. Endocrine Reviews. 2020;**41**(2):bnz003. DOI: 10.1210/endrev/bnz003

[68] Kobayashi T, Iwama S, Yasuda Y, Okada N, Okuji T, Ito M, et al. Pituitary dysfunction induced by immune checkpoint inhibitors is associated with better overall survival in both malignant melanoma and non-small cell lung carcinoma: A prospective study. Journal for Immunotherapy of Cancer. 2020;**8**(2):e000779. DOI: 10.1136/jitc-2020-000779

[69] Koo B, Choi B, Park H, Yoon KJ. Past, present, and future of brain

organoid technology. Molecules and Cells. 2019;**42**(9):617-627. DOI: 10.14348/molcells.2019.0162

[70] Acharya P, Choi NY, Shrestha S, Jeong S, Lee MY. Brain organoids: A revolutionary tool for modeling neurological disorders and development of therapeutics. Biotechnology and Bioengineering. 2024;**121**(2):489-506. DOI: 10.1002/bit.28606. Epub 2023 Nov 28

[71] Sun AX, Ng HH, Tan EK. Translational potential of human brain organoids. Annals of Clinical Translational Neurology. 2018;**5**(2):226-235. DOI: 10.1002/acn3.505

[72] Eichmüller OL, Knoblich JA. Human cerebral organoids—A new tool for clinical neurology research. Nature Reviews. Neurology. 2022;**18**(11):661-680. DOI: 10.1038/s41582-022-00723-9. Epub 2022 Oct 17

[73] Chambers SM, Fasano CA, Papapetrou EP, Tomishima M, Sadelain M, Studer L. Highly efficient neural conversion of human ES and iPS cells by dual inhibition of SMAD signaling. Nature Biotechnology. 2009;**27**(3):275-280. DOI: 10.1038/nbt.1529. Epub 2009 Mar 1. Erratum in: Nat Biotechnol. 2009 May;27(5):485

[74] Wang L, Meece K, Williams DJ, Lo KA, Zimmer M, Heinrich G, et al. Differentiation of hypothalamic-like neurons from human pluripotent stem cells. The Journal of Clinical Investigation. 2015;**125**(2):796-808. DOI: 10.1172/JCI79220. Epub 2015 Jan 2

[75] Rajamani U, Gross AR, Hjelm BE, Sequeira A, Vawter MP, Tang J, et al. Super-obese patient-derived iPSC hypothalamic neurons exhibit obesogenic signatures and hormone responses. Cell Stem Cell. 2018;**22**(5):698-712.e9. DOI: 10.1016/j.stem.2018.03.009. Epub 2018 Apr 19

[76] Dincer Z, Piao J, Niu L, Ganat Y, Kriks S, Zimmer B, et al. Specification of functional cranial placode derivatives from human pluripotent stem cells. Cell Reports. 2013;**5**(5):1387-1402. DOI: 10.1016/j.celrep.2013.10.048. Epub 2013 Nov 27

[77] Watanabe K, Kamiya D, Nishiyama A, Katayama T, Nozaki S, Kawasaki H, et al. Directed differentiation of telencephalic precursors from embryonic stem cells. Nature Neuroscience. 2005;**8**(3):288-296. DOI: 10.1038/nn1402. Epub 2005 Feb 6

[78] Eiraku M, Watanabe K, Matsuo-Takasaki M, Kawada M, Yonemura S, Matsumura M, et al. Self-organized formation of polarized cortical tissues from ESCs and its active manipulation by extrinsic signals. Cell Stem Cell. 2008;**3**(5):519-532. DOI: 10.1016/j.stem.2008.09.002

[79] Wataya T, Ando S, Muguruma K, Ikeda H, Watanabe K, Eiraku M, et al. Minimization of exogenous signals in ES cell culture induces rostral hypothalamic differentiation. Proceedings of the National Academy of Sciences of the United States of America. 2008;**105**(33):11796-11801. DOI: 10.1073/pnas.0803078105. Epub 2008 Aug 12

[80] Suga H, Kadoshima T, Minaguchi M, Ohgushi M, Soen M, Nakano T, et al. Self-formation of functional adenohypophysis in three-dimensional culture. Nature. 2011;**480**(7375):57-62. DOI: 10.1038/nature10637

[81] Ozone C, Suga H, Eiraku M, Kadoshima T, Yonemura S, Takata N, et al. Functional anterior pituitary generated in self-organizing culture of human embryonic stem cells. Nature

Communications. 2016;**7**:10351. DOI: 10.1038/ncomms10351

[82] Kodani Y, Kawata M, Suga H, Kaneko YS, Nakashima A, Kameyama T, et al. Characterization of hypothalamic MCH neuron development in a 3D differentiation system of mouse embryonic stem cells. eNeuro. 2022;**9**(2):ENEURO.0442-21.2022. DOI: 10.1523/ENEURO.0442-21.2022

[83] Yamamoto M, Ong ALC, Shinozuka T, Shirai M, Sasai N. Manipulation of signal gradient and transcription factors recapitulates: Multiple hypothalamic identities. Stem Cells. 2023;**41**(5):453-467. DOI: 10.1093/stmcls/sxad018

[84] Qian X, Nguyen HN, Song MM, Hadiono C, Ogden SC, Hammack C, et al. Brain-region-specific organoids using mini-bioreactors for modeling ZIKV exposure. Cell. 2016;**165**(5):1238-1254. DOI: 10.1016/j.cell.2016.04.032. Epub 2016 Apr 22

[85] Qian X, Jacob F, Song MM, Nguyen HN, Song H, Ming GL. Generation of human brain region-specific organoids using a miniaturized spinning bioreactor. Nature Protocols. 2018;**13**(3):565-580. DOI: 10.1038/nprot.2017.152. Epub 2018 Feb 22

[86] Kasai T, Suga H, Sakakibara M, Ozone C, Matsumoto R, Kano M, et al. Hypothalamic contribution to pituitary functions is recapitulated in vitro using 3D-cultured human iPS cells. Cell Reports. 2020;**30**(1):18-24.e5. DOI: 10.1016/j.celrep.2019.12.009

[87] Matsumoto R, Suga H, Aoi T, Bando H, Fukuoka H, Iguchi G, et al. Congenital pituitary hypoplasia model demonstrates hypothalamic OTX2 regulation of pituitary progenitor cells. The Journal of Clinical Investigation. 2020;**130**(2):641-654. DOI: 10.1172/JCI127378

[88] Huang WK, Wong SZH, Pather SR, Nguyen PTT, Zhang F, Zhang DY, et al. Generation of hypothalamic arcuate organoids from human induced pluripotent stem cells. Cell Stem Cell. 2021;**28**(9):1657-1670.e10. DOI: 10.1016/j.stem.2021.04.006. Epub 2021 May 6

[89] Blaess S, Szabó N, Haddad-Tóvolli R, Zhou X, Álvarez-Bolado G. Sonic hedgehog signaling in the development of the mouse hypothalamus. Frontiers in Neuroanatomy. 2015;**8**:156. DOI: 10.3389/fnana.2014.00156

[90] Corman TS, Bergendahl SE, Epstein DJ. Distinct temporal requirements for sonic hedgehog signaling in development of the tuberal hypothalamus. Development. 2018;**145**(21):dev167379. DOI: 10.1242/dev.167379

[91] Kapsimali M, Caneparo L, Houart C, Wilson SW. Inhibition of Wnt/Axin/beta-catenin pathway activity promotes ventral CNS midline tissue to adopt hypothalamic rather than floorplate identity. Development. 2004;**131**(23):5923-5933. DOI: 10.1242/dev.01453

[92] Ogawa K, Suga H, Ozone C, Sakakibara M, Yamada T, Kano M, et al. Vasopressin-secreting neurons derived from human embryonic stem cells through specific induction of dorsal hypothalamic progenitors. Scientific Reports. 2018;**8**(1):3615. DOI: 10.1038/s41598-018-22053-x

[93] Mitsumoto K, Suga H, Sakakibara M, Soen M, Yamada T, Ozaki H, et al. Improved methods for the differentiation of hypothalamic vasopressin neurons using mouse induced pluripotent stem cells. Stem Cell Research.

2019;**40**:101572. DOI: 10.1016/j.scr.2019.101572. Epub 2019 Sep 6

[94] Merkle FT, Maroof A, Wataya T, Sasai Y, Studer L, Eggan K, et al. Generation of neuropeptidergic hypothalamic neurons from human pluripotent stem cells. Development. 2015;**142**(4):633-643. DOI: 10.1242/dev.117978

[95] Kodani Y, Kawata M, Suga H, Kasai T, Ozone C, Sakakibara M, et al. EpCAM is a surface marker for enriching anterior pituitary cells from human hypothalamic-pituitary organoids. Frontiers in Endocrinology (Lausanne). 2022;**13**:941166. DOI: 10.3389/fendo.2022.941166

[96] Vierbuchen T, Ostermeier A, Pang ZP, Kokubu Y, Südhof TC, Wernig M. Direct conversion of fibroblasts to functional neurons by defined factors. Nature. 2010;**463**(7284):1035-1041. DOI: 10.1038/nature08797. Epub 2010 Jan 27

[97] Herb BR, Glover HJ, Bhaduri A, Colantuoni C, Bale TL, Siletti K, et al. Single-cell genomics reveals region-specific developmental trajectories underlying neuronal diversity in the human hypothalamus. Science Advances. 2023;**9**(45):eadf6251. DOI: 10.1126/sciadv.adf6251. Epub 2023 Nov 8

[98] Li J, Tang Y, Purkayastha S, Yan J, Cai D. Control of obesity and glucose intolerance via building neural stem cells in the hypothalamus. Molecular Metabolism. 2014;**3**(3):313-324. DOI: 10.1016/j.molmet.2014.01.012

[99] Zimmer B, Piao J, Ramnarine K, Tomishima MJ, Tabar V, Studer L. Derivation of diverse hormone-releasing pituitary cells from human pluripotent stem cells. Stem Cell Reports. 2016;**6**(6):858-872. DOI: 10.1016/j.stemcr.2016.05.005

[100] Kawata M, Kodani Y, Ohkuma M, Miyachi EI, Kaneko YS, Nakashima A, et al. Long-range axonal projections of transplanted mouse embryonic stem cell-derived hypothalamic neurons into adult mouse brain. PLoS One. 2022;**17**(11):e0276694. DOI: 10.1371/journal.pone.0276694

[101] Miwata T, Suga H, Kawaguchi Y, Sakakibara M, Kano M, Taga S, et al. Generation of hypothalamic neural stem cell-like cells in vitro from human pluripotent stem cells. Stem Cell Reports. 2023;**18**(4):869-883. DOI: 10.1016/j.stemcr.2023.02.006. 2023 Mar 23

[102] Sasaki H, Suga H, Takeuchi K, Nagata Y, Harada H, Kondo T, et al. Subcutaneous transplantation of human embryonic stem cells-derived pituitary organoids. Frontiers in Endocrinology (Lausanne). 2023;**14**:1130465. DOI: 10.3389/fendo.2023.1130465

[103] Taga S, Suga H, Nakano T, Kuwahara A, Inoshita N, Kodani Y, et al. Generation and purification of ACTH-secreting hPSC-derived pituitary cells for effective transplantation. Stem Cell Reports. 2023;**18**(8):1657-1671. DOI: 10.1016/j.stemcr.2023.05.002. Epub 2023 Jun 8

[104] Miranda-Angulo AL, Byerly MS, Mesa J, Wang H, Blackshaw S. Rax regulates hypothalamic tanycyte differentiation and barrier function in mice. The Journal of Comparative Neurology. 2014;**522**(4):876-899. DOI: 10.1002/cne.23451

[105] Mu W, Li S, Xu J, Guo X, Wu H, Chen Z, et al. Hypothalamic Rax^{+} tanycytes contribute to tissue repair and tumorigenesis upon oncogene activation in mice. Nature Communications. 2021;**12**(1):2288. DOI: 10.1038/s41467-021-22640-z

[106] Knobloch HS, Charlet A, Hoffmann LC, Eliava M, Khrulev S, Cetin AH, et al. Evoked axonal oxytocin release in the central amygdala attenuates fear response. Neuron. 2012;**73**(3):553-566. DOI: 10.1016/j.neuron.2011.11.030

[107] Hernández VS, Hernández OR, Perez de la Mora M, Gómora MJ, Fuxe K, Eiden LE, et al. Hypothalamic vasopressinergic projections innervate central amygdala GABAergic neurons: Implications for anxiety and stress coping. Frontiers in Neural Circuits. 2016;**10**:92. DOI: 10.3389/fncir.2016.00092

[108] Namba T, Kibe Y, Funahashi Y, Nakamuta S, Takano T, Ueno T, et al. Pioneering axons regulate neuronal polarization in the developing cerebral cortex. Neuron. 2014;**81**(4):814-829. DOI: 10.1016/j.neuron.2013.12.015

[109] Liu F, Pogoda HM, Pearson CA, Ohyama K, Löhr H, Hammerschmidt M, et al. Direct and indirect roles of Fgf3 and Fgf10 in innervation and vascularisation of the vertebrate hypothalamic neurohypophysis. Development. 2013;**140**(5):1111-1122. DOI: 10.1242/dev.080226

[110] van der Klaauw AA, Croizier S, Mendes de Oliveira E, LKJ S, Park S, Kong Y, et al. Human semaphorin 3 variants link melanocortin circuit development and energy balance. Cell. 2019;**176**(4):729-742.e18. DOI: 10.1016/j.cell.2018.12.009. 2019 Jan 17

Chapter 21

Stem Cell Therapy for Stroke: Mechanisms and Functional Recovery

Hongjian Pu

Abstract

Stem cell therapy has emerged as a promising approach for enhancing recovery following stroke, a leading cause of disability worldwide. This chapter provides a comprehensive overview of the mechanisms through which stem cell therapy exerts its effects on brain repair and functional recovery. We explore several types of stem cells, including mesenchymal stem cells, neural stem cells, and induced pluripotent stem cells, and their potential to promote neurogenesis, angiogenesis, and synaptic plasticity. Additionally, we discuss the paracrine effects of stem cells, highlighting their role in modulating inflammation and reducing apoptosis. Preclinical studies and clinical trials are reviewed to evaluate the efficacy and safety of different stem cell therapies. Furthermore, we address the challenges and limitations currently hindering the translation of stem cell therapy into routine clinical practice, such as optimal cell delivery methods, immune rejection, and ethical considerations. By elucidating the underlying mechanisms and summarizing the latest research advancements, this chapter aims to provide valuable insights into the potential of stem cell therapy as a viable treatment strategy for stroke patients, contributing to improved functional recovery and quality of life.

Keywords: stroke, stem cell, mechanism, brain repair, function recovery

1. Introduction

Stroke is a major global health concern, ranking as one of the leading causes of death and long-term disability [1]. The sudden interruption of blood supply to the brain results in neuronal death and the loss of crucial neurological functions, severely impacting the quality of life of millions of individuals. Current treatment strategies for stroke, such as thrombolysis and mechanical thrombectomy, are limited by narrow therapeutic windows and are not universally applicable to all patients [2]. As a result, there is an urgent need for novel therapeutic approaches that can enhance brain repair and functional recovery post-stroke.

Stem cell therapy has garnered significant attention as a potential treatment for stroke due to its capacity to repair and regenerate damaged brain tissue [3]. Stem cells are undifferentiated cells with the ability to self-renew and differentiate into various cell types. Their versatility and regenerative potential make them ideal candidates for treating neurological disorders. This chapter delves into the mechanisms by which stem cell therapy can facilitate brain repair and promote functional recovery after a stroke.

We will explore distinct types of stem cells, including mesenchymal stem cells (MSCs), neural stem cells (NSCs), and induced pluripotent stem cells (iPSCs), and their respective roles in neurogenesis, angiogenesis, and synaptic plasticity [4]. Additionally, we will discuss the paracrine effects of stem cells, particularly their ability to modulate inflammation and reduce apoptosis, which are crucial for creating a conducive environment for brain repair.

To provide a comprehensive understanding, we review both preclinical studies and clinical trials that investigate the efficacy and safety of stem cell therapies in stroke treatment. Despite promising results, the translation of stem cell therapy from research to clinical practice faces several challenges, including optimal delivery methods, immune rejection, and ethical considerations.

By elucidating the underlying mechanisms of stem cell therapy and summarizing the latest research advancements, this chapter aims to highlight the potential of stem cell therapy as a transformative treatment strategy for stroke. Our goal is to offer insights that could lead to improved functional recovery and a better quality of life for stroke patients.

2. Types of stem cells

Stem cell therapy for stroke involves several types of stem cells, each with unique characteristics and mechanisms of action. This chapter discusses mesenchymal stem cells (MSCs), neural stem cells (NSCs), and induced pluripotent stem cells (iPSCs), focusing on their potential for stroke treatment.

2.1 Mesenchymal stem cells

Mesenchymal stem cells (MSCs) are multipotent stromal cells that can differentiate into various cell types, including osteoblasts, chondrocytes, and adipocytes [5]. MSCs are sourced from bone marrow, adipose tissue, and umbilical cord blood. They are known for their immunomodulatory properties, making them an attractive option for treating inflammatory conditions like stroke [6, 7].

2.2 Neural stem cells

Neural stem cells (NSCs) are multipotent cells capable of differentiating into neurons, astrocytes, and oligodendrocytes. NSCs reside in specific regions of the adult brain, such as the subventricular zone and the hippocampus, where they contribute to neurogenesis throughout life [8].

2.3 Induced pluripotent stem cells

Induced pluripotent stem cells (iPSCs) are generated by reprogramming adult somatic cells to a pluripotent state, like embryonic stem cells (ESCs). iPSCs can differentiate into any cell type, offering enormous potential for regenerative medicine [9].

3. Functional recovery by stem cells

Functional recovery following a stroke involves the restoration of motor, sensory, cognitive, and emotional functions that were impaired due to the ischemic event [10].

Stem cell therapy has shown promising potential in enhancing functional recovery by targeting various pathological processes involved in stroke damage [11–13]. This section delves into the different mechanisms through which stem cells contribute to functional recovery, focusing on neurogenesis, synaptic plasticity, angiogenesis, and the modulation of inflammation and apoptosis.

3.1 Neurogenesis and synaptic plasticity

Stem cells, particularly neural stem cells (NSCs) and induced pluripotent stem cells (iPSCs), play a crucial role in neurogenesis, the process by which new neurons are formed in the brain [14, 15]. After a stroke, the brain's capacity for neurogenesis is significantly compromised. Transplanted stem cells can differentiate into neurons, replenishing the lost neuronal population. These newly formed neurons can integrate into existing neural networks, contributing to the restoration of synaptic connections and enhancing synaptic plasticity [16–18]. Synaptic plasticity, the ability of synapses to strengthen or weaken over time in response to increases or decreases in their activity, is essential for learning, memory, and recovery of motor skills post-stroke.

3.2 Angiogenesis

Angiogenesis, the formation of new blood vessels, is another critical process in stroke recovery. Mesenchymal stem cells (MSCs) are particularly effective in promoting angiogenesis [19–21]. MSCs secrete a variety of angiogenic factors such as vascular endothelial growth factor (VEGF), which stimulates the growth of new capillaries in the ischemic brain regions. This enhanced vascular network improves blood flow and oxygen delivery to the damaged areas, facilitating tissue repair and functional recovery. Improved perfusion in the penumbra, the area surrounding the ischemic core, helps to salvage at-risk neurons and glial cells, thereby limiting the extent of brain damage.

3.3 Modulation of inflammation and apoptosis

Inflammation and apoptosis (programmed cell death) are significant contributors to secondary brain damage following a stroke. Stem cells have demonstrated the ability to modulate the inflammatory response and reduce apoptosis, creating a more favorable environment for brain repair [22–24]. MSCs, for example, secrete anti-inflammatory cytokines and reduce the levels of pro-inflammatory cytokines, mitigating the inflammatory response [25]. Similarly, NSCs and iPSCs can modulate microglial activity, reducing neuroinflammation and promoting the release of neurotrophic factors that protect neurons from apoptosis [26, 27]. By reducing inflammation and preventing cell death, stem cells help to preserve the structural and functional integrity of the brain.

3.4 Enhancement of neuroplasticity

Neuroplasticity, the brain's ability to reorganize itself by forming new neural connections, is essential for recovery from stroke [28]. Stem cell therapy has been shown to enhance neuroplasticity by creating an environment conducive to the growth and strengthening of neural connections [29, 30]. This includes the upregulation of growth factors such as brain-derived neurotrophic factor (BDNF) and glial cell line-derived neurotrophic factor (GDNF), which support the survival, growth, and differentiation of neurons [31, 32]. Enhanced neuroplasticity contributes to the improvement of motor and cognitive functions, aiding patients in regaining lost abilities.

3.5 Clinical evidence and future directions

Preclinical studies and early-phase clinical trials have provided encouraging evidence regarding the efficacy of stem cell therapy in promoting functional recovery after stroke. For instance, studies have shown that intravenous administration of MSCs in rodent models leads to significant improvements in motor and cognitive functions, reduced infarct size, and enhanced neurogenesis [33]. Similarly, NSC and iPSC therapies have demonstrated potential benefits in animal models, including improved functional outcomes and reduced brain damage [14, 15].

Clinical trials in human stroke patients are ongoing, with some early-phase studies reporting promising results in terms of safety and functional improvement. However, several challenges remain in translating these findings into routine clinical practice. These include optimizing stem cell delivery methods, ensuring long-term safety, overcoming immune rejection, and addressing ethical considerations [34–37]. Continued research and well-designed clinical trials are essential to fully realize the potential of stem cell therapy in stroke recovery.

In summary, stem cell therapy offers a multifaceted approach to enhancing functional recovery after stroke by promoting neurogenesis, angiogenesis, and neuroplasticity, and by modulating inflammation and apoptosis. While considerable progress has been made, further research is needed to address the remaining challenges and bring these therapies from the bench to the bedside, improving the quality of life for stroke patients.

4. Mechanisms of stem cells' neuroprotection

4.1 Mesenchymal stem cells

4.1.1 Neuroprotection and neurogenesis

MSCs secrete a variety of neurotrophic factors, such as brain-derived neurotrophic factor (BDNF) and vascular endothelial growth factor (VEGF), which promote neuroprotection and neurogenesis [31, 38]. These factors help to protect neurons from apoptosis and stimulate the proliferation of neural progenitor cells, enhancing brain repair.

4.1.2 Angiogenesis

VEGF and other angiogenic factors secreted by MSCs promote the formation of new blood vessels in the ischemic brain, improving blood flow and nutrient delivery to damaged tissues [39]. This process is crucial for the recovery of the ischemic penumbra, the area surrounding the core of the stroke where cells are at risk but not yet dead.

4.1.3 Immunomodulation

MSCs modulate the immune response by secreting anti-inflammatory cytokines and reducing the levels of pro-inflammatory cytokines. This action helps to create a more favorable environment for brain repair by reducing inflammation-induced secondary damage [40, 41].

Preclinical studies have demonstrated the efficacy of MSCs in animal models of stroke. For instance, intravenous administration of MSCs in rodent models has shown significant improvements in functional recovery, reduced infarct size, and enhanced neurogenesis [20]. Clinical trials are ongoing to evaluate the safety and efficacy of MSCs in human stroke patients, with some early-phase trials reporting encouraging results.

4.2 Neural stem cells

4.2.1 Direct cell replacement

NSCs have the potential to replace lost neurons and glial cells in the ischemic brain. By differentiating into functional neural cells, NSCs can integrate into existing neural circuits, restoring lost functions [42].

4.2.2 Paracrine effects

Like MSCs, NSCs secrete neurotrophic factors and cytokines that promote neuroprotection, neurogenesis, and angiogenesis. These paracrine effects help to create a supportive microenvironment for brain repair [14, 42, 43].

4.2.3 Modulation of inflammation

NSCs can modulate the immune response by interacting with microglia and astrocytes, reducing the production of pro-inflammatory cytokines, and promoting the release of anti-inflammatory factors [42].

Preclinical studies have shown that transplantation of NSCs into animal models of stroke leads to improved functional outcomes, increased neurogenesis, and reduced infarct size [42]. Clinical trials investigating NSC therapy in stroke patients are in the initial stages, but preliminary results indicate potential benefits and warrant further investigation [43].

4.3 Induced pluripotent stem cells

4.3.1 Differentiation into neural lineages

iPSCs can be directed to differentiate into various neural cell types, including neurons, astrocytes, and oligodendrocytes [44]. These differentiated cells can potentially replace the lost or damaged cells in the ischemic brain.

4.3.2 Neuroprotection and neurogenesis

iPSCs and their derivatives can secrete neurotrophic factors that support neuroprotection and stimulate endogenous neurogenesis [45–47]. This helps to enhance the brain's intrinsic repair mechanisms.

4.3.3 Angiogenesis and immunomodulation

iPSC-derived cells can promote angiogenesis and modulate the immune response, creating a favorable environment for brain repair and recovery [27, 45].

Preclinical studies have demonstrated that iPSCs can improve functional outcomes in animal models of stroke. Transplantation of iPSC-derived neural progenitors has been shown to enhance neurogenesis, reduce infarct size, and improve behavioral recovery [45–47]. Clinical translation of iPSC therapy is still in the initial stages, with ongoing research focusing on optimizing differentiation protocols, ensuring safety, and minimizing the risk of tumorigenicity [44].

In conclusion, mesenchymal stem cells, neural stem cells, and induced pluripotent stem cells each offer unique advantages and mechanisms for promoting brain repair and functional recovery following stroke [48]. While preclinical studies provide convincing evidence for their potential efficacy [49], further research is needed to address the challenges and translate these therapies into clinical practice.

5. Conclusions

In this chapter, we explored the potential of stem cell therapy as a transformative treatment strategy for stroke, a leading cause of disability worldwide. Several types of stem cells, including mesenchymal stem cells (MSCs), neural stem cells (NSCs), and induced pluripotent stem cells (iPSCs), were discussed in detail, highlighting their unique characteristics and mechanisms of action. The chapter delved into the processes of neurogenesis, angiogenesis, synaptic plasticity, and the modulation of inflammation and apoptosis, all of which are critical for brain repair and functional recovery post-stroke.

MSCs, with their immunomodulatory properties and ability to secrete neurotrophic and angiogenic factors, offer significant potential in enhancing neuroprotection and promoting neurogenesis and angiogenesis [31, 38–41, 50]. NSCs contribute to direct cell replacement and exhibit paracrine effects that support neuroprotection and neurogenesis [14, 42, 43], while iPSCs provide a versatile source for generating various neural cell types and secreting factors that bolster neuroprotection and endogenous neurogenesis [27, 44–47].

Despite the promising results from preclinical studies and early-phase clinical trials, several challenges remain. These include optimizing delivery methods, ensuring long-term safety, overcoming immune rejection, and addressing ethical considerations. Continued research and well-designed clinical trials are essential to address these challenges and translate the potential of stem cell therapy into routine clinical practice.

In summary, stem cell therapy represents a multifaceted approach to enhancing recovery after stroke by promoting brain repair mechanisms and modulating detrimental processes such as inflammation and apoptosis. With further advancements, stem cell therapy holds the promise of improving functional recovery and the quality of life for stroke patients.

Acknowledgements

The author would like to express his gratitude to the individuals and institutions that contributed to this work. Special thanks to the research team at the University of Pittsburgh and Duquesne University for their invaluable contributions, including the laboratory staff and clinical collaborators who played a crucial role in conducting experiments and collecting data, especially Dr. Wenting Zhang and Dr. Rehana K Leak for their critical insights and contributions that significantly enhanced the quality of this research. The author also extends his appreciation to colleagues in the Department of Neurology and the Department of Orthopedic Surgery for their support and guidance throughout this study.

The author is particularly grateful to the University of Pittsburgh Medical Center Immune Transplant and Therapy Center for their financial support of the mesenchymal stem cell (MSC) research, provided through a grant awarded to Drs. Jun Chen and Lawrence R Wechsler. Additionally, the author acknowledges the patients and their families who participated in the clinical trials; their cooperation was instrumental in advancing our research.

The authors acknowledge the use of ChatGPT for the language polishing of the manuscript.

Conflict of interest

The authors declare no conflict of interest.

Notes

The author would like to extend sincere thanks to the administrative staff at the University of Pittsburgh, especially Pat Strickler for her excellent logistical and administrative support.

Author details

Hongjian Pu
University of Pittsburgh, Pittsburgh, USA

*Address all correspondence to: hop7@pitt.edu

References

[1] GBD 2021 Forecasting Collaborators. Burden of disease scenarios for 204 countries and territories, 2022-2050. A forecasting analysis for the global burden of disease study 2021. Lancet. 2024;**403**(10440):2204-2256. DOI: 10.1016/S0140-6736(24)00685-8

[2] Hurd MD, Goel I, Sakai Y, Teramura Y. Current status of ischemic stroke treatment: From thrombolysis to potential regenerative medicine. Regenerative Therapy. 2021;**18**:408-417. DOI: 10.1016/j.reth.2021.09.009

[3] Ya J, Pellumbaj J, Hashmat A, Bayraktutan U. The role of stem cells as therapeutics for ischaemic stroke. Cells. 2024;**13**(2):112. DOI: 10.3390/cells13020112

[4] Bhattacharya AK, Banik KK. Stem cells in stroke--appraisal of literature. Journal of the Indian Medical Association. 2009;**107**(6):400, 402

[5] Ding DC, Shyu WC, Lin SZ. Mesenchymal stem cells. Cell Transplantation. 2011;**20**(1):5-14. DOI: 10.3727/096368910X

[6] Ming J, Liao Y, Song W, Wang Z, Cui J, He L, et al. Role of intracranial bone marrow mesenchymal stem cells in stroke recovery: A focus on post-stroke inflammation and mitochondrial transfer. Brain Research. 2024;**1837**:148964. DOI: 10.1016/j.brainres.2024.148964

[7] Moñivas Gallego E, Zurita CM. Mesenchymal stem cell therapy in ischemic stroke trials. A systematic review. Regenerative Therapy. 2024;**27**:301-306. DOI: 10.1016/j.reth.2024.03.026

[8] Zhao X, Moore DL. Neural stem cells: Developmental mechanisms and disease modeling. Cell and Tissue Research. 2018;**371**(1):1-6. DOI: 10.1007/s00441-017-2738-1

[9] Ye L, Swingen C, Zhang J. Induced pluripotent stem cells and their potential for basic and clinical sciences. Current Cardiology Reviews. 2013;**9**(1):63-72. DOI: 10.2174/157340313805076278

[10] Mercier L, Audet T, Hébert R, Rochette A, Dubois MF. Impact of motor, cognitive, and perceptual disorders on ability to perform activities of daily living after stroke. Stroke. 2001;**32**(11):2602-2608. DOI: 10.1161/hs1101.098154

[11] Detante O, Legris L, Moisan A, Rome C. Cell therapy and functional recovery of stroke. Neuroscience. 2023;**550**:79-88. DOI: 10.1016/j.neuroscience.2023.11.027

[12] Ramos-Cabrer P, Justicia C, Wiedermann D, Hoehn M. Stem cell mediation of functional recovery after stroke in the rat. PLoS One. 2010;**5**(9):e12779. DOI: 10.1371/journal.pone.0012779

[13] Shyu WC, Lin SZ, Yang HI, Tzeng YS, Pang CY, Yen PS, et al. Functional recovery of stroke rats induced by granulocyte colony-stimulating factor-stimulated stem cells. Circulation. 2004;**110**(13):1847-1854. DOI: 10.1161/01.CIR.0000142616.07367.66

[14] Jiao Y, Liu YW, Chen WG, Liu J. Neuroregeneration and functional recovery after stroke: Advancing neural stem cell therapy toward clinical application. Neural Regeneration Research. 2021;**16**(1):80-92. DOI: 10.4103/1673-5374.286955

[15] Zents K, Copray S. The therapeutic potential of induced pluripotent stem cells after stroke: Evidence from rodent models. Current Stem Cell Research & Therapy. 2016;**11**(2):166-174. DOI: 10.2174/1574888 x10666150728121324

[16] Otero-Ortega L, Gutiérrez-Fernández M, Díez-Tejedor E. Recovery after stroke: New insight to promote brain plasticity. Frontiers in Neurology. 2021;**12**:768958. DOI: 10.3389/fneur.2021.768958

[17] Andres RH, Horie N, Slikker W, Keren-Gill H, Zhan K, Sun G, et al. Human neural stem cells enhance structural plasticity and axonal transport in the ischemic brain. Brain. 2011;**134**(Pt 6):1777-1789. DOI: 10.1093/brain/awr094

[18] Gutiérrez-Fernández M, Rodríguez-Frutos B, Ramos-Cejudo J, Otero-Ortega L, Fuentes B, Díez-Tejedor E. Stem cells for brain repair and recovery after stroke. Expert Opinion on Biological Therapy. 2013;**13**(11):1479-1483. DOI: 10.1517/14712598.2013.824420

[19] Tan N, Xin W, Huang M, Mao Y. Mesenchymal stem cell therapy for ischemic stroke: Novel insight into the crosstalk with immune cells. Frontiers in Neurology. 2022;**13**:1048113. DOI: 10.3389/fneur.2022.1048113

[20] Asgari Taei A, Nasoohi S, Hassanzadeh G, Kadivar M, Dargahi L, Farahmandfar M. Enhancement of angiogenesis and neurogenesis by intracerebroventricular injection of secretome from human embryonic stem cell-derived mesenchymal stem cells in ischemic stroke model. Biomedicine & Pharmacotherapy. 2021;**140**:111709. DOI: 10.1016/j.biopha.2021.111709

[21] Bian X, Ma K, Zhang C, Fu X. Therapeutic angiogenesis using stem cell-derived extracellular vesicles: An emerging approach for treatment of ischemic diseases. Stem Cell Research & Therapy. 2019;**10**(1):158. DOI: 10.1186/s13287-019-1276-z

[22] Anthony S, Cabantan D, Monsour M, Borlongan CV. Neuroinflammation, stem cells, and stroke. Stroke. 2022;**53**(5):1460-1472. DOI: 10.1161/STROKEAHA.121.036948

[23] Dabrowska S, Andrzejewska A, Lukomska B, Janowski M. Neuroinflammation as a target for treatment of stroke using mesenchymal stem cells and extracellular vesicles. Journal of Neuroinflammation. 2019;**16**(1):178. DOI: 10.1186/s12974-019-1571-8

[24] Stonesifer C, Corey S, Ghanekar S, Diamandis Z, Acosta SA, Borlongan CV. Stem cell therapy for abrogating stroke-induced neuroinflammation and relevant secondary cell death mechanisms. Progress in Neurobiology. 2017;**158**:94-131. DOI: 10.1016/j.pneurobio.2017.07.004

[25] Zhou L, Zhu H, Bai X, Huang J, Chen Y, Wen J, et al. Potential mechanisms and therapeutic targets of mesenchymal stem cell transplantation for ischemic stroke. Stem Cell Research & Therapy. 2022;**13**(1):195. DOI: 10.1186/s13287-022-02876-2

[26] Lee ST, Chu K, Jung KH, Kim SJ, Kim DH, Kang KM, et al. Anti-inflammatory mechanism of intravascular neural stem cell transplantation in haemorrhagic stroke. Brain. 2008;**131**(Pt 3):616-629. DOI: 10.1093/brain/awm306

[27] Spellicy SE, Hess DC. The immunomodulatory capacity of induced pluripotent stem cells in the post-stroke environment. Frontiers in Cell and

Development Biology. 2021;**9**:647415. DOI: 10.3389/fcell.2021.647415

[28] Aderinto N, AbdulBasit MO, Olatunji G, Adejumo T. Exploring the transformative influence of neuroplasticity on stroke rehabilitation: A narrative review of current evidence. Annals of Medicine and Surgery (Lond). 2023;**85**(9):4425-4432. DOI: 10.1097/MS9.0000000000001137

[29] Shyu WC, Lin SZ, Chiang MF, Su CY, Li H. Intracerebral peripheral blood stem cell ($CD34^+$) implantation induces neuroplasticity by enhancing beta1 integrin-mediated angiogenesis in chronic stroke rats. The Journal of Neuroscience. 2006;**26**(13):3444-3453. DOI: 10.1523/JNEUROSCI.5165-05.2006

[30] Ding DC, Shyu WC, Chiang MF, Lin SZ, Chang YC, Wang HJ, et al. Enhancement of neuroplasticity through upregulation of beta1-integrin in human umbilical cord-derived stromal cell implanted stroke model. Neurobiology of Disease. 2007;**27**(3):339-353. DOI: 10.1016/j.nbd.2007.06.010

[31] Wang C, Tian C, Cai D, Jiang H, Zhang W, Liu S, et al. BDNF-overexpressing MSCs delivered by hydrogel in acute ischemic stroke treatment. Annals of Translational Medicine. 2022;**10**(24):1393. DOI: 10.21037/atm-22-5921

[32] Chen L, Zhang G, Gu Y, Guo X. Meta-analysis and systematic review of neural stem cells therapy for experimental ischemia stroke in preclinical studies. Scientific Reports. 2016;**6**:32291. DOI: 10.1038/srep32291

[33] Sammali E, Alia C, Vegliante G, Colombo V, Giordano N, Pischiutta F, et al. Intravenous infusion of human bone marrow mesenchymal stromal cells promotes functional recovery and neuroplasticity after ischemic stroke in mice. Scientific Reports. 2017;**7**(1):6962. DOI: 10.1038/s41598-017-07274-w

[34] Bhasin A, Srivastava MV, Mohanty S, Bhatia R, Kumaran SS, Bose S. Stem cell therapy: A clinical trial of stroke. Clinical Neurology and Neurosurgery. 2013;**115**(7):1003-1008. DOI: 10.1016/j.clineuro.2012.10.015

[35] Bang OY. Clinical trials of adult stem cell therapy in patients with ischemic stroke. Journal of Clinical Neurology. 2016;**12**(1):14-20. DOI: 10.3988/jcn.2016.12.1.14

[36] Kawabori M, Shichinohe H, Kuroda S, Houkin K. Clinical trials of stem cell therapy for cerebral ischemic stroke. International Journal of Molecular Sciences. 2020;**21**(19):7380. DOI: 10.3390/ijms21197380

[37] Houkin K, Osanai T, Uchiyama S, Minematsu K, Taguchi A, Maruichi K, et al. Allogeneic stem cell therapy for acute ischemic stroke: The phase 2/3 treasure randomized clinical trial. JAMA Neurology. 2024;**81**(2):154-162. DOI: 10.1001/jamaneurol.2023.5200

[38] Liu C, Yang ZX, Zhou SQ, Ding D, Hu YT, Yang HN, et al. Overexpression of vascular endothelial growth factor enhances the neuroprotective effects of bone marrow mesenchymal stem cell transplantation in ischemic stroke. Neural Regeneration Research. 2023;**18**(6):1286-1292. DOI: 10.4103/1673-5374.358609

[39] Lai T, Li M, Zheng L, Song Y, Xu X, Guo Y, et al. Over-expression of VEGF in marrow stromal cells promotes angiogenesis in rats with cerebral infarction via the synergistic effects of VEGF and Ang-2. Journal of Huazhong University of Science and Technology. Medical Sciences. 2012;**32**(5):724-731. DOI: 10.1007/s11596-012-1025-3

[40] Song N, Scholtemeijer M, Shah K. Mesenchymal stem cell immunomodulation: Mechanisms and therapeutic potential. Trends in Pharmacological Sciences. 2020;**41**(9):653-664. DOI: 10.1016/j.tips.2020.06.009

[41] Shan XQ, Luo YY, Chang J, Song JJ, Hao N, Zhao L. Immunomodulation: The next target of mesenchymal stem cell-derived exosomes in the context of ischemic stroke. World Journal of Stem Cells. 2023;**15**(3):52-70. DOI: 10.4252/wjsc. v15.i3.52

[42] Baker EW, Kinder HA, West FD. Neural stem cell therapy for stroke: A multimechanistic approach to restoring neurological function. Brain and Behavior: A Cognitive Neuroscience Perspective. 2019;**9**(3):e01214. DOI: 10.1002/brb3.1214

[43] Hamblin MH, Lee JP. Neural stem cells for early ischemic stroke. International Journal of Molecular Sciences. 2021;**22**(14):7703. DOI: 10.3390/ijms22147703

[44] He JQ, Sussman ES, Steinberg GK. Revisiting stem cell-based clinical trials for ischemic stroke. Frontiers in Aging Neuroscience. 2020;**12**:575990. DOI: 10.3389/fnagi.2020.575990. Erratum in: Front Aging Neurosci. 2022 Mar 10; 14:875925

[45] Duan R, Gao Y, He R, Jing L, Li Y, Gong Z, et al. Induced pluripotent stem cells for ischemic stroke treatment. Frontiers in Neuroscience. 2021;**15**:628663. DOI: 10.3389/fnins.2021.628663

[46] Yuan T, Liao W, Feng NH, Lou YL, Niu X, Zhang AJ, et al. Human induced pluripotent stem cell-derived neural stem cells survive, migrate, differentiate, and improve neurologic function in a rat model of middle cerebral artery occlusion. Stem Cell Research & Therapy. 2013;**4**(3):73. DOI: 10.1186/scrt224

[47] Chang DJ, Lee N, Park IH, Choi C, Jeon I, Kwon J, et al. Therapeutic potential of human induced pluripotent stem cells in experimental stroke. Cell Transplantation. 2013;**22**(8):1427-1440. DOI: 10.3727/096368912X657314

[48] Yamaguchi S, Yoshida M, Horie N, Satoh K, Fukuda Y, Ishizaka S, et al. Stem cell therapy for acute/subacute ischemic stroke with a focus on intraarterial stem cell transplantation: From basic research to clinical trials. Bioengineering (Basel). 2022;**10**(1):33. DOI: 10.3390/bioengineering10010033

[49] Li Z, Dong X, Tian M, Liu C, Wang K, Li L, et al. Stem cell-based therapies for ischemic stroke: A systematic review and meta-analysis of clinical trials. Stem Cell Research & Therapy. 2020;**11**(1):252. DOI: 10.1186/s13287-020-01762-z

[50] Zhang W, Pu H, Hu X, Shi Y, Leak RK, Anne Stetler R, et al. Poststroke intravenous transplantation of human mesenchymal stem cells improves brain repair dynamics and functional outcomes in aged mice. Stroke. 2023;**54**(4):1088-1098. DOI: 10.1161/STROKEAHA.122.041507

Chapter 22

Stem Cell Use to Treat Dermatological Disorders

Hannah E. Doan, Mohammed S. Inayat and Vincent S. Gallicchio

Abstract

In recent advancements, stem cell therapy has emerged as a promising therapeutic approach. The development of stem cells to treat an ever-increasing list of clinical conditions focusing on restorative and/or regenerative applications has created significant expectations for their future use in medicine. This chapter presents a review of the potential therapeutic applications of stem cells in the field of dermatology. The review covers the historical and scientific origin of stem cells, followed by the laboratory processes involved in regulating stem cell derivation, cultivation, and propagation. Quality control is important in stem cell applications in medicine, starting from procurement to growth. The improvements for the use of media for maintaining optimal growth conditions for these cells is essential for controlling their differentiation. Among their use as dermatological treatments, they are applied as injections, topical applications, or through scaffolds. The growth in recent trials and improvements on stem cell knowledge improves the outlook on stem cell use as dermatological treatments. The review summarizes the therapeutic use of stem cells in dermatological disorders.

Keywords: stem cells, therapy, dermatological disorders, clinical use, regenerative medicine

1. Introduction

Stem cells are unspecialized cells and are the essential building blocks of all organisms. They can differentiate into any specialized cell within an organism [1]. In this capacity, stem cells possess the ability to self-renewal, in addition to differentiating into all cells within tissues and ultimately organ systems [2–4]. Stem cells exist from conception and remain functional through adulthood, with many regulatory factors responsible for their specialization. As stem cells mature, differentiation becomes more limited which is referred to as commitment to a specific lineage. This means a unipotent stem cell is restricted in differentiation compared to a pluripotential stem cell (PSC) that can produce a variety of lineage specific cells. Thus, PSCs are more restricted when compared to a totipotent stem cell (TSC) [5, 6].

TSCs are capable of cell division with the ability to differentiate into mature cells comprising all the physiological systems associated with an intact and complete organism

[6]. TSCs have unlimited potential to fully differentiate. This property allows TSCs to form both embryonic and extra-embryonic structures such as the placenta and the tissues associated with pregnancy [7, 8]. An example of a TSC is the zygote that forms after a sperm fertilizes an egg. TSCs will form a blastocyst which develops the inner cell mass (ICM). The ICM contains a unique population of stem cells known as embryonic stem cells (ESCs). ESCs are capable of remaining pluripotent *in vitro* [9, 10]. ESCs form the three germ layers associated with developmental biology, *i.e.*, ectoderm, mesoderm, and endoderm [10], thus providing the core foundation of an organism through each germ layer by providing all the anatomical and physiological systems of the organism [11].

Pluripotential stem cells (PSCs) form structures associated with only the germ layers [11]. Another example of stem cells possessing pluripotency was achieved following the reprogramming capability to produce induced pluripotent stem cells (iPSCs) [12]. iPSC pluripotency is a continuum, starting from totipotent cells to cells possessing less potency as in multi-, oligo- or unipotent cells. The independence of iPSCs allows for using improved methods that are more promising for therapeutic stem cell use now and for future applications as defined in regenerative medicine [13].

Within their respective lineages, multipotent stem cells can generate more specialized cells. It differentiates blood cell development to form a variety of diverse cells such as erythrocytes, leukocytes, and thrombocytes [14]. A myeloid stem cell is an example where a stem cell may differentiate into different types of leukocytes, *e.g.*, white blood cells such as granulocytes or monocytes, but never erythrocytes or platelets [15].

2. Stem cell biology

As mentioned above, during embryogenesis, stem cells form aggregates referred to as germ layers [16]. Once hESCs differentiate into a specific germ layer, they become multipotent stem cells and can only differentiate according to their respective layer. Pluripotent stem cells are present throughout the life of any organism existing as undifferentiated cells [17]. Regulatory signals influence stem cell specialization to create specific tissues that are produced via physical contact between cells through the microenvironment/stroma or as stimulators in the form of cytokines, interleukins, and/or tissue factors secreted by surrounding tissues. These factors from internal sources are controlled via the presence of the genome, *i.e.*, genes, thus DNA acting through transcription translation reactions [11]. Stem cells provide a mechanism designed to function as the body's internal repair system. For as long as an organism remains functional, its stem cells will continue to provide differentiation pathways to replace more mature cell lineages. This is the "repair" and "replenishment" aspect of stem cell vitality [11, 18].

3. Stem cell functional development

3.1 Whole-body development

The growth and development of an organism depends on the presence of stem cells. Overall, somatic stem cells such as ESCs can be distinguished based upon their characteristic lineage line of development. ESCs can be derived without isolating them from the inner cell mass; however, their growth potential is limited [11]. ESCs can be propagated *in vitro* using tissue culture conditions indefinitely without restriction if their growth requirements are maintained [19, 20]. ESCs can be propagated in culture with appropriate culture medium containing essential nutrients [19]. Passage of ESCs is an adequate method of sub-culturing stem cells to propagate their numbers over time. Because ESCs are totipotent, they can differentiate into every cell type required in any organ cell system [21]. However, because totipotent stem cells demonstrate immortality, ethical restrictions restrict the procurement of these cells. The origin of these totipotent stem cells is from the ICM of the blastocyst present in embryos. Thus, the procedure to obtain them destroys the viability of that embryo from further development. However, most ESCs are derived from fertilized eggs in an *in vitro* clinic rather than from eggs harvested from pregnant women [22].

Among the many stem cell types that exist are as follows:

- Hematopoietic stem cells have the potential to differentiate into many types of blood cells, e.g., erythrocytes, leukocytes, and thrombocytes.
- Mesenchymal stem cells are found in multiple types of tissues. They can differentiate into multiple lineages such as bone, adipose, vascular, and cartilage tissue. They can be harvested from sources including but not limited to the umbilical cord, bone marrow, and endometrial polyps [23].
- Neural stem cells develop into glial or neuronal cells such as nerve cells, oligodendrocytes, and astrocytes. These cells have been used in treatments regarding Parkinson's disease through transplants [24].
- Skin stem cells (SSCs) consist of several types that are separated into their own niches including hair follicle stem cells, melanocyte stem cells, and dermal stem cells. SSCs have greater potential to be used for stem cell therapies and treatments since these cells can differentiate into more cell lineages [25].

Human ESCs are involved in whole-body development and can eventually become pluripotent, multipotent, and unipotent stem cells. Compared to adult somatic stem cells, they also have a quicker proliferation time and greater range of differentiation causing them to be more ideal and preferred in therapies [26].

Stem cells can also be taken from the placenta. Placental fetal mesenchymal stem cells can differentiate into a wide variety of cells and are abundant, not requiring invasive procedures to procure. They are not surrounded with ethical issues that ESCs have since the placenta is usually considered medical waste after birth, making it favorable for use as treatment. They can produce ectodermal, endodermal, and mesodermal lineages *in vitro* and contain the same cell markers as ESCs, making them very similar. Placental stem cells are pluripotent and have low immunogenicity which allows them to be ideal for therapies and treatments [27].

4. Turning point in stem cell therapy

4.1 Induced pluripotential stem cells (IPSCs)

Differentiation was thought to be restricted and non-reversible. However, after several major experiments through cloning, even differentiated cells can be reprogrammed or "induced" to be pluripotent. Two major cloning-related discoveries were made in 1962 and 1987. The first was done by John Gurdon who cloned frogs through the process of somatic nuclear cell transfer (SNCT) into an enucleated frog egg [28]. This showed that the nucleus of a specialized somatic cell could be reverted and develop cells that could eventually produce an entirely new organism [29]. The specialized somatic cell became pluripotent which, before this experiment, was thought to be impossible [30, 31]. This technique was famously used successfully in the cloning of Dolly, the sheep [28]. The 1987 experiment focused on gene expression. The forced expression of one gene, known as myogenic differentiation 1 (Myod1), could cause fibroblasts to turn into myoblasts [32]. This was another example of transforming cells, but this was done through programming the cell in the DNA.

These discoveries provided the turning point in stem cell research by advancing the therapeutic application of stem cells when a Japanese team of scientists showed adult multipotent stem cells could be reverted into a pluripotent state. These cells functioned like ESCs but did not need to be acquired from embryos. This discovery created a process to avoid endangering the life of a fetus to obtain ESCs. The determining factor in the process using murine fibroblasts was incorporating a retrovirus-mediated transduction system containing four transcription factors found in ESCs known as Oct-3/4, Sox2, KLF4, and c-Myc [17]. These factors induced the fibroblasts to become pluripotent. The newly formed "reprogrammed" stem cells were named "induced pluripotent stem cells (iPSCs)". A later study succeeded using human cells [33]. This technological breakthrough created a new line of research in stem cell biology that coincided with the generation of iPSC cell lines. Importantly, as mentioned, iPSCs can be made biocompatible with any patient, thus dramatically improving the therapeutic potential of this newly created cellular therapy [13]. ESCs are still the only naturally occurring pluripotent cells, but from these experiments, terminally differentiated cells can be induced into pluripotency to become iPSCs. Still, reprogramming cells comes with risks to cellular development due to histone alteration. However, an experiment was done by sequencing DNA from murine iPSCs and confirmed that although mutations were introduced, reprogramming cells could create iPSCs that were not seriously genetically affected or produce ill-functioning cells [11, 34].

4.2 Quality control of iPSCs

As these cells are manufactured, controlling the quality of iPSC lines is necessary for use as treatments. Ways that they are controlled for their quality are as follows (**Table 1**) [35]:

4.3 Sourcing iPSCs

A common source for iPSCs includes fibroblasts. Especially in treatments, taking the patient's own fibroblasts for the treatment has shown to be beneficial as the autologous cells do not risk being rejected. However, at first, they were the only source that could be used, and obtaining these cells required a biopsy. Thus, further research was conducted to enhance the technique's efficiency. Other cell types have also been reprogrammed, but fibroblasts are still preferred since their stimulation can be fast and controlled [36, 37].

Stem cells are only potentially useful if they can be differentiated into specific lineages. If not, they can form a teratoma *in vivo.* However, this condition can be regulated; therefore, if the process can be controlled, it allows clinicians and researchers to improve their therapeutic use when using specific signaling pathways for differentiation. In regenerative medicine, it is important to ensure that these cells will

Characterization	To better identify unusual or abnormal cells, cells should be characterized. This can be done with immunotyping through flow cytometry.
Genetic fidelity and stability	Changes in iPSC genes pose the threat of harming the patient. To prevent this, residual vector testing is done to identify cells that may possess plasmids. Other tests that can be done include karyotyping, single nucleotide polymorphism arrays, and genetic and disease marker analysis. Though not all of them need to be applied to the cell lines, they provide information on possible genetic abnormalities that can occur in these cell lines.
Microbiological sterility	Even though it is not possible to sterilize living cells, bacterial, fungal, viral, mycoplasma, and endotoxin testing must be done to limit and control other microbiological growth.
Identity	Because there is a risk of cross contamination with other cell lines, short tandem repeat analysis or STR genotyping of source cells alongside iPSC cell lines can be used to confirm the source of the cells.
Potency	Measuring biological activity of the iPSC lines is done to ensure efficacy and reduce the risk of using poorly functioning cells for treatment. Phenotypic pluripotency assays are used to ensure the cell lines can differentiate by growing the cells into endoderm, mesoderm, and ectoderm cells.
Viability	After collecting cell samples, they should be tested with a cell viability test to ensure growth and patient dosage will be consistent.

Table 1.
Different ways that stem cells can be verified and tested during growth to ensure their quality and viability.

then differentiate in a timely and efficient manner. Directed differentiation exists to push the ESCs to differentiate. As cells develop, they send out signals within their surroundings [38]. Messages from the extracellular environment can also control the differentiation of stem cells which has been shown in *in vitro* cultures [39]. This can be done easily in *in vitro* cultures by controlling the conditions in culture. However, replicating such environments *in vivo*, has been challenging, requiring strict culture conditions [11].

For hESC treatments to be used on patients, the therapies must be "culture-free," meaning the stem cells are not contaminated with any feeder or animal cell components [40]. The FDA requires this pertaining to procurement and storage of any type of stem cells contemplated for human use [41]. A difficulty in procuring these treatments is that great amounts of these cells used for treatment must be cultivated in the absence of feeder cells.

Directed differentiation protocols replicate the development of the ICM during embryogenesis. Pluripotent stem cells differentiate into derived progenitors from each of the three germ layers, just as is observed *in vivo*. Specific molecules act as growth factors to induce stem cells to become specific progenitor cells eventually to develop into a specific cell type. Growth factors function as important regulatory molecules that affect germ layer development *in vivo*; examples include bone morphogenic proteins (BMP) [42, 43], fibroblast growth factors (FGFs) [44], transcription factors of the Wnt family [45], or transforming growth factors-beta (TGFβ). How each factor influences germ cell differentiation is unclear and research is ongoing.

The concentration levels and duration of action of a targeted signaling molecule such as a growth factor produces a variety of outcomes. However, the high cost of recombinant molecules currently restricts their routine use in therapy limiting their clinical application. A more promising approach is to focus on using small molecules, thereby activating or deactivating specific signaling pathways [46]. These methods are effective in improving reprogramming efficiency by helping to generate cells that are compatible with the target tissue type. Also, they offer a more cost-effective and non-immunogenic therapy method [47]. Endogenously generated small molecules, *e.g.,* retinoic acid is effective for patterning nervous system development *in vivo.* It functions effectively in embryonic development where it is used *in vitro* in culture systems to induce the differentiation of somatic cells [48, 49]. These cells can also induce retinal cell formation when hESCs are used [50]. Through the control of biochemical signals and the environment as important factors can be essential to achieve optimal hESC differentiation when culturing stem cells.

4.4 Stem cell manufacturing and ethical standards

Culture systems have been regulated by multiple agencies around the world including the Food and Drug Administration (FDA) and the European Medicine Agency (EMA). Initially, animal-derived products were utilized, however, that introduced possible animal pathogens. Some stem cell lines derived from embryos and human feeder cell lines have been established which include stem cell-derived cardiac progenitors and mesenchymal stem cells. Xeno-free culture systems also include the development of human foreskin fibroblasts (HFFs) [11, 51–53].

4.5 Stem cell use in medicine

Stem cells hold immense promise as an important therapeutic option for the future of medicine. Beyond their crucial role in regenerative medicine, stem cell research has demonstrated their intricate processes when involved in growth development. In stem cells, DNA is loosely organized, allowing genes to remain active. Differentiated cells differ in that these cells deactivate certain genes and activate others that are essential to the signals that the cell receives. This process is reversible, demonstrating that pluripotency can be induced through specific gene modifications. Several core transcription factors including Oct3/4, (SRY)-box 2, and Nanog genes have been found to keep these cells pluripotent [17, 54]. Nuclear transcription factors Oct3/4 and Sox2 are crucial for producing iPSCs [54].

Presently, various therapies using stem cells are offered as treatments for conditions like spinal cord injuries, heart failure, retinal and macular degeneration, tendon ruptures, and type 1 diabetes [52, 55–58]. Stem cell research improves our understanding of stem cell physiology, potentially leading to new treatments for presently untreatable diseases. Many of which are dermatological disorders which were previously thought to have no good solution. This chapter focuses on the application of stem cells treating various dermatological disorders and compliments recent reviews on the same topic [11, 59].

5. Therapeutic use of stem cells for dermatological conditions

5.1 Alopecia

Stem cell therapy has not been actively used as a solution for restoring hair growth, but current results are promising. One study used harvested autologous adipose-derived stromal vascular cells through injected into the scalp of 20 patients with alopecia areata (AA) [60]. At three and six months of follow-up, all patients produced statistically significant hair growth. Adipose-derived stem cell conditioned medium (ADS-CM) contains growth factors essential for hair follicle regrowth such as basic fibroblast growth factor, hepatocyte growth factor, platelet-derived growth factor, vascular endothelial growth factor, and transforming growth factor-beta (TGF-β) [61]. Another study isolated human adult stem cells by centrifuging human hair follicles obtained through punch biopsy and injected them into the scalps of 11 androgenetic alopecia (AGA) patients resulting in an increase in hair density and count compared to baseline and placebo [62]. In a larger study with 140 AGA patients, autologous cellular micrografts containing HFSCs were used as a treatment. Within one session, over two-thirds of the patients showed positive results while there was significant increase in their regrowth and thickness [63, 64].

A study randomly assigned 40 patients (20 with AGA and 20 with AA) to receive either autologous bone marrow-derived mononuclear cells or autologous follicular stem cell injections into the scalp, found significant improvement in hair loss with no significant difference between the two preparations [65]. An investigation introduced a novel stem cell method, termed "stem cell educator therapy" in which patient's mononuclear cells are separated from whole blood and allowed to interact with human cord blood–derived multipotent stem cells, thus "educating" these stem cells after returning them to patients [61]. In nine patients with severe AA, all but one experienced improved hair regrowth of varying degrees. Two patients (one with alopecia totalis and one with patchy AA) experienced complete hair regrowth at 12 weeks without relapse after two years. A combination of platelet-rich plasma and stem cell technology also showed promising results [61].

5.2 Atopic dermatitis

Numerous murine studies have demonstrated the progression of allergies in atopic dermatitis (AD) can be inhibited by using umbilical cord blood mesenchymal stem cells (UCB-MSCs), bone marrow mesenchymal stem cells (BM-MSCs), or adipose-derived mesenchymal stem cells (AD-MSCs) [66–69]. It is important to consider the type of stem cell used, the number of cells transplanted, the preconditioning of the cell preparation, the therapy's relevant targets, and the route and frequency of administration. One example highlighting the complexity of stem cell-based therapy was shown in a study where human UCB-MSCs were pre-treated with mast cell granules [68]. This pre-treatment method enhanced their therapeutic effectiveness, as evidenced by the reduced signs of AD in a NC/Nga mouse model. It was found that hUCB-MSCs primed with mast cell granules were more effective in suppressing the activation of mast cells and B lymphocytes compared to naïve MSCs, both *in vitro* and *in vivo* [70].

Despite promising results from murine studies in AD, only a few clinical trials have been conducted. In one study, a single subcutaneous administration of hUCB-MSCs was given to 34 adult participants with moderate-to-severe AD [66]. The improvement in AD symptoms was measured using the eczema area and severity index (EASI) score. Treatments for both low and high doses of hUCB-MSCs showed symptom improvement. In the higher dose group, six out of 11 subjects experienced a 50% reduction in EASI score, with no reported side effects. Additionally, typical biomarkers of AD, such as serum IgE levels and the number of eosinophils, decreased after treatment.

A later clinical trial had the injection of clonal mesenchymal stem cells (MSCs) into five patients with atopic dermatitis (AD) who had not responded to conventional treatments [71]. Patients received either one or two cycles of MSC treatment. Effective treatment was evaluated using cytokine biomarkers (CCL-17, CCL-22, IL-13, IL-18, IL-22, and IgE) and EASI scores. Results showed four out of five patients achieved more than a 50% reduction in EASI scores after one treatment cycle. Additionally, significant decreases in IL-13 and IL-22 levels were observed with other biomarkers showing decreasing trends during the studies.

In a more recent phase 1 clinical trial published in 2024, 20 subjects were treated intravenously with human clonal MSCs, given a low dose of cells in Arm 1 and a higher dose in Arm 2. There was an overall improvement for both arms, and the difference in dosage did not make a statistically significant effect. A phase 2 trial proceeded and was randomized, double-blind, and placebo controlled. In this, 72 subjects were tested. The half given the treatment were given the high dosage of hcMSCs originally tested in phase 1. Compared to the placebo group, the treated group had a statistically significant improvement response [72]. These findings suggest MSC administration might help normalize the immune system in AD patients. However, further studies are needed to understand the long-term mechanisms and effects of MSC treatment in this context.

5.3 Dermatomyositis

Dermatomyositis remains a mystery with its exact etiology still unknown. Research using stem cells to treat the disease is limited with few studies and case reports available. One report detailed successful autologous stem cell transplants for two patients with juvenile dermatomyositis who had not responded to initial treatments [73]. In the first patient, the procedure involved transferring CD3/CD19-depleted mobilized peripheral blood mononuclear cells (PBMCs), which included 7.5×10^6/kg $CD34^+$ stem cells and 2.9×10^4/kg T cells. Following a 26-month follow-up period, significant improvements were observed. The Childhood Myositis Assessment Scale (CMAS) score increased from 6 to 51and the manual muscle testing (MMT) score rose from 61 to 150. These results demonstrated a substantial improvement in symptoms with the patient regaining the ability to walk and showing significant reductions in inflammatory reactions after the autologous stem cell transplant.

In the second patient, a similar response was observed. The patient was treated with CD3/CD19-depleted autologous PBMC graft (7.51×10^6/kg CD34+; 1.6×10^4/kg CD3+). After three months of treatment, the patient had less muscle pain and contractures, and she began also regained the ability to walk [73].

An uncontrolled study in which 10 patients received allogenic mesenchymal stem cell therapy was reported where one or two MSC infusions were given to patients depending on whether they had disease recurrence within a short time after initial treatment. Out of the 10 patients, eight showed significant clinical improvement, with their symptoms improving after MSC therapy [74]. However, further research is required to evaluate the long-term effects of MSC treatment in patients with dermatomyositis.

5.4 Epidermolysis bullosa

Epidermolysis bullosa (EB) is a genetic condition that currently has no treatment, but stem cell therapy is one cell-based therapy under investigation that may be able to correct the skin and its underlying genetic component. Autologous or allogenic stem cells are options that can be used, with mesenchymal stem cell therapy showing potential; therefore, they may be more useful in alleviating some symptoms when tested in additional studies.

One study followed two patients with severe generalized recessive dystrophic epidermolysis bullosa (EB) treated with intradermal administration of allogenic mesenchymal stem cells from bone marrow showed complete healing of ulcers around the treated site by 12 weeks [75]. Type VII collagen was detected along the basement membrane zone and the dermal-epidermal junction was continuous in the treated site 1 week after treatment. Unfortunately, the clinical effect lasted for only 4 months in both patients.

In the case of junctional EB treated with primary cultured keratinocytes, it showed normal morphology and the absence of spontaneous and induced blisters or erosions at 21 months of follow-up [76]. Studies using BMSCs to treat recessive dystrophic EB have also shown promise [77, 78]. One study investigated 10 recessive dystrophic EB children treated with intravenous allogeneic bone marrow-derived mesenchymal stem cells and found that the procedure was well tolerated with minimal side effects over the nine-month period [79]. However, skin biopsies performed at the two-month time point showed no increase in type VII collagen and no new anchoring fibrils. While the initial clinical improvement was favorable, it was not maintained over time due to insufficient production of durable proteins like collagen and laminins. The current evidence for stem cell therapy in treating EB is limited because few patients have been treated. This underscores the need for additional research to assess the therapy's effectiveness and the balance of its risks and benefits [80].

5.5 Psoriasis

Despite significant progress in understanding psoriasis pathogenesis in recent years, it remains unclear what is the exact etiology. Current research suggests that dysfunction in certain types of stem cells might be a primary cause of the inflammatory response dysregulation in psoriasis [81]. This hypothesis came after observing long-term remission in psoriasis patients who underwent hematopoietic stem cell therapy [82, 83]. Conversely, there have been reports of acquired psoriasis in patients who received bone marrow transplants from donors with psoriasis, indicating a significant role of hematopoietic stem cells in disease pathogenesis [84, 85]. MSCs have also shown success in treatment likely due to their engraftment, paracrine, or immunomodulatory effects [86]. However, the availability of cost-effective and safe alternatives limits the use of stem cell transplantation as a practical option for treating psoriasis.

5.6 Scleromyxedema

Scleromyxedema is a chronic fibro-mucinous disorder that can result in respiratory complications. A study conducted on five patients who underwent high-dose chemotherapy followed by stem cell rescue led to durable remission in most cases, although it did not cure the disease [87]. Another study showed scleromyxedema was successfully treated with chemotherapy and autologous stem cell transplantation [88]. The patient achieved complete recovery within six months and remained in remission for 3 years post-transplantation. In a 2022 report, a male patient underwent an autologous hematopoietic stem cell (HSC) transplant after previous therapies failed to improve his symptoms. Improvements were seen in the patient's skin, but the renal and pulmonary complications required the use of steroids and plasmapheresis. Unfortunately, the patient contracted SARS-CoV-2 virus and died [89]. More studies still need to be done to determine if stem cell therapy might be useful alone or combined with other therapies to treat scleromyxedema.

5.7 Systemic sclerosis

Systemic sclerosis (SSc) is an autoimmune disease characterized by excess collagen in the internal organs and skin, causing ulcers and organ damage. HSC therapy and MSC therapy have been tested and found to improve pain, blood flow, lung function, among other symptoms of the disease [90]. Autologous hematopoietic stem cell therapy is preferred over allogeneic therapy due to its lower treatment-related mortality and absence of graft-vs.-host disease [91].

Stem cell therapy has been extensively studied in three randomized controlled trials: the American Scleroderma Stem Cell versus Immune Suppression Trial (ASSIST, phase 2, 19 patients), the Autologous Stem Cell Transplantation International Scleroderma Trial (ASTIS, phase 3, 156 patients), and the Scleroderma Cyclophosphamide or Transplantation study (SCOT, phase 3, 75 patients), with several pilot and case studies [92–94]. These studies have demonstrated autologous hematopoietic stem cell therapy is an effective and safe treatment for systemic sclerosis. However, patients with severe major organ involvement (pulmonary, cardiac, or renal) or serious comorbidities were excluded from all three trials due to contraindications [59].

MSC therapy has the ability to suppress innate and adaptive immunity and can differentiate into a wide variety of tissues, making it seem like an ideal choice for SSc [95]. However, if donors are not carefully chosen, there is the chance that collagen production can be increased, thus this therapy can worsen symptoms [96]. This research suggests that autologous MSCs from patients that have advanced stage SSc should not be used for treatment. On the other hand, allogenic MSC therapy has lived up closer to the promises of stem cell therapy. Allogenic MSCs were administered intravenously in a female patient, where her skin condition improved, reducing the appearance of ulcers and her pain score [95]. In a clinical trial, combining MSC therapy with plasmapheresis was shown to improve lung function and skin thickness shown in improved modified Rodnan Skin Scores. The current research suggests that MSC therapy may be most effective when paired with another therapeutic option, but research still needs to be done to explore this.

Stem cell therapy has been found to be more effective than conventional immunosuppressive drugs and is currently the only disease-modifying strategy that improves long-term survival, prevents organ deterioration, enhances skin and pulmonary function, and improves overall quality of life.

The European Society for Blood and Marrow Transplantation (ESBMT) and the British Society of Blood and Marrow Transplantation (BSBMT) classify autologous hematopoietic stem cell therapy in severe resistant cases as a "clinical option," requiring a risk-benefit assessment [97, 98]. Guidelines from the American Society for Blood and Marrow Transplantation (ASBMT) categorize this therapy as "standard of care, rare indication" for children (indicating it is an option for individual patients after careful risk-benefit evaluation) and "developmental" for adults [98]. Patients with acute onset rapidly progressive disease refractory to conventional therapy and mild initial organ damage carry a better prognosis after HSC therapy. Patients with long standing conditions, indolent course and/or irreversible organ damage are contraindications to this therapy [99]. Thus, the challenge is to identify patients who are likely to be benefitted with HSC therapy.

5.8 Systemic lupus erythematosus

HSC therapy has been tested in patients with refractory systemic lupus erythematosus (SLE). Many observational studies and clinical trials have been aimed at assessing the effectiveness and safety of this transplant approach [100–102]. In a long-term follow-up of a female patient who underwent allogenic BM-HSC treatment, her systemic lupus erythematosus disease activity index (SLEDAI) score was found to improve, pain improved, and engraftment remained functional [103]. Collectively, these reports show HSCs to be beneficial for patients with a shorter duration of refractory disease suggesting that earlier intervention might lead to better outcomes [104].

The therapeutic potential of MSCs has been investigated for various autoimmune diseases including SLE [105]. In a recent study, six refractory SLE patients were treated with an intravenous infusion of MSCs. Five of the patients reached the threshold for improvement, achieving an SLE Responder Index (SRI) of 4 [106]. In a separate long-term follow-up study done in 2021, 81 patients were treated with allogenic BM-MSC and/or UC-MSCs. After 5 years, 37 patients had achieved clinical remission. MSC therapy has been shown to improve patient survival and reduce the severity of the disease as it has been shown to be safe and effective in treatments [107]. MSCs have been shown to alleviate SLE severity, improve renal function, decrease autoantibody production, upregulate peripheral T-cells, and restore balance between Th1- and Th2-related cytokines [108]. These collective immunomodulatory and regenerative properties position MSCs as a promising treatment for SLE.

5.9 Vitiligo

Steroid topical treatment is the first line of therapy for vitiligo, but when it proves ineffective, surgical options may be viewed next [109, 110]. Cellular grafts using autologous non-cultured outer root sheath hair follicle cell suspension (NCORSHFS) have been tested as a method to treat vitiligo [111]. This method utilizes the regenerative capacity of hair follicle melanocytes, as they can repigment areas where vitiligo has caused depigmentation by allowing melanocyte precursors to proliferate into the areas that lack melanocytes, making them preferable over epidermal melanocytes for cell-based vitiligo treatments. One study reported NCORSHFS achieved an average repigmentation rate of 65.7%, with more than 75% repigmentation observed in nine out of 14 patients [112]. Another study investigated factors affecting therapeutic outcomes in 30 patients with 60 target lesions treated with NCORSHFS [111]. They found that 35% of the lesions achieved repigmentation greater than 75%. The study showed patients who achieved optimal repigmentation had significantly higher numbers of transplanted melanocytes and hair follicle stem cells. Also, the absence of dermal inflammation was a significant predictor of successful repigmentation. These results emphasize the importance of specific cellular components, and a favorable dermal environment is necessary for the effective treatment of vitiligo with NCORSHFS.

Another promising stem cell treatment for vitiligo is multilineage-differentiating stress-enduring (MUSE) cells [113]. In three-dimensional skin culture models, *ex vivo* studies have identified factors that encourage MUSE cells to differentiate into melanocytes. The melanocytes are integrated into the epidermis, promoting melanogenesis. However, the impact of MUSE cells *in vivo* remains to be determined [114].

5.10 Wound healing

Chronic or non-healing skin wounds present an ongoing challenge in advanced wound care. Current wound healing treatments remain insufficient. Stem cell therapy has emerged as a promising new approach for wound healing using MSCs [115]. MSCs are an attractive cell type for cell-based therapy due to their ease of isolation, vast differentiation potential, and immunomodulatory effects during transplantation. MSCs are known to play a key role in the wound healing process making them an obvious candidate for clinical use. When introduced into the wound bed, MSCs have been shown to promote fibroblast migration, stimulate extracellular matrix (ECM) deposition, facilitate wound closure, initiate re-epithelialization, enhance angiogenesis, and mitigate inflammation in preclinical animal models. MSC efficacy and safety use for the treatment of chronic wounds was further confirmed by several clinical studies involving human subjects which yielded similar positive results with no adverse side effects [116]. However, while MSCs appear to be a promising resource for chronic wound care, additional studies are needed to determine optimal cell source and route of delivery before this treatment can be recommended for clinical use.

MSCs for the treatment of chronic wounds has proven to be feasible, effective, and safe, reported through preclinical and clinical trials [117]. MSCs stimulate the healing process in chronic wounds through several biological and molecular mechanisms. One of the primary roles of MSCs is to promote the directional migration of fibroblast cells to the injury site where they can localize in the wound bed [115, 118]. Once localized fibroblasts facilitate wound closure and synthesize the necessary components of the ECM such as collagen. MSCs can also downregulate MMP-1, a type of collagenase primarily responsible for ECM degradation. MSCs function to preserve ECM and maintain dermal structure. MSC-treated wounds have increased elastin levels which provides recovering tissue with resiliency that is not typically seen in normal wound healing [116]. MSCs play a role in the re-epithelialization process by activating the proliferation, differentiation, and migration of keratinocytes that support the formation of a multi-layered and well-differentiated epidermis [117, 119].

MSCs are believed to stimulate the development of new hair follicles and sweat glands, which suggests these stem cells are capable of not only accelerating wound healing but also improving the quality of wound healing. MSC's use for chronic wounds supports angiogenesis by upregulating VEGF and Ang-1 increasing microvessels throughout the wound bed [120]. This allows the nutrient and oxygen transport to developing cells enhancing their longevity. Also, MSCs help to modulate the wound environment and in turn support proper healing by mitigating inflammation at the site of injury. Importantly, MSCs decrease infiltration of inflammatory cells and pro-inflammatory cytokines and initiate the polarization of M1 macrophages to anti-inflammatory M2 macrophages. MSCs also downregulate ICAM1, a protein involved in inflammation, and upregulate superoxide dismutase, an enzyme which breaks down harmful superoxide radicals [118, 121]. By supporting wound healing MSCs by optimizing the healing environment can produce efficient wound closure.

Several clinical trials in human subjects have generated positive results when MSCs were applied to chronic or non-healing wounds [122]. No adverse side effects have been observed which confirms the safety and feasibility of this cellular therapy for human application. However, further research is needed to determine the best cell source and route of delivery before this procedure can be recommended for human use clinically.

MSCs can be isolated from various tissue types including bone marrow, adipose tissue, cord blood, and placenta. MSCs demonstrate unique properties. Several comparative studies have reported MSCs as the most promise for cell therapy due to their abundance and ease of isolation as well as their regenerative and immunomodulatory properties [123]. How these MSCs are delivered into the wound is the critical question. MSCs can be delivered locally to the wound bed via injection, topical application, or incorporation into a 3D scaffold to avoid issues related to low engraftment efficiency observed following IV injection [124, 125]. Investigating local delivery methods, MSCs seeded into a biomaterial scaffold appears to hold promise as it allows for the localization of the cells into the wound bed and provides donor cells with protection and structure [126, 127]. Following additional research, the application of MSCs for chronic or non-healing wounds could provide a major development in advanced wound care.

Epidermal stem cells have potential to regenerate the epidermis and differentiate under appropriate stimuli into various skin cell types and tissues [128]. This property can be used to initiate and accelerate healing of chronic non-healing wounds. MSCs promote wound healing by decreasing inflammation, promoting angiogenesis, and decreasing scarring [129]. One study successfully applied human MSCs to non-healing and acute wounds using a specialized fibrin spray system [130]. Another study demonstrated the efficacy of stem cell therapy in diabetic foot ulcers [131].

6. Conclusion

This review addressed the potential of stem cells to function therapeutically for dermatological disorders highlighting the emphasis on their role in the treatment of these clinical conditions. Current evidence shows a diversified category of stem cells can be effective serving as a major contributor to treatment protocols for patients with a dermatological diagnosis. Stem cell diversification allows both effective functional and successful treatments and is of vital importance to understand the complex interactions involved in therapeutic responses. What will be the most efficacious stem cell type or perhaps in combination that would be best therapeutically will be confirmed following the performance of clinical trials. With the expected rise in dermatological disorders associated with the increase aging of the human population may allow enhance therapeutic utility of stem cells for continued human use for years to come.

Conflict of interest

The authors declare no conflict of interest.

Appendices and nomenclature

AA	alopecia areata
AD	atopic dermatitis
ADS-CM	adipose-derived stem cell conditioned media
AGA	androgenic alopecia
ASBMT	American society for blood and marrow transplantation
ASSIST	American scleroderma stem cell vs. immune suppression trial
ASTIS	autologous stem cell transplantation international scleroderma trial
BM-MSCs	bone marrow-derived mesenchymal stem cells
BMP	bone morphogenic protein
BSBMT	British society for blood and marrow transplantation
CAG	plasmid
CMAS	childhood myositis assessment scale
DNA	deoxyribonucleic acid
EASI	eczema area and severity index
EB	epidermolysis bullosa
EBNA	plasmid
ECM	extracellular matrix
EMA	European medical agency
ESBMT	European society for blood and marrow transplantation
ESC	embryonic stem cell
EULAR	European union labor and regulation
FDA	food and drug administration
FGF	basic fibroblast growth factor
GMP	good manufacturing practice
HFF	human foreskin fibroblasts
HSC	hematopoietic stem cell
hESC	human embryonic stem cell
hTERT	human telomerase reverse transcriptase
hUCB-MSC	umbilical cord blood mesenchymal stem cell
ICM	inner cell mass
IgE	immunoglobulin class E
ILGF	insulin like growth factor
iPSC	induced pluripotential stem cell
Klf4	kruppel-like factor 4
MMT	manual muscle testing
MSC	mesenchymal stem cell
Myod1	myogenic differentiation 1
MUSE	multilineage-differentiating stress-enduring cell
NCORSHFS	non-cultured outer root sheath hair follicle cell suspension
Oct3/4	octamer-minding transcription factor 3 & 4
PBMC	peripheral blood mononuclear cells
PSC	pluripotential stem cell
RCT	randomized control trial
SCOT	scleroderma cyclophosphamide or transplantation study
SLE	systemic lupus erythematosus
SNCT	somatic cell nuclear transfer
SRY-Box-2	sex determining region Y

SV	structural variation
TGFβ	transforming growth factor beta
TSC	totipotent stem cell
VEGF	variable epithelium growth factor

Author details

Hannah E. Doan[1], Mohammed S. Inayat[2] and Vincent S. Gallicchio[3]*

1 College of Medicine, University of Cincinnati, Cincinnati, OH, USA

2 Department of Internal Medicine, University of Cincinnati, Cincinnati, OH, USA

3 Department of Biological Sciences, College of Science, Clemson University, Clemson, SC, USA

*Address all correspondence to: vsgall@clemson.edu

References

[1] Lomberk G. Stem cells. Pancreatology. 2007;7(4):314-316

[2] Cascalho M, Ogle BM, Platt JL. The future of organ transplantation. Annals of Transplantation. 2006;**11**(2):44-47

[3] Wang Y et al. Cell differentiation lineage in the prostate. Differentiation. 2001;**68**(4-5):270-279

[4] Park SM. Stem cell research in gastroenterology. The Korean Journal of Gastroenterology. 2004;**43**(4):221-225

[5] Vaughan AE et al. Lineage-negative progenitors mobilize to regenerate lung epithelium after major injury. Nature. 2015;**517**(7536):621-625

[6] Wu J, Izpisua Belmonte JC. Dynamic pluripotent stem cell states and their applications. Cell Stem Cell. 2015;**17**(5):509-525

[7] Antoniadou E, David AL. Placental stem cells. Best Practice & Research. Clinical Obstetrics & Gynaecology. 2016;**31**:13-29

[8] Latos PA, Hemberger M. From the stem of the placental tree: Trophoblast stem cells and their progeny. Development. 2016;**143**(20):3650-3660

[9] Carpenter MK et al. Properties of four human embryonic stem cell lines maintained in a feeder-free culture system. Developmental Dynamics. 2004;**229**(2):243-258

[10] Thomson JA et al. Embryonic stem cell lines derived from human blastocysts. Science. 1998;**282**(5391):1145-1147

[11] Zakrzewski W et al. Stem cells: Past, present, and future. Stem Cell Research & Therapy. 2019;**10**(1):68

[12] Ye L, Swingen C, Zhang J. Induced pluripotent stem cells and their potential for basic and clinical sciences. Current Cardiology Reviews. 2013;**9**(1):63-72

[13] Chehelgerdi M et al. Exploring the promising potential of induced pluripotent stem cells in cancer research and therapy. Molecular Cancer. 2023;**22**(1):189

[14] Purton LE, Scadden DT. The hematopoietic stem cell niche. In: StemBook. Cambridge, MA: Harvard Stem Cell Institute; 2008

[15] Seita J, Weissman IL. Hematopoietic stem cell: Self-renewal versus differentiation. Wiley Interdisciplinary Reviews. Systems Biology and Medicine. 2010;**2**(6):640-653

[16] Oh SY et al. In vitro embryogenesis and gastrulation using stem cells in mice and humans. International Journal of Molecular Sciences. 2023;**24**(17):13655

[17] Takahashi K, Yamanaka S. Induction of pluripotent stem cells from mouse embryonic and adult fibroblast cultures by defined factors. Cell. 2006;**126**(4):663-676

[18] Bussolati B. Stem cells for organ repair: Support or replace? Organogenesis. 2011;7(2):95

[19] Lu J et al. Defined culture conditions of human embryonic stem cells. Proceedings of the National Academy of Sciences of the United States of America. 2006;**103**(15):5688-5693

[20] Hoffman LM, Carpenter MK. Characterization and culture of

human embryonic stem cells. Nature Biotechnology. 2005;**23**(6):699-708

[21] Morgani SM et al. Totipotent embryonic stem cells arise in ground-state culture conditions. Cell Reports. 2013;**3**(6):1945-1957

[22] Sills ES et al. Identification and isolation of embryonic stem cells in reproductive endocrinology: Theoretical protocols for conservation of human embryos derived from in vitro fertilization. Theoretical Biology & Medical Modelling. 2005;**2**:25

[23] Ding D-C, Shyu W-C, Lin S-Z. Mesenchymal stem cells. Cell Transplantation. 2011;**20**(1):5-14

[24] Zare S et al. Types of stem cells in regenerative medicine: A review. Journal of Skin and Stem Cell. 2014;**1**(3):e28471

[25] Díaz-García D et al. A Beginner's introduction to skin stem cells and wound healing. International Journal of Molecular Sciences. 2021;**22**(20):11030

[26] Altuntaş E, Gözütok K, Özkan B. Nanoengineering of stem cells for neural regenerative medicine. In: Razavi M, editor. Neural Regenerative Nanomedicine. London, UK: Academic Press; 2020. pp. 159-211

[27] Oliveira MS, Barreto-Filho JB. Placental-derived stem cells: Culture, differentiation and challenges. World Journal of Stem Cells. 2015;**7**(4):769-775

[28] Fulka J Jr et al. Cloning by somatic cell nuclear transfer. BioEssays. 1998;**20**(10):847-851

[29] Gurdon JB. The developmental capacity of nuclei taken from intestinal epithelium cells of feeding tadpoles. Journal of Embryology and Experimental Morphology. 1962;**10**:622-640

[30] Munsie MJ et al. Isolation of pluripotent embryonic stem cells from reprogrammed adult mouse somatic cell nuclei. Current Biology. 2000;**10**(16):989-992

[31] Kim JS et al. Reprogrammed pluripotent stem cells from somatic cells. International Journal of Stem Cells. 2011;**4**(1):1-8

[32] Davis RL, Weintraub H, Lassar AB. Expression of a single transfected cDNA converts fibroblasts to myoblasts. Cell. 1987;**51**(6):987-1000

[33] Takahashi K et al. Induction of pluripotent stem cells from adult human fibroblasts by defined factors. Cell. 2007;**131**(5):861-872

[34] Quinlan AR et al. Genome sequencing of mouse induced pluripotent stem cells reveals retroelement stability and infrequent DNA rearrangement during reprogramming. Cell Stem Cell. 2011;**9**(4):366-373

[35] Sullivan S et al. Quality control guidelines for clinical-grade human induced pluripotent stem cell lines. Regenerative Medicine. 2018;**13**(7):859-866

[36] Lowry WE et al. Generation of human induced pluripotent stem cells from dermal fibroblasts. Proceedings of the National Academy of Sciences of the United States of America. 2008;**105**(8):2883-2888

[37] Hilfiker A et al. Mesenchymal stem cells and progenitor cells in connective tissue engineering and regenerative medicine: Is there a future for transplantation? Langenbeck's Archives of Surgery. 2011;**396**(4):489-497

[38] Cohen DE, Melton D. Turning straw into gold: Directing cell fate for

regenerative medicine. Nature Reviews. Genetics. 2011;**12**(4):243-252

[39] Barthes J et al. Cell microenvironment engineering and monitoring for tissue engineering and regenerative medicine: The recent advances. BioMed Research International. 2014;**2014**:921905

[40] Hwang NS, Varghese S, Elisseeff J. Controlled differentiation of stem cells. Advanced Drug Delivery Reviews. 2008;**60**(2):199-214

[41] Tissue & Tissue Products, F.a.D. Administration. Available from: https://www.fda.gov/vaccines-blood-biologics/tissue-tissue-products [Accessed: March 1, 2024]

[42] Moustakas A, Heldin CH. The regulation of TGFbeta signal transduction. Development. 2009;**136**(22):3699-3714

[43] Jiang X et al. Otefin, a nuclear membrane protein, determines the fate of germline stem cells in drosophila via interaction with Smad complexes. Developmental Cell. 2008;**14**(4):494-506

[44] Teven CM et al. Fibroblast growth factor (FGF) signaling in development and skeletal diseases. Genes & Diseases. 2014;**1**(2):199-213

[45] Logan CY, Nusse R. The Wnt signaling pathway in development and disease. Annual Review of Cell and Developmental Biology. 2004;**20**:781-810

[46] Watt FM, Driskell RR. The therapeutic potential of stem cells. Philosophical Transactions of the Royal Society of London. Series B, Biological Sciences. 2010;**365**(1537):155-163

[47] Frank-Kamenetsky M et al. Small-molecule modulators of hedgehog signaling: Identification and characterization of smoothened agonists and antagonists. Journal of Biology. 2002;**1**(2):10

[48] Janesick A et al. ERF and ETV3L are retinoic acid-inducible repressors required for primary neurogenesis. Development. 2013;**140**(15):3095-3106

[49] Janesick A, Wu SC, Blumberg B. Retinoic acid signaling and neuronal differentiation. Cellular and Molecular Life Sciences. 2015;**72**(8):1559-1576

[50] Osakada F et al. In vitro differentiation of retinal cells from human pluripotent stem cells by small-molecule induction. Journal of Cell Science. 2009;**122**(Pt 17):3169-3179

[51] Shahdadfar A et al. In vitro expansion of human mesenchymal stem cells: Choice of serum is a determinant of cell proliferation, differentiation, gene expression, and transcriptome stability. Stem Cells. 2005;**23**(9):1357-1366

[52] Menasche P et al. Human embryonic stem cell-derived cardiac progenitors for severe heart failure treatment: First clinical case report. European Heart Journal. 2015;**36**(30):2011-2017

[53] Hovatta O et al. A culture system using human foreskin fibroblasts as feeder cells allows production of human embryonic stem cells. Human Reproduction. 2003;**18**(7):1404-1409

[54] Loh YH et al. The Oct4 and Nanog transcription network regulates pluripotency in mouse embryonic stem cells. Nature Genetics. 2006;**38**(4):431-440

[55] Nandoe Tewarie RS et al. Stem cell-based therapies for spinal cord injury.

The Journal of Spinal Cord Medicine. 2009;**32**(2):105-114

[56] Schwartz SD et al. Human embryonic stem cell-derived retinal pigment epithelium in patients with age-related macular degeneration and Stargardt's macular dystrophy: Follow-up of two open-label phase 1/2 studies. Lancet. 2015;**385**(9967):509-516

[57] Assady S et al. Insulin production by human embryonic stem cells. Diabetes. 2001;**50**(8):1691-1697

[58] Latres E et al. Navigating two roads to glucose normalization in diabetes: Automated insulin delivery devices and cell therapy. Cell Metabolism. 2019;**29**(3):545-563

[59] Khandpur S, Gupta S, Gunaabalaji DR. Stem cell therapy in dermatology. Indian Journal of Dermatology, Venereology and Leprology. 2021;**87**:753-767

[60] Anderi R et al. Cellular therapy with human autologous adipose-derived adult cells of stromal vascular fraction for alopecia areata. Stem Cell Research & Therapy. 2018;**9**(1):141

[61] Li Y et al. Hair regrowth in alopecia areata patients following stem cell educator therapy. BMC Medicine. 2015;**13**:87

[62] Gentile P et al. Stem cells from human hair follicles: First mechanical isolation for immediate autologous clinical use in androgenetic alopecia and hair loss. Stem Cell Investigation. 2017;**4**:58

[63] Zari S. Short-term efficacy of autologous cellular micrografts in male and female androgenetic alopecia: A retrospective cohort study. Clinical, Cosmetic and Investigational Dermatology. 2021;**14**:1725-1736

[64] Krefft-Trzciniecka K et al. Human stem cell use in androgenetic alopecia: A systematic review. Cells. 2023;**12**(6):951

[65] Elmaadawi IH et al. Stem cell therapy as a novel therapeutic intervention for resistant cases of alopecia areata and androgenetic alopecia. The Journal of Dermatological Treatment. 2018;**29**(5):431-440

[66] Kim HS et al. Human umbilical cord blood mesenchymal stem cell-derived PGE2 and TGF-beta1 alleviate atopic dermatitis by reducing mast cell degranulation. Stem Cells. 2015;**33**(4):1254-1266

[67] Na K et al. Bone marrow-derived clonal mesenchymal stem cells inhibit ovalbumin-induced atopic dermatitis. Cell Death & Disease. 2014;**5**(7):e1345

[68] Shin TH et al. Human adipose tissue-derived mesenchymal stem cells alleviate atopic dermatitis via regulation of B lymphocyte maturation. Oncotarget. 2017;**8**(1):512-522

[69] Hua C, Chen S, Cheng H. Therapeutic potential of mesenchymal stem cells for refractory inflammatory and immune skin diseases. Human Vaccines & Immunotherapeutics. 2022;**18**(6):2144667

[70] Lee BC et al. Disease-specific primed human adult stem cells effectively ameliorate experimental atopic dermatitis in mice. Theranostics. 2019;**9**(12):3608-3621

[71] Shin HT et al. Long-term efficacy and safety of intravenous injection of clonal mesenchymal stem cells derived from bone marrow in five adults with moderate to severe atopic dermatitis. The Journal of Dermatology. 2021;**48**(8):1236-1242

[72] Seo H-M et al. Phase 1/2 trials of human bone marrow–derived clonal mesenchymal stem cells for treatment of adults with moderate to severe atopic dermatitis. Journal of Allergy and Clinical Immunology. 2024;**154**(4):965-973

[73] Holzer U et al. Successful autologous stem cell transplantation in two patients with juvenile dermatomyositis. Scandinavian Journal of Rheumatology. 2010;**39**(1):88-92

[74] Wang D et al. Efficacy of allogeneic mesenchymal stem cell transplantation in patients with drug-resistant polymyositis and dermatomyositis. Annals of the Rheumatic Diseases. 2011;**70**(7):1285-1288

[75] Conget P et al. Replenishment of type VII collagen and re-epithelialization of chronically ulcerated skin after intradermal administration of allogeneic mesenchymal stromal cells in two patients with recessive dystrophic epidermolysis bullosa. Cytotherapy. 2010;**12**(3):429-431

[76] Hirsch T et al. Regeneration of the entire human epidermis using transgenic stem cells. Nature. 2017;**551**(7680):327-332

[77] El-Darouti M et al. Treatment of dystrophic epidermolysis bullosa with bone marrow non-hematopoeitic stem cells: A randomized controlled trial. Dermatologic Therapy. 2016;**29**(2):96-100

[78] Wagner JE et al. Bone marrow transplantation for recessive dystrophic epidermolysis bullosa. The New England Journal of Medicine. 2010;**363**(7):629-639

[79] Petrof G et al. Potential of systemic allogeneic mesenchymal stromal cell therapy for children with recessive dystrophic epidermolysis bullosa. The Journal of Investigative Dermatology. 2015;**135**(9):2319-2321

[80] Uitto J et al. Progress toward treatment and cure of epidermolysis bullosa: Summary of the DEBRA International research symposium EB2015. The Journal of Investigative Dermatology. 2016;**136**(2):352-358

[81] Owczarczyk-Saczonek A et al. Stem cells as potential candidates for psoriasis cell-replacement therapy. International Journal of Molecular Sciences. 2017;**18**(10):2182

[82] Kaffenberger BH et al. Remission of psoriasis after allogeneic, but not autologous, hematopoietic stem-cell transplantation. Journal of the American Academy of Dermatology. 2013;**68**(3):489-492

[83] Chen W et al. Complete remission of both immunoglobulin light chain amyloidosis and psoriasis after autologous hematopoietic stem cell transplantation: A case report. Medicine (Baltimore). 2018;**97**(50):e13589

[84] Gardembas-Pain M et al. Psoriasis after allogeneic bone marrow transplantation. Archives of Dermatology. 1990;**126**(11):1523

[85] Li X et al. Transmission of psoriasis by allogeneic bone marrow transplantation and blood transfusion. Blood Cancer Journal. 2015;**5**(3):e288

[86] Chen H et al. Treatment of psoriasis with mesenchymal stem cells. The American Journal of Medicine. 2016;**129**(3):e13-e14

[87] Lacy MQ et al. Successful treatment of scleromyxedema with autologous peripheral blood stem

cell transplantation. Archives of Dermatology. 2005;**141**(10):1277-1282

[88] Illa I et al. Steady remission of scleromyxedema 3 years after autologous stem cell transplantation: An in vivo and in vitro study. Blood. 2006;**108**(2):773-774

[89] Nathan S, Ustun C. Autologous stem cell transplantation in a patient with scleromyxedema. EJHaem. 2022;**3**(3):1084-1085

[90] Zare Moghaddam M, Mousavi MJ, Ghotloo S. Stem cell-based therapy for systemic sclerosis. Rheumatology Advances in Practice. 2023;**7**(3):rkad101

[91] Swart JF et al. Haematopoietic stem cell transplantation for autoimmune diseases. Nature Reviews Rheumatology. 2017;**13**(4):244-256

[92] Oyama Y et al. Autologous non-myeloablative hematopoietic stem cell transplantation in patients with systemic sclerosis. Bone Marrow Transplantation. 2007;**40**(6):549-555

[93] Nash RA et al. High-dose immunosuppressive therapy and autologous hematopoietic cell transplantation for severe systemic sclerosis: Long-term follow-up of the US multicenter pilot study. Blood. 2007;**110**(4):1388-1396

[94] Farge D et al. Autologous hematopoietic stem cell transplantation for autoimmune diseases: An observational study on 12 years' experience from the European group for blood and marrow transplantation working party on autoimmune diseases. Haematologica. 2010;**95**(2):284-292

[95] Christopeit M et al. Marked improvement of severe progressive systemic sclerosis after transplantation of mesenchymal stem cells from an allogeneic haploidentical-related donor mediated by ligation of CD137L. Leukemia. 2008;**22**(5):1062-1064

[96] Vanneaux V et al. Expression of transforming growth factor β receptor II in mesenchymal stem cells from systemic sclerosis patients. BMJ Open. 2013;**3**(1):e001890

[97] British society of blood and marrow transplantation indications table. Available from: https://bsbmtct.org/wp-content/uploads/2023/02/HSCT-Adult-Indications_final.pdf

[98] Majhail NS et al. Indications for autologous and allogeneic hematopoietic cell transplantation: Guidelines from the American society for blood and marrow transplantation. Biology of Blood and Marrow Transplantation. 2015;**21**(11):1863-1869

[99] Del Papa N et al. Autologous hematopoietic stem cell transplantation for treatment of systemic sclerosis. Frontiers in Immunology. 2018;**9**:2390

[100] Burt RK et al. Treatment of autoimmune disease by intense immunosuppressive conditioning and autologous hematopoietic stem cell transplantation. Blood. 1998;**92**(10):3505-3514

[101] Brunner M et al. Autologous blood stem cell transplantation in refractory systemic lupus erythematosus with severe pulmonary impairment: A case report. Arthritis and Rheumatism. 2002;**46**(6):1580-1584

[102] Gualandi F et al. Autologous stem cell transplantation for severe autoimmune diseases: A 10-year experience. Annals of the New York Academy of Sciences. 2007;**1110**:455-464

[103] Gladstone DE et al. Long-term systemic lupus erythematosus disease control after allogeneic bone marrow transplantation. Lupus. 2017;**26**(7):773-776

[104] de Silva NL, Seneviratne SL. Haemopoietic stem cell transplantation in systemic lupus erythematosus: A systematic review. Allergy, Asthma and Clinical Immunology. 2019;**15**:59

[105] Munir H, McGettrick HM. Mesenchymal stem cell therapy for autoimmune disease: Risks and rewards. Stem Cells and Development. 2015;**24**(18):2091-2100

[106] Kamen DL et al. Safety, immunological effects and clinical response in a phase I trial of umbilical cord mesenchymal stromal cells in patients with treatment refractory SLE. Lupus Science & Medicine. 2022;**9**(1):e000704

[107] Zare Moghaddam M, Mousavi MJ, Ghotloo S. Stem cell-based therapy for systemic lupus erythematous. Journal of Translational Autoimmunity. 2024;**8**:100241

[108] Xu J. Therapeutic applications of mesenchymal stem cells for systemic lupus erythematosus. Advances in Experimental Medicine and Biology. 2018;**1089**:73-85

[109] Gawkrodger DJ et al. Guideline for the diagnosis and management of vitiligo. The British Journal of Dermatology. 2008;**159**(5):1051-1076

[110] Kubelis-Lopez DE et al. Updates and new medical treatments for vitiligo (review). Experimental and Therapeutic Medicine. 2021;**22**(2):797

[111] Vinay K et al. Clinical and treatment characteristics determining therapeutic outcome in patients undergoing autologous non-cultured outer root sheath hair follicle cell suspension for treatment of stable vitiligo. Journal of the European Academy of Dermatology and Venereology. 2015;**29**(1):31-37

[112] Mohanty S et al. Noncultured extracted hair follicle outer root sheath cell suspension for transplantation in vitiligo. The British Journal of Dermatology. 2011;**164**(6):1241-1246

[113] Razmi TM, Afra TP, Parsad D. Vitiligo surgery: A journey from tissues via cells to the stems! Experimental Dermatology. 2019;**28**(6):690-694

[114] Fisch SC et al. Pluripotent nontumorigenic multilineage differentiating stress enduring cells (Muse cells): A seven-year retrospective. Stem Cell Research & Therapy. 2017;**8**(1):227

[115] Spence S, Gallicchio V. Mesenchymal stem cell therapy for chronic wound healing. Journal of Stem Cell Research. 2022;**3**:1

[116] Jeon YK et al. Mesenchymal stem cells' interaction with skin: Wound-healing effect on fibroblast cells and skin tissue. Wound Repair and Regeneration. 2010;**18**(6):655-661

[117] Nie C et al. Locally administered adipose-derived stem cells accelerate wound healing through differentiation and vasculogenesis. Cell Transplantation. 2011;**20**(2):205-216

[118] Smith AN et al. Mesenchymal stem cells induce dermal fibroblast responses to injury. Experimental Cell Research. 2010;**316**(1):48-54

[119] Hu MS et al. Mesenchymal stromal cells and cutaneous wound healing: A comprehensive review of

the background, role, and therapeutic potential. Stem Cells International. 2018;**2018**:6901983

[120] Li H et al. Adult bone-marrow-derived mesenchymal stem cells contribute to wound healing of skin appendages. Cell and Tissue Research. 2006;**326**(3):725-736

[121] Kemp K et al. Inflammatory cytokine induced regulation of superoxide dismutase 3 expression by human mesenchymal stem cells. Stem Cell Reviews and Reports. 2010;**6**(4):548-559

[122] Huang YZ et al. Mesenchymal stem cells for chronic wound healing: Current status of preclinical and clinical studies. Tissue Engineering. Part B, Reviews. 2020;**26**(6):555-570

[123] Teoh PL et al. Human mesenchymal stromal cells derived from perinatal tissues: Sources, characteristics and isolation methods. Malaysian Journal of Medical Science. 2023;**30**(2):55-68

[124] Wagner J et al. Optimizing mesenchymal stem cell-based therapeutics. Current Opinion in Biotechnology. 2009;**20**(5):531-536

[125] Chan BP, Leong KW. Scaffolding in tissue engineering: General approaches and tissue-specific considerations. European Spine Journal. 2008;**17**(Suppl. 4):467-479

[126] Formigli L et al. MSCs seeded on bioengineered scaffolds improve skin wound healing in rats. Wound Repair and Regeneration. 2015;**23**(1):115-123

[127] Bicer M. Revolutionizing dermatology: Harnessing mesenchymal stem/stromal cells and exosomes in 3D platform for skin regeneration. Archives of Dermatological Research. 2024;**316**(6):242

[128] Liang L, Bickenbach JR. Somatic epidermal stem cells can produce multiple cell lineages during development. Stem Cells. 2002;**20**(1):21-31

[129] Otero-Vinas M, Falanga V. Mesenchymal stem cells in chronic wounds: The Spectrum from basic to advanced therapy. Advances in Wound Care (New Rochelle). 2016;**5**(4):149-163

[130] Falanga V et al. Autologous bone marrow-derived cultured mesenchymal stem cells delivered in a fibrin spray accelerate healing in murine and human cutaneous wounds. Tissue Engineering. 2007;**13**(6):1299-1312

[131] Lu D et al. Comparison of bone marrow mesenchymal stem cells with bone marrow-derived mononuclear cells for treatment of diabetic critical limb ischemia and foot ulcer: A double-blind, randomized, controlled trial. Diabetes Research and Clinical Practice. 2011;**92**(1):26-36

Chapter 23

Stem Cells for Therapy of Eye Diseases: Current Status

Zala Lužnik Marzidovšek, Janina Simončič, Petra Schollmayer, Elvira Maličev, Primož Rožman and Marko Hawlina

Abstract

In recent years, the treatment of various ocular diseases using stem cells and stem cell-derived exosomes has rapidly evolved. In this chapter, we present the potential of different stem cells and their secreted extracellular vesicles for the treatment of ocular diseases based on a careful review of relevant pre-clinical and clinical studies. The regenerative and immunomodulatory capacity of stem cells is analyzed together with the complex role of extracellular vesicles in intercellular communication, regulation of inflammation, and tissue repair. In addition, the inevitable challenges in clinical translation and ethical considerations are presented. Thus, in this chapter, we highlight the importance of current advances in the field of stem cell-based therapy in ophthalmology, with a growing body of evidence confirming the potential of these therapeutic modalities not only to restore vision but also to inhibit the progression of various ocular diseases, promising a significant improvement in quality of patients' life.

Keywords: ophthalmology, stem cell therapy, regenerative medicine, extracellular vesicles, clinical trials

1. Introduction

The human eye is one of our most important sensory organs, which allows us to see. The anatomy is complex and can be divided into anterior and posterior structures (**Figure 1**). Anterior structures such as the cornea and conjunctiva are in direct contact with the external environment [1, 2]. The transparent cornea enables the transmission of light through the crystalline lens to the innermost light-sensitive neurosensory structures of the eye. The retina is accountable for the visual processing of light energy into an electrical signal that is transmitted via the optic nerve to the brain [1]. Several eye diseases and trauma can affect the normal function of these major ocular structures and can lead to progressive vision loss or blindness, severely affecting individual's mobility, independence, and quality of life [3]. Some of the most common and debilitating eye diseases are severe ocular surface disease, glaucoma,

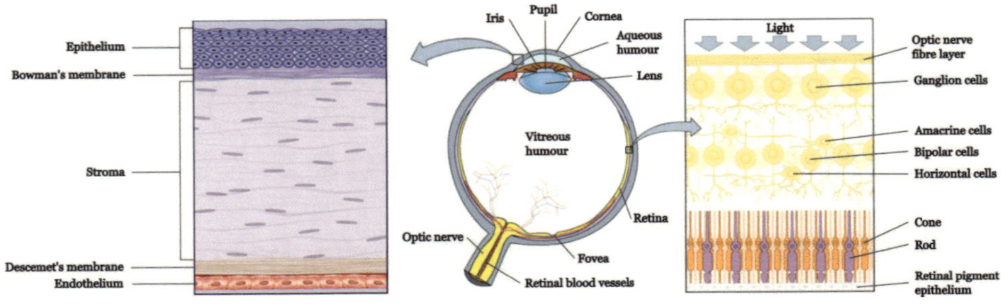

Figure 1.
Schematic cross-section of a human eye with an expanded view of the cornea (on the left) and retinal layers (on the right).

macular degeneration, autoimmune and genetic eye diseases [3], for which, there is currently available only symptomatic treatment.

Thus, extensive research effort is being put into the development of new preventive and causative therapeutic options such as advanced therapy medicinal products (ATMPs). ATMPs are medical products based on genes, tissues, or cells [4], including various stem cell (SC) and extracellular vesicle (EV) therapies [4, 5]. These new SC-based therapies offer the possibility of permanently restoring or replacing previously irreparable tissues or organs and have several well-described paracrine functions that can prevent or halt disease progression [5]. In recent years, the therapeutic potential of cell-free therapy using SC-secreted EVs has gained tremendous interest and considerable promise in regenerative medicine [6]. Due to its relative ease of accessibility, ocular immune privilege, and the accessibility of non-invasive diagnostic methods to follow-up and visualize eye structures after therapy (e.g., slit-lamp examination, optical coherence tomography, and *in vivo* confocal microscopy) predispose the human eye to be a prime target for SC-based therapy development [7, 8].

As presented in this chapter, we provide an overview of the current clinical advances in various SC-based therapies for treating several ocular diseases. Although published clinical studies using the therapeutic potential of EV secreted by SCs are currently still rare, with most reports in the pre-clinical stages, we additionally included some promising pre-clinical studies as this cell-free therapy may have several advantages over conventional SC therapies, as presented below.

2. Stem cell biology and types

Stem cells are undifferentiated or unspecialized cells present in the embryonic, fetal, and adult human body. The SC-specific characteristics are self-renewal and their ability to differentiate into different cell types, enabling embryonic development and, in adult life, tissue regeneration and repair after injury [9]. They can be classified into two types depending on their source. Embryonic or adult (somatic) SCs [10]. Another classification is based on their potency; the varying ability of SCs to differentiate into specialized cell types; and classifies SCs into totipotent (e.g., a zygote), pluripotent (e.g., embryonic SCs and induced pluripotent SCs), multipotent (e.g., hematopoietic SCs), oligopotent (e.g., myeloid SCs), or unipotent SCs (e.g., corneal/limbal epithelial SCs) [10]. Totipotent SCs can form embryonic and extra-embryonic tissues, whereas pluripotent SCs can form all germ layers but not extra-embryonic structures such as the placenta [10]. Another example of pluripotent SCs is artificially produced induced pluripotent SCs (iPSCs), which can be produced by reprogramming adult cells into a pluripotent state [11].

The differentiation potential is a continuum, starting from completely pluripotent SCs such as embryonic SCs or iPSCs and ending on representatives of adult SCs with much narrower spectrums of differentiation (e.g., multi-, oligo-, or unipotent SCs) [10]. Adult SCs can be found in various tissues and organs (e.g., skeletal muscles, brain, bone marrow, dental pulp, liver, spinal cord, cornea, and fat tissue) in specific microenvironments called SC niches [9], usually being able to differentiate into a limited number of terminated cell types allowing the maintenance/repairing of adult body tissues and organs.

Among these, one of the most important adult SC sources is multipotent mesenchymal stem cells (MSCs) that play a vital role in tissue repair and regeneration and can modulate immune responses via paracrine function [7, 12]. To date, they are the most commonly used SCs in clinical trials (https://clinicaltrials.gov/) and can be isolated from various adult as well as fetal tissues (such as placenta, Wharton's jelly, and umbilical cord blood) [12]. Thus, the International Society for Cellular Therapy (ISCT) has proposed a set of minimal *in vitro* criteria to define MSCs as cells that (a) can adhere to plastic, (b) express specific markers such as the cluster of differentiation (CD) 73, CD90, CD105, and lack the expression of CD14, CD34, CD45, and human leucocyte antigen-DR (HLA-DR), and (c) differentiate into adipocyte, chondrocyte, and osteoblast cell types [13]. Adult MSCs are most commonly isolated from bone marrow and adipose tissues [7].

Another interesting SC source for the development of SC-based therapies is tissue-specific adults SC. In the human eye, adult SCs have been identified in various regions such as the cornea, trabecular meshwork, crystalline lens, iris, ciliary body, retina, choroid, sclera, conjunctiva, eyelid, lacrimal gland, and the orbital fat [14]. However, to date, only cultured corneal epithelial SC transplantation revolutionized the current treatment options for patients with advanced painful and vision-threatening ocular surface disease, such as total limbal stem cell deficiency (LSCD) [15], as described in more detail below. A healthy corneal epithelium plays a vital role in the maintenance of corneal integrity and transparency, which is renewed by a small population of adult corneal epithelial SCs primarily found at the peripheral corneal area, the limbus (thus also named limbal epithelial SCs (LESC)) [2, 16]. Small populations of multipotent limbal mesenchymal or corneal stromal SCs have also been observed in the anterior limbal stroma [17] with increasing evidence suggesting a direct role of corneal stromal SCs in the provision of cells for corneal maintenance and regeneration [18]. Corneal stromal SCs could be used to regenerate the corneal stroma without a scar formation, thus preserving corneal transparency [19]. Although human corneal endothelium, the innermost corneal monolayer of neural crest-derived cells, has very low proliferative potential *in vivo* [20], recent *in vitro* studies also suggest the existence of corneal endothelial SCs located in the transition zone between the peripheral corneal endothelium and the beginning of the non-filtering portion of the trabecular meshwork called the Schwalbe's ring region, which might supply new cells for the corneal endothelium and trabecular meshwork [21]. Other eye-specific SC types were found also in the posterior segment of the eye in the retina [22, 23]. For example, Mueller cells display upon injury neural SC characteristics, capable of generating neurons and glial cells [22]. In addition, the retinal pigment epithelium also exhibits plasticity, with a subset of cells expressing SC features such as multipotency, potentially contributing to retinal repair [23].

3. Stem cell-derived extracellular vesicles

Extracellular vesicles (EVs) are cell-derived lipid membrane vesicles of varying sizes that can be secreted by many cell types [24–26]. Depending on the size and their biogenesis, they can be classified as (1) exosomes (30–150 nm in diameter), released into the extracellular space by the intracellular budding of endosomes; (2) microvesicles (100–1000 nm in diameter), formed by budding of the cell membrane; and into apoptotic bodies (1000–5000 nm in diameter), compartmentalization of the cell during programmed cell death [24]. EV carries important bioactive molecules, such as cytokines, growth factors, signaling lipids, mRNAs, and regulatory miRNAs, that are involved in intracellular communication and several signaling cascades [27]. The specific EV cargo depends on the cell type of origin and can influence a wide variety of biological functions, such as cell proliferation, regeneration, migration, apoptosis, and immunoregulation [27].

By transferring mRNA or miRNA, EVs can influence new protein synthesis and modulate gene expression [27]. MSCs are known to secrete EVs, which are supposed to have similar paracrine therapeutic effects as the original SCs, thus could be safer to use clinically due to their cell-free nature and can also be administered in higher concentrations [28]. Moreover, by genetic engineering, the original MSCs could be modified, and new therapeutic factors could be introduced into their EVs, avoiding safety concerns [28]. However, although the therapeutic potential of EVs is promising, the reproducibility, vesicle integrity, and maintenance of their biological activity to ensure the final product homogeneity remains challenging [25]. Thus, new Minimal Information for Studies of Extracellular Vesicles (MISEV) 2023 guidelines have been published to update the experimental requirements for EVs definition and their functions [29].

4. Clinical trials using stem cell therapy and stem cell-derived extracellular vesicles in various ocular disorders

The field of regenerative medicine using SC-based therapies holds immense promise for the treatment of debilitating ocular diseases. Based on encouraging results obtained from pre-clinical studies on animal models [30–38], several clinical trials utilizing various sources of SC therapies [6, 39–120] and a few studies using SC-derived EVs [121–125] emerged as novel approaches to address a wide range of ocular disorders. In this chapter, we focused on major pathologies with a large unmet medical need for new treatment strategies affecting the anterior segment (cornea), the retina (diabetic retinopathy, macular degeneration, and dystrophies), and the optic nerve (glaucoma).

4.1 Ocular surface and corneal regeneration

Corneal diseases represent the fifth leading cause of blindness worldwide, with approximately 4.5 million individuals being visually impaired due to loss of corneal clarity [126]. In case of treatable corneal blindness, the diseased or damaged corneal layers can be replaced by healthy donated corneal tissue (corneal transplantation or corneal graft). Corneal transplantation remains the most common form of solid tissue transplantation [127] with about 180,000 transplants being performed annually worldwide [128]. However, despite significant advances in surgery techniques and tissue storage, there are still major issues related to the availability and quality of donated tissues and postoperative corneal transplant survival [126–128]. Moreover, if patient's limbal area is severely compromised (e.g., LSCD disease), corneal transplantation is not curative in the long term as limbal epithelial SCs are absent in corneal grafts [129–131]. Thus, regenerative approaches using SC-based therapies provide a novel alternative treatment solution addressing severe corneal diseases such as LSCD, corneal stromal scarring, and endothelial dysfunction [132], as presented below.

4.1.1 Cultured epithelial stem cell transplantation for limbal stem cell deficiency disease

Severe ocular surface damage may lead to complete vision loss due to ocular surface failure as a result from insufficient corneal epithelial renewal and stromal scaring, which is clinically referred to as LSCD [16, 133, 134]. LSCD disease develops due to loss or dysfunction of LESC, the ultimate source of corneal epithelium, and can result from various causes including trauma, chemical burns, autoimmune diseases, infections, and genetic disorders [16, 133, 134]. LSCD can be partial or total and can affect one or both eyes.

Management of patients with total LSCD requires SC transplantation for corneal epithelial restoration. In unilateral cases, an autologous limbal graft from the fellow healthy eye can be used, which carries a hypothetical risk of inducing LSCD to the donor eye [135]. In bilateral cases, donor limbal tissue is needed. In case of no available or willing living related donor, the keratolimbal allograft (KLAL) transplantation can be performed, which uses cadaveric allogeneic limbal tissue as the source of corneal SCs and therefore allows for a larger SC supply [135]. However, the success rate of KLAL is limited by graft immune rejections that may develop despite intensive systemic immunosuppressive therapy in more than half of the cases [39]. As conventional surgical management of patients with total LSCD faces several challenges, a new approach using *ex vivo* cultured SC transplantation has been tried to be implemented into routine clinical practice since 1997 in many ophthalmological centers worldwide [40–73]. SCs are most commonly isolated and cultured from a small limbal biopsy of 1 to 2 mm^2 harvested from the peripheral corneal area (e.g., cultivated limbal epithelial transplantation (CLET)).

Autologous CLET was first documented by Pellegrini and colleagues in 1997 [53, 74] in two patients with LSCD caused by alkali burns and laid the foundation for several subsequent clinical studies. Until now, using several variants of autologous or allogeneic CLET technique, over 1250 eyes have been treated worldwide with promising clinical results including centers in Australia, Belgium, Germany, India, Iran, Italy, Japan, Norway, Spain, Taiwan, Thailand, the United States, and the United Kingdom [73, 135]. Moreover, in 2015, the European Medicines Agency (EMA) approved "Holoclar" as the first SC-based therapy for LSCD therapy in the world [75, 76], and in 2020, another autologous CLET product called "Nepic" was approved in Japan [77, 78]. In 2023, Jurkunas et al. published a case series of phase 1 clinical trial which enrolled five patients with unilateral LSCD using a new xenobiotic-, serum-, and antibiotic-free protocol for cultivated autologous limbal epithelial cell (CALEC) transplantation in the United States [73].

Long-term results of autologous CLET report improved corneal transparency and visual acuity, which is achieved in 50–86% [79], similar to conventional surgical approaches, but less invasive to the contralateral donor eye and with the ability to *ex vivo* expand or amplify the SC pool pre-transplantation [29, 53]. Similarly, in bilateral

LSCD cases, allogeneic CLET reported complete re-epithelialization of corneas and significant visual improvements postoperatively [80]. In addition, two meta-analyses found no differences in success rates between autologous and allogeneic CLET [81, 82]. The overall safety profile was favorable, with all side effects being amenable and transient to subsequent treatments [82, 83]; however, although allogeneic CLET is produced from small donor limbal biopsies with less immunogenic potential than the conventional KLAL transplantation (as fewer immune antigen-presenting cells are transplanted), systemic administration of immunosuppressive agents was needed to obtain good clinical results [81, 82].

Thus, other autologous SC sources such as oral mucosal cells or MSCs were sought for the *ex vivo* preparation of epithelial sheets in bilateral cases. Already in 2002, the first clinical study of autologous cultivated oral mucosal epithelial cell sheet transplantation (COMET) showed successful corneal surface reconstruction in four patients with bilateral total LSCD [84, 85]. In 2021, the first COMET product was approved in Japan, called "Ocural" [86]. In a recent comparative analysis of long-term results of three epithelial cell transplantation procedures comparing either CLET, COMET, or simple limbal epithelial transplantation, more than half of the patients achieved a stable ocular surface and visual acuity improvement up to 7 years postoperatively [74].

Another promising and potentially more sustainable SC source in ocular surface reconstruction is MSCs. Although MSCs have been widely studied in several animal disease models [127], to date, only a few research groups have reported clinical results with allogeneic bone marrow or autologous adipose-derived MSC transplantation in restoring corneal epithelium in LSCD [6, 87–89]. Calonge et al. reported the first prospective, randomized, double-masked clinical trial to treat patients affected by LSCD with cultured allogeneic bone marrow derived-MSC transplantation (MSCT), which was compared to CLET [87, 90]. The corneal epithelial phenotype improved in patients treated with MSCT without significant differences to CLET, with no reported adverse events related to cell products [87, 90].

4.1.2 Stem cell-based therapy in dry eye disease

Dry eye disease (DED) is a very common multifactorial disease characterized by a persistently unstable and/or deficient tear film causing discomfort and/or visual impairment, accompanied by variable degrees of ocular surface epitheliopathy, inflammation, and neurosensory abnormalities affecting millions of people [136]. Currently, there are only a few reported clinical trials in which severe DED was treated with MSC- or MSC-derived EV-based therapy [121, 122]. In an open-label, prospective, phase I clinical trial, seven patients affected by aqueous-deficient DED were treated with a single dose of allogeneic adipose-derived SCs, which were administered directly into the lacrimal gland through transconjunctival injection. Improvement in clinical signs and symptoms without adverse events was seen. Another study successfully used allogeneic bone-marrow-derived MSCs administered by intravenous injection to treat Graft-versus-host disease (GVHD)–associated DED [121]. Zhou et al. reported successful topical use of EV obtained from human umbilical cord-derived MSCs as eye drops in a prospective clinical trial to treat GVHD-associated DED [122].

4.1.3 Corneal stromal regeneration with stem cell-based therapy

The corneal stroma represents around 90% of corneal thickness and is composed of parallel collagen fibers (the extracellular matrix) and interspersed scarce keratocytes [137]. The precise spacing of fibers is essential for stromal transparency [137]. Stromal keratocytes remain quiescent throughout life and are derived from the embryonic periocular mesenchyme that originates from the neural crest [137]. Corneal stromal SCs (CSSC), a distinct cell population, are located in the anterior stroma close to limbal niche and serve as a reservoir of progenitor cells that can differentiate into functional keratocytes, which are responsible for producing and organizing the extracellular matrix [137]. CSSCs respond rapidly to corneal injury and also play a vital role in preserving corneal avascularity and immune privilege [137]. They can be characterized as MSCs according to the *in vitro* characteristics and ISCT guidlines, however, with more specific differentiation potentials [137]. After injury or severe corneal infections, the accompanying inflammatory response and fibrogenic growth factors activate keratocytes into fibroblasts and myofibroblasts initiating a wound-healing response which ultimately results in the formation of stromal scars [137]. In addition, many corneal diseases such as dystrophies or ectatic corneal disorders affect physiologic corneal stromal anatomy and transparency [137].

Thus, to find an alternative to classical corneal transplantation, CSSCs are of particular interest in regenerative medicine, offering a new therapeutic approach for corneal stromal scar therapy (regeneration instead of transplantation, enhancing the tissue's inherent but limited capacity to regenerate) [137].

CSSCs and MSCs have been widely studied in several animal disease models and were shown to restore corneal homeostasis and transparency. Although CSSCs might have greater differentiation potentials to differentiate into keratocytes as they are lineage committed compared to other MSCs sources [138], isolating CSSCs autologously is technically more demanding due to small amounts of obtained corneal tissues [6]. Previous studies also reported that extraocular sources of MSCs are capable to differentiate into keratocytes and produce new collagen *in vitro* and *in vivo* without inducing any inflammatory reaction [30] and can modulate pre-existing scars by stromal remodeling [31, 32]. For example, studies in rabbit LSCD models have shown that adipose-derived and dental pulp-derived MSCs improve corneal transparency, reduce inflammation, and promote re-epithelialization when transplanted to the cornea [33]. Similarly, intrastromal injections of umbilical cord-derived MSCs adopted the phenotype of corneal stromal keratocytes and improved corneal thickness and transparency [34]. Some research groups focused on iPSCs to generate healthy keratocytes *in vitro* [139].

However, to date, only Alió et al. [91–93] reported first promising results with autologous adipose-derived MSCs transplantation into a mid-stroma femtosecond laser-assisted lamellar pocket in patients with advanced keratoconus and poor visual function that were already candidates for corneal transplantation. Although they observed new collagen synthesis in the injected areas, it was quantitatively not enough to sufficiently restore corneal thickness [91–93]. Direct injection of MSCs might thus better provide a new treatment strategy for corneal scar modulation [6]. To treat severely thinned corneas, a better approach was to use decellularized corneal stroma as a carrier for MSCs [91]. During the 3-year follow-up, the authors noted no complications with a transparent cornea in all patients [94]. The authors also suggested that while SC-free therapy with MSC-derived EVs could potentially achieve a similar therapeutic effect, the *in vivo* production of extracellular matrix would probably still require the implantation of cells [6]. In India, a clinical trial is assessing the efficacy and safety of human donor cornea derived allogeneic CSSCs delivered to the ocular surface embedded into a fibrin gel for the prevention and treatment of corneal scars in pathologies such as corneal burns and ulcers with promising preliminary results [6].

4.1.4 Cultured human corneal endothelial cell transplantation

Corneal endothelium has very low proliferative potential *in vivo* and a gradual decline in endothelial cell density is observed during adulthood due to age-related cell death [140, 141]. In case of substantial endothelial cell loss or dysfunction, pathological corneal hydration (termed 'edema') occurs, leading to visual impairment [140, 141]. To date, the best treatment option is a corneal transplant [142]. Due to advancements in corneal transplantation surgery selective replacement of dysfunctional corneal endothelium enables fast visual rehabilitation in patients with corneal endothelial disease [142]. However, in many parts of the world, a shortage of donor corneas limits access to treatment, prompting a search for alternatives. One recent new strategy that can be performed only in patients with a healthy peripheral endothelium is central Descemetorhexis without endothelial transplantation, which necessitates migration of peripheral endothelial cells to cover the denuded stromal area. Thus, slower visual rehabilitation is expected [142]. Therefore, new clinical trials evaluating cultured human corneal endothelial cells, which aim to *ex vivo* multiply endothelial cells from a single donor, are studied [142].

Early attempts to culture human corneal endothelial cells began in the early 1980s with several published protocols [143]. As endothelial cells are growth arrested *in vivo*, it is extremely challenging to get them to proliferate in sufficient quantities *in vitro* without undergoing endothelial-mesenchymal transition or premature senescence [144]. In 2004, Amano and colleagues presented the potential of primary cultured human corneal endothelial cells isolated from explant cultures to adhere to human corneal stroma deprived of endothelial cells [145]. Refinements in protocols for *in vitro* endothelial cell expansion culminated in a pioneering clinical trial. In 2018, Kinoshita and colleagues achieved a significant milestone by restoring vision in patients with bullous keratopathy using cell injection with 1×10^6 passaged cultured endothelial cells supplemented with a Rock inhibitor [95]. At 5 years follow-up, normal corneal endothelial function was reported in 10 of 11 eyes [96]. Currently, there are a few other clinical trials studying the safety and efficacy of cultured human corneal endothelial cells with limited reports in the published literature: one in El Salvadore (NCT05309135) using human cultured corneal endothelial cells and Rho kinase Y-27632 inhibitor injection; one in Singapore (NCT04319848) using a tissue engineered graft consisting of a biological carrier and tissue engineered cells; and one in the United States (NCT04894110) using technology integrating biocompatible magnetic nanoparticles with cultured endothelial cells; similarly one other study in Mexico also using cell injection approach with human corneal endothelial cells laden with magnetic particles; and the last in India using an ultrathin carrier [96]. In addition, pre-clinical studies explore alternative sources for corneal endothelial cell differentiation from human embryonic SCs and iPSCs [146] or MSCs-derived EV with cytoprotective properties to induce proliferation and survival of damaged human corneal endothelial cells [123].

4.2 Retinal degenerative disorders

The neural retina is composed of several layers of neurons interconnected by synapses. The outermost cell layer consists of photoreceptor cells, which are supported by a layer of pigment epithelium cells. Photoreceptor cells are of two types, rods, and cones, and are essential components of the retina responsible for light perception. While rods function in low light conditions, cones are activated in bright light, particularly in the macula, the central region of the retina crucial for high acuity and color vision [147]. The innermost retinal layers (closest to the vitreous body) are formed by the ganglion cells (ganglion cell layer) and their axons that form the nerve fiber layer of the optic nerve [147]. Retinal degenerative disorders represent a major global health problem due to permanent vision loss characterized by pathological degeneration or loss of various retinal cells such as photoreceptors, retinal ganglion cells (RGC), and/or retinal pigment epithelium (RPE) with an inability to regenerate *in vivo* [148]. Therefore, cell replacement therapy emerges as a promising new therapeutic avenue for retinal degenerative disorders such as advanced retinitis pigmentosa (RP), age-related macular degeneration (AMD), and diabetic retinopathy (DR), as presented below.

4.2.1 Diabetic retinopathy

Diabetic retinopathy, a complication of diabetes that causes damage to the blood vessels in the retina, is one of the leading causes of blindness in the developed world in working-age population [97] DR can be classified into non-proliferative (NPDR) and proliferative DR (PDR). Early stages are characterized by inflammation, microglial activation, loss of endothelial cells and pericytes, loss of neuronal cells, and damage to the retinal microvasculature integrity [97]. With the progression of the disease, macular edema can lead to vision loss, and retinal ischemia stimulates new vessel growth and vision-threatening PDR [97]. Current treatment options include laser photocoagulation of ischemic peripheral retinal regions, intravitreal anti-VEGF, and glucocorticoid therapy, all of which can minimize vision loss if conducted promptly, but may have adverse side effects [97]. Thus, SC-based therapy is a promising new therapeutic option to replace lost endothelial cells or pericytes in DR, which are important for the maintenance of the inner blood-retinal barrier and vascular function, or to provide trophic factors and cytoprotective support to retinal neurons [97].

Several clinical trials (available at: https://www.clinicaltrials.gov) are currently underway to investigate the potential of SC-based therapies for DR. In one clinical study (NCT01736059), subjects with irreversible vision loss from retinal degenerative diseases or retinal vascular disease, including DR, are receiving intravitreal injections of autologous CD34+ bone marrow MSCs. The purpose of another trial (NCT03403283) is to determine if the reparative cells of blood vessels, called endothelial progenitor cells, are defective in people with diabetes. A third clinical trial (NCT03403699) is recruiting patients with diabetes and age-matched controls to generate iPSCs from the peripheral blood cells to be further used to generate mesoderm cells for injection into the vitreous cavity of diabetic rodents and primate eyes. In another recent pilot clinical trial, the safety and efficacy of intravenous autologous bone marrow-derived MSCs for the treatment of DR were evaluated [98]. Thirty-four eyes with non-proliferative or proliferative DR from 17 patients received one intravenous infusion of 3×10^6/kg cells. Treatment led to decreased fasting blood glucose and CRP levels, with improved visual acuity observed in the NPDR group [98]. No acute reactions or severe adverse events were reported, which suggests this treatment method is a safe and potentially effective option [98].

In addition, DR might also be improved after intravitreal injection of MSC exosomes, as demonstrated in a mouse model. In an animal model, Moisseiev et al. improved oxygen-induced retinopathy by the intravitreal injection of exosomes from human MSCs, as there was a suppression and amelioration of retinal ischemia and damage [35].

These trials thus represent significant steps toward understanding and potentially treating DR and related ocular conditions through innovative SC-based approaches.

4.2.2 Retinal dystrophies

Retinal dystrophies encompass a range of degenerative diseases affecting the retina with high genetic and clinical heterogeneity including syndromic or non-syndromic cases [149]. Typical symptoms may include impaired color or night vision and tunnel vision [149]. Vision impairment usually increases with age and eventually can progress to complete blindness [149]. Retinal dystrophies can be categorized depending on the type of photoreceptor affected into rod-dominated diseases, cone-dominated diseases, or generalized retinal degenerations involving both rod and cone photoreceptors [149]. One of the most common rod-cone retinal dystrophies is retinitis pigmentosa (RP), with high genetic heterogeneity [149]. It manifests as progressive night blindness (loss of the ability to see in dim light), followed by loss of peripheral vision that slowly progresses toward the center, resulting in tunnel vision [149]. Stargardt disease (STGD) is the most common cause of juvenile macular dystrophy [150]. It is a genetic disorder commonly inherited as an autosomal recessive trait that leads to the accumulation of lipofuscin in the retina, ultimately leading to retinal pigment epithelium (RPE) and photoreceptor cell death [150]. There are various other forms of retinal dystrophies; however, a more detailed description is beyond the scope of this chapter.

To date, there are limited treatment options for most retinal dystrophies, which usually progress to vision loss; thus, SC-based therapies might enable replacement or regeneration of the damaged cells (RPE or photoreceptor cells) in the near future.

The first clinical trial in humans was done by Schwarz et al. in 2012 [99] and 2015 [100]. In a study by Li et al., human ESC-derived retinal pigment epithelial (hESC-RPE) cell suspension was injected subretinally in seven early-stage Stargardt disease patients [101]. Over 60 months after transplantation, no systemic or local adverse reactions were observed, except for transiently high intraocular pressure in two cases, with improved or stable visual acuity [101]. Thus, the authors concluded that hESC-RPE transplantation in early-stage Stargardt disease was safe and well-tolerated [101]. Similarly, in another study, no serious adverse events over a 3-year period were found, suggesting long-term safety and feasibility of subretinal hESC-RPE transplantation, with the majority of patients remaining stable visual acuity [102]. However, in a study reported in 2023, which was conducted in 12 eyes with advanced Stargardt disease, no significant improvement in best-corrected visual acuity was observed during 1 year follow-up period after hESC-RPE transplantation [103]. A prospective clinical case series investigated the safety and efficacy of suprachoroidal adipose tissue-derived MSCs implantation in four patients with Stargardt disease with promising 6-month follow-up results [104]. Weiss and Levy reported recent results from the Stem cell ophthalmology treatment study (SCOTS) in which autologous bone marrow-derived MSCs in the treatment of Stargardt disease were used by various local ocular or systemic intravenous injections, concluding that the treatment might be beneficial after 1 year follow-up period [105]. Another study evaluated the therapeutic potential and safety of intravitreal use of a bone marrow mononuclear fraction containing CD34+ cells in patients with Stargardt-type macular dystrophy and showed discrete improvements in visual acuity and microperimetry in the treated eye compared to the untreated one [106].

In advanced RP, one study examined the safety of subretinal adipose-derived MSC implantation in 11 patients [107]. While systemic complications were absent, choroidal neovascular membrane and epiretinal membrane formation were observed [107]. Although no significant changes were noted in visual acuity or electroretinography recordings, one patient showed visual acuity and visual field improvement, and three reported enhanced brightness perception [107]. The findings highlight the short-term safety of this treatment but emphasize the need for caution in their application, suggesting further studies to optimize delivery techniques and evaluate long-term effects [107]. In a phase I clinical trial with 14 participants having advanced RP, intravitreal bone marrow-derived MSC injection was evaluated [108]. Mild, transient adverse events were noted during 12 months of follow-up [108]. While at the beginning, visual acuity improved significantly, it returned to baseline levels at 12 months; however, other tests did not indicate any disease progression. In one patient, diffuse vitreous hemorrhage required surgical intervention, which then resulted in restored vision [108]. Another phase 3 clinical study included 82 RP patients and reported promising short-term results (improvements in visual acuity, visual field, and electroretinography recordings with no serious complications) using suprachoroidal injection of umbilical cord-derived MSCs [109].

A recent three-level meta-analysis study identified seven cell-based therapy studies for Stargardt disease (including 40 eyes) and two studies for RP (including 44 eyes) until 2023, and their analysis highlighted that the application of cell therapy could enhance visual acuity in treated patients [148].

4.2.3 Age-related macular degeneration

Age-related macular degeneration (AMD) is a prevalent chronic retinal degenerative disorder characterized by pathological alterations in the macula and its adjacent vasculature (abnormalities in the photoreceptor/RPE/Bruch's membrane/choroidal complex), predominantly affecting the elderly population [151]. It leads to central vision loss, often resulting in geographic atrophy and/or neovascularization [151]. AMD can be classified into two types: dry and wet [151]. While dry AMD is more common, wet AMD typically leads to more severe vision loss and usually develops over a period of weeks to months [151].

Since Peyman et al. first described transplantation of RPE as an intact sheet in 1991 [152], tremendous progress has been made in the field of regenerative medicine, with various SC-based treatment options being clinically tested for AMD [148]. The first clinical trials in humans were done by Schwarz et al. in patients with AMD and Stargardt disease [99]. Successful integration of hESC-derived RPE cells into the host RPE layer was observed without abnormal growth or immune rejection [99]. Vision improved in both patients, highlighting the safety and therapeutic potential of hESC-derived RPE cells for retinal degenerative diseases [99, 100]. Since then, in a recent meta-analysis study, already 140 eyes in 12 clinical studies underwent various treatment strategies including suprachoroidal [104], intravitreal, subretinal, subfoveal [110] cell suspension injections or subretinal implantation of a bioengineered sheet of cells on human-vitronectin-coated polyester membrane or on a perylene membrane [111, 148]. Da Cruz et al. demonstrated safety and feasibility of subretinal transplantation of a manufactured hESC-RPE monolayer on a synthetic basement membrane (human-vitronectin-coated polyester membrane) for the treatment of wet AMD in two patients [112]. In most clinical studies, RPE cells were differentiated from ESCs

[110, 113–115] or included adipose-derived [104], umbilical cord-derived [116] or bone marrow-derived MSCs [105, 106]. Another promising SC source is iPSC, as they have the capacity to generate unlimited and readily available autologous RPE cells. These pluripotent SCs can proliferate indefinitely in laboratory settings while retaining their ability to differentiate into various cell types [117, 118]. Takagi et al. reported a case of iPSC-derived RPE cell transplantation in a patient with wet AMD [117]. Although the treatment did not improve vision, it remained stable for 4 years after transplantation without significant side effects [117].

Advancements are also observed in research focusing on EV use in retinal diseases. In China, a clinical trial is registered and currently ongoing (NCT03437759) to evaluate the safety and efficacy of MSC-derived exosomes in promoting the closure of macular holes (full-thickness defect in the macula). The trial involves 44 participants diagnosed with early phase 1 macular holes, who will receive MSC-derived exosomes via local intravitreal injection; however, the results have not been published yet.

4.3 Optic nerve

Retinal ganglion cells (RGCs) are central nervous system neurons located in the innermost layers of the retina, responsible for transmitting visual signals from the retina to the brain via their axons [119]. RGC axonal injury leads to neuronal death and loss of function with limited capacity to regenerate [119]. Successful optic nerve regeneration requires a source of healthy RGCs to replace damaged cells [119]. Research into nervous system regeneration is particularly intriguing and is still primarily conducted using animal models [119].

The first step in developing SC-based strategies for optic nerve regeneration requires a reliable and high-volume source of healthy RGC production [153]. Currently, there are two main methods for generating RGCs from pluripotent SC (hESC or iPSCs): organoid differentiation, which involves the development of self-organizing three-dimensional miniature organs *in vitro* from pluripotent SCs, and planar differentiation, where RGCs are trans-differentiated from SCs [153]. Unfortunately, simple intravitreal injection of cultured RGCs is not enough for their integration into proper retinal layers [153]. However, research by Wu et al. demonstrated that human iPSCs promote the survival of transplanted retinal ganglion cells (RGCs) and enhance neurite extension. Their *ex vivo* studies showed that RGCs cotransplanted with iPSCs onto adult rat retina explants had significantly better survival and longer neurite growth compared to RGC-only transplants [154].

4.3.1 Glaucoma

Glaucoma, a common optic neuropathy, is the leading cause of irreversible blindness [119]. Current glaucoma treatments focus on lowering intraocular pressure [119]. In the context of new glaucoma treatment strategy development, cell therapy has three primary goals: providing trophic and neuroprotective support, replacing lost cells and restoring function [119]. MSCs for glaucoma therapy have been widely investigated in animal models. Intravitreal injections of MSCs in a rodent model of glaucoma showed alignment of injected MSCs along the internal limiting membrane that survived for several weeks, promoting RGC survival [36]. A study compared human MSC treatment with their MSC-derived EVs alone in a rat model of optic nerve injury and concluded that cell treatment provided better RGCs neuroprotection than EVs alone, suggesting that cellular-based therapies may offer superior neurotrophic support [37]. In a study by Pan et al. [124], exosomes derived from umbilical cord MSCs were investigated in a rat model of optic nerve compression and revealed that while exosomes enhanced the survival of RGCs, they did not facilitate axonal regeneration. Results of another study showed that bone marrow MSC-derived EVs promoted significant neuroprotection of RGCs in a rat model of glaucoma and prevented thinning of the retinal nerve fiber layer [125]. Regular injections of EVs were effective, and RNA sequencing identified specific miRNA upregulated in these EVs, suggesting a potential mechanism for their neuroprotective function [125].

In another study, mouse iPSC and mouse ESC-derived RGCs were transplanted into both healthy and glaucomatous mouse retinas, with a remarkable survival and integration success rate exceeding 65% [38]. These transplanted RGCs exhibited prolonged survival for up to 12 months post-transplantation [38]. Notably, the transplanted cells demonstrated the ability to polarize within the host retina and form axonal processes [38].

To date, only one clinical study investigated the effect of a single intravitreal injection of bone marrow-derived MSCs in two patients with advanced-stage open-angle glaucoma (NCT02330978) [120]. Despite treatment, no improvements were observed in visual acuity or visual field [120]. ERG tests conducted at baseline and subsequent weeks showed stable responses in one patient, while the other developed retinal detachment with proliferative vitreoretinopathy; thus, further research is warranted [120].

5. Future directions and conclusions

Advances in regenerative medicine, including the development of new SC-based therapies, have revolutionized the treatment possibilities of some ocular disorders, as discussed in this chapter; however, several issues remain to be overcome before these methods can be used routinely in day-to-day clinical practice.

The utilization of pluripotent ESCs has been ethically and politically controversial since 1998 because it involves the use of human preimplantation embryos [10]. This hurdle can be overcome with the use of human iPSCs; however, their production is very costly and still prevents a more widespread use. Another concern with pluripotent SC transplantation is their growth control after transplantation and, thus, their possible role in malignant transformation [10]. It is, therefore, crucial to perform strict tumorigenicity testing before clinical application. Although a healthy eye is known as an immune-privileged site, it can be lost due to ocular disease states. Thus, transplantation of allogeneic SCs might need proper immunosuppression to be successful.

Another important and still unanswered challenge in treating retinal and optic nerve diseases is how to establish functional connections and proper integration of transplanted cells into the diseased tissue to enable the formation of a coherent visual image [155]. Although EV could overcome some of these issues and enable cell-free therapy, a better understanding of the precise therapeutic mechanisms is needed, as well as strict toxicological and microbiological studies, optimization of their therapeutic cargo, etc. [156].

As SC-based therapies are classified as advanced therapy medicinal products, the critical quality of the final product needs to be strictly defined with standardization and optimization of work protocols before clinical use. Thus, future directions might include research that would enable more widespread availability of the treatment by providing GMP-compliant, accessible, reliable, validated, pretested, and robust SC-based therapies for treating various advanced ocular diseases.

To sum up, cell therapy for retinal dystrophy develops rapidly; however, we should remain cautious with the completion of all necessary safety studies before treating patients. We should warn patients against practices that offer quick solutions on the Internet and are charging high prices for unapproved treatments for these potentially blinding diseases. In the coming years, results from well-conducted clinical trials will determine the best strategies of treatment for each of these diseases as well as the modes of immune suppression when required.

Acknowledgements

Funding was provided by the Slovenian Research Agency (project number J3-50107).

Conflict of interest

The authors declare no conflict of interest.

Author details

Zala Lužnik Marzidovšek[1*], Janina Simončič[2], Petra Schollmayer[1], Elvira Maličev[3], Primož Rožman[3] and Marko Hawlina[1,4]

1 Eye Hospital, University Medical Centre Ljubljana, Ljubljana, Slovenia

2 Medical Faculty, University of Ljubljana, Ljubljana, Slovenia

3 Blood Transfusion Centre of Slovenia, Ljubljana, Slovenia

4 Department of Ophthalmology, Medical Faculty, University of Ljubljana, Ljubljana, Slovenia

*Address all correspondence to: zala.luznik@kclj.si

References

[1] Purves D, Augustine GJ, Fitzpatrick D, Katz LC, LaMantia AS, McNamara JO, et al. Anatomy of the eye. In: Neuroscience [Internet]. 2nd ed. Sinauer Associates; 2001. Available from: https://www.ncbi.nlm.nih.gov/books/NBK11120/ [Accessed: May 05, 2024]

[2] Sridhar MS. Anatomy of cornea and ocular surface. Indian Journal of Ophthalmology. 2018;**66**:190-194

[3] Burton MJ, Ramke J, Marques AP, Bourne RRA, Congdon N, Jones I, et al. The Lancet Global Health Commission on Global Eye Health: Vision beyond 2020. The Lancet Global Health. 2021;**9**:e489-e551

[4] European Medicines Agency. Advanced therapy medicinal products: Overview [Internet]. Available from: https://www.ema.europa.eu/en/human-regulatory-overview/advanced-therapy-medicinal-products-overview [Accessed: May 05, 2024]

[5] Dzobo K, Thomford NE, Senthebane DA, Shipanga H, Rowe A, Dandara C, et al. Advances in regenerative medicine and tissue engineering: Innovation and transformation of medicine. Stem Cells International. 2018;**2018**:2495848

[6] Alió Del Barrio JL, De la Mata A, De Miguel MP, Arnalich-Montiel F, Nieto-Miguel T, El Zarif M, et al. Corneal regeneration using adipose-derived mesenchymal stem cells. Cells. 2022;**16**(11):2549

[7] Trounson A, McDonald C. Stem cell therapies in clinical trials: Progress and challenges. Cell Stem Cell. 2015;**2**(17):11-22

[8] Blenkinsop TA, Corneo B, Temple S, Stern JH. Ophthalmologic stem cell transplantation therapies. Regenerative Medicine. 2012;7:32-39

[9] Gonzalez MA, Bernad A. Characteristics of adult stem cells. Advances in Experimental Medicine and Biology. 2012;**741**:103-120

[10] Zakrzewski W, Dobrzyński M, Szymonowicz M, Rybak Z. Stem cells: Past, present, and future. Stem Cell Research & Therapy. 2019;**26**(10):68

[11] Takahashi K, Yamanaka S. Induction of pluripotent stem cells from mouse embryonic and adult fibroblast cultures by defined factors. Cell. 2006;**25**(126):663-676

[12] Ullah I, Subbarao RB, Rho GJ. Human mesenchymal stem cells - current trends and future prospective. Bioscience Reports. 2015;**35**:e00191

[13] Dominici M, Le Blanc K, Mueller I, Slaper-Cortenbach I, Marini F, Krause D, et al. Minimal criteria for defining multipotent mesenchymal stromal cells. The International Society for Cellular Therapy position statement. Cytotherapy. 2006;**8**:315-317

[14] Xiao Y, McGhee CNJ, Zhang J. Adult stem cells in the eye: Identification, characterisation, and therapeutic application in ocular regeneration – A review. Clinical & Experimental Ophthalmology. 2024;**52**:148-166

[15] De Luca M, Pellegrini G, Green H. Regeneration of squamous epithelia from stem cells of cultured grafts. Regenerative Medicine. 2006;**1**:45-57

[16] Lužnik Z, Hawlina M, Ferrari S, Ponzin D, Schollmayer P. Ocular

surface reconstruction in limbal stem cell deficiency: Current treatment options and perspectives. Expert Review of Ophthalmology. 2016;**2**(12):43-56

[17] Dziasko MA, Armer HE, Levis HJ, Shortt AJ, Tuft S, Daniels JT. Localisation of epithelial cells capable of holoclone formation in vitro and direct interaction with stromal cells in the native human limbal crypt. PLoS One. 2014;**9**:e94283

[18] Hashmani K, Branch MJ, Sidney LE, Dhillon PS, Verma M, McIntosh OD, et al. Characterization of corneal stromal stem cells with the potential for epithelial transdifferentiation. Stem Cell Research & Therapy. 2013;**24**(4):75

[19] Du Y, Funderburgh ML, Mann MM, Sundarraj N, Funderburgh JL. Multipotent stem cells in human corneal stroma. Stem Cells (Dayton, Ohio). 2005;**23**:1266-1275

[20] DelMonte DW, Kim T. Anatomy and physiology of the cornea. Journal of Cataract and Refractive Surgery. 2011;**1**(37):588-598

[21] Yu WY, Sheridan C, Grierson I, Mason S, Kearns V, Lo ACY, et al. Progenitors for the corneal endothelium and trabecular meshwork: A potential source for personalized stem cell therapy in corneal endothelial diseases and glaucoma. Journal of Biomedicine & Biotechnology. 2011;**2011**:412743

[22] Karl MO, Reh TA. Regenerative medicine for retinal diseases: Activating the endogenous repair mechanisms. Trends in Molecular Medicine. 2010;**16**:193-202

[23] Huang C, Albon JV, Ljubimov A, Grant MB. Chapter 60 - Stem cells in the eye. In: Lanza R, Langer R, Vacanti JP, Atala A, editors. Principles of Tissue Engineering (Fifth Edition) [Internet]. Cambridge, MA, USA: Academic Press; 2020. pp. 1115-1133. Available from: https://www.sciencedirect.com/science/article/pii/B9780128184226000629 [Accessed: May 05, 2024]

[24] Louie HH, Mugisho OO, Chamley LW, Rupenthal ID. Extracellular vesicles as biomarkers and therapeutics for inflammatory eye diseases. Molecular Pharmaceutics. 2023;**2**(20):23-40

[25] Kolenc A, Maličev E. Current methods for analysing mesenchymal stem cell-derived extracellular vesicles. International Journal of Molecular Sciences. 2024;**25**:3439

[26] Holcar M, Kandušer M, Lenassi M. Blood nanoparticles - influence on extracellular vesicle isolation and characterization. Frontiers in Pharmacology. 2021;**12**:773844

[27] Yáñez-Mó M, Siljander PRM, Andreu Z, Bedina Zavec A, Borràs FE, Buzas EI, et al. Biological properties of extracellular vesicles and their physiological functions. Journal of Extracellular Vesicles. 2015;**1**(4):27066

[28] Han Y, Yang J, Fang J, Zhou Y, Candi E, Wang J, et al. The secretion profile of mesenchymal stem cells and potential applications in treating human diseases. Signal Transduction and Targeted Therapy. 2022;**21**(7):92

[29] Welsh JA, Goberdhan DCI, O'Driscoll L, Buzas EI, Blenkiron C, Bussolati B, et al. Journal of Extracellular Vesicles. 2024;**13**:e12404

[30] Arnalich-Montiel F, Pastor S, Blazquez-Martinez A, Fernandez-Delgado J, Nistal M, Alio JL, et al. Adipose-derived stem cells are a source for cell therapy of the corneal stroma. Stem Cells (Dayton, Ohio). 2008;**26**:570-579

[31] Mittal S, Omoto M, Amouzegar A, Sahu A, Rezazadeh A, Katikireddy K, et al. Restoration of corneal transparency by mesenchymal stem cells. Stem Cell Reports. 2016;**7**:583-590

[32] Du Y, Carlson EC, Funderburgh ML, Birk DE, Pearlman E, Guo N, et al. Stem cell therapy restores transparency to defective murine corneas. Stem Cells (Dayton, Ohio). 2009;**27**:1635-1642

[33] Al-Jaibaji O, Swioklo S, Connon CJ. Mesenchymal stromal cells for ocular surface repair. Expert Opinion on Biological Therapy. 2019;**3**(19):643-653

[34] Liu H, Zhang J, Liu CY, Wang IJ, Sieber M, Chang J, et al. Cell therapy of congenital corneal diseases with umbilical mesenchymal stem cells: Lumican null mice. PLoS One. 2010;**19**(5):e10707

[35] Moisseiev E, Anderson JD, Oltjen S, Goswami M, Zawadzki RJ, Nolta JA, et al. Protective effect of intravitreal administration of exosomes derived from mesenchymal stem cells on retinal ischemia. Current Eye Research. 2017;**42**:1358-1367

[36] Johnson TV, Bull ND, Hunt DP, Marina N, Tomarev SI, Martin KR. Neuroprotective effects of intravitreal mesenchymal stem cell transplantation in experimental glaucoma. Investigative Ophthalmology & Visual Science. 2010;**51**:2051-2059

[37] da Silva-Junior AJ, Mesentier-Louro LA, Nascimento-dos-Santos G, Teixeira-Pinheiro LC, Vasques JF, Chimeli-Ormonde L, et al. Human mesenchymal stem cell therapy promotes retinal ganglion cell survival and target reconnection after optic nerve crush in adult rats. Stem Cell Research & Therapy. 2021;**19**(12):69

[38] Oswald J, Kegeles E, Minelli T, Volchkov P, Baranov P. Transplantation of miPSC/mESC-derived retinal ganglion cells into healthy and glaucomatous retinas. Molecular Therapy - Methods & Clinical Development. 2021;**21**:180-198

[39] Oh JY, Ko JH, Lee HJ, Kim MK, Lee JH, Wee WR. The antigenicity of ex vivo cultivated human corneal limbal epithelial and stromal cells: Temporal changes in vitro. Cornea. 2010;**29**:1302

[40] Baradaran-Rafii A, Ebrahimi M, Kanavi MR, Taghi-Abadi E, Aghdami N, Eslani M, et al. Midterm outcomes of autologous cultivated limbal stem cell transplantation with or without penetrating keratoplasty. Cornea. 2010;**29**:502-509

[41] Basu S, Ali H, Sangwan VS. Clinical outcomes of repeat autologous cultivated limbal epithelial transplantation for ocular surface burns. American Journal of Ophthalmology. 2012;**153**:643-650, 650.e1-2

[42] Bobba S, Chow S, Watson S, Di Girolamo N. Clinical outcomes of xeno-free expansion and transplantation of autologous ocular surface epithelial stem cells via contact lens delivery: A prospective case series. Stem Cell Research & Therapy. 2015;**12**(6):23

[43] Colabelli Gisoldi RAM, Pocobelli A, Villani CM, Amato D, Pellegrini G. Evaluation of molecular markers in corneal regeneration by means of autologous cultures of limbal cells and keratoplasty. Cornea. 2010;**29**:715-722

[44] Di Girolamo N, Bosch M, Zamora K, Coroneo MT, Wakefield D, Watson SL. A contact lens-based technique for expansion and transplantation of autologous epithelial progenitors for ocular surface reconstruction. Transplantation. 2009;**27**(87):1571-1578

[45] Di Iorio E, Ferrari S, Fasolo A, Böhm E, Ponzin D, Barbaro V. Techniques for culture and assessment of limbal stem cell grafts. The Ocular Surface. 2010;**8**:146-153

[46] Fasolo A, Pedrotti E, Passilongo M, Marchini G, Monterosso C, Zampini R, et al. Safety outcomes and long-term effectiveness of ex vivo autologous cultured limbal epithelial transplantation for limbal stem cell deficiency. The British Journal of Ophthalmology. 2017;**101**:640-649

[47] Ganger A, Vanathi M, Mohanty S, Tandon R. Long-term outcomes of cultivated limbal epithelial transplantation: Evaluation and comparison of results in children and adults. BioMed Research International. 2015;**2015**:480983

[48] Grueterich M, Espana EM, Touhami A, Ti SE, Tseng SCG. Phenotypic study of a case with successful transplantation of ex vivo expanded human limbal epithelium for unilateral total limbal stem cell deficiency. Ophthalmology. 2002;**109**:1547-1552

[49] Kawashima M, Kawakita T, Satake Y, Higa K, Shimazaki J. Phenotypic study after cultivated limbal epithelial transplantation for limbal stem cell deficiency. Archives of Ophthalmology (Chicago, Ill.: 1960). 1960;**2007**(125):1337-1344

[50] Kolli S, Ahmad S, Lako M, Figueiredo F. Successful clinical implementation of corneal epithelial stem cell therapy for treatment of unilateral limbal stem cell deficiency. Stem cells (Dayton, Ohio). 2010;**31**(28):597-610

[51] Nakamura T, Inatomi T, Sotozono C, Koizumi N, Kinoshita S. Successful primary culture and autologous transplantation of corneal limbal epithelial cells from minimal biopsy for unilateral severe ocular surface disease. Acta Ophthalmologica Scandinavica. 2004;**82**:468-471

[52] Pathak M, Cholidis S, Haug K, Shahdadfar A, Moe MC, Nicolaissen B, et al. Clinical transplantation of ex vivo expanded autologous limbal epithelial cells using a culture medium with human serum as single supplement: A retrospective case series. Acta Ophthalmologica. 2013;**91**:769-775

[53] Pellegrini G, Traverso CE, Franzi AT, Zingirian M, Cancedda R, De Luca M. Long-term restoration of damaged corneal surfaces with autologous cultivated corneal epithelium. The Lancet. 1997;**349**(9057):990-993

[54] Prabhasawat P, Ekpo P, Uiprasertkul M, Chotikavanich S, Tesavibul N. Efficacy of cultivated corneal epithelial stem cells for ocular surface reconstruction. Clinical ophthalmology (Auckland, N.Z.). 2012;**6**:1483-1492

[55] Rama P, Matuska S, Paganoni G, Spinelli A, De Luca M, Pellegrini G. Limbal stem-cell therapy and long-term corneal regeneration. The New England Journal of Medicine. 2010;**8**(363):147-155

[56] Rama P, Bonini S, Lambiase A, Golisano O, Paterna P, De Luca M, et al. Autologous fibrin-cultured limbal stem cells permanently restore the corneal surface of patients with total limbal stem cell deficiency. Transplantation. 2001;**15**(72):1478-1485

[57] Ramírez BE, Sánchez A, Herreras JM, Fernández I, García-Sancho J, Nieto-Miguel T, et al. Stem cell therapy for corneal epithelium

regeneration following good manufacturing and clinical procedures. BioMed Research International. 2015;**2015**:408495

[58] Sangwan VS, Vemuganti GK, Singh S, Balasubramanian D. Successful reconstruction of damaged ocular outer surface in humans using limbal and conjuctival stem cell culture methods. Bioscience Reports. 2003;**23**:169-174

[59] Sangwan VS, Matalia HP, Vemuganti GK, Ifthekar G, Fatima A, Singh S, et al. Early results of penetrating keratoplasty after cultivated limbal epithelium transplantation. Archives of Ophthalmology (Chicago, Ill.: 1960). 1960;**2005**(123):334-340

[60] Sangwan VS, Matalia HP, Vemuganti GK, Fatima A, Ifthekar G, Singh S, et al. Clinical outcome of autologous cultivated limbal epithelium transplantation. Indian Journal of Ophthalmology. 2006;**54**:29-34

[61] Sangwan VS, Basu S, Vemuganti GK, Sejpal K, Subramaniam SV, Bandyopadhyay S, et al. Clinical outcomes of xeno-free autologous cultivated limbal epithelial transplantation: A 10-year study. The British Journal of Ophthalmology. 2011;**95**:1525-1529

[62] Schwab IR. Cultured corneal epithelia for ocular surface disease. Transactions of the American Ophthalmological Society. 1999;**97**:891-986

[63] Schwab IR, Reyes M, Isseroff RR. Successful transplantation of bioengineered tissue replacements in patients with ocular surface disease. Cornea. 2000;**19**:421-426

[64] Sejpal K, Ali MH, Maddileti S, Basu S, Ramappa M, Kekunnaya R, et al. Cultivated limbal epithelial transplantation in children with ocular surface burns. JAMA Ophthalmology. 2013;**13**:731-736

[65] Sharma S, Tandon R, Mohanty S, Sharma N, Vanathi M, Sen S, et al. Culture of corneal limbal epithelial stem cells: Experience from benchtop to bedside in a tertiary care hospital in India. Cornea. 2011;**30**:1223-1232

[66] Sharma S, Tandon R, Mohanty S, Kashyap S, Vanathi M. Phenotypic evaluation of severely damaged ocular surface after reconstruction by cultured limbal epithelial cell transplantation. Ophthalmic Research. 2013;**50**:59-64

[67] Shimazaki J, Higa K, Morito F, Dogru M, Kawakita T, Satake Y, et al. Factors influencing outcomes in cultivated limbal epithelial transplantation for chronic cicatricial ocular surface disorders. American Journal of Ophthalmology. 2007;**143**:945-953

[68] Shortt AJ, Secker GA, Rajan MS, Meligonis G, Dart JK, Tuft SJ, et al. Ex vivo expansion and transplantation of limbal epithelial stem cells. Ophthalmology. 2008;**115**:1989-1997

[69] Subramaniam SV, Sejpal K, Fatima A, Gaddipati S, Vemuganti GK, Sangwan VS. Coculture of autologous limbal and conjunctival epithelial cells to treat severe ocular surface disorders: Long-term survival analysis. Indian Journal of Ophthalmology. 2013;**61**:202-207

[70] Tsai RJ, Li LM, Chen JK. Reconstruction of damaged corneas by transplantation of autologous limbal epithelial cells. The New England Journal of Medicine. 2000;**13**(343):86-93

[71] Vazirani J, Basu S, Kenia H, Ali MH, Kacham S, Mariappan I, et al. Unilateral

partial limbal stem cell deficiency: Contralateral versus ipsilateral autologous cultivated limbal epithelial transplantation. American Journal of Ophthalmology. 2014;**157**:584-590.e1-2

[72] Zakaria N, Possemiers T, Dhubhghaill SN, Leysen I, Rozema J, Koppen C, et al. Results of a phase I/II clinical trial: Standardized, non-xenogenic, cultivated limbal stem cell transplantation. Journal of Translational Medicine. 2014;**3**(12):58

[73] Jurkunas U, Yin J, Johns L, Li S, Negre H, Shaw K, et al. Cultivated autologous limbal epithelial cell (CALEC) transplantation: Development of manufacturing process and clinical evaluation of feasibility and safety. Science Advances. 2023;**9**:eadg6470

[74] Kengpunpanich S, Chirapapaisan C, Ngowyutagon P, Chotikavanich S, Sikarinkul R, Taetrongchit N, et al. Comparative analysis of long-term results of three epithelial cell transplantation procedures for treating limbal stem cell deficiency. The Ocular Surface. 2024;**32**:71-80

[75] Pellegrini G, Ardigò D, Milazzo G, Iotti G, Guatelli P, Pelosi D, et al. Navigating market authorization: The path Holoclar took to become the first stem cell product approved in the European Union. Stem Cells Translational Medicine. 2018;**7**:146-154

[76] European Medicines Agency. Holoclar [Internet]. 2015. Available from: https://www.ema.europa.eu/en/medicines/human/EPAR/holoclar [Accessed: May 05, 2024]

[77] Japan Tissue Engineering Co., Ltd. Medical device evaluation division pharmaceutical safety and environmental health bureau. Report on the Deliberation Results (nepic). [Internet]. Medical Device Evaluation Division Pharmaceutical Safety and Environmental Health Bureau Ministry of Health, Labour and Welfare. Report on the Deliberation Results (nepic). 2020; 2019. Available from: https://www.pmda.go.jp/files/000237511.pdf [Accessed: May 05, 2024]

[78] Oie Y, Sugita S, Yokokura S, Nakazawa T, Tomida D, Satake Y, et al. Clinical trial of autologous cultivated limbal epithelial cell sheet transplantation for patients with limbal stem cell deficiency. Ophthalmology. 2023;**130**:608-614

[79] Borderie VM, Ghoubay D, Georgeon C, Borderie M, de Sousa C, Legendre A, et al. Long-term results of cultured limbal stem cell versus limbal tissue transplantation in stage III limbal deficiency. Stem Cells Translational Medicine. 2019;**8**:1230-1241

[80] Campbell JDM, Ahmad S, Agrawal A, Bienek C, Atkinson A, Mcgowan NWA, et al. Allogeneic ex vivo expanded corneal epithelial stem cell transplantation: A randomized controlled clinical trial. Stem Cells Translational Medicine. 2019;**28**(8):323-331

[81] Le Q, Chauhan T, Yung M, Tseng CH, Deng SX. Outcomes of limbal stem cell transplant: A meta-analysis. JAMA Ophthalmology. 2020;**1**(138):660-670

[82] Mishan MA, Yaseri M, Baradaran-Rafii A, Kanavi MR. Systematic review and meta-analysis investigating autograft versus allograft cultivated limbal epithelial transplantation in limbal stem cell deficiency. International Ophthalmology. 2019;**39**:2685-2696

[83] Zhao Y, Ma L. Systematic review and meta-analysis on transplantation of ex

vivo cultivated limbal epithelial stem cell on amniotic membrane in limbal stem cell deficiency. Cornea. 2015;**34**:592-600

[84] Kohji N, Masayuki Y, Yasutaka H, Katsuhiko W, Kazuaki Y, Eijiro A, et al. Corneal reconstruction with tissue-engineered cell sheets composed of autologous oral mucosal epithelium. The New England Journal of Medicine. 2004;**351**:1187-1196

[85] Nakamura T, Inatomi T, Sotozono C, Amemiya T, Kanamura N, Kinoshita S. Transplantation of cultivated autologous oral mucosal epithelial cells in patients with severe ocular surface disorders. The British Journal of Ophthalmology. 2004;**88**:1280-1284

[86] Aketa N, Kasai M, Noda S, Asano J, Kunieda A, Kawanishi S, et al. Insights into the clinical development of regenerative medical products through a comparison of three cell-based products recently approved for limbal stem cell deficiency. The Ocular Surface. 2023;**29**:220-225

[87] Calonge M, Pérez I, Galindo S, Nieto-Miguel T, López-Paniagua M, Fernández I, et al. A proof-of-concept clinical trial using mesenchymal stem cells for the treatment of corneal epithelial stem cell deficiency. Translational Research: The Journal of Laboratory and Clinical Medicine. 2019;**206**:18-40

[88] Galindo S, de la Mata A, López-Paniagua M, Herreras JM, Pérez I, Calonge M, et al. Subconjunctival injection of mesenchymal stem cells for corneal failure due to limbal stem cell deficiency: State of the art. Stem Cell Research & Therapy. 2021;**13**(12):60

[89] de Los B, Bueis A, Vidal Arranz C, Del Hierro-Zarzuelo A, Díaz Valle D, Méndez Fernández R, et al. Long-term effects of adipose-derived stem cells for the treatment of bilateral limbal stem cell deficiency. Current Eye Research. 2024;**49**:345-353

[90] Calonge M, Nieto-Miguel T, de la Mata A, Galindo S, Herreras JM, López-Paniagua M. Goals and challenges of stem cell-based therapy for corneal blindness due to limbal deficiency. Pharmaceutics. 2021;**16**(13):1483

[91] Alió Del Barrio JL, El Zarif M, Azaar A, Makdissy N, Khalil C, Harb W, et al. Corneal stroma enhancement with decellularized stromal laminas with or without stem cell recellularization for advanced keratoconus. American Journal of Ophthalmology. 2018;**186**:47-58

[92] Alió Del Barrio JL, El Zarif M, de Miguel MP, Azaar A, Makdissy N, Harb W, et al. Cellular therapy with human autologous adipose-derived adult stem cells for advanced keratoconus. Cornea. 2017;**36**:952-960

[93] Alió JL, Alió Del Barrio JL, El Zarif M, Azaar A, Makdissy N, Khalil C, et al. Regenerative surgery of the corneal stroma for advanced keratoconus: 1-year outcomes. American Journal of Ophthalmology. 2019;**203**:53-68

[94] El Zarif M, Alió JL, Alió del Barrio JL, De Miguel MP, Abdul Jawad K, Makdissy N. Corneal stromal regeneration: A review of human clinical studies in keratoconus treatment. Frontiers in Medicine. 2021;**23**(8):650724

[95] Kinoshita S, Koizumi N, Ueno M, Okumura N, Imai K, Tanaka H, et al. Injection of cultured cells with a ROCK inhibitor for bullous keratopathy. The New England Journal of Medicine. 2018;**15**(378):995-1003

[96] Numa K, Imai K, Ueno M, Kitazawa K, Tanaka H, Bush JD, et al.

Five-year follow-up of first 11 patients undergoing injection of cultured corneal endothelial cells for corneal endothelial failure. Ophthalmology. 2021;**128**:504-514

[97] Gaddam S, Periasamy R, Gangaraju R. Adult stem cell therapeutics in diabetic retinopathy. International Journal of Molecular Sciences. 2019;**30**(20):4876

[98] Gu X, Yu X, Zhao C, Duan P, Zhao T, Liu Y, et al. Efficacy and safety of autologous bone marrow mesenchymal stem cell transplantation in patients with diabetic retinopathy. Cellular Physiology & Biochemistry. International Journal of Experimental Cellular Physiology, Biochemistry and Pharmacology. 2018;**49**:40-52

[99] Schwartz SD, Hubschman JP, Heilwell G, Franco-Cardenas V, Pan CK, Ostrick RM, et al. Embryonic stem cell trials for macular degeneration: A preliminary report. The Lancet. 2012;**25**(379):713-720

[100] Schwartz SD, Regillo CD, Lam BL, Eliott D, Rosenfeld PJ, Gregori NZ, et al. Human embryonic stem cell-derived retinal pigment epithelium in patients with age-related macular degeneration and Stargardt's macular dystrophy: Follow-up of two open-label phase 1/2 studies. The Lancet. 2015;**7**(385):509-516

[101] Li SY, Liu Y, Wang L, Wang F, Zhao TT, Li QY, et al. A phase I clinical trial of human embryonic stem cell-derived retinal pigment epithelial cells for early-stage Stargardt macular degeneration: 5-years' follow-up. Cell Proliferation. 2021;**54**:e13100

[102] Sung Y, Lee MJ, Choi J, Jung SY, Chong SY, Sung JH, et al. Long-term safety and tolerability of subretinal transplantation of embryonic stem cell-derived retinal pigment epithelium in Asian Stargardt disease patients. The British Journal of Ophthalmology. 2021;**105**:829-837

[103] Brant Fernandes RA, Lojudice FH, Zago Ribeiro L, Santos da Cruz NF, Polizelli MU, Cristovam PC, et al. Transplantation of Subretinal Stem Cell-Derived Retinal Pigment Epithelium for Stargardt Disease: A phase I clinical trial. Retina Phila Pa. 2023;**1**(43):263-274

[104] Oner A, Gonen ZB, Sevim DG, Smim Kahraman N, Unlu M. Suprachoroidal adipose tissue-derived mesenchymal stem cell implantation in patients with dry-type age-related macular degeneration and Stargardt's macular dystrophy: 6-month follow-up results of a phase 2 study. Cellular Reprogramming. 2018;**20**:329-336

[105] Weiss JN, Levy S. Stem cell ophthalmology treatment study (SCOTS): Bone marrow-derived stem cells in the treatment of Stargardt disease. Medicine. 2021;**3**(8):10

[106] Cotrim CC, Vieira Messias AM, Jorge R, Siqueira RC. Intravitreal use of a bone marrow mononuclear fraction (BMMF) containing CD34+ cells in patients with Stargardt type macular dystrophy. Stem Cells International. 2020;**2020**(10):8828256

[107] Oner A, Gonen ZB, Sinim N, Cetin M, Ozkul Y. Subretinal adipose tissue-derived mesenchymal stem cell implantation in advanced stage retinitis pigmentosa: A phase I clinical safety study. Stem Cell Research & Therapy. 2016;**1**(7):178

[108] Tuekprakhon A, Sangkitporn S, Trinavarat A, Pawestri AR, Vamvanij V, Ruangchainikom M, et al. Intravitreal autologous mesenchymal stem cell transplantation: A non-randomized

phase I clinical trial in patients with retinitis pigmentosa. Stem Cell Research & Therapy. 2021;**9**(12):52

[109] Kahraman NS, Oner A. Umbilical cord derived mesenchymal stem cell implantation in retinitis pigmentosa: A 6-month follow-up results of a phase 3 trial. International Journal of Ophthalmology. 2020;**13**:1423-1429

[110] Liu Y, Xu HW, Wang L, Li SY, Zhao CJ, Hao J, et al. Human embryonic stem cell-derived retinal pigment epithelium transplants as a potential treatment for wet age-related macular degeneration. Cell Discovery. 2018;**11**(4):50

[111] Kashani AH, Lebkowski JS, Rahhal FM, Avery RL, Salehi-Had H, Dang W, et al. A bioengineered retinal pigment epithelial monolayer for advanced, dry age-related macular degeneration. Science Translational Medicine. 2018;**4**(10):eaao4097

[112] da Cruz L, Fynes K, Georgiadis O, Kerby J, Luo YH, Ahmado A, et al. Phase 1 clinical study of an embryonic stem cell-derived retinal pigment epithelium patch in age-related macular degeneration. Nature Biotechnology. 2018;**36**:328-337

[113] Song WK, Park KM, Kim HJ, Lee JH, Choi J, Chong SY, et al. Treatment of macular degeneration using embryonic stem cell-derived retinal pigment epithelium: Preliminary results in Asian patients. Stem Cell Reports. 2015;**12**(4):860-872

[114] Sung Y, Ryul LD, Han SS, Chong SY, Choi SW, Song WK. Treatment of macular degeneration using human somatic cell nuclear transfer embryonic stem cell derived retinal pigment epithelium: 1-year results in an Asian patient. Investigative Ophthalmology & Visual Science. 2019;**22**(60):3938

[115] Yin ZQ, Liu Y, Li S, Xu HW, Wang Y, Qian C, et al. Clincal trial: Subretinal transplantation of CTS hESC derived RPE in the treatment of wet age-related macular degeneration (wAMD). Investigative Ophthalmology & Visual Science. 2016;**26**(57):3742

[116] Heier JS, Ho AC, Samuel MA, Chang T, Riemann CD, Kitchens JW, et al. Safety and efficacy of subretinally administered palucorcel for geographic atrophy of age-related macular degeneration: Phase 2b study. Ophthalmology Retina. 2020;**4**:384-393

[117] Takagi S, Mandai M, Gocho K, Hirami Y, Yamamoto M, Fujihara M, et al. Evaluation of transplanted autologous induced pluripotent stem cell-derived retinal pigment epithelium in exudative age-related macular degeneration. Ophthalmology Retina. 2019;**3**:850-859

[118] Thomson JA, Itskovitz-Eldor J, Shapiro SS, Waknitz MA, Swiergiel JJ, Marshall VS, et al. Embryonic stem cell lines derived from human blastocysts. Science. 1998;**282**(5391):1145-1147

[119] Miotti G, Parodi PC, Zeppieri M. Stem cell therapy in ocular pathologies in the past 20 years. World Journal of Stem Cells. 2021;**13**:366

[120] Vilela CAP, Messias A, Calado RT, Siqueira RC, Silva MJL, Covas DT, et al. Retinal function after intravitreal injection of autologous bone marrow-derived mesenchymal stromal cells in advanced glaucoma. Documenta Ophthalmologica. 2021;**143**:33-38

[121] Weng J, He C, Lai P, Luo C, Guo R, Wu S, et al. Mesenchymal stromal cells treatment attenuates

dry eye in patients with chronic graft-versus-host disease. Molecular Therapy is the Official Journal of the American Society of Gene Therapy. 2012;**20**:2347-2354

[122] Zhou T, He C, Lai P, Yang Z, Liu Y, Xu H, et al. miR-204–containing exosomes ameliorate GVHD-associated dry eye disease. Science Advances. 2022;**8**:eabj9617

[123] Buono L, Scalabrin S, De Iuliis M, Tanzi A, Grange C, Tapparo M, et al. Mesenchymal stem cell-derived extracellular vesicles protect human corneal endothelial cells from endoplasmic reticulum stress-mediated apoptosis. International Journal of Molecular Sciences. 2021;**6**(22):4930

[124] Pan D, Chang X, Xu M, Zhang M, Zhang S, Wang Y, et al. UMSC-derived exosomes promote retinal ganglion cells survival in a rat model of optic nerve crush. Journal of Chemical Neuroanatomy. 2019;**96**:134-139

[125] Mead B, Amaral J, Tomarev S. Mesenchymal stem cell-derived small extracellular vesicles promote neuroprotection in rodent models of glaucoma. Investigative Ophthalmology & Visual Science. 2018;**59**:702-714

[126] Flaxman SR, Bourne RRA, Resnikoff S, Ackland P, Braithwaite T, Cicinelli MV, et al. Global causes of blindness and distance vision impairment 1990-2020: A systematic review and meta-analysis. The Lancet Global Health. 2017;**5**:e1221-e1234

[127] Gain P, Jullienne R, He Z, Aldossary M, Acquart S, Cognasse F, et al. Global survey of corneal transplantation and eye banking. JAMA Ophthalmology. 2016;**134**:167-173

[128] Tan DTH, Dart JKG, Holland EJ, Kinoshita S. Corneal transplantation. The Lancet. 2012;**379**(9827):1749-1761

[129] Holland EJ. Management of limbal stem cell deficiency: A historical perspective, past, present, and future. Cornea. 2015;**34**(Suppl. 10):S9-S15

[130] Reinhard T, Spelsberg H, Henke L, Kontopoulos T, Enczmann J, Wernet P, et al. Long-term results of allogeneic penetrating limbo-keratoplasty in total limbal stem cell deficiency. Ophthalmology. 2004;**111**:775-782

[131] Burman S, Sangwan V. Cultivated limbal stem cell transplantation for ocular surface reconstruction. Clinical Ophthalmology (Auckland, N.Z.). 2008;**2**:489-502

[132] Kumar A, Yun H, Funderburgh ML, Du Y. Regenerative therapy for the cornea. Progress in Retinal and Eye Research. 2022;**87**:101011

[133] Shimmura S, Inagaki E, Hirayama M, Hatou S. The cornea: An ideal tissue for regenerative medicine. The Keio Journal of Medicine. 2024;**73**:1-7

[134] Ghareeb AE, Lako M, Figueiredo FC. Recent advances in stem cell therapy for limbal stem cell deficiency: A narrative review. Ophthalmology and Therapy. 2020;**1**(9):809-831

[135] Jurkunas U, Johns L, Armant M. Cultivated autologous limbal epithelial cell transplantation: New frontier in the treatment of limbal stem cell deficiency. American Journal of Ophthalmology. 2022;**1**(239):244-268

[136] Tsubota K, Pflugfelder SC, Liu Z, Baudouin C, Kim HM, Messmer EM, et al. Defining dry eye from a clinical

perspective. International Journal of Molecular Sciences. 2020;**4**(21):9271

[137] Volatier T, Cursiefen C, Notara M. Current advances in corneal stromal stem cell biology and therapeutic applications. Cells. 2024;**13**:163

[138] Dos Santos A, Balayan A, Funderburgh ML, Ngo J, Funderburgh JL, Deng SX. Differentiation capacity of human mesenchymal stem cells into keratocyte lineage. Investigative Ophthalmology & Visual Science. 2019;**1**(60):3013-3023

[139] Naylor RW, McGhee CNJ, Cowan CA, Davidson AJ, Holm TM, Sherwin T. Derivation of corneal keratocyte-like cells from human induced pluripotent stem cells. PLoS One. 2016;**11**(10):e0165464

[140] Tuft SJ, Coster DJ. The corneal endothelium. Eye. 1990;**4**:389-424

[141] Bourne WM. Biology of the corneal endothelium in health and disease. Eye. 2003;**17**:912-918

[142] Price MO, Mehta JS, Jurkunas UV, Price FW. Corneal endothelial dysfunction: Evolving understanding and treatment options. Progress in Retinal and Eye Research. 2021;**82**:100904

[143] Català P, Thuret G, Skottman H, Mehta JS, Parekh M, Ní Dhubhghaill S, et al. Approaches for corneal endothelium regenerative medicine. Progress in Retinal and Eye Research. 2022;**1**(87):100987

[144] Zhu C, Joyce NC. Proliferative response of corneal endothelial cells from young and older donors. Investigative Ophthalmology & Visual Science. 2004;**1**(45):1743-1751

[145] Amano S, Mimura T, Yamagami S, Osakabe Y, Miyata K. Properties of corneas reconstructed with cultured human corneal endothelial cells and human corneal stroma. Japanese Journal of Ophthalmology. 2005;**49**:448-452

[146] Zhao JJ, Afshari NA. Generation of human corneal endothelial cells via In vitro ocular lineage restriction of pluripotent stem cells. Investigative Ophthalmology & Visual Science. 2016;**57**:6878-6884

[147] Stern JH, Tian Y, Funderburgh J, Pellegrini G, Zhang K, Goldberg JL, et al. Regenerating eye tissues to preserve and restore vision. Cell Stem Cell. 2018;**1**(22):834-849

[148] Khaboushan AS, Ebadpour N, Moghadam MMJ, Rezaee Z, Kajbafzadeh AM, Zolbin MM. Cell therapy for retinal degenerative disorders: A systematic review and three-level meta-analysis. Journal of Translational Medicine. 2024;**2**(22):227

[149] Nash BM, Wright DC, Grigg JR, Bennetts B, Jamieson RV. Retinal dystrophies, genomic applications in diagnosis and prospects for therapy. Translational Pediatrics. 2015;**4**:139-163

[150] Kohli P, Tripathy K, Kaur K. Stargardt Disease. In: StatPearls [Internet]. Treasure Island (FL): StatPearls Publishing; 2024. Available from: http://www.ncbi.nlm.nih.gov/books/NBK587351/ [Accessed: May 05, 2024]

[151] Gheorghe A, Mahdi L, Musat O. Age-related macular degeneration. Romanian Journal of Ophthalmology. 2015;**59**:74-77

[152] Peyman GA, Blinder KJ, Paris CL, Alturki W, Nelson NC, Desai U. A technique for retinal pigment epithelium transplantation for age-related macular degeneration secondary to extensive

subfoveal scarring. Ophthalmic Surgery. 1991;**22**:102-108

[153] Gokoffski KK, Lam P, Alas B, Peng M, Ansorge H. Optic nerve regeneration: How will we get there? Journal of Neuro-ophthalmology: The Official Journal of the North American Neuro-Ophthalmology Society. 2020;**40**:234

[154] Wu S, Chang KC, Nahmou M, Goldberg JL. Induced pluripotent stem cells promote retinal ganglion cell survival after transplant. Investigative Ophthalmology & Visual Science. 2018;**59**:1571-1576

[155] Brianna, APK L, Wong YP. Applying stem cell therapy in intractable diseases: A narrative review of decades of progress and challenges. Stem Cell Investigation. 2022;**9**:4

[156] Mead B, Tomarev S. Extracellular vesicle therapy for retinal diseases. Progress in Retinal and Eye Research. 2020;**79**:100849

Section 4

Toxicity of SCT and Cellular Therapies

Chapter 24

CART Cell Toxicities and Management

*Nishanth Thalambedu, Sruthi Vellanki,
Anup Kumar Trikannad and Ankur Varma*

Abstract

Chimeric antigen receptor T-cell (CART) therapy is a novel therapeutic modality utilized for managing relapsed/refractory hematologic cancers. This approach entails the genetic modification of either autologous or allogeneic T-cells to incorporate chimeric antigen receptor (CAR) proteins, enabling them to target specific proteins present on the surface of cancer cells and causing cytotoxicity. Despite its promising therapeutic outcomes, there is a notable drawback wherein these modified cells may also attack normal tissue cells expressing similar surface proteins. This phenomenon of cross-reactivity often leads to adverse events (AEs), resulting in significant mortality and morbidity. This chapter focuses on the assessment and management of CART therapy-related toxicities.

Keywords: CART, lymphoma, myeloma, CRS, ICANS, HLH

1. Introduction

Chimeric antigen receptor T-cell (CART) therapy has revolutionized the treatment of hematologic malignancies, offering a new avenue of hope for patients with relapsed or refractory cancers. By genetically modifying a patient's T-cells to express a receptor specific to a tumor antigen, CART therapy allows for a targeted and potent immune response against cancer cells, ensuring precision and effectiveness. It is currently granted authorization by the Food and Drug Administration (FDA) for the treatment of acute lymphoblastic leukemia (ALL), large B-cell lymphoma (LBCL), follicular lymphoma (FL), mantle cell lymphoma (MCL), marginal zone lymphoma (MZL), and multiple myeloma (MM) [1]. The first CART therapy targeting CD-19, tisagenlecleucel (tisa-cel, ELIANA), was approved by the FDA in 2017 for the treatment of patients with ALL and was later approved for use in other hematologic malignancies like B-Cells lymphomas (JULIET). Around the same time, axicabtagene ciloleucel (axi-cel, ZUMA-1) was approved for relapsed/refractory (R/R) DLBCL and subsequently for R/R follicular lymphoma [2]. However, the nature of this immune activation can also lead to significant toxicities at various timelines, some of which

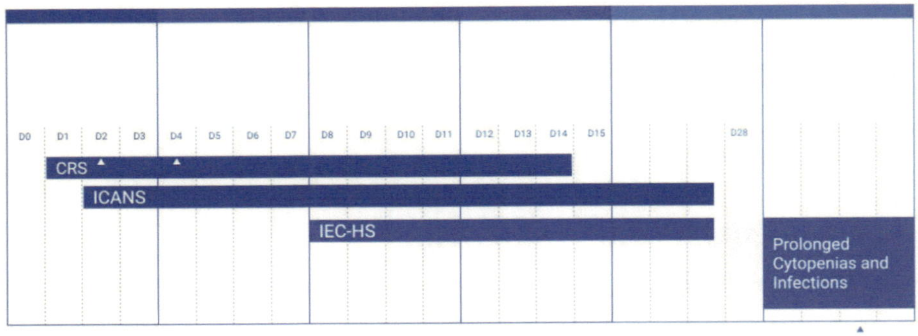

Figure 1.
Timeline of development of CART toxicities.

can be life-threatening (**Figure 1**). Understanding these toxicities, their underlying mechanisms, and appropriate management strategies is crucial for clinicians administering this therapy.

2. Overview of CART cell therapy

Currently, six FDA-approved CART cell therapies are available in the USA, with numerous ongoing clinical trials exploring the use of CART cell therapy for a wide range of diseases [1].

CART cells are generated through a multi-step process and depicted in **Figure 2** [4]:

i. Apheresis: T-cells are collected from the patient and are typically autologous but can also be derived from an allogeneic donor.

ii. Genetic modification: The T-cells are genetically modified to express a chimeric antigen receptor (CAR) that targets a specific tumor-associated antigen, such as CD19 in B-cell malignancies and B-cell maturation antigen (BCMA) in multiple myeloma.

iii. Expansion: The modified T-cells are expanded *ex vivo* to achieve the necessary cell dose.

iv. Preparation: The patient receives lymphodepletion chemotherapy to enhance the engraftment and efficacy of the CART cells.

v. Infusion: The CART cells are transfused into the patient, which will target and kill the cancer cells.

vi. Monitoring of the patient by the health care team for possible side effects and continued follow-up.

CART cell therapy is linked to significant side effects, with the most prominent toxicities being cytokine release syndrome (CRS) and immune effector cell-associated neurotoxicity syndrome (ICANS) [1].

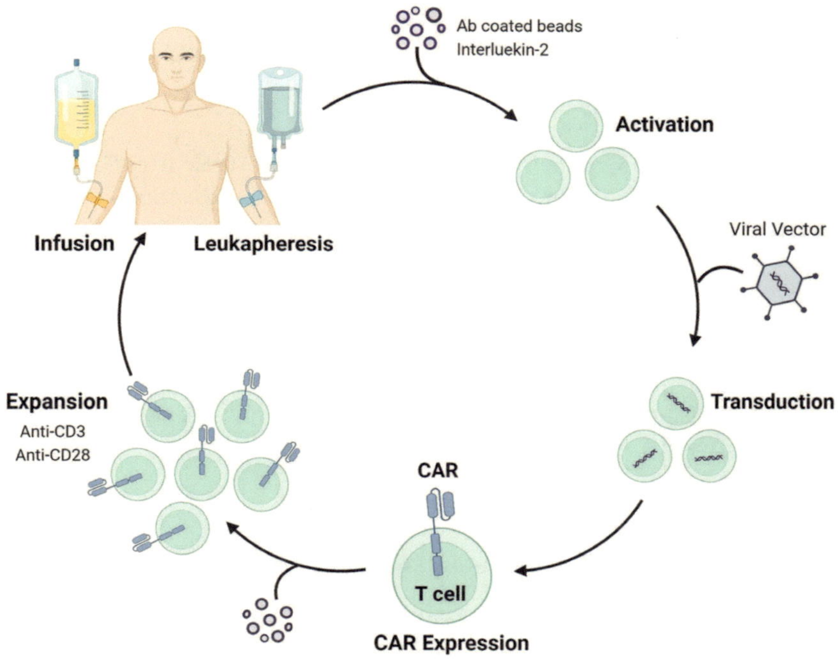

Figure 2.
CART cell manufacturing process [3].

3. Cytokine release syndrome (CRS)

3.1 Definition

CRS is the most common and well-characterized toxicity associated with CART cell therapy. It results from the widespread activation of immune cells, leading to the massive efflux of large amounts of pro-inflammatory cytokines [5, 6].

3.2 Epidemiology

CRS is generally considered a reversible complication of CART therapy, occurring in 42–100% of patients. Its severity can vary widely, with 0–46% of patients experiencing severe CRS and 0 to 9.1% of cases progressing to fatal outcomes [2].

3.3 Pathophysiology

CRS involves CART cells' rapid activation and proliferation upon encountering their target antigen in tumor cells, triggering the proliferation and release of various cytokines and chemokines by CART cells and other immune cells [7]. The most commonly elevated cytokines include interleukin (IL)-6, IL-8, IL-10, interferon (INF)-γ, granulocyte-macrophage colony-stimulating factor (GM-CSF), macrophage inflammatory protein-1β, and monocyte chemoattractant protein-1. IL-6 is a crucial mediator of CRS and, along with other immune mediators, activates macrophages and endothelial cells, further amplifying the inflammatory response [8]. The result is widespread endothelial activation and capillary leak, which can lead to hypotension, hypoxia, and organ dysfunction [8, 9].

3.4 Clinical presentation

CRS onset of symptoms usually occurs during the first week after CART cell infusion, with a typical duration of 7 to 8 days. It is a clinical diagnosis that typically presents with varying degrees of symptoms, including fever, tachycardia, hypotension, hypoxia, nausea, and vomiting [2]. Monitoring for CRS often includes regular measurement of C-reactive protein, ferritin, D-dimer, fibrinogen levels, and triglycerides [1]. Studies have indicated that disease characteristics such as tumor burden, Eastern Cooperative Oncology Group (ECOG) performance status, and levels of ferritin, lactate dehydrogenase (LDH), and C-reactive protein (CRP) prior to CART therapy can serve as predictors for the onset of CRS and ICANS, thereby helping to guide the outcomes of CART treatment [2].

3.5 Grading

Grading of CRS is crucial to accurately know its severity thereby guiding treatment decisions.

Grade 1: Fever, fatigue, headache, myalgias, and anorexia.

Grade 2: Persistent fever, hypoxia, and hypotension responsive to fluids or low-dose vasopressors.

Grade 3:

- fever,
- hypotension that requires one vasopressor, and/or
- hypoxia that necessitates high-flow O_2 (high-flow nasal cannula, mask, non-invasive ventilation) $\geq$40% FiO_2.

Grade 4: High fevers, severe hypotension requiring high-dose vasopressors, hypoxia requiring supplemental oxygen or mechanical ventilation, and multi-organ dysfunction.

3.6 Management

The management of CRS is guided by its severity and involves both supportive care and targeted therapies. A useful laboratory evaluation for CRS includes performing a complete blood count (CBC), renal and liver function tests, coagulation profile, lactate dehydrogenase (LDH), serum ferritin, IL-6, and C-reactive protein (CRP) which are monitored frequently to assess the severity and treatment response [2]. The treatment of CART-induced toxicity is primarily supportive. Still, when higher-grade adverse events occur, it is standard practice to administer toxicity-directed therapies such as tocilizumab, an IL-6 receptor antibody, and steroids [1]. For FDA-approved CART cell products, the Risk Evaluation and Mitigation Strategy (REMS) program mandates that at least two doses of tocilizumab per patient must be readily available and able to be administered within 2 hours of the CART infusion [2, 10].

Treatment of CRS is mainly supportive. Early recognition is the key. Management depends on the severity of CRS and is summarized in **Table 1**.

CRS grade	Treatment
Grade 1	• Supportive care with analgesics and antipyretics • Consider tocilizumab for persistent (lasting >3 days) and refractory fever
Grade 2	• IV fluids to maintain SBP > 90 mmHg. • Administer Tocilizumab early if persistent fever of ≥39°C, hypotension after initial fluid bolus or initiation of oxygen supplementation. • If persistent hypotension after two fluid bolus and Tocilizumab transfer to ICU for consideration of low-dose vasopressor therapy. • Add Dexamethasone 10 mg IV 6 hourly if hypotension persists after anti-IL-6 therapy, high risk for severe CRS, worsening hypoxia or clinical concern
Grade 3	• ICU admission • Administer Tocilizumab • Add steroids if unresponsive within 24 hours Dexamethasone 10 mg IV every 6 hours; If refractory increase to 20 mg IV every 6 hours. • If unresponsive CRS add Anakinra or Siltuximab
Grade 4	• Intensive care • Administer Tocilizumab • High dose methylprednisolone 1 g/day IV • If unresponsive, add Anakinra or Siltuximab
Grade 5	Death

Table 1.
Management of CRS.

If patients with CRS and ICANS continue to deteriorate despite treatment with tocilizumab and corticosteroids, siltuximab, a monoclonal antibody that binds to IL-6 and prevents its interaction with IL-6 receptors, can be considered as an alternative. Another option is anakinra, an IL-1 receptor antagonist, which has shown activity in cases of CRS and ICANS that are refractory to both corticosteroids and tocilizumab. Since IL-1 production precedes that of IL-6, anakinra may be effective in preventing CRS and ICANS, although further studies are needed to validate its efficacy [2].

The timing and sequence of these interventions are critical to managing CRS effectively. It's important to recognize CRS early and treat it promptly to prevent the progression to more severe forms of the syndrome.

4. Immune effector cell-associated neurotoxicity syndrome (ICANS)

4.1 Definition

ICANS, also known as CART cell-related encephalopathy syndrome (CRES), is the second most common toxicity associated with CART cell therapy after CRS, varying from mild confusion to severe cerebral edema [2].

4.2 Epidemiology

ICANS typically presents as a reversible complication in most patients after CART infusions. It occurs less frequently and often later than CRS, with an incidence ranging from 3–64% and 0–54% of cases showing signs of severe illness. Risk factors for ICANS include the presence of CRS, pre-existing neurologic dysfunction, high disease burden, elevated LDH levels, thrombocytopenia, and elevated ferritin within 72 hours after CART cell administration [2].

4.3 Pathophysiology

The pathophysiology of ICANS is not well understood. Theories suggest that it may involve the CART cells trafficking into the nervous system and disrupting the blood-brain barrier. Additionally, activation of myeloid cells in the central nervous system (CNS), leading to the secretion of IL-1 and IL-6 by monocytes, GM-CSF-mediated stimulation of monocytes after CART therapy, elevated levels of IL-15, and the presence of N-methyl-D-aspartate (NMDA) receptor agonists like glutamate and quinolinic acid, have all been proposed as potential contributors to the development of ICANS [11].

Another mechanism is thought to be immune-mediated endothelial activation, where CART cells rapidly proliferate and secrete cytokines that activate endogenous immune cells, such as macrophages. These macrophages release large amounts of cytokines, which in turn activate cerebral microvascular endothelial cells and disrupt tight junctions, causing the leaky blood-brain barrier (BBB). Consequently, high concentrations of serum cytokines passively diffuse into the BBB, and elevated levels of pro-inflammatory mediators in the cerebrospinal fluid (CSF) have been linked to ICANS, including IL-1β, IL-6, IL-8, IFN-γ, GM-CSF, MCP-1, and granzyme B. Additionally, it has been shown that T-cells and macrophages, including CART cells, can infiltrate the central nervous system (CNS) due to BBB disruption. These infiltrating immune cells and cytokines can activate microglia, which are brain-resident macrophages, further amplifying local inflammatory responses and leading to neurotoxicity. Because tocilizumab cannot cross the BBB, its efficacy in managing ICANS is limited. Given the enhanced CNS penetration of corticosteroids, they are recommended for the treatment of ICANS, as they do not impede the proliferation or anti-tumor effects of CART cells [12].

4.4 Clinical presentation

ICANS clinical and neuropsychiatric manifestation typically arises within days to 2-3 weeks after the administration of CART therapy but can also develop independently [13].

ICANS is diagnosed based on clinical symptoms such as headache, impaired attention and consciousness, lethargy, agitation, hallucinations, tremors, aphasia, encephalopathy, and seizures. The grading and surveillance of neurotoxicity are guided by the ICE criteria (**Table 2**). The total score ranges from zero to ten, with a higher score indicating better neurological function.

Laboratory evaluation for ICANS includes biomarkers similar to those used for CRS, along with additional tests such as lumbar puncture to analyze cerebrospinal fluid, neuroimaging, and EEG. These tests help determine the extent of ICANS-related damage and rule out other potential causes, including infections and other organic factors (**Table 3**) [14].

4.5 Management

The management of ICANS is based on the severity of the neurological symptoms and is summarized in **Table 4** [16].

ICE score component	Description	Score
Orientation	Oriented to year, month, city, hospital, and president/prime minister	4
Naming Objects	Name three objects (e.g., pen, watch, ball)	3
Following Commands	Follow simple commands (e.g., show two fingers)	1
Writing a Sentence	Ability to write a complete sentence	1
Attention	Count backward from 100 by 10s	1

Table 2.
ICE scoring criteria.

Grade	ICE score	Level of consciousness	Seizure	Motor findings	Elevated ICP/ cerebral edema
Grade 1	7–9	Awakens spontaneously	No	No abnormal motor findings	No
Grade 2	3–6	Awakens to voice	No	Mild motor weakness (e.g., hemiparesis)	No
Grade 3	0–2	Awakens only to tactile stimulation	Seizure(s) that responds to treatment	Moderate motor weakness	No
Grade 4	0	Unarousable or stupor	Life-threatening, prolonged seizures	Severe motor findings (e.g., decerebrate/ decorticate posturing)	Yes (any sign of increased ICP/ cerebral edema)

Table 3.
ICANS grading criteria.

ICANS grade	Treatment
Grade 1	• Close monitoring with supportive care including aspiration precautions and IV hydration and avoiding medications that depress the central nervous system. • Consult neurology, and imaging of the brain with and without contrast, is preferred • Tocilizumab if ICANS is accompanied by concurrent CRS.
Grade 2	• Investigations and supportive care as per grade 1 • Use dexamethasone 10 mg Intravenous(IV) every 6 hours or methylprednisolone 1 mg/kg IV every 12 hours
Grade 3–4	• ICU transfer is recommended. Consider mechanical ventilation for airway protection. • High-dose corticosteroids continued until improvement to grade 1 ICANS and then taper • In cases of seizures, anti-epileptic drugs are administered.
Grade 4	• Supportive care and Seizure management as per Grade 3 • Administer methylprednisolone 1000 mg/day for 3 days, followed by 250 mg every 12 hrs for 2 days, followed by 125 mg every 12 hrs for 2 days, followed by 60 mg every 12 hrs for 2 days • For patients with raised intracranial pressure, recommend acetazolamide 1000 mg IV, followed by 250–1000 mg IV every 12 h, along with head elevation, hyperventilation, and mannitol

Table 4.
Management of ICANS [15].

ICANS can be frequently confused with uncommon viral infections, stroke, chemotherapy-related toxicity, and the CNS involvement of malignancy. In severe cases of ICANS, early intervention is crucial to prevent irreversible neurological damage. The decision to use corticosteroids must balance the need to control inflammation with the potential impact on CART cell efficacy [17].

5. Cytopenias

5.1 Definition

Prolonged cytopenias are a common and significant complication following CART cell therapy. These can include neutropenia, thrombocytopenia, and anemia, which can persist for weeks or even months after treatment. Prior studies revealed that the rates of anemia, thrombocytopenia, and neutropenia of all grades among CD-19 CART therapy were 65, 55, and 78%, respectively. Risk factors thought to increase the risk of cytopenias include age, gender, type of disease, number of prior therapy lines, dose of CART cells infused, and the specific target and costimulatory domain used in the CART therapy [18, 19]. Understanding cytopenias' causes, risks, and management is crucial for minimizing complications and ensuring patient safety.

5.2 Pathophysiology

The pathophysiology of cytopenias after CART therapy is multifactorial and not fully understood, but some mechanisms have been proposed.

Severe CRS has been associated with cytopenias. There is a notable increase in the expression of CD40 on granulocytic progenitor/precursor cells, which also have low levels of CD40L during CRS, and the interaction between them enhances apoptosis, leading to cytopenias. Furthermore, inflammatory cytokines such as IL-1, TNF-α, and HMGB1 have the ability to inhibit the production of erythropoietin, and macrophages that are activated could potentially eliminate red blood cells. Reduced marrow reserve from prior lines of chemotherapy and hematopoietic stem cell transplantation (HSCT) can also lead to low blood cell counts. Additionally, certain target antigens for CART cells are found on normal hematopoietic stem or progenitor cells, enabling them to eliminate these cells directly [12].

5.3 Monitoring and diagnosis

Monitoring cytopenias in CART recipients involves regular complete blood counts (CBCs), typically performed multiple times per week in the immediate post-infusion period. The frequency of monitoring can be reduced as the patient stabilizes, but continued vigilance is necessary for several months after therapy. Cases of prolonged or unexplained cytopenias require a biopsy of bone marrow to identify any residual disease, marrow fibrosis, or other hematologic abnormalities. For febrile patients with neutropenia, an infection workup including blood cultures, chest X-ray, and possibly CT imaging is necessary to identify and treat potential sources of infection.

5.4 Management

The management of cytopenias after CART therapy requires a multifaceted approach. Cytopenias generally resolve over time, but in cases of prolonged or delayed cytopenias, a bone marrow biopsy is recommended to assess for secondary bone marrow abnormalities. Anemia, thrombocytopenia, and neutropenia are managed through blood product transfusions and granulocyte colony-stimulating factor (G-CSF), respectively. In some cases, persistent refractory cytopenia following CART therapy has been treated with the infusion of autologous stem cells [2].

Platelet and red blood cell transfusions are commonly used to manage thrombocytopenia and anemia, respectively. Patients should be monitored for transfusion-related complications, including allo-immunization and iron overload. The use of G-CSF stimulation and the optimal timing for its initiation remain topics of debate. There is no clear evidence linking G-CSF administration to higher grades of immunotoxicity. In the absence of randomized studies, early G-CSF use cannot be ruled out, but caution is advised if CRS or ICANS is present. In clinical practice, it is recommended to initiate G-CSF stimulation after the first week post-infusion. However, in cases of severe documented infection, the risk-benefit balance of earlier use to expedite neutrophil recovery should be carefully evaluated. Studies on erythropoietin or thrombopoietin analogs are much more limited, but they may be beneficial in cases of prolonged cytopenias several weeks after infusion [20].

A multi-center observational case series by Rejeski et al. reported on bone marrow findings and survival outcomes in 13 patients who received a stem cell boost (10 autologous and 3 allogeneic) for persistent cytopenia following CD19 CART therapy [21]. The data suggest that this approach is safe and clinically feasible. However, CRS remains a potential complication after infusion, possibly due to the presence of CD19+ progenitors within the CD34+ selected product. Autologous bone marrow harvest is currently not recommended when CART therapy is planned. Despite promising results, more data and longer follow-ups are necessary, as there have been reports of lymphoma contamination in the collected cells [20].

Prophylaxis with anti-bacterial, antifungal, and antiviral drugs during the periods of cytopenias is strongly encouraged. Patients with prolonged cytopenias require ongoing monitoring until hematologic recovery. In some cases, extended use of hematopoietic growth factors may be necessary to support bone marrow recovery. The decision to continue these agents should be based on individual patient responses and risk factors.

6. Infections

CART therapy poses several infection-related risks due to its impact on the immune system. Lymphodepleting chemotherapy, administered before CART infusion, enhances the therapy's effectiveness but weakens mucosal integrity and causes cytopenias, increasing early infection risks. Cytokine Release Syndrome (CRS), a common complication of CART therapy, requires immunosuppressive treatments that further elevate infection susceptibility. Prolonged cytopenias, especially neutropenia, leave patients vulnerable to bacterial and fungal infections for weeks or months post-therapy. Additionally, CART therapy can lead to hypogammaglobulinemia, reducing antibody levels and increasing the risk of infections from specific bacteria and viruses. All the risk factors are summarized in **Table 5**.

Risk factor	Description
Lymphodepleting therapy	Fludarabine and cyclophosphamide, prior to CART infusion, is essential for depleting regulatory T cells and enhancing the anti-tumor efficacy of the infused CART cells. However, this treatment also leads to a loss of mucosal integrity and cytopenias, increasing the risk of early infections.
Cytokine Release Syndrome (CRS) and ICANS	Immunosuppressive therapies like Tocilizumab and corticosteroids are often used to manage CRS/ICANS but can further suppress the immune system, thereby increasing the infectious risk.
Prolonged Cytopenias	Prolonged cytopenias, especially neutropenia, lasting weeks or months post-infusion, heightens the risk of bacterial and fungal infections.
Hypogammaglobulinemia	CART therapy can lead to B cell aplasia, causing hypogammaglobulinemia and an increased susceptibility to infections, particularly from encapsulated bacteria and viruses.

Table 5.
Risk factors leading to infections in CART recipients.

6.1 Timing of infections

The timing of infections in CART therapy can be broadly divided into three phases [22].

6.1.1 Early phase (day 0 to day 30)

In this phase, patients are most vulnerable to bacterial and fungal infections. This increased susceptibility is primarily due to the effects of lymphodepletion chemotherapy, which significantly suppresses the immune system, and the immediate toxicities of CART therapy, including cytokine release syndrome (CRS) and neurotoxicity. Febrile neutropenia is common during this period, increasing the risk of bacterial infections. The most common type of infection in the early phase, typically occurring within the first 2 weeks after infusion. These are often related to neutropenia and include bloodstream infections (BSIs), pneumonia, and urinary tract infections (UTIs). Invasive fungal infections (IFIs), particularly from Candida and Aspergillus species, are a concern during prolonged neutropenia. Patients with a history of prior fungal infections are at higher risk. Viral Infections such as reactivation of herpesviruses (e.g., herpes simplex virus (HSV), varicella-zoster virus (VZV), and cytomegalovirus [CMV]) can occur in the early phase, particularly in patients with a history of these infections or those with hypogammaglobulinemia.

6.1.2 Intermediate phase (day 31 to day 90)

The risk of infections during this phase is somewhat lower but remains considerable. Viral infections, especially the reactivation of latent viruses like cytomegalovirus (CMV), are more prevalent due to prolonged lymphopenia and incomplete T-cell recovery. The ongoing presence of CART cells and the associated immune system dysregulation further contribute to this vulnerability. As cytopenias persist and immune recovery remains incomplete, patients continue to be at risk for opportunistic infections, including *Pneumocystis jirovecii* pneumonia (PJP), fungal infections, and viral infections like CMV reactivation.

6.1.3 Late phase (beyond day 90)

In the late phase, infections are often associated with the chronic immunosuppressive state induced by CART cells. Hypogammaglobulinemia is a common complication, leading to an increased risk of bacterial infections, particularly encapsulated organisms such as *Streptococcus pneumoniae* and *Haemophilus influenzae*. Some patients may also develop secondary immunodeficiencies, leading to recurrent or opportunistic infections.

6.2 Monitoring and diagnosis

Early identification and treatment of infections are critical to improving outcomes in patients undergoing CART therapy. Key strategies include regular monitoring of complete blood counts (CBCs), inflammatory markers (such as C-reactive protein [CRP] and procalcitonin), and cultures from blood, urine, and other sites as indicated is essential for early detection of infections.

Chest X-rays or CT scans are often used to evaluate pneumonia or other deep-seated infections. Bronchoscopy may be required for suspected fungal infections or when a definitive diagnosis is needed. Polymerase chain reaction (PCR) assays for viral DNA or RNA are used to detect the reactivation of herpesviruses (e.g., HSV, VZV, and CMV) and respiratory viruses. This is particularly important for early identification of viral infections, allowing timely initiation of antiviral therapy.

Serum galactomannan and β-D-glucan are routinely checked to detect invasive fungal infections, particularly in patients with prolonged neutropenia or those with clinical signs suggestive of fungal disease.

6.3 Management

Prophylaxis and treatment of infections in CART recipients involve a comprehensive approach, including antimicrobial prophylaxis, prompt initiation of empiric therapy for suspected infections, and tailored therapy based on diagnostic findings. Antimicrobial prophylaxis for CART therapy is largely based on protocols used for hematopoietic stem cell transplants (HSCT). This includes the use of antivirals such as acyclovir or valacyclovir for herpes simplex virus (HSV) and varicella-zoster virus (VZV), beginning before conditioning chemotherapy and continuing for up to 1 year following CART therapy. For *Pneumocystis jirovecii* pneumonia (PJP) prophylaxis, trimethoprim-sulfamethoxazole or alternative agents are recommended, starting before conditioning chemotherapy and continuing until CD4 counts exceed 200 cells/μL. Prophylactic antifungal should be considered for high-risk patients such as those with prolonged steroid exposure or those with severe neutropenia (ANC < 0.5 × 10^9/L) lasting more than 14 days. Commonly used prophylaxis strategies are summarized in **Table 6** [1, 2].

The persistent B-cell aplasia and resultant hypogammaglobulinemia observed in CART cell therapy patients compromise humoral immunity, potentially leading to an infection risk akin to that seen in primary immune deficiencies. The role of immunoglobulin replacement therapy in mitigating this risk remains controversial due to the lack of robust clinical trial data. Consequently, current management practices vary widely, and the usual practice is to consider immunoglobulin replacement within the first 3 months after CART cell therapy in cases of severe hypogammaglobulinemia

Prophylaxis type	Recommendations
Antibacterial	Consider antibiotic prophylaxis with penicillin or fluoroquinolones for patients with ANC < 0.5 x 10^9/L.
Antifungal	• For patients with ANC < 0.5 x 10^9/L, consider fluconazole (200 mg/day) or micafungin (50 mg IV/day). • For those treated with tocilizumab or corticosteroids and with persistent ANC < 0.5 x 10^9/L for more than 4 weeks, consider posaconazole (300 mg/day).
Anti-pneumocystis	• Co-trimoxazole 80/400 mg once daily or 160/800 mg three times/week, starting from LD conditioning or after ANC > 0.5 x 10^9/L, until 6 months post-CART cell infusion, if CD4+ count >0.2 x 10^9/L. • In cases of prolonged cytopenias or co-trimoxazole allergy, consider pentamidine inhalation (300 mg once monthly), dapsone 100 mg daily, or atovaquone 1500 mg once daily.
Anti-viral	• Acyclovir 800 mg every 12 hours or valacyclovir 500 mg every 12 hours, starting from LD conditioning until 6 months post-CART cell infusion, if CD4+ count >0.2 × 10^9/L. • For hepatitis B infection, prophylaxis with entecavir or tenofovir is recommended for the first year after CART cell infusion. Consider monitoring viral load every 3–6 months during the first year after discontinuation.

Table 6.
Summary of prophylactic agents used in CART patients [20].

(IgG <400 mg/dL) or in moderate hypogammaglobulinemia (400–600 mg/dL) accompanied by recurrent and severe infections [23].

In patients presenting with fever and neutropenia or signs of infection, empiric broad-spectrum antibiotics should be initiated promptly, often covering both gram-positive and gram-negative bacteria. Broadening of antifungal therapy may be added in patients with persistent fever despite antibiotic therapy or those with risk factors for IFIs. Once specific pathogens are identified, therapy should be adjusted to target the identified organisms based on culture results, susceptibility testing, and clinical response. Duration of therapy is guided by the site of infection, the patient's response, and the recovery of immune function.

7. Immune effector cell-associated HLH-like syndrome (IEC-HS)

7.1 Definition

Hemophagocytic lymphohistiocytosis (HLH)/macrophage activation syndrome (MAS) is a serious and potentially toxic condition from aggressive over-activation of the immune system, leading to clinical signs and symptoms of extreme inflammation. It is caused by either a familial genetic disorder or a sporadic condition associated with a variety of triggers. HLH secondary to CART cell therapy is referred to as immune effector cell-associated HLH-like syndrome (IEC-HS).

7.2 Epidemiology

The incidence of IEC-HS from earlier studies was quite variable, ranging from 0 to 6% depending on the type of CART product used and the type of disease setting used. Some of the same CART products showed different incidence of IEC-HS when used to treat different diseases [24, 25]. However, a review of the FDA Adverse Events Reporting System (FAERS) and Vizient databases showed an incidence of IEC-HS was between 1 and 2% with no statistical differences reported between different CART products [24].

7.3 Pathogenesis

The exact pathogenesis of IEC-HS is poorly understood. Both severe CRS and IEC-HS develop fever, respiratory compromise, and deranged hemodynamics [26]. Overlap of laboratory findings was also noticed in CRS and IEC-HS, but the timing of IEC-HS is usually after the resolution of CRS, indicating a possible explanation of the prolonged inflammatory stage but involving different immunological mediators responsible for IEC-HS [27].

7.4 Diagnosis

The clinical presentation of IEC-HS often mirrors the exaggerated inflammatory response seen following chimeric antigen receptor T-cell (CAR) therapy. It is crucial not to be confused between severe CRA with multi-organ failure and IEC-HS. Usually, IEC-HS manifests later, following the initial CRS event. All documented cases of HLH-like toxicities associated with CART cell therapy have involved a prior or concurrent CRS [26].

The common scoring systems used for diagnosing non-CART-related secondary HLH were the HLH-2004 criteria or the H-score. According to the HLH-2004 criteria, HLH is diagnosed if a patient has five of the following: fever, splenomegaly, cytopenias in at least two cell lineages, hypertriglyceridemia or hypofibrinogenemia with an elevated D-dimer, hemophagocytosis in bone marrow, hyperferritinemia, high levels of soluble CD25 and low or absent NK cell activity [28, 29].

Using this criteria to diagnose IEC-HS is unreliable and challenging due to the fact that CART patients will meet the score for IEC-HS in the setting of other common CART complications like CRS. In order to mitigate this, the ASTCT defines IEC-HS as the development of hyperinflammatory syndrome independent from CRS and ICANS that manifests with features of MAS/HLH, is attributable to immune effector cell therapy, and is associated with cytopenias, hyperferritinemia, coagulopathy with hypofibrinogenemia and/or transaminitis [28]. The diagnostic criteria for HLH-2004 and ASTCT is listed in **Table 7**.

7.5 Management

Management of IEC-HS is broadly divided into supportive care and use of pharmacological agents, as summarized in **Table 8**.

Supportive care is fundamental in managing IEC-HS. It may involve admission into intensive care units for managing severe respiratory and hemodynamic compromise. The main aim is to stabilize the patient and address severe symptoms while closely monitoring for complications. Key aspects include regular monitoring of vital signs, volume status, mental status, and frequent laboratory tests such as complete blood counts, liver and renal function tests, and coagulation studies. Imaging techniques such as CT and MRI assess and characterize complications, guiding further management [16].

Diagnostic criteria	Components
HLH-2004	• Fever ≥38.5°C • Splenomegaly • Bicytopenia o Hemoglobin <9 g/dL o Platelets <100,000/uL o Absolute neutrophil count <1000/uL • Hypertriglyceridemia >265 mg/dL and/or hypofibrinogenemia <150 mg/dL • Hemophagocytosis in bone marrow, spleen, lymph node or liver • Hyperferritinemia >500 ng/mL • Low/absent NK-cell-activity • Elevated soluble interleukin-2-receptor levels
ASTCT	Most common manifestations • Elevated ferritin (>2 × ULN or baseline and/or rapidly rising) • Onset with resolving/resolved CRS or worsening inflammatory response after initial improvement with CRS-directed therapy • Hepatic transaminase elevation (>5 × ULN or > 5 × baseline) • Hypofibrinogenemia (<150 mg/dL or < LLN) • Hemophagocytosis in bone marrow or other tissue • Cytopenias Other manifestations that may be present • Lactate dehydrogenase elevations (>ULN) • Other coagulation abnormalities (e.g., elevated PT/PTT) • Direct hyperbilirubinemia • New-onset splenomegaly • Fever (new or persistent) • Neurotoxicity • Pulmonary manifestations (e.g., hypoxia, pulmonary infiltrates, pulmonary edema • Renal insufficiency (new onset) • Hypertriglyceridemia (fasting level, >265 mg/dL)

Table 7.
HLH-2004 and ASTCT diagnostic criteria.

Anti-inflammatory agents, especially systemic corticosteroids, are cornerstone treatment options for IEC-HS, reducing inflammation and modulating the immune response. Initial guidelines for dosing in IEC-HS suggest using corticosteroids in combination with cytokine-directed therapies, with the dosage adjusted based on the severity of concurrent CRS or other organ toxicities. For patients without severe end-organ damage, neurotoxicity, or grade 4 hypotension, a regimen of dexamethasone 10 mg IV every 6 hours, increasing to 20 mg IV every 6 hours if needed, followed by gradual tapering, has been recommended. In cases of secondary HLH or IEC-HS with severe CRS or end-organ damage that does not respond to standard treatments, pulse dose steroids have been considered, including methylprednisolone 1–2 mg/kg or dexamethasone 5–10 mg/m^2 every 6–24 hours [16, 30].

Treatment modality	Description
Supportive Care	*Clinical Manifestations*: Fever, cytopenias, respiratory distress, vasculopathy, coagulopathy, shock, high-grade organ damage (liver, kidneys, lungs, brain). • *Monitoring*: Vital signs, volume status, mental status, frequent laboratory tests (CBC, liver/renal function tests, coagulation studies). • *Management*: Close monitoring, CT/MRI imaging for complications.
Corticosteroids	• *Dosing*: Dexamethasone 10 mg IV every 6 h (up to 20 mg in refractory cases); higher doses in severe cases.
Anakinra (IL-1 receptor antagonist).	• *Dosing*: 100–200 mg IV/SQ every 6–12 h (adults); 10 mg/kg IV daily or 4 mg/kg IV every 6 h (pediatric). • Consider early use with or without corticosteroids.
Ruxolitinib (JAK1/JAK2 Inhibitors)	• *Dosing*: 10–20 mg daily • *Efficacy*: Effective in refractory HLH; recommended as a second-line option for IEC-HS with corticosteroids and anakinra.
Tocilizumab (IL-6 receptor antagonist)	• Mixed results in IEC-HS; potential for increased infections without significant benefit.
Emapalumab (monoclonal antibody against IFN-γ)	• *Dosing*: 1 mg/kg IV infusion twice weekly. • *Efficacy*: Promising in severe IEC-HS; used as salvage therapy in refractory cases
Etoposide, ATG, alemtuzumab (Cytotoxic Anti-T Cell Therapies)	• Used in refractory cases of IEC-HS. • Risk of further instability and end-organ damage; use cautiously

Table 8.
Summary of various management strategies for immune effector cell-hemophagocytic syndrome (IEC-HS).

Anakinra, an IL-1 receptor antagonist, which was originally approved for rheumatological conditions, is increasingly utilized in the management of hyperinflammatory states, including those with HLH/MAS. It is well tolerated and can be administered either intravenously or subcutaneously with dosing options of 100–200 mg every 6–12 hours. Given its potential benefits, the ASTCT IEC-HS committee recommends considering anakinra early in the treatment of IEC-HS, with or without corticosteroids [28].

Cytokine modulators are designed to target specific cytokines that contribute to the inflammatory response in IEC-HS. JAK1/JAK2 inhibitors, such as ruxolitinib, interfere with cytokine signaling pathways and have shown efficacy in treating refractory HLH and IEC-HS. Interleukin-6 (IL-6) inhibitors, like tocilizumab, disrupt IL-6 signaling but have had mixed results in IEC-HS, with concerns about increased infection rates and limited clinical benefit.

Targeted therapies focus on neutralizing specific inflammatory molecules or pathways directly. Emapalumab, an IFN-γ blockade agent, is approved for primary HLH and is being explored for IEC-HS, especially as a salvage therapy in cases resistant to other treatments. Its mechanism of action involves targeting IFN-γ, which plays a role in the inflammatory process associated with IEC-HS. The initial dosing is a twice-weekly IV infusion of 1 mg/kg. Currently, there is very limited data in the literature pertaining to its efficacy, specifically in IEC-HS [31].

Cytotoxic therapies are considered when other treatments fail to control IEC-HS. These therapies, such as topoisomerase inhibitors like etoposide, aim to reduce activated T-cells and manage severe inflammation [32]. They are used particularly in refractory cases and are known for their efficacy in primary HLH. Other anti-T-cell therapies, like antithymocyte globulin (ATG) and alemtuzumab, are also potential options. However, their use in IEC-HS is less well-documented and must be approached with caution due to the potential exacerbation of clinical instability [33, 34].

8. Conclusion

CART therapy is a revolutionary strategy and offers a promising approach to treating hematologic malignancies. Its remarkable efficacy comes with significant toxicities, which sometimes can be lethal. The most common toxicities are CRS and ICANS. Cytopenias and Infections will affect the morbidity of the patients. IEC-HS is rare but very lethal.

Early identification, careful monitoring, and timely intervention are crucial in managing toxicities, which will eventually improve patient and treatment outcomes. The need for a multidisciplinary approach to managing these toxicities is very important and cannot be overstated. Higher-grade toxicities need ICU-level care.

The future holds promise with novel CART constructs and innovative combination therapies. However, these will come at the expense of newer toxicities, and the need to mitigate them with newer agents with different mechanistic tools remains an area of active research.

Funding statement

This research received no external funding.

Conflict of interest

The authors declare no conflict of interest.

Author details

Nishanth Thalambedu*, Sruthi Vellanki, Anup Kumar Trikannad and Ankur Varma
University of Arkansas for Medical Sciences, USA

*Address all correspondence to: nthalambedu@uams.edu

References

[1] Chohan KL, Siegler EL, Kenderian SS. CAR-T cell therapy: The efficacy and toxicity balance. Current Hematologic Malignancy Reports. 2023;**18**(2):9

[2] Shaikh S, Shaikh H. CART Cell Therapy Toxicity. StatPearls Publishing; 19 Apr 2023

[3] Faeq MH, Al-Haideri M, Mohammad TAM, Gharebakhshi F, Marofi F, Tahmasebi S, et al. CAR-modified immune cells as a rapidly evolving approach in the context of cancer immunotherapies. Medical Oncology. 2023;**40**(5):155

[4] Squibb BM. Understanding the CAR T Cell Therapy Treatment Experience - Bristol Myers Squibb [Internet]. Bristol Myers Squibb; 2024. Available from: https://www.bms.com/media/media-library/scientific-media-resources/car-t-cell-therapy-treatment-experience.html [Accessed: August 4, 2024]

[5] Xiao X, Huang S, Chen S, Wang Y, Sun Q, Xu X, et al. Mechanisms of cytokine release syndrome and neurotoxicity of CAR T-cell therapy and associated prevention and management strategies. Journal of Experimental and Clinical Cancer Research: CR. 2021;**40**:367. DOI: 10.1186/s13046-021-02148-6

[6] Xiao X, Huang S, Chen S, Wang Y, Sun Q, Xu X, et al. Mechanisms of cytokine release syndrome and neurotoxicity of CAR T-cell therapy and associated prevention and management strategies. Journal of Experimental & Clinical Cancer Research. 2021;**40**(1):1-23

[7] Hay KA, Hanafi LA, Li D, Gust J, Liles WC, Wurfel MM, et al. Kinetics and biomarkers of severe cytokine release syndrome after CD19 chimeric antigen receptor-modified T-cell therapy. Blood. 2017;**130**(21):2295-2306

[8] Teachey DT, Lacey SF, Shaw PA, Melenhorst JJ, Maude SL, Frey N, et al. Identification of predictive biomarkers for cytokine release syndrome after chimeric antigen receptor T-cell therapy for acute lymphoblastic leukemia. Cancer Discovery. 2016;**6**(6):664-679

[9] Sheth VS, Gauthier J. Taming the beast: CRS and ICANS after CAR T-cell therapy for ALL. Bone Marrow Transplantation. 2021;**56**(3):552-566

[10] Booth JP et al. The pharmacist's role in chimeric antigen receptor T Cell Therapy. Journal of Oncology Pharmacy Practice: Official Publication of the International Society of Oncology Pharmacy Practitioners. 2020;**26**(7):1725-1731. DOI: 10.1177/1078155220948940

[11] Taraseviciute A, Tkachev V, Ponce R, Turtle CJ, Snyder JM, Liggitt HD, et al. Chimeric antigen receptor T cell-mediated neurotoxicity in nonhuman primates. Cancer Discovery. 2018;**8**(6):750-763

[12] Zhang X, Zhu L, Zhang H, Chen S, Xiao Y. CAR-T cell therapy in hematological malignancies: Current opportunities and challenges. Frontiers in Immunology. 2022;**10**(13):927153

[13] Holtzman NG et al. Immune effector cell-associated neurotoxicity syndrome after chimeric antigen receptor T-cell therapy for lymphoma: Predictive biomarkers and clinical outcomes. Neuro-oncology. 2021;**23**(1):112-121. DOI: 10.1093/neuonc/noaa183

[14] Chou CK, Turtle CJ. Assessment and management of cytokine release syndrome and neurotoxicity following CD19 CAR-T cell therapy. Expert Opinion on Biological Therapy. 2020;**20**(6):653-664

[15] Kröger N et al., editors. The EBMT/EHA CAR-T Cell Handbook. Springer; 2022. DOI: 10.1007/978-3-030-94353-0

[16] Neelapu SS, Tummala S, Kebriaei P, Wierda W, Gutierrez C, Locke FL, et al. Chimeric antigen receptor T-cell therapy - assessment and management of toxicities. Nature Reviews. Clinical Oncology. 2018;**15**(1):47-62

[17] Strati P, Ahmed S, Kebriaei P, Nastoupil LJ, Claussen CM, Watson G, et al. Clinical efficacy of Anakinra to mitigate CAR T-cell therapy-associated toxicity in large B-cell lymphoma. Blood Advances. 2020;**4**(13):3123-3127

[18] Xia Y et al. Cytopenias following anti-CD19 chimeric antigen receptor (CAR) T cell therapy: A systematic analysis for contributing factors. Annals of Medicine. 2022;**54**(1):2951-2965. DOI: 10.1080/07853890.2022.2136748

[19] Hayden PJ et al. Management of adults and children receiving CAR T-cell therapy: 2021 best practice recommendations of the European Society for Blood and Marrow Transplantation (EBMT) and the Joint Accreditation Committee of ISCT and EBMT (JACIE) and the European Haematology Association (EHA). Annals of Oncology: Official Journal of the European Society for Medical Oncology. 2022;**33**(3):259-275. DOI: 10.1016/j.annonc.2021.12.003

[20] Hernani R, Benzaquén A, Solano C. Toxicities following CAR-T therapy for hematological malignancies. Cancer Treatment Reviews. 2022;**111**(102479):102479

[21] Rejeski K, Burchert A, Iacoboni G, Sesques P, Fransecky L, Bücklein V, et al. Safety and feasibility of stem cell boost as a salvage therapy for severe hematotoxicity after CD19 CAR T-cell therapy. Blood Advances. 2022;**6**(16):4719-4725

[22] Hill JA, Seo SK. How I prevent infections in patients receiving CD19-targeted chimeric antigen receptor T cells for B-cell malignancies. Blood. 2020;**136**(8):925

[23] Bupha-Intr O, Haeusler G, Chee L, Thursky K, Slavin M, Teh B. CAR-T cell therapy and infection: A review. Expert Review of Respiratory Medicine [Internet]. 2021. Available from: https://www.tandfonline.com/doi/abs/10.1080/14787210.2021.1855143;**19**: 749-758 [Accessed: August 18, 2024]

[24] Priyadarshini S, Harris A, Treisman D, Cupac JN, Li N, Yan D, et al. Hemophagocytic lymphohistiocytosis secondary to CAR-T cells: Update from the FDA and Vizient databases. American Journal of Hematology. 2022;**97**(10):E374-E376

[25] Fugere T et al. Immune effector cell-associated HLH-like syndrome: A review of the literature of an increasingly recognized entity. Cancers. 26 Oct 2023;**15**(21):5149. DOI: 10.3390/cancers15215149

[26] Lichtenstein DA, Schischlik F, Shao L, Steinberg SM, Yates B, Wang HW, et al. Characterization of HLH-like manifestations as a CRS variant in patients receiving CD22 CAR T cells. Blood. 2021;**138**(24):2469-2484

[27] Diorio C, Shraim R, Myers R, Behrens EM, Canna S, Bassiri H, et al.

Comprehensive serum proteome profiling of cytokine release syndrome and immune effector cell-associated neurotoxicity syndrome patients with B-cell ALL receiving CAR T19. Clinical Cancer Research. 2022;**28**(17):3804-3813

[28] Hines MR, Knight TE, McNerney KO, Leick MB, Jain T, Ahmed S, et al. Immune effector cell-associated hemophagocytic lymphohistiocytosis-like syndrome. Transplantation and Cellular Therapy. 2023;**29**(7):438.e1-438.e16

[29] Henter JI, Horne A, Aricó M, Egeler RM, Filipovich AH, Imashuku S, et al. HLH-2004: Diagnostic and therapeutic guidelines for hemophagocytic lymphohistiocytosis. Pediatric Blood & Cancer. 2007;**48**(2):124-131

[30] La Rosée P, Horne A, Hines M, von Bahr GT, Machowicz R, Berliner N, et al. Recommendations for the management of hemophagocytic lymphohistiocytosis in adults. Blood. 2019;**133**(23):2465-2477

[31] Manni S, Del Bufalo F, Merli P, Silvestris DA, Guercio M, Caruso S, et al. Neutralizing IFNγ improves safety without compromising efficacy of CAR-T cell therapy in B-cell malignancies. Nature Communications. 2023;**14**(1):3423

[32] Johnson TS, Terrell CE, Millen SH, Katz JD, Hildeman DA, Jordan MB. Etoposide selectively ablates activated T cells to control the immunoregulatory disorder hemophagocytic lymphohistiocytosis. Journal of Immunology. 2014;**192**(1):84-91

[33] Moshous D et al. Alemtuzumab as first line treatment in children with familial lymphohistiocytosis. Blood. 2019;**134**(Supplement_1):80. DOI: 10.1182/blood-2019-124477

[34] Mahlaoui N, Ouachée-Chardin M, de Saint BG, Neven B, Picard C, Blanche S, et al. Immunotherapy of familial hemophagocytic lymphohistiocytosis with antithymocyte globulins: A single-center retrospective report of 38 patients. Pediatrics. 2007;**120**(3):e622-e628

Chapter 25

Oral Mucositis Management and Care

Stefano Botti, Doria Valentini and Stefania Grasselli

Abstract

One of the worst experiences for patients undergoing stem cell transplant is oral mucositis, caused by the toxicity of the treatments. It is a very common complication that causes patients pain and discomfort, increases their risk of infection, and raises hospitalization costs. As nurses are largely involved in mucositis management, they need to develop the necessary knowledge and skills as well as the competencies to educate patients on how to minimize its effects. Besides pain relief, we currently do not have clearly effective strategies to treat mucositis, and there are few preventive measures that have proven to be effective. Therefore, mucositis management is mostly focused on alleviating symptoms and relies on creating multidisciplinary protocols based on oral hygiene which must be applied systematically. Key steps in caring for patients include assessing risk, proactive intervention, monitoring, and addressing related issues such as vomiting, dietary problems, and oral infections. This review aims to provide useful elements for managing oral mucositis that go beyond the recommendations and suggestions of international guidelines.

Keywords: oral mucositis, stomatitis, mucosal toxicity, hematopoietic stem cell transplantation, conditioning regimen, risk assessment, oral hygiene, nursing skills, education, nursing management

1. Introduction

Oral mucositis (OM) is one of the most severe acute complications related to the toxicity of immuno-chemo-radiotherapy based conditioning regimens administered to patients undergoing hematopoietic stem cell transplantation (HSCT) [1].

OM is defined as a severe "pan-mucosal" inflammatory complication involving the mucous membranes in all anatomical structures, from the oropharyngeal tract to the anal area, including the entire gastrointestinal tract [2]. At the oral mucosa level, OM is characterized by inflammation and lesions of the mucous membranes, which can cause intense pain, ulcers, difficulty or inability to ingest food and water, significant patient discomfort, increased risk of infection, and increased costs [3, 4]. In patients undergoing HSCT, OM is associated with additional days of fever, increased time of artificial nutritional support, enhanced narcotic therapy needs, and increased risk of systemic infections [3] and bacteremia [5, 6] that may

IntechOpen

Age	Malnutrition
Sex	Metabolic diseases
Type of disease	Dehydration
Disease severity	Tachypnea
Type of conditioning regimen	Pre-existing or recurrent oral lesions
Conditioning regimen intensity	Poor oral hygiene
Other drugs	Lifestyle habits (alcohol, tobacco)

Table 1.
Risk factors leading to oral mucositis development in HSCT patients.

result in life-threatening or fatal consequences. Literature data showed that HSCT patients who experienced OM had a significantly higher mortality at day +100 post-transplant, and that the mortality rate increasing was associated with maximum severity score of OM [3].

The incidence of OM varies depending on the type of transplant. About 70% of the patients undergoing autologous HSCT are affected, while up to 98% of patients undergoing allogeneic transplantation receiving myeloablative conditioning regimens can be affected [7–9].

The pathobiological characteristics of OM are well recognized and divided into five features defined as initiation, signaling, amplification, ulceration, and healing [10]. However, despite the fact that the toxic effects of drugs and of their combinations are known and that the risk factors contributing to the development of OM (**Table 1**) are well described, the available literature is still unable to provide strong scientific evidence on OM prevention and treatment.

Healthcare professionals' actions in clinical practice are therefore guided by a plethora of interventions based on scarce or anecdotal knowledge, expert opinions, and long-standing procedures without any evidence of efficacy.

2. Healthcare professional training and documentation systems

2.1 Nursing skills

OM symptoms and features should be managed by nurses with appropriate skills and competencies [11]. Some studies have shown that healthcare providers assisting oncology patients often do not use the most up-to-date knowledge regarding oral care and OM management [12], and that even in settings with highly trained nurses, non-evidence-based practices are applied [13].

It is also known that where healthcare professionals are unfamiliar with the guidelines for the correct management of clinical problems, the evidence-based practice (EBP) process is poorly applied, and patient and family education pathways are deficient.

In terms of OM, routine training interventions should be provided to nurses that use the most recent knowledge and the best evidence with the aim to maintain and update their competencies over time [14]. Developing advanced competencies in OM management in terms of early intervention and symptom control can add value for transplant team nurses in their daily practice [15].

The effectiveness of training programs should be assessed through outcome comparisons with international benchmarking data. Reproducibility checks of collected data to test interprofessional competence levels for OM assessment and monitoring should be performed using validated grading scales with the aim to identify corrective actions based on educational needs. Nursing training should cover all main aspects of OM management, including assessment, oral care intervention, prevention and treatment strategies, and follow-up.

2.2 Documentation, data collection, and benchmarking

Healthcare professionals' competencies should be tracked, documented, and routinely assessed. The documentation in each transplant program should include specific standard operating procedures (SOPs) for OM management, including tools and procedures for data collection and recording. This will ensure that each intervention for the assessment, prevention, and treatment of a patient's oral cavity has been recorded in the clinical documentation.

Proper OM management means that a predefined set of benchmarking measures should be used to evaluate the effectiveness of the applied care strategies as well as the organization's performance over time. Data on OM prevalence, severity grading, duration, time of onset, pain levels, functional limitations (eating, drinking, swallowing), and patient-reported symptoms should be used to assess a center's performance on this clinical issue. In addition, data on antibiotics, antifungals, opioids, parenteral nutrition consumption, and length of hospital stay can be collected to assess the economic impact of OM [3, 4].

3. Patient and caregiver education

Before hospitalization for HSCT and routinely afterward, patients and their caregivers (CGs) should be informed about the risk of developing OM, its features, and associated complications [16].

Providing detailed information and education on potential conditioning regimen-related oral issues with the aim to promote the patient's timely recognition of the problem is fundamental [11, 17]. Patients should be educated on how to apply the local oral hygiene protocols by means of detailed explanations provided through in-person meetings, written tools (brochures, booklets, etc.) and/or digital resources (videos, tutorials), if available. The educational program should include methods for promoting oral hygiene and comfort, pain control, and dietary suggestions to maintain adequate oral intake of nutrients and water [18].

It should be considered that less than 20% of the information provided to patients during an educational meeting is normally retained [19]. The intervention's effectiveness can rise to 50% by adding written material and to 70–90% after learning verification strategies or practical exercises [20]. Therefore, nurses should pay attention to the context in which information are provided, to the quantity and quality of that information, and should provide the patient with options. The provided educational materials should apply principles of patient literacy, i.e., easily understandable to the patients and their CGs. This may allow their empowerment, which fosters the development of effective skills [21, 22].

Key topics of the educational interventions include the principles of oral care and a detailed description of the local policy by the SOPs for OM management, information on the correct use of the provided tools, and methods for self-assessment and care. Patients and CGs should be encouraged to report as early as possible any OM-related signs and symptoms as well as to maintain a balanced diet and adequate oral intake of water and food. To ensure optimal results from the educational process, strategies to assess patient's adherence to the provided recommendations should be implemented.

4. Assessment

4.1 Risk assessment

The implementation of all oral care, prevention, and treatment strategies must start by an effective assessment plan [10]. Patients' mouths should be assessed using validated tools by specifically trained nurses or other healthcare professionals. The assessment tools most used in the HSCT setting around the world are the World Health Organization (WHO) Oral Health Assessment Tool and the National Cancer Institute's Common Terminology Criteria for Adverse Events (CTCAE). The use of patient-reported tools (e.g., Oral Mucositis Daily Questionnaire – OMDQ) for OM evaluation are recommended for the early detection and scoring of OM-associated signs and symptoms [11]. Other tools such as the numeric rating scale (NRS) or the visual analog scale (VAS) can be associated for pain assessment. Tools to evaluate swallowing ability, salivation, and speech (e.g., the Tardieu scale) can also be considered.

From a proactive perspective, risk assessment before starting conditioning is fundamental. A thorough patient evaluation aimed to define the risk of developing OM should be performed to set a personalized prevention plan. The timing of risk assessment should consider the potential need of dental specialists (dentist or dental hygienist) to address any potential oral issues affecting patient's mouth and/or dentures before the transplant. At the end of the assessment process, a risk level should be attributed to the patient, and any preventive measure should be scheduled based on it [23].

In the absence of validated tools, risk assessment emerges as the result of a complex clinical evaluation that may yield different results depending on the context of care, the healthcare professional who performs the assessment, the availability of needed information, and the patient's compliance.

It is important to consider both direct and indirect risks.

Direct risks include age, sex, hygiene habits, baseline oral condition, including the presence of lesions, cavities, prostheses as well as any previous episodes of OM, and both patient and caregiver compliance.

Indirect risks are associated with the underlying disease and with treatment-related factors such as the intensity of conditioning regimen and its duration, the degree of immunosuppression, the presence of comorbidities (e.g., diabetes), previous antineoplastic drug exposure, other stomatotoxic drugs, and the development of complications (e.g., graft-versus-host disease – GvHD).

Each center should consider implementing tools and checklists to optimize the risk assessment process; the level of risk attribution must take into account the local expertise, interprofessional collaboration, and other local facilities and services, including organization of nursing activities and the availability of dental specialists. The following table (**Table 2.**) is to take in account as an example of risk stratification resulted by the analysis of literature reported data on incidence, severity and duration of OM [24, 25], and by the authors' opinion and experience on the field.

Risk level	Treatment	Factors increasing incidence and severity level
Higher risk	• MAC • High-dose Melphalan • Busulfan and/or TBI based	• Methotrexate (GvHD prophylaxis) • Female gender • Metabolic diseases (diabetes, renal insufficiency, etc..) • Severe malnutrition • Previous severe OM (grade 3–4 WHO) • Alcohol and tobacco consumption • Active mucosal lesions or infection loci • Elderly or pediatric patient • Poor adherence to oral hygiene
Moderate risk	• RIC or RTC	
Lower risk	• Autologous HSCT (RIC)	

HSCT = Hematopoietic Stem Cell Transplantation; MAC = Myeloablative Conditioning; TBI = Total Body Irradiation; OM = Oral Mucositis; WHO = World Health organization; RIC = Reduced Intensity Conditioning; RTC = Reduced Toxicity Conditioning.

Table 2.
OM risk stratification in patients undergoing HSCT.

4.2 Monitoring

A systematic evaluation of the oral cavity should be performed on all patients undergoing HSCT during and after their hospital stay. Each transplant center should have standardized protocols for OM assessment, prevention, and treatment as well as tools for data recording and their reporting among professionals [11]. Monitoring of a patient's oral condition during the hospital stay for HSCT should be frequent (at least daily) and scheduled based on the patient's clinical condition, time of OM onset and features, and any patient-reported symptoms.

Oral problems should be monitored until they have healed or until recovery is considered imminent. Routine assessments (at each hospital visit) of allogeneic HSCT patient's oral cavity are recommended after discharge and during follow-up to detect early signs and symptoms of oral GvHD to promptly establish an adequate intervention and monitoring plan [26].

5. Oral care

5.1 Multiprofessional approach

Various healthcare professionals should collaborate together with the aim to provide a holistic approach to patients affected by OM. However, an oral care team, which includes nurses, physicians, dentists, dental hygienists, dietitians, pharmacists, and other relevant figures, is still not widely implemented in oncology centers [16].

Although there is no strong scientific evidence regarding the adoption of professional oral hygiene programs versus daily hygiene in preventing OM [18], the pre-HSCT involvement of dentists and dental hygienists is recommended to reduce oral issues, especially infectious or traumatic issues, and to promote general oral well-being and comfort before transplantation [16, 18, 27].

A prophylactic evaluation performed by the dental specialist should be planned for all patients, and those who have oral conditions that increase the risk of infection or lesions during neutropenia should be treated to reduce that risk.

The presence of infectious foci, e.g., cavities, gingivitis, tooth fractures, excessive tartar, overly mobile prostheses with sharp elements or causing friction on the mucosa, must be detected and resolved or limited before HSCT [16, 18, 27].

5.2 Oral care: General concepts

The adoption of correct oral hygiene strategies and the implementation of standardized oral care protocols are fundamental to support all patients undergoing antineoplastic therapy [11, 16].

The oral care protocol (daily mouth care protocol) should include the following key elements: mouth cleansing after each meal, tooth brushing (and interdental cleansing), mouth rinses, and promotion of mucosal hydration [28].

This makes it possible to control the mouth microbial population, to limit the pathogenic bacterial species growth, and to prevent oral infection and/or systemic bacterial translocation as well as to reduce the risk of developing gingival or mucosal lesions.

5.3 Use of toothbrushes and other hygiene tools

Teeth, gums, and mucosa should be brushed at least after each meal and before bed [16] using regular fluoride toothpaste (1350–1500 ppm) and soft-bristle toothbrushes to limit the risk of mucosal and gingival trauma in pancytopenia patients [16, 18]. Brushing must be performed gently, with frequent rinsing with warm water.

Electric toothbrushes can be used instead of manually ones if patients accustomed to using them. Disposable oral swabs are not recommended due to their poor cleaning effectiveness [11, 16, 18]; however, they can be considered for use in severe thrombocytopenic patients with mucosal bleeding.

Tools for interdental cleansing (dental floss, interdental brushes) should be used with caution and limited to patients proficient in their use so as to avoid aggressive and potentially traumatic techniques [29]. Patients' competence in flossing must be evaluated by nurses before patient admission [11, 16, 18].

5.4 Oral rinses

A high number of mouth rinses have been studied, and probably many more have been used in clinical practice. Studied products include mixed solutions with antibacterial, antifungal, and anti-inflammatory action as well as various other mouthwashes (e.g., Magic solution). However, no recommendations are possible to support their use in daily practice due to the inconsistent and conflicting results [17, 27]. These solutions may potentially impair patients' comfort and/or safety if improperly stored or handled [27].

Bland saline solutions (0.9% normal saline, sodium bicarbonate solutions) are safe, inexpensive, and easy to use for frequent rinsing. They keep the mouth clean and moist [11, 18] without any adverse events.

There is no evidence to support the use of mouthwashes containing chlorhexidine in preventing or treating OM [11, 16, 30], but they can be useful as an oral antiseptic to treat specific lesions before or during neutropenia.

Moisturizing, emolliating, and barrier solutions belong to a very wide range of commercially available products. However, the use of many products has almost never supported by studies or scientific evidence on OM prevention and treatment; their use should be targeted to sign and symptom reduction (e.g., as a palliative treatment for mild/moderate pain) [31]. Wetting and emollient solutions as well as artificial saliva can be useful in cases of extreme dry mouth-related discomfort, and barrier and moisturizing products can be useful in reducing inflammation or burning.

5.5 Management of removable dentures

The fit of removable dentures should be assessed and managed before transplantation to avoid mucosal frictions and injuries [16].

During and after conditioning, it is advisable to minimize denture use. Carefully clean dentures with soap and/or specific products (consult the manufacturer's instructions), rinse with water, and carefully dry and store them in specific containers. Denture adhesives should be avoided during HSCT period until mucosal recovery [32].

Do not store dentures in antiseptic solutions for long periods of time to avoid solution contamination.

5.6 Managing dryness and bleeding

Saliva plays a crucial role in maintaining oral homeostasis, cleansing, and protection against infection.

The early treatment of dryness is considered one of the most important oral care interventions to delay the development of severe OM and to provide patients relief [31]. The use of moistening solutions and ensuring adequate fluid intake can help nurses to manage dryness.

Encourage patients to keep their mouths moist by sipping water or rinsing frequently with mild saline solutions (normal saline or bicarbonate solutions). It is possible to stimulate salivation by sucking ice chips, chewing sugar-free gum, and eating pieces of fruit such as pineapple or lemon (if tolerated).

Encourage the patient to avoid sweet and/or carbonated drinks, coffee, and smoking as well as crunchy foods, which can cause mucosal lesions. Dietary advice (see Section 6.3), a careful review of drugs (discontinuing those that cause dryness), and patient training to limit mouth breathing can be part of a dry mouth management protocol.

In cases of severe dryness, saliva substitutes or lubricating products can be considered, thick secretions can be removed with water rinses or aerosols, and dry lips can be treated with paraffin or petroleum jelly or oils, lip balms or ointments, or cocoa butter. Nurses should be aware that chronic use of Vaseline-based products may worsen the problem [23].

Regular assessment and appropriate intervention to manage gum and mucosal bleeding are necessary to reduce the risk of OM severity, which increases patient discomfort and complications [11, 16, 28]. Nurses must evaluate the sites and severity of bleeding in relation to any risk factors (coagulation disorders, thrombocytopenia, etc.). To remove blood and clots from the mouth, the patient may find rinsing with mild saline solutions gives relief, and cold solutions or ice chips may be used to reduce bleeding if tolerated. Rinses with tranexamic acid solutions can be provided as topical intervention to reduce oral bleeding.

WL (nm)	PD (mW/cm²)	TxS (s)	ED (j/cm²)	SS (cm²)	NS (n°)	Duration
632.8	31.25	40	1.0	0.8	18	For 5d after EoC
650	1000	2	2.0	0.04	54–70	SoC to d + 2 (7–13 d)

WL = Wavelength; PD = Power Density; TxS = Time per Spot; ED = Energy Density; SS = Spot Size; NS = Number of Spots; d = Day; EoC = End of Conditioning; SoC = Start of Conditioning.

Table 3.
Photobiomodulation protocols to prevent OM in HSCT patients.

5.7 Oral mucositis prevention and treatment

To date, the literature does not provide any strong evidence that can be applied in all settings. The evidence in the HSCT setting is very limited, and practice recommendations are often supported by studies with various methodological biases. The following recommendations for OM prevention and treatment are provided in the international guidelines [16, 18].

Preventive photobiomodulation by low-frequency oral laser applications is strongly recommended to prevent OM according to the protocols detailed in **Table 3**.

Cryotherapy through topical oral applications of ice chips has proven effective in preventing OM in autologous HSCT patients undergoing high-dose melphalan conditioning regimen. Ice should be applied during chemotherapy administration and for 30 minutes afterwards.

Keratinocyte growth factor (KGF-1) is strongly recommended for patients undergoing HSCT with myeloablative conditioning regimens who are expected to develop severe OM.

Not recommended preventive treatments include iseganan rinses, granulocyte growth factor rinses, systemic pentoxifylline and pilocarpine, topical and systemic sucralfate, and parenteral glutamine [16].

Patient-controlled morphine administration is indicated for severe OM-related pain management [16]. For mild to moderate pain, transdermal fentanyl patches and 0.5% doxepin oral rinses can be considered [16, 18].

Honey and photobiomodulation have shown encouraging results on OM treatment in pediatric patients, but further studies are needed to confirm their efficacy [33].

Numerous natural products were studied for the management of OM. Bioactive products from natural sources such as *Aloe vera, Glycyrrhiza glabra, Camellia sinensis, Calendula officinalis* and other herbal-derived remedies may be sources of bioactive compounds with pharmacological activities and have arisen as effective approaches for OM-related symptoms control. These activities are associated with the bioactive components of their matrix (such as flavonoids, salts and vitamins), which are associated with in vivo biological activities and minimal or absent toxicity [34, 35]. However, studies supporting recommendations on their use for prevention and/or treatment of OM in HSCT setting are lacking.

6. Further considerations

6.1 Oral infections

Prophylaxis of oral candidiasis in HSCT patients should be performed using drugs that are partially or fully absorbable by the gastrointestinal tract [36]. Various microorganisms contribute to the pathobiology of mucositis [10] and can cause mucosal infections.

Thus, it may be useful to perform culture swabs, increase the frequency of oral care interventions, use topical antimicrobial agents, and notify the physician to initiate systemic anti-infective therapy.

As recommended by the National Comprehensive Cancer Network (NCCN) Guidelines [37] antibiotic, antifungal and antiviral prophylaxis should be considered for patients undergoing autologous or allogeneic HSCT during neutropenia. Antifungal prophylaxis strategies with Fluconazole or other recommended drugs should be based on patient- and center-specific risk factors and could be avoided only in autologous HSCT patients who do not develop mucositis. A careful oral mucosal observation and the assessment of gastrointestinal symptoms can allow to early recognize mucositis-associated signs. Consequently, an appropriate diagnosis process and the timely starting of the antimicrobial treatment are mandatory. Considering that all patients with febrile neutropenia should receive broad-spectrum antibiotics, the assessment of mucosal lesions should leads the antimicrobial drugs review by the clinicians.

6.2 Emesis

Vomiting is an important factor that worsens mucosal conditions. Therefore, during conditioning regimen administration and the neutropenic phase, it is mandatory to prevent and treat emesis by adopting current guidelines. The assessment of taste alteration causes can lead to implementing intervention to limit patient's nausea, and providing fractionated meals composed of foods tolerated by the patient can help control the symptoms.

6.3 Dietary and nutritional aspects

The nutritional status of patients is correlated with functional recovery, infection development, and wound healing [38]. The development of OM leads to difficulties or inability to intake food and beverages orally, potentially altering the patient's nutritional status [16].

All HSCT patients should undergo routine screening to assess and monitor their nutritional status [38], especially those at higher risk of inflammatory gastrointestinal complications such as mucositis [16]. A multiprofessional approach to manage patients' nutritional issues can help to prevent or mitigate OM. Dietary advices should be provided by healthcare professionals (dietitian, nurses, etc....) to maintain a balanced food intake and avoid mucosal injuries and inflammation during treatment period and the follow-up.

However, it is necessary to consider various aspects, including taste alterations, nausea and vomiting, oral intake monitoring, nutritional needs calculation, and nutritional support options (enteral, parenteral).

6.4 Research

In 2019, the Multinational Association for Supportive Care in Cancer (MASCC) Mucositis Study Group (MSG) reviewed findings on established and emerging mediators of toxicity and potential insights from technological advances in mucositis research [39]. It concluded that technological advances had the potential to revolutionize the field of mucositis research, although focused effort was needed to move rationally targeted interventions to the clinical setting. However, exploring literature, less than 20 records reporting clinical trials findings on prevention and treatment of

N°	First author Year	Design	Population Setting	Study group Intervention	Control group Intervention	Major findings
1	Nakagaki et al. [40]	RCT	Adults. Auto/Allo HSCT	55 pts. Polaprezinc MW	53 pts. Sodium bicarbonate MW	No difference in grades 3–4 and grades 2–4 OM incidence and duration, patient-reported pain and functional limitations.
2	Hong et al. [41]	RCT	Adults. Auto HSCT	25 pts. Aciclovir PO prophylaxis	29 pts. No prophylaxis	OM incidence and HSV reactivations was significantly lower in SG
3	Yoshimatsu et al. [42]	RCT	Adults. Allo HSCT	15 pts. Hangeshashinto+Saline solution MW	15 pts. Saline solution MW	No difference in OM incidence. OM duration was significantly lower in SG
4	Çakmak and Nural [43]	RCT	Adults. HD-CHT/Auto HSCT	32 pts. Propolis mouthwashes + SP	32 pts. SP	OM incidence and onset time, and grade 2–3 time were significantly lower in SG.
5	Guberti et al. [44]	Non-randomized	Adults. Auto/Allo HSCT	59 pts. Colostrum & Aloe-Vera MW and gel	81 pts (historical cohort). SP	Grade 3–4 OM incidence, OM duration, FN incidence and duration were significantly lower in SG
6	Shahrabi et al. [45]	RCT	Pediatrics. Allo HSCT	29 pts. Mucosamin® spray	30 pts. Placebo	OM incidence, severity, duration were significantly lower in SG. OM onset time was longer in SG.
7	Walladbegi et al. [46]	RCT (Multicentric)	Adults. Auto HSCT	90 pts. Intraoral cooling device	92 pts. Ice Chips	No differences in OM incidence (whole sample). Significant lover OM incidence and severity were found in lymphoma patients who received ICD.
8	Parkhideh et al. [47]	RCT	>15 years old Auto/Allo HSCT	35 pts. Azithromycin oral suspension + SP	35 pts. SP	OM incidence and duration were significantly lower in SG
9	Baysal et al. [48]	Non-randomized (Multicentric)	Adults. Autologous HSCT	16 pts. Cryotherapy + SP	16 pts. SP	No differences in OM incidence, severity, and duration
10	Ramos-Pinto et al. [49]	RCT	Adults Auto/Allo HSCT	30 pts. Intra-oral PBM	30 pts. Extra-oral PBM	No differences between the groups.

N°	First author Year	Design	Population Setting	Study group Intervention	Control group Intervention	Major findings
11	Kitagawa et al. [50]	RCT (Multicentric)	Adults. Auto/Allo HSCT	41 pts. Polaprezinc lozenges by SoC	47 pts. Polaprezinc lozenges by G2-OM	Significant difference in G2-OM incidence. No other significant differences
12	Kamsvåg et al. [51]	RCT	Pediatrics. Auto/Allo HSCT	26 pts. Cryotherapy + SP	23 pts. SP	No differences in severe OM incidence, oral pain, and opioid use. Poor adherence.
13	Harman et al. [52]	Non-randomized	Adults. Auto/Allo HSCT	1. 28 pts. Ca-P solution MW 2. 25 pts. Black mulberry syrup MW	30 pts. SP	No differences in OM incidence (any grade)
14	Lu et al. [53]	RCT	Allo HSCT	1. 38 pts. Whole course cryotherapy + SP 2. 36 pts. Half course cryotherapy + SP 3. 36 pts. Twice daily cryotherapy + SP	35 pts. SP	Group 1 and 2 showed significant lower OM incidence, severity and duration. Highest OM incidence in group 3.
15	Ueno et al. [54]	Prospective	Adults. Allo HSCT & HNC	10 pts. Episil® MW	N/A	A significant decrease in the pain score was observed 5 minutes after intervention, and this effect lasted up to 120 min
16	Potting et al. [55]	RCT	Adults. Auto HSCT	23 pts. Powered tooth brushing	23 pts. Manual tooth brushing	Overall plaque scores were significantly lower in SG. No difference in OM incidence and severity.
17	Mubaraki et al. [56]	RCT	Pediatrics. Allo HSCT	1. 15 pts. Extra-soft toothbrushing + SP 2. 15 pts. Supersatured Ca oral spray	15 pts. SP	No difference on OM incidence. Lower OM severity in Supersatured Ca group.

RCT = Randomized Controlled Trial; Auto HSCT = Autologous Hematopoietic Stem Cell Transplantation; Allo HSCT = Allogeneic Hematopoietic Stem Cell Transplantation; pts. = patients; MW = Mouthwashing; OM = Oral Mucositis; PO = Per Os; HSV = Herpes Simplex Virus; SG = Study Group; HD-CHT = High-Dose Chemotherapy; SP = Standard Practice; FN = Febrile Neutropenia; ICD = Intraoral Cooling Device; PBM = Photobiomodulation; SoC = Start of Conditioning; G2-OM = Grade 2 Oral Mucositis; Ca-P Calcium-Phosphate; HNC = Head and Neck Cancer; N/A = Not Applicable; Ca = Calcium.

Table 4.
Summary of the trials on OM management in HSCT patient published in the last 5-years.

OM in HSCT patients are available in the last 5 years. In addition, by the 70 protocols actually registered in Clinicaltrials.gov, only 12 are reporting results, and less than one third are sustained by the Industry or Government facilities. This may reflects a poor appreciation for research in this field and a lack of financial appeal for seeking solutions to the problem. In **Table 4** are summarized the major findings of the most recent trials available in literature. It is evident that the major efforts are still provided by independent researcher who are exploring new strategies for OM prevention. However, it seems that the researchers are looking for the "miracle herb" for prevention instead having a comprehensive approach to the problem; thus the very limited trials on oral hygiene approaching strategies as well as the reduced availability of studies exploring new ways for the treatment of established OM.

7. Conclusions

Managing OM in HSCT patients requires an integrated, multidisciplinary approach based on up-to-date scientific evidence and advanced competencies. However, the research is slow in providing solutions to many of the problems affecting patients who experience this complication, including those related to the standardization of oral hygiene procedures. Continuous nurse training, patient and caregiver education, effective assessment and monitoring, and standardized oral hygiene protocols are still essential aspects to improve HSCT patients' clinical outcomes, safety and comfort.

Acknowledgements

The authors thank Dr. Francesco Merli, Medical Director of the Hematology Unit of Reggio Emilia. A special thank you to the GITMO (Italian Bone Marrow Transplantation) Nurses Group, and many thanks to Jacqueline M. Costa for the English language editing.

Conflict of interest

The authors declare no conflict of interest.

Author details

Stefano Botti*, Doria Valentini and Stefania Grasselli
Azienda USL-IRCCS of Reggio Emilia, Reggio Emilia, Italy

*Address all correspondence to: stefano.botti@ausl.re.it

References

[1] Bellm LA, Epstein JB, Rose-Ped A, Martin P, Fuchs HJ. Patient reports of complications of bone marrow transplantation. Supportive Care in Cancer. 2000 Jan;**8**(1):33-39

[2] Sonis ST. Mucositis: The impact, biology and therapeutic opportunities of oral mucositis. Oral Oncology. 2009 Dec;**45**(12):1015-1020

[3] Sonis ST, Oster G, Fuchs H, Bellm L, Bradford WZ, Edelsberg J, et al. Oral mucositis and the clinical and economic outcomes of hematopoietic stem-cell transplantation. Journal of Clinical Oncology. 2001;**19**(8):2201-2205

[4] Vera-Llonch M, Oster G, Ford CM, Lu J, Sonis S. Oral mucositis and outcomes of allogeneic hematopoietic stem-cell transplantation in patients with hematologic malignancies. Supportive Care in Cancer. 2007 May;**15**(5):491-496

[5] Blijlevens NM, Donnelly JP, de Pauw BE. Empirical therapy of febrile neutropenic patients with mucositis: Challenge of risk-based therapy. Clinical Microbiology and Infection. 2001;**7**(Suppl. 4):47-52. DOI: 10.1046/j.1469-0691.2001.00058.x

[6] Facchini L, Martino R, Ferrari A, Piñana JL, Valcárcel D, Barba P, et al. Degree of mucositis and duration of neutropenia are the major risk factors for early post-transplant febrile neutropenia and severe bacterial infections after reduced-intensity conditioning. European Journal of Haematology. 2012 Jan;**88**(1):46-51. DOI: 10.1111/j.1600-0609.2011.01724.x

[7] Filicko J, Lazarus HM, Flomenberg N. Mucosal injury in patients undergoing hematopoietic progenitor cell transplantation: New approaches to prophylaxis and treatment. Bone Marrow Transplantation. 2003;**31**(1):1-10

[8] Bhatt V, Vendrell N, Nau K, Crumb D, Roy V. Implementation of a standardized protocol for prevention and management of oral mucositis in patients undergoing hematopoietic cell transplantation. Journal of Oncology Pharmacy Practice. 2010 Sep;**16**(3):195-204. DOI: 10.1177/1078155209348721

[9] Vagliano L, Feraut C, Gobetto G, Trunfio A, Errico A, Campani V, et al. Incidence and severity of oral mucositis in patients undergoing haematopoietic SCT--results of a multicentre study. Bone Marrow Transplantation. 2011 May;**46**(5):727-732. DOI: 10.1038/bmt.2010.184

[10] Sonis ST. The pathobiology of mucositis. Nature Reviews. Cancer. 2004 Apr;**4**(4):277-284

[11] Quinn B, Potting CM, Stone R, Blijlevens NM, Fliedner M, Margulies A, et al. Guidelines for the assessment of oral mucositis in adult chemotherapy, radiotherapy and haematopoietic stem cell transplant patients. European Journal of Cancer. 2008 Jan;**44**(1):61-72

[12] Barker GJ, Epstein JB, Williams KB, Gorsky M, Raber-Durlacher JE. Current practice and knowledge of oral care for cancer patients: A survey of supportive health care providers. Supportive Care in Cancer. 2005 Jan;**13**(1):32-41

[13] Fulton JS, Middleton GJ, McPhail JT. Management of oral complications. Seminars in Oncology Nursing. 2002 Feb;**18**(1):28-35

[14] McGuire DB, Johnson J, Migliorati C. Promulgation of guidelines for mucositis

management: Educating health care professionals and patients. Supportive Care in Cancer. 2006 Jun;**14**(6):548-557

[15] FACT-JACIE International Standards for Hematopoietic Cellular Therapy Product Collection, Processing, and Administration Manual, Eight Edition (Version 8.1). December 14, 2021. Available from: https://www.ebmt.org/8th-edition-fact-jacie-standards [Accessed: August 10, 2022]

[16] Elad S, Cheng KKF, Lalla RV, Yarom N, Hong C, Logan RM, et al. Mucositis guidelines leadership Group of the Multinational Association of supportive Care in Cancer and International Society of Oral oncology (MASCC/ISOO). MASCC/ISOO clinical practice guidelines for the management of mucositis secondary to cancer therapy. Cancer. 2020;**126**(19):4423-4431. DOI: 10.1002/cncr.33100

[17] Clarkson JE, Worthington HV, Furness S, McCabe M, Khalid T, Meyer S. Interventions for treating oral mucositis for patients with cancer receiving treatment. Cochrane Database of Systematic Reviews. 2010;**8**:CD001973. DOI: 10.1002/14651858.CD001973.pub4

[18] Peterson DE, Boers-Doets CB, Bensadoun RJ, Herrstedt J. ESMO guidelines committee. Management of oral and gastrointestinal mucosal injury: ESMO clinical practice guidelines for diagnosis, treatment, and follow-up. Annals of Oncology. 2015 Sep;**26**(Suppl. 5):v139-v151. DOI: 10.1093/annonc/mdv202

[19] Sadler GR, Oberlee-Edwards L, Farooqi A, Hryniuk WM. Oral sequelae of chemotherapy: An important teaching opportunity for oncology health care providers and their patients. Support Care Cancer. 2000;**8**:209-214

[20] Arnold R et al. Educating for a Change. Toronto (Ontario, Canada), Between the Lines, Toronto, Ontario, Canada: Doris Marshall Institute for Education and Action; 1991

[21] Cooley ME, Moriarty H, Berger MS, Selm-Orr D, Coyle B, Short T. Patient literacy and the readability of written cancer educational materials. Oncology Nursing Forum. 1995;**22**:1345-1351

[22] Doak LG, Doak CC, Meade CD. Strategies to improve cancer education materials. Oncology Nursing Forum. 1996;**23**:1305-1312

[23] European Oral Care in Cancer group (EOCC): Oral Care Guidance and Support. First Edition. 2017. Available at: www.eocc.co.uk [Accessed: April 12, 2024]

[24] Chaudhry HM, Bruce AJ, Wolf RC, Litzow MR, Hogan WJ, Patnaik MS, et al. The incidence and severity of Oral mucositis among allogeneic hematopoietic stem cell transplantation patients: A systematic review. Biology of Blood and Marrow Transplantation. 2016 Apr;**22**(4):605-616. DOI: 10.1016/j.bbmt.2015.09.014

[25] Agholme MB, Dahllöf G, Törlén JK, Majorana A, Brennan MT, von Bültzingslöwen I, et al. Incidence, severity, and temporal development of oral complications in pediatric allogeneic hematopoietic stem cell transplant patients - a multicenter study. Supportive Care in Cancer. 2023;**31**(12):702. DOI: 10.1007/s00520-023-08151-1

[26] Kuten-Shorrer M, Woo SB, Treister NS. Oral graft-versus-host disease. Dental Clinics of North America. 2014 Apr;**58**(2):351-368. DOI: 10.1016/j.cden.2013.12.007. Review

[27] Hong CHL, Gueiros LA, Fulton JS, Cheng KKF, Kandwal A, Galiti D,

et al. Mucositis study Group of the Multinational Association of supportive Care in Cancer/International Society for Oral Oncology (MASCC/ISOO). Systematic review of basic oral care for the management of oral mucositis in cancer patients and clinical practice guidelines. Supportive Care in Cancer. 2019 Oct;**27**(10):3949-3967. DOI: 10.1007/s00520-019-04848-4

[28] Sonis ST, Elting LS, Keefe D, et al. Perspectives on cancer therapy-induced mucosal injury: Pathogenesis, measurement, epidemiology, and consequences for patients. Cancer. 2004;**100**:1995-2025

[29] Sambunjak D, Nickerson JW, Poklepovic T, Johnson TM, Imai P, Tugwell P, et al. Flossing for the management of periodontal diseases and dental caries in adults. Cochrane Database of Systematic Reviews. 2011;**12**:CD008829. DOI: 10.1002/14651858.CD008829.pub2

[30] Cardona A, Balouch A, Abdul MM, Sedghizadeh PP, Enciso R. Efficacy of chlorhexidine for the prevention and treatment of oral mucositis in cancer patients: A systematic review with meta-analyses. Journal of Oral Pathology & Medicine. 2017 Oct;**46**(9):680-688. DOI: 10.1111/jop.12549

[31] Sonis ST. Oral Mucositis. Anti-Cancer Drugs. 2011;**22**(7):607-612

[32] Duyck J, Vandamme K, Muller P, Teughels W. Overnight storage of removable dentures in alkaline peroxide-based tablets affects biofilm mass and composition. Journal of Dentistry. 2013 Dec;**41**(12):1281-1289. DOI: 10.1016/j.jdent.2013.08.002

[33] Miranda-Silva W, Gomes-Silva W, Zadik Y, Yarom N, Al-Azri AR, Hong CHL, et al. Mucositis study Group of the Multinational Association of supportive Care in Cancer / International Society for Oral Oncology (MASCC/ISOO). MASCC/ISOO clinical practice guidelines for the management of mucositis: Sub-analysis of current interventions for the management of oral mucositis in pediatric cancer patients. Supportive Care in Cancer. 2021;**29**(7):3539-3562. DOI: 10.1007/s00520-020-05803-4

[34] Ferreira AS, Macedo C, Silva AM, Delerue-Matos C, Costa P, Rodrigues F. Natural products for the prevention and treatment of Oral mucositis-a review. International Journal of Molecular Sciences. 2022;**23**(8):4385. DOI: 10.3390/ijms23084385

[35] Yarom N, Hovan A, Bossi P, Ariyawardana A, Jensen SB, Gobbo M, et al. Mucositis study Group of the Multinational Association of supportive Care in Cancer / International Society of Oral Oncology (MASCC/ISOO). Systematic review of natural and miscellaneous agents, for the management of oral mucositis in cancer patients and clinical practice guidelines - part 2: Honey, herbal compounds, saliva stimulants, probiotics, and miscellaneous agents. Supportive Care in Cancer. 2020;**28**(5):2457-2472. DOI: 10.1007/s00520-019-05256-4

[36] Clarkson JE, Worthington HV, Eden OB. Interventions for preventing oral candidiasis for patients with cancer receiving treatment. Cochrane Database Systematic Reviews. 24 Jan 2007;**2007**(1):CD003807. DOI: 10.1002/14651858.CD003807.pub3

[37] National Comprehensive Cancer Network. NCCN Clinical Practice Guidelines in Oncology. Prevention and Treatment of Cancer-Related Infection. Version 1. 2024. Available at: https://www.nccn.org/guidelines/

guidelines-detail?category=3&id=1457 [Accessed: August 12, 2024]

[38] Muscaritoli M, Arends J, Bachmann P, Baracos V, Barthelemy N, Bertz H, et al. ESPEN practical guideline: Clinical Nutrition in cancer. Clinical Nutrition. 2021;**40**(5):2898-2913. DOI: 10.1016/j.clnu.2021.02.005

[39] Bowen J, Al-Dasooqi N, Bossi P, Wardill H, Van Sebille Y, Al-Azri A, et al. Mucositis study Group of the Multinational Association of supportive Care in Cancer/International Society of Oral Oncology (MASCC/ISOO). The pathogenesis of mucositis: Updated perspectives and emerging targets. Supportive Care in Cancer. 2019 Oct;**27**(10):4023-4033. DOI: 10.1007/s00520-019-04893-z

[40] Nakagaki M, Kennedy GA, Gavin NC, Butler J, Clavarino A, Whitfield K. A randomised trial of topical polaprezinc to prevent oral mucositis in patients undergoing haematopoietic stem cell transplantation (ToPaZ study). Supportive Care in Cancer. 2023;**32**(1):44. DOI: 10.1007/s00520-023-08235-y

[41] Hong J, Park HK, Chang SH, Byun JM, Shin DY, Koh Y, et al. A randomized phase II study of acyclovir for the prevention of chemotherapy-induced oral mucositis in patients undergoing autologous hematopoietic stem cell transplantation. BMC Oral Health. 2023;**23**(1):1008. DOI: 10.1186/s12903-023-03623-6

[42] Yoshimatsu M, Kawashita Y, Soutome S, Murata M, Sawayama Y, Kurogi T, et al. Hangeshashinto for prevention of oral mucositis in patients undergoing hematopoietic stem cell transplantation: A randomized phase II study. Supportive Care in Cancer. 2023;**31**(12):707. DOI: 10.1007/s00520-023-08175-7

[43] Çakmak S, Nural N. Efficacy of propolis in the prevention of oral mucositis in patients undergoing high-dose chemotherapy: A randomized controlled trial. Cancer Nursing. 2024;**47**(4):E255-E268. DOI: 10.1097/NCC.0000000000001212

[44] Guberti M, Botti S, Caffarri C, Cavuto S, Savoldi L, Fusco A, et al. Efficacy and safety of a colostrum- and Aloe vera-based oral care protocol to prevent and treat severe oral mucositis in patients undergoing hematopoietic stem cell transplantation: A single-arm phase II study. Annals of Hematology. 2022;**101**(10):2325-2336. DOI: 10.1007/s00277-022-04934-4

[45] Shahrabi M, Solduzian M, Babaie MH, Mousavi SA, Goodarzi N, Ravari NS, et al. The effects of a combination oral spray (Mucosamin®) for the prevention of oral mucositis in pediatric patients undergoing hematopoietic stem cell transplantation: A double blind randomized clinical trial. Supportive Care in Cancer. 2022;**30**(10):7963-7972. DOI: 10.1007/s00520-022-07231-y

[46] Walladbegi J, Henriksson R, Tavelin B, Svanberg A, Larfors G, Jädersten M, et al. Efficacy of a novel device for cryoprevention of oral mucositis: A randomized, blinded, multicenter, parallel group, phase 3 trial. Bone Marrow Transplantation. 2022;**57**(2):191-197. DOI: 10.1038/s41409-021-01512-6

[47] Parkhideh S, Zeraatkar M, Moradi O, Hajifathali A, Mehdizadeh M, Tavakoli-Ardakani M. Azithromycin oral suspension in prevention and management of oral mucositis in patients undergoing hematopoietic

stem cell transplantation: A randomized controlled trial. Supportive Care in Cancer. 2022;**30**(1):251-257. DOI: 10.1007/s00520-021-06409-0

[48] Baysal E, Sari D, Vural F, Çağırgan S, Saydam G, Töbü M, et al. The effect of cryotherapy on the prevention of oral mucositis and on the oral pH value in multiple myeloma patients undergoing autologous stem cell transplantation. Seminars in Oncology Nursing. 2021;**37**(3):151146. DOI: 10.1016/j.soncn.2021.151146

[49] Ramos-Pinto MB, de Lima Gusmão TP, Schmidt-Filho J, Jaguar GC, Martins MD, Alves FA. Intraoral versus extraoral photobiomodulation therapy in the prevention of oral mucositis in HSCT patients: A Randomized, single-blind, controlled clinical trial. Supportive Care in Cancer. 2021;**29**(11):6495-6503. DOI: 10.1007/s00520-021-06228-3

[50] Kitagawa J, Kobayashi R, Nagata Y, Kasahara S, Ono T, Sawada M, et al. Polaprezinc for prevention of oral mucositis in patients receiving chemotherapy followed by hematopoietic stem cell transplantation: A multi-institutional randomized controlled trial. International Journal of Cancer. 2021;**148**(6):1462-1469. DOI: 10.1002/ijc.33316

[51] Kamsvåg T, Svanberg A, Legert KG, Arvidson J, von Essen L, Mellgren K, et al. Prevention of oral mucositis with cryotherapy in children undergoing hematopoietic stem cell transplantations-A feasibility study and randomized controlled trial. Supportive Care in Cancer. 2020;**28**(10):4869-4879. DOI: 10.1007/s00520-019-05258-2

[52] Harman M, Ovayolu N, Ovayolu O. The effect of three different solutions on preventing oral mucositis in cancer patients undergoing stem cell transplantation: A non-randomized controlled trial: A Turkish study - Non-randomised trial. The Journal of the Pakistan Medical Association. 2019;**69**(6):811-816

[53] Lu Y, Zhu X, Ma Q, Wang J, Jiang P, Teng S, et al. Oral cryotherapy for oral mucositis management in patients receiving allogeneic hematopoietic stem cell transplantation: A Prospective randomized study. Supportive Care in Cancer. 2020;**28**(4):1747-1754. DOI: 10.1007/s00520-019-04966-z

[54] Ueno T, Yatsuoka W, Ishiki H, Miyano K, Uezono Y. Effects of an oral mucosa protective formulation on chemotherapy- and/or radiotherapy-induced oral mucositis: A prospective study. BMC Cancer. 2022;**22**(1):90. DOI: 10.1186/s12885-021-09107-6

[55] Potting CMJ, van Leeuwen SJM, KurstjensMH,BronkhorstEM,ThomasRZ, Blijlevens NMA, et al. A randomized controlled trial of manual versus powered tooth brushing during haematopoietic stem cell transplantation. Oral Diseases. 2022;**28**(7):1987-1994. DOI: 10.1111/odi.13899

[56] Mubaraki S, Pani SC, Alseraihy A, Abed H, Alkhayal Z. The efficacy of two different oral hygiene regimens on the incidence and severity of oral mucositis in pediatric patients receiving hematopoietic stem cell transplantation: A prospective interventional study. Special Care in Dentistry. 2020;**40**(6):566-573. DOI: 10.1111/scd.12525

Chapter 26

Renal Complications Following Hematopoietic Stem Cell Transplantation

Alina Tanase, Andreea Andronesi, Lavinia Lipan, Mihaela Gherghiceanu and Gener Ismail

Abstract

Renal complications following hematopoietic stem cell transplantation (HSCT) pose substantial clinical challenges and impact patient outcomes. This chapter provides an in-depth analysis of the renal complications that may arise post-HSCT, including acute kidney injury (AKI), nephrotic syndrome, chronic kidney disease (CKD), and electrolyte imbalances. These complications occur from various factors such as nephrotoxic medications, graft-versus-host-disease (GVHD), infections, and pre-existing renal conditions. Early identification and management of renal dysfunction are crucial for optimizing patient care and prognosis. The chapter underscores the importance of vigilant monitoring, supportive measures, and timely interventions to mitigate renal complications post-HSCT. Enhanced understanding of these complications empowers clinicians to implement targeted preventive strategies and personalized therapeutic interventions, ultimately enhancing the overall outcomes and quality of life for HSCT recipients.

Keywords: hematopoietic stem cell transplantation, acute kidney injury, nephrotoxicity, nephrotic syndrome, chronic kidney disease

1. Introduction

The significance of renal complications post-stem cell transplantation (SCT) cannot be overstated, as they contribute substantially to morbidity, mortality, and long-term outcomes in transplant recipients. Renal complications can arise from various factors related to the transplantation process and subsequent treatment regimens, and their impact on patient health is multifaceted. Key concerns encompass acute kidney injury (AKI), chronic kidney disease (CKD), and electrolyte imbalances, each posing distinctive challenges with far-reaching implications. AKI, prevalent in the early transplant phase, often necessitates urgent intervention and can escalate mortality risks, exacerbating cardiovascular vulnerabilities and susceptibility to infections. Furthermore, the progression of CKD over time carries significant implications, potentially culminating in end-stage renal disease (ESRD), thereby limiting therapeutic options, and affecting overall longevity. Addressing renal dysfunction post-SCT requires a multifaceted approach, emphasizing vigilant monitoring, timely intervention, and ongoing management strategies to optimize patient outcomes and mitigate associated risks. Renal complications post-SCT pose significant challenges for transplant recipients and healthcare providers alike. Renal care emerges as a critical cornerstone in the comprehensive management of transplant recipients, underscoring its pivotal role in ensuring favorable post-transplant trajectories.

The chapter aims to provide an in-depth exploration of renal complications following stem cell transplantation (SCT), addressing their multifaceted nature and significant impact on patient outcomes. Through a comprehensive review, the chapter seeks to enhance understanding among healthcare providers involved in transplant care, facilitating early recognition, proactive intervention, and optimal management of renal complications in SCT recipients.

2. Etiology of renal complications

2.1 Pre-existing renal dysfunction

Patients with important comorbidities, including chronic kidney disease (CKD) (creatinine clearance <50 ml/min or creatinine level > 2 mg/ml), were usually excluded from SCT procedures mainly because of significant risk of drug toxicities due to impaired renal function, with exception of CKD secondary to multiple myeloma. However, many of these patients may benefit from SCT, especially patients with multiple myeloma, a malignant hematologic disease in which the overall prognosis is greatly influenced by the presence of CKD, but also by access to SCT. For MM patients, SCT together with a kidney transplant from a matched HLA donor may both treat the blood malignancy and end-stage renal disease [1, 2]. Besides multiple myeloma, patients with other diseases (e.g., sickle cell anemia) and CKD may also need an SCT. Abnormal kidney function is a factor in some risk scores for estimating the risk for post-transplant mortality [3, 4]. Cardiotoxicity of cyclophosphamide [5], as well as the mucosal and central nervous system toxicity of melphalan [6], may be exacerbated in the face of impaired GFR and dose adjustments are needed [2].

Nearly all the complications following SCT are more challenging in the presence of CKD. In a previous study published by our team, we found a prevalence of pre-existing CKD of 14% in 185 patients with multiple myeloma who suffered an autologous SCT, and this condition was an independent risk factor for acute kidney injury (AKI) during the first 30 days following SCT [7]. Dosing of antimicrobials must consider the kidney function. Amphotericin nephrotoxicity is higher in patients with pre-existing CKD. The use of aminoglycosides may also be problematic in these patients. Foscarnet, a well-known nephrotoxin, needs careful dose adjustments in CKD. This also applies to acyclovir and ganciclovir. Nephrotoxicity of calcineurin inhibitors (cyclosporine and tacrolimus), which are used for the prophylaxis of graft-versus-host-disease (GVHD), is more problematic if kidney impairment is already present. The toxicity of methotrexate, used for GVHD prophylaxis, is also enhanced by CKD.

2.2 Conditioning regimens and chemotherapy-induced nephrotoxicity

Conditioning therapy, while necessary for successful engraftment, can directly contribute to renal complications through its nephrotoxic effects.

Chemotherapeutic agents such as cyclophosphamide, busulfan, and melphalan, commonly used in conditioning regimens, can cause renal toxicity. These agents can damage the renal tubules, impair renal function, and lead to acute kidney injury (AKI) or chronic kidney disease (CKD).

Moreover, radiation therapy as part of conditioning can also affect renal function, especially when delivered to the abdominal or pelvic regions. Radiation nephropathy was first described in 1952 in 20 patients who suffered external beam irradiation for seminoma [8]. Renal shielding may reduce the incidence of radiation nephropathy and the incidence has dramatically declined with the advent of more effective chemotherapy (including myeloablative conditioning) [9]. Kidneys are organs with less radiosensitivity compared to organs with high cellular turnover (like the gastrointestinal tract or bone marrow). The risk dose is considered to be 10–14 Gy with a latent period of months to years after exposure [10].

Radiation affects all the compartments of the kidney. First, it encounters at the glomerular level, with endothelial dysfunction and mesangial injury, with thrombotic microangiopathy and mesangiolysis, which progress to glomerular scarring and eventually kidney failure. Tubular injury occurs later, with tubular atrophy and interstitial fibrosis. CKD due to radiation typically presents with severe hypertension and anemia. It seems that renin-angiotensin-aldosterone (RAA) system plays a major role in the development of fibrosis following radiation and treatment with angiotensin-converting enzyme inhibitors (ACE-I) or angiotensin receptor blockers (ARB) had a positive impact upon murine models of radiation nephritis [11].

2.3 Immunosuppressive therapies

The calcineurin inhibitors (CNI) available today are cyclosporine (CsA) and tacrolimus (TAC). Their immunosuppressive actions derive from binding to cyclophilin (for CsA) or FKBP12, and thus they inhibit calcineurin (a calmodulin and calcium-dependent phosphatase) with further suppression of transcription of interleukin-2 (IL-2). Calcineurin however is not specific to T cells, and thus inhibition of this pathway leads to toxicities beyond immunosuppression actions. The first report of chronic CsA nephrotoxicity comes from Myers et al. who described the chronic effects in heart transplant patients [12].

Pathogenic mechanisms of chronic CNI nephrotoxicity are only partially known. Chronic toxicity seems to derive from a combination of direct tubular effects and chronic hemodynamic changes due to endothelial dysfunction with repeated episodes of kidney ischemia [13]. Nox2 (the classical phagocytic NADPH oxidase) expression is increased in human and animal models of chronic CNI-induced fibrosis [14]. Prolonged vasoconstriction of arterioles was hypothesized to mediate CNI toxicity, although no direct evidence for this was found [15]. TGF-β is upregulated in the tubular epithelial cells because of CNI, and TGF-β promotes fibrosis through epithelial-mesenchymal-transition in these cells [16]. The drug transporter adenosine triphosphate-binding cassette protein B1 (ABCB1) is found especially in the brush border of the proximal tubular kidney cells and it was shown that lower ABCB1 expression is a risk factor for chronic nephrotoxicity in kidney transplantation; the reduced expression of CYP3A5 in kidney biopsies was also a risk factor for nephrotoxicity [17].

Histological lesions attributed to CNI (stripped interstitial fibrosis, fibrointimal thickening, nodular medial arteriolar hyalinosis, glomerulosclerosis, tubular atrophy, tubular microcalcifications, juxtaglomerular apparatus hyperplasia) are not specific for their nephrotoxicity and may be found in different other settings [15]. It seems that calcineurin-NFAT plays a major role in smooth muscle cells, and so arteriolar hyalinosis starts with an eosinophilic transformation of these cells in the afferent arterioles, followed by hyalinosis [18].

Calcium channel blockers are considered first-line therapy for lowering blood pressure in transplant recipients, since they may counterbalance the arteriolar vasoconstriction, although studies have failed to prove their superiority in preventing chronic nephrotoxicity. ACE-I and ARB, by suppressing the RAA system, also attenuate the vasoconstrictive effects of CNI. RAAS blockade attenuates CsA-induced interstitial fibrosis and arteriolopathy in murine models and decreases the circulating levels of TGF-β [19].

2.4 Infections and sepsis

Sepsis is an important cause of AKI, both in the general population as well as in patients with blood malignancies, including SCT. Patients with SCT are especially prone to develop sepsis due to the profound immunosuppression. Sepsis induces AKI through multiple mechanisms—systemic vasodilation with hypotension and decreased kidney perfusion, release of nephrotoxic inflammatory cytokines with direct tubular toxic effects, and use of nephrotoxic antibiotics [20].

BK virus reactivation and adenovirus may cause hemorrhagic cystitis with bladder obstruction due to clots and subsequent AKI or, rarely, acute tubulointerstitial nephritis [21]. Cidofovir, foscarnet, or leflunomide have not shown convincing data for the treatment of BK virus infection and immunosuppression reduction is the mainstay of treatment for both viral infections.

2.5 Sinusoidal obstruction syndrome (SOS)

Approximately, 20–40% of SCT patients will develop sinusoidal obstruction syndrome (SOS), especially those with allo-SCT. It is considered the main cause of severe AKI early after allo-SCT, with an incidence of AKI of up to 55%. AKI influences the prognosis of SOS and it is a type of hepatorenal syndrome, with normal kidney histology [22]. The underlying mechanisms of AKI in the context of SOS are not well understood; however, it seems to resemble a form of hepatorenal syndrome (HRS). HRS is multifactorial and primarily involves hemodynamic changes without causing structural kidney damage or tubular dysfunction [23].

Hepatic SOS occurs more frequently in recipients of myeloablative allogeneic transplants, busulfan, total body irradiation conditioning, or inotuzumab use before HSCT.

The most credible explanation suggests that the myeloablative conditioning regimen impacts sinusoidal endothelial cells and hepatocytes in zone 3 of the acinus, leading to the deposition of fibrin and other blood products beneath the endothelium of the affected venules. This accumulation results in the progressive narrowing and blockage of these venules, subsequently causing intrahepatic portal hypertension. The release of cytokines from tissue damage, along with glutathione depletion due to chemotherapeutic detoxification through the glutathione pathway, contributes to hepatocellular necrosis and fibrosis. Additionally, a reduction in nitric oxide has been linked to the onset of SOS, owing to the disruption of sinusoidal blood flow [24].

2.6 Thrombotic microangiopathy (TMA)

One of the causes for both AKI and CKD in SCT patients is thrombotic microangiopathy, which usually occurs 6–12 months after the procedure, when the immunosuppression is decreased. Up to 65% of all SCT patients will develop this serious complication which sometimes may be life-threatening [25, 26].

Dvorak et al. made a three-hit hypothesis for the pathogenesis of transplant-associated TMA (TA-TMA)—the first hit is complement activation and endothelial dysfunction, the second hit consists of a toxic conditioning regimen (either chemotherapy or TBI) leading to endothelial damage, and the third hit consists of alloreactivity, drugs, and/or infection caused damage [27]. Many studies have also confirmed the role of complement activation, as proved by the increase in serum C5b-9 and C3b [28]. Because of transplantation, the recipient may acquire complement regulatory protein variants from the donor, increasing the risk for TMA. In a study by Jodele et al., patients were found to have defective complement regulation [29]. Mutations of the genes responsible for complement activation were found in a greater percentage in patients with SCT who developed TA-TMA, compared to those without this complication [30]. Moreover, neutrophil extracellular traps (NET) may be an important link between endothelial dysfunction, complement activation, and TA-TMA pathogenesis. Neutrophils and thrombocytes aggregate at the site of the endothelial injury due to various factors in SCT (like infection, GVHD, conditioning chemotherapy, irradiation, CNI) and this process induces NET's formation, with subsequent release of complement factors and generation of terminal complement MAC. In pediatric SCT, a study showed that in patients who developed TA-TMA, NETs were significantly increased on day 14 following transplantation; damaged endothelial cells released IL8, which in turn activated neutrophils to release NETs, thus creating an amplification loop for complement activation [31].

Preventive measures, such as optimizing the conditioning regimen, avoiding infection through prophylactic anti-infective therapy, preventing GVHD, treating hypertension, and minimizing blood transfusions, are central to reducing endothelial damage. Other approaches include rituximab (especially if GVHD is present), discontinuation of CNI and mTOR inhibitors, therapeutic plasmapheresis, N-acetylcysteine, and complement inhibition through eculizumab [28]. The Food and Drug administration (FDA) has granted narsoplimab, a human monoclonal antibody targeting mannan-binding lectin-associated serine protease 2 (MASP2, the effector enzyme of the lectin pathway of complement system), priority review for SCT-associated TMA [32]. Daclizumab, an IL2 inhibitor, may replace other immunosuppressive drugs (especially when the suspected agent to induce TMA is a CNI) while promoting TA-TMA remission [33]. However, the best approach to patients with TA-TMA is still not very well known, since the efficiency of all these measures is still a matter of intense debate.

2.7 Graft-versus-host disease (GVHD)

Although chronic GHVD is a significant complication, involving 30–70% of SCT patients with a significant impact on their health status and quality of life, kidney involvement in cGVHD is a rare occurrence, being limited to only case reports or series of cases [20].

The pathogenesis of cGVHD-related NS is not clear. It is suspected that immune-mediated mechanisms may be involved. Bruijn demonstrated glomerular involvement through immune complexes in membranous nephropathy following a murine model of GVHD [34]. Moreover, membranous nephropathy is the most frequent histological pattern found in kidney biopsies from patients with SCT-associated NS (**Figure 1**). Other authors suggest that donor-activated T cells may induce podocytes injury [35], either through permeability factors (cytokines) with the development of minimal change disease (**Figure 2**) or focal segmental glomerulosclerosis, or by activating recipients' autoreactive B-cells clones, since treatment against B-cells, like rituximab, may play an important role in ameliorating cGVHD-associated NS. An argument in favor of the hypothesis that NS is a form of kidney cGVHD is the occurrence of NS soon after the reduction or withdrawal of CNI for the prevention or treatment of cGVHD and the complete response of NS promptly after reinitiating CNI in most of these patients.

However, there are cases of NS described in patients without any extrarenal signs of cGVHD [36, 37]. Whether this is a renal-limited type of GVHD or it is a different entity remains a matter of debate, and so is the pathogenesis of NS in this instance.

A careful renal monitoring of SCT patients, especially of those with allo-SCT with peripheral blood stem cells as the source (which is more often associated with cGVHD) together with kidney biopsy when there is a reliable suspicion of NS, is needed for proper diagnosis of this rare complication of SCT in order to establish the treatment.

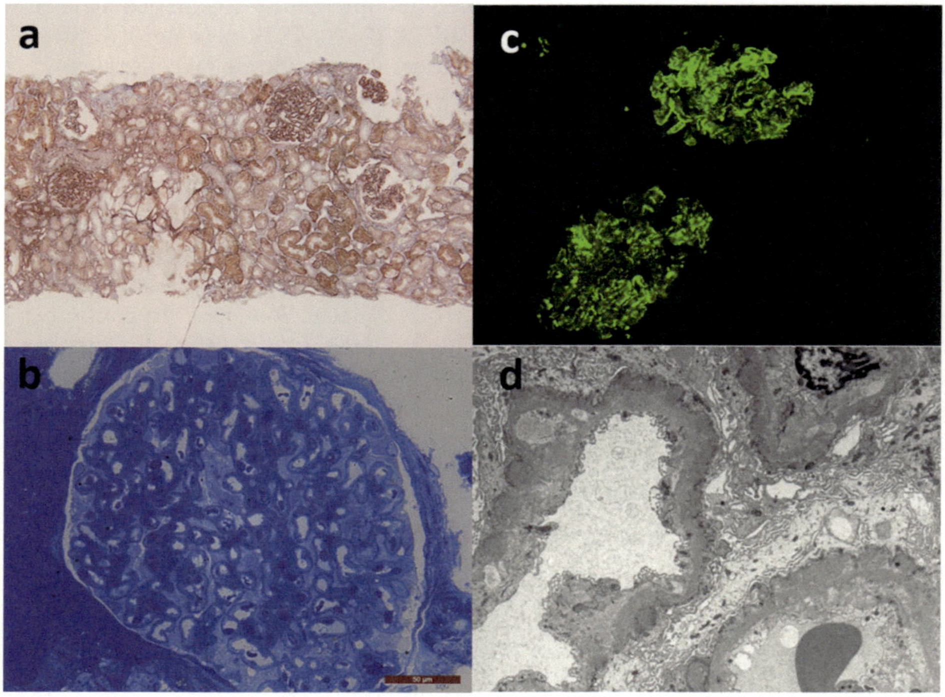

Figure 1.
Membranous nephropathy stage III in a patient with allo-SCT for acute myeloid leukemia. (a) Intense granular immune staining for IgG of the capillary wall (immunohistochemistry for IgG); (b) Significant thickening of the capillary wall (light microscopy, toluidine blue); (c) Moderate-intense granular immune staining for IgG of the capillary wall (immunofluorescence for IgG); (d) Subepithelial electron-dense deposits, thickening of the glomerular basement membrane, diffuse effacement of the foot processes of the podocytes (electron microscopy).

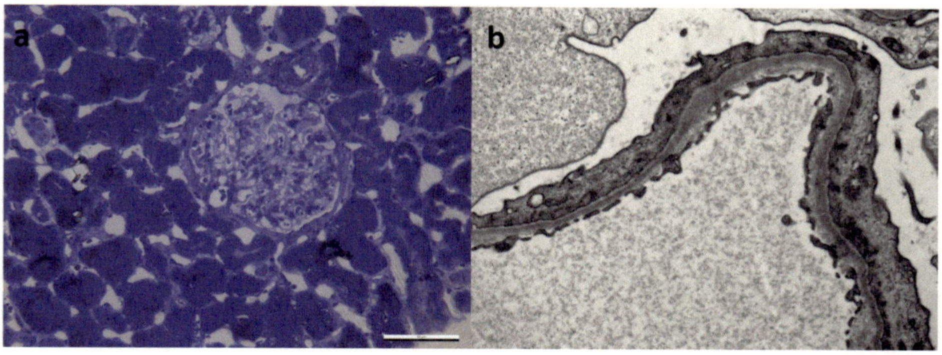

Figure 2.
Minimal change disease in a patient with allo-SCT for acute myeloid leukemia. (a) Normal glomerulus (light microscopy, toluidine blue); (b) Normal capillary wall, diffuse effacement of the foot processes of the podocytes (electron microscopy).

2.8 Nephrotoxic agents

Nephrotoxic medication, like antibiotics, antivirals, contrast dye, and immunosuppressants, may induce AKI. Among antibiotics, amphotericin, vancomycin, colistin, and aminoglycosides may have a direct tubular toxic effect, while beta-lactam antibiotics and cephalosporins may induce immune-allergic mediated acute interstitial nephritis. Acyclovir may precipitate in the renal tubules, especially after intravenous (iv) administration of high doses; a slower iv administration, together with proper hydration and dose adjustments, may be useful in prophylaxis of nephrotoxicity. CNI induces an acute reversible decrease of kidney function, in a dose-dependent manner, through arteriolar vasoconstriction and reversible tubular dysfunction. Vasoconstriction appears secondary to an imbalance between vasodilators (nitric oxide, prostacyclin, prostaglandin E2) and vasoconstrictors (RAA system, endothelin, thromboxane A2, together with COX-2 inhibition due to the inhibition of calcineurin-NFAT signaling) [38]. CNI may also induce endothelial dysfunction with thrombotic microangiopathy, which is a significant independent risk factor for AKI.

Marrow infusion toxicity may develop due to hemolysis because of exposure to cryoprecipitates (such as dimethyl sulfoxide). Heme is toxic on the epithelial cells of distal tubules, especially in highly acidic urine. Urine alkalinization and mannitol infusion are useful for the prevention of this complication [24].

3. Clinical manifestations

3.1 Acute kidney injury

Because acute kidney injury often occurs secondary to other conditions (such as SOS, sepsis, tumor lysis syndrome, TMA, GVHD, or nephrotoxic agents), the clinical manifestations overlap with those of the associated condition.

During the early post-SCT period, sepsis, nephrotoxic agents, sinusoidal obstruction syndrome, and acute GVHD may cause acute kidney injury. Its incidence is highly variable, depending on the definition used for AKI diagnosis but especially on the type of transplant; it ranges from 10% in auto-SCT [7] to around 50% in allo-SCT with reduced conditioning regimen, to up to 75% in allo-SCT with myeloablative conditioning [39, 40]. The median time to onset of AKI after SCT is 33 to 38 days and the prognosis of patients requiring dialysis is very poor, with 55–100% mortality [39]. Early AKI develops before cell engraftment and carries a poor prognosis, with very high mortality rates (since it occurs in patients with severe immunodepression, with significant risks of sepsis and bleeding complications), while late AKI develops after cell engraftment and is usually a milder form (**Table 1**) [22].

In the initial weeks following myeloablative hematopoietic cell transplantation (HCT), recipients face a heightened risk for various types of acute renal failure (ARF). This includes a *prerenal state* caused by vomiting and diarrhea, often due to conditioning regimens or acute graft-versus-host disease, or from the effects of calcineurin inhibitors. Hepatic SOS manifests with symptoms such as painful hepatomegaly, jaundice, and weight gain due to fluid accumulation. Prerenal-AKI frequently accompanies SOS, which is more common after allogeneic than autologous hematopoietic cell transplantation (HCT). Severe cases of SOS can progress to multiple organ failure. The development of SOS is linked to damage to sinusoidal endothelial cells caused by conditioning therapy, leading to hepatic portal hypertension, ascites, and elevated abdominal pressure [41]. Engraftment syndrome is a clinical condition of rash, fever, edema, weight gain, and liver/renal/pulmonary/cerebral dysfunction which occurs within four days of neutrophil recovery after SCT; it is included in the cytokine release syndromes [42]. Although it usually has a self-limited course, it may have a fulminant evolution in other patients progressing to multi-organ dysfunction and eventually death. Thrombocytopenia and neutropenia can also induce hemorrhagic or septic shock, respectively, increasing the risk of AKI.

Additionally, exposure to nephrotoxic agents such as amphotericin B, aminoglycosides, intravenous contrast, and calcineurin inhibitors can lead to acute tubular necrosis. TA-TMA patients may present with hypertension, progressive kidney insufficiency, and microangiopathic hemolytic anemia. Still, this typical triad may be absent in patients with renal-limited TMA or those with chronic TMA. In this instance, a kidney biopsy usually is the only diagnostic procedure that may establish a

Early onset (<30 days)	**Late onset (>3 months)**
Sinusoidal obstruction syndrome	Thrombotic microangiopathy
Sepsis	Calcineurin inhibitor toxicity
Hypotension (capillary leak syndrome)	
Hypovolemia (vomiting, diarrhea)	
Nephrotoxic agents	
Acyclovir	
Allopurinol	
Amphotericin B	
ACE-I/ARB	
Aminoglycosides	
Calcineurin inhibitors	
Cyclophosphamide	
Contrast dye	
Colistin	
Methothrexate	
NSAIDs	
Tumor lysis syndrome	
Marrow infusion toxicity (hemoglobinuria)	

Table 1.
Causes of AKI after SCT [22].

definite diagnosis (see Kidney biopsy below). Renal involvement is a prominent feature of SCT-associated TMA, patients may present with kidney insufficiency (either acute or chronic), proteinuria, hematuria, hypertension, and transitory neurological anomalies. Patients with a confirmed diagnosis of TMA were four times more likely to develop CKD. In patients who survived after acute TA-TMA, kidney function was 40% of normal two years after SCT [43]. Fulminating multifactorial TMA usually has a fatal course, it develops early after SCT, is associated with fungal or viral infections (including cytomegalovirus and herpes zoster), acute GVHD and presents with severe AKI, central nervous system disturbances.

Though less common, *obstructive uropathy* can arise from severe hemorrhagic cystitis triggered by cyclophosphamide, adenovirus, BK/polyomavirus infection, or fungal infections affecting the urinary collecting system [23].

3.2 Chronic kidney disease

The first report of CKD after SCT was in 1978 in a patient who developed radiation nephritis after total body irradiation [44]. Now CKD and ESRD are well-recognized complications following SCT, especially allo-SCT. It is estimated that around 17–35% of these patients will develop CKD during their lifetime, and this carries a significant negative impact on the patient's overall prognosis as well as cardiovascular prognosis, [45–47] since SCT patients who develop CKD have a reduced survival [48]. CKD at one year after SCT was also associated with worse GVHD-free/relapse-free survival [49]. CKD may eventually progress to ESRD, which is less common compared to other causes of ESRD but is 20 times more frequent compared to the gender and sex-matched general population [50]. It is considered that especially patients over 50 years of age are more vulnerable to renal toxicity of transplantation procedures [48]. CKD has important consequences for SCT patients. Besides increased cardiovascular risk, it may determine the reduction of immunosuppressive treatment (with a higher risk of GVHD), difficulties in choosing antiviral agents and antibiotics, and ineligibility for subsequent SCT or further cancer therapies in case of relapse.

Previous episodes of AKI, especially during the first 100 days after SCT, increase the risk of developing CKD; the same is true for the general population [47, 49]. In the study of Shimoi et al., each AKI stage was strongly associated with CKD. In this study, all the patients with CKD had previous episodes of AKI and almost half of all the patients who had AKI developed CKD [47]. Given the high incidence of AKI after allo-SCT and its proven relationship with subsequent CKD, as well as the significant decline of GFR especially during the first year after transplantation, all of these suggest that early intervention is the key in preventing the development of CKD in this population. However, in the study of Pelletier et al., although AKI incidence was 64% and CKD developed in 19% of patients, only 3% were referred to nephrology [49].

Patients with chronic kidney disease (CKD) stages 1–3, where the glomerular filtration rate (GFR) is above 30 mL/min/1.73 m^2, often do not exhibit symptoms. Specifically, at this level of GFR reduction, there are typically no noticeable clinical disturbances in water or electrolyte balance, nor any endocrine or metabolic abnormalities. However, such disturbances generally start to appear in stages 4–5 of CKD, when the GFR drops below 30 mL/min/1.73 m^2. Patients with conditions such as tubulointerstitial disease, cystic diseases, and nephrotic syndrome, which are linked with more "positive" symptoms like polyuria, hematuria, and edema, may show signs of the disease at earlier stages.

3.3 Hypertension

SCT is associated with either acute or chronic elevations of systemic blood pressure, which pose significant morbidity and mortality.

- *Acute exacerbations of blood pressure*

Acute hypertension may be seen in patients after SCT who were otherwise normotensive. This is either drug-related or may be associated with thrombotic microangiopathy (TMA). The overall prevalence of hypertension during the first 2 years after SCT is estimated to be around 70% [51]. Calcineurin inhibitors (CNI) and steroids, used for GVHD prophylaxis, are well-known drugs involved in the development of hypertension [52]. Stimulation of the renin-angiotensin-aldosterone (RAA) system with sodium and water retention, sympathetic overstimulation, vasoconstriction of the afferent arteriole, and nephrotoxicity due to CNI are the main pathogenic mechanisms involved. Dose reductions together with monitoring of CNI blood levels and initiation of antihypertensive drugs are the proper means of management.

In transplant-associated TMA (TA-TMA), hypertension due to kidney involvement is usually observed. Prompt diagnosis of TA-TMA is essential to improve patients' overall prognosis as well as kidney outcomes [30].

Posterior reversible encephalopathy syndrome (PRES) is a severe neurological disorder diagnosed in around 3% of all SCT recipients, [53] and hypertensive fluctuations were described in 61–80% of patients with PRES [54]. It is not known whether hypertension is a cause or consequence of PRES, but it was associated with CNI and mTOR inhibitors [55].

- *Chronic hypertension*

Cardiovascular events are a significant cause of morbidity and early premature death in SCT, especially after allogeneic procedures, death due to this complication being the second cause after GVHD in patients free from relapses of the blood disorder [56]. Overall, the prevalence of hypertension in long-term survivors after pediatric SCT is around 15% [57] and allo-SCT patients are more prone to develop hypertension compared to autologous recipients [58]. Intensive chemotherapy with cardiotoxic drugs may alter peripheral vascular hemodynamics. Chronic kidney disease (CKD) as well as radiation nephropathy are other causes of significant glomerular injury followed by secondary hypertension. There is a high prevalence of metabolic syndrome and endothelial dysfunction after SCT, and these may result in hypertension [59]. Hypertension was associated with cardiovascular outcomes in SCT survivors, independent of the conditioning regimen [60].

3.4 Hemorrhagic cystitis

Hemorrhagic cystitis (HC) is a potentially significant complication, especially after allo-SCT. BK virus reactivation and adenovirus may cause hemorrhagic cystitis with bladder obstruction due to clots and subsequent AKI. Urotoxic effects of conditioning regimen (especially cyclophosphamide, ifosfamide, busulfan, high doses of etoposide, and pelvic irradiation) or GVHD are another significant cause for HC.

DOI: http://dx.doi.org/10.5772/intechopen.1007385

Grade	Severity
I	Microscopic hematuria
II	Macroscopic hematuria
III	II + presence of blood clots
IV	III + renal impairment due to urinary obstruction

Table 2.
Clinical grading of hemorrhagic cystitis [61].

BK viral loads closely correlate to the occurrence of HC post-SCT and significant increases in BK viruria precede HC. The use of conditioning regimen damages the urothelium; the subsequent regeneration of epithelium creates a favorable milieu for BK replication. The immunosuppression state induced by the disease itself and by the chemotherapy favors BK reactivation. BK replication leads to cytopathic effects with BKV shedding in the urine and significant viruria. Finally, the return of anti-BKV-specific immunity due to hematopoietic reconstitution determines an immunological reaction against viral antigens expressed on the surface of urothelium, with the development of HC [61]. Intermittent reactivation with low level viruria ($<10^6$ virus copies/mL of urine) has no consequences. High level viruria ($>10^7$ virus copies/mL of urine) leads to HC +/− ureteral obstruction: obstructive AKI, or/and polyoma BKV nephropathy (very rare in SCT, mostly diagnosed in kidney transplant patients).

Diagnostic criteria of HC include microscopic or macroscopic hematuria with urinary symptoms (dysuria, pelvic pain, polakiuria), history of exposure to risk factors, and exclusion of other causes of painful hematuria (urinary neoplasia, stones, instrumentation, bleeding diathesis due to coagulopathy or thrombocytopenia, bacterial, fungal, or parasitic urinary infections) (**Table 2**).

3.5 Electrolyte disturbances

Electrolyte imbalances are frequently observed after stem cell transplantation (SCT), with particularly common occurrences of low electrolyte levels. Electrolyte disturbances add layers of complexity, potentially precipitating life-threatening conditions. Hypotonic hyponatremia is associated with edematous states because of nephrotic syndrome, renal insufficiency (either acute or advanced chronic one), heart failure (cardiomyopathy due to amyloidosis, thoracic radiation, drug-induced like anthracyclines, alkylating agents, immune checkpoint inhibitors), or cirrhosis (including Budd-Chiari syndrome due to thrombosis of suprahepatic veins or sinusoidal obstruction syndrome).

Hyperkalemia usually does not exhibit any symptoms, but when symptoms do occur, they are typically vague and primarily affect muscle or cardiac function. In some cases, patients may experience severe muscle weakness to the point of paralysis, shortness of breath, palpitations, chest pain, nausea, and vomiting.

4. Diagnostic evaluation

The diagnosis and evaluation of renal complications following stem cell transplantation necessitate a comprehensive approach, encompassing clinical assessment, laboratory tests, imaging studies, and occasionally renal biopsy.

Specialized tests, including serological tests for viral infections like cytomegalovirus (CMV) or BK virus (BKV) and autoimmune markers, may be conducted based on clinical suspicion.

4.1 Laboratory investigations

Laboratory tests, including serum creatinine, blood urea nitrogen (BUN), urinalysis, urine protein-to-creatinine ratio (UPCR), electrolyte levels, and complete blood count (CBC), aid in assessing renal function, proteinuria, electrolyte imbalances, and identifying signs of infection or hematologic abnormalities.

Although acute kidney injury (AKI) is a prevalent condition, it lacked internationally standardized criteria until recently. In 2004, the Acute Dialysis Quality Initiative (ADQI) introduced the Risk, Injury, Failure, Loss, and End-stage renal disease (RIFLE) criteria to define acute renal failure (ARF) based on increased serum creatinine (sCr), reduced glomerular filtration rate (GFR), and decreased urine output. Subsequently, in 2007, the Acute Kidney Injury Network (AKIN) refined the RIFLE criteria by including minor sCr elevations to broaden the definition to encompass early renal injury. This concept evolved into the AKI terminology. In 2012, the Kidney Disease Improving Global Outcomes (KDIGO) consolidated the RIFLE and AKIN criteria. The KDIGO criteria for AKI identify it based on any of the following: an increase in sCr by at least 0.3 mg/dl within 48 hours, an increase in sCr to at least 1.5 times the baseline within the last 7 days, or a urine output less than 0.5 ml/kg/h for 6 hours. AKI severity under KDIGO is staged from 1 to 3 based on sCr levels or urine output. According to recent studies, the KDIGO criteria are as predictive, if not more, of prognosis as the previous RIFLE or AKIN standards and are now commonly used in studies assessing the incidence of AKI post-hematopoietic cell transplantation (HCT) [20].

4.2 Electrolyte disturbances

Electrolyte abnormalities are quite common after SCT (**Table 3**). Low electrolyte levels are extremely common, with a prevalence of up to 80% for hypokalemia, 65% for hypomagnesemia, 50% for hypocalcemia, and 90% for hypophosphatemia, while high levels of electrolytes, especially hyperkalemia, occur less frequently, in up to 25% of all the recipients [62].

4.2.1 Acute electrolyte disturbances

Low levels of electrolytes may be explained by reduced intake of fluids and food due to nausea, increased gastrointestinal losses due to mucositis, abnormal bone metabolism, shifting of the electrolytes in the newly forming blood cells, or increased losses through the kidneys. Wollner et al. introduced the term "genesis syndrome" to define the increased transfer of electrolytes into the newly forming hematopoietic cells, and the nadir of low electrolyte levels is usually observed before the engraftment [62, 63].

Amphotericin is involved in magnesium and potassium losses, while aminoglycosides may act like loop diuretics. Low electrolyte levels are more obvious after SCT in multiple myeloma; in these patients, light chains may induce proximal tubulopathy with electrolyte losses, which become more evident when nutritional intake is also diminished [62]. Foscarnet, used for the treatment of cytomegalovirus in

Hypokalemia	Reduced intake (nausea) Increased gastrointestinal losses (mucositis) Increased renal losses (amphotericin, diuretics, aminoglycosides, foscarnet) Genesis syndrome
Hypomagnesemia	Increased renal losses (amphotericin, aminoglycosides, foscarnet) Genesis syndrome
Hypocalcemia	Foscarnet Genesis syndrome
Hypophosphatemia	Genesis syndrome
Hyponatremia	Syndrome of inappropriate antidiuretic hormone secretion Vomiting, diarrhea Renal losses (adrenal insufficiency, diuretics, cerebral salt-wasting syndrome
Hyperkalemia	Calcineurin inhibitors Reduced kidney function (acute kidney injury, chronic kidney disease)

Table 3.
Electrolyte disturbances after stem cell transplantation.

ganciclovir-resistant CMV disease, may be associated with hypocalcemia, hypomagnesemia, and hypokalemia since it is a chelating agent for divalent ions, and may also precipitate in the tubules with significant nephrotoxicity [64].

- *Hyponatremia*

Hyponatremia is the most common electrolyte abnormality in SCT patients, affecting almost 40% of all these patients [65]. It is an independent predictive factor for unfavorable outcomes after allo-SCT [66]. Symptoms of hyponatremia may be attributed by mistake to other causes, such as central nervous system (CNS) involvement in the underlying blood disorder or neutropenic sepsis. The main mechanisms involved are either excessive sodium (and potassium) loss, or, more often, water retention. It is important to exclude pseudohyponatremia due to hyperlipidemia (like in nephrotic syndrome) or hyperproteinemia (like in multiple myeloma or intravenous immunoglobulins). Hypertonic hyponatremia may be observed in diabetes mellitus (DM) associated with hyperglycemia, since SCT patients are at increased risk for developing type 2 DM [58].

Syndrome of inappropriate antidiuretic hormone secretion (SIADH) is one of the most frequent etiologies of euvolemic hyponatremia in SCT, which may develop insidious, or may progress rapidly with fatal outcomes [66]. SIADH is the leading cause of hyponatremia in hematology and its incidence is higher in the SCT setting compared to chemotherapy [67]. Among risk factors for SIADH post-SCT are HLA-mismatch, cyclophosphamide used for conditioning regimen, methylprednisolone for GVHD prophylaxis, age below 4 years, and cord blood as the source for stem cells (this case has been associated also with more severe symptoms of hyponatremia) [68]. IL-6 and TNF-alfa are elevated after allo-SCT and they have been implicated in the pathogenesis of SIADH. Pain, nausea, and stress are important factors that stimulate ADH secretion. Disseminated varicella zoster virus, CNS infections (like fungus, toxoplasmosis, or herpes simplex virus 6), or respiratory infections are other known causes of SIADH [68, 69]. Use of hypotonic fluids, which may aggravate hyponatremia, needs to be avoided and replaced by isotonic solutions [67].

Renal losses are another cause of hyponatremia in SCT. Adrenal insufficiency may occur due to sudden interruption of corticosteroids, while the hypothalamic adrenal axis is suppressed. Diuretics used to prevent or treat fluid overload in patients with AKI episodes are another cause for renal electrolyte losses. Vomiting and diarrhea, due to gastrointestinal toxicity of conditioning regimen or infectious enterocolitis, aggravate hyponatremia. Cerebral salt-wasting syndrome (CSWS) is a rare cause, often underdiagnosed, of renal sodium losses, first described by Peters JP in 1950 [70]. It was mainly identified in patients with CNS injuries, either due to hemorrhage (because of SCT-associated coagulopathy) or viral/bacterial/fungal encephalitis [71]. The diagnosis is made in the presence of a known risk factor for CSWS, associated with hyponatremia, dehydration, and high urine output. The exact mechanism involved in CSWS is unknown. It is presumed that the release of brain natriuretic peptide and interference with the afferent sympathetic tone of the kidney are involved [72]. Exclusion of other, more common causes of hyponatremia is essential for differential diagnosis. CSWS and SIADH share many biological and etiological similarities, and the key to differentiate one from another is the correct evaluation of hydration status since SIADH is associated with euvolemia and CSWS with signs of dehydration. Also, SIADH usually is accompanied by normaluria or oliguria, while polyuria is usually the rule for CSWS [71].

4.2.2 Chronic electrolyte disturbances

- *Hyperkalemia*

Hyperkalemia is not so common after SCT. It is usually diagnosed in patients with reduced renal function, either because of an AKI episode or due to CKD. It is encountered especially if these patients also receive drugs that potentiate potassium retention, like ACE-I, ARB, potassium sparring diuretics, amiloride, heparin, beta-blockers, or trimethoprim-sulfamethoxazole (TMP-SMX, which acts similarly to amiloride at the level of renal tubules).

CNI-induced hyperkalemia in patients with otherwise normal renal function is also quite common [73]. CNI has significant tubule toxicity which may manifest solely by hyperkalemia. Besides this, they may determine the inhibition of Na^{+}/K^{+} ATPase in the potassium-secreting cells of the collecting ducts and hypoaldosteronism by inhibition of aldosterone production by zona glomerulosa [74]. CNI-induced hyperkalemia is often accompanied by metabolic acidosis with hyperchloremia, hypertension, and hypercalciuria, similar to Gordon syndrome, probably by stimulating chloride reabsorption in the aldosterone-sensitive distal tubular cells [13].

4.3 Imaging studies

Imaging modalities such as renal ultrasound, CT scan, MRI, and renal Doppler ultrasound offer insights into renal structure, morphology, and vascular flow, facilitating the detection of structural abnormalities, masses, or vascular complications.

4.4 Renal biopsy indications and interpretation

Renal biopsy may be performed when the cause of renal dysfunction is uncertain or specific kidney diseases are suspected, allowing for direct examination of kidney tissue under a microscope to diagnose conditions such as acute tubular necrosis, interstitial nephritis, glomerulonephritis, or graft-versus-host disease (GVHD).

De novo glomerular diseases are an important complication following SCT, being encountered in 1–2% of all the recipients [75, 76]. However, there are only a few biopsy reports in SCT patients and most of them are case reports or case series, and it is estimated that it is performed in only 0.5% to 4.3% of these patients, mainly because of increased risk of bleeding complications [77]. Because the etiology of CKD or NS may be so diverse in SCT, a kidney biopsy (KB) is often necessary in this population to diagnose the etiology and to evaluate the prognosis of the treatment response depending on the chronicity of the lesions. While immunosuppression is unnecessary and maybe even contraindicated in TA-TMA, it is very useful in other glomerular pathologies associated with SCT. Recurrence of monoclonal gammopathies of renal significance after SCT may be another explanation for proteinuria in these patients and these diseases may require prompt clone-based therapy. Chronic GVHD may involve also a tubular compartment with mononuclear infiltration and collagen deposition with subsequent CKD [78]. CNI chronic nephrotoxicity may also be identified through KB. Disease recurrence after SCT, like in lymphoma, acute lymphocytic leukemia, or non-hematologic malignancies, may infiltrate kidneys. Indications for KB include AKI, CKD, NS, and sub-nephrotic proteinuria. The finding may be diverse—TMA, glomerular diseases (membranous nephropathy, minimal change disease, focal segmental glomerulosclerosis, membranoproliferative glomerulonephritis, monoclonal gammopathy of renal significance), tubulointerstitial disease (CNI nephrotoxicity, acute tubulointerstitial nephritis due to antibiotics or NSAIDs, myeloma kidney), malignant kidney infiltration [79, 80].

The most frequent pathological lesion identified is kidney TMA which may develop even in the absence of systemic signs for TMA, suggesting that the current criteria for diagnosis of TMA are not sensitive enough for kidney-limited TMA [76, 80–82]. KB will demonstrate intraluminal microthrombi and fibrin in the renal arterioles, widening of the subendothelial space, mesangiolysis and double contours of the glomerular basement membrane, and, sometimes, intraluminal schistocytes in the glomerular capillaries (**Figure 3**) [83].

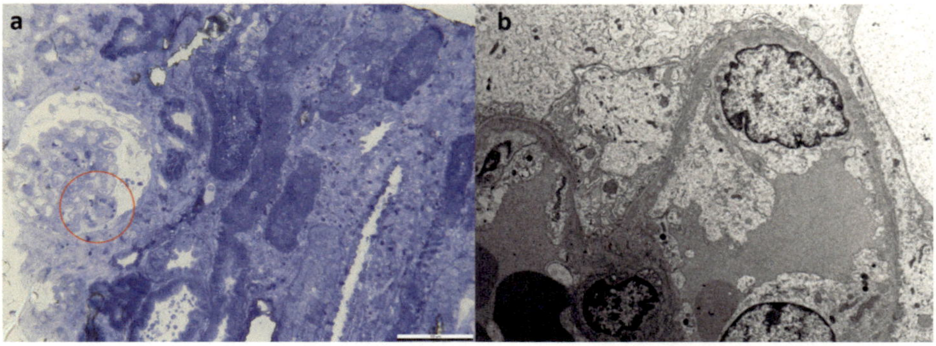

Figure 3.
Kidney-limited thrombotic microangiopathy in a patient with allo-SCT for acute myeloid leukemia. (a) Severe endothelial injury, fragmented red blood cells (light microscopy, toluidine blue); (b) Swollen endothelial cells (endotheliosis), loss of fenestration, detachment of endothelial cells from the glomerular basement membrane (electron microscopy).

Although quite common in kidney transplant recipients, BK virus nephropathy is a very rare occurrence after SCT. Usually, reactivation of the polyoma BK virus is associated with hemorrhagic cystitis. However, very few cases of BK nephropathy have been described in this population [84]. If suspected, a kidney biopsy is mandatory for definite diagnosis. There is no clear treatment, cidofovir was given with controversial results.

There are no criteria for diagnosis of kidney GVHD, but it seems that there is a strong correlation between TMA and GVHD. In the study of Roy et al., all the patients with kidney TMA had also signs of acute or chronic GVHD prior to KB, suggesting that kidney-limited TMA may be an equivalent of renal GVHD [81, 82]. Endothelial cells are a target for alloreactive T lymphocytes in chronic GVHD, and there is a close relationship between endothelial dysfunction and the development of TMA; this explanation may be the missing link between TA-TMA and renal GVHD. Some patients may also present immune deposits inside the kidney, suggesting humoral GVHD.

4.5 Novel biomarkers of kidney injury in SCT

Urinary biomarkers were studied in order to identify CNI-induced nephrotoxicity. In murine models, increased TNF-alfa, fibronectin, KIM-1, and microalbuminuria were associated with acute CsA nephrotoxicity, while in chronic toxicity, a delayed increase in urinary osteopontin and TGF-beta was observed [13].

Elafin is a protein produced by certain cells as a response to injury. Urinary elafin levels were associated with albuminuria, AKI, progression to CKD, and death in SCT [24].

Higher baseline urinary liver-type fatty acid-binding protein (L-FABP) was associated with an increased risk of developing AKI in allo-SCT recipients [85].

In the prospective study of Benoit et al. in children with allo-SCT, they evaluated weekly creatinine, cystatin C, and neutrophil gelatinase-associated lipocalin (NGAL). Patients with all these three biomarkers elevated were at the highest risk of adverse outcomes [86]. In a prospective study, Morito et al. found that urinary alpha 1 M was useful for predicting severe AKI after SCT [87].

However, there is a need for further research on useful biomarkers for the early identification of patients at risk of developing kidney injury after SCT and for better describing their prognosis.

5. Management strategies

The management of renal complications post-transplantation involves close monitoring of renal function, hydration, and the use of supportive care measures. In some cases, specific interventions such as reducing nephrotoxic medications, adjusting immunosuppressive therapy, or treating underlying infections may be necessary to preserve renal function. There are some general recommendations:

Optimization of Fluid and Electrolyte Balance means maintaining euvolemia and electrolyte balance through careful fluid management and monitoring of serum electrolytes, and using diuretics judiciously to manage volume overload while avoiding dehydration.

Medication Management: adjust or discontinue nephrotoxic medications, including chemotherapeutic agents, immunosuppressants, and antimicrobials, as appropriate; consider alternative medications with lower nephrotoxic potential when feasible.

In the case of HC, a reduction of immunosuppression and correction of hypogammaglobulinemia are important. Cidofovir is highly nephrotoxic and myelotoxic, leflunomide (a more potent suppressor of BKV), has myelosuppressive properties, and quinolones are potent inhibitors of BKV replication in vitro, but weak inhibitors in vivo. In patients at risk of HC, quinolone antibiotics are used for both their antibacterial and anti-BKV effects [61]. In grade I HC, patient observation is recommended. In grade II HC, analgesics together with bladder irrigation ameliorate symptoms and prevent obstruction by blood clots. Urology consultation is recommended for patients with grade III or persistent grade II >7 days, despite bladder irrigation [61].

Blood Pressure Control: manage hypertension aggressively to reduce the risk of further kidney damage. Regarding treatment, a personalized approach needs to be implemented taking into consideration the particularities of each patient and the current guidelines available for the treatment of hypertension. In GVHD, endothelial dysfunction plays a major role, so angiotensin-converting enzyme inhibitors (ACE-I) or angiotensin receptor blockers (ARB) might be useful. For CNI nephrotoxicity, vasodilators and calcium channel blockers (CCB) are a viable solution, patients with an AKI episode may benefit from CCB, while in patients with high-dose steroids, diuretics may be an option. Use antihypertensive medications, including angiotensin-converting enzyme inhibitors (ACEIs) or angiotensin II receptor blockers (ARBs), which may also confer renal protective effects.

Treatment of Underlying Causes and Supportive Care Measures. Address specific causes of renal complications, such as infections (e.g., antiviral therapy for CMV or BKV nephropathy), SOS (e.g., defibrotide), TMA (e.g., replaced calcineurin inhibitors by other immunosuppressives like mTOR inhibitors, Rituximab or Eculizumab), autoimmune disorders (e.g., corticosteroids for autoimmune-related nephritis), or graft-versus-host disease (e.g., immunosuppressive therapy for GVHD affecting the kidneys). Provide supportive care measures to manage symptoms and complications associated with renal dysfunction, such as anemia and bone mineral disorders.

Patient Education and Support: It is mandatory to educate patients and caregivers about the importance of medication adherence, lifestyle modifications (e.g., diet, exercise), and self-management strategies to optimize renal health and minimize complications. Also, provide psychosocial support and resources to address the emotional and practical challenges associated with renal complications and transplantation recovery.

Overall, the management of renal complications after stem cell transplantation requires a multidisciplinary approach, individualized treatment strategies, and close monitoring to optimize outcomes and quality of life for patients.

6. Conclusion

Renal complications after HSCT present significant clinical difficulties and affect patient outcomes, increasing non-relapse mortality. A range of factors contribute to the occurrence of renal failure: nephrotoxic medications, GVHD, infections, TA-TMA, and pre-existing renal conditions. Prompt recognition and early management of renal dysfunction are essential for improving patient care and prognosis. Although novel biomarkers are not standardized, future studies are needed to establish their role as diagnostic or monitoring tools. For the improvement of diagnosis, preventive measures, and personalized treatments, it is important that physicians have a better grasp of these issues.

Multidisciplinary collaboration among hematologists, nephrologists, radiologists, pathologists, and infectious disease specialists is essential for a comprehensive assessment and management approach of renal complications after stem cell transplantation.

Conflict of interest

The authors declare no conflict of interest.

Author details

Alina Tanase[1,2], Andreea Andronesi[2,3*], Lavinia Lipan[1], Mihaela Gherghiceanu[2,4] and Gener Ismail[2,3]

1 Department of Bone Marrow Transplantation, Fundeni Clinical Institute, Bucharest, Romania

2 University of Medicine and Pharmacy Carol Davila, Bucharest, Romania

3 Department of Nephrology, Fundeni Clinical Institute, Bucharest, Romania

4 “Victor Babes” National Institute of Pathology, Bucharest, Romania

*Address all correspondence to: andreea.andronesi@yahoo.com

References

[1] Bjorkstrand B et al. Allogeneic bone marrow transplantation versus autologous stem cell transplantation in multiple myeloma: A retrospective case-matched study from the European Group for Blood and Marrow Transplantation. Blood. 1996;**88**(12):4711-4718

[2] Heher EC, Spitzer TR. Hematopoietic stem cell transplantation in patients with chronic kidney disease. Seminars in Nephrology. 2010;**30**(6):602-614

[3] Sorror ML et al. Hematopoietic cell transplantation (HCT)-specific comorbidity index: A new tool for risk assessment before allogeneic HCT. Blood. 2005;**106**(8):2912-2919

[4] Parimon T et al. A risk score for mortality after allogeneic hematopoietic cell transplantation. Annals of Internal Medicine. 2006;**144**(6):407-414

[5] Gerrie A et al. Marrow transplantation for severe aplastic anemia with significant renal impairment. Bone Marrow Transplantation. 2007;**39**(5):311-313

[6] Sirohi B et al. Glomerular filtration rate prior to high-dose melphalan 200 mg/m(2) as a surrogate marker of outcome in patients with myeloma. British Journal of Cancer. 2001;**85**(3):325-332

[7] Andronesi AG et al. Incidence and risk factors for acute kidney injury following autologous stem cell transplantation for multiple myeloma. Cancer Medicine. 2019;**8**(6):3278-3285

[8] Luxton RW, Kunkler PB. RADIATION NEPHRITIS. Acta Radiologica: Therapy, Physics, Biology. 1964;**2**:169-178

[9] Cohen EP, Robbins ME. Radiation nephropathy. Seminars in Nephrology. 2003;**23**(5):486-499

[10] Breitz H. Clinical aspects of radiation nephropathy. Cancer Biotherapy and Radiopharmaceuticals. 2004;**19**(3):359-362

[11] Cohen EP, Fish BL, Moulder JE. Mitigation of radiation injuries via suppression of the renin-angiotensin system: Emphasis on radiation nephropathy. Current Drug Targets. 2010;**11**(11):1423-1429

[12] Myers BD et al. Cyclosporine-associated chronic nephropathy. The New England Journal of Medicine. 1984;**311**(11):699-705

[13] Farouk SS, Rein JL. The many faces of calcineurin inhibitor toxicity-what the FK? Advances in Chronic Kidney Disease. 2020;**27**(1):56-66

[14] Djamali A et al. Nox2 is a mediator of chronic CsA nephrotoxicity. American Journal of Transplantation. 2012;**12**(8):1997-2007

[15] Naesens M, Kuypers DR, Sarwal M. Calcineurin inhibitor nephrotoxicity. Clinical Journal of the American Society of Nephrology. 2009;**4**(2):481-508

[16] Rostaing L et al. Fibrosis progression according to epithelial-mesenchymal transition profile: A randomized trial of everolimus versus CsA. American Journal of Transplantation. 2015;**15**(5):1303-1312

[17] Joy MS et al. Cytochrome P450 3A5 expression in the kidneys of patients with calcineurin inhibitor nephrotoxicity. Nephrology Dialysis Transplantation. 2007;**22**(7):1963-1968

[18] Nieves-Cintrón M et al. Activation of NFATc3 down-regulates the beta1 subunit of large conductance, calcium-activated K+ channels in arterial smooth muscle and contributes to hypertension. Journal of Biological Chemistry. 2007;**282**(5):3231-3240

[19] Wang PX et al. Effect of losartan on slowing progression of chronic allograft nephropathy. Chinese Medical Sciences Journal. 2005;**20**(4):231-236

[20] Miyata M et al. Recent advances of acute kidney injury in hematopoietic cell transplantation. Frontiers in Immunology. 2021;**12**:779881

[21] Renaghan AD et al. Acute kidney injury and CKD associated with hematopoietic stem cell transplantation. Clinical Journal of the American Society of Nephrology. 2020;**15**(2):289-297

[22] Shingai N et al. Early-onset acute kidney injury is a poor prognostic sign for allogeneic SCT recipients. Bone Marrow Transplantation. 2015;**50**(12):1557-1562

[23] Humphreys BD, Soiffer RJ, Magee CC. Renal failure associated with cancer and its treatment: An update. Journal of the American Society of Nephrology. 2005;**16**(1):151-161

[24] Krishnappa V et al. Acute kidney injury in hematopoietic stem cell transplantation: A review. International Journal of Nephrology. 2016;**2016**:5163789

[25] Obut F, Kasinath V, Abdi R. Post-bone marrow transplant thrombotic microangiopathy. Bone Marrow Transplantation. 2016;**51**(7):891-897

[26] Wirtschafter E, VanBeek C, Linhares Y. Bone marrow transplant-associated thrombotic microangiopathy without peripheral blood schistocytes: A case report and review of the literature. Experimental Hematology & Oncology. 2018;7:14

[27] Dvorak CC, Higham C, Shimano KA. Transplant-associated thrombotic microangiopathy in pediatric hematopoietic cell transplant recipients: A practical approach to diagnosis and management. Frontiers in Pediatrics. 2019;7:133

[28] Qi J, Han Y. A narrative review of transplant-associated thrombotic microangiopathy: Pathogenesis and novel therapies. Annals of Blood. 2023;**8**:25

[29] Jodele S et al. Abnormalities in the alternative pathway of complement in children with hematopoietic stem cell transplant-associated thrombotic microangiopathy. Blood. 2013;**122**(12):2003-2007

[30] Jodele S et al. Diagnostic and risk criteria for HSCT-associated thrombotic microangiopathy: A study in children and young adults. Blood. 2014;**124**(4):645-653

[31] Gloude NJ et al. Circulating dsDNA, endothelial injury, and complement activation in thrombotic microangiopathy and GVHD. Blood. 2017;**130**(10):1259-1266

[32] Khaled SK et al. Narsoplimab, a Mannan-binding lectin-associated serine Protease-2 inhibitor, for the treatment of adult hematopoietic stem-cell transplantation-associated thrombotic microangiopathy. Journal of Clinical Oncology. 2022;**40**(22):2447-2457

[33] Wolff D et al. Replacement of calcineurin inhibitors with daclizumab in patients with transplantation-associated microangiopathy or renal insufficiency associated with graft-versus-host

disease. Bone Marrow Transplantation. 2006;**38**(6):445-451

[34] Bruijn JA et al. Murine chronic graft-versus-host disease as a model for lupus nephritis. The American Journal of Pathology. 1988;**130**(3):639-641

[35] Seconi J, Watt V, Ritchie DS. Nephrotic syndrome following allogeneic stem cell transplantation associated with increased production of TNF-alpha and interferon-gamma by donor T cells. Bone Marrow Transplantation. 2003;**32**(4):447-450

[36] Stevenson WS, Nankivell BJ, Hertzberg MS. Nephrotic syndrome after stem cell transplantation. Clinical Transplantation. 2005;**19**(1):141-144

[37] Reddy P et al. Nephrotic syndrome associated with chronic graft-versus-host disease after allogeneic hematopoietic stem cell transplantation. Bone Marrow Transplantation. 2006;**38**(5):351-357

[38] Issa N, Kukla A, Ibrahim HN. Calcineurin inhibitor nephrotoxicity: A review and perspective of the evidence. American Journal of Nephrology. 2013;**37**(6):602-612

[39] Hingorani S. Renal complications of hematopoietic-cell transplantation. The New England Journal of Medicine. 2016;**374**(23):2256-2267

[40] Andronesi A et al. Incidence and risk factors for acute kidney injury after allogeneic stem cell transplantation: A prospective study. Biomedicine. 2022;**10**(2):262

[41] Coppell JA et al. Hepatic veno-occlusive disease following stem cell transplantation: Incidence, clinical course, and outcome. Biology of Blood and Marrow Transplantation. 2010;**16**(2):157-168

[42] Cornell RF, Hari P, Drobyski WR. Engraftment syndrome after autologous stem cell transplantation: An update unifying the definition and management approach. Biology of Blood and Marrow Transplantation. 2015;**21**(12):2061-2068

[43] Glezerman IG et al. Chronic kidney disease, thrombotic microangiopathy, and hypertension following T cell-depleted hematopoietic stem cell transplantation. Biology of Blood and Marrow Transplantation. 2010;**16**(7):976-984

[44] Kamil E et al. Radiation nephritis following bone marrow transplantation. Kidney International. 1978;**14**(713):1336-1344

[45] Ellis MJ et al. Chronic kidney disease after hematopoietic cell transplantation: A systematic review. American Journal of Transplantation. 2008;**8**(11):2378-2390

[46] Cohen EP, Pais P, Moulder JE. Chronic kidney disease after hematopoietic stem cell transplantation. Seminars in Nephrology. 2010;**30**(6):627-634

[47] Shimoi T et al. The significant impact of acute kidney injury on CKD in patients who survived over 10 years after myeloablative allogeneic SCT. Bone Marrow Transplantation. 2013;**48**(1):80-84

[48] Jo T et al. Chronic kidney disease in long-term survivors after allogeneic hematopoietic stem cell transplantation: Retrospective analysis at a single institute. Biology of Blood and Marrow Transplantation. 2017;**23**(12):2159-2165

[49] Pelletier K et al. Chronic kidney disease, survival and graft-versus-host-disease-free/relapse-free survival in recipients of allogeneic hematopoietic

stem cell transplant. Clinical Kidney Journal. 2022;**15**(8):1583-1592

[50] Cohen EP, Drobyski WR, Moulder JE. Significant increase in end-stage renal disease after hematopoietic stem cell transplantation. Bone Marrow Transplantation. 2007;**39**(9):571-572

[51] Majhail NS et al. Hypertension and diabetes mellitus in adult and pediatric survivors of allogeneic hematopoietic cell transplantation. Biology of blood and marrow transplantation : Journal of the American Society for Blood and Marrow Transplantation. 2009;**15**(9):1100-1107

[52] Rossi GP et al. Drug-related hypertension and resistance to antihypertensive treatment: A call for action. Journal of Hypertension. 2011;**29**(12):2295-2309

[53] Chen Q et al. Posterior reversible encephalopathy syndrome (PRES) after haploidentical haematopoietic stem cell transplantation: Incidence, risk factors and outcomes. Bone Marrow Transplantation. 2020;**55**(10):2035-2042

[54] Fischer M, Schmutzhard E. Posterior reversible encephalopathy syndrome. Journal of Neurology. 2017;**264**(8):1608-1616

[55] Williams B et al. 2018 ESC/ESH guidelines for the management of arterial hypertension: The task force for the management of arterial hypertension of the European Society of Cardiology (ESC) and the European Society of Hypertension (ESH). European Heart Journal. 2018;**39**(33):3021-3104

[56] Tichelli A et al. Premature cardiovascular disease after allogeneic hematopoietic stem-cell transplantation. Blood. 2007;**110**(9):3463-3471

[57] Hoffmeister PA et al. Hypertension in long-term survivors of pediatric hematopoietic cell transplantation. Biology of Blood and Marrow Transplantation. 2010;**16**(4):515-524

[58] Baker KS et al. Diabetes, hypertension, and cardiovascular events in survivors of hematopoietic cell transplantation: A report from the bone marrow transplantation survivor study. Blood. 2007;**109**(4):1765-1772

[59] Borchert-Mörlins B et al. Factors associated with cardiovascular target organ damage in children after renal transplantation. Pediatric Nephrology. 2017;**32**(11):2143-2154

[60] Gavriilaki E et al. Hypertension in hematologic malignancies and hematopoietic cell transplantation: An emerging issue with the introduction of novel treatments. Blood Reviews. 2019;**35**:51-58

[61] Leung AY et al. Polyoma BK virus and haemorrhagic cystitis in haematopoietic stem cell transplantation: A changing paradigm. Bone Marrow Transplantation. 2005;**36**:929-937

[62] Philibert D et al. Incidence and severity of early electrolyte abnormalities following autologous haematopoietic stem cell transplantation. Nephrology Dialysis Transplantation. 2007;**23**(1):359-363

[63] Wollner A, Shalit M, Brezis M. Tumor genesis syndrome. Hypophosphatemia accompanying Burkitt's lymphoma cell leukemia. Mineral and Electrolyte Metabolism. 1986;**12**(3):173-175

[64] Ota R, Hirata A. Relationship between renal dysfunction and electrolyte abnormalities in

hematopoietic stem cell transplant patients treated with foscarnet. Journal of Chemotherapy. 2021;**33**(8):539-546

[65] Kobayashi R et al. Hyponatremia and syndrome of inappropriate antidiuretic hormone secretion complicating stem cell transplantation. Bone Marrow Transplantation. 2004;**34**(11):975-979

[66] Wei J et al. Early onset of syndrome of inappropriate antidiuretic hormone secretion (SIADH) after allogeneic haematopoietic stem cell transplantation: Case report and review of the literature. Journal of International Medical Research. 2010;**38**(2):705-710

[67] Lim Y-J et al. Syndrome of inappropriate secretion of antidiuretic hormone as a leading cause of hyponatremia in children who underwent chemotherapy or stem cell transplantation. Pediatric Blood & Cancer. 2010;**54**(5):734-737

[68] Koumpis E et al. Hyponatremia in patients with hematologic diseases. Journal of Clinical Medicine. 2020;**9**(11):3721

[69] Yoshida S et al. The preceding hyponatremia is a useful hallmark for the diagnosis of HHV-6 encephalitis after allogeneic hematopoietic stem cell transplantation. Bone Marrow Transplantation. 2023;**58**(1):97-99

[70] Peters JP et al. A salt-wasting syndrome associated with cerebral disease. Transactions of the Association of American Physicians. 1950;**63**:57-64

[71] Jeon YJ et al. Cerebral salt-wasting syndrome after hematopoietic stem cell transplantation in adolescents: 3 case reports. Annals of Pediatric Endocrinology & Metabolism. 2015;**20**(4):220-225

[72] Oh JY, Shin JI. Syndrome of inappropriate antidiuretic hormone secretion and cerebral/renal salt wasting syndrome: Similarities and differences. Frontiers in Pediatrics. 2014;**2**:146

[73] Caliskan Y et al. Cyclosporine-associated hyperkalemia: Report of four allogeneic blood stem-cell transplant cases. Transplantation. 2003;**75**(7):1069-1072

[74] Rizk JG et al. Mechanisms and management of drug-induced hyperkalemia in kidney transplant patients. Reviews in Endocrine and Metabolic Disorders. 2021;**22**(4):1157-1170

[75] Caza TN, Jhaveri KD. Glomerular diseases post-hematopoietic stem cell transplantation: Pathologic spectrum and plausible mechanisms. Clinical Kidney Journal. 2023;**16**(6):896-900

[76] Yap DYH et al. Clinico-pathological correlations and outcomes of de novo glomerular diseases in patients after haematopoietic stem cell transplantation. Clinical Kidney Journal. 2022;**16**(6):976-984

[77] Girsberger M et al. Kidney pathology after hematologic cell transplantation-a single-center observation study of indication biopsies and autopsies. Biology of Blood and Marrow Transplantation. 2018;**24**(3):571-580

[78] Bergijk EC et al. A histologic study of the extracellular matrix during the development of glomerulosclerosis in murine chronic graft-versus-host disease. American Journal of Pathology. 1992;**140**(5):1147-1156

[79] Chang A et al. Spectrum of renal pathology in hematopoietic cell transplantation: A series of 20 patients and review of the literature. Clinical

Journal of the American Society of Nephrology. 2007;**2**(5):1014-1023

[80] Brinkerhoff BT, Houghton DC, Troxell ML. Renal pathology in hematopoietic cell transplant recipients: A contemporary biopsy, nephrectomy, and autopsy series. Modern Pathology. 2016;**29**(6):637-652

[81] John EE et al. Patterns of renal dysfunction and profile of kidney biopsies in hematopoietic stem cell transplant recipients. Glomerular Diseases. 2023;**3**(1):98-115

[82] Roy G et al. Kidney biopsy findings among allogenic hematopoietic stem cell transplant recipients with kidney injury: A case series. Kidney Medicine. 2023;**5**(7):100674

[83] Young JA, Pallas CR, Knovich MA. Transplant-associated thrombotic microangiopathy: Theoretical considerations and a practical approach to an unrefined diagnosis. Bone Marrow Transplantation. 2021;**56**(8):1805-1817

[84] Lekakis LJ et al. BK virus nephropathy after allogeneic stem cell transplantation: A case report and literature review. American Journal of Hematology. 2009;**84**(4):243-246

[85] Shingai N et al. Urinary liver-type fatty acid-binding protein linked with increased risk of acute kidney injury after allogeneic stem cell transplantation. Biology of Blood and Marrow Transplantation. 2014;**20**(12):2010-2014

[86] Benoit SW et al. A novel strategy for identifying early acute kidney injury in pediatric hematopoietic stem cell transplantation. Bone Marrow Transplantation. 2019;**54**(9):1453-1461

[87] Morito T et al. Early identification of acute kidney injury after hematopoietic stem cell transplantation by the measurement of urinary biomarkers. Nihon Jinzo Gakkai Shi. 2011;**53**(8):1150-1158